Tissue Engineering Methods and Protocols

METHODS IN MOLECULAR MEDICINE™

John M. Walker, SERIES EDITOR

18. **Tissue Engineering Methods and Protocols,** edited by *Jeffrey R. Morgan and Martin L. Yarmush,* 1999
17. **HIV Protocols,** edited by *Nelson Michael and Jerome H. Kim,* 1999
16. **Clinical Applications of PCR,** edited by *Y. M. Dennis Lo,* 1998
15. **Molecular Bacteriology:** *Diagnostic and Experimental Applications,* edited by *Neil Woodford and Alan Johnson,* 1998
14. **Tumor Marker Protocols,** edited by *Margaret Hanausek and Zbigniew Walaszek,* 1998
13. **Molecular Diagnosis of Infectious Diseases,** edited by *Udo Reischl,* 1998
12. **Diagnostic Virology Protocols,** edited by *John R. Stephenson and Alan Warnes,* 1998
11. **Therapeutic Application of Ribozymes,** edited by *Kevin J. Scanlon,* 1998
10. **Herpes Simplex Virus Protocols,** edited by *S. Moira Brown and Alasdair MacLean,* 1998
9. **Lectin Methods and Protocols,** edited by *Jonathan M. Rhodes and Jeremy D. Milton,* 1998
8. ***Helicobacter pylori* Protocols,** edited by *Christopher L. Clayton and Harry L. T. Mobley,* 1997
7. **Gene Therapy Protocols,** edited by *Paul D. Robbins,* 1997
6. **Molecular Diagnosis of Cancer,** edited by *Finbarr Cotter,* 1996
5. **Molecular Diagnosis of Genetic Diseases,** edited by *Rob Elles,* 1996
4. **Vaccine Protocols,** edited by *Andrew Robinson, Graham H. Farrar, and Christopher N. Wiblin,* 1996
3. **Prion Diseases,** edited by *Harry F. Baker and Rosalind M. Ridley,* 1996
2. **Human Cell Culture Protocols,** edited by *Gareth E. Jones,* 1996
1. **Antisense Therapeutics,** edited by *Sudhir Agrawal,* 1996

METHODS IN MOLECULAR MEDICINE™

Tissue Engineering Methods and Protocols

Edited by

Jeffrey R. Morgan, PHD

and

Martin L. Yarmush, MD

*Center for Engineering in Medicine,
Massachusetts General Hospital/Harvard Medical School,
Boston, MA*

Humana Press ✻ Totowa, New Jersey

© 1999 Humana Press Inc.
999 Riverview Drive, Suite 208
Totowa, New Jersey 07512

All rights reserved. No part of this book may be reproduced, stored in a retrieval system, or transmitted in any form or by any means, electronic, mechanical, photocopying, microfilming, recording, or otherwise without written permission from the Publisher. Methods in Molecular Medicine™ is a trademark of The Humana Press Inc.

All authored papers, comments, opinions, conclusions, or recommendations are those of the author(s), and do not necessarily reflect the views of the publisher.

This publication is printed on acid-free paper. ∞
ANSI Z39.48-1984 (American Standards Institute) Permanence of Paper for Printed Library Materials.

Cover illustration: Fig. 3 from Chapter 27, "Micropatterning Cells in Tissue Enginering," by Sangeeta N. Bhatia, Martin L. Yarmush, and Mehmet Toner.

Cover design by Patricia F. Cleary.

For additional copies, pricing for bulk purchases, and/or information about other Humana titles, contact Humana at the above address or at any of the following numbers: Tel: 973-256-1699; Fax: 973-256-8341; E-mail: humana@humanapr.com or visit our website at http://www.humanapress.com

Photocopy Authorization Policy:
Authorization to photocopy items for internal or personal use, or the internal or personal use of specific clients, is granted by Humana Press Inc., provided that the base fee of US $8.00 per copy, plus US $00.25 per page, is paid directly to the Copyright Clearance Center at 222 Rosewood Drive, Danvers, MA 01923. For those organizations that have been granted a photocopy license from the CCC, a separate system of payment has been arranged and is acceptable to Humana Press Inc. The fee code for users of the Transactional Reporting Service is: [0-89603-516-6/99 $8.00 + $00.25].

Printed in the United States of America. 10 9 8 7 6 5 4 3 2 1

Library of Congress Cataloging in Publication Data

Main entry under title:

Methods in molecular medicine™.

Tissue engineering methods and protocols / edited by Jeffrey R. Morgan and Martin L. Yarmush.
    p. cm.— (Methods in molecular medicine; 18)
  Includes index.
  ISBN 0-89603-516-6 (alk. paper)
  1. Animal cell biotechnology. 2. Biomedical engineering.
3. Tissue culture. 4. Polymers in medicine. I. Morgan, Jeffrey Robert. II. Yarmush, Martin L.
III. Series.
  [DNLM: 1. Tissue Culture—methods. 2. Cell Culture—methods. 3. Cell Transplantation methods.
4. Genetic Engineering—methods. 5. Biomedical Engineering—methods. 6. Regeneration.
QS 525 T6164 1998]
TP248.27.A53T57 1998
571.5' 381—dc21
DNLM/DLC
for Library of Congress   98-11174

Preface

In recent years, the field of tissue engineering has begun, in part, to coalesce around the important clinical goal of developing substitutes or replacements for defective tissues or organs. These efforts are focused on many tissues including skin, cartilage, liver, pancreas, bone, blood, muscle, the vasculature, and nerves. There is a staggering medical need for new and effective treatments for acquired as well as inherited defects of organs/tissues. Tissue engineering is at the interface of the life sciences, engineering, and clinical medicine and so draws upon advances in cell and molecular biology, materials sciences, and surgery, as well as chemical and mechanical engineering. Such an interdisciplinary field requires a broad knowledge base as well as the use of a wide assortment of methods and approaches. It is hoped that by bringing together these protocols, this book will help to form connections between the different disciplines and further stimulate the synergism underlying the foundation of the tissue engineering field.

Tissue Engineering Methods and Protocols, a volume in the *Methods in Molecular Medicine* series, provides researchers with detailed techniques and protocols covering a comprehensive range of technologies and techniques used by leaders of the tissue engineering field. The first part of the book covers methods for the production and evaluation of materials used in tissue engineering, such as polymers, scaffolds, and composites. These materials, which are largely a product of expertise in materials science and polymer chemistry, are useful for multiple applications, including scaffolds for cells and the regeneration of tissues in vivo.

The next two parts of the book are concerned with methods for the isolation, culture, and analysis of cells, as well as the combination of cells with a variety of materials or devices. Methods covered in these two sections include media formulation, microencapsulation, microfabrication, and animal models of transplantation. The final part of the book is concerned with quantitative methods to assess cell performance and cell function. Such measurements are critical to our understanding of cellular processes, as well as the cellular response to the chemical and mechanical environment of a functional organ/tissue.

This collection of protocols from multiple disciplines should make *Tissue Engineering Methods and Protocols* an indispensable resource not only for undergraduate and graduate students, but also basic and clinical researchers in every area of the field.

Jeffrey R. Morgan
Martin L. Yarmush

Contents

Preface v
Contributors xi

PART I. MATERIALS

1 Preparation of Collagen–Glycosaminoglycan Copolymers for Tissue Regeneration
  Lila J. Chamberlain and Ioannis V. Yannas 3
2 Preparation and Use of Tethered Ligands as Biomaterials and Tools for Cell Biology
  Susan J. Sofia, Philip R. Kuhl, and Linda G. Griffith 19
3 Nondestructive Evaluation of Biodegradable Porous Matrices for Tissue Engineering
  Leoncio Garrido 35
4 Fabrication of Biodegradable Polymer Foams for Cell Transplantation and Tissue Engineering
  Peter X. Ma and Robert Langer 47
5 Formation of Highly Porous Polymeric Foams with Controlled Release Capability: *A Phase-Separation Technique*
  S. Kadiyala, H. Lo, and K. W. Leong 57
6 Magnetic-Induced Alignment of Collagen Fibrils in Tissue Equivalents
  Timothy S. Girton, Nerendra Dubey, and Robert T. Tranquillo 67
7 Preparation and Use of Thermosensitive Polymers
  Yi-Lin Cao, Clemente Ibarra, and Charles Vacanti 75
8 *In Situ* Photopolymerized Hydrogels for Vascular and Peritoneal Wound Healing
  Amarpreet S. Sawhney and Jeffrey A. Hubbell 85
9 Fabrication and Implantation of Gel-Filled Nerve Guidance Channels
  Ravi V. Bellamkonda and Robert F. Valentini 101
10 Small Animal Surgical and Histological Procedures for Characterizing the Performance of Tissue-Engineered Bone Grafts
  Kenneth S. James, Mark C. Zimmerman, and Joachim Kohn 121

11 Preparation and Use of Porous Poly(α-Hydroxyester) Scaffolds for Bone Tissue Engineering
Anna C. Jen, Susan J. Peter, and Antonios G. Mikos 133

PART II. CELLS

12 Quantitative Assessment of Autocrine Cell Loops
Gregory Oehrtman, Laura Walker, Birgit Will, Lee Opresko, H. Steven Wiley, and Douglas A. Lauffenburger 143

13 Measurement of Recovery of Function Following Whole Muscle Transfer, Myoblast Transfer, and Gene Therapy
John A. Faulkner, Susan V. Brooks, and Robert G. Dennis 155

14 Preparation of Immortalized Human Chondrocyte Cell Lines
James R. Robbins and Mary B. Goldring 173

PART III. CLINICAL APPLICATIONS

15 Isolation and In Vitro Proliferation of Chondrocytes, Tenocytes, and Ligament Cells
Sonya Shortkroff and Myron Spector 195

16 Culture and Identification of Autologous Human Articular Chondrocytes for Implantation
Ross Tubo and Francois Binette 205

17 Organogenesis of Skeletal Muscle in Tissue Culture
Herman Vandenburgh, Janet Shansky, Michael Del Tatto, and Joseph Chromiak 217

18 Hepatocyte Primary Cultures: *Currently Used Systems and Their Applications for Studies of Hepatocyte Growth and Differentiation*
George K. Michalopoulos 227

19 Formation and Characterization of Hepatocyte Spheroids
Julie R. Friend, Florence J. Wu, Linda K. Hansen, Rory P. Remmel, and Wei-Shou Hu 245

20 Isolation and Culture of Human Endothelial Cells
François Berthiaume 253

21 Isolation and Culture of Microvascular Endothelial Cells
Lisa Richard, Paula Velasco, and Michael Detmar 261

22 Initiation, Maintenance, and Quantification of Human Hematopoietic Cell Cultures
Paul C. Collins, Sanjay D. Patel, William M. Miller, and E. Terry Papoutsakis 271

Contents

23 Methods to Isolate, Culture, and Study Osteoblasts
   Mechteld V. Hillsley 293
24 Cryopreservation of Rat Hepatocytes in a Three-Dimensional Culture Configuration Using a Controlled-Rate Freezing Device
   Michael J. Russo and Mehmet Toner 303

PART IV. CELL MATERIAL COMPOSITES

25 Microencapsulation of Enzymes, Cells, and Genetically Engineered Microorganisms
   Thomas M. S. Chang 315
26 Methods for Microencapsulation with HEMA–MMA
   Shahab Lahooti and Michael V. Sefton 331
27 Micropatterning Cells in Tissue Engineering
   Sangeeta N. Bhatia, Martin L. Yarmush, and Mehmet Toner 349
28 Methods for the Serum-Free Culture of Keratinocytes and Transplantation of Collagen–GAG-Based Skin Substitutes
   Steven T. Boyce 365
29 Use of Skin Equivalent Technology in a Wound Healing Model
   Michael A. Vaccariello, Ashkan Javaherian, Nancy Parenteau, and Jonathan A. Garlick 391
30 Preparation and Transplantation of a Composite Graft of Epidermal Keratinocytes on Acellular Dermis
   Daniel A. Medalie and Jeffrey R. Morgan 407
31 Development of a Bioartificial Liver Device
   Linda K. Hansen, Julie R. Friend, Rory Remmel, Frank B. Cerra, and Wei-Shou Hu 423
32 Methods for the Implantation of Liver Cells
   Stephen S. Kim, Hirofumi Utsunomiya, and Joseph P. Vacanti 433
33 Isolation and Long-Term Maintenance of Adult Rat Hepatocytes in Culture
   François Berthiaume, Ronald G. Tompkins, and Martin L. Yarmush 447
34 Design and Fabrication of a Small Caliper Hybrid Arterial Bioprosthesis
   Gilbert J. L'Italien and William M. Abbott 457
35 Methods for the Immunoisolation and Transplantation of Pancreatic Cells
   Anthony M. Sun 469
36 Methods for the Study of Nerve Cell Migration and Patterning
   Helen M. Buettner and Hsin-Chien Tai 483

PART V. MEASUREMENT TECHNIQUES

37 Estimating Number and Volume of Islets Transplanted Within a Planar Immunobarrier Diffusion Chamber
   Kazuhisa Suzuki, Clark K. Colton, Susan Bonner-Weir, Jennifer Hollister, and Gordon C. Weir 497

38 Quantitative Measurement of Cell–Cell Adhesion Under Flow Conditions
   Carroll L. Ramos and Michael B. Lawrence 507

39 Quantitative Measurement of the Biological Response of Cartilage to Mechanical Deformation
   R. Gregory Allen, Solomon R. Eisenberg, and Martha L. Gray 521

40 Measuring Receptor-Mediated Cell Adhesion Under Flow: Cell-Free Systems
   Daniel A. Hammer and Debra K. Brunk 543

41 In Vitro and In Vivo Quantification of Adhesion Between Leukocytes and Vascular Endothelium
   Rakesh K. Jain, Lance L. Munn, Dai Fukumura, and Robert J. Melder 553

42 Quantitative Measurement of Shear-Stress Effects on Endothelial Cells
   Maria Papadaki and Larry V. McIntire 577

43 Quantitative Modeling of Limitations Caused by Diffusion
   Athanassios Sambanis and Sanda A. Tan 595

Index 607

Contributors

WILLIAM M. ABBOTT • *Vascular Surgery Laboratory, Division of Vascular Surgery, Massachusetts General Hospital and Harvard Medical School, Boston, MA*
R. GREGORY ALLEN • *Harvard-MIT Division of Health Sciences and Technology, Cambridge, MA*
RAVI V. BELLAMKONDA • *Biomaterials, Cells and, Tissue Engineering Laboratory, Department of Biomedical Engineering, Case Western Reserve University, Cleveland, OH*
FRANÇOIS BERTHIAUME • *Center for Engineering in Medicine/Surgical Services, Massachusetts General Hospital, Department of Surgery, Harvard Medical School and Shriners Burns Hospital, Boston, MA*
SANGHETA N. BHATIA • *Center for Engineering in Medicine/Surgical Services, Massachusetts General Hospital, Department of Surgery, Harvard Medical School and Shriners Burns Hospital, Boston, MA*
FRANCOIS BINETTE • *Genzyme Tissue Repair, Inc., Framingham, MA*
SUSAN BONNER-WEIR • *Joslin Diabetes Center and Harvard Medical School, Cambridge, MA*
STEVEN T. BOYCE • *Department of Surgical Research, University of Cincinnati, OH*
SUSAN V. BROOKS • *Institute of Gerontology, University of Michigan, Ann Arbor, MI*
DEBRA K. BRUNK • *Department of Chemical Engineering, University of Pennsylvania, Philadelphia, PA*
HELEN M. BUETTNER • *Department of Chemical and Biochemical Engineering, Rutgers University, Piscataway, NJ*
YI-LIN CAO • *Department of Anesthesia, University of Massachusetts Medical School, Worcester, MA*
FRANK B. CERRA • *Department of Surgery and the Biomedical Engineering Center, University of Minnesota, Minneapolis, MN*
LILA J. CHAMBERLIN • *Fibers and Polymers Laboratory, Department of Mechanical Engineering, Massachusetts Institute of Technology, Cambridge, MA*
THOMAS M. S. CHANG • *Artificial Cells and Organs Research Center, McGill University, Montreal, Quebec, Canada*

JOSEPH CHROMIAK • *Department of Pathology, Brown University School of Medicine and the Miriam Hospital, Providence, RI*
PAUL C. COLLINS • *Department of Chemical Engineering, Northwestern University, Evanston, IL*
CLARK K. COLTON • *Department of Chemical Engineering, Massachusetts Institute of Technology, Cambridge, MA*
MICHAEL DEL TATTO • *Department of Pathology, Brown University School of Medicine and the Miriam Hospital, Providence, RI*
ROBERT G. DENNIS • *Institute of Gerontology, University of Michigan, Ann Arbor, MI*
MICHAEL DETMAR • *Departments of Pathology and Dermatology, Beth Israel Deaconess Medical Center, Boston, MA*
N. DUBEY • *Department of Chemical Engineering and Materials Science, University of Minnesota, Minneapolis, MN*
SOLOMON R. EISENBERG • *Department of Biomedical Engineering, Boston University, Boston, MA*
JOHN A. FAULKNER • *Institute of Gerontology, University of Michigan, Ann Arbor, MI*
JULIE R. FRIEND • *Department of Chemical Engineering and Computer Science, University of Minnesota, Minneapolis, MN*
DAI FUKUMURA • *Edwin L. Steele Laboratory, Department of Radiation Oncology, Massachusetts General Hospital and Harvard Medical School, Boston, MA*
JONATHAN GARLICK • *School of Dental Medicine, Department of Oral Biology and Pathology, SUNY at Stony Brook, NY*
LEONCIO GARRIDO • *Biomaterials Laboratory, NMR Center, Department of Radiology, Massachusetts General Hospital and Harvard Medical School, Charlestown, MA*
T. S. GIRTON • *Department of Chemical Engineering and Materials Science, University of Minnesota, Minneapolis, MN*
MARY B. GOLDRING • *Arthritis Research Laboratory, Medical Services, Massachusetts General Hospital and Division of Medical Sciences, Harvard Medical School, Boston, MA*
MARTHA L. GRAY • *Harvard-MIT Division of Health Sciences and Technology, Cambridge, MA*
LINDA G. GRIFFITH • *Department of Chemical Engineering, Massachusetts Institute of Technology, Cambridge, MA*
DANIEL A. HAMMER • *Department of Chemical Engineering, University of Pennsylvania, Philadelphia, PA*

Contributors

LINDA K. HANSEN • *Departments of Laboratory Medicine and Pathology, and the Biomedical Engineering Center, University of Minnesota, Minneapolis, MN*

MECHTELD V. HILLSLEY • *Department of Chemical Engineering, Pennsylvania State University, University Park, PA*

JENNIFER HOLLISTER • *Joslin Diabetes Center and Harvard Medical School, Boston, MA*

WEI-SHOU HU • *Departments of Chemical Engineering and Materials Science, University of Minnesota, Minneapolis, MN*

JEFFREY A. HUBBELL • *Division of Chemistry and Chemical Engineering, California Institute of Technology, Pasadena, CA*

CLEMENTE IBARRA • *Department of Anesthesia, University of Massachusetts Medical School, Worcester, MA*

RAKESH K. JAIN • *Edwin L. Steele Laboratory, Department of Radiation Oncology, Massachusetts General Hospital and Harvard Medical School, Boston, MA*

KENNETH S. JAMES • *Department of Chemistry, Rutgers University, Piscataway, NJ*

ASHKAN JAVAHERLAN • *School of Dental Medicine, Department of Oral Biology and Pathology, SUNY at Stony Brook, NY*

ANNA C. JEN • *Cox Laboratory for Biomedical Engineering, Department of Chemical Engineering, Rice University, Houston, TX*

S. KADIYALA • *Osiris Therapeutics, Inc., Baltimore, MD*

STEVEN S. KIM • *Laboratory for Transplantation and Tissue Engineering, Children's Hospital, Boston, MA*

JOACHIM KOHN • *Department of Chemistry, Rutgers University, Piscataway, NJ*

PHILIP R. KUHL • *Department of Chemical Engineering, Massachusetts Institute of Technology, Cambridge, MA*

S. LAHOOTI • *Department of Chemical Engineering and Applied Chemistry, University of Toronto, Ontario, Canada*

ROBERT LANGER • *Department of Chemical Engineering, Massachusetts Institute of Technology, Cambridge, MA*

DOUGLASS LAUFFENBURGER• *Department of Chemical Engineering, Massachusetts Institute of Technology, Cambridge, MA*

MICHAEL B. LAWRENCE • *Department of Biomedical Engineering, University of Virginia, Charlottesville, VA*

KAM W. LEONG • *Department of Biomedical Engineering, Johns Hopkins University, Baltimore, MD*

GILBERT J. L'ITALIEN • *Vascular Surgery Laboratory, Division of Vascular Surgery, Massachusetts General Hospital and Harvard Medical School, Boston, MA*

H. LO • *Gliatech, Cleveland, OH*

PETER X. MA • *Departments of Biologic and Materials Sciences and Biomedical Engineering, Macromolecular Sciences and Engineering Center, University of Michigan School of Dentistry, Ann Arbor, MI*

LARRY V. MCINTIRE • *Cox Laboratory for Biomedical Engineering, Institute of Bioscience and Bioengineering, Rice University, Houston, TX*

DANIEL A. MEDALIE • *Center for Engineering in Medicine/Surgical Services, Massachusetts General Hospital, Department of Surgery, Harvard Medical School and Shriners Burns Hospital, Boston, MA*

ROBERT J. MELDER • *Edwin L. Steele Laboratory, Department of Radiation Oncology, Massachusetts General Hospital and Harvard Medical School, Boston, MA*

GEORGE MICHALOPOULOS • *Department of Pathology, University of Pittsburgh Medical School, Pittsburgh, PA*

ANTONIOS G. MIKOS • *Cox Laboratory for Biomedical Engineering, Department of Chemical Engineering, Rice University, Houston, TX*

WILLIAM M. MILLER • *Department of Chemical Engineering, Northwestern University, Evanston, IL*

JEFFREY R. MORGAN • *Center for Engineering in Medicine/Surgical Services, Massachusetts General Hospital, Department of Surgery, Harvard Medical School and Shriners Burns Hospital, Boston, MA*

LANCE L. MUNN • *Edwin L. Steele Laboratory, Department of Radiation Oncology, Massachusetts General Hospital and Harvard Medical School, Boston, MA*

GREGORY OERHTMAN • *Department of Chemical Engineering and Center for Biomedical Engineering, Massachusetts Institute of Technology, Cambridge, MA*

L. OPRESKO • *Department of Pathology, University of Utah, Salt Lake City, UT*

MARIA PAPADAKI • *Cox Laboratory for Biomedical Engineering, Institute of Bioscience and Bioengineering, Rice University, Houston, TX*

E. TERRY PAPOUTSAKIS • *Department of Chemical Engineering, Northwestern University, Evanston, IL*

NANCY PARENTEAU • *Organogenesis Inc., Canton, MA*

SANJAY D. PATEL • *Department of Chemical Engineering, Northwestern University, Evanston, IL*

SUSAN J. PETER • *Cox Laboratory for Biomedical Engineering, Department of Chemical Engineering, Rice University, Houston, TX*

CAROLL L. RAMOS • *Department of Biomedical Engineering, University of Virginia, Charlottesville, VA*
RORY P. REMMEL • *College of Pharmacy, University of Minnesota, Minneapolis, MN*
LISA RICHARD • *Department of Pathology, Beth Israel Deaconess Medical Center, Boston, MA*
JAMES R. ROBBINS • *Arthritis Research Laboratory, Medical Services, Massachusetts General Hospital and Division of Medical Sciences, Harvard Medical School, Boston, MA*
MICHAEL J. RUSSO • *Center for Engineering in Medicine/Surgical Services, Massachusetts General Hospital, Department of Surgery, Harvard Medical School and Shriners Burns Hospital, Boston, MA*
ATHANASSIOS SAMBANIS • *School of Chemical Engineering and Parker H. Petit Institute for Bioengineering and Bioscience, Georgia Institute of Technology, Atlanta, GA*
AMARPREET S. SAWHNEY • *Focal Inc., Lexington, MA*
M. SEFTON • *Department of Chemical Engineering and Applied Chemistry, University of Toronto, Ontario, Canada*
SONYA SHORTKROFF • *Orthopedic Research Laboratory, Brigham and Woman's Hospital and Harvard Medical School, Boston, MA*
JANET SHANSKY • *Department of Pathology, Brown University School of Medicine and the Miriam Hospital, Providence, RI*
SUSAN J. SOFIA • *Department of Chemical Engineering, Massachusetts Institute of Technology, Cambridge, MA*
MYRON SPECTOR • *Orthopedic Research Laboratory, Brigham and Woman's Hospital and Harvard Medical School, Boston, MA*
ANTHONY M. SUN • *Department of Physiology and Applied Chemistry, University of Toronto, Ontario, Canada*
KAZUHISA SUZUKI • *Joslin Diabetes Center and Harvard Medical School, Boston, MA*
HSIN-CHIEN TAI • *Department of Chemical and Biochemical Engineering, Rutgers University, Piscataway, NJ*
SANDA A. TAN • *School of Chemical Engineering and Parker H. Petit Institute for Bioengineering and Bioscience, Georgia Institute of Technology, Atlanta, GA*
RONALD G. TOMPKINS • *Center for Engineering in Medicine/Surgical Services, Massachusetts General Hospital, Department of Surgery, Harvard Medical School and Shriners Burns Hospital, Boston, MA*
MEHMET TONER • *Center for Engineering in Medicine/Surgical Services, Massachusetts General Hospital, Department of Surgery, Harvard Medical School and Shriners Burns Hospital, Boston, MA*

ROBERT T. TRANQUILLO • *Department of Chemical Engineering and Materials Science, University of Minnesota, Minneapolis, MN*
ROSS TUBO • *Genzyme Tissue Repair, Inc., Framingham, MA*
HIROFUMI UTSUNOMIYA • *Laboratory for Transplantation and Tissue Engineering, Children's Hospital, Boston, MA*
CHARLES VACANTI • *Department of Anesthesia, University of Massachusetts Medical Scool, Worcester, MA*
JOSEPH P. VACANTI • *Laboratory for Transplantation and Tissue Engineering, Children's Hospital, Boston, MA*
MICHAEL VACCARIELLO • *School of Dental Medicine, Department of Oral Biology and Pathology, SUNY at Stony Brook, NY*
ROBERT F. VALENTINI • *Department of Molecular Pharmacology, Physiology, and Biotechnology, Brown University, Providence, RI*
HERMAN VANDENBURGH • *Department of Pathology, Brown University School of Medicine and the Miriam Hospital, and Department of Molecular Pharmacology, Physiology, and Biotechnology, Brown University, Providence, RI*
PAULA VELASCO • *Department of Pathology, Beth Israel Deaconess Medical Center, Boston, MA*
L. WALKER • *Department of Chemical Engineering and Center for Biomedical Engineering, Massachusetts Institue of Technology, Cambridge, MA*
GORDON C. WEIR • *Joslin Diabetes Center and Harvard Medical School, Boston, MA*
H. WILEY • *Department of Pathology, University of Utah, Salt Lake City, UT*
B. WILL • *Department of Pathology, University of Utah, Salt Lake City, UT*
FLORENCE J. WU • *Department of Chemical Engineering and Materials Science, University of Minnesota, Minneapolis, MN*
IOANNIS V. YANNAS • *Fibers and Polymers Laboratory, Department of Mechanical Engineering, Massachusetts Institute of Technology, Cambridge, MA*
MARTIN L. YARMUSH • *Center for Engineering in Medicine/Surgical Services, Massachusetts General Hospital, Department of Surgery, Harvard Medical School and Shriners Burns Hospital, Boston, MA*
MARK C. ZIMMERMAN • *Johnson and Johnson Professional, Inc., Corporate Biomaterials Center, Somerville, NJ*

I

Materials

1

Preparation of Collagen–Glycosaminoglycan Copolymers for Tissue Regeneration

Lila J. Chamberlain and Ioannis V. Yannas

1. Introduction

Certain analogs of the extracellular matrix (ECM) have been shown to possess surprising morphogenetic activity during healing of lesions in various anatomical sites. This chapter describes methods for synthesis of the two ECM analogs that have been studied most extensively. The reader is referred to descriptions of these methods in the original literature *(1–3)*. The biological activity of ECM analogs has been reviewed elsewhere *(4)*.

One of these analogs, referred to as the skin regeneration template (SRT), has induced regeneration of dermis in full-thickness skin wounds in the guinea pig model *(2,5–7)*, the porcine model *(8)*, and in humans *(5,9–11)*. Since it is well known that the dermis of the adult mammal does not regenerate spontaneously *(12,13)*, the SRT is required for dermal regeneration in all commonly encountered skin wounds that are sufficiently deep to have compromised the dermis. The SRT is currently used as a dermal regeneration treatment for patients who have sustained deep burns or deep mechanical trauma, including trauma from elective surgery, and who would otherwise have been treated with autografts *(10)*. In the clinical setting or in animal models, the SRT is applied on wounds as a bilayer graft; the proximal layer is the highly porous ECM analog and the distal layer is a silicone film (**Fig. 1**). The latter has no biological activity, but serves as a temporary dressing that protects the proximal layer from dehydration and bacterial invasion, and also converts the bilayer into a mechanically competent sheet, capable of being handled conveniently and sutured on the patient's tissues.

Another ECM analog, referred to as the nerve regeneration template (NRT), has induced regeneration of a functional peripheral nerve across a 15-mm gap

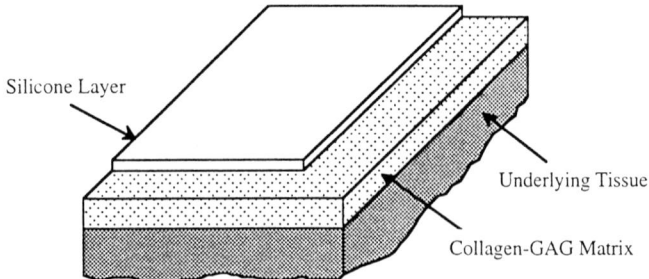

Fig. 1. SRT as it would be implanted in the skin wound model.

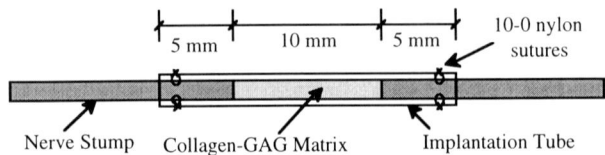

Fig. 2. NRT, ensheathed by an implantation tube, as it would be implanted in the peripheral nerve wound model.

in the rat sciatic nerve model *(14)*. In this model, the highly porous ECM analog is used to fill the lumen of a tube, made either of silicone (nondegradable) or collagen (biodegradable); the nerve stumps are inserted into the tube and are prevented from being displaced further by two sutures at each stump (**Fig. 2**). If the silicone tube is used without an ECM analog, the maximum gap distance that can be bridged by a functional peripheral nerve in this animal model is only 10 mm *(15)*. The structure of the NRT has been determined by selection of the network structure that resulted in maximum regenerative activity in the sciatic nerve model, using a gap of 10 mm *(3,16)*. The selection of the collagen tube instead of the silicone tube, based on superior regenerative activity, has also been made in the same animal model *(17)*.

Another ECM analog, similar in structure to the SRT and the NRT, has been shown capable of regenerating the canine knee meniscus *(18)*.

The structure of these biologically active ECM analogs has been characterized on the scale of the nanometer, as well as on the scale of the micrometer. In the former scale, both ECM analogs referred to above are graft copolymers of type I collagen and chondroitin 6-sulfate, which are crosslinked covalently, and can therefore be described as insoluble macromolecular networks. On the larger scale, the analogs are highly porous matrices that are characterized in terms of the pore volume fraction, the average pore diameter and the average orientation of pore channel axes. Being insoluble, the ECM analogs cannot be isolated and characterized structurally using common biochemical techniques

Table 1
Structural Properties of Two Regeneration Templates

Design parameter of ECM analog	Skin regeneration template (SRT)	Nerve regeneration template (NRT)
Type I collagen/chondroitin 6-sulfate, (w/w)	98/2	98/2
Degradation half-life, wk	1.5	6–8
Average pore diameter, μm	20–125	5–10
Pore channel orientation	Random	Axial

for the structural analysis of proteins. However, structural methodology that has been used to characterize synthetic polymeric networks, including infrared and Raman spectroscopy, rubber elasticity analysis of network structure and various forms of microscopy, have been employed in the characterization of the ECM analogs . The structures of the SRT *(2)* and the NRT *(3,16)* (**Table 1**) have been identified by selecting the analogs of maximum activity from a large number of ECM analogs with related structure. Inspection of **Table 1** shows that the NRT is significantly different from the SRT regarding network cross link density, average pore diameter, and average orientation of pore channel axes (**Fig. 3**).

2. Materials
2.1. Preparation of Collagen–GAG Suspension

1. Type I collagen, from bovine tendon, (Integra LifeSciences, Plainsboro, NJ), in the form of hydrated fibrillar granules, is divided into 14-g aliquots and stored at 0°C. Freeze–thaw cycles during storage should be avoided. If dry collagen is used, it should be kept refrigerated at 4°C.
2. Cooled overhead blender (Granco overhead blender, Granco, Kansas City, MO), including a cooling system (Brinkman cooler model RC-2T, Brinkman, Westbury, NY). The blender is used to mix the collagen–glycosaminoglycan (GAG) suspension, which must be kept at 4°C during the entire preparation.
3. 0.05 M acetic acid solution: Add 8.7 mL glacial acetic acid (Mallinckrodt Chemical, Paris, KY) to 3 L dH_2O. This solution has a shelf life of approx 1 wk.
4. Peristaltic pump (Manostat Cassette Pump, cat. no. 75-500-0.00, Manostat, New York).
5. 0.11% w/v chondroitin 6-sulfate solution: Dissolve 275 mg chondroitin 6-sulfate (from shark cartilage, cat. no. C-4384, Sigma, St. Louis, MO) in 250 mL 0.05 M acetic acid solution. The chondroitin 6-sulfate solution is stored at 4°C and has a shelf life of 1 d. The chondroitin 6-sulfate powder is stored in a desiccator at 4°C.

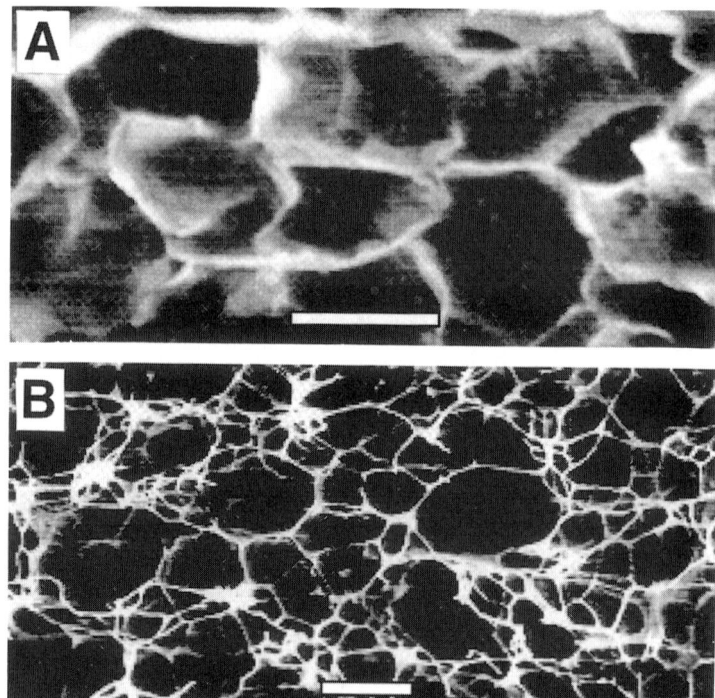

Fig. 3. Environmental scanning electron micrographs of the (**A**) skin regeneration template (scale bar = 100 μm) and the (**B**) nerve regeneration template (scale bar = 20 μm).

2.2. Formation of Matrix Pore Structure

2.2.1. Skin Regeneration Template

1. Freeze dryer (VirTis Genesis, VirTis, Gardiner, NY). Required to freeze the suspension and to sublimate the ice crystals, leaving behind a highly porous matrix structure. The freeze dryer is equipped with trays that are pressed against the chamber shelves when placed in the freeze dryer. These trays ensure good contact between the cooled shelf and the product, and are important for proper pore formation in the skin regeneration template.

2.2.2. Nerve Regeneration Template

1. Polyvinylchloride (PVC) tubing (0.125 in. id, 0.25 in. od), cut into 12-cm lengths.
2. Silicone processing tubes (model 602-235 medical grade Silastic, 0.058 in. id, 0.077 in. od, Dow-Corning, Midland, MI) cut into 15-cm lengths.
3. Silicone adhesive (Medical GradeSilastic, Dow-Corning, MI).
4. Liquid nitrogen: 160-L canister.
5. Axial freezing bath: This custom-made device (**Fig. 4**) is required to freeze the suspension for a nerve regeneration template *(19)*. To achieve the appropriate

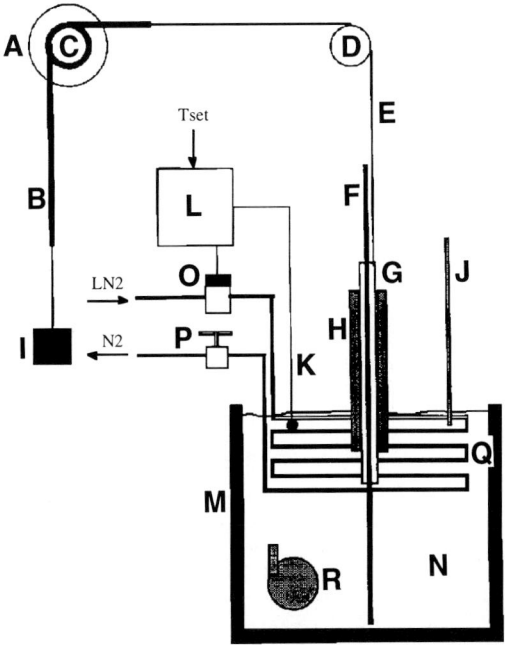

Fig. 4. Schematic of the custom-made axial freezing bath *(19)*. Not to scale. (**A**) Electric timing motor. (**B**) Gear drive chain. (**C**) Drive gear. (**D**) Idler pulley. (**E**) String. (**F**) Guide rod. (**G**) PNS Graft tube carrier. (**H**) PNS Graft tubes. (**I**) Counterweight. (**J**) Thermometer. (**K**) Thermocouple. (**L**) Temperature controller. (**M**) Insulate glass vessel. (**N**) Silicone heat transfer fluid (**O**) LN2 Solenoid valve. (**P**) N2 Throttle valve. (**Q**) One-quarter inch copper cooling (**R**) Circulator.

pore structure, the suspension is injected into tubes and lowered into a freezing bath. The freezing apparatus consists of a liquid nitrogen-controlled cooling system and a gear train arrangement, which allows for variable lowering velocities. The cooling system uses liquid nitrogen, traveling through coiled copper tubing, to cool the heat transfer fluid inside the bath (Silicone Oil, Syltherm XLT Heat Transfer Liquid, Dow Corning, MI). A simple temperature controller is used to regulate the flow of liquid nitrogen. The freezing bath is insulated with hard styrofoam and capped with an acrylic disk.

6. Freeze dryer (VirTis Genesis). Required to sublimate the ice crystals, leaving behind a highly porous matrix structure.

2.3. Crosslinking, Sterilization, and Hydration

2.3.1. Skin Regeneration Template

1. Vacuum oven (Fisher Isotemp Vacuum Oven, Fisher Scientific, Boston, MA; VacTorr 150 Vacuum Pump, GCA/Precision Scientific, Chicago, IL).

2. Silicone adhesive (Silastic, Dow-Corning, MI); sterilize by autoclaving.
3. Sterile implements: 5-L plastic tub (approx W11 × L14 × D4 in.) with Teflon cover (does not need to seal, only cover the tub), gauze, Teflon working surface, forceps, metal spatulas, rulers, scalpel blade holder, and scalpel blades. Sterilize by autoclaving.
4. Laminar flow bench (Relialab, Tenney Engineering, Union, NJ). All sterile procedures are performed in the laminar flow bench.
5. 0.05 M acetic acid solution: Add 2.9 mL glacial acetic acid (Mallinckrodt) to 1000 mL dH_2O. Sterilize by filtration using a 0.2-μm filter (cat. no. 8310, Costar Scientific, Cambridge, MA). This solution has a shelf life of approx 1 wk.
6. 0.25% glutaraldehyde in 0.05 M acetic acid: Combine 10 mL of 25% glutaraldehyde and 3 mL glacial acetic acid. Add distilled water to 100 mL. Add an additional 900 mL of dH_2O. This solution has a shelf life of about 1 wk, and is stored in a dark container at room temperature. Sterilize by filtration using a 0.2-μm filter.
7. 4000 mL dH_2O: Sterilize by filtration using a 0.2-μm filter.
8. Teflon cutting template: Make a matrix-cutting template by cutting a piece of Teflon the size and shape of the desired matrix sheet. Using this type of template the matrix is cut without tearing, and is ensured the proper size matrix sheet. Sterilize template by autoclaving.
9. Phosphate buffered saline (PBS) (cat. no. P-3813, Sigma), 1000 mL: Sterilize by filtration using a 0.2-μm filter.
10. 70% isopropanol in dH_2O, 1000 mL: Sterilize by filtration using a 0.2-μm filter.

2.3.2. Nerve Regeneration Template

1. Implantation tubes (see **Note 1**): For implantation, the nerve regeneration template is ensheathed by an implantation tube (**Fig. 2**). Tubes that can be used include porous collagen tubes (1.5 mm id, 3.0 mm od, Integra), nonporous collagen tubes (1.5 mm id, 1.8 mm od, Integra), and silicone tubes (model 602-235 medical grade Silastic, 0.058 in. id, 0.077 in. od, Dow-Corning).
2. Vacuum oven (Fisher Isotemp Vacuum Oven, Fisher Scientific; VacTorr 150 Vacuum Pump, GCA/Precision Scientific).
3. Sterile implements: several pair of forceps, scalpel blade holder, scalpel blades, ruler, specimen jars, and a Teflon working surface. Sterilize by autoclaving.
4. PBS (cat. no. P-3813, Sigma), 1000 mL: Sterilize by filtration using a 0.2-μm filter.
5. 70% isopropanol in dH_2O, 1000 mL: Sterilize by filtration using a 0.2-μm filter.

3. Methods
3.1. Preparation of Collagen–GAG Suspension

The technique for preparing the collagen–GAG suspension is identical for the SRT and NRT. It is important that the collagen and GAG components remain refrigerated; therefore, the entire suspension preparation must take place at 4°C.

1. Defrost a 14-g aliquot of frozen hydrated tendon collagen for 30–60 min at room temperature.

Collagen–Glycosaminoglycan Copolymers

2. Turn on cooling system for blender and cool to 4°C (takes about 30 min).
3. Add 13.69 g of defrosted hydrated tendon collagen (or 3.6 g of dry collagen), all at once, to 600 mL of 0.05 M acetic acid in one blender, and blend at high speed setting (approx 20,000 RPM) for 90 min (*see* **Note 2**).
4. Calibrate the peristaltic pump to 40 mL/5 min.
5. Add 120 mL of 0.11% w/v chondroitin 6-sulfate solution dropwise to the blending collagen dispersion over 15 min, using the peristaltic pump (maintain blender at 4°C and high-speed setting).
6. Blend the mixture for an additional 90 min on high-speed setting (approx 20,000 RPM).
7. Pour out the collagen–GAG suspension and store in a capped bottle at 4°C. The suspension has a shelf life of about 4 mo (*see* **Note 3**). If stored more than 4 wk, reblend for 15 min at low speed (approx 10,000 RPM), in cooled blender (4°C), before using.

3.2. Formation of the Matrix Pore Structure
3.2.1. Skin Regeneration Template

1. Remove the air from the collagen–GAG suspension by placing it into a 1500-mL Erlenmeyer flask under vacuum for 10 min with agitation, or until bubbles are no longer visible.
2. Set the shelf temperature of the freeze-dryer to –45°C.
3. Turn on the condenser of the freeze-dryer.
4. Allow at least 1 h for the shelf temperature to reach –45°C.
5. Pour the collagen–GAG suspension into an aluminum VirTis freeze-dryer tray. The depth of the suspension can be varied to change the thickness of the resulting dry matrix.
6. Place the suspension-filled tray on the freeze-dryer shelf, and close the chamber door. Be sure that the tray and the shelf are in good contact.
7. Wait for approx 1 h (or longer as necessary) until the collagen–GAG suspension is frozen (*see* **Note 4**).
8. Check the condenser temperature. It must be at –50°C or below before proceeding to the next step.
9. After the suspension is frozen, turn on the freeze-dryer vacuum pump. Make sure the chamber door makes a good seal.
10. Once the vacuum is below 200 mtorr, increase the shelf temperature to 0°C.
11. Leave overnight (at least 15 h).
12. Increase the shelf temperature to 20°C.
13. When the chamber reaches 20°C, turn off the vacuum pump and condenser. Release the vacuum in the chamber and remove the dry collagen–GAG matrix in the form of a white, highly porous sheet.

3.2.2. Nerve Regeneration Template

1. Prepare vented PVC jackets by heating 12-cm sections of flexible PVC tubing at 105°C for 2 h, to straighten. Puncture each tube with a 25 gage needle at 90-degree intervals around the tube, spaced 1 cm apart for the length of the tube.

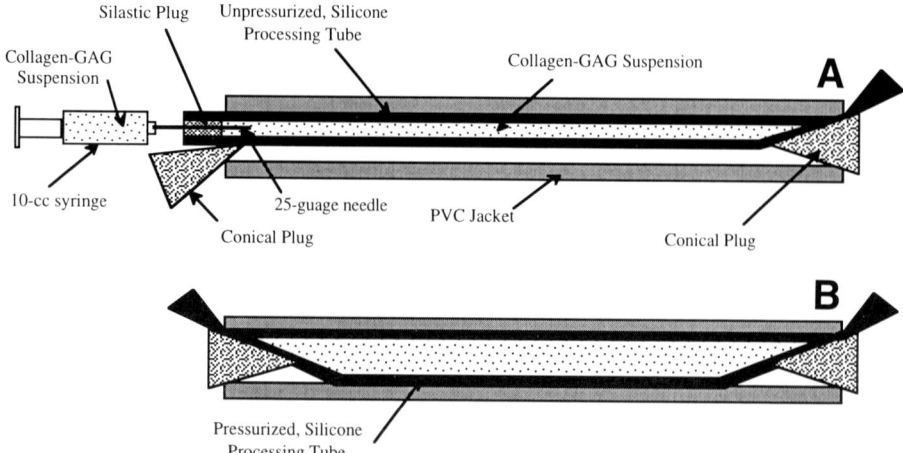

Fig. 5. Preparation of the PVC jacket assemblies for freezing the NRT. (**A**) Diagram of PVC assembly during injection of the collagen–GAG suspension. (**B**) Final PVC jacket assembly, ready for freezing. Not to scale.

2. Flush silicone processing tubes (15 cm in length) with dH_2O, and let dry.
3. Seal one end of each silicone processing tube with silicone adhesive. Inject a cylindrical plug of adhesive, approx 5 mm in length, into the end of silicone tube and allow the excess to stay on the outside of the tube. The excess is important for adhesion and can be cut off later. Let cure for 24 h at room temperature to a tack-free, elastomeric state.
4. Prepare for use a 160-L liquid nitrogen tank for the bath cooling system.
5. Remove the air from the collagen–GAG suspension by placing into a 1500-mL Erlenmeyer flask under vacuum for 10 min with agitation, or until bubbles are no longer visible.
6. Turn on the cooling system of the axial freezing bath and set the bath temperature to –80°C (*see* **Note 5**). It will take approx 45 min of liquid nitrogen cooling for the bath to reach this temperature.
7. Insert each plugged silicone processing tube into a prepared PVC jacket.
8. Draw collagen–GAG suspension into a 10-cc syringe (Becton Dickinson model 5604, Becton Dickinson, Rutherford, NJ) and expel all the air bubbles. Attach a 25-gage needle (Becton Dickinson model 25G5/8, Becton Dickinson) to the syringe and insert the needle carefully into the plugged end of the silicone tube. The needle should be inserted far enough so that a needle length of about 3–5 mm extends beyond the Silastic plug into the tube (**Fig. 5A**).
9. Inject collagen–GAG suspension until the tube is full and no air remains in the tube. Pinch the free end of the silicone processing tube against the wall of the PVC jacket using a conical, plastic plug (the end of a pipet tip works well). Insert the plug far enough so that the silicone processing tube is sealed, and no suspen-

Collagen–Glycosaminoglycan Copolymers

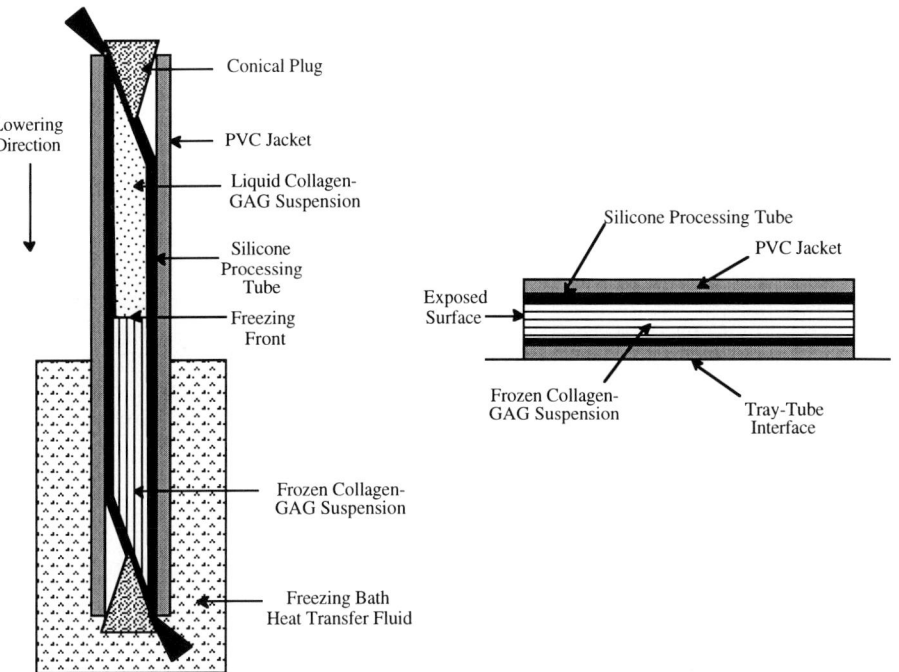

Fig. 6. Manufacturing assembly diagram for the peripheral nerve regeneration template. (**A**) Freezing orientation. (**B**) Freeze-drying (sublimation) orientation. Not to scale.

sion can leak out. Insert another conical, plastic plug into the needle end of the tube. The plug at the needle end should not block the flow of the suspension into the silicone tube via the needle (**Fig. 5A**).

10. Inject additional suspension until the silicone processing tube becomes pressurized and expands to fill the entire PVC jacket (**Fig. 5B**). The silicone tube will inflate because of pressure from the injection of additional suspension. The end of the needle should be inside the PVC jacket to help prevent pressure build up at the needle tip. When the silicone tube has completely filled the PVC jacket, carefully remove the needle; simultaneously, press the conical plug into the end of the tube until the silicone processing tube is pinched against the PVC jacket and sealed. Pressure should be kept on the syringe plunger until the needle is completely out of the tube. Check to make sure the silicone processing tube is still filling the entire PVC jacket (**Fig. 5B**).

11. Attach the drive gear to the electric timing motor on the axial freezing apparatus (**Fig. 4**). Place prepared PVC jackets, up to four at a time, on the tube carrier (**Fig. 4**). Place the tube carrier on the gear train and manually lower until the bottom of the PVC jacket assembly is just touching the freezing bath. Start the motor and let the tubes lower into the bath at a velocity of 10^{-4} m/s (*see* **Note 6**; **Fig. 6A**).

Monitor the process of lowering to ensure that the tubes do not stick to the copper tubing in the freezing bath.
12. Turn on the freeze dryer and set the shelf temperature to $-20°C$.
13. Turn on the condenser of the freeze dryer.
14. When the PVC jackets are fully immersed in the freezing bath, turn off the timing motor and remove the tubes from the bath. Quickly separate the tubes and remove the conical plugs. Cut off the plugged end of the silicone tube and cut each PVC jacket assembly approximately in half with a sharp razor blade. This process provides more exposed surface for sublimation of the ice crystals. Lay the PVC jacket assemblies on a freeze dryer tray and place the tray in the $-20°C$ freeze dryer (**Fig. 6B**). This step must be done as quickly as possible (within a minute) to ensure that the tubes stay completely frozen.
15. Seal the chambers on the freeze dryer and close the vacuum outlet tube. Check to be sure the condenser temperature is below $-45°C$ (if not, wait for the condenser temperature to reach $-45°C$ before proceeding to the next step).
16. Turn on the vacuum pump and wait for the vacuum to reach 200 mtorr. Make sure that the chamber door is sealed.
17. Once the vacuum reaches 200 mtorr, increase the shelf temperature to $0°C$. Leave the PVC jacket assemblies in the freeze dryer for 17 h at this temperature and pressure.
18. Increase the temperature to $25°C$, then turn off the vacuum pump and the condenser. Release the vacuum and remove the PVC jacket assemblies, which contain the dry, white, highly porous matrix inside the silicone processing tubes.

3.3. Crosslinking, Sterilization, and Hydration
3.3.1. Skin Regeneration Template
1. After removing the dry collagen–GAG matrix from the freeze-dryer, inspect the matrix for any irregularities; using a scalpel blade, remove any regions that appear to be distinctly different in appearance from that expected of a very highly porous solid of uniform thickness. Usually these regions will be located near the pan edges. Take note of the difference between the pan side (the side that was in contact with the horizontal pan surface) and the air side (the side that was in contact with the environment) of the dry matrix. The pan side has a much smoother surface. Future steps will require distinguishing between the pan and air sides of the dry matrix.
2. Make an aluminum foil pouch large enough to fit the sheet of dry matrix. Take a large piece of foil and fold it in half. The folded edge is now the bottom of the pouch. Take the left edges and fold, at least twice, to form a sealed side. Repeat on the right side of the pouch. Insert the dry matrix into the pouch (one sheet of matrix per pouch) and leave the top open.
3. Place the matrix-filled pouch (top open) in the vacuum oven for dehydrothermal (DHT) treatment (*see* **Note 7**). The conditions of treatment in the vacuum oven are: 30 mtorr, $105°C$, 24 h.
4. After 24 h, remove the pouch and immediately seal the top by folding the top edges of the foil pouch at least twice. If the matrix is not being prepared for

immediate use, it can be stored in the foil pouch in a desiccator (up to 1 yr) until needed. The matrix is now sterile and must be handled using sterile procedure from this point.

5. After the DHT process, or after storage, sterile silicone adhesive is placed on the dry matrix. Prepare a sterile field in the laminar flow hood and place in it all sterile implements, including the sterile silicone adhesive and the dry matrix (dropped in the sterile area out of the foil pouch; the pouch can be discarded). Prepare operator for sterile work in gown, cap, coat, and sterile gloves. Pour 1000 mL of 0.05 M sterile acetic acid solution into the sterile 5-L plastic tub. Cover tub loosely with the sterile Teflon cover.

6. Place the matrix on the sterile Teflon working surface with the air side up (the silicone adhesive is placed on the air side of the dry matrix). Squeeze a long bead of the viscous silicone fluid along one edge of the dry matrix sheet. Holding the matrix with one finger on the edge, spread the silicone in a thick layer, using straight sweeps of the spatula toward the opposite edge of the sheet. Use only one sweep of the spatula for each portion of the dry matrix. Wipe the spatula clean with sterile gauze. Another bead of silicone adhesive may be necessary in the center of the dry matrix to reach the opposite edge, if the matrix sheet is large. Spread along the pore channels, if possible, so the matrix does not tear.

7. Next, use the spatula at a 90-degree angle to remove the excess silicone and create a thin layer of approximately constant thickness (approx 1 mm). Use single strokes to cover each area. Exercise care, since the dry matrix tears easily. If tearing occurs, it will be necessary to discard the torn portion of the coated sheet. Wipe the spatula often with the gauze to remove excess silicone adhesive.

8. Immerse the matrix, silicone side up, in the acetic acid solution for 15 s. Flip the matrix over, silicone side down, and leave it for 20 h in the covered tub filled with acetic acid.

9. After 20 h in the acetic acid solution, the air bubbles must be removed from the hydrated matrix. Prepare operator for sterile procedure (as in **step 5**). Remove the air in the hydrated matrix by holding one edge of the matrix (wearing sterile gloves) and gently rubbing the hydrated matrix with one finger until air bubbles come out. Be careful not to tear the hydrated matrix.

10. Turn the matrix over so that the silicone side faces up; allow it to rehydrate in the acetic acid for 4 h. Remove any trapped air bubbles. After the 4-h rehydration, remove the acetic acid from the tub with suction. If the matrix is allowed to dehydrate, its activity will be lost.

11. Within 1 min or less after removing the acetic acid, begin pouring 1000 mL of sterile 0.25% glutaraldehyde in 0.05 M acetic acid into the tub. Pour carefully, to avoid air bubbles. Remove any air bubbles by gently rubbing the foam (*see* **step 9**). Soak the matrix in the glutaraldehyde solution for 24 h in the covered tub at ambient temperature (20–22°C) (*see* **Note 8**).

12. Remove the excess glutaraldehyde solution with suction. Rinse the matrix 3× by adding three 1000-mL sterile water rinses to the tub, allowing the matrix to soak for 10 min in each rinse, then remove the rinse water with suction. Add a fourth

rinse of sterile water (1000 mL) to the tub, cover, and soak the matrix for 24 h. Remove any air bubbles (*see* **step 9**).
13. After the 24-h period, the matrix must be cut and stored. Remove the hydrated, porous matrix from the water and place it, silicone side down, on the Teflon sheet. Be careful not to let the matrix fold. Cut away any jagged edges with a scalpel. Place the Teflon cutting template on the matrix and cut around it gently with a scalpel. Do not press down on the template while cutting. Hold only the edges of the matrix. This step must be completed within less than 1 min to prevent dehydration (which leads to deactivation of the matrix).
14. For immediate use, place the cut matrix in a storage container filled with sterile PBS. The matrix can be stored in this medium for 1 d. For longer term storage, place the cut matrix into a storage container with 70% sterile isopropanol. Store at 4°C in this medium up to 30 d. Place the matrix in PBS for 12–24 h prior to use as an implant.

3.3.2. Nerve Regeneration Template

1. Make an aluminum foil pouch for each PVC jacket assembly (containing the silicone processing tube and the matrix). Take a large piece of foil and fold it in half. The folded edge is now the bottom of the pouch. Take the left edges and fold, at least twice, to form a sealed side. Repeat on the right side of the pouch. Insert the PVC jacket assembly into the pouch (one tube per pouch) and leave the top open. The PVC jacket is left in place to protect the matrix from damage during handling and storage.
2. Prepare foil pouches (*see* **step 1**) for the implantation tubes. Place each implantation tube in a pouch, leaving the top open.
3. Place the matrix-filled pouches (top open) and the implantation tube pouches (top open) in the vacuum oven for DHT treatment (*see* **Note 7**). The conditions of treatment in the vacuum oven are: 30 mtorr, 105°C, 24 h.
4. After 24 h, remove each pouch and immediately seal the top by folding the top edges of the foil pouch, at least twice. If the matrix is not being prepared for immediate use, it can be stored in the foil pouch in a desiccator (up to 1 y) until needed. The matrix is now sterile and must be handled using sterile procedure from this point.
5. After the DHT process is complete, the matrix can be cut and hydrated for use. Prepare a sterile field and place in it all sterile implements, including the PVC jacket assemblies and implantation tubes. Prepare the operator for sterile work in gown, cap, coat, and sterile gloves.
6. Under sterile conditions, trim each implantation tube to a length of 20 mm, using a scalpel.
7. Remove the matrix from the silicone processing tube by making a careful slit with the scalpel down the length of the silicone tube and gently pulling out the matrix with forceps. Discard the silicone processing tube.
8. Trim off any crushed or otherwise damaged pieces of the dry matrix. Cut the remaining portion of the dry matrix into 10-mm segments. The exact length depends on the experimental design for use of the implant.

Collagen–Glycosaminoglycan Copolymers

9. Insert each 10-mm segment of matrix into the center of a trimmed implantation tube (**Fig. 2**).
10. Place each implant into a specimen jar filled with sterile PBS for hydration and short term (less than 2 d) storage. If implants will not be used immediately, store at 4°C in 70% isopropanol for up to 30 d. Transfer the implants to sterile PBS solution 1 d prior to implantation.

4. Notes

1. Using a degradable, collagen implantation tube yields a superior regenerated nerve than when using a nondegradable, silicone tube for regeneration across a 10-mm gap in the rat sciatic nerve. This finding is based on histological data at 30 wk *(17)*.
2. The blending time is critical because this is the step that induces swelling of the collagen fibrils and conversion of about 90% of banded collagen fibers to unbanded structures *(20)*. The blending procedure does not, however, remove the triple helical structure of collagen *(1)*. Banded collagen induces platelet aggregation; by removing the banding, the collagen fibrils do not aggregate platelets *(21)*. Blending for shorter times, or at slower speeds, does not eliminate the banding to the desired extent.
3. Collagen–GAG suspension that is stored for over 4 mo may be contaminated with fungus. It is recommended to prepare the suspension shortly before matrices are to be manufactured to avoid possible contamination.
4. Freezing the collagen–GAG suspension on a cool, flat surface creates randomly oriented pore channels with approximately circular cross sections. The average pore diameter can be manipulated by varying the shelf temperature. For optimum dermal regeneration the pores must be between 20 and 125 µm *(2)*. Freezing at –45°C typically results in pores with an average channel diameter of 70 ± 30 µm (**Fig. 3A**).
5. The temperature of the axial freezing bath controls the average pore diameter. In the case of NRTs, pores with an average diameter of 5–10 µm were determined to be optimal for the regeneration of axons *(3,16,22)*. Freezing at –80°C results in pore channels that average 5–10 µm in diameter (**Fig. 3B**). Higher freezing temperatures result in larger pore diameters *(19)*.
6. By slowly lowering the suspension-filled tubes into the freezing bath, the pores form as axially aligned channels. The degree of axial alignment can be modified by changing the lowering velocity. For example, quenching the tubes in the freezing bath (V = 1 m/s) creates radial pores, while lowering the device into the bath much more slowly, say at 10^{-4} m/s, results in highly aligned axial pores *(19,23)*, which were optimal for regeneration *(3,16,22)*.
7. DHT treatment in a vacuum oven at 105°C for 24 h serves as a method of crosslinking *(24–25)* and sterilizing *(1)* the prepared matrices. Treatment at 105°C does not affect the triple-helical structure of the collagen, provided the moisture content at the beginning of DHT treatment is below 10 wt%, which is achieved by freeze-drying the matrices *(1)*.

8. Glutaraldehyde is used to covalently crosslink the matrix in addition to crosslinking introduced by DHT treatment and may be omitted if a lower level of crosslink density (leading to a more rapid degradation rate in vivo) is desired *(26)*. *(1)*. For the SRT, a 24-h treatment in glutaraldehyde resulted in a degradation rate favorable for dermal regeneration. The maximum allowable degradation rate is 140 enzyme units, as determined by using an in vitro collagenase digestion assay *(2)*. Glutaraldehyde treatment is not used for the NRT, since a rapidly degrading collagen–GAG matrix was found to be optimal for peripheral nerve regeneration *(3,16,22)*. Although the SRT has a higher crosslink density because of glutaraldehyde treatment, the degradation half-life of the SRT is much shorter than the NRT probably owing to a significantly lower collagenalytic activity in a nerve lesion compared to a skin lesion (**Table 1**).

References

1. Yannas, I. V., Burke, J. F., Gordon, P. L., Huang, C., and Rubenstein, R. H. (1980) Design of an artificial skin. II. Control of chemical composition. *J. Biomed. Mater. Res.* **14,** 107–131.
2. Yannas, I. V., Lee, E., Orgill, D. P., Skrabut, E. M., and Murphy, G. F. (1989) Synthesis and characterization of a model extracellular matrix that induces partial regeneration of adult mammalian skin. *Proc. Natl. Acad. Sci. USA* **86,** 933–937.
3. Chang, A. S., Yannas, I. V., Perutz, S., Loree, H., Sethi, R., Krarup, C., et al. (1990) Electrophysiological study of recovery of peripheral nerves regenerated by a collagen–glycosaminoglycan copolymer matrix, in *Progress in Biomedical Polymers* (Gebelein, C. G. and Dunn, R. L., eds.), Plenum, New York, pp. 107–120.
4. Yannas, I. V. (1995) Regeneration templates, in *The Biomedical Engineering Handbook* (Bronzino, J. D., ed.), CRC, Boca Raton, FL, pp. 1619–1635.
5. Yannas, I. V., Burke, J. F., Warpehoski, M., Stasikelis, P., Skrabut, E. M., Orgill, D., and Giard, D. J. (1981) Prompt, long-term functional replacement of skin. *Trans. Am. Soc. Artif. Intern. Organs* **27,** 19–22.
6. Yannas, I. V., Burke, J. F., Orgill, D. P., and Skrabut, E. M. (1982) Wound tissue can utilize a polymeric template to synthesize a functional extension of skin. *Science* **215,** 174–176.
7. Murphy, G. F., Orgill, D. P., and Yannas, I. V. (1990) Partial dermal regeneration is induced by biodegradable collagen–glycosaminoglycan grafts. *Lab. Invest.* **63,** 305–313.
8. Orgill, D. P., Butler, C. E., and Regan, J. F. (1996) Behavior of collagen–GAG matrices as dermal replacement in rodent and porcine models. *Wounds* **8,** 151–157.
9. Burke, J. F., Yannas, I. V., Quinby, W. C., Jr., Bondoc, C. C., and Jung, W. K. (1981) Successful use of a physiologically acceptable artificial skin in the treatment of extensive skin injury. *Ann. Surg.* **194,** 413–428.
10. Heimbach, D., Luterman, A., Burke, J., Cram, A., Herndon, D., Hunt, J., et al. (1988) A multi-center randomized clinical trial: artificial dermis for major burns. *Ann. Surg.* **208,** 313–320.

11. Stern, R., McPherson, M., and Longaker, M. T. (1990) Histologic study of artificial skin used in the treatment of full-thickness thermal injury. *J. Burn Care Rehabil.* **11**, 7–13.
12. Billingham, R. E. and Medawar, P. B. (1951) The technique of free skin grafting in mammals. *J. Exp. Biol.* **28**, 385–394.
13. Billingham, R. E. and Medawar, P. B. (1955) Contracture and intussusceptive growth in the healing of extensive wounds in mammalian skin. *J. Anat.* **89**, 114–123.
14. Yannas, I. V., Orgill, D. P., Silver, J., Norregaard, T. V., Zervas, N. T., and Schoene, W. C. (1987) Regeneration of sciatic nerve across 15mm gap by use of a polymeric template, in *Advances in Biomedical Polymers* (Gebelein, C. G., ed.), Plenum, New York, pp. 1–9.
15. Lundborg, G., Dahlin, L. B., Danielsen, N., Gelberman, R. H., Longo, F. M., Powell, H. C., and Varon, S. (1982) Nerve regeneration in silicone chambers: influence of gap length and distal stump components. *Exp. Neurol.* **76**, 361–375.
16. Chang, A. S. and Yannas, I. V. (1992) Peripheral nerve regeneration, in *Neuroscience Year* (Supplement 2 to the *Encyclopedia of Neuroscience*), (Smith, B. and Adelman, G., eds.), Birkhaüser, Boston, pp. 125–126.
17. Chamberlain, L. J. (1996) Long term functional and morphological evaluation of peripheral nerves regenerated through degradable collagen implants. S. M. Thesis, Massachusetts Institute of Technology.
18. Stone, K. R., Rodkey, W. G., Webber, R. J., McKinney, L., and Steadman, J. R. (1990) Future directions, collagen-based prostheses for meniscal regeneration. *Clin. Orthop.* **252**, 129–135.
19. Loree, H. (1988) A freeze-drying process for fabrication of polymeric bridges for peripheral nerve regeneration. S. M. Thesis, Massachusetts Institute of Technology.
20. Forbes, M. J. (1980) Cross-flow filtration, transmission electron micrographic analysis and blood compatibility testing of collagen composite materials for use as vascular prosthesis. S. M. Thesis, Massachusetts Institute of Technology.
21. Sylvester, M., Yannas, I. V., and Salzman, E. W. (1989) Collagen banded fibril structure and the collagen platelet reaction. *Thromb. Res.* **55**, 135.
22. Chang, A. S.-P. (1988) Electrophysiological recovery of peripheral nerves regenerated by biodegradable polymer matrix. S. M. Thesis, Massachusetts Institute of Technology.
23. Loree, H. M., Yannas, I. V., Mikic, B., Chang, A. S., Perutz, S. M., Norregaard, T. V., and Krarup, C. (1989) A freeze drying process for fabrication of polymeric bridges for peripheral nerve regeneration. *Proc. NE. Bioeng. Conf.* 53–54.
24. Yannas, I. V. and Tobolsky, A. V. (1967) Cross-linking of gelatine by dehydration. *Nature* **215**, 509–510.
25. Yannas, I. V. (1972) Collagen and gelatin in the solid state, *J. Macromol. Sci. Rev. Macromol. Chem.* C7, 49–104.
26. Yannas, I. V., Burke, J. F., Huang, C., and Gordon, P. L. 91975) Correlation of in vivo collagen degradation rate with in vitro measurements. *J. Biomed. Mater. Res.* **8**, 623–628.

2

Preparation and Use of Tethered Ligands as Biomaterials and Tools for Cell Biology

Susan J. Sofia, Philip R. Kuhl, and Linda G. Griffith

1. Introduction

The immobilization of biomolecules to various supports has been an important research area for many years. Molecules such as heparin *(1,2)*, as well as various enzymes *(3,4)*, antibodies *(5,6)*, and adhesion ligands *(7)*, have been bound to such supports as silicon or glass, agarose gels, polyethylene oxide (PEO) gels, poly(vinyl alcohol) (PVA) gels, and polymer surfaces. These studies have had important applications in the creation of antithrombogenic surfaces for blood contact, affinity chromatography, and tissue growth and regeneration.

Most applications of surface-bound biomolecules in biotechnology and medicine have involved soluble ligands or substrate interactions with the bound molecule (e.g., affinity chromatography, biosensors, immobilized enzymes), and have thus employed immobilization techniques in which these molecules are indeed essentially immobilized (covalently bound directly to the substrate or linked via short spacers). A major concern in such applications is orientation of the molecule when immobilized *(4)*, particularly with proteins, since multiple reactive sites (typically amine or carboxyl side chains) are usually present.

Applications of surface-bound biomolecules in biomaterials and tissue engineering, in contrast to those in biotechnology, usually involve interactions of the bound molecules with receptors in the cell membrane. In biotechnology applications, the density of bound biomolecule may affect the rate of reaction or the efficiency of a packed-column process, but variations in density of surface-bound ligands for cell receptors can have profound effects on cellular responses, e.g., engendering cell adhesion and spreading as threshold phenomena *(8)*, or enabling formation of specialized cell adhesion structures *(9)*.

From: *Methods in Molecular Medicine, Vol. 18: Tissue Engineering Methods and Protocols*
Edited by: J. R. Morgan and M. L. Yarmush © Humana Press Inc., Totowa, NJ

Increasing the density of substrate-bound ligand presumably allows receptors, which bind the ligands, to aggregate or cluster together because of the decreased distance between receptor–ligand complexes. In adhesion, clustering of ligand-bound integrins is associated with activation of signaling pathways *(10,11)*, and controlling the spatial organization of receptors is thus of interest. It seems likely that, for some aspects of cell behavior mediated by adhesion, the effects of increased ligand concentration on cell processes arise mainly from altering the ability of receptor–ligand complexes in the cell membrane, and not because of greater number of receptor–ligand bonds binding. For example, by increasing the length of the spacer linking the ligand to the substrate, cell behavior, such as spreading, can be observed at a lower ligand density *(12)*.

The ability to control ligand spatial organization independent of ligand density—and thus, presumably, the spatial organization of receptor–ligand complexes—could be a useful tool in studying receptor-mediated processes, including interactions with cytokines and growth factors. Receptor dimerization has been suggested as a necessary event in signal transduction by several growth factors, but remains controversial for some growth factors, including epidermal growth factor (EGF) *(13–15)*. We thus focus on procedures to tether, rather than immobilize, biomolecules on substrates suitable for cell culture, and present procedures for creating clusters of tethered ligands, using EGF as an example.

In the following procedures, we use PEO star molecules as linkers to tether the biomolecule to the surface. PEO star molecules have a poly(divinyl benzene) core, with PEO arms extending radially outward from the core *(16–18)*. The length (M_{arm}) and number (functionality) of arms can vary. The multiple-arm nature of stars allows a single molecule to have several ligands bound (binding occurs at the free ends of the PEO arms), and to retain sites for linking the molecule to the substrate. PEO is perhaps ideally suited as a tether since it is extremely flexible and mobile when in aqueous solution *(19)*. It therefore gives a very mobile presentation of the ligand to the surrounding environment. PEO is also well known to be protein- and cell-adhesive *(20)*. This results in the ligand and cell interacting in a specific manner, as desired, without nonspecific interactions with the PEO tether. The advantages of using PEO star molecules over linear PEO to tether the ligand are twofold. First, because of steric hindrance in the core region, the PEO arms of a star molecule are in an extended conformation, thus making the arm chain ends more accessible to the outer regions of the star molecule and surrounding environment *(21,22)*. This is unlike the chain ends of a linear molecule, which can exist anywhere within the volume of the molecule. Second, because of the larger number of chain ends per star molecule than per linear (depending on the star functionality), star molecules have a higher probability of binding to a surface via these arm chain ends, as well as leaving a large number of free chain ends, once the star is bound, for the attachment of a ligand.

We present here methods for binding EGF to glass surfaces, using PEO star molecules as tethers. With this method, the ligand is bound first to the stars, and the ligand–star conjugate is then bound to the surface. The stars are also reacted with ethylene diamine, which functionalizes some of the arm ends with an amine, which are then used to bind the star to the surface. This approach minimizes nonspecific adsorption of protein on the surface and provides a facile means for independently varying the ligand density and the ligand spatial organization.

Two similar procedures are presented. Procedure I describes a tethering scheme more specific to EGF; in this scheme, each star molecule bound to the surface has only one EGF molecule bound to it. Procedure II is more general; we use EGF as an example, but the procedure can be applied generically to biomolecules that can be linked via a free amine. Also, the number of molecules bound to a star molecule is not limited to one.

We use murine EGF for tethering, because it possesses only a single amine, the amine terminus, to which the PEO tether can bind. The amine terminus is not involved in receptor binding, and thus EGF has the potential to be fully active when tethered, as confirmed by its ability to stimulate DNA synthesis in cultured hepatocytes *(23)*.

2. Materials
2.1. Tresylation of PEO Star Molecules

1. PEO star molecules (Shearwater Polymers, Huntsville, AL).
2. Oven-dried, glass, round-bottom (rb) flask (100 mL) and cap with stir bar.
3. Triethyl amine (TEA) (Aldrich, Milwaukee, WI).
4. Tresyl chloride (TrCl, trifluoroethanesulfonylchloride) (Aldrich). **Caution:** Use extreme care when using TrCl, as it is very toxic and corrosive. TrCl is also moisture sensitive. Unused material can be stored in an oven-dried glass vial with plastic cap in a desiccated jar at 4°C.
5. Dichloromethane (CH_2Cl_2); methanol (MeOH); concentrated hydrochloric acid (HCl, ~37%); liquid nitrogen.
6. Molecular sieves, maintained in convection oven at 120°C, cooled to room temperature (RT) before use.

2.2. Preparation of Glass Surface

1. Glass slides, 1 × 3 in., with Teflon-lined wells (Cel-Line, Newfield, NJ).
2. Concentrated sulfuric acid, concentrated hydrochloric acid, glacial acetic acid, methanol.
3. 18 *M*Ohm purified water.

4. Trimethoxysilylpropyldiethylene triamine, stored desiccated away from light (Hüls Silicon Compounds catalog, United Chemical Technologies, Bristol, PA).
5. Phosphate buffer: 0.1 M NaH_2PO_4, pH 7 (Procedure I only).
6. Glutaraldehyde: 70% aqueous solution (Sigma, St. Louis, MO); dilute to 2.5% with phosphate buffer (Procedure I only).
7. Reductant solution (**Caution:** Prepare in fume hood): Dissolve 12.8 mg sodium cyanoborohydride (Aldrich) per 1 mL 10 mM NaOH (Procedure I only).
8. Succinic anhydride, stored under nitrogen (Aldrich) (Procedure II only).
9. ECD (1-ethyl-3-[3-dimethylaminopropyl] carbodiimide hydrochloride), stored desiccated at –20°C. NHS (N-hydroxysuccinimide), stored desiccated at –20°C (Pierce, Rockford, IL) to be mixed in dioxane in chemical Fume hood (Procedure II only).

2.3. Preparation of ^{125}I-EGF (Iodination)

1. Epidermal growth factor (mouse, Collaborative Biomedical, Bedford, MA): Resuspend lyophilized material in 100 mM NaH_2PO_4, pH 7.0, to concentration of 200 µg/mL. Aliquot in 0.25 mL portions; store at –20°C.
2. 1.5 mL screw-cap microcentrifuge tubes.
3. Iodo-beads (Pierce), stored desiccated at 4°C.
4. Iodination reaction buffer: 100 mM NaH_2PO_4, pH 7.0.
5. Iodination column buffer: 1 mg/mL poly(ethylene oxide); mol wt 1000 g/mol, dimethyl ether (DMPEO) (Polysciences, Warrington, PA) in iodination reaction buffer.
6. Chase solution: one crystal KI (potassium iodide, Aldrich) in 20 mL column buffer.
7. Carrier protein solution: Dissolve 100 mg bovine serum albumin (BSA) and one crystal KI in 20 mL Dulbecco's phosphate buffered saline (IX, without Ca^{2+} and Mg^{2+} ions, Sigma).
8. 12 mg/mL sodium metabisulfite in iodination reaction buffer, made up on the day of use, 1 mL.
9. KwikSep Excellulose disposable gel filtration columns (5 mL) (Bio-Rad; Hercules, CA): Equilibrate with 2 column volumes of iodination column buffer.
10. 1% phosphotungstic acid in 1 N HCl.
11. Na^{125}I; 2–8 mCi, depending on the specific activity desired (NEN Life Science, Boston, MA). 2 mCi reacted with 10 µg EGF results in specific activity 30–40 µCi/µg EGF.

2.4. Preparation and Purification of Amine-Star-EGF

2.4.1. Preparation Using ^{125}I-EGF

2.4.1.1. Procedure I

1. ^{125}I-EGF.
2. Tresylated star PEO.
3. Star PEO dissolution buffer; 10 mM sodium acetate, pH 4.

Tethered Ligands

4. Ethylene diamine (EDA) (Sigma).
5. Econo-column, 2.5-cm diameter, 10-cm bed height (Bio-Rad).
6. Gel filtration column packing: Bio-gel P-60, fine (mol wt cutoff 60 k, Bio-Rad) (*see* **Note 2**)
7. GuHCl gel filtration buffer: 10 mM NaH$_2$PO$_4$, 0.5 M guanidine HCl (purest grade, Sigma), with 1 mg/mL dimethoxy PEO (mol wt 1000 g/mol, Shearwater Polymers), pH 6.2.
8. Anion exchange column. Hi-Trap Q, 1 mL (Pierce).
9. Anion exchange start buffer. 10 mM piperidine, pH 10.8 (Sigma).
10. Anion exchange regeneration buffer. 10 mM *bis*-Tris propane, 0.5M NaCl, pH 7.
11. Grafting buffer: Dissolve 1 mg/mL dimethoxy PEO in 0.1 M NaH$_2$PO$_4$; adjust pH to 7.0.

2.4.2. Preparation using ^{125}I-EGF and Noniodinated EGF

2.4.2.1. PROCEDURE II

1. ^{125}I-EGF and noniodinated EGF.
2. Tresylated PEO stars.
3. Ethylated diamine (EDA), ethylene diamine dihydrochloride (EDA-HCl), sodium azide (NaN$_3$) (Aldrich).
4. Dulbecco's phosphate buffered saline (DPBS), 10z (0.1 M phosphate, without Ca^{2+} and Mg^{2+} ions), titrated to pH 7.4 (Sigma). (When diluted to 1X, no need for titration—already at pH 7.4).
5. Econo-column 2.5-cm diameter, 10-cm bed height (Bio-Rad).
6. Bio-gel P-30, fine, gel-filtration packing (mol wt cutoff 45k, Bio-Rad), prepared with 1X DPBS with 0.04% (w/v) NaN$_3$ (*see* **Note 2**).

3. Methods
3.1. Tresylation of Star PEO
3.1.1. Preparation

1. Dry 100-mL rb flask, with stir bar, in convection oven, at 120°C, for at least 24 h.
2. Dissolve 1 g star PEO in 12 mL CH$_2$Cl$_2$ (**Caution:** in fume hood). Once fully dissolved, add 1 g molecular sieves, wait ~ 30 min for bubbling to cease, seal the vessel, and store at 4°C for 24 h.
3. Add 1g molecular sieves to 10 mL of CH$_2$Cl$_2$ and to 10 mL TEA. Wait ~30 min for bubbling to cease, tightly seal the vial, and store at 4°C for 24 h.

3.1.2. Reaction (**Caution:** in fume hood)

1. Cool the rb flask and stir bar, while capped, to room temperature.
2. Decant the PEO star solution into the rb flask. Decant the excess 10 mL CH$_2$Cl$_2$ into the star/sieves vessel and swirl to recover most of the residual star solution left in the container. Decant this into the rb flask. Only uncap the rb flask for as long as necessary to add the required solutions; otherwise, keep it capped (*see*

Note 1). Set the stir bar rotating at a slow to intermediate speed, to create a small vortex.
3. Pipet dropwise into the rb flask the required volume of TEA, to be 1.5× the number of mol of -OH groups in the 1 g star sample, i.e., there should be a 50% excess number of moles of TEA as -OH groups. (The groups of moles of -OH groups is determined by dividing the mass of star to be tresylated (1 g) by the mol wt of an arm, M_{arm}.)
4. Pipet dropwise into the rb flask the required volume of TrCl, so that it is also at 50% excess over the number of moles of OH groups.
5. Cap tightly and let react at room temperature for 90 min.

3.1.3. Purification of Tresyl-PEO Star

Caution: Except when centrifuging, work in a chemical fume hood.

1. Evaporate off the CH_2Cl_2 using a liquid nitrogen trap.
2. Redissolve the dried residue in 50 mL acidified MeOH (250 µL conc. HCl in 50 mL MeOH).
3. Place the solution into centrifuge tubes and maintain at –20°C for at least 3 h, to allow the star PEO to precipitate.
4. Centrifuge at 7000 rpm, at –20°C, for 25 min. Immediately decant off the top solvent.
5. Redissolve the precipitated residue with moderately acidified MeOH (50 µL conc. HCl in 50 mL MeOH).
6. Reprecipitate at –20°C for at least 3 h, centrifuge, and immediately decant off the top solvent.
7. Repeat **steps 5** and **6** 3× for a total of five precipitations.
8. With final precipitate, evaporate off the solvent using a liquid nitrogen trap. Either use the tresyl-PEO immediately, or store under nitrogen, desiccated, at –70°C.

3.2. Iodination of EGF

Several methods for iodination of proteins are available. The use of Iodobeads from Pierce is a mild method applicable for proteins containing tyrosine residues, like EGF. The only departure from the standard technique is that no protein stabilizer (usually BSA) is added to the protein mixture in order to avoid interference with the later coupling and grafting reactions. This procedure should, of course, be performed only in authorized areas by researchers trained in the use of radioactive materials. Care is taken to maximize recovery of EGF, in order to increase the accuracy of the SA measurement. The following procedure is given for iodinating 10 µg of protein. More or less protein can be iodinated as needed, with the specific activity increased or decreased by varying the amount of $Na^{125}I$ used in the iodination reaction.

Tethered Ligands

3.2.1. Iodination Reaction

1. Equilibrate KwikSep column with 2 column volumes (10 mL) of column buffer.
2. Place an Iodo-bead in a microcentrifuge reaction tube and rinse with 1 mL reaction buffer. Pour off the buffer.
3. Dilute the Na^{125}I with 20 µL of reaction buffer, then add the solution to the reaction tube.
4. Let sit for 5 min.
5. Add 100 µL EGF solution (0.1 mg/mL) to reaction tube. Save pipet tip and EGF tube (if now empty).
6. Wait 15 min (during this time, prepare the two precolumn tubes; see **step 14**).
7. Using the saved tip, transfer the contents of the reaction vial (~120 µL) to the original EGF tube (or fresh tube, if EGF tube not empty). Save tip and reaction tube.
8. Add 80 µL sodium metabisulfite solution to the reaction tube, and swirl gently.
9. With saved tip, transfer the 80 µL to the EGF tube. Save tip and reaction tube again.
10. Add 60 µL chase solution to the reaction tube; swirl gently.
11. With saved tip, transfer the 60 µL to the EGF tube. Save tip and reaction tube.
12. Repeat steps 10 and 11 once.
13. Mix the contents of the EGF tube.
14. Take two precolumn samples 3 µL added to 997 µL carrier protein solution.
15. Load the labeled protein mixture onto the Kwik Sep column, but do not open outlet yet.
16. Rinse the EGF tube twice with 150 µL column buffer each rinse, and add the rinses to the sample on the column before opening the outlet.
17. Open the outlet and collect fractions in screw cap microcentrifuge tubes. Collect 5 min in the first tube, then take 2-min fractions until 25 min, adding column buffer to the top of the column once the sample fully enters the column gel.
18. Count the fractions (a pancake-type, hand-held probe works well for this purpose, or an extremely low-efficiency NaI detector) and combine the hottest 3–5 fractions in one tube. Mix well.
19. Take two postcolumn samples: Add 2 µL combined fractions solution to 998 µL carrier protein solution. Mix well.

3.2.2. Determination of Specific Activity

Specific activity of the iodinated protein is determined by precipitation of the protein with phosphotungstic acid.

1. Vortex the two precolumn and two postcolumn samples.
2. Nonprecipitated samples.
   a. Put 995 µL reaction buffer into each of eight gamma tubes.
   b. Transfer 5 µL from each pre- and postcolumn sample into each of the gamma tubes; do in duplicate.

3. Precipitated samples.
   a. Put 20 μL carrier protein solution into each of eight screw-cap tubes.
   b. Add 5 μL from each of the pre- and postcolumn samples to the carrier protein tubes, do in duplicate.
   c. Add 1 μL phosphotungstic acid to each tube, vortex.
   d. Centrifuge 2–3 min at high speed on desktop centrifuge.
   e. Discard supernatant and add 1 mL 1 N NaOH.
   f. Vortex to dissolve pellet.
   g. Transfer to labeled gamma tubes.
4. Count all nonprecipitated and precipitated tubes in gamma counter.
5. To calculate specific activity in counts/min/ng (CPM/ng), multiply the CPM average value obtained for the precipitated precolumn samples by (200/3) × (sample volume loaded on column)/(ng EGF iodinated). To convert CPM to μCi, divide CPM by the efficiency of the gamma counter, then divide by 2.22 × 10^6. The concentration of protein in the postcolumn combined fractions in CPM/mL can be determined by multiplying the average nonprecipitated postcolumn CPM by 10^5. The specific activity can then be used to convert from CPM/mL to ng/mL.

3.3. Activation of Glass Slide Surfaces

In these first two steps, the glass slides are cleaned and reacted with an aminosilane to obtain reactive primary amines, $-NH_2$, on the surface. These amines are used in subsequent chemistry to bind the ligand–star conjugate to the surface. The Si–O bond that binds the aminosilane to the glass surface is sensitive to degradation by base. Therefore, whenever using the aminosilane-treated samples, the pH should be maintained below pH 8.0.

3.3.1. Cleaning of Glass Slides

1. Immerse the slides in a 40% conc. H_2SO_4 in water solution. Let soak for 30–60 min.
2. Rinse slides well with purified water, then immerse in a 1:1 conc. HCl:methanol solution. Let soak for 30–60 min.
3. Rinse slides well with purified water, then rinse once in methanol.
4. React slides with aminosilane immediately after cleaning.

3.3.2. Reaction with Aminosilane

1. Prepare a 1 mM acidified methanol solution, using glacial acetic acid.
2. Prepare a 95% (v/v) acidified methanol, 1% (v/v) aminosilane, 4% (v/v) water solution, adding first the aminosilane, mixing for several seconds, then immediately adding the water and mixing.
3. Immediately add the cleaned glass slides and let react for 15 min at room temperature.
4. Rinse the slides well with 3–4 washes of methanol. Cure in a convection oven, 120°C, for 15 min, or let sit covered, in air, overnight to cure.

3.3.3. Conversion of Surface –NH$_2$ to –COH

3.3.3.1. Procedure I

1. Rinse slides with grafting buffer.
2. Apply glutaraldehyde solution to each Teflon-lined well and incubate at RT in a humid box overnight.
3. Remove glutaraldehyde from wells and rinse thoroughly with grafting buffer.
4. React immediately with amine-star-EGF conjugate (*see* **Subheading 3.4.3.**)

3.3.4. Conversion of Surface –NH$_2$ to –COOH

3.3.4.1. Procedure II

1. Dissolve 0.3 g succinic anhydride in 100 mL 5X DPBS (0.05 M phosphate) titrated to pH 6.0. It takes several minutes of stirring for the succinic anhydride to fully dissolve.
2. Place cured slides in the solution and let react, at room temperature, for 15 h.
3. Rinse well with water, then rinse once with methanol.

3.3.5. Conversion of -COOH to Activated Ester

Surfaces activated with EDC/NHS are moisture- and UV-sensitive. They should be stored desiccated and sealed from light and used no more than 1 wk from when first made. They can also be stored in dried dioxane for 1–2 mo. Use care when running the activation step, as dioxane is extremely toxic. **Caution:** Work in a chemical fume hood.

1. Place slides in the appropriate volume of dioxane (e.g., 100 mL), and set stir bar spinning at intermediate speed.
2. Slowly add in NHS and EDC, each to a concentration of 0.1 M (e.g., in 100 mL dioxane, add 1.2 g of NHS and 1.9 g of EDC). A white slurry results.
3. Let react at room temperature for 90 min. By the end of the reaction, some residue may adhere to the slides.
4. Rinse the slides once in dioxane, then 3–4× in methanol. Place immediately in desiccated container until ready for reaction with amine–star–ligand conjugate (*see* **Subheading 3.4.7.**).

3.4. Formation and Grafting of Amine-Star-EGF Conjugate

Of key importance in the formation of amine-star-ligand conjugates is the number of ligand molecules that covalently bind to a single star molecule. The number of ligands/star is mainly governed by the ratio of moles of tresylated star to moles of ligand when binding the ligand to the star molecules (*see* **Note 3**).

Procedure I provides a method of binding only one EGF molecule to a star molecule; Procedure II gives a method of binding one or more EGF (or other ligand) molecules to a star molecule. Procedure II also gives a means for estimating the number of ligand molecules bound to each star. It is written with a variable mole ratio of ligand to stars not fixed, so that the experimenter has the freedom to choose this parameter. The resulting number of ligand molecules per star molecule for Procedure II is a calculated average; in actuality, there is a distribution in the number of protein molecules bound to the star molecules.

3.4.1. Synthesis of Amine-Star-EGF

The amount of EGF to be reacted should be such that there is 1000× molar excess of star -OH groups over moles of EGF.

1. Add 104 × molar excess EDA (excess over moles EGF) to reaction vial containing iodinated EGF and vortex (e.g., add 42 µmol EDA to 4.2 nmol 125I-EGF).
2. Dissolve tresyl-activated star PEO in small volume (300 µL for 4.2 µmol star OH groups for the example given in **step 1**) 10 mM acetate (pH 4). Take care that the star PEO completely dissolves (*see* **Note 4**).
3. Add completely dissolved tresyl–star solution to the reaction vial and vortex again.
4. Incubate at 4°C overnight.

3.4.2. Purification of Amine-Star-EGF

1. Equilibrate gel filtration (GF) column with GuHCl gel filtration buffer.
2. Dilute reaction mixture 1:2 with GuHCl gel filtration buffer.
3. Load diluted reaction mixture on GF column. Wait for sample to completely enter the gel, then load GF column buffer to the top of the column.
4. Collect 3-min fractions in gamma tubes for 70–90 min.
5. Count the fractions on gamma counter and pool the fractions from the first peak. These are the PEO star fractions.
6. Dialyze pooled PEO fractions against anion exchange start buffer at 4°C, using, for example, small volume dialysis cassettes (Pierce). Load pooled fractions into cassettes using syringe and needle, and dialyze against 1–2 L anion exchange buffer. Change buffer twice.
7. Anion exchange chromatography:
   a. Equilibrate anion exchange column with 5 mL start buffer, followed by 5 mL regeneration buffer, and then 5 mL start buffer again.
   b. Load dialyzed star fractions onto column and wash through with 5 mL start buffer.
   c. Elute the amine-star-EGF conjugate with 5 mL regeneration buffer.
   d. Dialyze the amine-star-EGF against grafting buffer at 4°C (1–2 L); replace buffer three times.

3.4.3. Grafting of Amine-Star-EGF to Glass Surface

1. Add amine-star-EGF conjugate in grafting buffer to -COH activated wells and incubate at 4°C for 2 h.
2. Remove conjugate solution (it may be reused).
3. Rinse wells thoroughly with grafting buffer to remove nonspecifically adsorbed conjugate.

Tethered Ligands 29

4. To quantify star-EGF binding to solids, place them face up in thin-walled, flat-bottom plastic containers (to maintain sterility).
5. Place the samples, along with standards of known ^{125}I activity, on a phosphor-imager plate (Phosphorimager with reusable phosphor screens, Molecular Dynamics, Sunnyvale, CA). The isotope emits radiation of sufficient energy to penetrate the glass and plastic and expose the plate on the order of hours to days. Star–EGF slides should be stored at 4°C.

3.4.4. Synthesis of Amine-Star-EGF

3.4.4.1. PROCEDURE II

1. Dissolve the appropriate amount of EGF in 1 mL purified water. Use 500 µL to first dissolve the EGF, then rinse the EGF product vial(s) with the remaining 500 µL to recover as much of the EGF as possible.
2. Add 500 µL of the ^{125}I-EGF solution to the unlabeled EGF solution. Mix well.
3. Dissolve the appropriate amount of tresyl–star in 1 mL of 10X DPBS, titrated to pH 7.4. The solution will bubble as the PEO dissolves.
4. Immediately upon dissolving all the tresyl–PEO, add it to the EGF solution. Mix well and let react for 15 min.
5. Add 100 times molar excess of EDA over the total number moles of –OH groups, with 75% of the added moles of EDA from EDA-HCl, and the remaining 25% from EDA, with the EDA-HCl being added first (to prevent the pH from becoming too basic). Mix well, and let react overnight at 4°C.

3.4.5. Purification of Amine-Star-EGF

1. Prepare gel filtration column with Bio-Gel P-30 packing, bed volume 50–70 mL, with 1X DPBS buffer with 0.04% NaN$_3$.
2. Run the sample through the column, collecting fractions every 2 min in gamma tubes, for 70–90 min. If possible, wrap the column in lead while the sample is running through, as well as using lead to shield the collected samples, for the safety of those working in the area.
3. Count the fractions on a gamma counter and pool the highest fractions from the first peak.

3.4.6. Calculation of EGF Functionality on Stars

1. Weigh an empty glass vial and cap. Put the pooled fractions in the vial and weigh the vial again to obtain the approximate volume. Lyophilize to recover the dried amine-star-EGF conjugate and buffer salts.
2. Reweigh the vial and cap with solids. Subtracting the weights of the vial, cap, and buffer salts gives the mass of conjugate recovered.
3. Knowing the total radioactivity from the counts of the pooled fractions allows a calculation of the total moles of EGF present. A calculated estimate of the number of EGF molecules per star molecule can then be obtained.

$$\text{No. mol EGF per star} = (\text{moles EGF ion pooled fractions})(M_w \text{ star})/(\text{mass conjugate recovered}).$$

3.4.7. Grafting of Amine-Star-EGF to Glass Surfaces

1. Dissolve the lyophilized amine-star-EGF product with purified water to the desired concentration; generally, between 1 and 10 mg/mL should be sufficient (*see* **Note 6**). Titrate the solution of pH 7.4, if needed, using 0.1 M NaOH or HCl.
2. Pipet the solution onto the activated slides and let react for 4 h at room temperature. (For a 4-cm^2 activated surface area, 250–300 µL of solution are needed to fully cover the surface for reaction.)
3. Rinse the slides thoroughly with 1X DPBS, using either a gentle stream pipeted onto the surface (e.g., a dozen pipet rinses with 1–1.5 mL buffer per rinse), or swirled gently in 3–4 washes of 1X DPBS. If pipeting, the rinses can be counted in a gamma counter to check when radioactivity coming off the surfaces reaches a minimum.
4. To determine the amount of EGF present on the surface, radioactivity can be measured using phosphorimaging, as described in **Subheading 3.4.3. step 5**, or sample slides not to be further used can be carefully cut into pieces and the counts on each individual piece measured in a gamma counter. Knowing the surface area treated, a surface density of EGF can then be calculated.

4. Notes

1. When tresylating PEO, it is important that all glassware and reagents be dried, free of moisture, and that the reaction solution be exposed to air as little as possible, because the extent of tresylation is very dependent on the absence of water. Of course, more extensive measures could be taken, such as running the reaction under inert gas, but with the method outlined here, yields of 75–80% can be obtained when starting with equimolar quantities of -OH groups, TrCl, and TEA.
2. Using different gels to separate stars and free ligand depends on the relative sizes of two species.
3. In binding ligands to star molecules, the relative amounts of both star and ligand to be used need to be taken into consideration. In the case of EGF, because EGF is a rather expensive protein, smaller quantities of materials were generally used. As an example, for a star molecule with a functionality of 70 and arm mol wt of 5200 g/mol (therefore, M_w = 350,000 g/mol), if 20 mg of tresylated star material is used to react with EGF, then an equimolar amount of EGF would be ~350 µg. Considering that the yield of this particular reaction (i.e., the fraction of starting EGF that binds to the stars and recovered from the GF column) is about 30%, then roughly three times this amount of EGF, ~1 mg, would be needed to ensure that all star molecules, on average, have one EGF molecular bound to them. If 10 mg tresyl–star were reacted with 1 mg EGF, then on average ~2 EGF molecules would bind to each star. For other ligand/star molecule systems, these amounts would probably change.
4. If tresylated PEO has been stored for a period of time, one way to determine if the stored material has become unusable (because of crosslinking) is by watching for the presence of tiny microgels when dissolved in aqueous solution. In dissolving

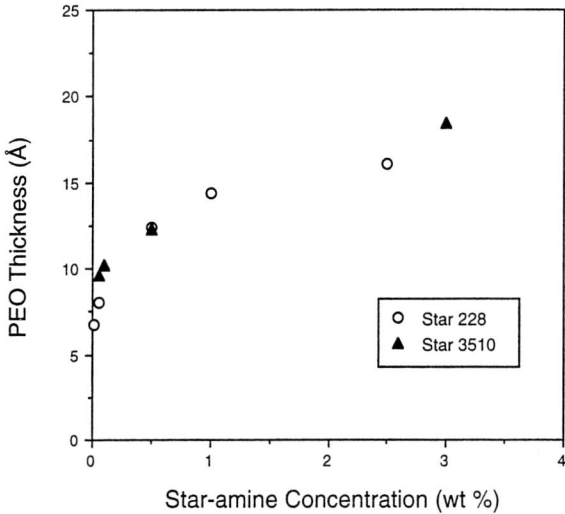

Fig. 1. Dependence of PEO dry layer thickness on star–amine concentration.

the material, once the bubbles rise to the top, check the solution to *see* that it is clear, free of floating microgel particles.
5. Another issue to consider is the particular star type to use, i.e., star functionality and arm mol wt. As to the former, with functionalities greater than ~18–20, star molecules in good solvent behave as hard spheres *(18)*. They have very strong excluded volume effects, so that they are very difficult to overlap, both in solution and on a surface. Larger functionalities provide for more sites per star to which EGF and EDA can bind. As stated in the introduction, the star arms are in an extended conformation as a result of steric hindrance in the core region. This arm extension increases with functionality. The accessibility of the arm ends, however, decreases with increasing arm mol wt, because of a return to more randomly coiling characteristics. We therefore recommend using shorter arm mol wt (ca. 5000–10,000 g/mol) and larger functionalities (20 or more) for use in tethering ligands to surfaces.
6. When working with the protein, it is of course important to maintain the pH of solutions around neutral pH, to prevent the protein from denaturing. Generally, a range of pH that is usually safe is pH 6–8, but the stability of the specific ligand to be used should be investigated before use.
7. To give an idea of the dependence of star–amine coupling to an EDC/NHS activated surface on coupling concentration (i.e., concentration of star–amine in solution), **Fig. 1** shows a plot of the dependence of PEO dry layer thickness, as measured by XPS, as a function of coupling concentration for two specific star molecules. Both star types have functionalities in the hard-sphere regime, with arm mol wt in the range of 5000–10,000 g/mol. The data was obtained for star molecules reacted with EDA only, no ligand.

References

1. Tay, S. W., Merrill, E. W., Salzman, E. W., and Lindon, J. (1989) Activity towards thrombin-antithrombin of heparin immobilized on two hydrogels. *Biomaterials* **10,** 11–15.
2. Lindberg, B., Maripuu, R., Siegbahn, K., Larsson, R., Gölander, C.-G., and Eriksson, J. C. (1983) ESCA studies of heparinized and related surfaces. *J. Coll. Int. Sci.* **95,** 308–321.
3. Trevan, M., ed. (1980) *Immobilized Enzymes: An Introduction and Applications in Biotechnology,* Wiley: New York.
4. Leckband, D. and Langer, R. (1991) An approach for the stable immobilization of proteins. *Biotechnol. Bioeng.* **37,** 227–237.
5. Huang, S.-C., Caldwell, K. D., Lin, J.-N., Wang, H.-K., and Herron, J. N. (1996) Site-specific immobilization of monoclonal antibodies using spacer-mediated antibody attachment. *Langmuir* **12,** 4292–4298.
6. von Sommern, A. P. G., Machielsen, P. A. G. M., and Gribnau, T. C. J. (1993) Comparison of three activated agaroses for use in affinity chromatography. *J. Chromatog.* **639,** 23–31.
7. Massia, S. P. and Hubbell, J. A. (1990) Covalent surface immobilization of Arg-Gly-Asp and Tyr-Ile-Gly-Ser-Arg-containing peptides to obtain well-defined cell-adhesive substrates. *Analyt. Biochem.* **187,** 292–301.
8. Weigel, P. H., Schnaar, R. L., Kuhelnschmidt, M. S., Schmell, E., Lee, R. T., Lee, Y. C., and Roseman, S. (1979) Adhesion of hepatocytes to immobilized sugars: a threshold phenomenon. *J. Biol. Chem.* **254,** 10,830–10,838.
9. Massia, S. P. and Hubbell, J. A. (1991) An RGD spacing of 440 nm is sufficient for integrin alpha V beta 3-mediated fibroblast spreading and 140 nm for focal contact and stress fiber formation. *J. Cell Biol.* **114,** 1089–1100.
10. Kornberg, L. J., Earp, H. S., Turner, C. E., Prockop, C., and Juliano, R. L. (1991) Signal transduction by integrins: Increased protein tyrosine phosphorylation caused by clustering of beta-1 integrins. *Proc. Natl. Acad. Sci. USA* **88,** 8392–8396.
11. Plopper, G. E., McNamee, H. P., Dike, L. E., Bojanowski, K., and Ingber, D. E. (1995) Convergence of integrin and growth factor receptor signaling pathways within the focal adhesion complex. *Mol. Biol. Cell* **6,** 1349–1365.
12. Lopina, S. T., Wu, G., Merrill, E. W., and Griffith-Cima, L. G. (1996) Hepatocyte culture on carbohydrate-modified star polyethylene oxide hydrogels. *Biomaterials* **17,** 559–569.
13. Sorkin, A. and Carpenter, G. (1991) Dimerization of internalized epidermal growth factor receptors. *J. Biol. Chem.* **266,** 23453–23460.
14. Mohammadi, M., Honegger, A., Sorokin, A., Ullrich, A., Schlessinger, J., and Hurwitz, D. R. (1993) Aggregation-induced activation of the epidermal growth factor receptor protein tyrosine kinase. *Biochemistry* **32,** 8742–8748.
15. Carraway, K. L. and Cerione, R. A. (1993) Inhibition of growth factor receptor aggregation by an antibody directed against the epidermal growth factor receptor extracellular domain. *J. Biol. Chem.* **268,** 23,860–23,867.

16. Merrill, E. W. (1993) Poly(ethylene oxide) star molecules: synthesis characterization, and applications in medicine and biology. *J. Biomater. Sci., Polymer Ed.*, in press.
17. Gnanou, Y., Lutz, P., and Rempp, P. (1988) Synthesis of star-shaped poly(ethylene oxide). *Makromol. Chemie* **189,** 2885–2892.
18. Lutz, P. and Rempp, P. (1988) New developments in star polymer synthesis: star-shaped polystyrenes and star-block copolymers. *Makromol. Chemei* **189,** 1051–1060.
19. Wong, J. Y., Kuhl, T. L., Israelachvili, J. N., Mullah, N., and Zalipsky, S. (1997) Direct measure of a tethered ligand-receptor interaction potential. *Science* **275,** 820–823.
20. Harris, J. M., ed. (1992) *Poly(ethylene glycol) Chemistry. Topics in Applied Chemistry* (Katritzky, A. R. and Sabongi, G. J., eds.), Plenum, New York, p. 385.
21. Forni, A., Ganazzoli, F., and Vacatello, M. (1996) Local conformation of regular star polymers in a good solvent: a Monte Carlo study. *Macromolecules* **29,** 2994–2999.
22. Grest, G. S., Kremer, K., and Witten, T. A. (1987) Structure of many-arm star polymers: a molecular dynamics simulation. *Macromol.* **20,** 1376–1383.
23. Kuhl, P. R. and Griffith-Cima, L. G. (1996) Tethered epidermal growth factor as a paradigm for growth factor-induced stimulation from the solid phase. *Nature Med.* **2,** 1022–1027.

3

Nondestructive Evaluation of Biodegradable Porous Matrices for Tissue Engineering

Leoncio Garrido

1. Introduction

Significant advances in transplantation therapy for end stage organ failure or tissue repair have led to a shortage of donor organs and reconstructive tissue. Alternatives to remedy this deficiency are being investigated. In particular, the engineering of tissue using natural or synthetic polymers and isolated cells from the organ or tissue of interest has received increasing attention in the past few years *(1–9)*. The idea behind this approach is that the extracellular matrices made of biodegradable and biocompatible polymers can provide initial structural support for cell adhesion and growth, and gradually reabsorb during organ/tissue generation without any detrimental effect to formed tissue.

In this chapter, we describe the preparation of bioabsorbable extracellular matrices of poly(3-hydroxybutyrate-co-3-hydroxyvalerate) (PHBHV) copolymers by solvent casting *(10,11)*. This family of polymers degrades slowly, and, in the process, produces a less acidic environment than other polymers currently used for tissue scaffolding *(12)*. Because of these properties, PHBHV copolymers may be alternatives to polylactides and polyglycolides for those applications in which prolonged cell support to maintain cell viability and differentiated function is required.

The porous matrix for tissue engineering should have high porosity to be able to accommodate a large number of cells, as well as large interconnected pores to facilitate a uniform distribution of cells and the diffusion of oxygen and nutrients to them *(13,14)*. In this context, it would be advantageous for designing and optimizing extracellular matrices to be able to study the proliferation, viability, and function of cells in well-characterized matrices. To

From: *Methods in Molecular Medicine, Vol. 18: Tissue Engineering Methods and Protocols*
Edited by: J. R. Morgan and M. L. Yarmush © Humana Press Inc., Totowa, NJ

implement this approach, the use of nondestructive methods of analysis is required.

The ability of nuclear magnetic resonance (NMR) to nondestructively determine the composition of a given compound is well known. Moreover, the NMR signal can be spatially encoded using magnetic field gradients *(15)*. NMR imaging (MRI) techniques can produce pictures in which the visual contrast, i.e., difference in brightness between separate regions, is related to the NMR properties of the sample (i.e., spin–lattice, T_1, and spin–spin, T_2, relaxation times) and the concentration of the nuclei being observed (NMR is chemically specific), and it depends on the parameters of the pulse sequence used for the imaging experiment (e.g., flip angle, echo time, repetition time, and so on). The nondestructive and noninvasive nature of the MRI experiment (it does not require sectioning or ultrahigh vacuum, and does not cause radiation damage, as does, for example, scanning electron microscopy) offers a unique opportunity to characterize samples prior to being used, and to study *in situ* time-dependent chemical and morphological processes.

In this chapter, we describe the use of proton MRI to determine the porosity of bioabsorbable extracellular matrices of PHBHV copolymers prepared by solvent casting *(10,11)*. The proton MRI method works by imaging the content of a fluid, in this case water, introduced into the pore space of the matrix *(16,17)*. For the application discussed here, this does not represent a problem, since the extracellular matrices are immersed in aqueous media at several stages during the preparation process, i.e., to remove the salt particles, and, often, prior to cell seeding or implantation.

The size of the features, e.g., pores, agglomerates, on so on, detected by the MRI method depends on the spatial resolution achieved in the image. The acquisition of images with high spatial resolution, or small volume elements (voxels), requires long imaging times. This is because the mass of material giving NMR signal in a voxel decreases as the cube of the increase in linear resolution (increasing resolution means voxels with smaller dimensions). Since the signal-to-noise ratio obtainable from a fixed voxel is proportional to the square root of the time spent signal averaging, there is a sixth-power dependence between the linear resolution and the time required to obtain a fixed signal-to-noise ratio per image voxel. Here, we present a two-steps MRI approach for efficiently screening extracellular matrices to assess the uniformity in porosity and the pore structure. First, a fast MRI protocol (total imaging time less than 30 min), for mapping gradual variations in porosity and macroscopic (>100 µm) heterogeneities or defects in the matrix caused by salt agglomerates and nondissolved polymer particles, is used. Second, high-resolution MR images, 200 × 40 × 40 µm, are acquired (total imaging time of 4.5 h) to illustrate the pore size in those matrices free of defects.

The correlation of the MRI findings with the set of conditions used to prepare the matrices allows us to determine, in a reproducible manner, the values of experimental parameters that yield matrices with the desired micromorphology. Although the specific details are not included in this chapter, the proton MRI methods described here can be used to characterize morphological changes in the bioabsorbable matrices *in situ* during aging. In addition, in vitro cellular proliferation and in vivo tissue growth in the matrices can be monitored with MRI.

2. Materials

2.1. Poly(3-Hydroxybutyrate-co-3-Hydroxyvalerate) (PHBHV) Porous Matrices

1. Poly(3-hydroxybutyrate-co-3-hydroxyvalerate) with 24% hydroxyvalerate; mol wt average, M_w: 297,000 g/mol; mol wt number average, M_n: 104,000 g/mol, and a melting temperature, T_c, of 112°C (Aldrich, Milwaukee, WI).
2. Chloroform Optima (Fisher Scientific, Pittsburgh, PA).
3. Whatman No 5 filter paper (Fisher Scientific).
4. Methanol, ACS reagent (Fisher Scientific).
5. Sodium acetate anhydrous, 99%, cell culture reagent (Sigma, St. Louis, MO).
6. Carbon tetrachloride, 99.9%, ACS reagent (Aldrich).
7. USA standard testing sieves, ASTM-11 specification (Fisher Scientific) with openings 180 μm (No. 80) and 212 μm (No. 70).
8. Phosphate-buffered saline (PBS): 138 mM NaCl, 2.7 mM KCl, 10 mM phosphate buffer (Sigma).

2.2. Nuclear Magnetic Resonance: Instrumentation

1. 2.0 T superconducting magnet (Nalorac, Martinez, CA), horizontal bore of 18.3 cm in diameter, operating at a proton frequency of 84.742 MHz, and interfaced with a SIS spectrometer/imager (Varian, Palo Alto, CA; formerly SISCO, Fremont, CA). The system is equipped with Oxford Instruments (Osney Mead, UK) gradient coils with id 11 cm and a maximum gradient amplitude of 12 g/cm in each of the three orthogonal axis.
2. Home-built NMR probe consisting of a four-turns solenoidal coil, 14.5 mm in diameter, 23 mm in length, tuned and matched to the proton frequency at 2.0 T, mounted on a Lucite® tube (id 10 mm, od 14 mm), with its axis perpendicular to the main static magnetic field (E. I. du Pont de Nemours, Wilmington, DE).
3. The sample holder consists of a 10 mm od NMR tube, 6.5 cm long, with a four-legged plastic insert to hold the sample in the desired position (*see* **Fig. 1**).

2.3. Nuclear Magnetic Resonance: Pulse Sequence

The pulse sequence used in the present experiments consists of a conventional two-dimensional multislice spin–echo acquisition, with short

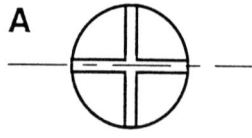

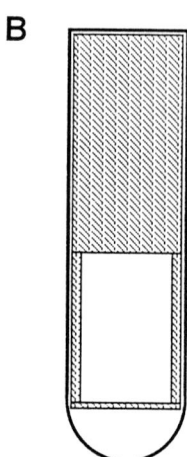

Fig. 1. Sketch illustrating the sample holder used to maintain the matrix in the desired position during imaging. **(B)** shows a cross-section of the holder, as indicated by dashed line in **(A)**.

echo time to minimize the effects of magnetic susceptibility-induced signal attenuation *(16,17)*.

1. For low resolution images, the pulse sequence consists of two sinc-shaped, 2 ms long, 90- and 180-degree radiofrequency (rf) pulses. The echo time is 11 ms and the repetition time is 3 s. The slice thickness is 1 mm, and gap between slices is 0.1 mm. Total number of slices varies between 13 and 17 (sufficient to cover the entire sample). The digital resolution in the images is 128 X by 128 Y pixels, corresponding to 86 X by 86 Y µm. The total imaging time is 26.5 min.
2. For high-resolution images, the same pulse sequence is used, with the following changes: the length of the rf pulses is 5 ms, the echo time is 35 ms, the repetition time is 2 s, and the slice thickness is 200 µm, with no gap. The digital resolution in the images is 256 X by 256 Y pixels, corresponding to 39 X by 39 Y µm. The total imaging time is 4.5 h min.

2.4. Image Processing and Display

1. Spectrometer/imager software, VNMR.
2. NIH image, version 1.55, National Institutes of Health, Bethesda, MD.

3. Methods
3.1. Purification of PHBHV

Prior to being used for the preparation of matrices, PHBHV copolymer, with 24% HV, is purified (*see* **Note 1**).

1. Dissolve 4 g of the copolymer in 200 mL of hot chloroform, in a Soxhlet apparatus, for 4 h under nitrogen atmosphere.
2. Filter the solution to remove impurities.
3. Precipitate the copolymer in the filtered solution with 300 mL of methanol.
4. Filter and wash the precipitate with cold (4°C) dH_2O at least 2×.
5. Dry the precipitate under vacuum to constant weight for approx 48 h. Store the copolymer in a desiccator under vacuum at room temperature (22°C) until use.

3.2. Preparation of Salt Particles

1. Sieve sodium acetate particles with sieve no. 70 (nominal opening of 212 μm).
2. Sieve the fraction having a particle size <212 μm with sieve no. 80 (nominal opening of 180 μm), and collect the salt fraction retained.

3.3. Preparation of PHBHV Porous Matrices

PHBHV porous matrices are prepared by a solvent casting technique described elsewhere *(10,11)*. To obtain highly porous matrices (>90% porosity) with large interconnected pores, high concentration of salt particles (size ~200 μm) in the final mixture is required, which leads to several problems. The method was optimized to improve the mechanical stability of the porous matrix and to reduce the agglomeration and sedimentation of salt particles during casting (*see* **Note 2**).

1. To prepare the PHBHV porous matrices, dissolve 2 g of purified copolymer in 20 mL of a mixture of carbon tetrachloride (42% v/v) and chloroform (58% v/v) (*see* **Note 2**), and mix with 8 g of salt (sodium acetate anhydrous, with particle size of 196 ± 16 μm).
2. Stir and cast the mixture on a polypropylene mold with dimensions of 5.2 × 4.2 × 1 cm.
3. Cover with tissue paper to reduce the evaporation rate of the solvent. Let the mixture dry for 72 h at room temperature. After drying, the thickness of the resulting cake is ~3.5 mm.
4. Cut the cake in pieces with dimensions of approx 15 × 5 × 3.5 mm^3, and immerse the samples in dH_2O (change every 12 h) at 37°C in a shaking water bath to remove the salt, until constant weight is reached (3–5 d).
5. Store dry matrices in a desiccator under vacuum at room temperature until use.

3.4. MR Imaging of Water Content

To map the internal empty volume of a matrix, using proton MRI, the sample must be filled with water. Thus, the signal intensity in the MR images repre-

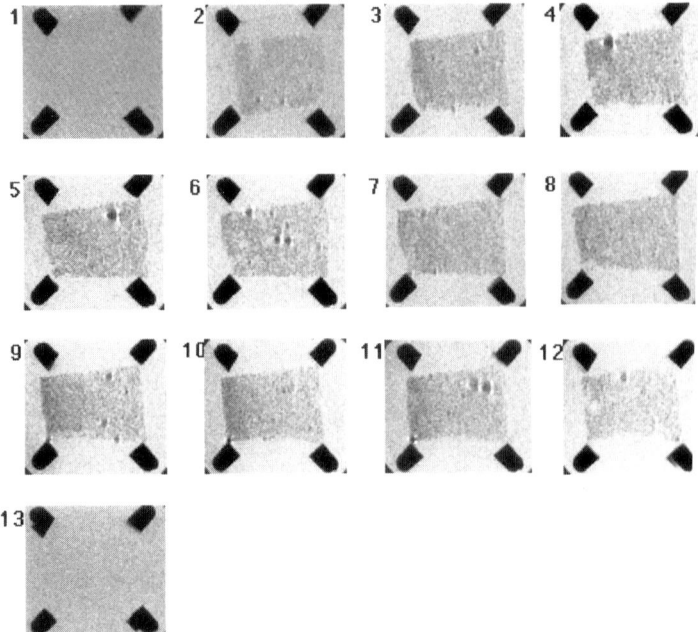

Fig. 2. Multislice proton MR images corresponding to a PHBHV (24% HV) matrix acquired with the pulse sequence parameters described in **Subheading 2.3.1.** Regions of low NMR signal attributed to nondissolved polymer particles are observed in images 4, 5, 6, 9, 11, and 12. Also, large voids are observed in images 2, 4, 11, and 12. The three bright spots surrounding each of the polymer particles is an artifact caused by the magnetic susceptibility difference in the water concentration.

sents the concentration of water at each location in the matrix. The porosity at a specific location in the sample is given by the fractional image intensity at that location normalized to the intensity of pure water, which represents 100% porosity.

3.4.1. Detection of Defects

1. Place the sample with the plastic sample holder in the NMR tube and add PBS until the sample is fully immersed.
2. Fill the open space in the sample by using vacuum to remove the air and facilitate the penetration of water in the porous structure (*see* **Note 3**).
3. Place the NMR probe with the sample in the imager and proceed to acquire a low-resolution MR image, using the pulse sequence with the set of imaging parameters described in **Subheading 2.3.1.**
4. After the acquisition is completed, Fourier-transform the data using the imager software (*see* **Note 4**). **Figure 2** shows an example of a multislice data set corre-

Biodegradable Porous Matrices

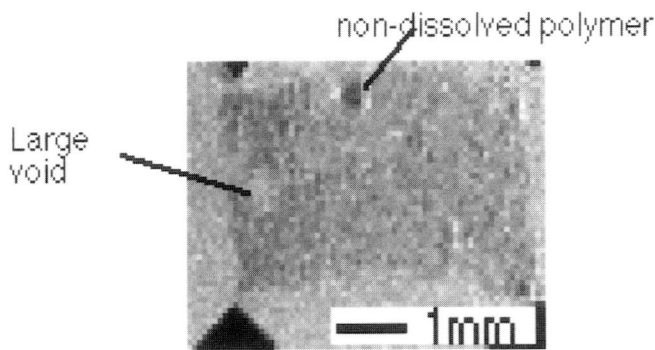

Fig. 3. Detail of image #12 shown in **Fig. 2**, illustrating two classes of defects observed in PHBHV matrices prepared by the solvent casting method described in **Subheading 3.3.**

sponding to a PHBHV porous matrix. The presentation of the data in two dimensions, using a gray scale, allows visual discrimination of signal voids caused by nondissolved polymer particles and very large pores formed by trapped air or solvent vapor during casting (see **Fig. 3**).

5. At this stage, it can be decided if samples are accepted or discarded, depending on the size and distribution of the defects observed.

3.4.2. Measurement of Porosity

After screening the samples for macroscopic defects, the quality of a sample is further assessed by measuring the total porosity and its variation across the sample.

1. Read image data with NIH Image software and smooth sharp variations in signal intensity (see **Note 5**).
2. To determine the open porosity of the sample, normalize the signal intensity in the image by the mean signal intensity measured in a region with only water (100% porosity). The information obtained is presented in two ways to facilitate the evaluation of this parameter.
   a. Histograms (see **Note 6**). After selecting a region of interest, defined by the sample contour, a histogram of this region is used to construct the distribution function for porosity (see **Fig. 4**).
   b. Surface plots (see **Note 6**). A surface plot allows easy visualization of gradual variations of porosity across the sample (see **Fig. 5**).

3.4.3. In-Plane Pore Size and Connectivity

1. High-resolution MR images are acquired using the pulse sequence with the set of imaging parameters described in **Subheading 2.3.2.**

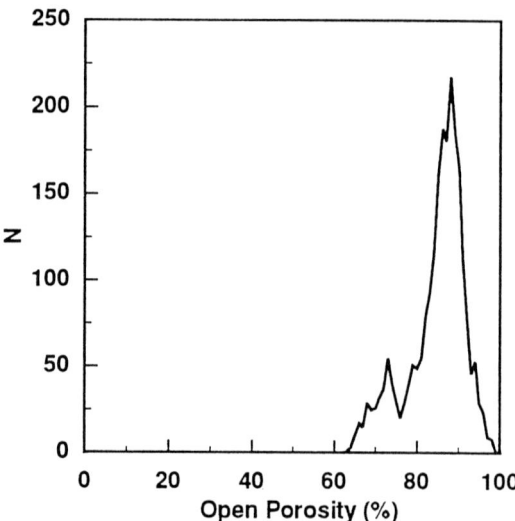

Fig. 4. Histogram representation of the signal intensity data corresponding to a region of interest containing 95% of the total cross-section of the sample shown in image 10 (*see* **Fig. 2**). The horizontal axis is the signal intensity normalized to the mean intensity of the water distribution and, therefore, represents the NMR-derived porosity distribution function.

Fig. 5. Surface plot representation of cross-section of the sample shown in image 10 (**Fig. 2**). The plot illustrates the gradual variation in porosity across the sample.

Biodegradable Porous Matrices

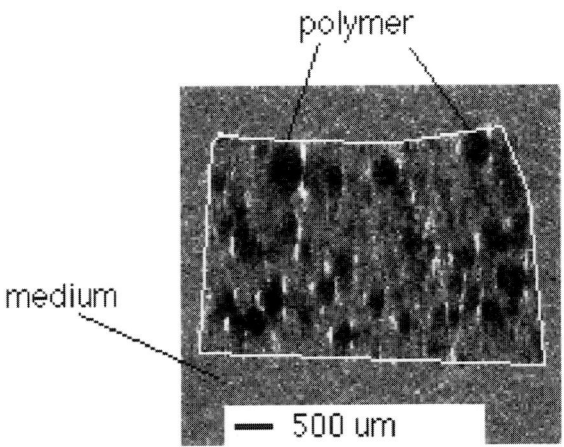

Fig. 6. High-resolution MR image (200 × 39 × 39 μm^3 per voxel) of water in a PHBHV (24% HV) porous matrix acquired with the pulse sequence imaging parameters described in **Subheading 2.3.2.**

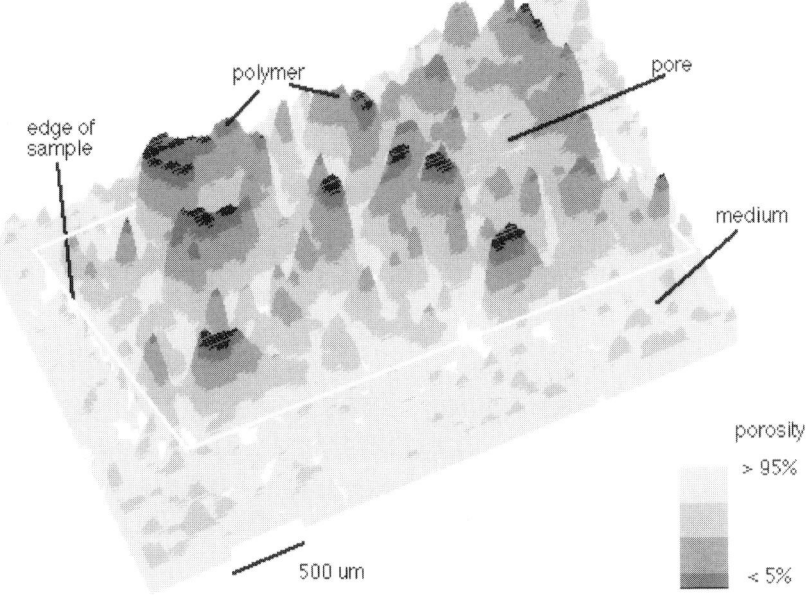

Fig. 7. Surface plot representation of cross-section of the sample shown in **Fig. 6**. The plot illustrates the distribution and connectivity of empty volume in an image plane with a thickness of 200 μm.

2. After Fourier transforms the raw data using the imager software (*see* **Fig. 6**), the images are processed and displayed using NIH Image software, as described in **Subheading 3.4.2.** (*see* **Fig. 7**).

4. Notes

1. Poly(hydroxybutyrate-hydroxyvalerate) copolymers are of bacterial origin. Generally, polymer samples as received from supplier contain cellular debris from synthesis, which can be removed by the treatment described.
2. The agglomeration of salt particles was reduced using sodium acetate, instead of sodium chloride. To minimize the sedimentation of salt particles during casting, the density of the organic phase was adjusted to match that of the sodium acetate anhydrous (1.528 g/cm^3) by mixing carbon tetrachloride and chloroform with densities of 1.594 and 1.472 g/cm^3, respectively, in a ratio of 42:58. In addition, we found that the amount of polymer should be at least 20% of the total solids present in the initial mixture, in order to obtain matrices dimensionally stable after removing the salt.
3. Instead of vacuum, the impregnation of the PHBHV matrices with water is facilitated by filling the samples first with a fluid having low interfacial tension with the microscopic surfaces of the matrix, such as methanol or ethanol. Then, the alcohol in the matrix is replaced by immersing the sample in water, allowing sufficient time for the exchange between the two fluids to occur. The immersion of samples in alcohol does not add unnecessary steps to the preparation of porous matrices for cell implantation, since this is a usual procedure for sterilization of the matrices.
4. The data-processing software varies between NMR spectrometers, depending on the manufacturer, but the outcome should be independent of the instrument used, assuming that the experimental conditions and processing parameters used are the same. Similarly, instead of the NIH Image software, other image processing software commercially available may be used to process and display the MRI data with comparable results, provided that the same algorithms and procedures are used.
5. The smoothing of the image data will reduce the artifacts caused by regional differences in magnetic susceptibility, and will allow a better visualization of the variations of signal intensity attributed to the changes in the sample geometry.
6. The width and shape of the porosity distribution function, as well as the percentage of regional variation in porosity with respect to the mean value of porosity may be used to set a criteria for quality control of the extracellular matrices.

References

1. Langer, R. and Vacanti, J. (1993) Tissue engineering. *Science* **260,** 920–926.
2. Vacanti, J. P., Morse, M., Saltzman, M. A., Domb, W. M., Perez-Atayde, A., Langer, R. (1988) Selective cell transplantation using bioabsorbable artificial polymers as matrices. *J. Pediatr. Surg.* **23,** 3–9.
3. Mooney, D. J., Cima, L., Langer, R., Johnson, L., Hansen, L. K., Ingber, D. E., Vacanti, J. P. (1992) Principles of tissue engineering and reconstruction using polymer-cell constructs. *Mater. Res. Soc. Symp. Proc.* **252,** 345–352.
4. Cima, L. G., Ingber, D. E., Vacanti, J. P., and Langer, R. (1991) Hepatocyte culture on biodegradable polymeric substrates. *Biotechnol. Bioeng.* **38,** 145–158.

5. James, K. and Kohn, J. (1996) New biomaterials for tissue engineering. *MRS Bull.* **21,** 22–26.
6. Lu, L. and Mikos, A. G. (1996) The importance of new processing techniques in tissue engineering. *MRS Bulletin* **21,** 28–32.
7. Ripamonti, U. and Duneas, N. (1996) Tissue engineering of bone by osteoinductive biomaterials. *MRS Bull.* **21,** 36–39.
8. Dunn, M. G. (1996) Tissue-engineering strategies for ligament reconstruction. *MRS Bull.* **21,** 43–46.
9. Borkenhagen, M. and Aebischer, P. (1996) Tissue-engineering approaches for central and peripheral nervous-system regeneration. *MRS Bull.* **21,** 59–61.
10. Mikos, A. G., Wald, H., Sarakinos, G., Leite, S., and Langer, R. (1992) Biodegradable cell transplantation devices for tissue regeneration. *Mat. Res. Soc. Symp. Proc.* **252,** 353–358.
11. Mikos, A. G., Thorsen, A. J., Czerwonka, L. A., Bao, Y., Langer, R., Winslow, D. N., Vacanti, J. P. (1994) Preparation and characterization of poly(L-lactic acid) foams. *Polymer* **35,** 1068–1077.
12. Holland, S. J., Jolly, A. M., and Tighe, B. J. (1987) Polymers for biodegradable medical devices. II. Hydroxybutyrate-hydroxyvalerate copolymers: hydrolytic degradation studies. *Biomaterials* **8,** 289–295.
13. Mikos, A. G., Sarakinos, G., Lyman, M. D., Ingber, D. E., Vacanti, J. P., and Langer, R. (1993) Prevascularization of porous biodegradable polymers. *Biotechnol. Bioeng.* **42,** 716–723.
14. Rotem, A., Toner, M., Sangeeta, B., Foy, B. D., Tompkins, R. G., and Yarmush, M. L. (1994) Oxygen is a factor determining in vitro tissue assembly: effects on attachment and spreading of hepatocytes. *Biotechnol. Bioeng.* **43,** 654–660.
15. Mansfield, P. and Morris, P. G. (1982) NMR imaging in biomedicine, in *Advances in Magnetic Resonance.* (Waugh, J. S., ed.) Suppl 2. Academic, New York.
16. Ackerman, J. L., Garrido, L., Ellingson, W. A., and Weyand, J. D. (1987) The use of nmr imaging to measure porosity and binder distribution in green-state and partially sintered ceramics, in *American Ceramic Society Proceedings Joint Conference on Nondestructive Testing of High-Performance Ceramics* (Vary, A. and Snyder, J., eds.), American Ceramic Society, Boston, pp. 88–113.
17. Ackerman, J. L., Garrido, L., Moore, J. R., Pfleiderer, B., and Wu, Y. (1992) Fluid and solid state mri of biological and nonbiological ceramics, in *Magnetic Resonance Microscopy. Methods and Applications in Materials Science, Agriculture and Biomedicine* (Blümich, B. and Kuhn, W., eds.), VCH, Weinheim, Germany, pp. 237–260.

4

Fabrication of Biodegradable Polymer Foams for Cell Transplantation and Tissue Engineering

Peter X. Ma and Robert Langer

1. Introduction

Organ transplantation has been successful since the early 1960s as a result of the success in immunologic suppression in the clinical setting *(1)*, and has saved, and is continuing to save, countless lives, but is far from a perfect solution to tissue losses or organ failures. By far the most serious problem facing transplantation is donor scarcity. Approximately 30,000 Americans need liver transplantation each year, but only about 10% of the patients have the chance to receive a donated liver transplant *(2)*. There is a total of approx 100,000 people in the United States with transplants, but there are more than 1 million with biomedical implants *(3)*. Tissue engineering and cell transplantation are fields emerging to resolve the missing tissue and organ problems.

There are three strategies in cell transplantation and tissue engineering:

1. The use of isolated cells or cell substitutes to replace those cells that supply the needed function, including various manipulations before the cell infusion *(4)*;
2. The delivery of tissue-inducing substances, such as growth factors, to targeted locations *(5)*;
3. Growing cells on three-dimensional matrices or devices *(2)*.

The advantage of the use of isolated cells is to avoid surgical complications. The disadvantages include cell death and loss of function because many cells are substrate-dependent, i.e., they can grow and function only when they are attached on an appropriate substrate. The uses of tissue-inducing substances are dependent on the development of delivery systems that can deliver bioactive molecules and the economical large-scale production of the bioactive agents. Therefore, there has been very active research aimed at the regeneration of new tissues and organs by culturing mammalian cells on three dimensional matrices or devices *(2)*. Some of these devices work as an immunoprotective barrier, which allows nutrients and metabolic wastes to diffuse through, but not antibodies or immune cells. The core of this technology is a semipermeable membrane with a well-defined mol wt cutoff or size cutoff *(6,7)*. These technologies are leading to the development of immobilized cell systems, either as implants or extracorporeal devices *(6,8–10)*. Some other three-dimensional matrices work as templates to guide cells to grow, to synthesize biologically functional molecules and extracellular matrices in three dimensions *(11,12)*. The unique feature in this approach is the control over cell adhesion and function by choosing or developing the right substrate materials, and the control over new tissue or organ structure by scaffold design and processing.

Matrix materials can be further classified as natural materials or synthetic materials. They can also be classified as degradable or nondegradable. The most frequently used matrix materials are either natural macromolecules, such as extracellular matrix components and polysaccharides *(13,14)*; engineered macromolecules from natural origins, such as collagen–glycosaminoglycan copolymers *(15)*; and synthetic polymers, either degradable *(11,12)* or nondegradable *(16)*. However, nondegradable materials carry a risk of infection and permanent connective-tissue reaction *(17)*. Natural extracellular matrix materials provide the advantage of cellular recognition. However, there is less control over their mechanical properties, biodegradability, and batch-to-batch consistency. They might exhibit immunogenicity *(18)*. Many of them are also limited in supply and can therefore be costly. Synthetic biodegradable polymers offer control over structure and properties such as chemical and physical structure, crystallinity, hydrophobicity, degradation rate, and mechanical properties. They can be processed into various shapes and microstructures, such as desired surface area, porosity, pore size, and pore size distribution. Their surface properties can be altered to adapt to the biologic requirements for cell adhesion, growth, and function. Therefore, synthetic biodegradable polymers have been widely used as vehicles for cell transplantation and templates for tissue engineering *(2)*.

There are several basic requirements for biodegradable polymer scaffolds. They need to be highly porous, with the proper pore size distribution to give cells room to grow, and paths for nutrients and metabolites to permeate. They should have a high surface area (surface:volume ratio) to give cells sufficient surface to adhere. They should have the right degradation rate to match the tissue regeneration process. They should have the structural integrity and mechanical strength to maintain the desired shape before the new tissue takes over. They should be biocompatible and their degradation products should not be cytotoxic. Their surface properties should be adaptable to the requirements for cell attachment, growth, and function.

Biodegradable Polymer Foams

Various biodegradable polymers have been processed into scaffolds to engineer new tissues and organs *(11,12,19–23)*. Biodegradable foam processing via salt-leaching technology has been reported earlier *(11)*. The foams have also been used in bone- and liver-tissue engineering *(24,25)*. The fabrication processes have been further modified, improved, and standardized to better satisfy scaffolding requirements, multi-specimen preparation, and reproducibility. The poly(L-lactic acid) (PLLA) foams fabricated using the improved procedure are currently being used for liver tissue engineering *(26–28)*. In this chapter, the improved biodegradable polymer foam processing will be described in detail, using PLLA as an example. However, the method should be of general use for other biodegradable polymer or nondegradable polymers.

2. Materials

2.1. Salt Crystals

1. NaCl crystals: Mallinckrodt, Paris, KY (*see* **Note 3**).
2. Grinder: Scienceware Micro-Mill Grinder, Fisher Scientific, Pittsburgh, PA.

2.2. Polymer Solution

1. Polymer: poly(L-lactic acid) with an inherent viscosity of approx 1.6, Boehringer Ingelheim, Ingelheim, Germany (*see* **Notes 1** and **5**).
2. Solvent: chloroform, EM Science, Gibbstown, NJ.
3. 20-mL glass scintillation vials: Cole-Parmer Instrument, Niles, IL.

2.3. Surface-Coating Solution

1. Poly(vinyl alcohol) (PVA) partially hydrolyzed from poly(vinyl acetate): 75% hydrolyzed, average mol wt 3,000, Aldrich, Milwaukee, WI (*see* **Note 6**).
2. Membrane filter: Schleicher and Schuell (Keene, NH) nylon membrane filters, pore size 0.45 μm, diameter 47 mm, VWR Scientific, Boston, MA.

2.4. Fabrication

1. Clean room ionizing air gun: model M-1205wc, Terra Universal, Anaheim, CA.
2. Teflon PFA vials: 18 mm id, 5 mL, Cole-Parmer.
3. Falcon polystyrene tissue culture dishes: 35 × 10 and 60 × 15 mm, Becton Dickinson, Lincoln Park, NJ.
4. Blotting paper: Gel Blot Paper, size 20 × 20 cm, Schleicher and Schuell.

2.5. Sterilization and Packing

1. Ethylene oxide sterilization agent: Anprolene gaseous sterilant, H. W. Andersen, Haw River, NC.
2. Sterilization pouches: Chex-all II instant sealing pouch, Propper Manufacturing, Long Island, NY.
3. Aluminum sealing pouch: Kapak Pouches, 6.5 × 8 in., Kapak, Minneapolis, MN.

3. Methods

3.1. Salt Particles

NaCl crystals are ground into smaller particles using a micro-mill grinder (*see* **Note 3**). The mixture of the ground particles is separated with sieves into different size ranges. The salt particles of the desired size range are collected and are sealed with parafilm in a glass bottle to prevent moisture adsorption and particle aggregation.

3.2. Polymer Solution

The biodegradable polymers, contained in glass bottles and sealed with parafilm, are stored in freezers (<-20°C) to prevent hydrolysis. They are taken out and put on the lab bench for at least 1 h to equilibrate with ambient temperature, before opening the bottles to prevent water from condensing on the polymers, which could lead to weight change (water adsorption) and molecular weight decrease (hydrolysis). The PLLA is weighed accurately into a scintillation vial. In a fume hood, the calculated amount of chloroform is added into the glass vial to make a 5% solution (w/v) (*see* **Note 6**). The vials are kept on an orbital shaker (approx 100 rpm) at 25°C, and are vortexed from time to time to facilitate polymer dissolution. It takes about 2 d to dissolve the polymer well.

3.3. Preparation of Surface Coating Solution

PVA is weighed into a 500-mL bottle with deionized water, and stirred with a magnetic stirrer on a stirring hot plate, to make a 1% (w/v) solution. The heater is periodically switched between off and low to speed up the dissolution. However, the temperature is kept lower than 60°C. High temperature can cause decomposition and color change (yellow or brown). It takes about 2 d to dissolve the PVA well. The PVA solution is then filtered through a nylon membrane filter (pore size 0.45 μm), using a regular lab vacuum filtration setup.

3.4. Fabrication of Foam

The whole process to fabricate one batch of samples takes approx 2 wk after the polymer solutions and salt particles have been prepared. A schematic illustration of the fabrication procedure is shown in **Fig. 1**. The detailed fabrication processes are described below:

3.4.1. Preparation

Teflon vials are washed with solvent to remove residual polymer and stuck salt particles. Dry vials are blown with high-pressure air to clean away loose dust. An ionizing air gun is connected with a nitrogen gas tank. The gas pressure is set at 10 psi with a gas regulator. The vials are ionized thoroughly by aiming and circling the nozzle at the entire surface area of both the inside and outside of each vial, to eliminate static.

Biodegradable Polymer Foams

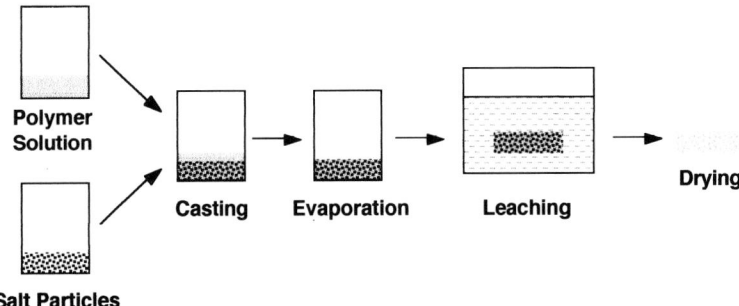

Fig. 1. Schematic illustration (not proportional) of basic fabrication processes of biodegradable polymer foam with salt-leaching technique.

3.4.2. Casting

0.400 ± 0.005 g of salt of the desired size are added into each Teflon vial (*see* **Note 3**). The salt surface is leveled by first tapping the vial on edge to roughly spread out salt, then pressing it down onto a thin plate adapted on top of a vortex shaker. The vibration setting of the vortex shaker should be adjusted to shake individual grains of salt without disturbing already leveled salt. A flat metal plate with three adjustable legs (homemade) is set in a fume hood, and is leveled with a lab level. The vials are transferred carefully onto the leveled metal plate and are placed as far apart from each other as possible. 0.24 mL prepared PLLA solution is pipeted onto the salt. In order to evenly distribute the polymer solution in the salt, it is better to drip a few drops equally apart on the salt instead of one large drop (five drops usually work fine, one in the center and four at corners of an imaginary square); or allow the polymer solution to drip slowly onto salt, using the free hand to rotate the vial, so that the solution covers nearly all surface area. The polymer solution should not drip when transferring from the stock solution to the vials, nor should the solution drip down sides of vials.

3.4.3. Evaporation

The vials are covered with their caps to slowly evaporate the solvent for 24 h. The covers are removed carefully, without disturbing the salt and polymer distribution. The solvent continues to evaporate from the mixture of polymer solution and salt in the fume food for another 24 h. The metal plate should be covered to prevent foreign matter from settling on the polymer salt mixture (**Fig. 2**).

3.4.4. Leaching

The polymer–salt composite disks are removed from the vials onto a clean sheet of paper. Handling carefully with forceps, the disks are placed

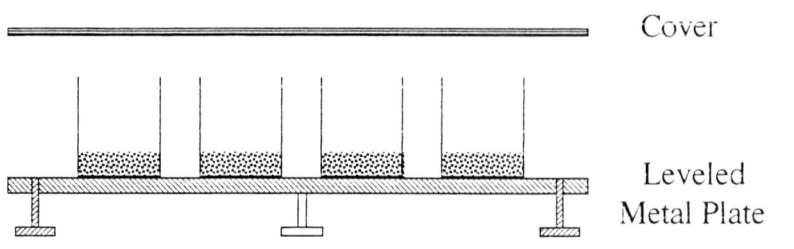

Fig. 2. Second-stage evaporation setup.

into 35 × 10 mm culture dishes (one piece per dish), and five dishes are banded together with elastic bands. An 1-L beaker is filled with distilled water. Holding a banded group of dishes under the water surface with gloves on, each dish is opened slightly to allow air bubbles to escape. The foams at the edges should not be squeezed by the covers. Then the banded dishes are allowed to sink to the bottom. When the beaker is full, a clean glass plate is put on top of the dishes to prevent floating and all the dishes are submerged in the water. The beaker is placed on an orbital shaker and the shaker is turned on (setting roughly at 75 rpm). The foams are leached for 2 d, and the water is changed 3×/d. The foams are taken out one by one (handling lightly and carefully) and placed on blotting paper, smooth side down. The paper absorbs water, then the foams are placed in 60 × 15-mm culture dishes, rough side down (five pieces per dish). The covers are placed on, and air drying under a hood occurs for 24 h (*see* **Note 4**).

3.4.5. Surface Coating

The dishes containing the half-dry foams (rough side up) are placed on a flat surface. 5 mL 1% PVA solution is slowly added to each dish by pipeting over the foams (*see* **Note 6**). No piece of foam should be allowed to float over others or to overlap. The dish is covered and coating occurs for 24 h without disturbing.

3.4.6. Drying

The PVA-coated foams are removed and placed on blotting paper, smooth side down. The paper is allowed to absorb the excess PVA solution. The foams are placed in clean 60 × 15-mm culture dishes, rough side down (five pieces a dish). The dishes are covered and air-drying occurs for 24 h. The dishes are banded with elastic bands and transferred in freeze-drying flasks. The foams are lyophilized for 72 h. Porous PVA-filtrated PLLA disks, 1-mm thick and 18 mm in diameter, with approx 95% porosity, are obtained (*see* **Notes 2–4**).

3.5. Sterilization and Packaging

Defective foams are discarded. The remaining foams are rearranged in dishes, based on convenience for future use (such as five in a dish), and are covered. The foams (in their dishes) are sealed in gas-permeable sterilization pouches (Chex-all II instant-sealing pouches). The pouches are placed in a plastic liner bag. A plastic gas-release bag containing one ethylene oxide ampule is unrolled and placed in the liner bag alongside the seam. Excess air is pressed out of the liner bag and the bag is tied shut with a twist-tie. The ampule is broken inside the bag and the bag is placed in a sterilization metal box under a fume hood. The lock lid to the box is closed; ethylene oxide sterilizes the foams in the bag for 24 h. The bags are untied and the contents are shaken into the sterilization box; ethylene oxide is allowed to diffuse away for 24 h. All materials are left in the hood until the end of the sterilization period. **Caution:** The operator should have protective gloves and a face mask on while handling the sterilization agent.

The sterilized foams (in the gas-permeable pouches) are placed on a sheet of dry filter paper into a Kapak aluminum sealing pouch, and the pouches are thermally sealed with an iron or a pouch sealer to prevent moisture adsorption and degradation of the fabricated biodegradable polymer foams.

4. Notes

1. The methods described are of general use for fabricating porous structures, either from degradable polymers or from nondegradable polymers, as long as the polymers can be dissolved in a good solvent, but not in water. Poly(lactic acid), poly(glycolic acid), and their copolymers are a series of polymers that are approved by FDA for human use as suture materials and some implantable devices. They are considered biocompatible, with minimal tissue reaction *(29,30)*. This chapter is aimed at the need for cell transplantation devices and porous scaffolds for tissue engineering, therefore, a biocompatible and biodegradable polymer PLLA is used as an example, and the sterilization and packaging procedures are included. With these techniques, highly porous foams can be reproducibly fabricated, and multiple samples can be fabricated in one batch. These techniques do not require complicated equipment or large lab space, and therefore are useful for research sample preparations. These methods can be easily extended to large scale productions. The methods can also be extended for melt processing, as well.
2. The geometry and size of porous devices can also be altered to satisfy specific needs. In this chapter, a cylindrical vial is used as an example to make round porous disks. In order to fabricate a device of a particular size and shape, a mold of that particular size and shape is required. Lamination techniques can be used to fabricate very thick devices by use of many layers of thin-contour, shaped porous sheets *(31)*. The lamination technique also permits the fabrication of more complex scaffolds for complex tissue and organ engineering.
3. The porosity, pore shape, pore size, and pore-size distribution can be controlled with the salt:polymer ratio, particle shape, particle size, and size distribution. For example, increasing the initial salt:polymer ratio can increase the porosity of the

foams to be fabricated. Increasing the salt particle size can increase the pore size of the foam *(11)*. NaCl is easily available, biocompatible and economical, and therefore is used in this example. However, other salts and substances can also be used. The porosity, surface:volume ratio, pore size, and pore-size distribution can be measured with mercury-intrusion porosimetry.
4. The microstructure of the devices can also be manipulated by various treatments used in polymer science and engineering. For example, heat treatment can be used to control the crystallinity of the porous polymer foams *(11)*, and thereby their mechanical properties and even degradation rate. It is known that water has easier access to the amorphous regions. Therefore, polymer chains in the amorphous region are believed to degrade first *(12)*. By manipulating the crystallinity with thermal treatment, degradation rate and mechanical properties can be manipulated to a certain extent.
5. However, in order to drastically change the properties, such as degradation rate, it would be wise to modify the chemical structure or to choose a different polymer. For example, the degradation rate can be changed an order of magnitude by just changing from poly(glycolic acid) to poly(L-lactic acid). A whole range of degradation rates can be achieved by choosing poly(glycolic acid-co-lactic acid) with varying monomer ratios. There are also other synthetic biodegradable polymers, such as polyanhydrides *(32,33)* and poly(ortho ethers) *(34)*, to choose from. There are endless possibilities in modifying existing polymers and synthesizing new polymers.
6. There are countless choices for surface coating materials. PVA is used here as a hydrophilic coating for easier filtration of cell suspensions, for cell loading in cell transplantation, and cell seeding in tissue engineering. There are many other hydrophilic materials that can be used for the same purpose. There are also biologically active components, such as peptides, proteins, and growth factors, that can be potentially attached on the surfaces to control cell attachment and differentiation.

References

1. Couch, N., Wilson, R., Hager, E., and Murray, J. (1966) Transplantation of cadaver kidneys: experience with 21 cases. *Surgery* **59**, 183–188.
2. Langer, R. and Vacanti, J. (1993) Tissue engineering. *Science* **260**, 920–926.
3. Nerem, R. M. and Sambanis, A. (1995) Tissue engineering: from biology to biological substitutes. *Tissue Eng.* **1**, 3–13.
4. Brittberg, M., Lindahl, A., Nilsson, A., Ohlsson, C., Isaksson, O., and Peterson, L. (1994) Treatment of deep cartilage defects in the knee with autologous chondrocyte transplantation. *N. Engl. J. Med.* **331**, 889–895.
5. Aebischer, P., Salessiotis, A., and Winn, S. (1989) Basic fibroblast growth factor released from synthetic guidance channels facilitates peripheral nerve regeneration across long nerve gaps. *J. Neurosci. Res.* **23**, 282–289.
6. Colton, C. (1995) Implantable biohybrid artificial organs. *Cell Transplant* **4**, 415–436.

7. Darquy, S. and Reach, G. (1985) Immunoisolation of pancreatic B cells by microencapsulation. An in vitro study. *Diabetologia* **28,** 776–780.
8. Nyberg, S., Shirabe, K., Peshwa, M., Sielaff, T., Crotty, P., Mann, H., et al. (1993) Extracorporeal application of a gel-entrapment, bioartificial liver: demonstration of drug metabolism and other biochemical functions. *Cell Transplant* **2,** 441–452.
9. Reach, G. (1993) Bioartificial pancreas. *Diabet. Med.* **10,** 105–109.
10. Colton, C. and Avgoustiniatos, E. (1991) Bioengineering in development of the hybrid artificial pancreas. *J. Biomech. Eng.* **113,** 152–170.
11. Mikos, A. G., Thorsen, A. J., Czerwonka, L. A., Bao, Y., Langer, R., Winslow, D. N., and Vacanti, J. P. (1994) Preparation and characterization of poly(l-lactic acid) foams. *Polymer* **35,** 1068–1077.
12. Ma, P. X. and Langer, R. (1995) Degradation, structure and properties of fibrous nonwoven poly(glycolic acid) scaffolds for tissue engineering, in *Polymers in Medicine and Pharmacy* (Mikos, A. G., Leong, K. W., Radomsky, M. L., Tamada, J. A., and Yaszemski, M. J., eds.), MRS, Pittsburgh, pp. 99–104.
13. Bell, E., Rosenberg, M., Kemp, P., Gay, R., Green, G., Muthukumaran, N., and Nolte, C. (1991) Recipes for reconstituting skin. *J. Biomech. Eng.* **113,** 113–119.
14. Krewson (née Beaty), C. E., Chung, S. W., Dai, W., and Saltzman, W. M. (1994) Cell aggregation and neurite growth in gels of extracellular matrix molecules. *Biotechnol. Bioeng.* **43,** 555–562.
15. Yannas, I. V. (1994) Applications of ECM analogs in surgery. *J. Cell. Biochem.* **56,** 188–191.
16. Pongor, P., Betts, J., Muckle, D., and Bentley, G. (1992) Woven carbon surface replacement in the knee: independent clinical review. *Biomaterials* **13,** 1070–1076.
17. Gristina, A. (1987) Biomaterial-centered infection: microbial adhesion versus tissue integration. *Science* **237,** 1588–1595.
18. Cima, L., Vacanti, J., Vacanti, C., Ingber, D., Mooney, D., and Langer, R. (1991) Tissue engineering by cell transplantation using degradable polymer substrates. *J. Biomech. Eng.* **113,** 143–151.
19. Ma, P. X., Schloo, B., Mooney, D. and Langer, R. (1995) Development of biomechanical properties and morphogenesis of in vitro tissue engineered cartilage. *J. Biomed. Mater. Res.* **29,** 1587–1595.
20. Freed, L. E., Marquis, C. J., Nohria, A., Emmanual, J., Mikos, A. G., and Langer, R. (1993) Neocartilage formation *in vitro* and *in vivo* using cells cultured on synthetic biodegradable polymers. *J. Biomed. Mater. Res.* **27,** 11–23.
21. Vacanti, C., Kim, W., Upton, J., Vacanti, M., Mooney, D., Schloo, B., and Vacanti, J. (1993) Tissue-engineered growth of bone and cartilage. *Transplantation Proc.* **25,** 1019–1021.
22. Organ, G., Mooney, D., Hansen, L., Schloo, B., and Vacanti, J. (1993) Enterocyte transplantation using cell-polymer devices to create intestinal epithelial-lined tubes. *Transplantation Proc.* **25,** 998–1001.
23. Shinoka, T., Ma, P. X., Shum-Tim, D., Breuer, C. K., Cusick, R. A., Zund, et al. (1996) Tissue-engineering heart valves: autologous valve leaflet replacement study in a lamb model. *Circulation* **94 (Suppl.),** II-164–II-168.

24. Thomson, R., Yaszemski, M., Powers, J., and Mikos, A. (1995) Fabrication of biodegradable polymer scaffolds to engineer trabecular bone. *J. Biomater. Sci. Polym. Ed.* **7,** 23–38.
25. Mooney, D., Park, S., Kaufmann, P., Sano, K., McNamara, K., Vacanti, J., and Langer, R. (1995) Biodegradable sponges for hepatocyte transplantation. *J. Biomed. Mater. Res.* **29,** 959–965.
26. Cusick, R. A., Lee, H., Sano, K., Pollok, J. M., Utsunomiya, H., Ma, P. X., Langer, R., and Vacanti, J. P. (1997) The effect of donor and recipient age on engraftment of tissue engineered liver. *J. Pediat. Surg.* **32(2),** 357–360.
27. Lee, H., Cusick, R. A., Browne, F., Kim, T. H., Ma, P. X., Utsunomiya, H., Langer, R., and Vacanti, J. P. Increased angiogenesis by local delivery of bFGF increases survival of transplanted hepatocytes, to be published.
28. Lee, H., Cusick, R. A., Utsunomiya, H., Ma, P. X., Langer, R., and Vacanti, J. P. Effect of implantation site on hepatocytes heterotopically transplanted on biodegradable polymer scaffolds, to be published.
29. Matlaga, B. and Salthouse, T. (1983) Ultrastructural observations of cells at the interface of a biodegradable polymer: Polyglactin 910. *J. Biomed. Mater. Res.* **17,** 185–197.
30. Craig, P., Williams, J., Davis, K., Magoun, A., Levy, A., Bogdansky, S., and Jones, J. J. (1975) A biologic comparison of polyglactin 910 and polyglycolic acid synthetic absorbable sutures. *Surg. Gynecol. Obstet.* **141,** 1–10.
31. Mikos, A., Sarakinos, G., Leite, S., Vacanti, J., and Langer, R. (1993) Laminated three-dimensional biodegradable foams for use in tissue engineering. *Biomaterials* **14,** 323–330.
32. Leong, K., Brott, B., and Langer, R. (1985) Bioerodible polyanhydrides as drug-carrier matrices. I: Characterization, degradation, and release characteristics. *J. Biomed. Mater. Res.* **19,** 941–955.
33. Domb, A. and Langer, R. (1987) Polyanhydrides: I. Preparation of high molecular weight polyanhydrides. *J. Polymer Science.* **25,** 3373–3386.
34. Choi, N. S. and Heller, J. (1978) US Patent **4,093,**709.

5

Formation of Highly Porous Polymeric Foams with Controlled Release Capability

A Phase-Separation Technique

S. Kadiyala, H. Lo, and K. W. Leong

1. Introduction

The success of many tissue engineering applications depends on a scaffold with the suitable physical properties, one of which might be a macroporous structure that allows cellular ingrowth. Such a porous implant further raises the possibility of delivering chemotactic or growth factors to influence the course of cell proliferation and differentiation *in situ*. The scaffolds can also be preseeded with cells to accelerate tissue growth or repair. Even in the absence of these payloads, they still provide the benefit of introducing the minimal amount of foreign material into the tissue. Furthermore, by making the porous scaffold, or foam, from biodegradable polymers, the regenerated tissue would be rid of any synthetic component, leading to a more functional biological equivalent, and eliminating concerns of long-term tissue compatibility.

Because of the inherent limitations of polymer strength and the requirement for high porosity, it is not feasible to use the foams in conditions in which they are subjected to large mechanical loads. However, it is important that the foams can maintain their form under physiologic conditions and do not collapse prior to any tissue ingrowth. For cellular infiltration to occur, the pore sizes would have to be larger than the cells—actually, substantially larger, considering the effect of tortuosity on cell migration. It is unlikely that pore sizes below 40 mm can support tissue ingrowth, though such foams may still be used to deliver growth factors or to act as tissue barriers. The optimal pore size range of the foam is probably tissue-dependent. For instance, it has been hypothesized that, in orthotopic sites, pore sizes below 400 mm lead to bone formation and pore sizes above 400 mm lead to fibrous tissue ingrowth *(1–4)*. In addition to pore size, porosity, which more reflects the interconnectivity of the foam, is also important. High porosity maximizes the volume of tissue ingrowth and minimizes the amount of polymer used. It also facilitates transport of nutrients and cellular waste products. Another parameter is the pore morphology, which may be meaningful in favoring the ingrowth of certain cell types.

There exist a large number of methods for producing a macroporous structure from polymers *(5)*, many of which are developed for non-biological applications. We have studied the technique of foam fabrication by phase separation from a homogenous polymer solution *(6,7)*, a versatile method that appears capable of addressing the important issues of a desirable porous scaffold.

1.1. Theory of Foam Fabrication by Phase Separation

The foam fabrication technique is based on the phase separation of the polymer from a homogenous solution in the spinodal region of the phase diagram. As the polymer solution is cooled down to the solidification temperature of the solvent, the homogenous polymer solution goes through a phase-separation region, called the spinodal region where two distinct liquid phases emerge: a polymer phase and a solvent phase. The size of the phase domains depend on the temperature and the duration at which the solution is maintained at that temperature *(8–10)*. The closer the temperature to the homogenous region and the shorter the duration, the finer is the phase structure obtained. The morphology at any stage can be preserved by rapid quenching of the solution to prevent any coarsening of the microstructure. When the solvent is subsequently removed without disturbing this morphology, a highly porous structure will remain. For our purpose, the key to success is that the solvent can be removed without aqueous washing, because that would leach out any imbedded bioactive agents. This can be accomplished by using a solvent that has a high vapor pressure at room temperature, and hence can be removed by sublimation.

In addition to the factors of temperature and time mentioned, there are other factors that can be varied to change the pore structure of the foam, including the solution concentration of the polymer, the mol wt of the polymer, the chemical composition of the polymer, and the choice of the solvent. To predict the microstructure of the foam made from a particular polymer-solvent, detailed phase diagrams for each system need to be constructed by studying the phase behavior under different temperature, pressure, and annealing periods. In the absence of these phase diagrams, the foam microstructure can be optimized empirically, and a variety of foams can be fabricated by changing some of the factors described above *(6,7,11)*.

The choice of solvents for this technique of foam fabrication is based on the following considerations: The polymer of interest should be soluble in the solvent; the melting point of the solvent should be low enough to minimize the thermal degradation of the polymer and the incorporated bioactive agent; the solvent should have a low energy of sublimation, and its freezing point should be high enough to allow ease of sublimation; the solvent should not pose a toxicity risk at the residual amounts left at the end of the fabrication process. After an extensive search of most of the organic solvents, we have tested naphthalene and phenol, two solvents that satisfy most of these criteria. The melting points of naphthalene and phenol are 80°C and 42°C, respectively. Although the temperatures are higher than those that have been shown to denature many bioactive proteins in aqueous conditions, if the fabrication can be performed under anhydrous conditions, much bioactivity, in principle, can still be retained *(12,13)*. Thus, any protein added to the molten polymer needs to be in an anhydrous form, easily achieved, for instance, by lyophilization.

2. Materials

1. Naphthalene and phenol (99% pure): Aldrich, Milwaukee, WI.
2. (PLA): Zimmer, Warsaw, IN.
3. Atomizer: General Glassblowing, Richmond, CA.
4. Mercury porosimetry: Porous Materials, Ithaca, NY.

3. Methods

The foam fabrication methodology will be described here with specific examples *(6,7,11)* to illustrate the points:

3.1. Method A: Foam Fabrication by Atomization (Naphthalene System)

Chromatography sprayers are used to provide a fine, uniform spray. The sprayer consists of an atomizer top attached with a screw top and O ring to glass reservoir flasks. The sprayer is connected to a nitrogen gas tank with a pressure regulator. The spray pattern, which in turn influences the morphology of the foam, can be adjusted by the spraying pressure and the coverage of the vent hole. In the following procedure, the vent hole is fully covered.

1. Into a 125-mL reservoir flask containing a magnetic stirrer, 57 g of solid anhydrous naphthalene was added, followed by addition of another 3 g of anhydrous (PLLA) (mol wt 500,000) over the naphthalene solid.

    Note that the order of addition will reduce the chance that the polymer adheres to the surface of the flask, thereby facilitating dissolution of the polymer.

2. Stopper the flask loosely. Heat the flask containing the solid naphthalene and polymer to 90°C by a stir plate. Stir the contents at 200 rpm for 20 min, or until a homogeneous melt is obtained.

    Wrapping the flask with a heating tape preheated at 90°C would ensure an uniform temperature of the molten mixture and facilitate polymer dissolution. If

necessary, add fresh naphthalene to compensate for any loss caused by sublimation and evaporation of naphthalene. If a drug or any excipient is to be incorporated into the foam, the agent should be added into the melt at this time. Most hydrophilic drugs will not be soluble in the naphthalene melt. Stir another 5–10 min to achieve uniform suspension of drugs for each addition. If decomposition of material is noted (e.g., yellowing), the stock should be remade.
3. Wrap the atomizer top with a heating tape at a temperature of 90°C.
   This is a critical step. Allow plenty of time to heat the atomizer tip. Insufficient heating will cause deposition of naphthalene on the sprayer nozzle and clog the spray. Should that happen, a heat gun can be used to dissolve the naphthalene.
4. Cut a Teflon sheet to the desired shape for the deposition of naphthalene mixture. The Teflon surface allows easy removal of samples following the spray. Connect the heated atomizer top to the reservoir flask. Connect warm nitrogen (40–50°C) to the sprayer at a pressure 5 psig, and begin to spray by closing the vent hole with thumb. A disk of 3 cm in diameter and 3 mm in thickness can be obtained in less than a minute.
5. Place samples into a vacuum oven heated to 50°C. Apply vacuum overnight at less than 1 torr to remove the naphthalene.

3.2. Method B: Foam Fabrication Method by Casting (Naphthalene System)

Sample cups made of tinplate with slip-on covers (VWR Scientific, Boston, MA) were used for casting naphthalene mixture.

1. A stock solution of 5 wt% PLLA in naphthalene is prepared according to **step 1** of Method A.
2. Heat a 1-oz tin cup to 90°C on a heat plate.
3. Charge 10 g of stock solution into the heated tin cup from **step 2**. Close the cover immediately.
4. Immediately transfer the tin cup to liquid nitrogen. It will only take less than 10 s to solidify the naphthalene melt.
5. Place samples into a vacuum oven heated to 50°C. Apply vacuum overnight at less than 1 torr to remove the naphthalene.

3.3. Method C: Foam Fabrication Method by Casting (Phenol System)

The procedures are similar to Method B, with the following exceptions. The temperature of the stock solution of the mold should be heated to 50°C, instead of to 90°C. The sublimation of phenol should be done at room temperature, instead of 50°C in the oven.

3.4. Other Considerations

The principle of foam fabrication has been demonstrated by the work described in this review, but the relationship between the thermodynamics of phase separation and the final porous architecture of the foam remains to be established. A detailed study of the phase diagram of a given polymer–solvent system will be required to allow prediction of the microstructure of the foam, based on the chosen parameters. Also, additional work is required to evaluate the biocompatibility of the foams.

One might argue that if a controlled-release function is desired, one can inject drug-loaded microspheres into the foam. However, it might be difficult to uniformly distribute the microspheres in the foam and might compromise the goal of leaving the most volume for tissue ingrowth. Even not considering the merit of drug incorporation, the phase-separation technique is still attractive, compared to some of the other methods, in terms of simplicity and versatility.

4. Notes

1. Characterization of Foams. Two methods that can be utilized to characterize the foams are mercury porosimetry and scanning electron microscopy (SEM). Mercury porosimetry can yield data on the porosity and pore-size distribution. In tissue-engineering applications, only pore sizes above a certain threshold (10 μm) should be used to calculate the porosity, since pores below that threshold may not contribute to the healing response. SEM is an effective tool to determine the pore morphology, orientation, and, to a limited extent, pore sizes.
2. Modifications to the system. The effect of changes of mol wt of the polymer, concentration of the polymer in the fabrication mix, the solvent used, and the polymer used are shown in **Fig. 1A–H**. As can be seen in these SEM micrographs, the foam microstructure can be readily changed by changing any of the aforementioned parameters.
3. Change of foam morphology. **Fig. 1E** illustrates the result of directional cooling on the foam morphology. To obtain this kind of morphology, the polymer solution is poured into a mold preheated to the temperature of the polymer solution, and cooling is applied from one surface of the mold, while the other surfaces of the mold are kept insulated. Such an architecture would be useful in cases like tendon regeneration, in which the fibers are aligned in a longitudinal direction.
4. Inclusion of bioactive moieties. The bioactive agent to be included in the foam is added to the polymer mix in the molten state. As mentioned previously, when operating at high temperatures, it is essential that the system be maintained in an anhydrous state to retain the bioactivity of the protein. Sustained release of small molecules, such as bromothymol blue and sulphorhodamine *(6)*, or large proteins, such as alkaline phosphatase *(7)* or bone morphogenetic protein (BMP) *(11)*, have been achieved. The release kinetics of the embedded agent is governed by many parameters. The hydrophobicity/hydrophilicity of the agent relative to the solvent and the polymer determines its partition or dissolution in the polymer phase. Hydrophilic compounds are likely to just adsorb on the polymer phase of the foam and will not be released in a sustained manner. The loading level (wt% of agent in the polymer phase) determines the fraction of the agent that is dis-

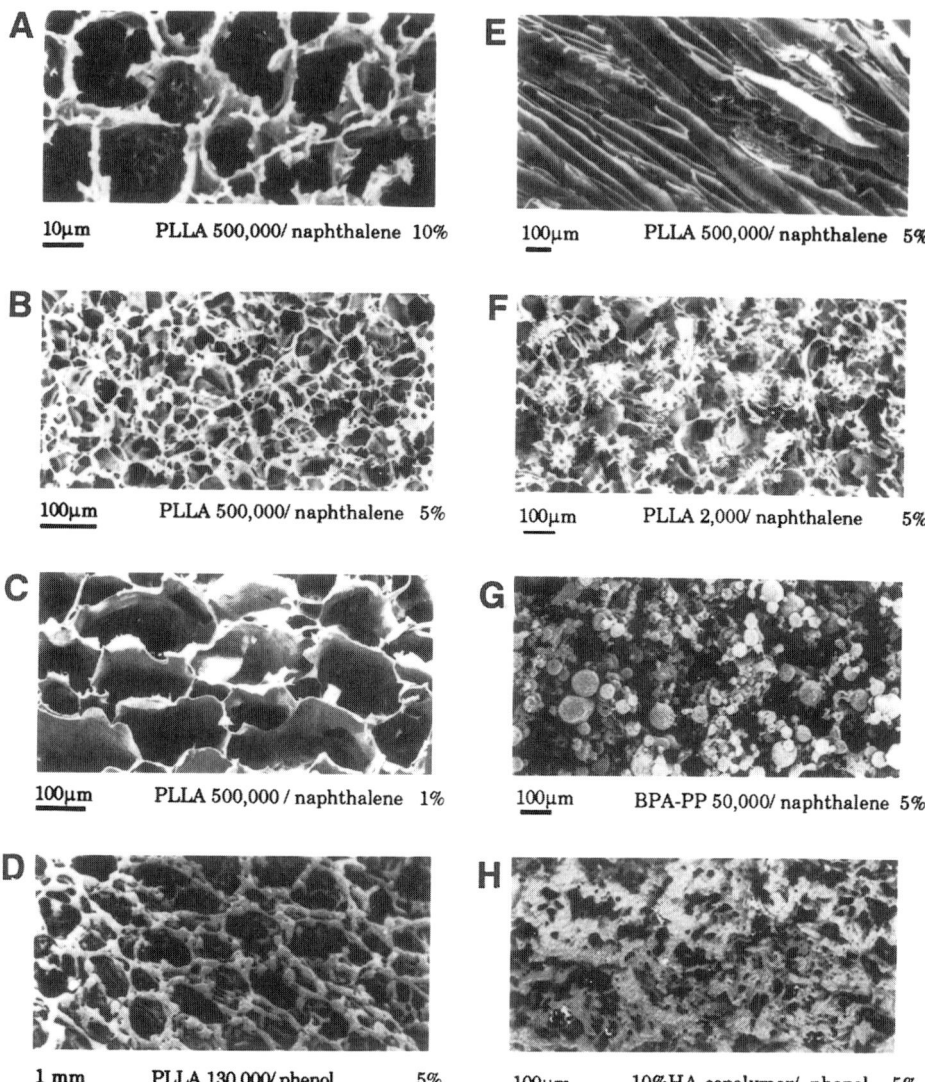

Fig. 1. Scanning electron micrographs of assorted biodegradable foams made by phase separation. Foams are identified by polymer type and mol wt, solvent used, and percentage of polymer in solvent. Foams A, B, C, and G were fabricated by atomization; D, E, F, and H were fabricated by casting. Reproduced with permission from **ref. 7**.)

solved in the polymer phase and the fraction that is adsorbed on the surface. A biphasic release profile can be expected in this situation, the first phase being release of the surface exposed drug, and the second phase controlled by diffusion and matrix degradation. Morphology of the foam—not so much the pore size and

Highly Porous Polymeric Foams

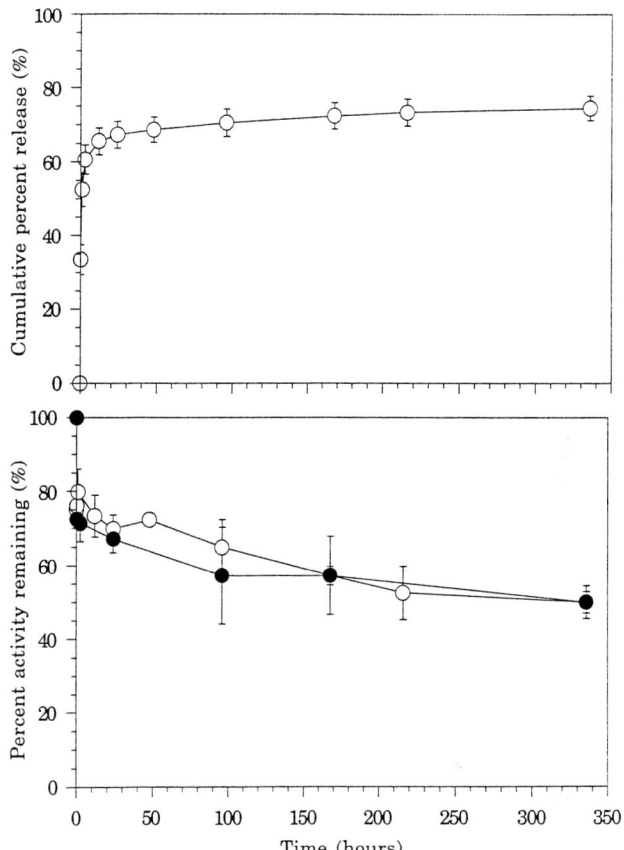

Fig. 2. **(A)** In vitro release curve of FITC-labeled alkaline phosphatase (FITC-AP) from 10 wt% FITC-AP-loaded PLLA foams (–○–, % AP released) in 0.1 M pH 7.4 phosphate buffer at 37°C (upper curve). **(B)** Changes in bioactivity of FITC-AP after being released from PLLA foams (–○–) with pure FITC-AP placed under the same conditions as a control (lower curve) –●–, AP control. Reproduced with permission from Lo **ref. 7**.)

porosity, but the thickness of the individual cells, affects the diffusional distance for release. Finally, degradation of the polymer may overshadow many of these factors if the degradation rate is high.

The release profile of alkaline phosphatase from a PLA foam fabricated with a PLA–naphthalene system, along with assessment of the bioactivity of the released alkaline phosphatase is shown in **Fig. 2**. The large burst and biphasic release seen in this case is probably caused by the 10% loading level, as explained above. A sustained and steady release, however, can be obtained with a lower drug loading, as illustrated in the delivery of BMP *(11)*.

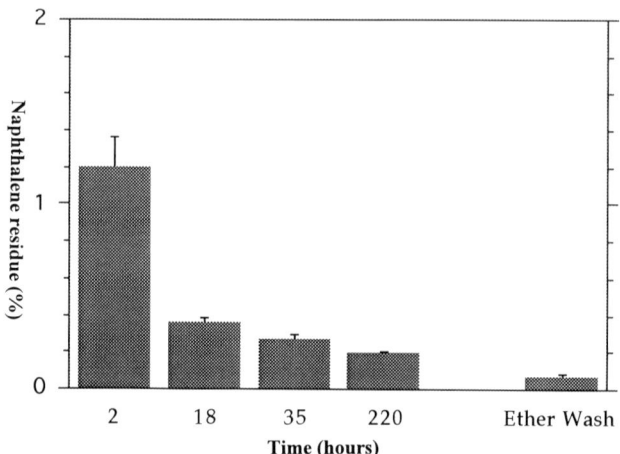

Fig. 3. Levels of residual naphthalene in the foams after sublimation and ether wash.

5. Biocompatibility issue. A major concern for this fabrication technique, as is the case with all other techniques that might require contact with organic solvents, is the residual levels of solvents that might influence the biocompatibility of the implant. The residual amounts of naphthalene in the foam as a function of the sublimation time are shown in **Fig. 3**, and translate to µg quantities in a typical implant. Moreover, the hydrophobicity of naphthalene would mean that it would be released only when the polymer degrades, over a long period of time. In vitro culturing of cells in the foam *(6)* and in vivo implantation *(11)* have revealed no acute problems of biocompatibility. However, long-term studies will need to be conducted. One way to reduce the amount of residual naphthalene might be to include a secondary vacuum exposure at a higher temperature and higher vacuum. Naturally, this may not be advisable if delicate bioactive agents are embedded in the foam.

References

1. Ducheyne, P. (1985) Success of prosthetic devices fixed by ingrowth or surface interaction. *Acta Orthop. Belg.* **51,** 144–161.
2. Schliephake, H., Neukam, F. W., and Klosa, D. (1991) Influence of pore dimensions on bone ingrowth into porous hydroxyapatite blocks used as bone graft substitutes. A histometric study. *Int. J. Oral. Maxillofac. Surg.* **20,** 53–58.
3. Eggli, P. S., Muller, W., and Schenk, R. K. (1988) Porous hydroxyapatite and tricalcium phosphate cylinders with two different pore size ranges implanted in the cancellous bone of rabbits. A comparative histomorphometric and histologic study of bony ingrowth and implant substitution. *Clin. Orthop. Rel. Res.* **232,** 127–138.
4. Collier, J. P., Mayor, M. B., Chae, J. C., Surprenant, V. A., Surprenant, H. P., and Dauphinais, L. A. (1988) Macroscopic and microscopic evidence of prosthetic fixation with porous-coated materials. *Clin. Orthop.* **235,** 173–180.

5. Kadiyala, S., Lo, H., and Leong K. W. (1994) Biodegradable polymers as synthetic bone grafts, in *Bone Formation and Repair* (Brighton, C. T., Friedlander, G., Lane, J. M., eds.), AAOS, Boca Raton, FL.
6. Lo, H., Kadiyala, S., Guggio, S. E., and Leong, K. W. (1996) Poly(L-lactic acid) foams with cell seeding and controlled-release capacity. *J. Biomed. Mater. Res.* **30**, 475–484.
7. Lo, H., Ponticiello, M. S., and Leong, K. W. (1996) Fabrication of controlled release biodegradable foams by phase separation. *Tissue Eng.* **1**, 15–28.
8. Tanaka, H. and Nishi, T. (1987) Direct determination of the probability determination of concentration in polymer mixtures undergoing phase separation. *Phys. Rev. Lett.* **59**, 692–695.
9. Siggia, E. D. (1979) Late stages of spinodal decomposition in binary mixtures. *Phys. Rev.* **A20**, 595–605.
10. Aubert, J. H. (1988) Interfacial tension of demixed polymer solutions. *Polymer.* **29**, 118–122
11. Kadiyala, S., Lo, H., Ponticiello, M. S., Reddi, A. H., and Leong, K. W. (1996) Bone induction achieved by controlled release of BMP from PLA/Hydroxyapatite foams, in *Transactions of the Fifth World Biomaterials Congress,* Toronto, p. 289.
12. Zaks, A. and Klibanov, A. M. (1984) Enzymatic catalysis in organic media at 100 degrees C. *Science* 1249–251.
13. Zaks, A. and Klibanov, A. M. (1988) Enzymatic catalysis in nonaqueous solvents. *J. Biol. Chem.* **263**, 3194–3201.

6

Magnetic-Induced Alignment of Collagen Fibrils in Tissue Equivalents

Timothy S. Girton, Narendra Dubey, and Robert T. Tranquillo

1. Introduction

The use of reconstituted type I collagen gel as a scaffold for engineered soft tissues is a highly attractive prospect, given that collagen is the principal component of the extracellular matrix (ECM) in vivo, providing a mechanically suitable and information-rich scaffold for cell–ECM interactions. It has the advantage that cells can be directly entrapped within the comprising collagen fibrils as they grow into an entangled network from a cell containing solution of monomeric type I collagen. These tissue equivalents have the further advantage that the collagen fibrils can be aligned by applying a magnetic field during fibrillogenesis. Then, through a process termed "contact guidance," the cells align with the fibrils by directing their motility. Such alignment is characteristic of many tissues, and may provide microstructural and mechanical cues for regulation of cell phenotype and function, as well as influence the gross mechanical properties of the tissues. Recent research in our laboratory has used magnetic-induced alignment in the fabrication of tissue-equivalents, notably circumferential alignment in tubes, and longitudinal alignment in rods (patent pending). The former is aimed at reproducing the architecture of the arterial media; the latter is aimed at providing a bridge that promotes directed axonal growth between severed nerve ends. These tissue engineering applications exploit the finding of Torbet and Ronziere *(1)* in their cell-free studies that forming fibrils tend to align in the plane normal to the direction of the field (because of the negative diamagnetic anisotropy of collagen) and parallel to the mold surfaces (because of an uncharacterized interfacial effect). The eventual entanglement with other growing fibrils mechanically stabilizes the

From: *Methods in Molecular Medicine, Vol. 18: Tissue Engineering Methods and Protocols*
Edited by: J. R. Morgan and M. L. Yarmush © Humana Press Inc., Totowa, NJ

magnetic-induced alignment, in that the alignment is sustained after the formed network is removed from the magnetic field.

There is significant potential for using magnetic processing of type I collagen for engineering the alignment in tissue equivalents. It is simple: The mold is merely placed with appropriate orientation in the magnetic field during fibrillogenesis; it is reproducible, as documented in our previous studies of axially aligned tissue-equivalent slabs *(2)*; and it is controllable, via the rate of fibrillogenesis (e.g., via temperature) and the magnetic field strength *(2)*. The method is suitable for mass production, since it is essentially scale-independent, the aligning effect of the magnetic field being independent of the mold dimensions, except for the gap size, which must be small, so that the interfacial effect, which selects the alignment direction from the plane, is significant. In practice, "small" is on the order of a few millimeters, which, fortunately, is suitable for many tissue-engineering applications.

The only practical method reported outside of our group for imposing alignment in tissue equivalents is stretch-induced alignment *(3)*. Other methods reported for (acellular) collagen gel that might be adapted include ordered convection, driven by a collagen concentration gradient *(4)*, or temperature gradient *(5)* and weak electric currents *(6)*; however, the fibril alignment in these methods was very nonuniform.

Although unnecessary for exploiting magnetic-field-induced collagen fibril alignment, knowledge of the mechanism is of fundamental interest. A collagen molecule exhibits negative diamagnetic anisotropy primarily because of its peptide bonds. In a magnetic field, a molecule possessing a negative diamagnetic anisotropy experiences a torque that tends to orient the molecule normal to the field direction. However, the ratio of aligning energy to randomizing thermal energy (Brownian motion) is only 5×10^{-3} for a single collagen molecule at room temperature in the highest magnetic field available (about 20 T). It has been proposed that this ratio should exceed 6, in order to have more than 75% of the maximum magnetic alignment *(1)*. It was therefore concluded that the high degree of alignment observed is a result of the summation of diamagnetic anisotropies of the collagen molecules (monomers) that comprise the fibril (polymer). For a rotationally symmetric, rigid polymer comprised of monomers whose symmetry axes are parallel, the diamagnetic anisotropy of the polymer can be considered to a good approximation as the sum of the diamagnetic anisotropies of the monomers. Thus, for such monomers that self-assemble in a linear fashion (fibrin, actin, and tubulin being other biopolymer examples), the additivity of the diamagnetic anisotropies causes the polymer to experience an aligning torque in the presence of a strong magnetic field. However, once extensive fibrillogenesis has occurred, the fibrils are too constrained within the entangled network to be further aligned. Alignment, therefore, is restricted to a window of time during fibrillogenesis in which the growing fibrils are sufficiently large to experience a torque able to overcome Brownian motion, yet small enough that they are not too constrained by neighboring fibrils. This time-window can thus be extended by conditions that slow the rate of fibrillogenesis, such as decreased temperature.

In this chapter, we detail our methods for generating magnetic-induced alignment in tissue equivalents, in particular, circumferentially aligned tubes and longitudinally aligned rods. We discuss the preparation of the collagen gel, generation of the required geometry, and optimization and control of the desired alignment.

2. Materials

2.1 Monomeric Collagen Solution

1. Vitrogen 100 monomeric type I collagen (Collagen, Palo Alto, CA) (*see* **Note 1**).
2. 1 M HEPES buffer solution (Gibco-BRL, Grand Island, N.Y.).
3. 0.1 N endotoxin free NaOH solution (Sigma, St. Louis, MO.).
4. 10X minimum essential medium (MEM) (Gibco-BRL).
5. Fetal bovine serum (FBS) (HyClone, Logan, UT).
6. Penicillin–streptomycin solution (5000 U of penicillin and 5000 µg streptomycin/mL in 0.85% saline solution) (Gibco-BRL).
7. L-Glutamine solution: 29.2 mg/mL of L-glutamine in 0.85% saline solution (Gibco-BRL).

2.2. Hardware

1. Molds (comprised entirely of autoclavable parts)
   a. Tubes: an 8-mm diameter Teflon rod, two 12-mm diameter 4-mm-tall rubber cylinders, and a 12-mm id clear polypropylene tube (**Fig. 1A**).
   b. Rods: a 3-mm id, 2-cm-long Teflon tube, and two 3-mm id, 5-mm-long Teflon caps (**Fig. 1B**).
2. 6-cc syringe with 21-gage needle.
3. 21-gage needle.
4. A horizontal bore (preferably 6-cm diameter or larger), high-strength (preferably 4 T or higher) magnet (*see* **Note 2**).
5. A watertight acrylic tube, whose dimensions allow it to be placed in the bore of the magnet and hold the desired number of molds (*see* **Note 3**).
6. A large-capacity hot-water heater and recirculator (Fisher, Springfield, NJ).
7. A filled icebox large enough to hold all of the molds.

3. Methods

3.1. Collagen Solution Preparation

The preparation of the collagen solution is rather straightforward and simple. However, the temperature of the reagents and the order of mixing, as well as

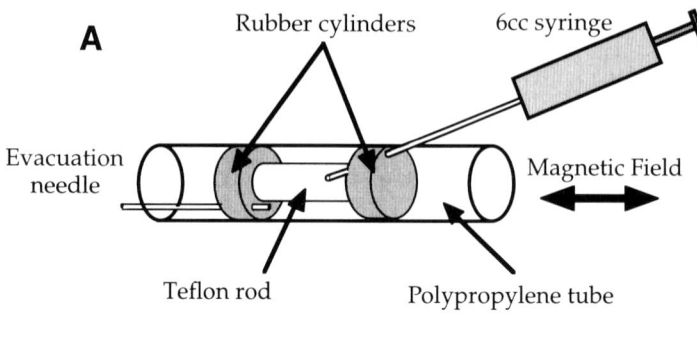

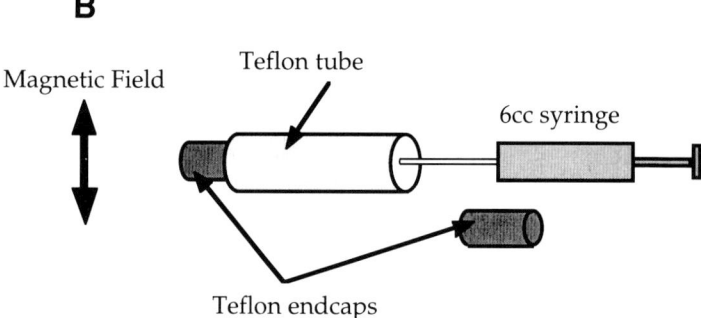

Fig. 1. (A) The tubular mold comprised of two rubber cylinders, a Teflon rod, and a polypropylene tube. Shown are the two needles used to fill the mold, as well as the necessary orientation of the mold relative to the magnetic field for circumferential alignment. (B) The rod mold comprised of two Teflon cylindrical endcaps and a Teflon tube. Shown is the needle and syringe used to fill the mold, as well as the necessary orientation of the mold relative to the magnetic field for longitudinal alignment.

the source of the collagen, all prove to be important. (Vitrogen 100 is preferred, because its fibrillogenesis rate is relatively slow *(7)*, thereby enhancing the degree of magnetic-induced alignment, as previously explained.) If the temperature of the reagents and the resulting collagen solution are kept close to 5°C, turbidity experiments (a good indication of both the rate and extent of fibrillogenesis) indicate that no significant fibrillogenesis occurs. This allows time for the injection of the solution into the molds and their transfer to the magnetic field before significant fibrillogenesis. The temperature maintained during fibrillogenesis is discussed subsequently.

To make 1 mL of 2 mg/mL collagen solution, thoroughly mix the following cold (5°C) ingredients, in the order listed below, in a sterile container that is kept on ice until fibrillogenesis is desired:

1. 20 µL HEPES-buffer solution (helps maintain the correct pH).
2. 132 µL 0.1 N NaOH (raises the pH of the solution from 2 to about 7.6).
    Vitrogen 100 collagen fibrillogenesis begins once both the pH and the temperature are raised. The ideal pH for fibrillogenesis is about 7.6; increasing temperature increases the rate of fibrillogenesis. Note that Vitrogen 100 is 3 mg collagen/mL and shipped at 5°C and pH 2.
3. 50 µL 10X MEM (used primarily as a source of nutrients for cells entrapped or postseeded (*see* **Note 4**), but also provides a phenol red pH indicator).
4. 60 µL FBS (used primarily for cell survival; however, FBS may alter the diameter and structure of the collagen fibrils formed).
5. 1 µL Penicillin–streptomycin solution (used to ensure sterility of the gel).
6. 10 µL L-glutamine solution (provides cells with needed nutrients, so can be omitted if there are no cells present).
7. 677 µL Vitrogen 100 (*see* **Note 5**).

3.2. Preparation of Molds

After the preparation of the collagen solution, the autoclaved mold pieces are assembled and the mold filled.

1. Tube mold: Fit a sterile rubber cylinder onto each end of the Teflon rod and slide the resulting piece into the clear polypropylene tube (**Fig. 1A**). Pierce one of the rubber cylinders with a 21-gage needle, so that the needle tip is exposed in the tubular cavity between the Teflon and the polypropylene outer cylinder. Fill the 6-cc syringe with the collagen solution, and pierce the other rubber cylinder with the syringe needle as before. Fill the tubular cavity with the collagen solution in the syringe (the other needle provides an evacuation hole for the air).
2. Rod mold: Insert a Teflon endcap into one end of the Teflon tube (**Fig. 1B**). Fill the tube with the collagen solution and cap the end with the remaining Teflon endcap.

After filling, wrap the molds with Parafilm and place them on ice.

3.3. Magnetic-Induced Alignment

Once the molds are full, they are placed in the temperature control chamber, and with the use of a hot-water recirculator, the temperature of the samples is maintained during fibrillogenesis. As noted previously, the degree of alignment obtained increases with decreasing temperature; however, for Vitrogen 100 fibrillogenesis to occur within 2 h when prepared as described, the temperature must typically be 33–37°C.

Placement of the molds in the proper orientation in the magnetic field is essential for achieving the desired fibril alignment (cf. **Figs. 1A,B**). Collagen will align in the direction parallel to the surfaces that define the gap that is in the plane normal to the magnetic field direction; thus, for circumferential alignment of the collagen in the tubular mold, the axis of the mold must be placed parallel to the direction of the magnetic field; for longitudinal alignment in the collagen rod, the axis of the mold must be placed normal to the direction of the magnetic field.

1. Connect the temperature control chamber to the hot water recirculator with Tygon tubing.
2. Place the molds in the appropriate orientation in the chamber. Seal the chamber, and start the hot-water recirculator. Place the chamber immediately in the bore of the magnet.
3. Leave the chamber in the magnet until fibrillogenesis is complete.
4. Remove the chamber from the bore, and then the molds from the chamber (*see* **Note 6**).

After removal of the tissue equivalents from the molds, it is often desirable to verify that the collagen fibrils are aligned (*see* **Note 7**). Viewing the samples between crossed polarizing elements is a simple and noninvasive method of checking for alignment. The tissue equivalents are too delicate right after fabrication to be removed from the medium without disruption; therefore, crosslinking of the collagen by the addition of 3.7% formalin is first recommended. Rotate the slide holding the sample, and verify that the intensity of transmitted light varies with the angle of rotation (there should be alternating maxima and minima every 90 degrees). The associated birefringence can be measured to quantify the degree of alignment *(2)*, but this is substantially more complicated and unnecessary for the simple purpose of verification of alignment.

4. Notes

1. Vitrogen 100 is a pepsin-solubilized collagen from bovine dermis. Depending on the source of the collagen and the manner of extraction , the rate of gel formation (collagen fibrillogenesis) may vary drastically. Acid-extracted type I collagen from rat tail tendon can form a gel in a few minutes, in comparison to 30–60 min typically required for Vitrogen 100.
2. Safety should be a primary concern when using such high-strength magnets: Be sure that no metallic parts are used in the molds or temperature-control chamber.
3. A temperature-control chamber should be used to keep the samples at the optimal fibrillogenesis temperature in the magnetic field. Our chamber is a watertight acrylic tube with a removable endcap possessing an entrance and exit port for the warm water from the recirculating water bath. There are few restrictions on the type of device, as long as it maintains the temperature and can fit in the bore of the magnet.
4. Cells can be entrapped within the collagen gel by the addition of a small amount of a concentrated suspension of cells in culture medium to the collagen solution. Cells are entrapped within the fibrils of the gel during fibrillogenesis, and subsequently spread and migrate during incubation of the tissue equivalent, with an associated cell-induced compaction of the network of collagen fibrils.

5. The collagen concentration may be altered, although this may also affect the rate of fibrillogenesis.
6. Immediately after fibrillogenesis (i.e., before cell-induced compaction), the tissue equivalents are easily disrupted, and must therefore be removed carefully from the molds. For the tubular molds, simply insert a plunger into one end of the polypropylene tube, making contact with the rubber cylinder, and force the rubber cylinders, the Teflon rod, and the tissue equivalent out into warm cell-culture medium. For the rod molds, simply remove both Teflon caps and allow the tissue equivalent to slide out of the Teflon tube into the medium. All of these manipulations must be done in a sterile environment.
7. Collagen alignment can be inferred from cell alignment, since contact guidance appears to be a property of all motile blood and tissue cells, but a period of incubation is first required for cell spreading to occur, so that cell orientation can be measured.

References

1. Torbet, J. (1984) Magnetic alignment of collagen during self-assembly. *Biochem. J.* **219**, 1057–1059.
2. Guido, S. and Tranquillo, R. T. (1993) A methodology for the systematic and quantitative study of cell contact guidance in oriented collagen gels: Correlation of fibroblast orientation and gel birefringence. *J. Cell Sci.* **105**, 317–331.
3. Kanda, K. and Matsuda, T. (1993) Behavior of arterial wall cells cultured on periodically stretched substrates. *Cell Transplantation* **2**, 475–484.
4. Ghosh, S. and Comper, W. D. (1988) Oriented fibrillogenesis of collagen in vitro by ordered convection. *Conn. Tissue Res.* **17**, 33–41.
5. Hughes, K. E., Fink, D. J., Hutson, T. B., and Veis, A. (1988) High-strength collagen biomaterials. *JALCA* **83**, 372–378.
6. Becker, R. O., Bassett, C. A., and Bachmann, C. H. (1964) Bioelectrical factors controlling bone structure, in *Bone Biodynamics*, (Frost, H. M., ed.), Little, Brown, Boston, pp. 209–230.
7. Bell, E., Ivarson, B., and Merrill, C. (1979) Production of a tissue-like structure by contraction of collagen lattices by human fibroblasts of different proliferative potential in vitro. *Proc. Natl. Acad. Sci. USA* **76**, 1274–1278.

7

Preparation and Use of Thermosensitive Polymers

Yi-Lin Cao, Clemente Ibarra, and Charles Vacanti

1. Introduction

Restoration of organ structure and function, utilizing tissue engineering technologies, often requires the use of a temporary porous scaffold. The function of the scaffold is to direct the growth of cells migrating from the surrounding tissue (tissue conduction), or of cells seeded within the porous structure of the scaffold. The scaffold must therefore provide a suitable substrate for cell attachment, differentiated function, and, in certain cases, cell proliferation *(1–3)*. These critical requirements may be met by the choice of an appropriate material from which to construct the scaffold, although the suitability of the scaffold may also be affected by the processing technique. There are many biocompatible materials that could potentially be used to construct scaffolds, however, a biodegradable material is normally desirable, since the role of the scaffold is usually only a temporary one. Many natural and synthetic biodegradable polymers, such as collagen, poly(α-hydroxyesters), and poly(anhydrides), have been widely and successfully used as scaffolding materials because of their versatility and ease of processing. Many researchers have used poly(α-hydroxyesters) as starting materials from which to fabricate scaffolds, using a wide variety of processing techniques. These polymers have proven successful as temporary substrates for a number of cell types, allowing cell attachment, proliferation, and maintenance of differentiated function *(10)*. Poly(α-hydroxyesters) such as the polyoxamers are a family of more than 30 different nontoxic, nonionic surface active agents. These compounds are made at elevated temperature and pressure by the sequential addition of propylene oxide and then ethylene oxide to neutralize the salt that is generally retained in the final product. The polyoxamer series cover a wide range of liquids, pastes, and solids, with mol wt varying from 1100 to about 14,000. The ethylene oxidepropylene oxide weight ratios range from about 1:9 to about 8:2 *(4)*.

Concentrated aqueous solutions of the polyoxamers form gels. A unique physical property of these gels is their reversible nature. Upon lowering the temperature, the gels revert to a liquid state. Upon rewarming, the gels reform. This change in physical state can occur innumerable times and is temperature-dependent.

Polyoxamer 407 was found to form a gel at only 20% concentration by weight at 25°C, which is less than that of the other members of the polyoxamer series. Because of this, polyoxamer 407 has been selected for further study as an innocuous carrier for medications to treat burn wound injuries, as well as other dermatologic or surgical disorders *(4–9)*. A desirable feature of a polyoxamer gel for the clinical treatment of burns, is its ability to adhere to the surface to which it is applied. This is a physical interaction between the aqueous polyoxamer gel and the surface to which it adheres.

A novel approach to the formation of a layer of tissue engineered cartilage on host bone using synthetic polymers, is described. The ability of thermosensitive, biocompatible, and biodegradable liquid polymer, a synthetic copolymer of ethylene and propylene oxide, which transforms to gels at physiologic temperatures to support cartilaginous tissue formation when mixed with isolated chondrocytes on a viable osseous surface, was investigated.

2. Materials

1. Chondrocyte culture medium (complete medium). The medium used to culture chondrocytes is composed of Ham's F-12 nutrient mixture (Gibco-BRL, Grand Island, NY, # 21250–089), supplemented with 10% fetal bovine serum (Gibco-BRL, # #0000), penicillin (100 U/mL, [Sigma, St. Louis, MO], #A-9909), gentamycin (100 µg/mL, [BioWhittaker, MD], #17–518Z), streptomycin (100 µg/mL [Sigma, Ann Arbor, MI], #A9909), amphotericin B (0.25 µg/mL [JRH Biosciences, Lenexa, KS], #59–604–076), L-glutamine (292 µg/mL [Gibco], #0000), ascorbic acid (5 µg/mL [Sigma], #A-0278). After sterilization through a 0.2 µm filter (Gelman Sciences, Ann Arbor, MI, #66234), the culture media is stored at 4°C in 500-mL aliquots.
2. Dulbecco's phosphate buffered saline (PBS) (Gibco, # 21600–010) Ca^{2+}- and Mg^{2+}-free. Chondrocyte wash solution: 1X solution is made by mixing 9.6 g of powder with 1 L deionized ddH_2O. pH was adjusted to 7.0 with $NaHCO_3$ if necessary, and filter-sterilized through a 0.2-µm filter (Gelman Sciences). pH will rise 0.1–0.3 units after filtering. Store at room temperature.
3. PBS/antibiotic solution is made by mixing 1X PBS (described above) with 1 g/L cefazolin sodium (Apothecon, Princeton, NJ, #NDC 0015–7339–12). Filter through a 0.2 µm filter (Gelman Sciences), and store at 4°C.
4. Collagenase solution is made with 300 mg collagenase class 2 powder (Worthington Biochemicals, Freehold, NJ, #LSO4177), mixed in 100 mL plain Ham's F-12 medium (Gibco-BRL), prefiltered through Fisher P4 paper disks (Fisher Scientific, Pittsburgh, PA, #09–803–6H), and finally filtered through a 0.2-µm filter (Gelman Sciences), and stored frozen at −5 to −8°C.

Thermosensitive Polymers

5. Pluronic gels. Pluronic F127 NF (Polyoxamer407NF) surfactant (BASF, Mount Olive, NJ, #549926) will form a gel at concentrations greater than 20% w/v (g/mL) in plain Ham's F-12 medium at temperatures above 15°C. Gels will reliquefy when cooled below 15°C. Concentrations of 30–40% (w/v) are ideal for tissue engineering.
6. Trypsin 0.25%/EDTA solution. Gibco-BRL, #27250-042/11267-010.
7. Trypan blue vital dye (Sigma, # T-8154).
8. Tissue-culture dishes (100 × 20 mm). Falcon, Becton Dickinson, Lincoln Park, NJ, #3003.
9. 50-cc Conical (centrifuge) tubes. VWR Scientific, Westchester, PA, # 21008-178.
10. 0.2-µm Filter disks (47-mm diameter) (Gelman Sciences, Ann Arbor, MI).
11. 250 µm Polypropylene spectra mesh (autoclavable). Fisher Scientific#08-670-250c, Pittsburgh, PA..

3. Methods
3.1. Pluronic (Polyoxamer) Gels

The technique used for making aqueous polyoxamer gels is very simple. A weighed amount of the polyoxamer 407 is slowly added to a known weight of cold water (less than 10°C). Stirring rate should be controlled to maintain a slight vortex in the liquid; Too rapid a stirring rate will cause aeration and the formation of foam. When all the polymer has been added, stirring can be continued (while keeping the solution cool), until a clear solution is formed, or the container can be placed in a refrigerator and left undisturbed for several hours, at which time solution is complete. Known burn medications, such as silver salts, antibiotics, and so on, can be added to the cold solution and dissolved or suspended therein, while warming to induce gel formation. The temperature at which the liquid gels will vary, depending on the poloxmar concentration and on the type and amount of additives. Thus, the addition of a humectant, such as glycerin, will increase the gel strength more than the addition of the same amount of propylene glycol. The polyoxamer gels can be autoclaved and are stable on storage **(Fig. 1)**.

Protocol to prepare Pluronic (polyoxamer) gels:

1. On a magnetic stirrer, place an autoclavable bottle with a stir bar in a beaker with ice and water to keep the mixture cold. Alternatively, this procedure can be performed in a cold room at 5°C.
2. Add cold Ham's F-12 medium to the bottle. Use a bottle that is large enough to allow 30% headspace above the mixture.
3. Add Pluronic F127 powder slowly through a dry funnel, while stirring, gradually adding 1–2 g at a time every 10–15 min. Avoid allowing large clumps to form.
4. Keep the mixture stirring until clumps of powder are smaller than 2–3 mm. This may take up to 16 h. The mixture must be kept cold throughout this process.

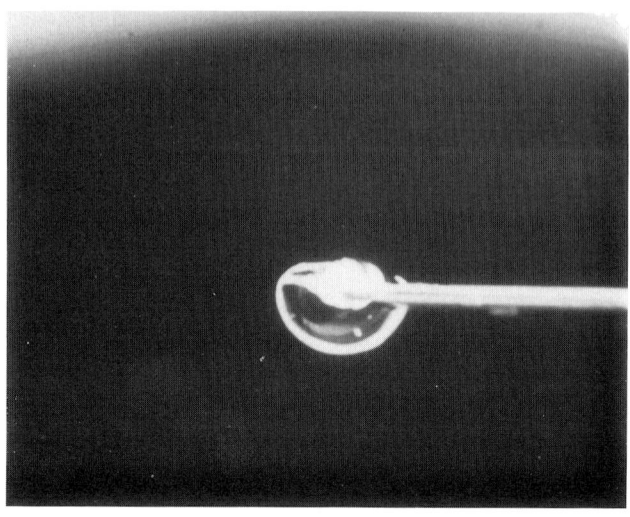

Fig. 1. Photographic close-up view of a drop of Pluronic gel in liquid state at the distal end of an 18-gage needle.

When the Pluronic powder has completely dissolved, remove the stir bar and autoclave the mixture in a pan of water for 20 min. Some researchers report successful results sterilizing these gels with UV light and filtering through a 0.25-µm filter in the liquid (cooled) stage. Store at 4–8°C.

3.2. Isolation of Chondrocytes

Newborn calf forelimbs were obtained from a local abattoir within 6 h of animal sacrifice. Cartilage was harvested by sharp dissection under sterile conditions from the articular surfaces of the humerus and scapula. Tissue fragments were washed twice with 1X PBS solution with antibiotics in 50 cc conical tubes. PBS was then replaced with a 3% solution of class 2 collagenase to digest the tissue at 37°C for 12 h in a shaker. The solution and tissue debris was filtered through a 250-µm-pore nylon mesh, and was centrifuged to obtain a cell pellet, which was then washed twice with PBS with antibiotics solution. The cells were finally suspended in 10 cc of complete culture media and were counted in a hemocytometer. Trypan blue vital dye was used for counting to assess cell viability.

3.3. Cell–Polymer Suspension

After counting the chondrocytes, the cell–polymer suspension is centrifuged again and the supernatant is removed. Chilled 30% (w/v) copolymer solution is added to obtain a cellular concentration of 5×10^7 cells/mL. The conical tube containing the cell pellet must be maintained in ice at the moment of adding the chilled polymer solution, to prevent it from setting before the cells can be evenly suspended in it by vortexing. The cell–polymer suspension is kept at 4°C to maintain the liquidity of the copolymer.

3.4. Tissue-Engineering Applications

3.4.1. Cartilage Growth on Bone by Painting

3.4.1.1. Methods

Under general anesthesia with inhaled methoxyflurane, the calvariae of 12 male athymic mice were exposed through a longitudinal incision on the head, and the surface of the bone was abraded using a drill bit. With a sterile fine brush (0.5 × 1 cm), each calvarium was painted with several overlying applications of the chondrocyte–copolymer admixture described above. After each application, the liquid was allowed to polymerize over the calvarial surface as it was exposed to body temperature. A total of five layers were applied to each animal. Subsequently, the incision was closed with 5-0 nonabsorbable sutures. Four animals were sacrificed at 4, 6, and 8 wk. Twelve additional animals were used as controls, applying the liquid copolymer without chondrocytes or chondrocytes suspended in isotonic saline, in similar fashion. The animals were euthanized by anesthetic overdose at the different time intervals and the calvariae were harvested for gross and microscopic examination.

3.4.1.2. Results

The painted chondrocyte–copolymer solution gave rise to a semisolid sticky gel a few minutes after application, and showed evidence of cartilage formation in all the animals in the experimental groups at all time-points. The new cartilage had been incorporated to the surface of the underlying bone, as a 3–5 mm thick layer of a white, opaque coating that was difficult to dislodge mechanically. Histologic sections stained with hematoxylin and eosin (H&E) demonstrated the presence of hyaline-like cartilage. Remarkably, the painted cartilage infiltrates the underlying osseous substrate, forming bone–cartilage interface. None of the control animals demonstrated any cartilage formation either grossly or on histologic examination. The calvariae of animals painted with either chondrocytes, suspended in isotonic saline or liquid copolymer alone, showed no changes in their bony contour.

3.4.1.3. Conclusions

This study demonstrated the formation of new cartilage using isolated chondrocytes suspended in a thermopolymerizable gel, applied with a brush on a bony surface. A relatively thick surface of articular-like cartilage with a near-normal, strong, and stable bone–cartilage interface was formed. This technique shows potential usefulness in joint resurfacing, with the unique characteristic of allowing the formation of new cartilage in the desired shape and location, with potential minimal invasiveness.

3.4.2. Injectable Cartilage

3.4.2.1. METHODS

Twelve athymic mice were anesthetized with inhaled methoxyflurane. Five 100-μL aliquots of a chondrocyte-gel suspension of 30% (w/v) polyoxamer, with a cell concentration of 5×10^7 cells/mL were injected subcutaneously, through a 22-gage needle, into five different locations on the dorsum of each mouse. Twelve additional mice were injected with the same amount of cells suspended in isotonic saline, and 12 mice were injected with gel without cells, in similar locations, as controls. Three animals from each group were euthanized by anesthetic overdose after 2, 4, 6, and 8 wk respectively. The presence of areas of induration under the skin was assessed previous to euthanasia in all animals. Specimens harvested at this time were evaluated grossly and histologically in search of evidence of new cartilage formation.

3.4.2.2. RESULTS

All the animals in the experimental groups presented with five subcutaneous nodules at the moment of euthanasia. Upon palpation through the skin of the mice, the nodules had firm rubbery consistency. There were no apparent nodules or sites of induration under the skin of any of the control animals. After surgical exposure, the nodules on the animals in the experimental groups were white, opalescent, amorphous masses of varying sizes (7–10 × 5–7 × 2–5 mm) **(Fig. 2)**. The nodules had maintained the size and shape that the gel adopted right after injection, as it dissected its way through the subcutaneous space. They were all enclosed within a well-vascularized capsule that could be easily dissected from the underlying fascia and overlying skin. After removal of the vascular capsule, the texture and consistency of the nodules resembled that of cartilage. Specimens harvested after 2 wk were relatively softer at palpation, compared to specimens harvested at later time-points. Specimens harvested after 8 wk presented the greatest apparent resistance to manual compression. Histologic examination of all specimens, stained with H&E, demonstrated the presence of new cartilage formation **(Fig. 3)**. Specimens at the earlier time-points showed the presence of very cellular immature-appearing cartilage. After 8 wk, the histologic appearance of the nodules was that of more mature-appearing, hyaline-like cartilage, with chondrocytes in lacunae, surrounded by extracellular matrix with a higher matrix:cell ratio. New carti-

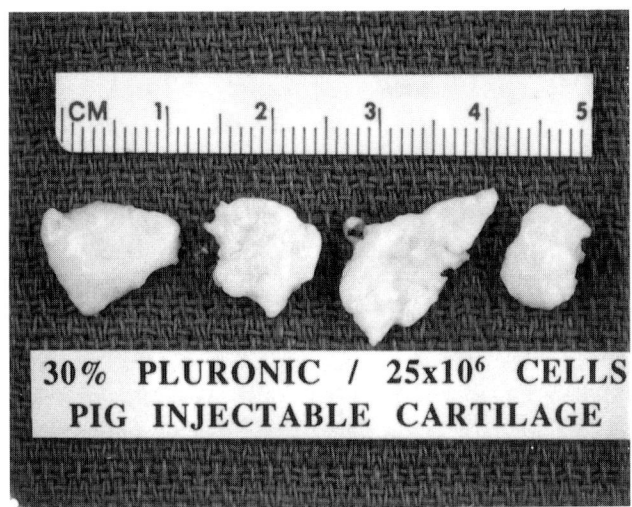

Fig. 2. Gross appearance of tissue-engineered cartilage specimens, harvested after 4 wk of subcutaneous injection of porcine chondrocytes suspended in Pluronic gel in nude mice.

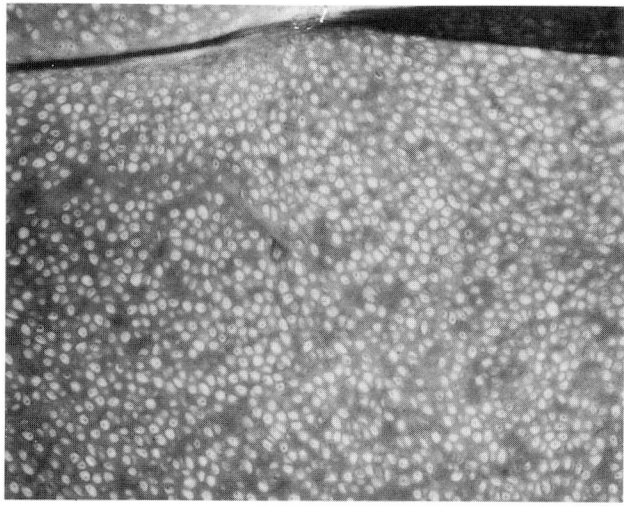

Fig. 3. Microphotograph (×10) of histologic section stained with H&E of tissue-engineered cartilage, after 8 wk of subcutaneous injection of porcine chondrocytes suspended in Pluronic gel in nude mice.

lage was not found in the subcutaneous space after dissection of animals in which plain polymer gel was injected. Small remnants of polymer scattered in some areas could only be found in animals in this group at the earlier timepoints (2–4 wk). Very small nodules of cartilage-appearing tissue were found during dissection of animals in which chondrocytes suspended in saline were injected. Such small nodules could not be appreciated through the skin of these mice. Histologic appearance of this nodules resembled that of the new tissue formed in the animals in the experimental groups.

3.4.2.3. Conclusions

New cartilage can be obtained by injecting isolated chondrocytes suspended in a thermopolymerizable gel as a carrier in the subcutaneous space of athymic mice. The volume of new tissue formed was significantly greater than that obtained by the simple injection of cells suspended in saline. The new tissue matured over time, but by 2 wk, the presence of new cartilage formation was evident both grossly and histologically with this technique. The copolymer gel utilized is degraded and disappears almost completely after 2 wk in the subcutaneous space, if it is not combined with cells that can replace its presence with newly synthesized extracellular matrix, and thus, new tissue.

In summary, this work demonstrates the ability of these relatively new thermosensitive polymers to act as a cell-delivery device that can be utilized in a number of different ways to form new cartilage. This system shows great promise in the fields of orthopedics and reconstructive surgery as a means to form new cartilage, perhaps utilizing autologous cells, for the repair or reconstruction of cartilage surfaces and structures employing minimally invasive techniques.

References

1. Langer, R. and Vacanti, J. P. (1993) Tissue engineering. *Science* **260**, 920–926.
2. Vacanti, J. P. and Vacanti, C. A. (1996) The challenge of tissue engineering, in *Text Book of Tissue Engineering* (Lanza, R., Langer, R., and Chick, W., eds.), R. G. Landes, Austin, TX, pp. 1–4.
3. Vacanti, C. A., Langer, R., Schloo, B., and Vacanti, J. P. (1991) Synthetic polymers seeded with chondrocytes provide a template for new cartilage formation. *Plastic Reconstruct. Surg.* **88**, 753–759.
4. Wong, W. H. and Mooney, D. J. (1997) Synthesis and properties of biodegradable polymers used as synthetic matrices for tissue engineering, in *Synthetic Biodegradable Polymer Scaffolds* (Atala, A., Mooney, D. J., Vacanti, J. P., Langer, R., eds.), Birkhauser, Boston, pp. 51–82.
5. Schmolka, I. (1972) Artificial Skin I. Preparation and properties of Pluronic F-127 gels for treatment of burns. *J. Biomed. Mater. Res.* **6**, 571–582.

6. Faulkner, D. M., Sutton, S. T., Hesford, J. D., Faulkner B. C., Major, D. A., Hellewell, T. B., Laughon, M. M., Rodeheaver, G. T., and Edlich, R. F. (1997) A new stable Pluronic F68 gel carrier for antibiotics in contaminated wound treatment. *Am. J. Emerg. Med.* **15,** 20–24.
7. Miyazaki, S., Tobiyama, T., Takada, M., and Attwood, D. (1995) Percutaneous absorption of Indomethacin from Pluronic F127 gels in rats. *J. Pharm. Pharmacol.* **47,** 455–457.
8. Paustian, P. W., McPherson, J. C., III, Haase, R. R., Runner, R. R., Plowman, K. M., Ward, D. F., Nguyen, T. H., and McPherson, J. C. (1993) Intravenous Pluronic F-127 in early burn wound treatment in rats. *Burns* **19,** 187–191.
9. Rice, V. M., Shanti, A., Moghissi, K., and Leach, R. E. (1993) A comparative evaluation of Polyoxamer 407 and oxidized regenerated cellulose (intercede TC7) to reduce postoperative adhesion formation in the rat uterine horn model. *Fertil. Steril.* **59,** 901–906.
10. Thomson, R. C., Yaszemski, M. J., and Mikos, A. G. (1996) Polymer scaffold processing. In *Text Book of Principles of Tissue Engineering.* (Lanza, R., Langer, R., and Chick, W., eds.), R. G. Landes, Austin, TX, pp. 263–272.

8

In Situ Photopolymerized Hydrogels for Vascular and Peritoneal Wound Healing

Amarpreet S. Sawhney and Jeffrey A. Hubbell

1. Introduction
1.1. Wound Healing in the Peritoneal Cavity

The process of wound healing in internal organs lined with mesothelial cells is different in some respects from healing in cutaneous injuries. Hertzler *(1)* noted that cutaneous wound re-epithelialization takes place from the wound borders, but peritoneal defects become re-mesothelialized simultaneously. Most organs in the pelvic cavity possess a mesothelial lining that can be injured as a result of trauma from surgical intervention or handling. It is generally agreed that the remesothelialization of the peritoneum can take place over 5–8 d *(2)*. During this period, events of inflammation and mesothelial repair occur that can set the course for postsurgical adhesion formation. Postsurgical adhesion formation, or the formation of scar tissue bridges between organs in the pelvic cavity, is an undesirable result of a natural healing process. Ryan et al. *(3)* showed experimentally that the surface of bowel is subject to demesothelialization and loss of native fibrinolytic activity by surgical manipulation, whether it is allowed to dry or is kept moist with saline. This denuded surface could allow blood clots to form and act as precursors to adhesion formation before remesothelialization can occur. The formation of adhesions can distort the normal anatomy and natural positioning of organs, and has often been implicated in postoperative pelvic pain and bowel obstructions, in addition to other complications. Ryan recommended that to minimize adhesion formation, tissue exposure and manipulation should be minimized *(3)*. However, this is often not possible, and so the use of barriers to isolate injured organs from their surroundings has been proposed as a way to minimize adhesion formation.

Barrier materials, such as Gore-Tex™ surgical membrane, have been shown to reduce the instance of adhesion formation *(4)*. However, this material requires retrieval, because it is not absorbable. The barrier material should also be extremely noninflammatory, so as not to further enhance the already active inflammatory processes. Therein lies a challenge for absorbable biomaterials to degrade within the time frame of a few weeks, and yet be inert and noninflammatory during this degradation process. Fabrics such as those made from oxidized regenerated cellulose absorb within the required time frame but were seen by Haney et al. *(5)* to cause peritoneal injury.

1.2. Vascular Wound Healing

The process of balloon angioplasty of arteries that have been narrowed because of artheroscelerotic plaque is used to restore patency of the vessels. The balloon dilation process is traumatic and causes denudation of the endothelium, as well as injury of the intimal and medial tissues as a result of dilation. The clinical significance of the ensuing events of wound healing that occur are the renarrowing (or restenosis) of approx 30–40% of coronary arteries dilated by balloon angioplasty within 6 mo of treatment *(6)*. Wound healing in this case is a complex series of events involving both luminal and medical events. It has been observed that smooth-muscle cell migration, and proliferation and elaboration of matrix proteins, contribute to intimal thickening and consequent luminal narrowing. Several growth factors, cytokines, and chemoattractants play a role in smooth muscle-cell stimulation. Some of these are luminal in origin, such as platelet-derived growth factor and thrombin, and may be deposited locally as a result of the thrombosis at the angioplasty site *(7,8)*. This mural thrombus, which forms because of the denudation of the artery and the exposure of underlying thrombogenic medial tissues, may also serve as a matrix for smooth muscle cell proliferation. Other factors, such as basic fibroblast growth factor, are intrinsic to the vessel wall and are also mitogens for smooth-muscle cell proliferation *(9,10)*.

1.3. Photopolymerized Bioresorbable Hydrogels

Hydrogels are materials that have water as a significant component of their structure and have the capacity to swell in an aqueous surrounding. Because of the high water content of these materials, and their low interfacial tension with water, they are often extremely biocompatible, and can serve as barriers and scaffolds over injured tissue to help direct the wound healing process. Nonionic hydrogels, such as the poly(ethylene glycol) (PEG)-based gels discussed herein, contain large amounts of water, and have hydrophilic, nonionic surfaces. This lowers the driving force for protein adsorption onto these surfaces from the physiological surroundings *(11)*. Cell adhesion is mediated by protein

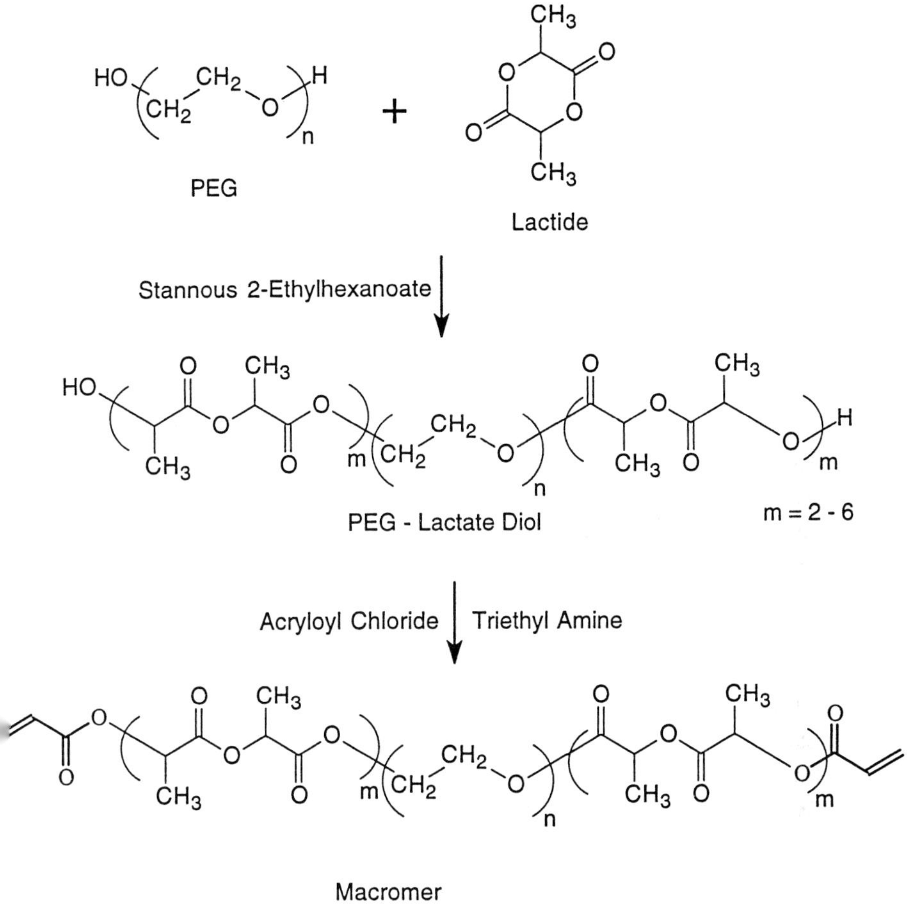

Fig. 1. Schematic showing the synthesis of the photosensitive macromers.

adsorption and/or proteoglycan interaction with charged surfaces *(12,13)*. The use of PEG to increase biocompatibility is well documented in literature. PEG has been used in numerous studies to increase biocompatibility of blood-contacting materials *(14)* and soft-tissue contacting materials *(15)*.

The precursor molecules (called "macromers" here), that can be used to form the crosslinked hydrogels by polymerization, are novel molecules that have three domains along the polymer chain. The synthesis and structure of these macromers is illustrated in **Fig. 1**. The water-soluble central domain enables the entire polymer to be water soluble. Short α-hydroxyester oligomeric domains that flank the central water-soluble domain are terminated with acrylates as polymerizable end groups. The presence of hydrophobic end regions on the central PEG chain has been seen to result in the formation of micellar structures in aqueous solutions *(16)*. The formation of such preorganized configurations in aqueous solutions of these precursors results in the rapid polymerization and gelation of these multifunctional precursors.

The bioresorbability of the hydrogels described above is afforded by hydrolysis of the well-known poly(α-hydroxy acids), such as poly(lactic acid) and poly(glycolic acid). The use of a light-sensitive initiation mechanism to provide the rapid and spatially controllable gelation of a liquid precursor, in direct contact with the treated tissues, provides a conformal lubricious and adherent hydrogel film on the tissues. This is presumably caused by the flow of the liquid precursor into crevices on the surface of the tissue, with subsequent mechanical interlocking upon conversion into a gel. In this chapter, we describe the synthesis and use of such hydrogels, and their use in guiding wound healing in the peritoneal cavity and in vascular tissues.

Photopolymerized bioresorbable hydrogels have been shown to be effective in guiding wound healing and in prevention of postsurgical adhesion formation in several animal models *(17–19)*. These gel barriers are resorbed over a time frame of less than 2 wk in vivo. Studies conducted in rats, using only the ungelled viscous precursor solution, have failed to prevent the formation of adhesions. This is quite expected, because the use of viscous macromolecular solutions, despite the coating and lubricating effects, has been documented to be insufficient to prevent the formation of adhesions *(20)*. The gel degrades into water-soluble substances, primarily PEG, lactic acid and its oligomers, and oligomers of acrylic acid. These components are all water-soluble, and are either naturally existent in the body or are rapidly cleared by excretion through the kidneys. Photoinitiation is used to induce gelation. This reaction takes place under very mild conditions, with no excessive liberation of heat or toxic factors to cause tissue necrosis. This is because of the relatively low acrylic group concentration, compared to the larger nonreactive central PEG segment, which forms the bulk of the material. The photoinitiator used in this work presents a very low toxicological burden. The oral LD_{50} in rats is in excess of 6 g/kg *(21)*, and, typically, e.g., in a rabbit model, the total dose is about 0.6 mg/kg. The photoinitiator used in this study is sensitive to long-wavelength ultraviolet (LWUV) light, which is essentially lacking in cytotoxic or mutagenic potency. The total dose of light used is about four orders of magnitude less than that at which an effect would be expected to be observable *(22)*.

The approach of applying conformal barrier layers in the liquid state, and then gelling the liquid by application of light, provides several conceptual advantages over the application of preformed barriers, such as fabrics. First, the liquid may be readily deployed through a laparoscope, whereas a fabric must be rolled, pushed through the laparoscope, unrolled, and placed on the tissue to be treated. Moreover, the liquid is inherently conformal, i.e., it adapts its shape precisely to that of the tissues; by contrast, fabric barriers are not conformal to the macrotopology of the tissue surface. The methods described here can be used to study the formation and prevention of the adhesions between the injured tissue and neighboring tissues.

Isolation of the injured tissues after angioplasty, to prevent local thrombosis, may be a way to change the course of events leading to restenosis. In this chapter, a nonpharmacological approach to prevent intimal thickening, by inhibiting contact between blood and subendothelial tissues with a resorbable polymeric barrier, is demonstrated. A method is described to synthesize a thin nonthrombogenic, conformal hydrogel barrier on the arterial wall by interfacial photopolymerization of a macromolecular precursor *in situ*. This novel barrier has been previously shown to virtually eliminate thrombosis in injured arteries. The transient presence of the barrier has also been shown to preserve patency and reduced intimal thickening long after the disappearance of the barrier *(23)*. It is critical to achieve this coating process with careful control of the thickness and growth of the coatings, e.g., with minimal loss in luminal diameter of a treated artery. Polymerization in contact with living tissue in a physiological environment necessitates that the chemical entities used and generated during the polymerization process be nontoxic, and that the polymerization conditions be mild. In this chapter, we describe techniques to reproducibly and accurately deposit hydrogel films on the surface of vascular tissue.

2. Materials

All chemical, reagents, and solvents can be purchased from Aldrich (Milwaukee, WI), unless otherwise mentioned.

2.1. Peritoneal Wound Healing

2.1.1. Synthesis of PEG-co-Poly(α-Hydroxy Acid) Copolymers

1. Reagents: α, ω-dihydroxy PEG with mol wt 10,000 (PEG 10kDa), DL-lactide, stannous 2-ethylhexanoate, triethyl amine, acryloyl chloride.
2. Solvents: benzene, ethyl acetate, dichloromethane, anhydrous diethyl ether.
3. Glassware: round bottomed flasks, tubing, Y-connectors, gas manifold, beakers, crystallization dishes, 1-cc syringe with needle, fritted glass filter.
4. Other supplies: heating mantle, magnetic stirrer and stirbars, vacuum pump.

2.1.2. Synthesis and Characterization of PEG Macromers

1. Polyethylene films for infrared spectra, NMR tubes.
2. Tetramethyl silane, $CDCl_3$.

2.1.3. LWUV-Induced Photopolymerization

1. Phosphate-buffered saline (PBS): 0.2 g/L KCl, 0.2 g/L KH_2PO_4, 8 g/L NaCl, 1.15 g/L Na_2HPO_4, pH 7.4.
2. Initiator solution: 300 mg 2,2 dimethoxy 2-phenyl acetophenone dissolved in 1 mL of N-vinyl pyrrolidinone.
3. LWUV-emitting lamp (Blak-Ray, model B-100A with Flood, VWR Scientific, Boston, MA).
4. Sterilized aluminum foil (autoclaved).
5. Polymerizable solution: A 10% w/v solution of macromer in PBS is used. This solution can be filtered through a 0.45-μm filter (Millipore, Bedford, MA) to sterilize it. To 1 mL of this solution, 3 μL of initiator solution is added, and the solution is vortexed briefly. This solution is then drawn up in a 10 cc syringe that is shielded from light using the aluminum foil.

2.1.4. Rabbit Uterine Horn Model

1. New Zealand white rabbits between 2.5 and 3.5 kg in weight (HRP, Denver, PA).
2. Anesthesia: A cocktail containing equal parts of ketamine ("Ketaset" 100 mg/mL in 10-mL vials Fort Dodge Labs, Fort Dodge, IA, dose at 100 mg/kg), Xylazine (100 mg/mL 50-mL vials, Fermenta Animal Health Care, Kansas City, MO, dose at 5 mg/kg) and Acepromazine (10 mg/mL in 50-mL vials, Fermenta, Kansas City, MO, dose at 0.5 mg/kg).
3. 21-gage intravenous catheter (Baxter Healthcare, Mc Gaw Park, IL), 1-, 5-, and 10-cc syringes, 21-gage hypodermic needles.
4. Betadine solutions (Perdue Fredericks, Norwalk, CT) for scrubbing and preparing, and hair clipper.
5. Standard abdominal surgery pack containing abdominal scissors, self-retaining retractors, surgical blade (#10) and handle, suture scissors, needle holder, towel clamps, tissue forceps. (Johnson & Johnson Professional, Raynham, MA).
6. Sutures: 3-O Vicryl and 2-O Vicryl (Ethicon, Sommerville, NJ).
7. Monopolar electrocautery (Hyfrecator Plus, Birtcher Medical Systems, Irvine, CA).

2.2. Interfacially Formed Thin Hydrogels for Vascular Wound Healing

2.2.1. Interfacial Photopolymerization of Thin Hydrogel Films

1. Reagents: eosin-Y, triethanolamine, N-vinylpyrrolidinone.
2. Argon ion laser (American Laser, Salt Lake City, Utah) and 200-μm core optical fiber (Spectra-Physics, Cranbury, NJ).
3. Solution 1: eosin-Y is dissolved in PBS at a concentration of 0.05% w/w. Solution 2: to 50 mL of 0.67 g vacuum-distilled triethanolamine is added, mixed well, and pH is adjusted to 7.4 using 0.6 N hydrochloric acid. The photopolymerizable macromonomer is dissolved in the PBS-triethanolamine solution at a concentration of 10% w/v. N-vinyl pyrrolidone is distilled over argon, and then added at a concentration of 0.5 μL/1 mL, to form the macromonomer solution.

4. Inverted microscope (Nikon TMS, Avon, MA), mechanical sizer (Boeckeler Instruments, Tucson, AZ), and microscope slides.

2.2.2. Rabbit Balloon Angioplasty Injury Model

1. Rabbits, anesthesia, betadine scrub solution, and hair clippers, as in **Subheading 2.1.4**.
2. Polyethylene tubing (PE-50).
3. 4F arterial embolectomy catheter (Baxter Healthcare).
4. Silk suture ties (4-O silk, Ethicon).
5. Vascular Clamps (Johnson & Johnson Professional).
6. HEPES buffered saline solution (HBSS): 10 mM HEPES and 0.9% NaCl (w/w) in deionized water, pH adjusted to 7.4 using HCl and NaOH.
7. Neutral buffered formalin (10% w/w).
8. Laser, solutions, and microscopy setup are as described in **Subheading 2.2.1**. All solutions are filter-sterilized in a laminar flow hood, and drawn up in a sterile fashion in syringes.
9. Scanning electron microscope, critical point drier, and graded series of ethanol.
10. Heparin solution: 1000 U/mL in 1 mL vials (Elkins-Sinn, Cherry Hill, NJ).

3. Methods
3.1. Peritoneal Wound Healing
3.1.1. Synthesis of Photosensitive Macromers

3.1.1.1. Synthesis of PEG-co-Poly(α-hydroxy acid) Copolymers (24)

1. The synthesis of the PEG-co-poly(α-hydroxy acid) copolymer is illustrated below, and the synthesis scheme is outlined in **Fig. 1**. PEG is dried by azeotropic distillation with benzene as a 10% solution. DL-Lactide is recrystalized from ethyl acetate prior to use.
2. 50 g of dry PEG 10 kDa, 3.60 g of DL-lactide (5 mol DL-lactide per mol of PEG), and 15 mg of stannous 2-ethylhexanoate are charged into a 100-mL round-bottom flask under a nitrogen atmosphere. The reaction mixture is stirred under vacuum at 200°C for 4 h, and at 160°C for 2 h, and is subsequently cooled to room temperature.
3. The resulting copolymer is dissolved in dichloromethane, precipitated in anhydrous ether, filtered, and dried (yield typically 95%). The α- and ω-hydroxyl end groups of PEGs with various mol wt can be used as ring opening reagents to initiate the polymerization of either DL-lactide or glycolide, to similarly form several other copolymers (*see* **Note 1**).

3.1.1.2. Synthesis and Characterization of PEG Macromers

1. The above-described PEG-co-poly(α-hydroxy acid) copolymers, which are themselves α- and ω-terminated by hydroxyl groups, are end-capped with acrylate groups to form a polymerizable macromer; the synthesis of the macromer with the PEG 10K is illustrated. 50 g of the aforementioned intermediate copolymer is dissolved in 500 mL of dichloromethane in a 500-mL round-bottom flask, and is cooled to 0°C in an ice bath.

2. 1.31 mL of triethylamine and 1.58 mL of acryloyl chloride are added to the flask, and the reaction mixture is stirred for 12 h at 0°C, and for 12 h at room temperature. The reaction mixture is filtered to remove triethanolamine hydrochloride, and the macromer is obtained by pouring the filtrate in a large excess of dry diethyl ether. It is further purified by dissolution and reprecipitation once, using dichloromethane and hexane, respectively. Finally, it is dried at 70°C under vacuum for 1 d. This PEG macromer is called 10KL5. This indicates that the acrylate-terminated macromer is synthesized from PEG segments of mol wt 10,000 (10KL5), using DL-lactide as the extension of the PEG α- and ω-hydroxy end groups (10KL5), and the degree of polymerization of the lactoyl repeats (not lactidyl repeats, the cyclic dimer of lactic acid, which is actually polymerized) is, on average, 5 per hydroxy end group (10KL5). Several other macromers can be similarly synthesized by terminating PEG-co-poly(α-hydroxy acid) copolymers with acrylate or methacrylate groups (*see* **Note 2**).

3. The structure of the synthesized macromers can be verified using IR and NMR spectroscopy. NMR spectra can be recorded in $CDCl_3$, and tetramethyl silane is used as internal standard. IR spectra can be run on a polyethylene film on which a drop of chloroform solution of the macromer is placed and dried. The disappearance of the -OH absorbance at 3510/cm confirms the complete acrylation of the macromer. The presence of lactate ester is confirmed by an absorbance at about 1750/cm. The degree of incorporation of lactate moieties can be calculated using the integral NMR intensities of the $-CH_3$ (1.3 ppm) and the $-CH_2O$ (3.63 ppm) of PEG.

3.1.2. LWUV Induced Photopolymerization

The macromer, synthesized as described above, is used to prepare a photopolymerizable macromer solution, as described in **Subheading 2.1.3**. 0.1 mL of this photopolymerizable macromer solution is poured onto an 18 × 18-mm glass cover slip, and irradiated using a low-intensity, portable, LWUV lamp, until gelation occurs. The time required to induce nontacky gelation is noted. This can be accomplished by continuously scratching a film of the macromer solution with a sharp instrument, until the scratched film retains the scratch mark. The chemical reactions occurring during this photopolymerization process are outlined in a schematic in **Fig. 2**. Visible light can be used for the photopolymerization process, also. (*see* **Note 3**, **Fig. 3**).

3.1.3. Rabbit Uterine Horn Model

The animals are placed in two groups: 7 for control (injured, but not treated), and treatment (injured and treated) for the evaluation of the wound healing in the absence and presence of the photopolymerized hydrogels.

1. Anesthesia: The rabbits are anesthetized with an intramuscular injection of the anesthetic cocktail. An intravenous catheter is placed in the ear vein and the ani-

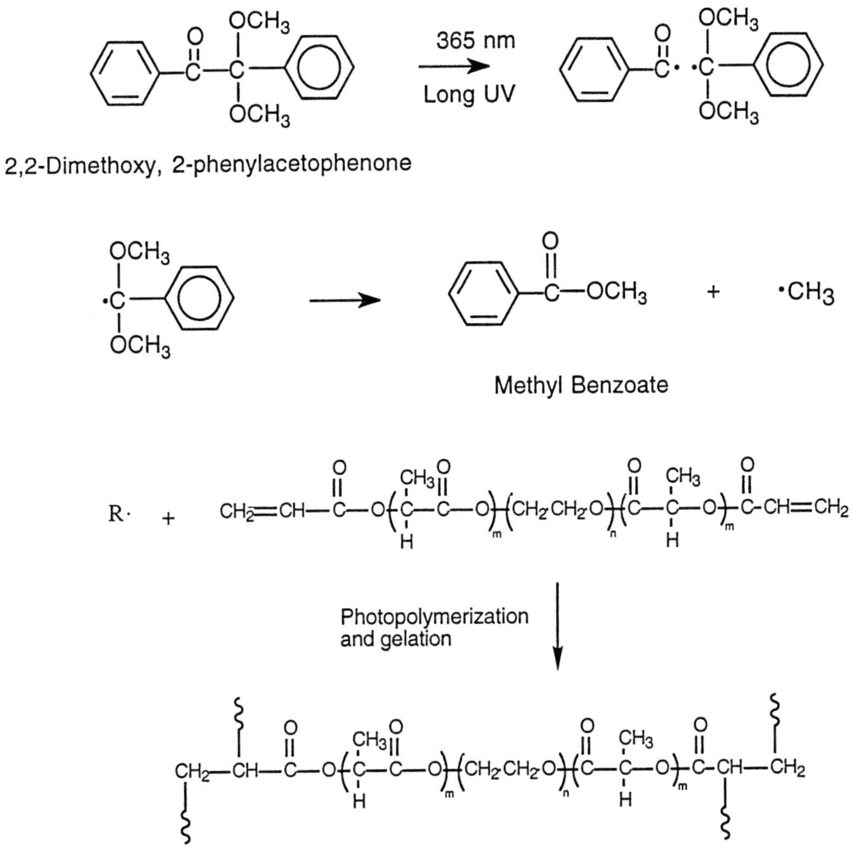

In Situ *Photopolymerized Hydrogels*

2,2-Dimethoxy, 2-phenylacetophenone

Methyl Benzoate

Fig. 2. Schematic showing the initiation and photopolymerization of the photosensitive macromers using LWUV light.

mals are maintained under anesthesia on periodic infusions of ketamine and Xylazine, as necessary.

2. Surgery: The abdomen is shaved and prepared with Betadine. A midline incision is made in the lower abdominal region, using the scalpel. Any bleeding encountered during muscle incision is controlled using electrocautery. The abdominal retractor is placed to allow good visualization of the visceral organs. The uterine horns are located and the vasculature to both horns is systematically cauterized using monopolar electrocautery (setting at 30), to induce an ischemic injury to the horns. After cauterization, 1.0 mL of the polymerizable solution is applied along the obverse side of both horns, and also along the cauterized injury area, by dripping the solution through the syringe and gently rubbing the solution into the tissue, using a gloved finger. After uniform application of the solution, the horns are exposed to the LWUV lamp, at an irradiance of 8–10 mW/cm^2, for 1 min, to

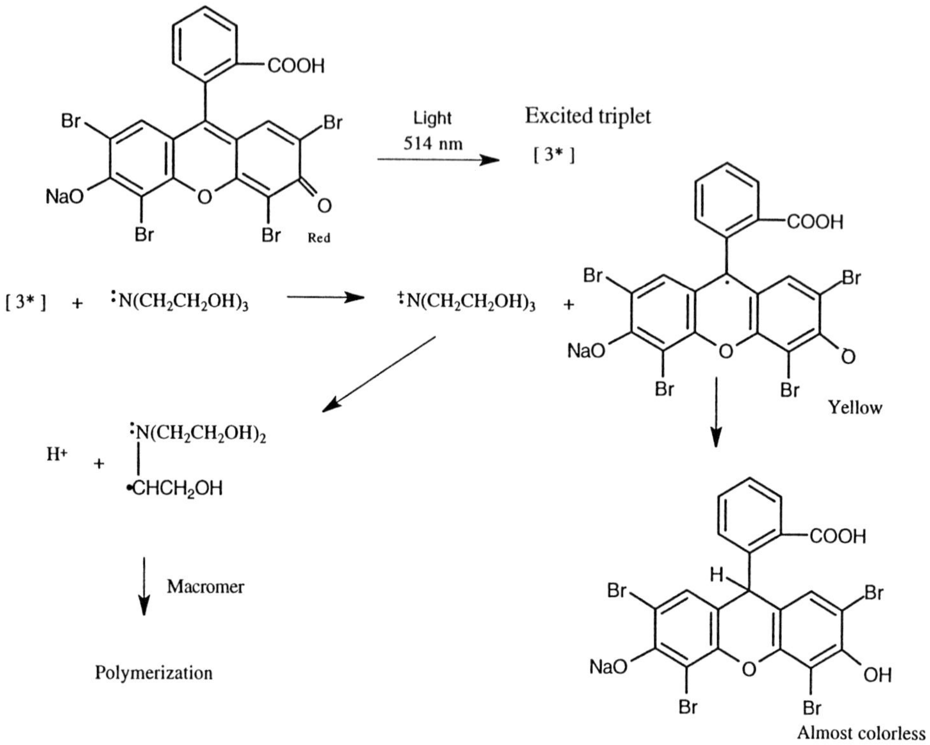

Fig. 3. Schematic showing the visible-light-initiated photopolymerization process.

induce gelation. The procedure is repeated on the reverse side of both horns and the cauterized area. The incisions are then closed using a continuous 2-0 Vicryl suture for the musculoperitoneal layer, and a 3-0 Vicryl suture for the cutaneous layer. No prophylactic antibiotics are administered. For the control group, the ischemic injury is made as described above, and the incision is closed without the application of the polymerizable solution or illumination with the LWUV light. All techniques should be identical between the treatment group and the control group, and all animals should be operated on by the same researcher.

3. Evaluation: The animals should be reoperated at the end of 2 wk under ketamine anesthesia (35 mg/kg, administered intramuscularly) to evaluate adhesion formation, and are sacrificed by iv injection of euthanasia solution. Gross observations are recorded as to sites of adhesion formation and organs involved in adhesion formation. Presence or absence of hydrogel material should be noted, as well as any gross evidence of inflammation, necrosis, hematoma or seroma formation, and presence of cysts or infection. Microscopic confirmation of any gross changes should be carried out by excision of site involved and histological processing. Quantitative assessment of adhesion formation is evaluated using scoring sys-

Table 1
Grading of Postoperative Adhesions

Grade of Adhesion	Type of Adhesion Formed
Grade 0	No adhesions
Grade 1	Tentative transparent adhesions that frequently separated on their own
Grade 2	Adhesions that give some resistance, but can be separated by hand
Grade 3	Adhesions that require blunt instrument dissection to separate
Grade 4	Dense, thick adhesions that required sharp instrument dissection in the plane of the adhesion to separate

tems for extent and tenacity. Extent of adhesion is evaluated by measuring the length of the uterine horn that has formed adhesions, and/or measuring the total length of the uterine horn after lysis of adhesions. Adhesions are described as a fraction of the length of the horn that is involved in adhesion formation. Tenacity of adhesion is classified according to the numerical scale depicted in **Table 1**.

At least two researchers should be present for independent scoring of all animals. Results are analyzed using ANOVA, with the Fisher PLSD post hoc test used to assess significance, comparing the control group (injured, but not treated) and treatment group (injured and treated). The sites of adhesion formation can be excised for histology and fixed in 10% neutral buffered formalin. Gomori's or Masson's trichrome stain are used for the visualization of collagen in the adhesions. The sites of dense adhesion will stain green and show well-organized, collagen-rich tissue.

3.2. Interfacially Formed Thin Hydrogels for Vascular Wound Healing

The interfacial photopolymerization process takes advantage of the fact that the eosin-Y is anionic in nature, and thus can adsorb ionically to the surface of tissue through interactions with positively charged functional groups present. The polymerizable solutions are prepared so as to isolate the photoinitiator and adsorb it on to the vessel wall. Thereafter, the polymerizable solution containing the photosensitive macromer, but lacking the photoinitiator, is infused, and the tissue is illuminated. All components needed for the photopolymerization process, i.e., light, photoinitiator, and the polymerizable solution, are present only at the vessel wall. This causes the interfacial growth of a thin hydrogel at the vessel wall. As the photoinitiator is depleted (*see* **Fig. 3** for reaction schematic), the reaction becomes self-limiting and ceases after the hydrogel has grown to a certain thickness.

3.2.1. Interfacial Polymerization of Thin Hydrogel Films on Aortic Tissue

The luminal tissue surface of a 1-cm^2 piece of porcine aortic tissue is patted dry with a paper towel. Solution 1 is applied to the entire luminal surface, and then copiously rinsed with PBS. The eosin-Y-stained tissue is submerged in 0.5 mL of solution 2 and illuminated with an argon ion laser operated at multiline, multimode through an optical fiber at an illumination intensity of 70–100 mW/cm^2. The interfacially polymerized piece of tissue is placed in PBS for at least 1 h to allow the hydrogel to completely hydrate. A variety of formulation and photopolymerization conditions can be changed to manipulate the thickness of the resulting hydrogel. A summary of the photopolymerization conditions that can be employed to produce films of various thicknesses appears in **Table 2** (*see* **Note 4**).

Sections, less than 0.5-mm thick, of fully hydrated, interfacially formed hydrogel–porcine aortic tissue laminate, prepared as described above, are cut with a razor blade applied roughly to the center of the sample and moving from the hydrogel side downward to the tissue side. The cross-section is placed on a glass microscope slide, flooded with PBS, and observed under an inverted-stage microscope equipped with a mechanical sizer with a resolution of 1 μm. Five thickness measurements are taken along the length of each section, using the mechanical sizer.

3.2.2. Rabbit Balloon-Angioplasty Injury Model

A balloon denudation injury in the carotid artery of the rabbit is employed as a model more closely related to balloon angioplasty. Male New Zealand white rabbits are used, and no anticoagulant is administered prior to surgery.

1. Anesthesia: Induction and maintenance of anesthesia is identical to that in **Subheading 3.1.3.**
2. Surgery and injury process: The neck region of the rabbits is shaved and prepared using a Betadine scrub solution. The left carotid artery is surgically exposed through a midline incision. Any bleeding from the skin incision or the muscle layers is controlled using electrocautery and a zone is isolated by placing atraumatic vascular clamps on the internal, external, and common carotid arteries. The PE-50 tubing is inserted and the isolated zone is flushed with HBSS several times, to remove blood from the isolated zone. The tubing is removed, and a 4F arterial embolectomy catheter is inserted through the internal carotid artery. The balloon on the catheter is inflated and dragged through the common carotid artery. This process is performed a total of three times by a surgeon who is blinded to treatment grouping.
3. Hydrogel deposition: The balloon catheter is removed, the tubing is replaced, and the isolated zone is flushed with solution 1 (containing the eosin-Y). This solution is allowed to contact the vessel for about 30 s, and is then flushed away

Table 2
Thickness Values of Interfacially Photopolymerized Hydrogels for a Variety of Conditions

Formulation	Component concentration[a]				Illumination intensity (mW/cm^2)	Plateau thickness (μm)
	Eosin Y (mg/mL)	TEOA (mM)	VP (μL/mL)	Macromonomer (w/v%)		
1	0.02	90	2	10	100	81 ± 13
2	0.05	90	2	10	100	107 ± 12
3	0.10	90	2	10	100	191 ± 24
4	0.02	90	2	23	100	140 ± 27
5	0.05	90	2	23	100	181 ± 12
6	0.10	90	2	23	100	257 ± 24

[a]TEOA, triethanolamine; VP, N-vinylpyrrolidone.

using the HBSS. The zone is then filled with solution 2. Excess solution is allowed to flow out of the vessel and the external surface of the vessel is lightly massaged with a gloved finger to contact the solution well with the vessel wall. Illumination is then performed externally, using the argon ion laser, as described above, for about 30 s, at an intensity of 70–100 mW/cm^2. The tubing is removed, the internal carotid artery is ligated, and the clamps are removed to allow blood flow. The unpolymerized precursor is rinsed into the vasculature. The incisions are then closed using a continuous 2-0 Vicryl suture for the musculoperitoneal layer, and a 3-0 Vicryl suture for the cutaneous layer. No prophylactic antibiotics are administered. For the control group, the balloon injury is made as described above, and the incision is closed without the application of the polymerizable solution or illumination. All techniques should be identical between the treatment group and the control group, and all animals should be operated on by the same researcher.

4. Re-exploration: The sites of the injured segment of the common carotid artery are re-explored either 2 h or 14 d after injury, under ketamine anesthesia (35 mg/kg). An intravenous dose of heparin (100 U/kg) is given to the animal about 10 min before re-exploration. On visualization of the injury sites, the patency of the arteries is established by clamping them distal to the site of injury and deposition, gently milking them to remove blood present, and then releasing the clamps to visualize refill of the vessels with blood.

5. Histology: The arteries are then clamped proximally and distally, and flushed clean with saline and fixed using neutral-buffered formalin (10% w/w) under 1.5 m of aqueous solution column pressure. The tissues are allowed to fix in this fashion for 2–3 h and are then harvested and allowed to additionally fix for 3–5 d prior to processing for histology. Slides are stained with either Verhoeff's elastin stain or Masson's trichrome stain. Microscopy using the mechanical sizer is used to measure the cross-sectional areas of the thrombus and lumen (2 h), and of the intima and media (14 d).

Half of each of three vessel segments from each group, randomly selected, are reserved for scanning electron microscopy to determine the presence of endothelial cells at the 14-d time-point. The tissue is cut open longitudinally, dehydrated in a graded ethanol series, and dried by critical-point drying. Samples can be coated with gold and imaged by scanning electron microscopy. A region in each vessel segment, corresponding to the approximate center of the injured zone in the animal, is examined.

4. Notes

1. In addition to using DL-lactide, several other α-hydroxy ester linkages can be used to control the rate of absorption of such hydrogels. For example, glycolide can be used to give faster absorbing hydrogels, and ε-caprolactone can be used to form slower absorbing hydrogels.
2. In addition to using acrylate moieties as the polymerizable end groups, other end groups, such as methacrylate or cinnamate, can also be used.

3. Visible light from a xenon arc lamp or an argon ion laser can be used to initiate the polymerization of these solutions. For example, a 10% w/v solution of macromer in PBS with eosin Y (0.5 mM) and triethanolamine (0.5 M) containing 0.5 µL/mL N-vinyl pyrrolidinone can be irradiated with an argon ion laser. Light output should be in the range of 400–600 mW. 0.1 mL of this macromer solution is poured onto an 18 × 18-mm glass cover slip and illuminated. The fiber is held an appropriate distance from the sample, to get a 2–3-cm spot size. Gelation is noted as described.
4. The thickness of the resulting interfacially polymerized thin hydrogels can be controlled by varying several components in the formulation. A sampling of some of these formulation conditions is given in **Table 2**.

References

1. Hertzler, A. E. (1919) *The Peritoneum,* Mosby, St. Louis, MO.
2. DiZerega, G. S. and Rogers, K. E. (1992) *The Peritoneum,* Springer-Verlag, New York.
3. Ryan, G. B., Grobety, J., and Majno, G. (1973) Mesothelial injury and recovery, *Am J. Pathol.*, **71,** 93–112.
4. Surgical Membrane Study Group (1992) Prophylaxis of pelvic sidewall adhesions with Gore-Tex surgical membrane: a multicenter clinical investigation, *Fertil. Steril.* **57,** 921–923.
5. Haney, A. F. and Doty, E. (1992) Murine peritoneal injury and de novo adhesion formation caused by oxidized regenerated cellulose (INTERCEED TC7) but not expanded polytetrafluoroethylene (Gore-Tex surgical membrane), *Fertil. Steril.* **57,** 202–208.
6. Schwartz, R. S., Holmes, D. R., and Topol, E. J. (1992) The restenosis paradigm revisited: an alternative proposal for cellular mechanisms. *J. Am. Coll. Cardiol.* **20(5),** 1284–1293.
7. Fuster, V., Badimon, L., Badimon, J. J., and Chesebro, J. H. (1992) The pathogenesis of coronary artery disease and the acute coronary syndromes. *N. Eng. J. Med.* **32(5),** 242–250.
8. Ross, R., Raines, E. W., and Bowen-Pope, D. F. (1986) The biology of platelet-derived growth factor. *Cell* **46,** 155–169.
9. Edelman, E. R., Nugent, M. A., Smith, L. T., and Karnovsky, M. J. (1992) Basic fibroblast growth factor enhances the coupling of intimal hyperplasia and proliferation of vasa vasorum in injured rat arteries. *J. Clin. Invest.* **89,** 465–473.
10. Linder, V., Lappi, D. A., Baird, A., Majack, R. A., and Reidy, M. A. (1991) Role of basic fibroblast growth factor in vascular lesion formation. *Circ. Res.* **68,** 106–113.
11. Lee, S. H. and Ruckenstein, E. (1988) Adsorption of proteins onto polymeric surfaces of different hydrophilicities: a case study with bovine serum albumin, *J. Colloid. Interface Sci.* **125,** 365–379.
12. Buck, C. A. and Horwitz. A. F. (1987) Cell surface receptors for extracellular matrix molecules, *Ann. Rev. Cell Biol.* **3,** 179–205.
13. Andrade, J. D. and Hlady, V. (1986) Protein adsorption and materials biocompatibility: a tutorial review and suggested hypotheses, *Adv. Polymer. Sci.* **79,** 1–63.

14. Desai, N. P. and Hubbell, J. A. (1991) Solution technique to incorporate polyethylene oxide and other water-soluble polymers into surfaces of polymeric biomaterials, *Biomaterials* **12,** 144–153.
15. Desai, N. P. and Hubbell, J. A. (1992) Tissue response to intraperitoneal implants of polyethylene oxide-modified polyethylene terepthalate, *Biomaterials* **13,** 505–510.
16. Ito, K., Tanaka, K., Tanaka, H., Imai, G., Kawaguchi, S., and Itsuno, S. (1991) Poly(ethyleneoxide) macromonomers. 7. Micellar polymerization in water, *Macromolecules* **24,** 2348–2354.
17. Sawhney, A. S., Pathak, C. P., van Rensburg, J. J., Dunn, R. C., and Hubbell, J. A. (1994) Optimization of photopolymerized bioerodible hydrogel properties for adhesion prevention, *J. Biomed. Mater. Res.* **28,** 831–838.
18. Hill-West, J. L., Chowdhury, S. M., Sawhney, A. S., Pathak, C. P., Dunn, R. C., and Hubbell, J. A. (1994) Prevention of postoperative adhesions in the rat by in situ photopolymerization of bioresorbable hydrogel barriers, *Obstet. Gynecol.* **83,** 59–64.
19. Hill-West, J. L., Chowdhury, S. M., Sawhney, A. S., Pathak, C. P., Dunn, R. C., and Hubbell, J. A. (1996) Efficacy of a novel hydrogel and conventional hydrogel barriers, *J. Reprod. Med.* **41,** 149–154.
20. Verreet, P. R., Fakir Ohmann, C., and Roher, H. D. (1989) Preventing recurrent postoperative adhesions: an experimental study in rats, *Eur. Surg. Res.* **21,** 267–273.
21. Ciba-Geigy Corporation, Material Safety Data Sheet for Irgacure **651,** Hawthorne, NY.
22. Coohill, T. P., Peak, M. J., and Peak, J. G. (1987) The effects of the ultraviolet wavelengths present in sunlight on human cells *in vitro, Photochem. Photobiol.* **46,** 1043–1050.
23. Hill-West, J. L., Chowdhury, S. M., Slepian, M. J., and Hubbell, J. A. (1994) Inhibition of thrombosis and intimal thickening by in situ photopolymerization of thin hydrogel barriers, *Proc. Natl. Acad. Sci. USA* **91,** 5967–5971.
24. Sawhney, A. S., Pathak C. P., and Hubbell, J. A. (1993) Bioerodible hydrogels based on photopolymerized poly(ethylene glycol)-co-poly(α-hydroxy acid) diacrylate macromers, *Macromolecules* **26,** 581–587.

9

Fabrication and Implantation of Gel-Filled Nerve Guidance Channels

Ravi V. Bellamkonda and Robert F. Valentini

1. Introduction

In human adults, the peripheral nervous system (PNS) is capable of healing and regeneration. In order to reestablish function, nerve tissue must heal by true regeneration of a functional structure, since healing by simple scar will not reestablish electrical connectivity. Nerve guidance systems have been used experimentally to enhance regeneration through the use of functionalized gels, the delivery of growth-promoting molecules, and the use of neuronal support cells or genetically engineered cells. The objectives of this chapter are to overview the methods used to construct gels for nerve stimulating regeneration and to outline the surgical techniques to implant nerve guidance systems.

The behavior of cells is influenced by both their intrinsic genetic programs and their extracellular environment. Humoral factors, cells, and the extracellular matrix (ECM) are the main components of a cell's external environment. Extracellular matrices modulate the organization of intracellular cytoskeleton, cell differentiation, and the spatial architecture of cells and tissues. The protein and proteoglycan macromolecules of the extracellular matrix are highly regulated and are thought to be organized to provide permissive and nonpermissive three-dimensional (3-D) pathways for neural cell migration and axon guidance during development *(1,2)*.

1.1. The Problem of Regeneration

It has been estimated that more than 200,000 nerve repair procedures are performed annually in the United States alone *(3)*. However, complete recovery after damage to the nervous system is rare. Various treatments to prevent nerve degeneration have been attempted *(4,5)*. These include cooling of injured nerves in an attempt to avoid ischemia *(6)*; and the application of agents like naloxone *(7)* to avoid the consequences of trauma. In most nerve injuries, however, degeneration of the distal neural elements cannot be prevented. Neurons in adults are not capable of proliferating. Therefore, recovery of the injured neuron depends on its ability to survive and regenerate. Nerve regeneration thus may be defined as the regrowth of injured axons, followed by restoration of their original synaptic connections and the recovery of their physiological functions. Neurons of differing species, and different neurons within the mammalian species, have differing abilities to regenerate. Older animals regenerate less well than younger ones. In mammals, the neurons of the PNS have a relatively greater ability to regenerate, compared to neurons of the central nervous system (CNS), which seldom regenerate *(8,9)*.

1.2. Lack of Appropriate Environment Contributes to Regenerative Failure

One of the main factors that contributes to the differential regenerative abilities of peripheral and central neurons is the environment in the adult PNS, compared to the environment in the adult CNS. The environment, in particular, the extracellular matrix (ECM), plays a critical role in both development and regeneration of the PNS and CNS *(1,10)*.

One of the major differences in the PNS and CNS environments is in the cells that myelinate neurons in the two systems. The PNS is myelinated by Schwann cells, which, in response to injury, secrete growth factors and neurite promoting basement membrane proteins, such as laminin (LN) *(1,11–13)*. Studies show that the use of PNS nerve grafts (which carry Schwann cells) may facilitate limited regeneration of lesioned nerves in the CNS *(14,15)*. In contrast to the PNS, CNS neurons are myelinated by oligodendrocytes, which express molecules that inhibit neurite growth and are thought to contribute to the regenerative failure in the CNS *(16)*. Even within the PNS, the spinal dorsal roots, which lie close to the junction of PNS and CNS, do not regenerate *(17)*, and present a challenge to regeneration efforts. These studies indicate that the environment of regeneration is a determining factor influencing the regenerative fate of the particular neural system.

It thus has been hypothesized that when polymer scaffolds containing ECM proteins, such as laminin and collagen, are presented to a severed nerve in vivo, the regeneration of injured or transected nerves would be significantly enhanced. Hydrogels are popular materials for 3-D polymer scaffolds for soft tissues such as nerves, because their physical properties match well with soft tissues in vivo. Permissive, hydrogel-based 3-D scaffolds are often functionalized with neurite-promoting proteins/agents, such as Laminin *(18)* or Schwann cells *(9)* to enhance the scaffold's ability to support and promote

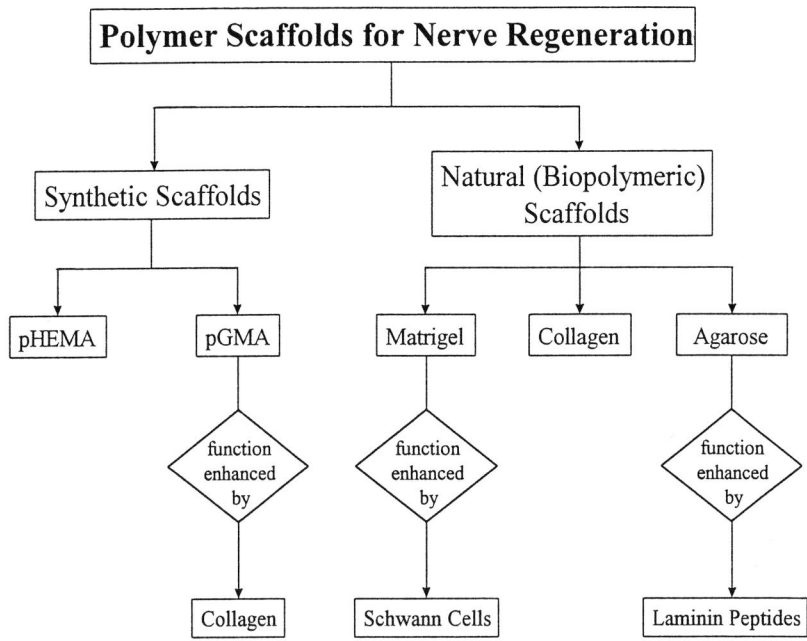

Fig. 1. Listing of 3-D polymer scaffolds covered in this chapter.

axonal regrowth. Small, synthetic peptides mimicking the function of larger ECM proteins are also used.

The important 3-D polymeric scaffolds used to support neurite extension and growth in vitro and in vivo are shown in **Fig. 1**. The following sections will detail the synthesis and functionalization methods.

2. Materials
2.1. Polyhydroxyethylmethacrylate (pHEMA)

1. Tris-(hydroxymethyl) aminomethane, sodium chloride, ammonium persulphate, sodium bisulphite, and sodium bicarbonate (Sigma, St. Louis, MO).
2. Penicillin–streptomycin (Gibco-BRL, Grand Island, NY).
3. 2-Hydroxyethylmethacrylate (Polysciences, Warrington, PA).
4. Ethylene glycol, sodium hydroxide, and hydrochloric acid (Fisher Scientific, Fair Lawn, NJ).
5. Untreated 24-well tissue-culture plastic trays (Corning, Corning, NY).

2.2. Polyglycerylmethacrylate (pGMA)

1. Glycidyl methacrylate (GdMA, 2,3-epoxypropyl methacrylate, Aldrich, Milwaukee, WI).
2. (Dimethylamino) ethyl methacrylate (DMAEMA, Aldrich).

3. Methacrylic acid (MAA) (Aldrich).
4. Ethylene glycol dimethacrylate (EGDMA, Aldrich Chemical Co., Milwaukee, WI).
5. Ammonium persulphate and sodium metabisulphite (Aldrich Chemical Co., Milwaukee, WI).

2.3. MATRIGEL®

1. Matrigel (Collaborative Research, Bedford, MA). Store at –20°C, thaw overnight at 4°C. Gels rapidly and irreversibly at room temperature of 22°C.
2. Dulbecco's modified Eagle's medium (DMEM, Gibco-BRL). Store at 4°C. Must be kept cold (on ice) before using to dilute Matrigel.
3. Calf Serum (FCS, Gibco-BRL).
4. Pencillin–streptomycin (Gibco-BRL).

2.3.1. Isolation of Schwann Cells (9).

1. Adult male inbred Fisher rats (250–300 g, Taconic, Germantown, NY).
2. Vitrogen 100 (Collagen, Palo Alto, CA). Must be stored at 4°C. Gels irreversibly at room temperature of 22°C.
3. Ca^{2+}/Mg^{2+}-free Hanks' balanced salt solution (HBSS, Sigma).
4. Trypsin (Sigma).
5. Collagenase.
6. Hyaluronidase.
7. Forskolin (2 M, Sigma).
8. Pituitary extract (10 g/mL, Sigma).
9. EDTA (Sigma).

2.4. Collagen Type I (Vitrogen 100)

1. Sodium chloride, sodium hydroxide, and monophosphate (Sigma).
2. Type I collagen (Vitrogen 100) is purchased from Collagen. Store at 4°C.

2.5. Agarose Gels

1. SeaPrep® agarose powder (FMC Bioproducts, Rockland, ME).
2. Dulbecco's phosphate buffered saline (Gibco-BRL).

2.5.1. Peptide and Laminin Immobilization to Agarose Gels Using Imidazole-Based Solution Chemistry

1. Agarose (SeaPrep, FMC).
2. Dulbecco's PBS.
3. Laminin (mouse, Gibco-BRL) or laminin peptides CDPGYIGSR, SIKVAV, or GRGDSP (Anawa, Wagen, Switzerland). Store at –20°C.
4. 1 carbonyldiimidazole (CDI), albumin (bovine), acetone, and sodium bicarbonate (Sigma). Avoid exposure to moisture during storage. Moisture-sensitive.
5. Molecular sieves (4 Å, Fisher Scientific). Dry sieves by heating at 350°C for 2 h.

2.5.2. Peptide and Laminin Immobilization to Agarose Gels Using Benzophenone-Based Photo Chemistry

1. Agarose (SeaPrep, FMC).
2. Dulbecco's PBS.
3. Laminin (mouse; Gibco-BRL).
4. Albumin (bovine), acetone, and sodium bicarbonate (Sigma).
5. Molecular sieves and 1,4 - dioxane (Fisher Scientific).
6. 4-benzoylbenzoic acid, succinimidyl ester (Molecular Probes, Eugene, OR). Light-sensitive. Store in dark.
7. Gate MR-4 UV lamp (Glo-Mark Systems, Saddle River, NJ).

2.6. In Vitro Analysis of Polymer Scaffolds: Visualization of 3-D Neurite Extension

It is important to quantify neurite extension in 3-D polymer scaffolds, to optimize their design and functionalization with neurite promoting agents, such as laminin or Schwann cells. This is achieved by capturing images of neurite growing in 3-D polymer scaffolds, using the system described below.

2.6.1. Digital Image Analysis System

1. Nikon inverted light microscope model TMS-F (Nikon, Tokyo, Japan).
2. Javelin Electronics JE7862 high-resolution monochrome CCD camera (Javelin, Japan).
3. Scion LG-3 frame grabber (Scion, Frederick, MD).
4. Power Macintosh 7200 computer (Apple Computer, Cupertino, CA).
5. NIH *Image* Software Package (developed at the U.S. National Institutes of Health, and available on the Internet at http://rsb.info.nih.gov/nih-image/).

2.6.2. Other Materials

1. Hemacytometer (Hausser Scientific, Horsham, PA).
2. Nerve guidance channels can be fabricated from a range of commercially available, tubular materials, as listed in **Table 1**.

3. Methods
3.1. Hydroxyethylmethacrylate Gel Synthesis (see Note 1) (19,20)

1. 2-hydroxyethylmethacrylate, ethylene glycol, and an aqueous buffer, usually 0.05 M Tris-HCl, 0.15 M NaCl, are mixed at 1:1:1 (by vol, so that the final HEMA concentration is 33% w/v).
2. After degassing under low pressure, and while on ice, 0.1 mL of 12% sodium bisulphite and 0.1 mL of 6% ammonium persulphate solution are added to the mixture.
3. Gels in the shape of flat sheets are prepared by using a form that consists of a sandwich of two precleaned glass microscope slides spaced by two no. 1 coverslips.

Table 1
Materials Used for Nerve Guidance Channels

Nonporous
    Polytetrafluoroethylene (PTFE)
    Polyethylene (PE)
    Silicone elastomers (SE)
    Polyvinyl chloride (PVC)
Microporous
    "Gortex," expanded polytetrafluoroethylene (ePTFE)
    "Millipore" (cellulose filter)
Semipermeable
    Polyacrylonitrile (PAN)
    Polyacrylonitrile/polyvinyl chloride (PAN/PVC)
    Polysulfone (PS)

4. This sandwich is held in position with an alligator clip, and the separation of the slides is approx 0.5 mm.
5. The polymer mixture is injected between the glass slides with a Pasteur pipet, and is drawn into the space by capillary action.
6. Following injection of the slides with polymer, the slides are polymerized at 37°C for 2 h.
7. When polymerization is complete, the sandwich form is separated and the sheet of polymer is peeled from the form.
8. These sheets are then exhaustively dialyzed against 0.05 M Tris-HCl and 0.15 M NaCl buffer solution, pH 7.4.
9. When all of the ethylene glycol had been removed from the polymerized gel, the gels became crystal clear and are then ready to be prepared for cell-culture experiment.
10. Disks or buttons of polymer are cut from the polymerized sheets with a no. 10 cork borer; the radius of the resultant buttons is 140 mm.
11. The buttons are then placed, 1/well, into untreated Corning 24-well tissue culture trays.
12. To sterilize the gels, 1 mL of a 10% penicillin–streptomycin antibiotic solution, made in Puck's Ca- and Mg-free saline, is added, and the trays are exposed for 2 h to UV germicidal irradiation.
13. The antibiotic solutions are removed and the gels are incubated for 30 min with media, prior to cell seeding.

3.2. Polyglyceryl Methacrylate Gels (21,22)

1. GMA is obtained by hydrolysis of glycidyl methacrylate in aqueous sulphuric acid, according to the method described by Refojo (1965).
2. The GMA is further purified by distillation under dynamic vacuum (1.33×10^{-3} pa) to remove epoxy-type residues and traces of polymer.
3. Hydrogels based on pGMA are prepared by radical polymerization with the crosslinking agent EGDMA in dH_2O.

4. Polymerization is initiated in a small glass tube, with the redox initiator ammonium persulphate (6% v/v) and sodium metabisulphite (12% v/v) added in a 0.37 wt% ratio to GMA, at 60°C for 12 h.
5. The hydrogels are washed in dH$_2$O for a minimum of 4 wk and kept in water solution.
6. The hydrogels are sterilized by boiling in dH$_2$O.

3.2.1. Copolymeric Poly(glyceryl methacrylate)–Collagen Hydrogels (see Note 2) (21–23)

1. GMA is obtained by hydrolysis of glycidyl methacrylate in aqueous sulphuric acid, according to the method described by Refojo (1965).
2. The GMA is further purified by distillation under dynamic vacuum (1.33×10^{-3} pa) to remove epoxy-type residues and traces of polymer.
3. Hydrogels based on pGMA are prepared by radical polymerization with the crosslinking agent EGDMA in dH$_2$O.
4. Polymerization is initiated in a small glass tube with the redox initiator ammonium persulphate (6% v/v) and sodium metabisulphite (12% v/v) added in a 0.37 wt% ratio to GMA, at 60°C for 12 h.
5. The hydrogels are washed in dH$_2$O for a minimum of 4 wk, and are kept in water solution.
6. The hydrogels are sterilized by boiling in dH$_2$O.
7. Collagen is mixed with a concentrated salt solution of 1.3 M NaCl in 0.2 M monophosphate, and with 0.1 M NaOH, in the proportion of 8:1:1 (v/v), and the final concentration of collagen is 2.4 mg/mL.
8. The pH of the collagen is adjusted to 7.2.
9. The solution is kept on ice to prevent spontaneous gelation.
10. Dehydrated pGMA are re-equilibrated by swelling in the collagen solution for 2 d at 4°C.
11. The gels are then maintained at 37°C in a humidified incubator, to allow polymerization of collagen.

3.3.1. Loading of pHEMA and pGMA Gels into PAN–PVC Guidance Channels

1. Because pHEMA and pGMA both involve in situ free-radical polymerization, the authors think it best to polymerize the gels after loading the PAN–PVC guidance channels with the monomeric reagents (**step 4** of **Subheading 3.2.**).
2. For pGMA-collagen gels, guidance channels loaded with pGMA may be lyophilized and reequilibrated with collagen solution, as described above.

3.4. Matrigel® Loading into Nerve Guidance Channels (see Note 3) (9)

Matrigel is derived from an Engelbreth-Holm-Swarm (EHS) mouse sarcoma. All pipets and tissue-culture materials that contact Matrigel must be cold, to prevent premature gelation.

3.4.1. Schwann Cell Isolation

1. Matrigel is diluted with cold DMEM, so that the ratio of DMEM:Matrigel is 7:3 (v/v).
2. The 30% Matrigel solution is then loaded into 10-cm long-nerve guide tubes by attaching a syringe with a 22-gage needle to one end and drawing the Matrigel into the tube. Care should be taken to avoid air bubbles.
3. The tubes are then cut into 10-mm-long tubes, their ends sealed with polymer glue, and stored for 24 h at 37°C in an incubator to allow for Matrigel gelation.
4. Sciatic nerves are collected into DMEM, stripped of their epineurium, and chopped into 1-mm² pieces.
5. The nerve pieces are placed on Vitrogen 100-coated Petri dishes in DMEM supplemented with 10% fetal calf serum and penicillin streptomycin (1000 U/mL).
6. Every 5 d, the nerve pieces are transferred to new Petri dishes, leaving behind the fibroblasts that migrate out onto the Vitrogen 100.
7. After four such transfers, almost all of the fibroblasts leave the nerve explant and the explants are then dissociated by incubating the nerve chunks at 37°C for 2 h in Ca^{+2}- and Mg^{+2}-free HBSS containing 0.3% trypsin, 0.1% collagenase, and 0.1% hyaluronidase.
8. The cells are then triturated, washed, and cultured in DMEM with 10% FCS for 1 d.
9. The following day, the culture medium is replaced with motogenic medium containing DMEM, FCS (10%), forskolin (2 mM), and pituitary extract (10 mg/mL). For the next 4–6 d, Schwann cells are cultured in mitogenic medium at 37°C in a humidified atmosphere with 5% CO_2.

3.5. Schwann Cell Seeding onto Matrigel Scaffolds: Schwann Cell Functionalized Matrigel: Loading into Nerve Guidance Channels (see Note 4)

1. Schwann cells are used to functionalize and enhance the neurite-promoting properties of Matrigel. Here, Matrigel is used as a scaffolding material that presents Schwann cells to the regeneration environment.
2. Schwann-cell-seeded Matrigel scaffolds are introduced into nerve guidance channels using a method similar to the one described above. Except, instead of using a mixture of DMEM:Matrigel at 70:30 (v/v), 80 10^6 to 120×10^6/mL of Schwann cells are suspended in DMEM, and this cell–DMEM suspension is used to dilute Matrigel in a ratio of 70:30 (v/v).
3. The cell–Matrigel suspension is then loaded into 10-cm tubes, as described above.

3.6. Type I Collagen Gels (see Note 5) (24)

1. Collagen solution (2.0 mg/mL) is mixed with a concentrated salt solution of 1.3 M NaCl in 0.2 M monophosphate, and with 0.1 M NaOH, in the proportion of 8:1:1 (v/v), and the final concentration of collagen is 2.4 mg/mL.
2. The pH of the collagen is adjusted to 7.2.
3. The solution is kept on ice to prevent gelation until cells are entrapped.

3.7. Loading Collagen Gels into PAN-PVC Tubes

1. Collagen gels may be loaded into nerve guidance channels in a procedure similar to the loading of Matrigel into tubes, using a 22 gage needle and syringe to draw liquid collagen into the tube (at 4°C).
2. The guidance channels' ends are sealed with polymer glue and the channels placed at 37°C for 24 h.
3. The channels are then cut to desired length to bridge a nerve gap.

3.8. Functionalizing Agarose Gels with Laminin or Laminin-Peptides, Using Imidazole Chemistry (see Note 6) (25,26)

1. Four mL gels of 1% agarose in PBS are dehydrated by repeated washes in acetone, followed by dry acetone (dried under 4 Å molecular sieves).
2. A CDI solution prepared in dry acetone (5 mg/mL) is added to the acetone-washed agarose gels (5 mL/4 g gel).
3. The activation reaction is allowed to proceed for 5 min with gentle agitation.
4. Gels are then washed 5× with dry acetone for 5 min per wash to remove unbound CDI.
5. CDI-activated gels are made into 2% solution in 2 mL, 100 mM sodium bicarbonate buffer (pH = 8.5) by heating and agitation.
6. Laminin (0.5 mg) or peptide (2.0 mg) is dissolved into 2 mL of the sodium bicarbonate buffer, and the solution mixed with the CDI-activated gel solution.
7. The coupling reaction is allowed to proceed for 36 h under gentle agitation at room temperature.
8. The gels are further quenched in sodium bicarbonate for 6 h. at room temperature, and then washed thoroughly with PBS for another 48 h, lyophilized, and redissolved to the desired gel concentration, typically 1.0% in PBS.

3.9. Functionalizing Agarose Gels with Laminin or Laminin-Peptides, Using Benzophenone-Based photochemistry (see Note 7) (27,28)

1. Four mL gels of 1% agarose in PBS are dehydrated by repeated washes in acetone, followed by dry acetone (dried under 4 Å molecular sieves).
2. Benzophenene (BP) solution prepared in dioxane (1 mg/mL) is added to the acetone-washed agarose gels (2 mL/4 g gel), and the mixture is exposed under UV light (wave length: 250 nm) for 1 h.
3. Gels are then washed 5× with dioxane for 5 min per wash to remove unbound BP.
4. Laminin (0.5 mg) or peptide (2.0 mg) is dissolved into 3 mL of 100 mM sodium bicarbonate buffer (pH = 8.5), and this solution is added to the BP-activated gels (4 g gel).
5. The coupling reaction is allowed to proceed for 36 h under gentle agitation at room temperature.
6. The gels are further quenched in sodium bicarbonate for 6 h at room temperature, and then washed thoroughly with PBS for another 48 h, lyophilized, and redissolved to the desired gel concentration of 1.0% in PBS.

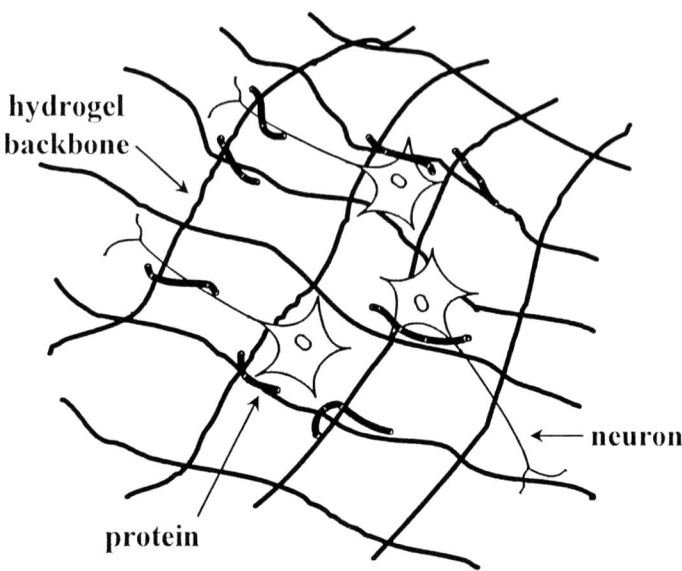

Fig. 2. Schematic depiction of neural cell entrapment in peptide/protein-modified 3-D hydrogels.

3.10. Preparation of Agarose Gels and Loading of Neural Cells into Agarose Gels for In Vitro Analysis of Hydrogel Scaffold (see *Fig. 2* for schematic; see Note 8)

3.10.1. Preparation of Agarose (for 10 mL of 1.0% Agarose Solution)

1. Add 0.10 g of SeaPrep agarose powder to 10 mL PBS.
2. Stir the mixture over medium heat until the agarose is completely dissolved (usually 10–15 min).
3. Sterilize the liquid agarose through a 0.2 µm syringe filter into a sterile container.

3.10.2. Cell Entrapment in Agarose Gels

1. Coat the bottom of each well of a sterile 24 well cell culture dish with 200 µL of the sterilized liquid agarose.
2. Swirl the dish, evenly distributing the liquid over the bottom of the well.
3. Chill the dish at 4°C until the liquid agarose gels (usually about 15 min, depending on the concentration).
4. Draw 200 µL of the sterilized liquid agarose into a 1-cc syringe; set aside.
5. Take up the desired volume of the cell and medium mixture into a micropipet (typically 4–5 dorsal root ganglia pieces or 50,000–100,000 embryonic cortical cells from rodent brains in 10 µL or less of sterilized cell culture medium).
6. Insert the tip of the micropipet inside the bevel of the syringe, creating a seal.

Gel-Filled Nerve Guidance Channels

7. Quickly pull back on the syringe plunger until the cell and medium mixture is entirely transferred into the syringe.
8. Carefully pump the plunger in and out to thoroughly mix the cells and the liquid agarose.
9. Add the contents of the syringe to the previously gelled agarose in the 24-well cell-culture dish.
10. As described above, swirl the dish and chill until the entire well has completely gelled.
11. Add 1 mL of cell-culture medium to each well.
12. Add 80 μL NGF to each well (when required).
13. Every 48 h, the cell-culture medium and the NGF should be replaced and the cells should be visualized/quantified.

3.11. In Vitro Analysis and Quantification of Neurite Extension in Hydrogel Scaffolds (29–31)

3.11.1. Calibration of the Digital Image Analysis System

1. Place the hematocytometer under the light microscope.
2. Focus on the smallest area of the grid under the desired magnification (one side of the smallest squares is 50 μm long).
3. Switch the optical path of the microscope to the CCD camera
4. Start the NIH *Image* software package.
5. Select **Start Capturing** from the **Special** menu (an image of the hematocytometer will appear on the monitor).
6. Adjust the light and focus the microscope.
7. Trace along a known length of a hematocytometer grid line with the straight line tool (typically, a 600 μm length).
8. Select **Set Scale** from the **Analyze** menu.
9. In the set scale dialog box, change the units to micrometers and the known length to 600.0.
10. Click **OK** and the calibration is complete.

3.11.2. Measuring the DRG Neurite Length

1. Calibrate the system.
2. Place the gel to be imaged under the microscope.
3. Using the binocular eyepieces, focus on a DRG.
4. Switch the optical path of the microscope to the CCD camera.
5. Start the NIH *Image* software package.
6. Select **Start Capturing** from the **Special** menu.
7. Adjust the light power and the focus of the microscope.
8. Trace along the length of one of the longest neurites with the freehand line tool.
9. Select **Measure** from the **Analyze** menu.
10. Select **Show Results** form the **Analyze** menu (the length is given in the units specified during the calibration).

11. Repeat this procedure for seven more of the longest neurites (the average of these eight lengths is the average neurite length for that particular DRG).
12. Repeat the entire process for 2–3 more DRGs.

3.11.3. Measuring the DRG Spread

1. Calibrate the system.
2. Place the gel to be imaged under the microscope.
3. Using the binocular eyepieces, focus on a DRG.
4. Switch the optical path of the microscope to the CCD camera.
5. Start the NIH *Image* software package.
6. Select **Start Capturing** from the **Special** menu.
7. Adjust the light power and the focus of the microscope (at this point, image enhancement procedures can be performed).
8. Trace along the outer edge of the cell body with the freehand selection tool.
9. Select **Measure** from the **Analyze** menu.
10. Trace along the outer edge of the ganglionic area with the freehand selection tool (the ganglionic area is defined as the area covered by the cell body and all of the neurites).
11. Select **Measure** from the **Analyze** menu.
12. Select **Show Results** form the **Analyze** menu (the area is given in the square of the unit specified during the calibration).
13. The spread is defined as the difference of the total ganglionic area and the cell-body area divided by the cell body area.
14. Repeat this process for 2–3 more DRGs.

3.11.4. Computing the Percent of Cells Extending Neurites

(NOTE: The digital image analysis system is not required for this procedure)

1. Place the gel to be imaged under the microscope.
2. Focus on a typical area of cells.
3. With two counters, simultaneously count the total number of cells present in the gel and the number of cells that are extending neurites.
4. Repeat this procedure on several different areas of the gel, keeping a running total (typically, at least 900 total cells are counted).
5. The percent of cells extending neurites is calculated by dividing the number of cells extending neurites by the total number of cells.

3.12. Implantation of Nerve Guides (see Note 9)

In repair procedures using nerve guidance channels, the free or mobilized ends of a severed nerve are positioned within the lumen of a tube and anchored in place with sutures (**Fig. 3**). Tubulation repair provides prevention of scar tissue invasion into the regenerating environment, provision of directional guidance for elongating axons, and maintenance of endogenous trophic or growth factors *(32,33)*.

thigh. The underlying fascia is incised along the whitened border indicating the juncture of the vastus lateralis and biceps femoris muscles. The femur lies deep to this avascular plane and can be exposed without cutting into the body of any muscle group, thus minimizing blood in the field. The sciatic nerve can be exposed by pushing the hamstring muscle group posteriorly to expose the nerve. A retractor or suture can be used to retract the muscle while surgery is underway. The nerve is encased in a mass of loose connective tissue, fat, and small blood vessels.

3.12.6. Nerve Guide Placement

1. At this point, an operating or dissecting microscope should be used. At least 1 cm of nerve is isolated by carefully dissecting away everything except the outer connective tissue sheath (e.g., epineurium), which is glistening white in appearance. Care should be exercised so as not to cut any blood vessels or to handle the nerve except by the epineurium using 5-0 forceps. The nerve should be handled gently, because axonal damage can occur.
2. The exposed nerve is divided with microscissors at the level of the mid-thigh and the nerve guide is positioned. Sutures (10-0) are placed as in **Fig. 3**. The first suture is placed 1 mm in from the proximal end of the nerve guide, pulled through the lumen, placed through the epineurium 1 mm in from the end of the proximal nerve stump, placed back into the lumen, and carried through to the outer wall, close to the original entry point. Pull the suture up fully after each placement. The nerve is pulled into the tube by gently pulling on the suture/nerve until it moves into the tube. The knot is tied using three ties. It may be easier to place the sutures loosely in the channel and nerve at one end before cutting the nerve. The tube is flushed with sterile saline to remove any trapped air bubbles. A segment of distal nerve is resected, to correct for the length of nerve guide used, and the same procedure is used to position and anchor the distal stump through outer tube to lumen to epineurium to lumen through tube.

3.12.7. Wound closure

The field is inspected for bleeding, disrupted sutures, air bubbles in lumen, foreign material, and so on. The fascia is closed using 3–4 interrupted sutures (5-0). The skin is closed using interrupted 5-0 sutures, or running, subcuticular 5-0 bioresorbable sutures. Wound clips may also be used.

4. Notes

1. HEMA Gels. When the HEMA gel concentration was dropped below 33%, in our experience, HEMA gels lost their ability to become optically clear after the ethylene glycol was washed away. This of course poses problems for visualization of cells in vitro via optical microscopy. On the other hand, 33% HEMA gels may be too stiff (or their porosity too low) for these gels to allow neurite extension when neural cells are trapped within the gel matrix. This may be a problem with HEMA gel synthesis by this method.

2. pGMA gels. pGMA gels, along with other acrylic gels, such as HEMA, are polymerized at the site of final use (or inside nerve guidance channels), because they are irreversible gels and cannot be reformed into another shape once polymerized. This may pose practical difficulties for their use and application. Also, the polymerized gels should be washed thoroughly to remove any remaining unpolymerized monomeric compounds to ensure biocompatibility, because acrylic monomers are often toxic.
3. Matrigel. Although Matrigel has been shown to be a potent promoter of neurite extension from neural cells in vitro, its application in clinical medicine is in doubt, because it is derived from a mouse sarcoma, rendering difficult its FDA approval. Also, in some instances it has been shown to inhibit nerve regeneration. In addition, Matrigel, because of its source, may have great variations in batch to batch lots at the manufacturer, and this factor should be considered in the experimental design using Matrigel. Care also should be taken while handling Matrigel, because it is an irreversible gel that gels rapidly at room temperature, and hence it should always be handled on ice and thawed at 4°C overnight, not at room temperature.
4. Schwann cell loading onto Matrigel scaffolds. Care should be taken that fibroblast contamination is avoided in the Schwann cell culture. Also, sufficient numbers of Schwann cells are critical to observe enhanced regeneration. A minimum of 80×10^6 cells/mL may be required to observe an effect in vivo when Schwann cells are seeded onto 3-D Matrigel scaffolds.
5. Collagen gels. Collagen gels at 2.4 mg/mL sometimes precipitate in the process of gelation, and this may diminish optical transmission. It is the authors experience that the rate of warming of the gel may influence this. Also, it is the authors' experience that Vitrogen 100 from Collagen is less likely to have this problem, compared to collagen type I from other sources.
6. CDI modification of agarose gels. Modification of agarose gels with CDI chemistry, and subsequent binding of proteins (e.g., laminin) or laminin peptides, may decrease the solubility of agarose in PBS. If problems with solubulization occur, the authors suggest that the concentration of CDI solution in acetone be decreased in step **Subheading 3.8, step 2.**, appropriately. Also, the time of exposure of the above CDI solution to acetone-washed agarose may be decreased appropriately. This would impact the yield (amount of protein bound to the gel), but CDI has a high protein-binding efficiency, and the decreased yields may still be effective in promoting neurite growth from neural cells in vitro and in vivo.
7. Benzophenone (BP) modification of agarose gels. BP chemistry yields of protein conjugation to agarose gels are generally lower than that of CDI chemistry. However, the big advantage with BP chemistry is the possibility of spatially controlling peptide conjugation in different areas of the agarose gel. This may be useful for 3-D axon guidance in vitro and in vivo. Care should be taken at the UV exposure step, so that the agarose polymeric chains are not broken because of prolonged exposure. This may result in loss of gelation properties of agarose gels.

8. In vitro visualization of neurite extension In 3-D polymeric scaffolds (e.g., agarose gels with embryonic dorsal root ganglia). For optimal visualization (via optical microscopy) of neurite extension of cells entrapped in 3-D matrices, the authors recommend that Nomarski optics (for glass culture dishes) or Hoffman optics (for tissue-culture plastic dishes) be used. Phase contrast microscopy does not yield the best results because of a halo of light obscuring neurite extension imaging close to the neurons. Also, it should be remembered that the neurite length measure in **Subheading 3.11., step 2.** is a 2-D projection of 3-D neurite extension, and hence underestimates the actual neurite length. Measurement of neurite extension in the 'Z' direction requires the use of specially design culture dishes *(30)*.
9. Nerve guide placement. Care should be taken to perform all procedures aseptically and with the aid of an operating microscope. When isolating the nerve, the nerve should be handled with caution; cutting of blood vessels, especially those near the knee, should be avoided. The sciatic nerve bifurcates into the tibial and peroneal branches above the level of the knee, and the sural nerve branches off at midthigh, necessitating its sacrifice. Care should be taken to remove any air bubbles from the nerve guide by rinsing with syringes containing bubble-free sterile saline solution.

Acknowledgments

The authors thank Xiaojun Yu and Laura Okun for their technical assistance in the preparation of this manuscript. The authors thank George P. Dillon for his technical assistance and for assistance, in the preparation of figures used in this manuscript.

References

1. Sanes, J. R. (1989) Extracellular matrix molecules influence neural development. *Ann. Rev. Neurosci.* **12,** 491–516.
2. Purves, D. and Lichtman, J. (1985) *Principles of Neural Development.* Sinauer Associates, Sunderland, MA, pp. 81–130.
3. Medical Devices and Diagnostic Industry. (1985) August P. 3.
4. De La Torre, J. C. (1981) Spinal cord injury, review of basic and applied research. *Spine* **6,** 315.
5. Puchala, E. and Windle, W. F. (1977) The possibility of structural and functional restitution after spinal cord injury, a review. *Exp. Neurol.* **55,** 1.
6. Beggs, J. L. and Waggener, J. D. (1979) The acute microvascular responses to spinal cord injury. *Adv. Neurol.* **22,** 179.
7. Faden, A. I., Jacobs, T. P., and Holaday, J. W. (1981) Endorphins in experimental spinal injury, therapeutic effect of naloxone. *Ann. Neurol.* **10,** 326.
8. Hausmann, B., Sievers, J., Hermanns, J., and Berry, M. (1989) Regeneration of axons from adult rat optic nerve: influence of fetal brain grafts, laminin and artificial basement membrane. *J. Comp. Neurol.* **281,** 447–466.

9. Guenard, V., Kleitman, N., Morrissey, T. K., Bunge, R. P., and Aebischer, P. (1992) Syngeneic Schwann Cells derived from adult nerves seeded in semipermeable guidance channels enhance peripheral nerve regeneration. *J. Neurosci.* **12,** 3310–3320.
10. Martini, R. (1994) Expression and functional roles of neural cell surface molecules and extracellular matrix components during development and regeneration of peripheral nerves. *J. Neurocytol.* **23,** 1–28.
11. Le Beau, J. M., Liuzzi, F. J., Depto, A. S., and Vinik, A. I. (1995) Up-regulation of laminin B2 gene expression in dorsal root ganglion neurons and nonneuronal cells during sciatic nerve regeneration. *Exp. Neurol.* **134,** 150–155.
12. Cornbrooks, C. J., Carey, D., McDonald, J. A., Trimpl, R., and Bunge, R. P. (1983) *In vivo* and in vitro observations on laminin production by Schwann cells. *Proc. Natl. Acad. Sci.* **80,** 3850.
13. Bunge, R. P. and Bunge, M. B. (1983) Interelationship between Schwann cell function and extracellular matrix production. *Trends Neurosci.* **6,** 499.
14. Aguayo, A. J., Vidal-Sanz, M., Villegas-Perez, M. P., Bray, G. M. (1987) Growth and connectivity of axotomized retinal neurons in adult rats with optic nerves substituted by PNS grafts linking eye and the midbrain. *Ann. NY Acad. Sci.* **495,** 1–9.
15. Villegas-Perez, M. P., Vidal-Sanz, M., Bray, G. M., Aguayo, A. J. (1988) Influences of peripheral nerve grafts to enhance neuronal survival, promote growth and permit terminal reconnections in the central nervous system of adult rats [review]. *J. Exp. Biol.* **132,** 5–19.
16. Schwab, M. E. (1990) Myelin associated inhibitors of neurite growth and regeneration in the CNS. *Trends Neurosci.* **13,** 452–456.
17. McCormack, M. L., Goddard, M., Guenard, V., and Aebischer, P., (1991) Comparison of dorsal and ventral spinal root regeneration through semipermeable guidance channels. *J. Comp. Neurol.* **313,** 449–456.
18. Manthorpe, M., Engvall, E., Ruoslahti, E., Longo, F. M., Davis, G. E., and Varon, S. (1983) Laminin promotes neuritic regeneration from cultured peripheral and central neurons. *J. Cell Biol.* **97,** 1882–1890.
19. Civerchia-Perez, L., Faris, B., Lapointe, G., Beldekas, J., Leibowitz, H., and Franzblau, C. (1980) Use of collagen hydroxyethylmethacrylate hydrogels for cell growth. *Proc. Natl. Acad. Sci. USA* **77,** 2064–2068.
20. Bergethon, P. R., Trinkaus-Randall, V., and Franzblau, C. (1989) Modified hydroxyethylmethacrylate hydrogels as a modeling tool for the study of cell–substratum interactions. *J. Cell Sci.* **92,** 111–121.
21. Woerly, S., Marchand, R., and Lavallee, C. (1991) Interactions of copolymeric poly (glyceryl methacrylate)-collagen hydrogels with neural tissue : effects of structure and polar groups. *Biomaterials* **12,** 197–203.
22. Refojo, M. F. (1965) Glyceryl methacrylate hydrogels. *J. Appl. Polymer. Sci.* **9,** 3161–3170.
23. Elsdale, T. and Bard, J. (1972) Collagen substrate for studies on cell behavior. *J. Cell Biol.* **54,** 626–637.
24. Collagen Corporation, Technical Product Notes on Vitrogen 100®, # 3001-11-0395.

25. Hearn MTW (1987) 1,1 - carbonyldiimidazole- mediated immobilization of enzymes and affinity ligands. *Methods. Enzymol.* **135**, 102–117.
26. Bellamkonda, R., Ranieri, J. P., and Aebischer, P. (1995) Laminin oligopeptide derivatized agarose gels allow three-dimensional neurite extension in vitro. *J. Neurosci. Res.* **41**, 501–509.
27. Parker, J. M. R. and Hodges, R. S. (1985) Photoaffinity probes provide a general method to prepare peptide-conjugates from native protein fragments. *J. Protein Chem.* **3**, 479–489.
28. Bellamkonda, R., Dillon, G. P., and Xiaojun, Yu. (1996) A hydrogel based 3D biopolymeric matrix for Nerve Regeneration. *Soc. Neurosci. Abstracts* **22**, 316.
29. Lennard, P. R. (1990) Image analysis for all, *Nature* **347**, 103–104.
30. Bellamkonda, R., Ranieri, J. P., Bouche, N., and Aebischer, P. (1995) A hydrogel based three dimensional matrix for neural cells. *J. Biomed. Mater. Res.* **29**, 663–671.
31. Shaw, E. D., Salmon, and Quatrano, R. S., (1995) Digital photography for the light microscope: Results with a gated, video-rate CCD Camera and NIH-*Image* software. *Biotechniques* **19**, 946–955.
32. Valentini, R. F., Aebischer, P., Winn, S. R., and Galletti, P. M. (1987) Collagen- and laminin-containing gels impede peripheral nerve regneration through semipermeable nerve guidance channels. *Exp. Neurol.* **98**, 350.
33. Valentini, R. F., Sabatini, A. M., Dario, P., Aebischer. (1989) Polymer electret guidance channels enhance peripheral nerve regeneration in mice. *Brain Res.* **48**, 300.

10

Small Animal Surgical and Histological Procedures for Characterizing the Performance of Tissue-Engineered Bone Grafts

Kenneth S. James, Mark C. Zimmerman, and Joachim Kohn

1. Introduction

Developing effective tissue-engineered constructs for bone regeneration requires careful assessment of the in vivo bone response to novel biomaterials, scaffold architectures, and biologically augmented, tissue-engineered constructs. Both the implant material and scaffold architecture are known to significantly effect the local tissue response *(1–3)*. Consequently, in characterizing the performance of new bone implants, it is prudent to establish material-dependent and scaffold-architecture-dependent bone-growth phenomena, in addition to the effect of biological augmentation, e.g., preseeded cells, growth factors, and cell-attachment proteins. Here we describe rabbit transcortical pin and trephine defect models, which, in combination, yield a method to investigate such variables on bone regeneration. The necessary histological and histomorphometry procedures are also detailed.

The transcortical bone-pin model is adapted from ASTM standard F981-91 *Standard Practice for Assessment of Compatibility for Surgical Implants with Respect to Effect of Materials on Muscle and Bone* (American Society for Testing of Materials, West Conshohocken, PA). Fabricated 2.0-mm diameter pins of the investigational material are implanted in the rabbit distal femur and proximal tibia to assess the cortical and cancellous bone response *(2)*. The method is designed to specifically assess material effects on bone. For example, the material may elicit a strong inflammatory response, osteoclastic bone resorption at the bone–implant interface, or a significant fibrous layer may form around the implant, preventing direct bone apposition to the material. In the case of a degradable implant material, the effects of degradation products and material resorption are also readily documented with this model. Note that rabbits have life spans over 5 yr, allowing for short- and long-term experiments that can be performed relatively inexpensively.

The rabbit skull trephine defect model is designed to examine architectural and biological augmentation effects on bone growth. It can be used to analyze a number of quantitative variables, including the rate of bone formation and volume of tissue ingrowth (bone, cartilage, or soft tissue) *(4–6)*. Two 8-mm-diameter defects are created in the rabbit skull, allowing for a paired comparison between implant configurations or to a positive (autograft) or negative (empty) control defect. Note that this is not a critical size defect model (defect size = 15 mm), but a delayed healing model *(7)*. Kramer et al. *(4)* studied the normal healing process of unfilled 8.0-mm trephine defects in rabbit crania. Defects were observed at 1, 2, 4, 8, 16, 20, and 24 wk: 1 wk, areas of the original inner and outer tables were filled with fibrous tissue and spindle shaped cells; 2 wks, tongues of newly formed trabecular bone jutted inward; 4 wk, new bone formed tapering tongues, which extended into the center of the defect 1–3 mm, along with a thin layer of cellular fibrous tissue between the advancing bone fronts; 8 wk, increased bone formation into the center of the defect site; 16–20 wks, bone bridging of the 8-mm defect was observed. In contrast, materials shown to be osteoconductive or osteoinductive have induced defect bone bridging by 2–4 wk *(5,6)*. Consequently, time-points of 2, 4, and 8 wk are typically investigated, with longer time periods providing insights into possible material degradation effects and long term bone remodelling.

2. Materials
2.1. Surgical Procedures

1. Implants: pins 2 mm in diameter × 2 cm, or porous scaffolds 8 mm in diameter × 2.5 mm.
2. Male New Zealand white rabbits weighing approx 3.5 kg.
3. Disinfecting solutions: 70% ethyl alcohol in dH_2O, Betadine (Henry Schein, Port Washington, NY), and 0.2% nitrofurazone ointment.
4. Anesthesia: ketamine hydrochloride (55 mg/kg), acepromazine maleate (1.0 mg/kg), xylazine (5 mg/kg), bupivacaine hydrochloride (1.25 mg/site).
5. Antibiotics: enrofloxacin (10 mg/kg).
6. Analgesia: buprenorphine HCl (0.1 mg/kg).
7. General surgical supplies: lubricant eye ointment (Vedco, St. Joseph, MO), disposable surgical drapes, sterile surgical gloves, sterile surgical gowns, scalpel handles, #10 and #15 blades, suture holders, scissors, small self-retaining retractor, small Sein retractors, towel clips, small battery operated cautery, forceps, hemostats, rongeur, osteotome, periosteal elevator, 3-0 and 4-0 resorbable sutures (Vicryl, Ethicon, Somerville, NJ), 3-0 nylon sutures, 4x4 in. and sterile gauze rolls, sterile saline, plastic spray sealant (Bard, Murray Hill, NJ).

8. Transcortical pin protocol surgical supplies: manual hand held drill, 2-mm stainless steel drill bit, hammer, awl.
9. Trephine protocol surgical supplies: pneumatic drill, universal T-handle chuck, two 8-mm stainless steel trephine bits (one with a center post and one without).
10. Euthanisa solution: sodium pentobarbital.

2.2. Histological Processing

1. Dissection instruments: scalpel, forceps, scissors, osteotome, rongeur, bone saw.
2. Medium square glass bottles (Fisher, Pittsburgh, PA, 03-325BB).
3. 10% buffered formalin (Fisher, SF100-4).
4. 100% (200 proof), 95, 80, 70, and 40 ethyl alcohol solutions in distilled water.
5. HEMO-DE (Fisher, 15-182-507A) clearing solution.
6. Dry benzoyl peroxide (Fisher, B-274-1). Benzoyl peroxide, shipped wet, is dried with the following procedure:
   a. Place the required amount in a beaker of 100% ethyl alcohol.
   b. Mix and let stand for 15 min.
   c. Filter solution to recover the benzoyl peroxide, using a vacuum filtering system.
   d. Place benzyol peroxide in a shallow dish to allow ethyl alcohol to evaporate
   e. When dry, weigh out the amount of catalyst needed and store remaining benzoyl peroxide in a closed metal container away from heat.

   Caution: Note that dry benzoyl peroxide is highly explosive.
7. Methylmethacrylate (Fisher, 0–3629) PMMA solution I; PMMA solution II (100 mL PMMA solution I and 1.0 g dry benzoyl peroxide); PMMA solution III (100 mL PMMA solution I and 2.5 g dry benzoyl peroxide). PMMA solution II and III are mixed for at least 3 h on a stirring plate prior to use.
8. Plexiglas microscope slides and staining dishes.
9. Cyanoacrylate glue, such as Permabond Industrial Grade 910 Adhesive (Permabond International, Englewood, NJ).
10. Diamond-tipped electric engraver.
11. Stevenel's blue stain, prepared by mixing solutions A (1 g of methylene blue [Fisher, M-291] in 75 mL of distilled water) and B (1.5 g of potassium permanganate [Fisher, P279-500]) in a boiling water bath, until the precipatate dissolves. Allow the stain to cool to room temperature and filter through medium-grade filter paper. Stain stored at room temperature is stable for many months and can be reused several times.
12. Van Geison picro-fuschin stain prepared by first mixing 0.1 g acid fuschin (Fisher, F-97) with 10 mL of dH_2O, followed by the addition of 100 mL of saturated picric acid (Sigma, St. Louis, MO, 925-40). Stain stored at room temperature is stable for many months and can be reused several times.
13. Specimen cutting, grinding, polishing stations, and accessories equivalent to the Buehler Isomet (Lake Bluff, IL) 11-1180 low-speed diamond saw and blade; Buehler Handimet II Roll Grinder, 240, 320, 400, and 600 grit Buehler Carbimet abrasive paper rolls, Buehler Ecomet III Polishing Wheel, and 1.0 and 0.3 µ Buehler alumina alpha micropolish II.

2.3. Histological Interpretation and Histomorphometry

1. Hardware: a stereomicroscope equipped with CCD camera and computer interface permitting digital image capturing. We have found a system consisting of an Olympus BH-2 microscope, Apple Macintosh IIci computer, 17-in. color monitor, and a MTI CCD72 video camera with adaptor box and video card (Dage-MTI, Michigan City, IN) works well.
2. Software: image analysis software comparable to NIH Image 1.60 (available for free download from the NIH at http://rsb.info.nih.gov/nih-image/download.html).

3. Methods
3.1. Surgical Procedures

The transcortical pin and trephine defect models require approx 1–2 h to complete, including animal preparation, surgery, and recovery. A team of two (a surgeon and an anesthesiologist) can efficiently complete these procedures. The rabbits tolerate the procedures very well and are typically found to be alert, weight bearing, eating, urinating, defecating, and free of significant discomfort 1 d postoperation.

3.1.1. Hard Tissue Transcortical Pin Biocompatibility Model

1. Induce anesthesia with an intramuscular (im) injection of ketamine hydrochloride and acepromazine maleate (doses given in **Subheading 2.1.**).
2. Administer preoperative antibiotics (enrofloxacin) and analgesia (buprenorphine HCl) im.
3. Administer eye lubricant.
4. Shave the legs from the ankle to hip and bring animal into the surgical suite.
5. Per normal sterile surgical procedure guidelines, scrub the legs with betadine, rinse with 70% ethyl alcohol, wrap the feet, and drape the legs (*see* **Note 1**).
6. Administer an im injection of xylazine just prior to initiating the procedure. Supplemental injections of xylazine at one-quarter the initial dose is administered as needed to maintain a surgical plane of anesthesia.
7. Monitor and record the heart rate at 15-min intervals, either by manually palpating the pulse in an ear vessel or with an EKG machine.
8. Create two 1-cm medial incisions along the knee to expose the proximal tibia and distal femur. Displace the periosteum anteriorly and posteriorly with the periosteal elevator.
9. The implant sites are approx 7 mm from the distal end of the femur and the proximal tibia. In practice, the tibia implant site is midway between the joint space and tibial tuberosity. The femur implant site is circa the proximal aspect of the medial condyle. To help guide the drill bit, create a small indention in the bone at the implant sites by lightly tapping on the awl with a hammer. Drill 2.0-mm diameter holes in the tibia and femur perpendicular to the long axis of the femoral-tibial shaft. Penetration of the medial and lateral cortices is easily felt with the manual hand drill.

Tissue-Engineered Bone Grafts

10. Press-fit the implants into place and carefully trim any excess material with scissors.
11. Close the soft tissue in separate layers with resorbable (subcutaneous layer) and nylon interrupted sutures (skin).
12. Perform implant procedure on the contralateral leg.
13. Administer local anesthesia (bupivacaine HCl) subcutaneously at each implant site.
14. Wash wounds of blood, dress with antibiotic ointment, and coat with a plastic spray sealant.
15. For pain management, administer, as needed, a subcutaneous injection of buprenorphine HCl twice daily until the third postoperative day.
16. Antibiotics (such as enrofloxacin, 10 mg/kg) should be administered im daily until the third postoperative day.
17. Remove the nylon skin sutures 10–14 d postoperation.

3.1.2. Trephine Model

1. Animal preparation, anesthesia, and monitoring are identical to steps 1–7 in the *Transcoritical Pin Protocol,* except for the surgical site. Prepare a surgical site the size of a silver dollar on the rabbit skull, taking care to protect the eyes of the animal from the disinfecting solutions. The rabbit skull can be stabilized with rolled towels and placed on either side of the rabbit skull. Otherwise, an assistant can physically stabilize the head when necessary.
2. Create a midline incision approx 2.5 cm in length through the skin along the sagittal suture of the skull, caudal to the eyes. Continue the incision through the peripheral muscles and periosteum over the sagittal suture.
3. With a small osteotome, carefully elevate the muscles and periosteum. Use small retractors to expose an area large enough for two 8.0-mm trephine holes to be respectively made on either side of the sagittal suture of the skull.
4. Using the trephine bit with the post and the pneumatic drill, start the trephine hole and penetrate the other diploë (*see* **Note 2**). Use the trephine bit without the post and universal hand chuck to slowly penetrate the inner diploë. Take care to continually check for full penetration by pushing on the the trephine bone with forceps. Once the trephine bone is free, pry the circular bone fragment from the surrounding bone with forceps to expose the intact dura with viable blood vessels (*see* **Note 3**).
5. Place the test material into the defect (*see* **Note 4**).
6. Repeat procedure for the second defect.
7. Close the soft tissue in separate layers with resorbable (subcutaneous layer) and nylon interrupted sutures (skin).
8. Postoperative care is in accordance with steps 13–16 in the *Transcoritical Pin Protocol* (*see* **Note 5**).

3.2. Histological Processing

It is critical that the fixation, dehydrating, clearing, and embedding procedures described below be performed on the pins and scaffolds prior to implantation, to document the effects of these procedures on implant morphology.

Here we describe a fixation, dehydration, clearing, and polymethylmethacrylate-embedding technique that has been successfully employed in investigating the histological response of bone to degradable polymers *(2,3)*. However, it is possible that some implant materials may swell significantly or be soluable in one of these solutions. In such a case, it may be necessary to chose alternative procedures, and employ, for example, water soluable embedding media such as JB4 (Polysciences) *(8)*.

Caution: The fixation, dehydration, and embedding steps should all be performed under a fume hood.

3.2.1. Specimen Retrieval and Fixation

1. Expose the implant sites through routine dissection procedures.
2. With a bone saw, grossly isolate the implant, taking care not to disturb the implant or surrounding bone. Dissect away any remaining soft tissues.
3. Using a low-speed diamond saw with saline as the lubricating fluid, make additional bone cuts to isolate the implant, while retaining approx 5 mm of peripheral bone.
4. Place the implant/bone in an appropriately labeled storage vial (medium glass bottle) containing a volume of fixative (Formalin; *see* **Note 6**) equal to or greater than 10 times the specimen volume. In most cases, allowing the sample to remain in fixative overnight is adequate. If necessary, change the fixative solution several times over the next few days, until the solution remains clear.

3.2.2. Dehydration, Clearing, and Embedding

1. Sequentially place specimens in 40% ethyl alcohol for at least 8 h, and in 70% ethyl alcohol for at least 16 h. If required, the specimen can be safely left in these alcohol solutions over the weekend.
2. Replace dehydrating solution with 80% ethyl alcohol for 24 h. Refresh the solution after the first 8 h. For this and the following dehydration steps, it is increasingly important that the specimens soak in the specified alcohol solution for the stated duration.
3. Replace dehydrating solution with 95% ethyl alcohol for 24 h. Refresh the solution after the first 8 h.
4. Replace dehydrating solution with 100% ethyl alcohol for 24 h. Refresh the solution after the first 8 h.
5. Replace dehydrating solution with Hemo-De clearing agent for 8 h. Refresh the solution after 4 h.
6. Replace the clearing agent with PMMA solution I. Allow to infiltrate for 3–7 d.
7. Replace embedding solution with PMMA solution II. Allow to infiltrate for 3–7 d.
8. Replace embedding solution with PMMA solution III. Allow to infiltrate for 3–7 d.
9. Place bottle in 32°C water bath to polymerize methylmethacrylate solution (*see* **Notes 7** and **8**). The solution should polymerize in 1–3 d. Check if block is polymerized by removing the cap and touching the top of the block with forceps. If it is soft (sticky), leave the bottle in the waterbath until the material has hardened.

3.2.3. Preparing Histological Slides

1. Break the embedded specimen out of glass bottle with a hammer. To prevent the glass from scattering, break the bottle within a thick polyethylene bag. Placing the bottle in the freezer for about an hour prior to breaking helps the glass to fall away from the block. Rinse the blocks in water to remove any small fragments of glass. Immediately engrave necessary identification information into the block.
2. Carefully orient the specimen and cut 0.5-mm-thick sections with the diamond saw. Use pure ethyl alcohol as the lubricating fluid. For both the transcortical pin and trephine models, saggital and coronal sections, respectively, are routinely made through the implants. For the implanted pins, cut three sequential sections, starting at either the medial or lateral cortex. This yields cortical and cancellous bone histological sections. For the embedded trephine defects, first cut the specimen in half, along the diameter in the coronal plane, followed by 2–3 coronal serial sections. The remaining pieces of the embedded specimens are cut in the transverse plane to visualize the implant sites from this perspective.
3. Clean the section to be glued and the Plexiglas microscope slide in a Petri dish filled with 100% ethyl alcohol and dry with a Kimwipe. Put a drop of cyanoacrylate on the slide. Place the specimen on the slide by holding the specimen with forceps at a 45-degree angle and slowly dropping it down on the glue and slide (similar to the way you would coverslip a slide). Put a piece of plastic wrap underneath the slide and fold it over to cover the slide. Clamp slide between two scrap Plexiglas slides. Leave at room temperature overnight to make sure that the glue is set.
4. Manually grind the glued section on the Buehler Handimet II Roll Grinder (or equivalent), using abrasive paper and water as a lubricant, until it becomes approximately one cell-layer thick (*see* **Note 9**). Begin by grinding on the lower grit papers and subsequently move to higher grit papers as the section becomes thinner. Take care to assure that equal pressure is applied to section, so that it grinds evenly. Turning the slide occasionally helps to resolve this problem. Check slide frequently under the microscope to determine when the one cell layer thickness is reached, i.e., only one distinct cell focal plane exists.
5. Polish the ground slides on the Buehler Ecomet III Polishing Wheel (or equivalent) until all the scratches are no longer visible under the microscope. Initially use a suspension of 1.0 alumina polishing particles, followed by 0.3 alumina particles, for the final polished finish.

3.2.4. Staining

1. Fill staining dishes with Stevenel's blue and Van Geison's picro-fuschin. Place staining dish with Stevenel's blue and a beaker of dH_2O in a 60°C water bath and bring up to temperature.
2. Soak slides in Stevenel's blue for 10 min, rinse in dH_2O, and blot dry.
3. Soak slides in Van Geison's picro-fuschin for 2 min (room temperature), rinse in a beaker of 100% ethyl alcohol, and blot dry.

3.3. Histological Interpretation and Histomorphometry

The Stevenel's Blue and Van Geison picro-fuschin combination distinctively stains tissues over a wide color spectrum: extracellular structures and nonmineralized tissues, shades of blue; collagen fibers, green to green-blue; bone, red-orange or purple; osteoid, yellow-green; muscle fibers, blue to blue-green. In examining histological sections of the transcortical pins (**Fig. 1**) or trephine defects (**Fig. 2**), we routinely scan the slides for changes in pin dimensions, surface roughness and scaffold morphology, evidence of direct bone–implant contact, the presence of intervening fibrous tissue between the implant and bone, the quality of the bone surrounding and/or penetrating the implant (is it woven or lamellar), the number of osteoclasts at the bone-implant interface and whether significant osteoclastic bone resorption is taking place, and the presence of inflammatory cells (**Fig. 1**). A calibrated histomorphometry system, i.e., calibrate the system, using a 1.0-mm calibration bar microscope slide, is used to accurately measure attributes such as the average thickness of any fibrous layer. Histomorphometric analysis of the the trephine defect slides also includes measuring the linear extent of bone growth across the defect and the area of the defect site occupied by bone. These values are normalized to the length or area of the original defect *(6)*.

4. Notes

1. We have found it convenient to place the rabbit's feet within sterile surgical gloves, which are subsequently wrapped with sterile gauze.
2. The trephine cutter with a post is a key component in starting the trephine defect. We have also found that it is sometimes easier to start the trephine with the drill rotating in reverse. The trephine teeth are large and by rotating the cutter in reverse, the cutter will simply start the trephine and cut very slowly. The drill is then reversed back once the trephine has been initiated.
3. One possible major complication is the disruption of the dura. If the trephine cutter is used too aggressively, the dura may be disrupted, resulting in bleeding and local necrosis. The implant material will reduce the bleeding and the animals should be carefully observed for neurological deficits and/or complications.
4. An autograft bone-control implant is readily made by morselizing the trephine bone with a rongeur and carefully packing the defect. Likewise, a negative control site is created by leaving the site empty.
5. Some have found it useful to incorporate florescent bone label injections, i.e., oxytetracycline, 2-4-*bis*(N,N'-dicarboxymethyl aminomethyl) fluorescein (DCAF), and so on, to identify areas of bone formation at intermediate time points *(6)*.
6. If florescent bone markers are being used, e.g., oxytetracycline, the specimens should be fixed in 70% ethyl alcohol, because Formalin causes the florescent dyes to fade.
7. Close the cap tightly, because oxygen inhibits the PMMA solution polymerization reaction. Do not shake or tilt the bottle. The PMMA solution will dissolve

Tissue-Engineered Bone Grafts

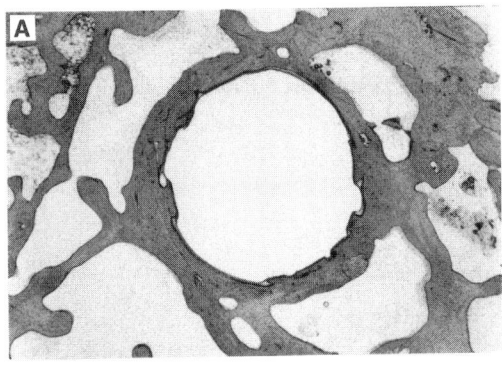

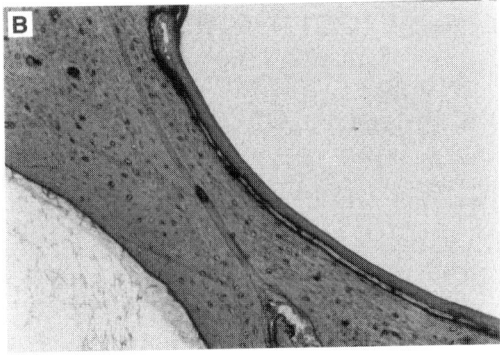

Fig. 1. A representative sagittal section through a polymer pin (clear circular area) implanted in the rabbit distal femur for 9 mo. A ring of lamellar bone (red) surrounds the implant, which approximates the original shape and dimensions of the original 2-mm diameter pin (**A**). A very thin (2–3 cell layers) fibrous layer (blue) is found between the implant and bone, and very few osteoclasts are found at the bone–implant interface (**B**). The surrounding marrow cavities are free of inflammatory cells. In summary, the material elicited a very mild inflammatory response. However, it appears the implant does not promote direct bone bonding.

the bottle cap if it comes in contact with it. Place the bottles in a 32°C water bath under a fume hood. Besides warming the methacrylate to facilitate polymerization, the water serves as a heat sink for the energy generated by the polymerization reaction, that if not dissipated, will cause the the PMMA to bubble. Make sure that all four sides of each bottle are surrounded by water and that they are not touching other bottles. Check the water level and the temperature of the water bath often. Keep the water level as high as it can be without reaching the neck of the shortest bottle. The PMMA solution should polymerize in 1–3 d.

8. If the PMMA bubbles significantly, the specimen can be re-embedded by first dissolving in PMMA I, and subsequently moving on to PMMA II and PMMA III

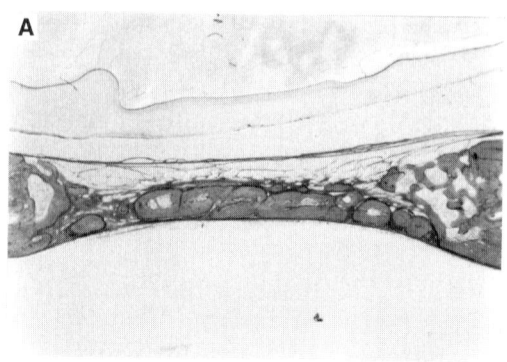

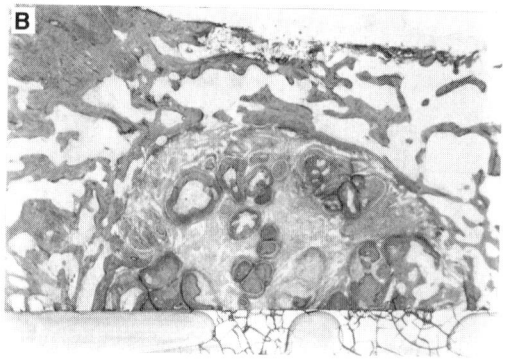

Fig. 2. Coronal (**A**) and transverse (**B**) sections of negative control (empty) 8-mm trephine defects at 12 wk. Bone (red) has successfully bridged the defect, though the inner and outer diploë have yet to fully form. Bone fragments are interspersed with fibrous tissue (blue) and osteoid (blue-green). Note that measurements of linear bone ingrowth are most conveniently made on the coronal sections, and bone area measurements on the transverse sections.

per the normal embedding schedule. Recheck or lower the temperature of the water bath to prevent further bubbling.
9. To assist in gripping the slide during grinding, we machined a piece of Plexiglas that could be comfortably gripped and consisted of a number of 1.5-mm diameter holes in the bottom surface that spanned the width and length of a typical microscope slide. A vacuum is drawn through the holes by an attached vacuum line, thus gripping the slide.

References

1. Mikos, A. G., Sarakinos, G., Lyman, M. D., Ingber, D. E., Vacanti, J. P., and Langer, R. (1993) Prevascularization of porous biodegradable polymers. *Biotechnol. Bioeng.* **42,** 716–723.

2. Ertel, S. I., Kohn, J., Zimmerman, M. C., and Parsons, J. R. (1995) Evaluation of poly(DTH carbonate), a tyrosine-derived degradable polymer, for orthopaedic applications. *J. Biomed. Mater. Res.* **29,** 1337–1348.
3. Choueka, J., Charvet, J. L., Koval, K. J., Alexander, H., James, K. S., Hooper, K. A., and Kohn, J. (1996) Canine bone response to tyrosine-derived polycarbonates and poly(L-lactic acid). *J. Biomed. Mater. Res.* **31,** 35–41.
4. Kramer, I., Killey, H., and Wright, H. (1968) A histological and radiological comparison of the healing of defects in the rabbit calvarium with and without implanted heterogenous inorganic bone. *Arch. Oral Biol.* 1095–1106.
5. Damien, C. J., Parsons, J. R., Benedict, J. J., and Weisman, D. S. (1990) Investigation of hydroxyapatite and calcium sulfate composite supplemented with an osteoinductive factor. *J. Biomed. Mater. Res.* **24,** 639–654.
6. Damien, C. J., Parsons, J. R., Prewett, A. B., Rietveld, D. C., and Zimmerman, M. C. (1994) Investigation of an organic delivery system for demineralized bone matrix in a delayed-healing cranial defect model. *J. Biomed. Mater. Res.* **28,** 553–561.
7. Frame, J. (1980) A convenient model for testing bone substitute materials. *J. Oral Surg.* **38,** 176–180.
8. Burg, K. J. L., Jenkins, L., Powers, D. L., and Shalaby, S. W. (1996) Special considerations in embedding a lactide absorbable polymer. *J. Histotechnol.* **19,** 39–43.

11

Preparation and Use of Porous Poly(α-Hydroxyester) Scaffolds for Bone Tissue Engineering

Anna C. Jen, Susan J. Peter, and Antonios G. Mikos

1. Introduction

Skeletal defects resulting from tumor resection, congenital abnormalities, or trauma often require surgical intervention to restore function. Current options for bone replacement include autografts, allografts, metals, ceramics, and polymeric bone cements. However, all of these materials have drawbacks, and their selection usually requires some degree of compromise. Autografts represent the ideal repair material, but are limited by availability and donor site morbidity. Allografts may be potential transmitters of disease, and also solicit immune response if not sufficiently pretreated. Ceramics suffer from slow integration and remodeling, and wear-debris from nondegradable polymeric implants may evoke chronic inflammation. Finally, metallic implants may cause atrophy of surrounding tissue through stress shielding, requiring corrective procedures.

Recent investigations into tissue engineering implants offer alternatives to the current methods of skeletal repair. Synthetic biodegradable polymers are promising as extracellular matrix analogs to facilitate tissue development and growth *(1)*. Poly(α-hydroxyesters) are among the few synthetic biodegradable polymers approved by the Food and Drug Administration for human clinical use. These include poly(glycolic acid) (PGA), poly(L-lactic acid) (PLLA), poly(DL-lactic acid) (PLA), and copolymers of poly(DL-lactic-co-glycolic acid) (PLGA). All are biocompatible, degrading to products that can be eliminated from the body through either metabolic pathways or by direct renal excretion *(2)*. In addition, the PLGA copolymers have the distinct advantage of being capable of degrading from weeks to years, as required for a specific application, by altering their copolymer ratio *(3)*.

The ideal bone substitute should be osteoconductive, osteoinductive, structurally sound, and, if possible, provide an osteogenic cell population *(4)*. Cell presence is believed to enhance healing and promote bone formation. A tissue engineering approach using polymer–cell constructs attempts to improve upon previous methods by satisfying all such requirements *(5)*. Advances in cell culture also allow for in vitro expansion of bone forming cells, which reduces harvest requirements.

We have studied the function of primary rat calvarial osteoblasts on different poly(α-hydroxyesters) and showed that these polymers support cell adhesion, proliferation, and migration *(6,7)*. In addition, we have successfully cultured rat marrow stromal osteoblasts on three-dimensional (3-D) PLGA foam scaffolds and showed the formation of bone-like tissue in vitro as guided by the polymer *(8)*. Moreover, we have demonstrated both ectopic and orthotopic bone formation in vivo by transplantation of polymer–cell constructs in rat models *(9,10)*.

In this chapter, we illustrate the fabrication of 3-D poly(α-hydroxyester) foam scaffolds by a solvent-casting and particulate-leaching method in combination with heat-compression molding. We also discuss techniques for seeding and culturing osteoblasts in 3-D polymer scaffolds.

2. Materials

2.1. Polymer Processing

2.1.1. Chemicals

1. Polymers: poly(L-lactic acid) and poly(DL-lactic-co-glycolic acid)
   The polymers come sealed under vacuum. Once the original packaging has been opened, the polymers must be stored either under N_2 in a freezer at –4°C, or under vacuum in a desiccator. This precaution reduces premature degradation by hydrolysis caused by the moisture in the air. The inherent viscosity of the polymer is provided by the company, but exact mol wt and polydispersity can be determined using gel permeation chromatography. The physical properties of the polymers, including the glass transition temperature (T_g) and melting temperature for the semicrystalline polymers (T_m), are listed in ref. *11* and *12*. The PLGA copolymers are usually available at lactic:glycolic acid ratios of 85:15, 75:25, 65:35, and 50:50. The polymers can be purchased from: Birmingham Polymers (Birmingham, AL); Boehringer Ingelheim KG (Ingelheim, Germany); Medisorb® (Cincinnati, OH); Purasorb® (Lincolnshire, IL)
2. Solvents: dichloromethane (Sigma, St. Louis, MO); chloroform (Sigma) (*see* **Note 1**).
3. Sodium chloride (Sigma): Store in a dry, cool place, especially after size separation.

2.1.2. Laboratory Supplies

1. 50-mm glass Petri dishes (with covers) (Fisher Scientific, Pittsburgh, PA).
2. Teflon-foil protective overlay (Cole-Parmer, Niles, IL).

3. USA standard testing sieves, ASTM E-11 Specification (Fisher Scientific).
4. Heat-jacketed Teflon molds (custom-designed).
5. Press (Carver Laboratory Equipment, Wabash, IN).

2.2. Cell Culture

The culture solutions should be sterilized and kept refrigerated when not in use. To prevent temperature shock, heat up to 37°C in water bath before contact with cells.

1. Primary media:
   a. Dulbecco's modified eagles medium (DMEM: Gibco-BRL, Gaithersburg, MD).
   b. Fetal bovine serum (FBS) (HyClone, Logan, UT).
   c. 25 mg/mL Gentamicin (GS) (Sigma).
2. Complete media:
   a. DMEM (Gibco BRL).
   b. 10–15% FBS (HyClone) or Calf Serum (Gibco-BRL).
   c. 8 mg/mL GS (Sigma).
   d. 10 mM Na β-glycerol phosphate (Sigma).
   e. 50 mg/mL L-ascorbic acid (Sigma).
   f. 10^{-8} M dexamethasone (Sigma) (*see* **Note 2**).
3. Phosphate buffered saline (PBS) (Sigma): The buffer should be magnesium- and calcium-free at pH 7.4.
4. Trypsin solution: The trypsin solution is kept in the freezer and can be thawed–frozen up to 3× before activity is lost.
   a. 0.0625% Trypsin (Sigma).
   b. 0.0125% Ethylene glycol-*bis*(b-aminoethyl ether) N,N,N',N'-tetraacetic acid (EGTA) (Sigma).
   c. 0.625% Polyvinylpyrrolidone (Sigma).
   d. 26 mM N-(2-hydroxyethyl)piperazine-N'-(2-ethanesulfonic acid) (HEPES) (Sigma).
   e. 1.125% NaCl (Sigma).

3. Methods
3.1. Polymer Processing

The solvent-casting and particulate-leaching method described here is adapted from ref. *13*. The addition of the heat-compression molding step is adapted from ref. *14*.

3.1.1. Solvent-Casting

1. Sieve the salt to desired size range. Grinding, for instance, with a micromill, may be helpful if small salt size is desired (*see* **Note 3**).
2. Remove polymer from storage. If frozen, let equilibrate 1 h and wipe off condensation before opening container. Meanwhile, cover Petri dish bottoms with Teflon foil. This is necessary for removal of the polymer–salt composite later.

3. In a glass vial, dissolve polymer in 10× (g/mL) dichloromethane for 30 min, or until no polymer particles are visible.
4. Fill Petri dish with desired wt% of NaCl (*see* **Note 4**).
5. In the fume hood, cast the polymer solution into the Petri dish with the salt. Quickly cover the Petri dish to slow down evaporation of the solvent. Fast evaporation may cause uneven distribution of the salt and polymer mixture, and result in nonuniform foams. Swirl around until salt appears wetted and is evenly distributed. Leave on a leveler platform overnight.
6. The following day, the polymer–salt composite should be further dried in the hood with the lids removed, then vacuum-dried overnight.

3.1.2. Heat-Compression Molding

A Teflon mold of desired shape should be previously constructed, with a heating jacket attached to a temperature control unit. A press is also needed to apply the compression (*see* **Note 5**).

1. Remove the dry polymer–salt composites from the Petri dishes using a strong spatula, and place into the mold in small chunks. Fill the mold as much as possible, but leave room for the end plugs and enough indention for the pressing bar to be properly aligned.
2. If possible, design a heating program so as not to overshoot the desired internal temperature of approx 20°C above the T_g or T_m of the polymer.
3. After heating cycle (1 h), allow cooling. Note that pressure will drop a little with the temperature decrease. If crystalline properties are important, one can adjust the cooling cycle to quench quickly or anneal slowly.
4. Remove from mold.
5. A diamond saw (Isomet 11-1180, Buehler, Evanston, IL) can be used to cut the molded samples into individual sizes. We generally use foams 2-mm thick to minimize diffusion limitations of nutrients to the cells. When cutting, avoid the use of cutting fluid, which can contaminate the samples. Water is the best lubricant, although one has to be careful not to leach the samples prematurely.

3.1.3. Particulate Leaching

1. Both large casted foams and smaller precut foams are leached in deionized-dH_2O for 24 h to remove the salt porogen. Use of excess water is recommended to prevent saturation between water changes (6 h intervals).
2. The leached foams should first be dried in air and then in vacuum. They can be stored in a desiccator until use.

3.2. Cell Culture

3.2.1. Prewetting and Sterilization of Foams

1. Soak foams at least 30 min in 100% EtOH. Keep the foams in individual containers (welled plates) during this step, because the ethanol softens the polymer, and aggregates may otherwise form (*see* **Note 6**).

2. Maintaining sterile conditions, transfer foams into fresh wells. Wash for 1 h in sterile PBS, changing the solution every 15–20 min. This rinses out the EtOH, which is harmful to cells.
3. Immerse foams in media for 1.5 h. Use of complete media is preferred, because this solution is to be added after seeding.
4. Change into fresh media and begin preparing cell suspension.

3.2.2. Cell Seeding and Culture

1. After rinsing primary culture with PBS, add enough trypsin solution to cover the bottom of the culture flask (about 1.5 mL for a T-75 flask) and incubate for 8–10 min.
2. Make sure cells are detached from the bottom of the flask. If not, tap lightly on countertop to dislodge stubborn cells. Add 5 mL of complete media, pipeting up and down to create an uniform cell suspension.
3. Remove a sample amount to determine cell number, either by a hemacytometer or Coulter counter (Hialeah, FL).
4. Centrifuge for 10 min at 400g and 4°C (*see* **Note 7**).
5. Aspirate the supernatant and resuspend cells in desired dilution, according to seeding density and cell counts (*see* **Note 8**).
6. Seed cells (*see* **Note 9**).
7. Six hours after seeding, add more complete media to sufficiently cover sample (*see* **Note 10**).
8. Media should be changed every 1–3 d, depending on the seeding density and media volume. Regular changes will reduce the effects of polymer degradation products, reflected in a decrease in pH, as well as ensure sufficient nutrient supply.

4. Notes

1. PGA is insoluble in both dichloromethane and chloroform; it is only soluble in hexafluoroisopropanol, which is highly toxic *(12)*.
2. This potent glucocorticoid is often added to induce osteoblastic phenotype of marrow stromal cells.
3. Salt size is directly correlated to pore diameter *(13,14)*. Pore diameters ranging from 150 to 700 µm were found to have a negligible effect on marrow stromal osteoblast proliferation and function in vitro *(8)*. The size of the pores, however, is important for vascular ingrowth into implants/transplants *(15,16)*. Larger pores promote faster vascular penetration into the foam interior.
4. Porosity is directly related to the initial salt-weight fraction *(13)*. The actual porosity can be determined with mercury porosimetry, or calculated from total volume and properties of the polymer. Porosity of 90% is recommended to create an interconnected pore network and facilitate nutrient transport.
5. We have used cylindrical molds to create templates of either 6 or 12 mm diameter for cell culture. **Figure 1** presents a diagram of the heat-compression-molding system used in our laboratory.
6. Passing the polymer foams through alcohol serves as the sterilization step before cell seeding. This prewetting step also enhances permeation of water (and

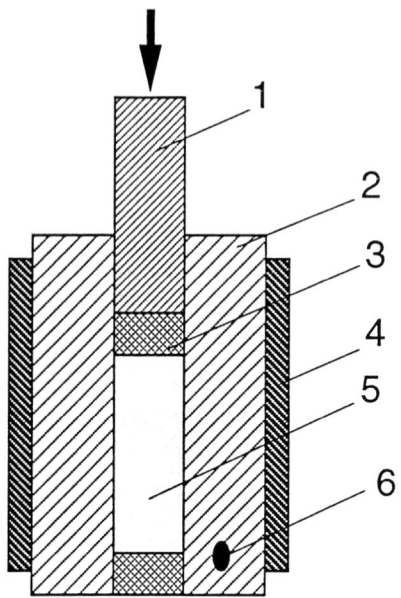

Fig. 1. Diagram of the custom-made Teflon mold with heating jacket: (1) compression bar, (2) Teflon mold, (3) Teflon end plug, (4) heating jacket, (5) center of mold, and (6) thermocouple insert. The polymer is first placed into the center of the Teflon mold. Then, with the end plugs in place, the compression bar is inserted into the top of the mold and the unit is placed in a press. An inserted thermocouple is connected to a temperature controller for the heating jacket.

medium) into the core of the foam *(17)*. Use of a shaker or rotor mixer is recommended to enhance this effect. Ethylene oxide sterilization is another method we have used successfully.
7. Generally, to avoid leaving cells in suspension longer than necessary, adjust centrifuge to 4°C in advance. Also, do the cell counting while the rest of the suspension is being centrifuged.
8. We generally seed cells on 3-D polymer foam scaffolds at approx 53,000 cells/cm^2, based on the top surface area. This corresponds to confluency of osteoblasts on an equivalent two-dimensional substrate.
9. We seed by micropipeting the desired amount of cell suspension onto of the scaffold. The seeding volume should be adjusted according to the size of the foams. Avoid using too much volume, because excess cell suspension will flow out of the foams and populate the well-plate instead of the scaffold. For 6-mm disks, we use about 100–200 μL. We also suggest changing wells before analyses are performed, so results reflect population on the foam and not of the surrounding well.
10. Addition of media less than 4 h after seeding is not recommended, because the cells require time to attach onto the scaffold so they will not be washed off during

handling. Waiting too long is also discouraged, because the culture may not have enough nutrients for the seeded cells to properly attach and function.

5. Final Comments

The methods presented in this chapter demonstrate one method of fabricating biodegradable polymer foam scaffolds for 3-D cell culture. Other methods of polymer processing are available for the manufacture of scaffolds *(18,19)*. The development of new polymer-cell constructs in tissue engineering is integrally related to advances in cell culture engineering through the development of new bioreactors for enhanced nutrient supply to overcome diffusion limitations and form clinically useful bone.

References

1. Langer, R. and Vacanti, J. P. (1993) Tissue engineering. *Science* **260,** 920–926.
2. Hollinger, J. O. and Battistone, G. C. (1986) Biodegradable bone repair materials: synthetic polymers and ceramics. *Clin. Orthop. Rel. Res.* **207,** 290–305.
3. Suggs, L. J. and Mikos, A. G. (1996) Synthetic biodegradable polymers for medical applications, in *Physical Properties of Polymers Handbook* (Mark, J. E., ed.), AIP, Woodbury, NY, pp. 615–624.
4. Gazdag, A. R., Lane, J. M., Glaser, D., and Forster, R. A. (1995) Alternatives to autogenous bone graft: efficacy and indications. *J. Am. Acad. Orthop. Surg.* **3,** 1–8.
5. Crane, G. M., Ishaug, S. L., and Mikos, A. G. (1995) Bone tissue engineering. *Nature Med.* **1,** 1322–1324.
6. Ishaug, S. L., Payne, R. G., Yaszemski, M. J., Aufdemorte, T. B., Bizios, R., and Mikos, A. G. (1996) Osteoblast migration on poly(α-hydroxy esters). *Biotechnol. Bioeng.* **50,** 443–451.
7. Ishaug, S. L., Yaszemski, M. J., Bizios, R., and Mikos, A. G. (1994) Osteoblast function on synthetic biodegradable polymers. *J. Biomed. Mater. Res.* **28,** 1445–1453.
8. Ishaug, S. L., Crane, G. M., Miller, M. J., Yasko, A. W., Yaszemski, M. J., and Mikos, A. G. (1997) Bone Formation by three-dimensional stromal osteoblast culture in biodegradable polymer scaffolds. *J. Biomed. Mater. Res.*, **36,** 17–28.
9. Ishaug-Riley, S. L., Crane, G. M., Gurlek, A., Miller, M. J., Yasko, A. W., Yaszemski, M. J., and Mikos, A. G. (1997) Ectopic bone formation by marrow stromal osteoblast transplantation using poly(DL-lactic-co-clycolic acid) foams implanted into the rat mesentery. *J. Biomed. Mater. Res.*, **36,** 1–8.
10. Smith, M. L., Miller, M. J., Crane, E., Khoo, A. K. M., Gulek, A., and Mikos, A. G. (1997) Cranial defect repair with osteoblast transplantation. Abstracts of Plastic Surgery Research Council, Galveston, TX, p. 45.
11. Lu, L. and Mikos, A. G. Poly(lactic acid), in *Polymer Data Handbook* (Mark, J. E., ed.), Oxford University Press, New York, in press.
12. Lu, L. and Mikos, A. G. Poly(glycolic acid), in *Polymer Data Handbook* (Mark, J. E., ed.), Oxford University Press, New York, in press.
13. Mikos, A. G., Thorsen, A. J., Czerwonka, L. A., Bao, Y., and Langer, R. (1994) Preparation and characterization of poly(L-lactic acid) foams. *Polymer* **35,** 1068–1077.

14. Thomson, R. C., Yaszemski, M. J., Powers, J. M., and Mikos, A. G. (1995) Fabrication of biodegradable polymer scaffolds to engineer trabecular bone. *J. Biomater. Sci.—Polym. Ed.* **7,** 23–38.
15. Mikos, A. G., Sarakinos, G., Lyman, M. D., Ingber, D. E., Vacanti, J. P., and Langer, R. (1993) Prevascularization of porous biodegradable polymers. *Biotechnol. Bioeng.* **42,** 716–723.
16. Wake, M. C., Patrick, C. W., and Mikos, A. G. (1994) Pore morphology effects on the fibrovascular tissue growth in porous polymer substrates. *Cell Transplant.* **3,** 1–5.
17. Mikos, A. G., Lyman, M. D., Freed, L. E., and Langer, R. (1994) Wetting of poly(L-lactic acid) and poly(DL-lactic-co-glycolic acid) foams for tissue culture. *Biomaterials* **15,** 55–58.
18. Thomson, R. C., Yaszemski, M. J., and Mikos, A. G. (1997) Polymer scaffold processing, in *Principles of Tissue Engineering* (Lanza, R. P., Langer, R., and Chick, W. L., eds.), R. G. Landes, Austin, TX, pp. 263–272.
19. Lu, L. and Mikos, A. G. (1996) The importance of new processing techniques in tissue engineering. *Mater. Res. Soc. Bull.* **21,** 28–32.

II

Cells

12

Quantitative Assessment of Autocrine Cell Loops

Gregory Oehrtman, Laura Walker, Birgit Will, Lee Opresko, H. Steven Wiley, and Douglas A. Lauffenburger

1. Introduction

Regeneration of functioning tissue essentially involves recapitulating relevant aspects of organogenesis, so that the starting composite of cells, matrix, and molecular factors develops into the desired structure and physiology. A crucial aspect of development is local cell–cell communication; that is, molecular regulatory factors are more typically paracrine and autocrine than endocrine in nature. Autocrine loops were originally thought of predominantly as being involved in pathological behavior, but it is becoming increasingly clear that a large portion of normal physiological behavior—and a tremendous portion of development—is strongly regulated by autocrine factors *(1)*. Thus, continuing progress of the field of tissue engineering will require increased understanding of how autocrine loops operate, so that they can be designed or manipulated systematically. We have made an effort in this direction, and some early experimental and modeling results can be found in the literature *(2–5)*. In this chapter, we describe the methods we have used for creating autocrine cell loops and quantitatively assessing their operation.

A crucial assessment issue is how much of the ligand synthesized by autocrine cells is captured by the secreting cells themselves, and how much is permitted to escape into the bulk extracellular medium. A perhaps counterintuitive principle of autocrine loops is that they might be providing the most useful physiological function when one cannot experimentally find any autocrine ligand in the bulk medium, because that is when the system should be most sensitively regulated by what is captured; when one can experimentally find substantial levels of autocrine ligand in bulk medium, that may be when pathological behavior arises. Another important principle of autocrine loops is that proper regulation can only occur when the ligand is synthesized in its transmembrane precursor form for proteolytic cleavage at the cell surface. When a purely secreted form of the ligand is synthesized, receptor binding can take place along the secretory pathway even before the ligand reaches the cell surface; this short-circuiting probably compromises aspects of receptor trafficking dynamics and interactions of ligand with extracellular matrix that affect cell regulation by autocrine loops.

Our assessment methods are ELISA, for determination of the amount of autocrine ligand escaping into bulk extracellular medium, and the Molecular Devices Cytosensor, for determination of the amount of autocrine ligand captured by secreting cells. The Cytosensor measures extracellular acidification rate (ECAR) in real time, which can be correlated to receptor-mediated signaling processes *(6,7)*. Our model experimental system is of B82 mouse L-cells, transfected with the genes for human epidermal growth factor receptor (EGFR) and transforming growth factor alpha (TGF-α) using the two-plasmid tetracycline-controlled gene-expression system. This system allowed us to calibrate readouts from the Cytosensor with ligand binding, because we could quantify Cytosensor readouts from nonautocrine cells challenged with exogenous ligand, and to vary ligand synthesis rate by changing the concentration of tetracycline in the medium.

2. Materials

2.1. Creation of Autocrine Cell Loops

2.1.1. Reagents for Construction of pUHD10.3-TGF-α Plasmid

1. pUHD15.1neo and pUHD10.3 (first and second plasmids of tetracycline controlled, inducible expression system) obtained from M. Gossen and H. Bujard *(8)* (Heidelberg University, Germany).
2. pMTE4 (transmembrane TGF-α cleavable protein) obtained from R. Derynck *(9)* (University of California at San Francisco).
3. Bluescript II KS+ (Strategene, La Jolla, CA).
4. Restriction enzymes: *Hind*III, *Nco*I, *Rsa*I, *Xho*I, *Eco*RI, *Bam*HI, *Pst*I (Boehringer Mannheim, Indianapolis, IN).
5. Modifying enzymes: Klenow, T4 ligase (Boehringer Mannheim).
6. Low melting point agarose (Sigma, St. Louis, MO) and β-agaraseI (New England Biolabs, Beverly, MA)
7. Wizard Maxiprep (Promega, Madison, WI)

2.1.2. Reagents for Creation of Autocrine Clones

1. Histidinol resistance plasmid (pREP8) (Invitrogen, San Diego, CA). pREP is modified into pR8 with the removal of the EBNA-1 and OriP segments (*Sac*I to *Xba*I deletion and religation) to prevent episomal replication, and to allow for plasmid incorporation into chromosomal DNA.

2. CaPO$_4$ transfection reagents:
   a. Salt solution: dissolve 3.7 g KCl (Sigma), 10 g D(+) glucose (Sigma), and 1 g Na$_2$HPO$_4$ (Sigma) into 50 mL final volume ddH$_2$O, filter sterilize with 0.2-μm filter, and store at –20°C.
   b. 2X HBS (HEPES buffered saline): Add 1 mL salt solution, 1 g HEPES (Sigma), and 1 g NaCl (Sigma) to 85 mL of ddH$_2$O, pH to 7.1 with 1 N NaOH, and bring volume up to 100 mL with ddH$_2$O. Filter sterilize with 0.2-μm filter; store in 20-mL aliquots at 4°C.
   c. 2.5 M CaCl$_2$: dissolve 36.8 g of CaCl$_2$·2H$_2$O (Sigma) into 100 mL final volume H$_2$O. Autoclave sterile. Store at room temperature or 4°C.
3. 8 × 8 glass cylinders (Bellco, Vineland, N.J.)

2.2. Determination of Autocrine Ligand in Bulk Extracellular Medium

2.2.1. Reagents for Measurement of Autocrine Ligand

1. Monoclonal anti-EGFR antibodies 528 and 225 (hybridomas; ATCC, Rockville, MD). Antibodies purified from hybridoma supernatant, 50% (NH$_4$)$_2$SO$_4$ (Sigma)–ddH$_2$O cut at 4°C, spin at 7600g for 10 min at 4°C. Dialysis pellet overnight in PBS with 30,000 MWCO dialysis tubing (Spectrum, Houston, TX).
2. TGF-α ELISA kit (Calbiochem, San Diego, CA).
3. 1X PBS buffer; 0.2 g KCl (Sigma), 0.2 g KH$_2$PO$_4$ (Sigma), 8 g NaCl (Sigma), 1.15 g Na$_2$HPO$_4$ (Sigma) dissolved in 1 L ddH$_2$O, pH 7.0, sterile-filter through 0.2-μm filter and store at 4°C.

2.2.2. TGF-α Cellular Processing

1. Protein A-conjugated Sepharose beads (Sigma).
2. Amicon Concentrators (YM3000 membrane)/Centricons (3000 MWCO) (Amicon, Beverly, MA).
3. Protein standards: albumin, carbonic anhydrase, cytochrome, aprotinin (Sigma).
4. Glycerol (Sigma).
5. G-50 Sephadex (Pharmacia, Piscataway, NJ).

2.3. Determination of Autocrine Ligand Captured by Secreting Cells

2.3.1. Molecular Devices Cytosensor Material

1. Transwells: 12 mm diameter, 3 μm pore size (Corning, Cambridge, MA).
2. Transwell inserts: 8 mm diameter, 3 μm pore size (Molecular Devices, Menlo Park, CA).
3. Spacer: 50 μm height/6 mm id (Molecular Devices).
4. DV-cyto buffer: On the day of a Cytosensor experiment, make 1 L of Dulbecco's volt modified Eagle's media (DMEM, 4500 mg/L glucose, L-glutamine, Sigma) without sodium bicarbonate. Instead, add 2.59 g of NaCl (Sigma), pH to 7.4, and add 100 mg of BSA (Sigma). Filter-sterilize with 0.2-μm filter. Sometimes, DV-cyto can be stored at 4°C, but re-pH to 7.4 and filter.

2.3.2. Calibration of Cytosensor Readout Reagents

1. hEGF (Gibco BRL): EGF at 1 mg/mL in 50 mM sodium phosphate, stored at –20°C.
2. 1X WHIPS buffer: 1 g polyvinyl pyrrolidone mol wt 40,000 (Sigma), 7.6 g NaCl (Sigma), 0.373 g KCl (Sigma), 0.102 g $MgCl_2 \cdot 6H_2O$ (Sigma), 0.147 g $CaCl_2 \cdot 2H_2O$ (Sigma), 4.76 g HEPES-acid (Sigma), dissolve in 1 L of ddH_2O at a pH of 7.4, store at 4°C.
3. I^{125} EGF: Add together 1 iodobead (Pierce, Rockport, IL), 100 mM 80 µL Tris (Sigma), 30 µL 100 mCi/mL I^{125} (NEN, Boston, MA), 10 µL of 10 ng/µL EGF (Gibco-BRL). Wait 15 min and remove protein to new tube. Made fresh day of use, add 40 µL of 12 mg/mL sodium metabisulfite (Sigma) in PBS to iodobead tube, remove, and add to EGF tube. Add 40 µL of BSA chase solution (20 mL ddH_2O, 100 mg BSA-RIA grade (Sigma), 1 crystal KI (Sigma), sterile-filtered with 0.2-µm filter, and store 1-mL aliquots at –20°C). Separate I^{125}-EGF from I^{125} over a G-10 Sepharose column (Pharmacia, Piscataway, NJ) with PBS. Store I^{125}-EGF at 4°C.
4. D/H/B buffer: make 1 L of Dulbecco-V-modified Eagle media (DMEM, 4500 mg/L glucose, 1 mM L-glutamine, Sigma) without sodium bicarbonate. Add 5.95 g/L HEPES (Sigma) and 1 mg/mL BSA (Sigma), pH to 7.2.

2.3.3. Cytosensor Readout of Antibody Inhibition Reagents

1. Polyclonal anti-TGF-α antibody (R & D Systems, Minneapolis, MN).
2. Monoclonal anti-EGFR antibodies 528 and 225 (hybridomas - ATCC) (*see* **Note 2**).
3. Antibodies used in Cytosensor experiments.
    a. Because of the sensitivity of Cytosensor measurements, it has been noted that noise and baseline changes occur when adding a high concentration of antibodies in glycerol or a buffer with high salt concentrations, such as PBS.
    b. To avoid this problem, antibodies or any other additives used at a high concentration need to be dialyzed in DV-s (DV-cyto buffer without BSA added) overnight at 4°C. Aliquot protein into small volumes to prevent cyclic freeze–thaw problems. Suggestion for antibodies and small volumes is Slide-a-Lyzer <0.5 mL capacity, 10K MWCO (Pierce, Rockford, IL).

2.4. B82 Fibroblast Tissue Culture Medium

B82 cells are a mouse L-cell line that lacks the EGFR chromosomal segment and does not express detectable levels of epidermal growth factor (EGF) and TGF-α. B82 receptor minus and receptor positive cells were obtained from G. Gill (University of California at San Diego) *(10)*. All B82 cells are grown at 37°C, 5% CO_2 and 98% relative humidity.

1. R⁻ media (Parental cells (B82R⁻) tissue-culture medium):
    a. DMEM: 4500 mg/L-glucose, 1 mM L-glutamine, Sigma); add 3.7 g/L sodium bicarbonate (Sigma), filter-sterilize with a 0.2-µm filter (Gelman Sciences, Ann Arbor, MI) under 12 psig positive pressure (N_2 gas), and store at 4°C until use.

b. L-glutamine 1 mM (Sigma). 100 mM glutamine stock solution; dissolve 3.65 g into 250 mL 1X PBS buffer, filter sterilize through 0.2-µm filter, and store at –20°C in 10-mL aliquots.
   c. Penicillin–streptomycin 100 IU/mL-100 µg/mL (Sigma).
   d. Bovine calf serum 10% (Sigma).
2. R$^+$ media (B82 with pXER (EGFR, B82R$^+$) tissue culture media).
   a. DMEM (Sigma).
   b. L-glutamine 1 mM (Sigma).
   c. Penicillin–streptomycin 100 IU/mL–100 µg/mL (Sigma).
   d. 10,000 MWCO dialyzed bovine calf serum 10% (Sigma).
   e. 1 µM methotrexate (Sigma). 1 mM 1000X stock solution; dissolve 25 mg methotrexate into 55 mL of 0.1 N HCl, sterile-filter with 0.2-µm filter, and store at –20°C in 10-mL aliquots.
3. R$^+$/1st plasmid media (B82R$^+$ cells with pUHD15.1neo (B82R$^+$/1st plasmid) tissue-culture media).
   a. R$^+$ media plus 600 µg active G418/mL (Sigma). 60 mg/mL-100X geneticin sulfate (G418) stock solution; 5 g (70% active ingredient) dissolved in DMEM, final volume 58.3 mL, sterile-filtered using a 0.2-µm filter, store at –20°C in 10-mL aliquots.
4. Autocrine media (B82 EGFR cells with pUHD15.1neo, pUHD10.3/TGF-α, pR8 (B82R$^+$/TGF-α) tissue-culture media).
   a. R$^+$/1st plasmid media plus 2.4 mM histidinol (Sigma). 2.4 M histidinol stock solution; 5 g dissolved in 9.7 mL ddH$_2$O final volume, sterile-filtered using a 0.2-µm filter, store at 4°C.
   b. 1 µg/mL tetracycline (repression of TGF-α expression, Sigma). 20 mg/mL tetracycline superstock, dissolve tetracycline powder into ddH$_2$O, and store at –20°C. Dilute tetracycline 1:40 in ddH$_2$O, sterile-filter using 0.2-µm filter, and store at –20°C in 1-mL aliquots.

3. Methods
3.1. Creation of Autocrine Cell Loops
3.1.1. Construction of pUHD10.3/TGF-α

First, Bluescript II KS+ (pBS)/TGF-α plasmid is constructed for ease of molecular cloning because of Bluescript's multiple cloning region.

1. The entire 800 bp TGF-α sequence is removed from pMTE4 with a *Hind*III digest (37°C, 2 h) and gel purified (1% low melting point agarose, 2.5 hours, 50 mAmps). Isolate the TGF-α gene fragment from the agarose using β-agaraseI and ligate into *Hind*III digested Bluescript II KS$^+$ plasmid (T4 ligase, overnight, 16°C). Minipreps can be checked via *Nco*I (only cuts if insert is present) digests and large scale preps of pBS/TGF-α purified with Wizard Maxiprep. Orientation of TGF-α insert (need T7 to T3 orientation for part 2) in Bluescript can be determined by *Rsa*I digests (if T7 to T3, bands are 577, 1754, 1430 bps; if T3 to T7, bands are 348, 1754, 1659 bps) and sequencing.

2. The construction of pUHD10.3/TGF-α. Remove TGF-α sequence from pBS/TGF-α by digesting the plasmid with *Xho*I and *Eco*RI (37°C, 2 h each). Before digesting the plasmid with *Eco*RI, however, the linear plasmid (cut with *Xho*I) must be blunted with Klenow enzyme. Prepare pUHD10.3 by digesting it with *Bam*HI (37°C, 2 h), blunting with Klenow enzyme (same method as with pBS/TGF-α), and digesting the linear plasmid with *Eco*RI (37°C, 2 h). Gel purify the DNA fragments with 1% low-melting-point agarose, 2.5 h, 50 mAmps, and isolate the fragments from the agarose using β-agaraseI. Ligate the two DNA fragments together (T4 ligase, overnight, 16°C); check resulting plasmid preps for the correct pUHD10.3/TGF-α via *Pst*I linearization (*Pst*I only in TGF-α insert, no sites in negative pUHD10.3 plasmids) and sequencing.

3.1.2. Creation and Selection of Autocrine Cells

The transfection of the B82 cells were accomplished using the $CaPO_4$/DNA precipitation method *(11)*.

1. Day 1: Seed one million B82 R^+/1st plasmid cells into 60-mm tissue-culture dishes in R^+/1st plasmid media.
2. Day 2: Prepare a 500-mL aliquot of pUHD10.3/TGF-α and pR8-$Ca_3(PO_4)_2$ precipitate by mixing 30 μg of pUHD 10.3/TGF-α and 10 μg of pR8, up to 450 μL ddH_2O (depending on plasmid concentrations), 500 μL of 2X HBS, and 50 μL of 2.5 M $CaCl_2$. Add the precipitate to the cells' media plated on d 1. On d 3, rinse the cells with PBS and add back fresh R^+/1st plasmid media.
3. Day 4: Dilute cells 1:4 into autocrine media, keeping all of the cells. Note: Sometimes, if a large number of clones continue to grow in the selective media, and isolating individual clones is difficult, dilute the cells serially 1:2, 1:10, 1:100, 1:1000, 1:10,000. Refresh the selective media every third d until individual cell colonies appear (approx 200–1000 cells per colony), typically appearing 1–2 wk after beginning selection. Isolate the colonies using autoclaved Bellco 8 × 8 mm glass cloning cylinders dabbed into autoclaved Vaseline, and pass the colony into individual well of a 24-well plate containing selective media.

3.2. Determination of Autocrine Ligand in Bulk Extracellular Medium

3.2.1. ELISA Measurement of Autocrine Ligand

1. Plate autocrine cells in four sets (one for induced TGF-α expression and one for uninduced expression, plus and minus antireceptor blocking antibodies) at a cell concentration allowing for 2–3 d of growth before cell confluence is reached in normal autocrine growth media. For example, B82 cells double once a day, thus plate 125,000 cells into a 35-mm tissue culture dish and confluence is 3 d later at 1.5 million cells.
2. Two days before cell confluence, remove tetracycline containing autocrine media, rinse the cells with PBS, and, to half of the cell wells, add tetracycline-free

autocrine media, inducing TGF-α expression. To the other cell wells, add tetracycline containing autocrine media, keeping TGF-α expression off. NOTE: Protein synthesis takes about 6 h before detection at the surface and in the media, thus TGF-α expression is induced 24 h before beginning the experiment to ensure a steady-state expression of TGF-α.
3. Next day, with TGF-α expression at steady state, rinse the cells with PBS and add fresh autocrine media (tetracycline-free for induced cells, and tetracycline-containing media for uninduced cells). To half of the induced and uninduced autocrine cells' media, add 10 µg/mL anti-EGFR-blocking antibody 225 or 528. A comparison of TGF-α concentrations, plus and minus antibody, allows the analysis of TGF-α uptake by its receptor.
4. On the last day, remove 1 mL of conditioned media, spin at 13,000g for 10 min at 4°C to remove cell debris. Store the supernatant at –20°C, if measuring TGF-α concentration another day, otherwise on ice. Measure TGF-α concentration via TGF-α ELISA. Determine cell density by trypsinizing the cells, and count cell number using a hemocytometer or Coulter counter (*see* **Note 1**).

3.2.2. Tetracycline Concentration Effect on TGF-α Secretion

1. Plate autocrine cells at a similar cell density at a cell concentration that allows 2–3 d before cell confluence is reached in normal autocrine growth media. Plate enough cell dishes for 6 or more different tetracycline concentrations, with replicates.
2. Two days before cell confluence is reached, rinse the cells with PBS and replace the media with fresh autocrine media containing a gradient of tetracycline concentrations. For example, TGF-α expression is mostly inhibited by 1 µg/mL tetracycline, thus a 10, 1, 0.1 µg/mL to 10, 1, 0.1, 0 ng/mL tetracycline concentrations is an appropriate range.
3. Next day, with TGF-α expression at steady state, rinse cells with PBS and replace the media with a fresh autocrine media containing the tetracycline gradient. Also, include in the media, 10 µg/mL antireceptor blocking antibody 225 or 528, to prevent receptor TGF-α uptake and accurate measurement of TGF-α expression.
4. On the last day, remove 1 mL of conditioned media, spin at 13,00g for 10 min at 4°C to remove cell debris. Store the supernatant at –20°C, if measuring TGF-α concentration another day, otherwise on ice. Measure TGF-α concentration via TGF-α-ELISA. Determine cell density by trypsinizing the cells and count cell number using a hemocytometer or Coulter Counter.

3.2.3. Determining Cellular Processing of Secreted TGF-α Protein

1. Plate a large number of either transfected B82R$^-$/TGF-α cells or with autocrine B82R$^+$/TGF-α cells into 100-mm dishes, T75 flasks, or roller bottles in tetracycline-free growth media. Once cells have grown confluent and expressing TGF-α, replace the media with serum/protein-free media. Allow the media to condition for 1–2 d.

a. If using autocrine cells, one must have a high TGF-α expressor, overcoming ligand uptake by EGFR, or use antireceptor-blocking antibodies 225 or 528. If using antibodies, they should be removed before adding the conditioned media to the sizing column, because they might effect the mol wt standards elution or mask protein detection.
   b. The antibody can be removed using rabbit antimouse antibodies with protein A-conjugated Sepharose beads. Incubate the conditioned media with rabbit antimouse antibody, protein A-beads for a minimum of 2 h, at 4°C on a rocker. Spin beads down at 130g for 2 min and continue with the supernatant.
2. A high concentration of TGF-α (>100 ng/mL) will be required to run on the sizing column; thus, it may be necessary to concentrate the conditioned media using Amicon concentrators (large volumes, >100 mL) or centricons (small volumes, <10 mL) with 3000 mol wt cutoffs (MWCO) done at 4°C.
3. Mix 100 ngs of TGF-α with protein standards (albumin, 66 kDa, 2.5 mgs), carbonic anhydrase (29 kDa, 1 mg), cytochrome c (12.4 kDa, 1 mg), aprotinin (6.5 kDa, 3 mg), and 50 μL of glycerol, to a total volume of 1 mL.
4. Overlay the TGF-α/protein standards mixture on a 1-m G-50 fine Sephadex column (equilibrated with PBS) with a small buffer head (<2 mL). The glycerol in the protein mixture will help keep the proteins from diluting in the buffer head.
5. Elute the protein mixture into the beads, and then add a larger buffer head. Elute the proteins through the column, collecting 5 minute fractions.
6. Measure the fractions' 280 nm to determine when the protein standards eluted off the column. Measure TGF-α concentrations in the elute fractions by a TGF-α ELISA kit.

3.3. Determination of Autocrine Ligand Captured by Secreting Cells

3.3.1. Molecular Devices Cytosensor Operation/Setup

1. Plate cells in normal growth media at 250,000 cells/transwell the day before an experiment (4 transwells/experiment), 1 mL of cells/media in the top portion of the transwell and 2 mL of media in the bottom.
2. On the day of the experiment, add the spacer and transwell insert to each transwell. To avoid air bubbles, add the spacer or insert into the media at an angle.
3. Purge the Cytosensor's lines with DV-cyto and equilibrate the cells for approx 2 h on the Cytosensor at a 100 μL/min flow rate, 50% pump speed.
4. Pump cycle for all experiments is 1 min cycles to measure cells' quick response to EGF. Thus, the pump cycle is 30 s at 100 μL/min (50% pump speed); 8 s, pumps off delay; 20 s, pumps off, get extracellular acidification rate (ECAR); 2 s, pumps off delay.

3.3.2. Calibration of Cytosensor Readout to Ligand/Receptor Binding

A correlation of ECAR to EGF/TGF-α-receptor complexes can be found by taking the best fit line determined in **Subheading 3.3.2.1., step 4** and **Sub-**

heading 3.3.2.2., step 6, entering in a free-ligand concentration, and plotting the resulting number (ECAR and complexes, respectively) as EGF-receptor complexes vs ECAR. Thus, ECAR output from the Cytosensor can be converted into receptor–ligand complex numbers.

3.3.2.1. Cytosensor Readout to Exogenous Ligand

1. Plate and set up the Cytosensor as described in **Subheading 3.3.1.** with B82 R$^+$/1st plasmid cells.
2. After equilibrating the cells with DV-cyto, do a stepwise EGF concentration gradient exposure to the B82R$^+$/1st plasmid cells. For example: one EGF gradient series may be 0.05, 0.1, 0.5, 1, 2, 5, 10, 20, and 100 ng/mL EGF. Since the Cytosensor runs four transwells (channels) per day's experiment, a suggestion would be to leap-frog EGF additions in each channel to minimize previous EGF additions effects. Thus, the run would be Channel A: 0 ng/mL EGF run continuously as blank/control; Channel B: 0.05, 1, 10 ng/mL EGF; Channel C: 0.1, 2, 20 ng/mL EGF; and Channel D: .5, 5, and 100 ng/mL EGF. EGF is diluted in the running buffer, DV-cyto.
3. Expose cells for 10 min (or until peak ECAR is reached) to desired EGF concentration, before switching cells back to DV-cyto media and allowing the cells to re-equilibrate their ECAR to baseline. Once cells have re-established their baseline, a new EGF concentration can be introduced to each channel. NOTE: There is a 300-μL dead volume between the inlet and line switch; thus, the EGF solution must be flowing at least 6 min before switching the lines to EGF (Line 1, DV-cyto; Line 2, EGF solution).
4. Analyze data by recording the peak ECAR, usually occurring between 6 and 10 min, subtract the baseline, and plot as a function of EGF concentration (free EGF). A best-fit equation can be determined as:

$$\text{ECAR response} = \text{free EGF} * \text{ECAR coefficient} * \text{Receptor Number}/(\text{Free EGF} + K_d)$$

where K_d is the EGF/EGFR disassociation rate constant. Use ECAR coefficient * receptor number combined as one parameter, such as Receptor-ECAR, for the purposes of obtaining a parameter fit.

3.3.2.2. Binding of Exogenous Ligand

1. Plate R+/1st plasmid cells into 35-mm tissue-culture dishes in normal growth media and grow cells to confluence.
2. Two h before beginning the experiment, switch the cells' media to D/H/B.
3. Add desired I^{125} EGF gradient for 10 min at 37°C. Suggested EGF concentrations are 0.1, 0.3, 0.6, 1.2, 2.5, 5, 10, 20, 30, 40, 50 ng/mL.
4. After 10 min, immediately move the cells to 4°C, remove, and save the media. Rinse the cells with 1X WHIPS, and add the rinse to the corresponding media fractions. Count the combined solution fractions on a gamma counter as free EGF. Lyse the cells with 1 M NaOH for 10 min at 4°C. Add 1 mL rinse of 1X

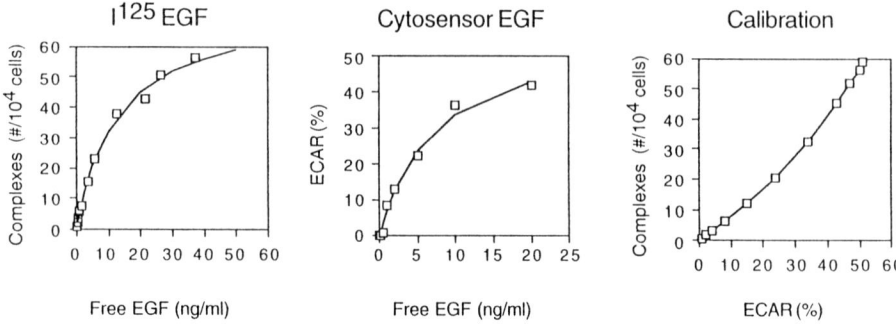

Fig. 1. Calibration of receptor–ligand complex numbers with ECAR.

WHIPS, and count the lyse cells/WHIPS rinse in a gamma counter as bound ligand–receptor complexes.
5. Count an extra, similarly plated cell dish, using a Coulter counter or hemocytometer to determine cell concentration.
6. Analyze data by converting free EGF (cpm to ng/mL) and bound EGF–receptor complexes (cpm to complexes/cell, EGF's MW 6045). Plot as complexes vs free EGF. A best-fit equation can be determined for complexes = Free EGF * receptor number/(Free EGF + K_d), where K_d is the EGF/EGFR disassociation rate constant.

3.3.2.3. CALIBRATION CURVE OF B82 EGFR POSITIVE CELLS

1. Use plots of ECAR vs EGF and Complexes vs EGF, as described in the two previous subheadings, to obtain the best-fit equations and parameter values. Obtain an expected value for ECAR and complex numbers as a function of the same EGF gradient concentration. Plot the expected values as complex numbers vs ECAR. Thus, an ECAR reading from the Cytosensor can be directly correlated to receptor–ligand complex number. See **Fig. 1**.

3.3.3. Cytosensor Readout of Autocrine Ligand

1. Plate and set up the Cytosensor as described in **Subheading 3.3.1.** with autocrine cells in normal growth media containing tetracycline. Also, equilibrate the cells in DV-cyto buffer containing 1 µg/mL tetracycline.
2. Upon obtaining steady-state ECAR, impose the following conditions (one condition per Cytosensor channel - cell):
   a. DV-cyto buffer.
   b. DV-cyto buffer, 1 µg/mL antireceptor antibody 225 or 528.
   c. DV-cyto buffer, 10 µg/mL tetracycline and 1 µg/mL anti-EGFR antibody 225 or 528.
   d. DV-cyto buffer, 10 µg/mL tetracycline.
3. Run these conditions for 24 h and measure TGF-α's expression induction by the increase in Channel B's ECAR.

3.3.4. Cytosensor Readout of Antibody Inhibition

1. Plate cells and set up the Cytosensor as described in **Subheading 3.3.1.** with autocrine cells in tetracycline-free (initiates TGF-α expression) growth media.
2. Check to make sure cells are responding similarly by exposing the cells to a 10-ng/mL EGF pulse. After cells' ECAR peaks, remove the EGF and allow the cells to re-equilibrate their ECAR to baseline.
3. Leaving one Cytosensor channel as control (DV-cyto buffer only), introduce a desired antibody concentration (anti-TGF-α decoy antibody or anti-EGFR blocking antibodies) to each channel. Steady-state ECAR response should occur within 30 min. Analysis of data can be performed by recording the minimum ECAR value on addition of antibody, or integrating the ECAR response for a 30-min period. For the conversion of ECAR to complexes, use the highest antibody concentration (antibody concentration at which further increases in antibody concentration did not decrease the cells' ECAR) as a baseline equal to zero receptor–ligand complexes. Complexes in the presence of other antibody concentrations can then be determined from this baseline (*see* **Note 3**).

4. Notes

1. Cell density will have an effect on ligand concentrations in the extracellular media. Thus, all TGF-α concentrations must be measured at similar final cell densities. If necessary, because of slight differences in growth rates, dilute the initial cell concentrations accordingly.
2. When using high concentrations of an additive to DV-cyto (i.e., antibodies), accurate results will only be obtained by dialyzing the additive in DV-cyto to remove any additional components (i.e., salts, glycerol, and so on).
3. Theoretical predictions indicate that antiligand decoy antibody concentrations will need to be 100–1,000 times higher than antireceptor blocking antibody's because of transport phenomena. Suggested concentrations for blocking antibodies is in the low μg/mL concentration range (i.e., 0.1–10 μg/mL).

Acknowledgments

This work was supported by a grant from the NSF Biotechnology Program to DAL and HSW.

References

1. Sporn, M. B. and Roberts, A. B. (1992) Autocrine secretion—10 years later. *Ann. Intern. Med.* **117,** 408–414.
2. Forsten, K. and Lauffenburger, D. A. (1992a) Autocrine ligand binding to cell receptors: mathematical analysis of competition by solution "decoys." *Biophys. J.* **61,** 518–529.
3. Forsten, K. and Lauffenburger, D. A. (1992b) interrupting autocrine ligand-receptor binding: comparison between receptor blockers and ligand decoys. *Biophys. J.* **63,** 857–861.

4. Lauffenburger, D. A., Chu, L., French, A., Oehrtman, G., Reddy, C., Wells, A., Niyogi, S., and Wiley, H. S. (1996) Engineering dynamics of growth factors and other therapeutic ligands. *Biotechnol. Bioeng.* **52,** 61–80.
5. Will, B. H., Lauffenburger, D. A., and Wiley, H. S. (1995) studies on engineered autocrine systems: requirements for ligand release from cells producing an artificial growth factor. *Tissue Eng.* **1,** 81–94.
6. McConnell, H. M., Owicki, J. C., Parce, J. W., Miller, D. L., Baxter, G. T., Wada, H. G., Pitchford, S. (1992) the cytosensor microphysiometer: biological applications of silicon technology. *Science* **257,** 1906–1912.
7. Hirst, M. A., Pitchford, S. (1993) use of a single assay system to assess functional coupling of a variety of receptors. *J. NIH. Res.* **5,** 69.
8. Gossen, M. and Bujard, H. (1992) tight control of gene expression in mammalian cells by tetracycline-responsive promoters. *PNAS* **89,** 5547–5551.
9. Derynck, R., Roberts, A. B., Winkler, M. E., Chen, E. Y. and Goeddel, D. V. (1984) Human transforming growth factor-α: precursor structure and expression in *E. coli. Cell* **38,** 287–297.
10. Chen, W. S., Lazar, C. S., Poenie, M., Tsien, R. Y., Gill, G. N., and Rosenfeld, M. G. (1987) requirement for intrinsic protein tyrosine kinase in the immediate and late actions of the EGF receptor. *Nature* **328,** 820–823.
11. Kriegler, M. (1990) *Gene transfer and expression* (Kriegler, M., ed.), H. Freeman, New York, pp. 96–98.

13

Measurement of Recovery of Function Following Whole Muscle Transfer, Myoblast Transfer, and Gene Therapy

John A. Faulkner, Susan V. Brooks, and Robert G. Dennis

1. Introduction

For a skeletal muscle tissue engineer, the most important issue following an experimental intervention is the evaluation of the recovery of the functional capabilities of the tissue, relative to those of the control tissue. Whether investigators perform whole muscle transfers with spontaneous *(1)* or surgical *(2)* vascular and nerve repair, myoblast transfers *(3)*, or manipulations of muscle-specific genes *(4,5)*, the question remains the same: Has the intervention impaired, maintained, or enhanced the functional capabilities of the skeletal muscles involved? In each case, determining structure–function relationships is of vital importance because structure–function relationships are frequently disrupted following an intervention, so that muscle mass and total muscle fiber cross-sectional area (CSA) are not different from control values, but function is impaired, or both are impaired, but with different magnitudes of impairment.

The contractile properties provide quantitative measurements of the functional capabilities of skeletal muscles. A muscle contraction is defined as the initiation of crossbridge cycling by increasing intracellular calcium concentration with an attempt of the muscle to shorten. Whether the muscle fibers shorten, remain at the same length, or are stretched depends on the interaction between the force developed by the muscle and the load on the muscle. A force greater than the load results in the muscle fibers shortening in a miometric contraction. When the force and the load are equal, or the length is fixed, an isometric contraction occurs. Finally, a load greater than the force developed stretches the fibers in a pliometric contraction. Force and displacement are measured directly, and the interactions between force and the rate and magnitude of displacements allow the determination of velocity, work, power, and stiffness. Muscle mass and muscle-fiber length provide the data to normalize the contractile properties appropriately. Additional insightful measurements of structure are: fiber-type composition *(6)*, the dry mass:wet mass ratio *(7)*, protein and connective-tissue content *(7)*, and myofibrillar-protein content *(8)*.

Making accurate measurements of even basic contractile properties of skeletal muscles is laborious and costly. Experienced investigators make only those measurements necessary to establish the functional status of skeletal muscles unequivocally. Contractile properties of skeletal muscles may be measured at the level of a muscle group *(9,10)*, single whole muscles *(11,12)*, single motor units *(6)*, small bundles of muscle fibers *(13–15)*, or single fibers *(16–20)*, and measurements are made either in vivo *(9,10)*, *in situ* *(11)* or in vitro *(12,14,15,17–20)*. Properties of single motor units *(6)* and single intact mammalian fibers *(16)* provide important data, but the operative and measurement requirements of these preparations are extremely demanding and will not be dealt with in this chapter. Since the correct selection of the most appropriate preparation determines both the cost and success of the outcome of experiments, the pros and cons of each preparation will be discussed in detail.

2. Materials
2.1. Instrumentation

Vital to the valid measurement of contractile properties is correct instrumentation. The simplest apparatus includes a temperature-controlled fluid bath, a fixed post attached to a linear positioner for adjusting muscle length, a force transducer, electrodes, and a stimulus wave-form generator. Stimulus pulses may be initiated manually and force signals may be recorded directly on a strip-chart recorder or oscilloscope. With this system, only isometric contractions can be evaluated, but these are the most commonly reported data and the easiest to analyze and interpret. The addition of a servo motor to the system permits measurements of velocity, work, power, and stiffness. With the addition of a servo motor, the integration of a full computer-based data-acquisition system for data collection and analysis is required, and software programs must be written.

2.1.1. Stimulators

Stimulators are readily available (Grass Instruments, West Warlock, RI). Muscle groups, single muscles, and bundles of intact fibers may be activated by the application of electrical fields. For muscles or bundles of muscle fibers in vitro, these fields are applied via parallel plate electrodes, which will provide a uniform and well-defined stimulus to the entire muscle. Platinum is the

Measurement of Recovery of Function 157

material of choice for in vitro plate electrodes, but stainless steel (type 18-8) is also acceptable. Plate electrodes should be parallel and close to, but not touching, the muscle, with the muscle entirely between the plates. Since the current drawn is a function of the plate area, plates should be no larger than necessary. With an underpowered amplifier, the voltage setting on the stimulator can be increased without actually increasing the stimulus intensity, resulting in an erroneous measurement of maximum force. Such errors are prevented by verification of the voltage directly across the electrodes in the bath with an oscilloscope. Inadequate current output may be corrected by reducing the electrode plate area, placing the electrodes closer together, or upgrading the amplifier. The current draw of muscles stimulated directly with needle electrodes or for nerve stimulation is much less than in an in vitro bath, but verification is still warranted.

2.1.2. Force Transducers

1. Types of force transducers: A wide variety of force transducers are commercially available, and force transducers with performance characteristics well beyond those of most commercially available units have been reported in detail *(21–23)*. The principle of operation of every force transducer is a sensitive displacement transducer that measures the displacement, or strain, of an elastic element when subjected to an external force. A number of different types of force transducers provide effective measurements of the mechanics of muscles during isometric contractions.
   a. The most common type of force transducer is a bonded-resistance-type force transducer, which uses a small foil resistor to measure small strains in the load element (Kulite Semiconductor Products, Leonia, NJ). Bonded-resistance force transducers are robust and are available in a wide range of load capacities, but are generally not suitable for high-frequency applications, and are subject to noise because of the small output signal prior to amplification. Although some manufacturers claim infinite resolution, in practice the resolution is limited to ~0.05% of full scale.
   b. Piezoelectric force transducers operate on the principle that certain materials generate a voltage when subjected to strain. Piezoelectric transducers tend to have wide bandwidths and generate large signals, but they are easily damaged and often leak charge rapidly. These latter characteristics limit their usefulness for many experiments on muscle mechanics.
   c. Capacitance-based force transducers usually employ an air gap between metallized glass plates. A small change in the separation between the two plates results in a capacitance change. While capacitance-based transducers generally perform well, they tend to be physically large, extremely fragile, quite expensive, and subject to atmospheric changes and radio interference.
   d. The general principle of optical force transducers *(24)* is that a light source (an [LED], laser diode, or fiber optic cable) illuminates one or two photo

detectors. A small opaque vane attached to the load element of the force transducer interrupts the light. The output of the detectors is amplified and calibrated directly into units of force. Two detectors can be physically positioned so that the signal from one increases while the signal from the other decreases. The resulting two signals can be differentially amplified to improve the linearity of the device. Such devices are generally robust and can be designed in a very wide range of sensitivities and physical sizes.

2. Performance characteristics of force transducers: For muscle mechanics, the important performance characteristics of force transducers are: mechanical robustness, range, linearity, sensitivity, resolution, noise, signal type, compliance, drift, off-axis sensitivity, and bandwidth.

   a. Force transducers with metallic load elements are generally more mechanically robust than those with glass or ceramic load elements. When damage to the force transducer is detected, all data taken since the last calibration is suspect.

   b. Range, the simplest and most common specification for a force transducer, is usually defined by the maximum force value for which the transducer is linear. Sensitive transducers, such as those with glass or ceramic load elements, are likely to be destroyed by forces only slightly beyond the full range.

   c. Linearity describes how well the output of the force transducer follows a straight line over the full range of applied forces, and passes through zero when zero force is applied. Nonlinearity may result from the mechanical load element configuration, the electromechanical transduction of the load element deflection, or the preamplification electronics. Linearity may be improved by use of less than the full range of the transducer, but this procedure will reduce the signal-to-noise ratio.

   d. Sensitivity is defined as the output of the device divided by the input, usually expressed in units of voltage divided by force.

   e. The resolution of the force measurements is defined by both the sensitivity of the force transducer and the voltage resolution of the data-acquisition system recording the force. The manufacturer of the data acquisition system will specify the smallest voltage change detectable by their system. Divide this number by the sensitivity of the force transducer to determine the force resolution of the total system. The resolution is reduced by noise.

   f. Noise is frequently the limiting factor in the performance of an experimental apparatus. If the frequency of the noise is ~60 Hz, the source is probably the line current or a device near the experimental setup. Fluorescent lights, fans, compressors, video terminals, and power supplies introduce noise. To minimize the noise in a force signal, first eliminate from the area sources of electromagnetic noise, such as motors and spark gaps. Make sure the force transducer is receiving a clean source of power. Some power supplies are noisy, resulting in noise being superimposed on the force signal. Batteries, the most noise-free source of power, can be used with some force transducers. The length of the cables from the power supply to the transducer, and from the transducer to the amplifier, should be minimized, and instrument-quality-shielded twisted-pair cable should be used. Ground loops should be eliminated by grounding the device at only one point (usually done by the manufacturer), or the device will act as a loop antenna. Preamplify the signal as close to the transducer as possible to minimize any amplification of the noise. Some manufacturers build a preamp right into the transducer housing. Shield the force transducer in a tight-fitting, electrically grounded metal box. The entire setup may also be shielded within a Faraday cage. These noise-reduction techniques are applicable to any electrical signal and should be considered early in the design of the experimental system. Noise can be further reduced by filtering the signal, using either analog circuitry or software. Filtering always removes information from the signal, in addition to noise, and may introduce significant distortions in the force signal. The original data should always be retained, because filtered data cannot be unfiltered. Highly sensitive force transducers may also be subject to mechanical noise. Building vibrations and acoustic coupling to the setup can be eliminated by using an antivibration table.

g. Most force transducers provide analog signals. The type of amplifier and data-acquisition system required depends on whether the analog signal is differential (two voltages) or single-ended (one voltage referenced to ground). Some transducers provide digital signals, which may need to be converted to analog for the data-acquisition system. For basic contractile properties, force signals should be sampled at least at several hundred Hz, and for rapidly changing forces, such as for stiffness measurements, the sampling rate should be at least 20× the oscillation frequency to accurately detect peak forces to within 1–2%.

h. The compliance of the force transducer is critical, since a large compliance allows the muscle to shorten when activated. An ideal force transducer would have zero compliance, but in practice this is not possible to attain. The compliance of the force transducer should be less than 1% of the muscle fiber length for the full range of forces.

i. Both zero drift and sensitivity drift should be considered when selecting an appropriate force transducer. Zero drift is the change in the intersection, not the slope, of the calibration curve. Zero drift is not a problem if forces are measured relative to the resting baseline force determined immediately prior to each contraction. Sensitivity drift is a change in the slope of the calibration curve. Sources of sensitivity drift include damage to the force transducer, thermal instability (which usually also affects zero drift), or other environmental conditions, such as humidity. The operating principles of different types of force transducers determine their susceptibility to drift. Sensitivity drift occurs gradually, so force transducers must be calibrated regularly. Some manufacturers recommend daily calibration.

j. Off-axis sensitivity is the tendency of force transducers to respond to forces applied in directions other than that intended during the experiment. To reduce the effects of off-axis sensitivity, the line of force should be aligned with the transducer, and side-loading of the specimen should be eliminated.

k. Bandwidth is perhaps the most important and often overlooked performance specification of a force transducer. The force transducer bandwidth must include DC, so that steady, unchanging forces can be measured. Some transducers, especially those based on piezoelectric elements, may be AC-coupled. When using an AC-coupled device, a steady force will appear to decay to zero. If the time-constant of the decay is long, the transducer may be suitable for some muscle mechanics experiments, provided that the forces are not to be maintained for long periods of time. In addition to the steady-state performance of the force transducer, in some experiments the force transducer must be able to measure rapidly changing forces. Rapid force changes occur during stiffness measurements and step-release experiments, in which the muscle specimen is subjected to a rapid length change and the force response is recorded. To ensure that rapid force changes are detected with minimal distortion, the resonant frequency and damping factor of the force transducer must be determined. The specifications provided by manufacturers can only be used in the initial selection of a force transducer, because the actual performance of the transducer will change significantly in each individual system with a muscle specimen attached. When measuring rapidly oscillating forces, the oscillation frequency must be restricted to <30% of the resonant frequency. Details on resonance and mechanical damping are beyond the scope of this chapter, but are available in standard mechanical engineering and dynamic systems texts.

2.1.3. Servo Motors

1. Types of servo motors: For muscle mechanics, linear and rotary servo motors are commercially available (Cambridge Technology, Watertown MA; Aerotech, Pittsburgh, PA) and complete setups may be purchased (Scientific Instruments GMBH, Heidelberg, Germany). The complete setups are costly, have limited flexibility, and may have unreported or unknown technical limitations.
2. Performance characteristics of servo motors: The important performance characteristics of servo motors for muscle mechanics experiments are signal type, load capacity, hysteresis, compliance, range, speed, and force-measurement capability.
    a. All closed-loop servo mechanisms require electronics to continuously monitor and control position, compensate for position errors, provide position output data, and interface with the data-acquisition system. Monitoring the actual servo position—not only the control signal from the computer to the servo—is critical, because some error between the desired and actual position is present in any servo mechanism. Some controllers provide an analog signal proportional to servo position that can be measured directly. In contrast, modern industrial servo controllers may have digitally-resolved position information. To increase the servo speed by reducing demand on the controller, a quadrature decoder to read the digital position signal directly may be built or

provided by the manufacturer. The digital signal can be collected by the digital I/O ports of the computer, or converted to analog and collected in parallel with the force signal.

b. During pliometric contractions, forces may reach 3× the maximum isometric force, and the servo must be capable of bearing loads at this level.

c. Depending on the type of controller, the servo will exhibit varying degrees of hysteresis, which will introduce directionally dependent errors in the muscle length signal. Hysteresis errors can usually be eliminated by properly tuning the controller gains. Tuning information for reducing hysteresis should be supplied by the manufacturer.

d. The mechanical compliance of the motor introduces errors into the muscle-length measurement, because muscle length is measured indirectly by the servo-position feedback signal. Therefore, the compliance in the connection of the muscle specimen to the motor must be minimized by short, rigid attachments. An additional source of compliance results from the limitations of the controller electronics, which may tolerate larger position errors at larger loads. This compliance may give large errors during experiments when muscle length must be maintained during force transients. A compliance compensation circuit reduces the compliance in Cambridge servo motors significantly *(25)*, and the general principle may be employed with other servo controllers that have force-feedback capability. Total compliance (mm/N) of a servo is measured by hanging weights near, but not exceeding, its rated load capacity, and measuring the deviation of the position-output signal from the unloaded position. No change in the servo position signal during compliance measurements suggests that the position input signal, rather than position output from the servo, is being monitored.

e. The range of displacements through which the servo must operate for measurements of velocity, work, power, and stiffness varies with the size of the muscle. In general, servos should be capable of displacements equal to at least one fiber length of the specimen.

f. All servos have a limited speed, also known as the slew rate. The maximum slew rate is the maximum velocity of the servo under specific loading conditions. In general, larger loads result in slower slew rates. Therefore, when determining the servo speed, the loading conditions should approximate those under which the servo will operate. The load on the servo actuator includes inertial and viscous components, as well as the components of interest originating from the passive and active properties of the muscle. Viscous forces, related directly to velocity and arising from the movement of the motor actuator and muscle specimen, may be ignored, except for very small specimens *(26)*. Inertial forces, which result from acceleration of the servo effector mechanism and the muscle specimen, usually limit the speed of the servo. Therefore, the mass of the moving portion of the servo mechanism should be made as small as possible, and the mass distribution in rotating servo mechanisms is concentrated as close to the axis of rotation as possible.

g. Servo motors accomplish the measurement of force by monitoring the coil current of the motor. In a constant magnetic field, the torque (or force) is, in principle, directly proportional to the coil current. Some manufacturers allow the coil current to be used in a feedback loop, enabling the specimen to be subjected to constant-force length changes. Although convenient, the coil-current signal must be used with caution. The caution is required because, for linear motors, or rotary motors with large angular displacements, the magnets may have variable strengths over their working lengths, which result in varying relationships between coil current and force at different servo positions. The variation in the coil current–force relationship causes systematic errors in the force measurement. For rotary servos with small angular displacements (less than ± 20 degrees), coil current may be used with reasonable confidence, but the motor force should still be calibrated for the full range of expected loads at several servo positions. The primary disadvantage of using coil currents for measuring force is that viscous and inertial force components of the system are included, as well as the passive and active forces of the muscle. Consequently, a stationary force transducer positioned at the end of the muscle opposite to the servo motor is preferable.

2.1.4. The Data-Acquisition System

Many data-acquisition systems are available commercially. Once force and position signals have been generated by the servo motor and force transducer, and verified with an oscilloscope, the collection and analysis of the data becomes relatively straightforward. In general, a programmer will be required to develop customized software for a given experiment.

2.2. Experimental Apparatus

2.2.1. Apparatus for In Vivo Preparations *(3,9,27,28)*

When repeated assessments of muscle function in a single animal are required over time, muscle function must be assessed noninvasively without disruption of the health and condition of the animal. Such experiments include studies of the effects of age or an experimental intervention on muscle function, with or without changes induced by programs of muscle conditioning, dietary alterations, or pharmacologic dosing. Various monitoring devices have been developed to evaluate hindlimb muscle groups in vivo in small animals. These devices are essentially rotary servo motors, with shoe fixtures for securing the foot of a human being *(29)*, rabbit *(27)*, rat *(9)*, or mouse *(28)*. The shoe is attached to the rotary servo via a torque transducer. Body temperature of the experimental animal is maintained at ~37°C through the use a heating pad, or by circulating warmed water through a Plexiglas platform, and the platform must allow for fixation of the experimental limb to minimize compliance in the system.

2.2.2. Apparatus for In Situ Preparations *(2,11,30)*

When maintenance of blood flow is required during the assessment of the function of single muscles, the *in situ* preparation is appropriate. Such experiments include studies of large muscles that cannot be maintained in vitro *(2)*, or assessments of function during prolonged contraction protocols *(11)*. As with evaluations in vivo, body temperature must be maintained with a temperature-controlled platform. The platform must also permit rigid stabilization of the experimental limb, and the distal end of a single muscle is attached either to an isometric force transducer or to the tip of the lever arm of a servo motor.

2.2.3. Muscle Bath for In Vitro Preparations *(12,14,15)*

The in vitro preparation is essentially the same for evaluating isolated whole muscles *(12)*, small bundles of intact fibers *(13,14)*, or single permeabilized-fiber segments *(19)*. The major difference between the preparations is the sensitivity requirements of the force and displacement transducers. The muscle or muscle fiber is securely attached to a displacement transducer or a fixed post at one end, and to a force transducer at the other end, and suspended in either a horizontal or vertical orientation in a muscle bath. The advantages of the horizontal orientation are easier access to the muscle for adjustments of length and less volume for the bathing solution. Temperature control of the bath is also critical and is usually achieved by circulating water at a thermostatically controlled temperature through the jacket of the bath.

2.3. Solutions

For all in vitro preparations, solutions are required to maintain the viability of the specimen.

2.3.1. Single Muscles and Small Bundles of Fibers

Small single muscles and small bundles of parallel fibers up to ~8 mm^2 in total fiber cross-sectional area may be studied in vitro in Krebs-Ringer solution *(7)*. The single muscles and bundles of fibers are suspended in a Krebs-Ringer bicarbonate solution: 137 mM NaCl, 5 mM KCl, 2 mM CaCl$_2$, 1 mM MgSO$_4$, 1 mM NaH$_2$PO$_4$, 24 mM NaHCO$_3$, 11 mM glucose, and 0.025 mM tubocurarine chloride. The pH of the bath is maintained by continuously bubbling the solution with 95% O$_2$–5% CO$_2$.

2.3.2. Single Permeabilized Muscle Fibers

Bundles of muscle-fiber segments may be obtained from muscles by needle or open biopsy. Permeabilized single muscle fibers are obtained and studied by the following techniques.

1. Skinning solution for permeabilized fibers *(19)*: Small bundles of fiber segments are permeabilized in a skinning solution (in mM): K-propionate, 125;

ethyleneglycol-*bis* (β-aminoethyl ether) tetraacetic acid (EGTA), 5; ATP, 2; $MgCl_2$, 2; imidazole, 20; and glycerol 50% (v/v). The solution is adjusted to pH 7 with 4 M KOH. Bundles of fibers will remain viable stored in this solution at $-20°C$ for over 6 mo.

2. Relaxing solution for single permeabilized fibers *(20)*: The sarcolemmal barrier is disrupted in permeabilized fiber segments, making it possible to control directly the composition of the solution that surrounds the myofibrils. For attaching fibers to the apparatus and measuring fiber dimensions, fibers are placed in a low-calcium relaxing solution (in mM): EGTA, 7; $MgCl_2$, 5.4; ATP, 4.74; CrP, 14.5; imidazole, 20; $CaCl_2$, 0.016; and KCl, 79, to give a final ionic strength of 180 mM. The solution is adjusted to pH 7 with KOH.

3. Activating solutions for single permeabilized fibers *(20)*: Permeabilized fibers are activated by high calcium-activating solutions. Similar to the relaxing solution, the maximal activating solution of pCa 4.5 contains (in mM): EGTA, 7; $MgCl_2$, 5.3; ATP, 4.81; CrP, 14.5; imidazole, 20; $CaCl_2$, 7; and KCl, 64, to give a final ionic strength of 180 mM, and is adjusted to pH 7 with KOH. The pCa 4.5 activating solution is mixed with various volumes of relaxing solution to obtain solutions with a range of pCa values. The final concentrations of all metals and ligands in the solutions are calculated using the computer program of Fabiato and Fabiato *(31)*, based on the stability constants for each metal–ligand complex reported by Godt and Lindley *(17)*. The apparent stability constant for the Ca^{2+}–EGTA complex must be corrected for ionic strength and temperature *(31)*.

3. Methods

Animal care procedures and all experimental and operative procedures must be performed in compliance with the guidelines set forth in the United States Public Health Service, National Institutes of Health Publication No. 85–23. For the measurement of contractile properties in vivo, *in situ,* or in vitro, commonly accepted forms of anesthesia for mice and rats include inhaled ether *(1)* or intraperitoneal injections of sodium pentobarbital (30–60 mg/kg for old and young rats *[7]*, respectively, and 80 mg/kg for mice *[11]*) , or Avertin *(28)*. Supplemental anesthetic is administered, as required, to maintain a depth of anesthesia that prevents response to tactile stimuli. Following removal of muscles, the animals are euthanized. Acceptable methods for euthanasia include administration of an overdose of the anesthetic or an intravenous injection of 1 M KCl, and successful euthanasia is ensured by the immediate induction of a pneumothorax.

3.1. Types of Skeletal Muscle-Fiber Preparations

3.1.1. In Vivo Preparations *(3,9,27–29)*

The contractile properties of the muscle groups have been measured in vivo on the upper and lower limb flexor and extensor muscle groups of many mammalian species, including human beings. An in vivo apparatus actually measures the moment developed by the muscle group about a joint. The use of an apparatus of this type requires that one of the two articulating bones be immobilized without injuring the animal. For rodents, the most easily studied muscle groups are the ankle flexors and extensors, using a shoe apparatus, with the knee held stationary at a 90 degree angle by a screw clamp fixed at the femoral condyle. Care must be taken not to impair blood flow to the muscle group. The foot is positioned and strapped in the shoe so that the centers of rotation of the ankle and the shoe are coaxial. Torque is converted to muscle force by the relationship $T = r \times F$, where T is the torque, r is the moment arm length, F is the force generated by the muscle group, and x denotes the cross product. In general, the moment arm length is not constant over the range of motion of a joint. The effect of ankle-joint angle on moment arm length for ankle flexors of mice has been quantified and modeled *(32)*, and this general method can be applied to many different species and muscle groups. If moment arm length as a function of joint position is known, muscle force is directly calculated from the joint angle and torque.

Typically, percutaneous needle electrodes are used to stimulate the motor nerve for activation of the muscle group. Complete nerve stimulation without damage to the nerve and minimization of contraction of antagonistic muscle groups are considerations with this type of stimulation. Additional forces are also generated because of surrounding tissue, such as the joint capsule. Finally, because of the rotation of the torque transducer, inertial forces are included in any dynamic experiment. In some cases, dynamic effects can be measured and subtracted from the data.

3.1.2. In Situ Preparations *(2,11,30)*

The contractile properties of a wide variety of single muscles in mice, rats, rabbits, cats, dogs, sheep, and goats can be measured *in situ*. For measuring contractile properties *in situ,* animals are anesthetized, and an incision is made over the experimental muscle. The muscle is freed from surrounding tissues, and the distal tendon is ligated, sectioned distal to the tie, folded back on itself to form a loop, and ligated again. The experimental limb is immobilized, and the ligature is used to tie the distal tendon to the force transducer, or to the tip of the lever arm of a servo motor. The muscle is activated by isolation and stimulation of the nerve with needle or cuff electrodes or by direct stimulation of the muscle with needle electrodes. Nerves to synergistic and antagonistic muscles may be transected to avoid artifacts from contractions of these muscles. Any exposed portion of the muscle or nerve is bathed regularly with warmed saline, or a skin pouch is filled with warmed mineral oil to maintain muscle temperature at ~35°C.

3.1.3. In Vitro Preparations

The in vitro preparations include whole muscles, bundles of intact muscle fibers or fiber segments, or single permeabilized muscle fibers.

1. Whole muscles: The limitation for the measurement of contractile properties in vitro is that all of the fibers in a muscle remain viable throughout the period of measurement. For skeletal muscles in vitro, the calculated critical radius for oxygen diffusion decreases almost linearly from 1.19 mm at 20°C to 0.51 mm at 40°C *(33)*. Based on evaluations of slow soleus and fast extensor digitorum longus (EDL) muscles of rats over a period of 60 min, and measurements of fiber glycogen content in muscle cross sections, in vitro measurements are preferably made at 25°C, to ensure viable fibers in the core of the muscle *(33)*. The largest muscles on which valid measurements of contractile properties can be made in vitro are ~200 mg. To remove a muscle, an incision is made through skin and fascia, and the muscle is isolated with blood flow intact. A tie is placed securely around the distal and proximal tendons. The distal tendon is cut, and then the nerves and blood vessels are transected. The muscle is lifted gently by the distal tie and the proximal tendon is cut. The muscle is placed in the bath and tied to the transducer(s).
2. Bundles of intact fibers or fiber segments: Bundles of intact fibers *(14)* or fiber segments *(15)* may be obtained from many muscles and from many species. One very useful application has proven to be the strips of intact fibers obtained from diaphragm muscles *(4,14)*. To obtain the diaphragm strips, animals are anesthetized deeply and the complete diaphragm muscle is excised rapidly and immersed in aerated Krebs-Ringer solution. Under a dissection microscope, the diaphragm muscle is pinned with fibers at approximately resting length, and strips 2–4 mm wide are cut from the costal portion of the hemidiaphragm with straight-edged dissection scissors. Cuts are made parallel to the fibers on the cephalad surface of the muscle. Fibers on the caudal surface of the diaphragm muscle run at an angle to these fibers, preventing the possibility of a pure intact-fiber preparation. The diaphragm strips include a small section of a single rib and a portion of the central tendon. Ties of 5-0 suture are placed firmly around the rib and the central tendon, and the diaphragm strip is suspended in a muscle bath between force and displacement transducers.
3. Single permeabilized fiber segments: Single permeabilized fibers segments may be obtained from any muscle and from any species, including human muscles *(18)*. Small fiber bundles are tied at slack length with 5-0 suture to glass capillary tubes. The bundles are placed in skinning solution and stored at 4°C overnight. Fiber bundles are subsequently transferred to –20°C. For a given experiment, single fiber segments are pulled from the bundle with fine forceps and transferred to a 15°C bath containing relaxing solution. Single fibers are mounted between a force transducer and the lever arm of a servo motor. Single permeabilized fibers are activated by exposing the fiber segment to high calcium-activating solutions. Throughout an experiment, fiber segments are typically

cycled between an isometric contraction, and short periods of rapid isovelocity shortening, followed by a rapid restretch back to the initial fiber-segment length while maximally activated. This cycling protocol allows the maintenance of a stable sarcomere striation pattern and a constant maximum isometric force for more than 10 min of continuous activation *(20)*.

3.2. Measurement of Contractile Properties

The measurements of contractile properties require the setting of optimum voltage and optimum length for the development of maximum force *(14)* and power *(11)*.

3.2.1. Optimum Voltage

Muscle fibers in vivo *(3,10)* or *in situ* *(11)* are activated by nerve stimulation, and those in vitro directly by field stimulation between two platinum plate electrodes *(14)*. The duration of a square wave pulse is set at 200 µs. The optimum voltage is determined by increasing the voltage during single twitches until twitch force plateaus or decreases. The voltage of the stimulator is then set for supramaximum stimulation, slightly above the value that produced maximum twitch force.

3.2.2. Optimum Length (L_o)

The development of maximum force by a muscle fiber requires maximum overlap of thick myosin filaments and thin actin filaments in a maximum number of half sarcomeres. To optimize sarcomere length for force development, the muscle or bundle of fibers is placed at a length just below the threshold for the development of passive force. The length of the muscle or bundle of muscle fibers is increased until the force developed during a single twitch reaches a maximum. Optimum muscle length (L_o) is set using twitches to prevent fatigue of the muscle fibers during this determination. The conditions that maximize the twitch response are the same as those for tetanic contractions *(12)*. With the muscle at L_o, the length of the muscle fibers is defined as optimum fiber length (L_f). For mammalian muscles at L_o, average sarcomere length is between 2.5 and 3.0 µm. Consequently, the length of single permeabilized fiber segments (L_f) is adjusted to give an average sarcomere length in this range. All subsequent measurements of isometric force are made with the muscles at L_o and muscle fibers at L_f.

3.2.3. Measurement of Maximum Isometric Tetanic Force (P_o)

For groups of muscles, single muscles, and small bundles of intact fibers, a frequency–force curve is obtained by stimulating the muscle with trains of 200 µs square-wave pulses for a duration so that force plateaus during the train.

The frequency of stimulation is increased from 10 Hz at ~40 Hz increments, until the force plateaus at P_o. The duration of the train of pulses that results in a force plateau, and the stimulation frequency that results in P_o depend on the fiber type composition of the muscle and temperature. For muscles of mice and rats composed of fast fibers at 25°C, a train duration of ~300 ms is appropriate, and P_o is observed at a stimulation frequency between 120 and 150 Hz. For single permeabilized fiber segments, the analogous relationship to the frequency–force relationship is the pCa–force relationship, which is established by immersing fiber segments in activating solutions of varying free-calcium concentration. Single fiber P_o is achieved when a higher calcium concentration does not produce a higher force.

3.2.4. Measurement of Maximum Power Output During a Single Contraction (11)

Measurements of power are made during isovelocity shortening contractions with equal portions of the displacement above and below L_o. Very high forces can be maintained for all fiber types during shortening through displacements from 105 to 95% of L_f. The servo motor is programmed to move the lever arm at a selected constant velocity through a prescribed displacement. Stimulation is initiated simultaneously with beginning of the shortening ramp and ceases at the end of the ramp. Despite the termination of stimulation at the end of the ramp, the maintenance of strongly bound crossbridges in muscle fibers during relaxation causes an increase in force development as the muscle is returned to resting length. During the shortening ramp, the average force is calculated from the integrated area under the force–time curve, and power is calculated as the product of the average force and velocity. At any stimulation frequency, as velocity is increased from low to high values, power increases, peaks, and then declines. The peak is defined as the maximum power, and at each stimulation frequency, the velocity of shortening that results in maximum power is defined as the optimum velocity for the development of power (V_{opt}). V_{opt} is ~1/3 of the maximum velocity of shortening (V_{max}) determined from a force–velocity relationship *(11)*. Although an important property of fibers *(18)* and muscles *(12)*, V_{max}, need not be known for the measurement of power. The values of V_{opt} and maximum power will vary, depending on the species and the fiber type composition of the muscle being tested.

3.2.5. Derived Data

Following the measurements of contractile properties in whole muscles, L_o is measured with calipers, the muscle is removed from the animal, or from the bath, and the tendons are trimmed. The muscle is blotted and weighed to obtain the muscle wet mass (M_m). The L_f is evaluated by partial dissection and direct measurements of fibers with calipers, with the muscles pinned at L_o. The measurements by calipers are in good agreement with the $L_f:L_o$ ratios determined by nitric acid digestion *(12)*. Total fiber CSA (mm²) is calculated by dividing M_m (mg) by the product of L_f (mm) and the density of skeletal muscle, 1.06 (mg/mm³) *(12)*. Force development is a function of the total number of crossbridges per half sarcomere developing force in parallel. The total fiber CSA provides the best estimate of this value. Consequently, for whole muscles, P_o (N) is divided by the total fiber CSA (mm²) to obtain an estimate of the specific P_o (N/mm²). For single permeabilized fibers, L_f, width, and depth are measured using a stereomicroscope, high-power objective, and a camera system (Wild-Heerbrugg, Heerbrugg, Switzerland). Assuming an elliptical cross-section, the CSA of single fibers and specific forces are calculated.

The velocity of shortening is a function of the number of half sarcomeres that exist in series, and L_f is the best estimate of this value. Because of the dependence of velocity on the number of sarcomeres in series, in order to compare velocities between muscles, the velocities must be normalized by L_f (L_f/s). Work and power are dependent on both total fiber CSA and L_f making muscle mass the most appropriate normalization of work (J/kg) and power (W/kg) *(14)*.

4. Notes

Although the measurement of skeletal muscle contractile properties appears straightforward, even experienced investigators make critical errors that result in the generation of invalid data.

1. Injury, severe enough to cause decreases in P_o and power, may occur during the isolation of muscles for measurements of contractile properties *in situ*. Injury is much more likely to occur during the removal of muscles for measurements in vitro. Force is impaired by passive stretch, even when the remainder of the operative technique is sound.
2. If slippage or loss of the attachment at either end of the muscle fibers occurs, as it usually does, during a contraction, the contracting muscle fibers are injured by the excessive shortening. Under these circumstances, the decreased P_o invalidates the whole experiment.
3. To obtain specific P_o, the P_o is best normalized by the total fiber CSA. Normalization of P_o by muscle mass is incorrect, because muscle mass can change, independent of total fiber CSA, through changes in L_f. Normalization by dry mass/L_f *(34)* or by myofibrillar protein content/L_f *(8)* have been suggested to avoid changes in specific P_o resulting from changes in water content or noncontractile protein content, respectively. These normalization techniques are useful if specific P_o (kN/m²) is reduced, compared with the control value, but force is dependent on the number of cross bridges per half sarcomere in parallel, and neither

alternate technique assures that the dry mass or the myofibrillar protein mass represent masses of viable sarcomeres.

4. Between 25 and 35°C, P_o of soleus muscles from rats varies by ~13%, and that of EDL muscles by 7% *(33)*. The highest value for the soleus muscle is at 35°C; for EDL muscles, the differences are not significant between 25 and 35°C. Although P_o is not highly dependent on temperature, the isometric twitch characteristics, the rate of force development (dP/dt), the velocity of shortening, and, consequently, power are extremely sensitive to temperature *(35)*. The temperature of the muscle bath must be measured and controlled rigorously with a thermostatic feedback system.

5. Data obtained through muscle mechanics experiments are validated best by comparisons of control data with data published by highly reputable laboratories. Once the validity and reliability of control data have been established, contractile properties can be used to trouble-shoot both experimental equipment and methods. Valid control data must always be presented to demonstrate that any differences in structure, function, or structure–function relationships observed following an experimental or therapeutic intervention reflect real differences, as well as to support the biological relevance of the result.

Acknowledgment

Supported by NIA through Program Project AG10821 and the Nathan Shock Center for the Basic Biology of Aging AG13283.

References

1. Carlson, B. M. and Faulkner, J. A. (1996) The regeneration of non-innervated muscle grafts and marcaine-treated muscles in young and old rats. *J. Gerontol.* **51,** B43–B49.
2. Guelinckx, P. J., Faulkner, J. A., and Essig, D. A. (1988) Neurovascular-anastomosed muscle grafts in rabbits: functional deficits result from tendon repair. *Muscle Nerve* **11,** 745–751.
3. Mendell, J. R., Kissel, J. T., Amato, A. A., King, W., Signore, L., Prior, T. W., et al. (1995) Myoblast transfer in the treatment of Duchenne's muscular dystrophy. *N. Engl. J. Med.* **333,** 832–838.
4. Cox, G. A., Cole, N. M., Matsumura K., Phelps, S. F., Hauschka, S. D., Campbell, K. P., Faulkner, J. A., and Chamberlain, J. S. (1993) Overexpression of dystrophin in transgenic MDX mice eliminates dystrophic symptoms without toxicity. *Nature* **364,** 725–729.
5. Phelps, S. F., Howser, M. A., Cole, N. M., Raphael, J. A., Hinkle, R. T., Faulkner, J. A., and Chamberlain, J. S. (1995) Prevention of muscular dystrophy by full length and internally truncated dystrophins. *Human Mol. Genet.* **4,** 1251.
6. Burke, R. E. and Edgerton, V. R. (1975) Motor unit properties and selective involvement in movement. *Exerc. Sport Sci.* **3,** 31–81.
7. Segal, S. S., White, T. P., and Faulkner, J. A. (1986) Architecture, composition and contractile properties of rat soleus muscle grafts. *Am. J. Physiol.* **250** (*Cell Physiol.* **19**): C474–C479.

8. Taylor, J. A. and Kandarian, S. C. (1994) Advantage of normalizing force production to myofibrillar protein in skeletal muscle cross-sectional area. *J. Appl. Physiol.* **76**, 974–978.
9. Ashton-Miller, J. A., He, Y., Kadhiresan, V. A., McCubbrey, D., and Faulkner, J. A. (1992) An apparatus to measure in vivo biomechanical behavior of dorsi- and plantarflexors of the mouse ankle. *J. Appl. Physiol.* **72**, 1205–1211.
10. Miller, S. W., Hassett, C. A., White, T. P., and Faulkner, J. A. (1994) Recovery of medial gastrocnemius muscle grafts in rats: implications for the plantarflexor group. *J. Appl. Physiol.* **77**, 2773–2777.
11. Brooks, S. V. and Faulkner, J. A. (1991) Forces and powers of slow and fast skeletal muscles in mice during repeated contractions. *J. Physiol. (London)* **436**, 701–710.
12. Brooks, S. V. and Faulkner, J. A. (1988) Contractile properties of skeletal muscles from young, adult, and aged mice. *J. Physiol. (London)* **404**, 71–82.
13. Mutungi, G. and Ranatunga, K. W. (1996) Tension relaxation after stretch in resting mammalian muscle fibers: stretch activation at physiological temperatures. *Biophys. J.* **70**, 1432–1438.
14. McCully, K. K., and Faulkner, J. A. (1983) Length-tension relationship of mammalian diaphragm muscles. *J. Appl. Physiol.* **54**, 1681-1686.
15. Faulkner, J. A., Claflin, D. R., McCully, K. K., and Jones, D. A. (1982) Contractile properties of bundles of fiber segments from skeletal muscles. *Am. J. Physiol.* **243** (*Cell Physiol.* **12**): C66–C73.
16. Lännergren, J. and Westerblad, H. (1987) The temperature dependence of isometric contractions of single, intact fibres dissected from a mouse foot muscle. *J. Physiol. (London)* **390**, 285–293.
17. Godt, R. E. and Lindey, B. D. (1982) Influence of temperature upon contractile activation and isometric force production in mechanically skinned muscle fibers of the frog. *J. Gen. Physiol.* **80**, 279–297.
18. Larsson, L. and Moss, R. L. (1993) Maximum velocity of shortening in relation to myosin isoform composition in single fibres from human skeletal muscles. *J. Physiol. (London)* **472**, 595–614.
19. Lynch, G. S., Duncan, N. D., Campbell, S. P., and Williams, D. A. (1995) Endurance training effects on the contractile activation characteristics of single muscle fibres from the rat diaphragm. *Clin. Exp. Pharmacol. Physiol.* **22**, 430–437.
20. Sweeney, H. L., Corteselli, S. A., and Kushmerick, M. J. (1987) Measurements on permeabilized skeletal muscle fibers during continuous activation. *Am. J. Physiol.* **252** (*Cell Physiol.* **21**), C575–C580.
21. Huxley, A. F. and Lombardi, V. (1980) A sensitive force transducer with resonant frequency 50 kHz. *J. Physiol (London) Proc.* **305**, 14–16P.
22. Cecchi, G., Colombo, F., and Lombardi, V. (1979) A capacitance-gauge force transducer for isolated muscle fibers. *J. Physiol. (London) Proc.* 1–2P.
23. Hellam, D. C. and Podolsky, R. J. (1969) Force measurement in skinned muscle fibers. *J. Physiol. (London)* **200**, 807–819.

24. Fearn, L. A., Bartoo, M. L., Myers, J. A., and Pollack, G. H. (1993) An optical fiber transducer for single myofibril force measurement. *IEEE Transactions on Biomed. Eng.* **40,** 1127–1132.
25. Cole, N. M. (1992) Mechanism of skeletal muscle sounds: Acoustic measures of resonant frequency and tension. Doctoral Dissertation, University of Michigan.
26. Ford, L. E., Huxley, A. F., and Simmons, R. M. (1977) Tension responses to sudden length change in stimulated frog muscle fibers near slack length. *J. Physiol. (London)* **269,** 441–515.
27. Lieber, R. L., Schmitz, M. C., Mishra, D. K., and Friden, J. (1994) Contractile and cellular remodeling in rabbit skeletal muscle after cyclic eccentric contractions. *J. Appl. Physiol.* **77,** 1926–1934.
28. Miller, R. A., Bookstein, F., van der Meulen, J. H., and Faulkner, J. A. (1997) Candidate biomarkers of aging: age-sensitive indices of immune and muscle function co-vary in genetically heterogeneous mice. *J. Gerontol.:Biol. Sci.,* **52,** B39–B47.
29. Marsh, E. D., Sale, D., McComas, A. J., and Quinlan, J. (1981) Influence of joint position on ankle dorsiflexion in man. *J. Appl. Physiol.* **51,** 160–167.
30. Caiozzo, V. J., Ma, E., McCue, S., Smith, E., Herrick, R. E., and Baldwin, K. M. (1992) A new animal model for modulating myosin isoform expression by altered mechanical activity. *J. Appl. Physiol.* **73,** 1432–1440.
31. Fabiato, A. and Fabiato, F. (1979) Calculator programs for computing the composition of solutions containing multiple metals and ligands used for experiments in skinned muscle cells. *J. Physiol. (Paris)* **75,** 463–505.
32. Miller, S. W. and Dennis, R. G. (1996) A parametric model of muscle moment arm as a function of joint angle: application to the dorsiflexor muscle group in mice. *J. Biomech.* **29,** 1621–1624.
33. Segal, S. S. and Faulkner, J. A. (1985) Temperature dependent physiological stability of rat skeletal muscle in vitro. *Am. J. Physiol.* **248** (*Cell Physiol.* **17**), C265–C270.
34. Elzinga, G., Howarth, J. V., Rall, J. A., Wilson, M. G. A., and Woledge, R. C. (1989) Variation in the normalized tetanic force of single frog muscle fibres. *J. Physiol. (London)* **410,** 157–170.
35. Ranatunga, K. W. and Wylie, S. R. (1983) Temperature-dependent transitions in isometric contractions of rat muscle. *J. Physiol. (London)* **339,** 87–95.

14

Preparation of Immortalized Human Chondrocyte Cell Lines

James R. Robbins and Mary B. Goldring

1. Introduction

The chondrocyte is responsible for synthesis of cartilage matrix proteins, and, thereby, the specialized mechanical properties of articular cartilage, including tensile strength and resistence to mechanical loading *(1)*. The limited repair response by chondrocytes accounts for a major component of the loss of articular cartilage in joint diseases such as osteoarthritis, a progressive disease associated with normal wear and tear of joints, aging, or trauma. Although research has been directed primarily toward developing therapeutic strategies that prevent degradation of cartilage matrix, recent work has also focused on promoting cartilage repair. Success of either strategy depends on the development of reliable cell culture models of human origin.

Primary cultures of human chondrocytes in monolayer culture have been used extensively to study the synthesis of cartilage matrix proteins and proteolytic enzymes that degrade them *(2)*, but their sources, numbers, and metabolic states are difficult to control. In these systems, chondrocytes briefly retain the capacity to synthesize cartilage-specific matrix components, including collagens II, IX, and XI, and the large aggregating proteoglycan, aggrecan. However, synthesis of type II collagen, for example, is replaced by collagens I and III as the chondrocytes dedifferentiate during long-term primary culture, or after subculture *(3–5)*, as first described in chick chondrocyte cultures *(6)*. When cultured in suspension over or within agarose gels, or encapsulated in alginate, they assume the normal nonproliferative state and rounded morphology, and retain or regain expression of chondrocyte-specific phenotypic markers *(7,8)*.

Investigators have utilized several different approaches to develop stable chondrocyte lines that retain the capacity to express chondrocyte-specific markers. Introduction of viral genes into chondrocytes from nonhuman sources has generated several immortalized chondrocytes lines that demonstrate high proliferative capacities while retaining at least some differentiated chondrocyte properties *(9–13)*. Although all of the lines retain the capacity to synthesize sulfated proteoglycans, synthesis of type II collagen appears to be more fragile, being observed in only a few of the lines. Introduction of simian virus 40 (SV40) large T antigen (TAg) into human articular chondrocytes has resulted in generation of type II collagen-negative clonal lines that express types I and III collagens in monolayer culture *(14)*. In contrast, chondrocytes of mouse origin appear to be more easily immortalized and retain the ability to synthesize type II collagen, when SV40-TAg is introduced either in vitro *(13)* or in vivo *(15)*. Thus, the tissue source from which the chondrocytes are derived is an important factor. The developmental stage of the tissue is also important. For example, SV40-immortalized rib chondrocytes from newborn mice express type X collagen, a hypertrophic chondrocyte marker found in OA, but not normal, articular cartilage, but not type II collagen *(16)*. In contrast, human juvenile rib chondrocytes immortalized with SV40-TAg express type II, but not type X collagen *(17)*. Therefore, success in establishing immortalized human chondrocyte lines that can be used as models to study cartilage breakdown and repair depends on the source and developmental state of the tissue from which the cell lines are established, and the ability of the culture system to support the differentiated phenotype.

Expression of SV40-TAg is known to stabilize proliferative capacity by inactivating cellular proteins such as p53 and Rb, which are responsible for maintaining quiescence or retarding progression through G1 of the cell cycle *(18)*. Immortalization with SV40 does not always proceed by the same genetic mechanism, even within a single cell population *(19)*, and the phenotype of the immortalized cell type is therefore not necessarily frozen. For example, matrix protein synthesis and secretion are greatly reduced in proliferating monolayer cultures of immortalized fibroblasts *(20,21)*, and other SV40-immortalized cell types are known to dedifferentiate following prolonged culture *(22)*. SV40-TAg-immortalized human chondrocytes are not fully transformed, since they have not been transfected with a cooperating oncogene such as *src* and *ras*, which have been shown to markedly suppress phenotypic collagen gene expression in fibroblasts and chondrocytes, and they do not form tumors in nude mice *(17)*. Since the suppression of chondrocyte phenotype by immortalizing agents, such as *myc* and SV40-TAg, has been attributed to the secondary effects of their mitogenic activities *(9,23)*, it is essential to consider the culture conditions that will promote differentiated chondrocyte phenotype.

In this chapter, we focus on adult human articular chondrocytes as target cells for immortalization, since articular cartilage is the primary joint tissue requiring replacement or reconstruction after it is damaged in arthritic conditions. We also focus on cells derived from embryonic rib and vertebral tissues, which yield chondrocytes at different stages of differentiation. We describe the use of a retroviral vector encoding a temperature-sensitive (ts) mutant of SV40-TAg *(24)*, which has the advantage that the immortalized chondrocytes can be expanded in monolayer culture at the permissive temperature, and then transferred to the nonpermissive temperature for TAg expression at which cell proliferation ceases and matrix synthesis increases. Furthermore, this system allows us to overcome potential experimental problems caused by SV40-TAg-associated disruption of normal cellular functions *(25–27)*. Similar to normal human chondrocytes, immortalized chondrocytes can be transferred to, and maintained in, culture systems such as fluid suspension culture *(8,28,29)*, in micromass or pellet cultures *(30–33)*, or within a three-dimensional gel or matrix, such as agarose *(7,34)*, alginate *(35,36)*, collagen gels *(37)*, or synthetic polymers *(38–40)*, in which chondrocytes assume the normal spherical morphology, and synthesize and deposit cartilage matrix *(41)*. In this chapter, we focus on the alginate culture system.

Studies on experimental engraftment of chondrocyte-laden collagen scaffolds for resurfacing cartilage defects in horses *(42)*, and on autologous chondrocyte transplantation in traumatic defects of knee cartilage in young adult humans *(43)*, indicate that more work is required to understand how to promote a fully regenerative phenotype in chondrocytes that would participate in the repair of advanced cartilage lesions in older adults. Availability of phenotypically stable immortalized human chondrocyte culture models will permit prior testing in vitro of potential modes of gene therapy that would affect metabolism of cartilage in vivo *(44–46)*. Understanding mechanisms that control chondrocyte differentiation and chondrogenesis will provide important insights into the regulation of cartilage formation during development and growth, and will also aid in development of rational strategies for promoting repair of cartilage defects.

2. Materials
2.1. Isolation and Culture of Adult Human Chondrocytes

1. Dulbecco's modified Eagle's medium (high glucose, without pyruvate).
2. Ham's F12 (DMEM/F12; 50/50, v/v), containing 10% fetal calf serum (FCS).
3. Dulbecco's phosphate-buffered Ca^{2+}- and Mg^{2+}-free saline (PBS).
4. Trypsin–EDTA solution: 0.05% trypsin and 0.02% EDTA in Hanks' balanced salt solution without Ca^{2+} and Mg^{2+}.

These solutions have shelf lives as recommended by the supplier. If DMEM is prepared from powder, high quality distilled and deionized water (i.e., using a Milli-Q apparatus), dedicated sterilized bottles, and 0.22-µm filters should be used. FCS and trypsin–EDTA are stored at –20°C, but should not be refrozen after thawing for use.

2. Enzymes for cartilage digestion:
   a. Hyaluronidase (Sigma, St. Louis, MO), 1 mg/mL in PBS: Prepare freshly and filter through a sterile 0.22-µm filter.
   b. Trypsin (Gibco-BRL): 0.25% in Hanks' balanced salt solution without Ca^{2+} and Mg^{2+}.
   c. Collagenase (bacterial, clostridiopeptidase A, Worthington), 3 mg/mL in DMEM with 10% FCS for articular cartilage or serum-free for costal cartilage: Prepare freshly in ice-cold DMEM, and filter immediately through a sterile 0.22-µm filter.

2.2. Isolation and Culture of Human Embryonic Chondrocytes

1. Cell culture reagents (*see* **Subheading 2.1., step 1.**).
2. Enzymes for cartilage digestion:
   a. Collagenase D (Boehringer Mannheim): 3 mg/mL in DMEM with 10% FCS.
   b. Pronase (*Streptomyces griseus*, Boehringer Mannheim): 2 mg/mL in PBS. Prepare fresh by adding the Pronase to cold PBS, and filter through a sterile 0.22-µm filter.

2.3. Retrovirus Production and Infection of Chondrocytes

1. Polybrene (Sigma): 10 mg/mL in PBS (1000X stock), sterile-filtered, and stored at –20°C. Add to freshly thawed virus-containing medium to a final concentration of 10 µg/mL.
2. Geneticin (G418, Sigma): 30 mg/mL in PBS (100X stock), sterile-filtered, freshly prepared or stored in aliquots at –20°C (avoid freeze–thawing). Added to a final concentration of 300 µg/mL in DMEM/F12, 10% FCS.

2.4. Alginate Culture System

1. Alginate (Keltone LVCR, Kelco, Chicago, IL): 1.2% (w/v) solution of Alginate in 0.15 M NaCl. (We use the low-viscosity (LV) alginate, but others have successfully used high-viscosity (HV). Request LVCR for more highly purified preparation.): Dissolve alginate in a 0.15 M NaCl solution, heating the solution in a microwave until it just begins to boil and swirl, and repeat this process 2–3× until the alginate is completely dissolved. Allow the solution to cool to about 37°C, and sterile-filter. (**Caution:** Do not autoclave. Filtering when warm permits the viscous solution to pass through the filter.)
2. 102 mM $CaCl_2$: Prepare in a tissue culture bottle and autoclave.
3. 0.15 M NaCl: Prepare in a tissue culture bottle and autoclave.
4. 55 mM Na citrate, 0.15 M NaCl, pH 6.0: Prepare in a tissue culture bottle, sterile filter, and store at 4°C. Make fresh weekly.

2.5. Western Blotting Analysis of TAg

1. Gel sample buffer (GSB): 0.1 M Tris-HCl, pH 7.6, 3% (w/v) SDS, and 16% (v/v) glycerol.

2. Loading dye: 1% (w/v) bromophenol blue (sodium salt, Sigma).
3. Tris-glycine/SDS-10% polyacrylamide precast gels (Bio-Rad) and Laemmli buffer system.
4. Nitrocellulose, nylon-supported: Nitroplus 2000 (0.45 μm, Micron Separations).
5. Transfer buffer: 25 mM Tris-HCl, pH 7.6, 192 mM glycine, 20% (v/v) methanol.
6. Tris-buffered saline/Tween (TBST): 20 mM Tris-HCl, 137 mM NaCl, 0.1% (v/v) Tween-20, pH 7.6. Add 5% (w/v) nonfat dry milk (Carnation) as required.
7. SV40-TAg monoclonal antibody (Ab-2, Oncogene Science) *(47)*: Dilute 1:500 in TBST containing 5% (w/v) nonfat dry milk.
8. ACL Western blotting analysis system (Amersham).

2.6. Analysis of ^{35}S-Labeled Proteoglycans

1. Biosynthetic labeling of proteoglycans: [^{35}S]sodium sulfate (2 mCi/mL; specific activity >1000 Ci/mmol). Add to culture medium at 50 μCi/mL.
2. Guanidine extraction buffer: 8.0 M guanidine-HCl, buffered with 0.01 M sodium acetate, and containing 0.02 M disodium EDTA, 0.20 M 6-aminocaproic acid. Add immediately before use 5.0 mM benzamidine HCl, 10 mM N-ethylmaleimide, and 0.5 mM (PMSF) (diluted at 100× in absolute ethanol).
3. SDS-PAGE: Precast Tris-glycine SDS-polyacrylamide 4–20% gradient gels (Bio-Rad). Use the Laemmli buffer system *(48)*.

2.7. Alcian Blue Staining of Alginate Beads

1. 2.5% glutaraldehyde (diluted from 50% solution, Sigma) in 0.4 M MgCl$_2$ and 25 mM sodium acetate, pH 5.6.
2. Alcian blue 8GX (Sigma or Aldrich). Dissolve in the 2.5% glutaraldehyde solution to give final concentration of 0.05%. Filter the solution through Whatman paper (or coffee filter).
3. Washes: 3% acetic acid solutions without and with 25 and 50% ethanol.

2.8. Extraction of RNA

1. Trizol reagent (Gibco-BRL).
2. 75% ethanol: Dilute bottled absolute ethanol with DEP-treated water in sterile tube. Store at −20°C.
3. 10 mM HEPES: Dilute 0.5 M HEPES stock solution in DEP-treated H$_2$O in RNase-free bottle, and autoclave.

2.9. Analysis of mRNA by RT-PCR

1. Total RNA: Dilute to 100 ng/μL in DEP-treated H$_2$O.
2. Antisense (3') and sense (5') oligonucleotide primers (15 μM). See **Table 1** for some primer sequences used to detect chondrocyte-specific mRNAs.
3. (AMV) reverse transcriptase and 5X buffer (Promega), rTaq polymerase and 10X buffer (Fisher Biotech.), dNTPs (10 mM; Promega).
4. RT mix (for 20 μL rxn):

Table 1
Oligonucleotides Used for the Detection of Chondrocyte-Specific mRNAs by RT-PCR

Primer[a]	S/AS[b]	Sequence	Amplified fragment bp[c]
Type II collagen	S	5'-CTGGCTCCCAACACTGCCAACGTC-3'	414
	AS	5'-TCCTTTGGGTTTGCAACGGATTGT-3'	
Aggrecan	S	5'-TGAGGAGGGCTGGAACAAGTACC-3'	346
	AS	5'-GGAGGTGGTAATTGCAGGGAACA-3'	
Type X collagen	S	5'-AGCCAGGGTTGCCAGGACCA-3'	387
	AS	5'-TTTTCCCACTCCAGGAGGGC-3'	
GAPDH	S	5'-GCTCTCCAGAACATCATCCCTGCC-3'	350
	AS	5'-CGTTGTCATACCAGGAAATGAGCTT-3'	

[a]The primer sequences for type II collagen, aggrecan core protein, and glyceraldehyde-3-phosphate dehydrogenase (GAPDH) are adapted with permission from ref. 53, and those for type X collagen are adapted with permission from ref. 54.
[b]AS, antisense primer; S. sense primer.
[c]bp, base pairs.

	Per rxn	Final concn.
5X RT buffer	4 µL	1X
dNTPs (10 mM)	1 µL (each)	200 µM
AMV reverse transcriptase (10 U/µL)	0.5 µL	5 U

5. PCR Mix (for 50 µL rxn):

	Per rxn	Final concn.
sterile H$_2$O	29 µL	
10X PCR buffer	5 µL	1X
MgCl$_2$ (25 mM)	3 µL	1.5 mM
dNTPs (10 mM)	2 µL (each)	200 µM
5' oligo (15 µM)	1 µL	0.2 µM
3' oligo (15 µM)	1 µL	0.2 µM
rTaq DNA polymerase (5 U/µL)	1 µL	2.5 U

6. Gel loading buffer: 50% (v/v) glycerol (Sigma), 1% (w/v) bromophenol blue (Sigma), 1% (w/v) xylene cyanol (Sigma) in 1X TAE.

3. Methods

3.1. Isolation and Culture of Adult Human Chondrocytes (see Note 1)

1. Human adult articular cartilage is obtained from knee joints or hips after surgery following joint replacement, or at autopsy, and dissected free from underlying bone and any adherent connective tissue.

2. Place slices of cartilage in a 10-cm dish and wash several times with PBS. Incubate slices at 37°C in hyaluronidase for 10 min, followed by 0.25% trypsin for 30–45 min with 2–3 washes in PBS after each enzyme treatment. Use ~10 mL of proteinase solution for digestion of each g of tissue.
3. Add collagenase solution, chop the cartilage in small pieces using a scalpel blade, and incubate at 37°C overnight (18–24 h), until the cartilage matrix is completely digested and the cells are free in suspension. Break up any clumps of cells by repeated aspiration of the suspension through a 10-mL pipet or a 12-cc syringe without a needle.
4. Transfer cell suspension to a sterile 50-mL conical polypropylene tube, and wash the plate with PBS to recover remaining cells and combine in tube. Centrifuge cells at ~1000g in a benchtop centrifuge for 10 min at room temperature and wash the cell pellet three times with PBS, resuspending cells each time, and centrifuging.
5. Resuspend the final pellet in DMEM/F12 containing 10% FCS, perform cell count with a Coulter counter or hemacytometer, and bring up to volume with culture medium to give 1×10^6 cells/mL. Plate cells at 2×10^6 cells/10-cm dish already containing some culture medium, and agitate without swirling to evenly distribute the cells. Incubate at 37°C in an atmosphere of 5% CO_2 in air with medium changes after 2 and 4 d and every 3 d thereafter. Retroviral infection should be performed, once the chondrocytes are actively dividing, within 24 h after a medium change, but prior to ~d 10 of culture, when they are actively expressing the differentiated phenotype (*see* **Note 2**).

3.2. Isolation and Culture of Human Embryonic Chondrocytes

1. Place dissected vertebral column and rib cage in a 10-cm dish, wash several times with PBS to remove tissue debris and blood, and dissect as much of the loose connective tissue from the developing cartilage as possible with a sterile scalpel blade.
2. Digest the remaining loose connective tissue from the cartilage with Pronase (~10 mL of enzyme solution per g wet wt of tissue), incubating at 37°C for 30 min. Following the digestion, aspirate the Pronase repeatedly over the cartilage with a 10-mL pipet to facilitate the removal of extraneous tissue, and wash several times with PBS. At this point, the cartilage should be free of loose connective tissue. Place the ribs and vertebrae in separate 10-cm dishes and proceed with the collagenase digestion.
3. Add the collagenase solution (~10 mL of enzyme solution per g wet wt of tissue) to each of the dishes, and chop the tissue into small pieces with a sterile scalpel blade. Incubate at 37°C for 24 h. Following the digestion, the vertebrae should be completely solubilized. The ribs are somewhat more resistant to collagenase digestion, and remnants of the tissue will probably remain. Remove each of the resulting cell suspensions, including any remaining undigested tissue, to sterile 50-mL centrifuge tubes, washing the plates with an additional 10 mL of PBS to recover remaining cells, and combine in the tube. Briefly allow any remaining

tissue pieces to settle to the bottom of the tubes and remove the cell suspensions to fresh 50-mL centrifuge tubes. Break up any remaining clumps of cells by repeated aspiration of the suspensions with a 10-mL pipet. Wash the cells several times with PBS.

4. Resuspend the cell pellets in 10–20 mL of growth medium and count with a hemacytometer. For retroviral infection, plate the chondrocytes at a density of 2×10^6 cells in 10-cm dishes, place in a 37°C incubator, and culture for 3–4 d, or until the cells have begun to divide rapidly (*see* **Note 3**).

3.3. Retrovirus Production and Infection of Chondrocytes

1. Packaging cell line: The retrovirus packaging cell line PA/tsA58-3 was derived by stable transfection of the amphotropic PA317 cell line *(49)*, with the plasmid form of the retroviral vector tsA58-3 encoding a temperature-sensitive mutant of SV40-large T antigen (tsTAg) and the neomycin-resistance marker *(24)* (*see* **Note 4**).
   a. Culture the PA317 cell line in DMEM containing 10% FCS at 37°C in 75-cm^2 flasks. Passage the cultures twice weekly with a split ratio of 1:15 (*see* **Note 5**).
   b. Collect virus-containing conditioned medium in 15-mL centrifuge tubes from subconfluent, actively dividing cultures. Filter through a 0.45-µm filter and quick freeze on dry ice. Store the conditioned media at –70°C in 5-mL aliquots until needed.
   c. Liquid nitrogen storage of packaging cell line: Expand the cell line in several 10-cm dishes. When the cultures are nearly confluent (50–70% confluency), trypsinize and wash the cells in PBS, and count with a hemacytometer. Pellet the cells, and resuspend at a concentration of 2×10^6 cells/mL in DMEM/F12 containing 10% FCS. To the suspension, slowly add a one-tenth vol of DMSO, gently swirling, and aliquot 1.5 mL of the suspension into 1.8-mL cryo-vials. Freeze the vials at –70°C for 24 h in a styrofoam rack with lid or a Cryo 1°C freezing chamber (Nalgene). Transfer the vials to liquid nitrogen for long-term storage. Subsequent experiments should be carried out with cultures derived from freshly thawed cells.
2. Retroviral infection of chondrocytes:
   a. Three h prior to infection, remove the spent medium from chondrocyte cultures (in 10-cm plates), and add 8 mL of fresh growth medium (DMEM/F12 containing 10% FCS) to stimulate cell division.
   b. Begin infection by removing the medium and adding 5 mL of freshly thawed virus-containing medium supplemented with polybrene (10 µg/mL). Replace the dishes in the 37°C incubator and incubate for 6 hr to overnight (*see* **Note 6**).
   c. Remove the conditioned medium, replace with 8 mL of fresh growth medium, and transfer the cultures to a 32°C incubator. Incubate for 42 h, or until the cultures become confluent.
3. Selection and culture of immortalized chondrocytes:
   a. Passage confluent cultures with a split ratio of 1:2, and begin selection with G418 the following day.

b. Remove the medium and add fresh growth medium containing G418 (geneticin) at 300 µg/mL. Continue selection for 3–4 wk, feeding the cultures 3× per wk with fresh growth medium supplemented with G418. During selection a percentage of the cells will die, substantially reducing the culture density. Following 3–4 wk, however, G418-resistant cells should repopulate the dishes, and when the cultures become confluent, they should be passaged into fresh growth medium containing G418.

c. Passaging of immortalized chondrocytes: Remove culture medium by aspiration with a sterile Pasteur pipet attached to a vacuum flask, and wash with PBS. Add trypsin–EDTA (1 mL/10-cm dish) and incubate at room temp for 10 min, with periodic gentle shaking of dish, and observation through a microscope to assure that all cells have come off the plate (*see* **Note 7**). If significant numbers of cells remain attached, continue the incubation for a longer time (<20 min), or at a higher temperature (37°C), and/or scrape the cell layer with a sterile plastic scraper, Teflon™ policeman, or syringe plunger. Repeatedly aspirate and expel the cell suspension into the plate using a 5- or 10-mL pipet containing culture medium, and then transfer to a sterile conical 15- or 50-mL polypropylene tube. Perform cell counts or determine the split ratio required, distribute equal volumes of the cell suspension in dishes or wells that already contain culture medium, and agitate the culture plate immediately after each addition to assure uniform plating density on the culture surface (*see* **Note 8**).

4. Growth kinetics: Trypsinize the cells from a 10-cm dish and wash with PBS. Plate cells in two sets of 12-well plates at a concentration of 25,000 cells/well in 2 mL of growth medium. Culture one set at the permissive temperature for tsTAg expression (32°C), and the other at the nonpermissive temperature (39°C). At intervals of 48 h, trypsinize duplicate wells from each plate and determine the perform cell counts. Cell counts should be obtained until the cultures reach stationary phase (~12 d for cultures maintained at 32°C). Representative growth curves from tsTAg-immortalized chondrocytes cultured at the permissive and nonpermissive temperatures are shown in **Fig. 1**.

3.4. Alginate Culture System

1. Trypsinize several 10-cm plates and wash the cells with PBS. Determine the cell count with a hemacytometer and pellet the cells. Resuspend the pellet in a 1.2% solution of alginate, 0.15 M NaCl at a concentration of 4×10^6 cells/mL (*see* **Note 9**).

2. Express the alginate suspension in a dropwise manner, through a 10-mL syringe equipped with a 25-gage needle, into a 50-mL centrifuge tube containing 40 mL of 102 mM CaCl$_2$. Allow the beads to polymerize in the CaCl$_2$ solution for 10 min and wash twice with 25 mL of 0.15 M NaCl. **Caution:** The alginate beads should not be washed in PBS, because they will become cloudy.

3. Resuspend the beads (7–15 beads/mL) in 20 mL of growth medium supplemented with 25 µg/mL Na ascorbate, and decant the beads to a 10-cm dish. As many as

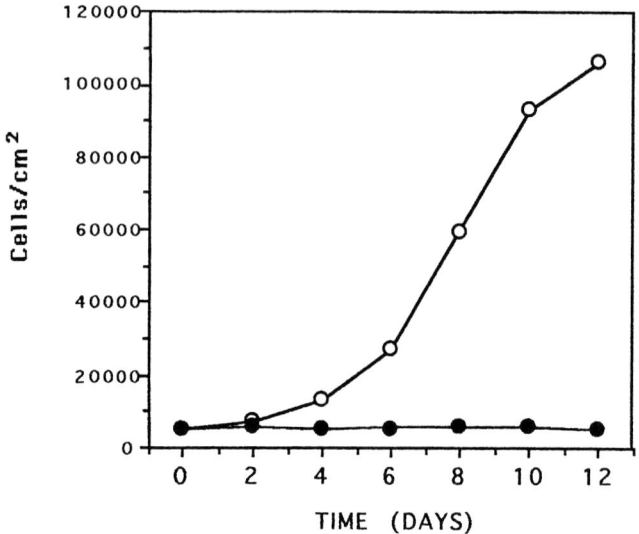

Fig. 1. Growth characteristics of tsTAg-immortalized human articular chondrocytes at the permissive (32°C) and nonpermissive (39°C) temperatures. Cells were seeded into multiwell plates at a density of 6.6×10^3 cells/cm^2, and cultured in DMEM/F12 supplemented with 10% FCS for the indicated time-periods, with medium changes every 3 d. Open circles, cultures maintained at 32°C; closed circles, cultures maintained at 39°C.

150 beads (~ 9×10^6 cells) can be cultured in a single dish, with medium changes every 3 d.
4. To recover cells from alginate, carefully aspirate the medium from the cultures and wash twice with PBS. Depolymerize the alginate by adding 3 vol of a solution of 55 mM Na citrate/0.15 M NaCl, and incubate at 37°C for 10 min. Aspirate the solution over the surface of the dish several times to dislodge adherent cells (the cells are sticky), and transfer the suspension to a 50-mL centrifuge tube.
5. Because the solution is quite viscous, centrifuge the cells at 2000g for a minimum of 10 min to completely pellet the cells. Wash the cells twice with PBS before using them for any further analysis.

3.5. Western Blotting Analysis of TAg

1. Trypsinize a 10-cm dish of immortalized cells and seed them into two 6-well tissue culture plates at a density of 0.25×10^6 cells/well. Culture both plates at 32°C until nearly confluent (~3 d).
2. Change both cultures to fresh DMEM/F12 containing 10% FCS, and transfer one plate to a 39°C incubator, while leaving the other plate at 32°C. After 12, 24, and 72 h, remove the medium from two wells of each plate and wash the cell layers with 3 mL of PBS. Add 1 mL of trypsin–EDTA to each well, and incubate at

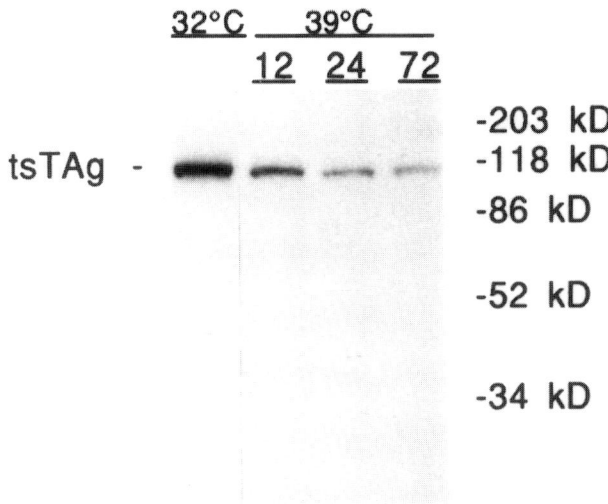

Fig. 2. Western blotting analysis of SV40 large T antigen in tsTAg-immortalized human articular chondrocytes. Cells were seeded into multiwell plates at a density of 2.6×10^4 cells/cm^2 in medium supplemented with 10% FCS and cultured at 32°C until nearly confluent. Cultures were kept at 32°C or switched to 39°C for 12, 24 and 72 h, as indicated, and TAg was detected using the Ab-2 monoclonal antibody specific for the SV40 large T antigen.

room temperature for ≤10 min to allow the cells to detach. Transfer the suspension to a 15-mL centrifuge tube, and wash the cells twice with 5 mL of PBS.
3. Pellet the cells, and extract each pellet with 100 µL of gel sample buffer.
4. Electrophorese 7 µL of each extract on a Tris-glycine/SDS-10% polyacrylamide precast gel, using the Laemmli buffer system, and transfer to a nylon-supported nitrocellulose membrane by electroblotting with a transfer current of 100 V for 1 h.
5. Block membrane with 10 mL TBST containing 5% nonfat powdered milk at 4°C for 1 h, with rocking, and wash three times (10 min/wash) in 10 mL TBST.
6. Incubate membrane with the TAg monoclonal antibody (1:500 dilution) in 5 mL of TBST containing 5% powdered milk for 1 h at room temperature, with rocking.
7. Wash the membrane as before (**step 5**), and detect bound TAg antibody with the ACL Western blotting analysis system (Amersham). As shown in **Fig. 2**, TAg is significantly reduced in immortalized cells after a 72 h incubation at 39°C.

3.6. Alcian Blue Staining of Alginate Beads

1. Using a 25-mL pipet, transfer five beads to a 12 × 15 cm tube, and wash twice with 2 mL PBS. Add 1 mL Alcian blue stain, and 50 µL of a 50% glutaraldehyde solution. Store for 24 h at 4°C.
2. Aspirate the stain from the beads, and wash twice with 2 mL of 3% (v/v) acetic acid. Destain the beads with rocking at room temp for 5 min, sequentially, with

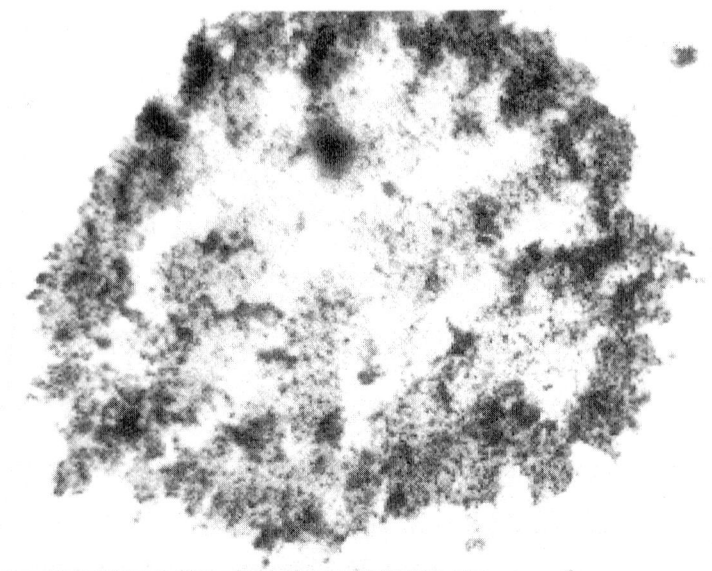

Fig. 3. Alcian blue staining of tsTAg-immortalized human articular chondrocytes in alginate bead culture. The cells were encapsulated in alginate beads, cultured at 37°C in medium supplemented with 10% FCS and 25 µg/mL ascorbate for 2 wk, and stained with Alcian blue. Note that a single bead is shown in the field. Dense metachromatic staining is present within and around clumps of cells, and diffuse staining can be observed between the cell clumps. Magnification ×40.

    2 mL of each of the following solutions: 3% acetic acid; 3% acetic acid/25% ethanol; and 3% acetic acid/50% ethanol. Store the beads in 70% ethanol.
3. Before photographing the stained beads, gently flatten beneath a glass cover slip, taking care not to disrupt the alginate matrix (**Fig. 3**). Alternatively, the beads may be embedded and sectioned prior to photography.

3.7. Analysis of [^{35}S]Sulfate-Labeled Proteoglycans

1. Monolayer cultures:
    a. Aspirate the medium from a 10-cm dish and wash the cell layer with PBS. Add 4 mL of DMEM/F12 containing 10% FCS, supplemented with [^{35}S]sulfate at 50 µCi/mL. Culture at 32°C (or other temperature as desired) for 12 h.
    b. Aspirate the medium from the culture and wash the cell layer 3× with 5 mL of PBS (5 min/wash).
    c. Extract radiolabeled PGs from a 10-cm dish by adding 2 mL of guanidine extraction buffer to the dish and extracting at 4°C for 24 h, with rocking. Transfer extract to a 2-mL screw-cap microcentrifuge tube and spin at 12,000 rpm in an Eppendorf microcentrifuge at 4°C to pellet particulate material from

the sample. Remove supernatant to a fresh 2-mL tube and count 20 µL in a liquid scintillation counter.
2. Alginate suspension cultures:
   a. Aspirate the medium from a culture containing 50 beads in a 25-cm² flask cultured on end and wash the beads with 5 mL of PBS. Add 4 mL of growth medium, supplemented with [^{35}S]sulfate at 50 µCi/mL, and culture at 32°C (or other temperature, as desired) for 12 h.
   b. Aspirate the medium from the culture and wash the beads 3× with 5 mL of PBS (5 min/wash).
   c. Add 2 mL of guanidine extraction buffer to the flask and extract radiolabeled PGs from the beads at 4°C for 24 h, with rocking. Transfer extract (leaving the beads behind) to a 2-mL screw-cap microcentrifuge tube, and spin at 12,000 rpm in an Eppendorf microcentrifuge at 4°C to remove particulate material. Remove supernatant to a fresh 2-mL tube and count 20 µL in a liquid scintillation counter.
   d. Transfer ~200,000 cpm of extract (from either monolayer or alginate culture) to a 2-mL screw-cap tube, and precipitate the radiolabeled PGs by adding 3 vol of absolute ethanol, and leave at –20°C for 1 h. Spin at 12,000 rpm for 15 min, remove supernatant to a fresh tube, and wash pellet with 1 mL of cold 75% ethanol; shake vigorously or vortex to break up the pellet. Centrifuge at 12,000 rpm for 15 min at 4°C. Dry pellet at room temp for 30 min.
3. Redissolve pellet in 50 µL of GSB, add 1 µL of 10 mM DTT, heat to 100°C for 5 min, and count 5 µL in a liquid scintillation counter.
4. Electrophorese ~20,000 cpm in a 4–20% polyacrylamide gradient gel. Fix the gel in acetic acid/methanol for 1 h, soak in scintillant, and dry. Visualize radiolabeled PGs by fluorography, as shown in **Fig. 4** *(50,51)* (*see* **Note 10**).

3.8. Extraction of RNA

1. Alginate cultures (50 beads, or ~ 3 × 10⁶ cells): Aspirate the medium from the culture and wash the beads twice with 10 mL of PBS. Depolymerize the alginate as described in **Subheading 3.4.4.**
2. Monolayer cultures (10-cm dishes): Aspirate the medium and wash the cell layer with 10 mL of PBS. Trypsinize cells in 0.5 mL of trypsin–EDTA (add 1.5 mL, then remove 1 mL immediately) at room temp, ≤10 min. Resuspend cells in 2 mL of medium containing 10% FCS. Transfer to sterile polypropylene 10 × 75 mm tubes (15 mL). Wash plates with an additional 2 mL of medium and combine in tube with cell suspension. Place tube on ice. Remove aliquot for cell count, if required. Spin down cells at 1200–1500g and wash with cold PBS. Resuspend and combine pellets from 1–4 dishes (1–5 × 10⁶ cells) for each extraction.
3. Pellet the cells (from either monolayer or alginate cultures), add 1 mL of Trizol reagent, vortex vigorously, and transfer to RNase-free 1.2-mL centrifuge tube.
4. Add 200 µL of chloroform and let stand at room temperature for 5 min. Centrifuge at 12,000 rpm (in Eppendorf centrifuge) for 15 min at 4°C. Remove upper aqueous phase to fresh sterile 1.2-mL tube, taking care not to take the interphase. Discard lower organic phase with interphase containing DNA and protein.

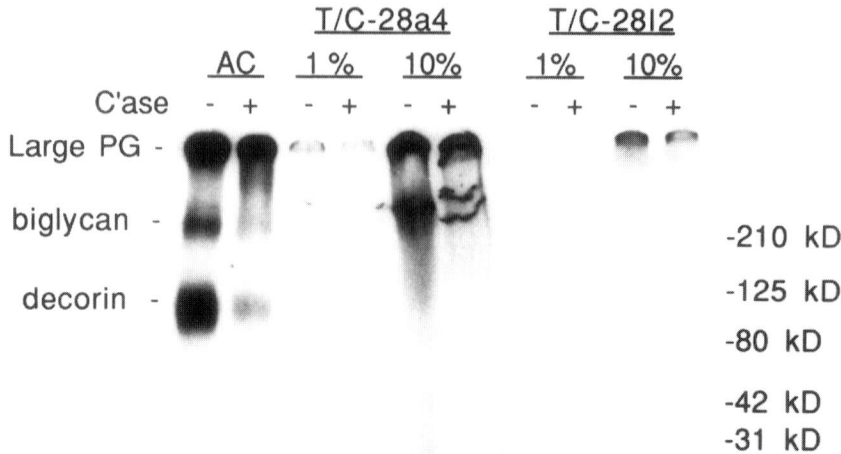

Fig. 4. SDS-PAGE and fluorography of radiolabeled proteoglycans synthesized by TAg-immortalized juvenile human costal chondrocytes. Explant cultures of adult human articular cartilage (AC), or alginate bead cultures of the T/C-28a4 and C-28/I2 immortalized juvenile costal chondrocytes *(17)* were radiolabeled with [^{35}S]sulfate for 12 h in DMEM/F12 supplemented with either 1 or 10% FCS, as indicated. Radiolabeled proteoglycans were purified by DEAE ion-exchange chromatography from the medium of the explant cultures, or extracted from the alginate bead cultures prior to electrophoresis on SDS-polyacrylamide 4–20% gradient gels and visualization by fluorography. The migration of large proteoglycan (PG), biglycan, and decorin is indicated to the left of the panel. Treatment with chondroitinase ABC (C'ase) resulted in the expected loss of sulfation. Mol wt standards are indicated to the right of the panel.

5. To precipitate the RNA, add an equal volume of isopropanol (500 μL) and incubate at room temperature for 10 min. Centrifuge 30 min at 12,000 rpm for 15 min at 4°C. Discard supernatant.
6. Wash pellet with 800 μL of 75% ethanol and shake tube, or vortex to break up pellet. Centrifuge 30 min at 12,000 rpm for 15 min at 4°C. (Note: Samples can be left in ethanol at –20°C, if necessary.)
7. Speed vac final pellet, but not to dryness. Dissolve pellets in 30 μL of 10 mM HEPES.
8. Read ODs at 240, 260, and 280. Use OD 260 to calculate RNA concentration. The final preparations should give yields of approx 10 μg of RNA per 1×10^6 cells, with the appropriate A_{260}:A_{280} ratio of approx 2.0. Store at –20°C in non-self-defrosting freezer or at –80°C.

3.9. Analysis of mRNA by RT-PCR

1. Combine 5 μL (100 ng/μL dilution) of total RNA with 5 μL of RNase-free, filtered Millipore water and 1 μL (15 μM) downstream oligonucleotide primer.

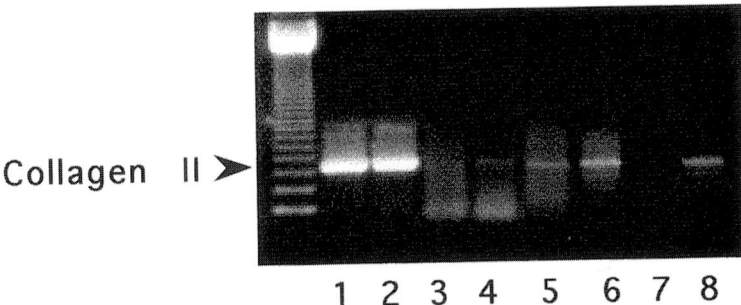

Fig. 5. RT-PCR analysis of type II collagen mRNA in immortalized human chondrocytes. Total RNAs were extracted from cultures of normal human primary costal chondrocytes (lanes 1 and 2), tsTAg-immortalized adult articular chondrocytes incubated at 32°C (lane 3) and 37°C (lane 4), and SV40-immortalized juvenile costal chondrocytes that were untreated (lanes 5–7) or treated with bFGF (lane 8). RT-PCR was performed as described in **Subheading 3.9.** using 125 ng (lane 1), 250 ng (lanes 2–5), 500 ng (lanes 6 and 8), or 1000 ng (lane 7) of total RNA in the RT reaction. Note that high levels of total RNA in the RT reaction prevent detection of the specific mRNA.

2. Denature by heating to 65°C for 5 min and cool (2 min) to room temp. Perform reverse transcription by adding 8.5 µL of RT-mix and incubating at 42°C for 60 min.
3. Aliquot 48 µL of PCR-mix into a fresh 0.5-mL PCR tube and add 2 µL of the reverse transcription product.
4. Amplify DNA by denaturing at 95°C for 2 min, followed by 40 cycles of 95°C for 30 s, and 60°C for 30 s, and a final extension at 60°C for 7 min. (These conditions are for the Perkin-Elmer 9600 Thermocycler, and should be adjusted according to the machine.)
5. Combine 30 µL of PCR product with loading dye and run in a 1.2% agarose gel containing 2 µL of ethidium bromide (10 mg/mL), and visualize the amplified bands by UV transillumination (**Fig. 5**; *see* **Note 11**).

4. Notes

1. Ideally, batches of serum should be tested on primary cultures of normal chondrocytes and selected on the basis of the capacity to support expression of chondrocyte-specific matrix expression. High capacity to induce cell proliferation is not necessarily associated with the ability to maintain phenotype. Further extensive notes on isolation and culture of primary human chondrocytes may be found in ref. *41*.
2. After initial plating of the primary cultures, the chondrocytes require 2–3 d before they have settled down and spread out completely. Although the cultures may continue to express chondrocyte phenotype (e.g., type II collagen and aggrecan)

for several weeks, expression of nonspecific collagens I and III may begin as early as d 7 after isolation. Thus, it is recommended that infection with the tsSV40-TAg retrovirus be carried out between d 4 and 7. Since cells must be actively dividing in order to be infected efficiently with retroviruses, the culture medium should be changed 24 and 3 h prior to addition of the retroviral supernatants.
3. The embryonic vertebral and rib chondrocyte preparations are expected to have a mixture of cell types at different stages of chondrocyte differentiation, as well as contaminating muscle, fat, and bone cells. Therefore, cultures with a higher proportion of cells with chondrocyte-like morphology should be selected for immortalization. Furthermore, once immortalized cultures have been established, it is necessary to clone and select for collagen II- vs collagen X-producing phenotypes. The method for isolating cells from embryonic rib and vertebral specimens has been adapted from Lefebvre et al. *(52)*.
4. Human chondrocytes, similar to other human cell types, are not as easily immortalized with TAg-containing retroviruses as their mouse counterparts. Furthermore, the original φ2 packaging cell line and its derivatives, which package viral RNA into ecotropic particles and have the capacity to infect rat and mouse cells only, must be substituted with an amphotropic packaging line such as PA317 for the production of helper virus-free stocks that will infect human cells.
5. Safety considerations in handling viral stocks: **Caution:** Special care should be taken to prevent skin contamination and aerosol of virus particles. Spent medium, pipets, vacuum flasks, and any other materials contaminated with retroviral containing media should be treated with a virucidal detergent such as LpHse (Calgon Vestal Laboratories, Inc., St Louis, MO).
6. Since viral activity is depleted after 4–6 h of incubation at 37°C, it is not necessary to leave the viral supernatants on the chondrocytes overnight, except as a convenience.
7. Since chondrocytes are strongly adherent to tissue-culture plastic, possibly because of the strongly charged glycosaminoglycans in their matrix, a trypsin–EDTA solution, rather than trypsin alone, is used for full release of chondrocytes from tissue culture plastic during passaging. We have found that tsTAg-immortalized adult articular chondrocytes adhere more strongly to the culture dish than the immortalized juvenile costal or embryonic chondrocytes, which also proliferate more rapidly.
8. The expansion of immortalized chondrocytes in monolayer culture in medium containing 10% FCS favors a loss of phenotype. Established cultures should not be plated at densities lower than those resulting from a split ratio of 1:10, and should be passaged frequently, but not more than twice weekly, just prior to reaching confluence. Since prolonged passage of immortalized lines will select for rapidly growing cells, immortalized lines should be expanded as early as possible and frozen down in liquid nitrogen.
9. The method for culture of chondrocytes in alginate beads has been adapted from previously published methods *(35,36)*.

10. Various methods are available for analysis and characterization of proteoglycans. We have found the described method to be a convenient and rapid approach for screening the relative amounts and molecular sizes of newly synthesized proteoglycans. Specific antibodies are available for more precise identification by either Western blotting or immunocytochemistry. Collagen typing may also be performed, as described previously *(17,41)*.
11. The RT-PCR method described here represents a rapid, sensitive, and semiquantitative approach for screening immortalized cell lines for the expression of chondrocyte-specific mRNAs. Northern blotting may also be used, as described previously *(17,41)*.

References

1. Poole, A. R. (1993) Cartilage in health and disease, in *Arthritis and Allied Conditions: A Textbook of Rheumatology* (McCarty, D. J. and Koopman, W. P., eds.), Lea and Febiger, Philadelphia, pp. 279–333.
2. Goldring, M. B. (1993) Degradation of articular cartilage in culture: regulatory factors, in *Joint Cartilage Degradation: Basic and Clinical Aspects* (Woessner, J. F., Jr. and Howell, D. S., eds.), Marcel Dekker, New York, pp. 281–345.
3. Goldring, M. B., Sandell, L. J., Stephenson, M. L., and Krane, S. M. (1986) Immune interferon suppresses levels of procollagen mRNA and type II collagen synthesis in cultured human articular and costal chondrocytes. *J. Biol. Chem.* **261**, 9049–9056.
4. Goldring, M. B. and Krane, S. M. (1987) Modulation by recombinant interleukin 1 of synthesis of types I and III collagens and associated procollagen mRNA levels in cultured human cells. *J. Biol. Chem.* **262**, 16,724–16,729.
5. Goldring, M. B., Birkhead, J., Sandell, L. J., Kimura, T., and Krane, S. M. (1988) Interleukin 1 suppresses expression of cartilage-specific types II and IX collagens and increases types I and III collagens in human chondrocytes. *J. Clin. Invest.* **82**, 2026–2037.
6. von der Mark, K., Gauss, V., von der Mark, H., and Muller, P. (1977) Relationship between cell shape and type of collagen synthesised as chondrocytes lose their cartilage phenotype in culture. *Nature* **267**, 531–532.
7. Benya, P. D. and Shaffer, J. D. (1982) Dedifferentiated chondrocytes reexpress the differentiated collagen phenotype when cultured in agarose gels. *Cell* **30**, 215–224.
8. Castagnola, P., Moro, G., Descalzi-Cancedda, F., and Cancedda, R. (1986) Type X collagen synthesis during in vitro development of chick embryo tibial chondrocytes. *J. Cell Biol.* **102**, 2310–2317.
9. Alema, S., Tato, F., and Boettiger, D. (1985) Myc and src oncogenes have complementary effects on cell proliferation and expression of specific extracellular matrix components in definitive chondroblasts. *Mol. Cell. Biol.* **5**, 538–544.
10. Gionti, E., Pontarelli, G., and Cancedda, R. (1985) Avian myelocytomatosis virus immortalizes differentiated quail chondrocytes. *Proc. Natl. Acad. Sci. USA* **82**, 2756–2760.
11. Horton, W. E., Jr., Cleveland, J., Rapp, U., Nemuth, G., Bolander, M., Doege, K., Yamada, Y., and Hassell, J. R. (1988) An established rat cell line expressing chondrocyte properties. *Exp. Cell Res.* **178**, 457–468.

12. Thenet, S., Benya, P. D., Demignot, S., Feunteun, J., and Adolphe, M. (1992) SV40-immortalization of rabbit articular chondrocytes: alteration of differentiated functions. *J. Cell. Physiol.* **150,** 158–167.
13. Mallein-Gerin, F. and Olsen, B. R. (1993) Expression of simian virus 40 large T (tumor) oncogene in chondrocytes induces cell proliferation without loss of the differentiated phenotype. *Proc. Natl. Acad. Sci. USA* **90,** 3289–3293.
14. Benoit, B., Thenet-Gauci, S., Hoffschir, F., Penformis, P., Demignot, S., and Adolphe, M. (1995) SV40 large T antigen immortalization of human articular chondrocytes. *In Vitro Cell. Dev. Biol.* **31,** 174–177.
15. Mataga, N., Tamura, M., Yanai, N., Shinomura, T., Kimata, K., Obinata, M., and Noda, M. (1996) Establishment of a novel chondrocyte-like cell line derived from transgenic mice harboring the temperature-sensitive simian virus 40 large T-antigen. *J. Bone Miner. Res.* **11,** 1646–1654.
16. Lefebvre, V., Garofalo, S., and deCrombrugghe, B. (1995) Type X collagen gene expression in mouse chondrocytes immortalized by a temperature-sensitive simian virus 40 large tumor antigen. *J. Cell Biol.* **128,** 239–245.
17. Goldring, M. B., Birkhead, J. R., Suen, L.-F., Yamin, R., Mizuno, S., Glowacki, J., Arbiser, J. L., and Apperley, J. F. (1994) Interleukin-1β-modulated gene expression in immortalized human chondrocytes. *J. Clin. Invest.* **94,** 2307–2316.
18. Fanning, E. and Knippers, R. (1992) Structure and function of simian virus 40 large T antigen. *Annu. Rev. Biochem.* **61,** 55–85.
19. Duncan, E. L., Whitaker, N. J., Moy, E. L., and Reddel, R. R. (1993) Assignment of SV40-immortalized cells to more than one complementation group for immortalization. *Exp. Cell Res.* **205,** 337–344.
20. Rowe, D. W., Moen, R. C., Davidson, J. M., Byers, P. H., Bornstein, P., and Palmiter, R. D. (1978) Correlation of procollagen mRNA levels in normal and transformed chick embryo fibroblasts with different rates of procollagen synthesis. *Biochemistry* **17,** 1581–1590.
21. Berman, A. E. and Morozevich, G. E. (1990) Secretion and intracellular degradation of collagen in cultures of normal and SV-40-transformed human fibroblasts. *FEBS Lett.* **263,** 285–263.
22. Woodworth, C. D., Kreider, J. W., Mengel, L., Miller, T., Meng, Y. L., and H. C., I. (1988) Tumorigenicity of simian virus 40-hepatocyte cell lines: effect of in vitro and in vivo passage on expression of liver-specific genes and oncogenes. *Mol. Cell Biol.* **8,** 4492–4501.
23. Iwamoto, M., Yagami, K., LuValle, P., Olsen, B. R., Petropoulos, C. J., Ewert, D. L., and Pacifici, M. (1993) Expression and role of c-*myc* in chondrocytes undergoing endochondral ossification. *J. Biol. Chem.* **268,** 9645–9652.
24. Jat, P. S. and Sharp, P. A. (1989) Cell lines established by a temperature-sensitive simian virus 40 large-T-antigen gene are growth restricted at the nonpermissive temperature. *Mol. Cell. Biol.* **9,** 1672–1681.
25. Mitchell, P. J., Wang, C., and Tjian, R. (1987) Positive and negative regulation of transcription in vitro: enhancer binding protein AP-2 is inhibited by SV40 T antigen. *Cell* **50,** 847–861.

26. Hansell, E. J., Frisch, S. M., Tremble, P. M., Murnane, J. P., and Werb, Z. (1995) Simian virus 40 transformation alters the actin cytoskeleton, expression of matrix metalloproteinases and inhibitor of metalloproteinases, and invasive behavior of human skin fibroblasts. *Biochem. Cell Biol.* **73,** 373–389.
27. Logan, S. K., Hansell, E. J., Damsky, C. H., and Werb, Z. (1996) T-Antigen inhibits metalloproteinase expression and invasion in human placental cells transformed with temperature-sensitive simian virus 40. *Matrix Biol.* **15,** 81–89.
28. Glowacki, J., Trepman, E., and Folkman, J. (1983) Cell shape and phenotypic expression in chondrocytes. *Proc. Soc. Exp. Biol. Med.* **172,** 93–98.
29. Reginato, A. M., Iozzo, R. V., and Jimenez, S. A. (1994) Formation of nodular structures resembling mature articular cartilage in long-term primary cultures of human fetal epiphyseal chondrocytes on hydrogel substrate. *Arthritis Rheum.* **37,** 1338–1349.
30. Paulsen, D. F. and Solursh, M. (1988) Microtiter micromass cultures of limb-bud mesenchymal cells. *In Vitro Cell. Dev. Biol.* **24,** 138–147.
31. Kato, Y., Iwamoto, M., Koike, T., Suzuki, F., and Takano, Y. (1988) Terminal differentiation and calcification in rabbit chondrocyte cultures grown in centrifuge tubes: Regulation by transforming growth factor β and serum factors. *Proc. Natl. Acad. Sci. USA* **85,** 9552–9556.
32. Ballock, R. T. and Reddi, A. H. (1994) Thyroxine is the serum factor that regulates morphogenesis of columnar cartilage from isolated chondrocytes in chemically defined medium. *J. Cell Biol.* **126,** 1311–1318.
33. Denker, A. E., Nicoll, S. B., and Tuan, R. S. (1995) Formation of cartilage-like spheroids by micromass cultures of murine C3H10T1/2 cells upon treatment with transforming growth factor-β1. *Differentiation* **59,** 25–34.
34. Sun, S., Aydelotte, M. B., Maldonaldo, B., Kuettner, K. E., and Kimura, J. H. (1986) Clonal analysis of the population of chondrocytes from the Swarm rat chondrosarcoma in agarose culture. *J. Orthopaed. Res.* **4,** 427–436.
35. Guo, J., Jourdian, G. W., and MacCallum, D. K. (1989) Culture and growth characteristics of chondrocytes encapsulated in alginate beads. *Connect. Tiss. Res.* **19,** 277–297.
36. Hauselmann, H. J., Fernandes, R. J., Mok, S. S., Schmid, T. M., Block, J. A., Aydelotte, M. B., Kuettner, K. E., and Thonar, E. J. (1994) Phenotypic stability of bovine articular chondrocytes after long-term culture in alginate beads. *J. Cell Sci.* **107,** 17–27.
37. Gibson, G. J., Schor, S. L., and Grant, M. E. (1982) Effects of matrix macromolecules on chondrocyte gene expression: synthesis of a low molecular weight collagen species by cells cultured within collagen gels. *J. Cell Biol.* **93,** 767–774.
38. Freed, L. E., Marquis, J. C., Nohria, A., Emmanual, J., Mikos, A. G., and Langer, R. (1993) Neocartilage formation *in vitro* and *in vivo* using cells cultured on synthetic biodegradable polymers. *J. Biomed. Mater. Res.* **27,** 11–23.
39. Nicoll, S. B., Denker, A. E., and Tuan, R. S. (1995) *In vitro* characterization of transforming growth factor-β1–loaded composites of biodegradable polymer and mesenchymal cells. *Cells Materials* **5,** 231–244.

40. Mizuno, S. and Glowacki, J. (1996) Chondroinduction of human dermal fibroblasts by demineralized bone in three-dimensional culture. *Exp. Cell Res.* .
41. Goldring, M. B. (1996) Human chondrocyte cultures as models of cartilage-specific gene regulation, in *Methods in Molecular Biology: Human Cell Culture Protocols* (Jones, G. E., ed.), Humana, Totawa, NJ, pp. 217–231.
42. Sams, A. E. and Nixon, A. J. (1995) Chondrocyte-laden collagen scaffolds for resrufacing extensive articular cartilage defects. *Osteoarthritis Cartilage* **3,** 47–59.
43. Brittberg, M., Lindahl, A., Nilsson, A., Ohlsson, C., Isaksson, O., and Peterson, L. (1994) Treatment of deep cartilage defects in the knee with autologous chondrocyte transplantation. *N. Engl. J. Med.* **331,** 889–895.
44. Crystal, R. G. (1995) Transfer of genes to humans: Early lessons and obstacles. *Science* **270,** 404–410.
45. Bandara, G., Mueller, G. M., Galea-Lauri, J., Tindal, M. H., Georgescu, H. I., Sucharek, M. K., et al. (1993) Intraarticular expression of the interleukin-1 receptor antagonist protein by ex vivo gene transfer. *Proc. Natl. Acad. Sci. USA* **90,** 10,764–10,768.
46. Geiler, T., Kriegsmann, J., Keyszer, G. M., Gay, R. E., and Gay, S. (1994) A new model for rheumatoid arthritis generated by engraftment of rheumatoid synovial tissue and normal human cartilage into SCID mice. *Arthritis Rheum.* **37,** 1664–1671.
47. Harlow, E., Crawford, L. V., Pim, D. C., and Williamson, N. M. (1981) Monoclonal antibodies specific for simian virus 40 tumor antigens. *J. Virol.* **39,** 861–869.
48. Laemmli, U. K. (1970) Cleavage of structural proteins during the assembly of the head of bacteriophage T4. *Nature* **227,** 680–685.
49. Miller, A. D. and Buttimore, C. (1986) Redesign of retrovirus packaging cell lines to avoid recombination leading to helper virus production. *Mol. Cell Biol.* **6,** 2895–2902.
50. Vogel, K. G., Sandy, J. D., Pogany, G., and Robbins, J. R. (1994) Aggrecan in bovine tendon. *Matrix Biol.* **14,** 171–179.
51. Robbins, J. R. and Vogel, K. G. Mechanical loading and TGF-β regulate proteoglycan synthesis in tendon. *Arch. Biochem. Biophys.* submitted.
52. Lefebvre, V., Garofalo, S., Zhou, G., Metsaranta, M., Vuorio, E., and deCrombrugghe, B. (1994) Characterization of primary cultures of chondrocytes from type II collagen/β-galactosidase transgenic mice. *Matrix Biol.* **14,** 329–335.
53. Lum, Z.-P., Hakala, B. E., Mort, J. S., and Recklies, A. D. (1996) Modulation of the catabolic effects of interleukin-1β on human articular chondrocytes by transforming growth factor-β. *J. Cell. Physiol.* **166,** .
54. Bonaventure, J., Kadhom, N., Cohen-Solal, L., Ng, K. H., Bourguigno, J., Lasselin, C., and Freisinger, P. (1994) Reexpression of cartilage-specific genes by dedifferentiated human articular chondrocytes cultured in alginate beads. *Exp. Cell Res.* **212,** 97–104.

III

Clinical Applications

15

Isolation and In Vitro Proliferation of Chondrocytes, Tenocytes, and Ligament Cells

Sonya Shortkroff and Myron Spector

1. Introduction

In recent years it has become possible to grow large numbers of selected cell types in vitro from relatively small tissue samples. This capability has served as the foundation for using autogenous, as well as allogeneic, cells expanded in culture for engineering tissues that display little potential for spontaneous regeneration. Cartilage, tendons, and ligaments are connective tissues that have common attributes that limit their capabilities of regeneration (**Table 1**). These tissues are relatively avascular, sparsely populated with cells, and their cell populations have a low mitotic activity. Because the use of allogeneic grafts brings with it concerns regarding transfer of disease and immunological rejection of the foreign material, the availability of an autologous system for regeneration of these tissues is of paramount importance for restoring joint function and preventing further degeneration.

Articular cartilage is composed of a highly organized extracellular matrix, which consists mainly of type II collagen and proteoglycan (chondroitin and keratan sulfate glycosaminoglycans) and a relatively sparse highly specialized cell population (articular chondrocytes), that accounts for less than 5% of the tissue by volume. Cartilage is described as having three zones, in accordance with the organization of the collagen fibrils. The layer closest to the joint space is the superficial tangential zone, which comprises approx 10% of the cartilage; the middle zone is composed of 40–60%, and the deep zone adjacent to the calcified cartilage contains approx 30–40% of the cartilage *(1,2)*. The superficial zone displays minimal glycosaminoglycans and the cells in the zone appear flattened and elongated parallel to the surface. The collagen fibrils are oriented tangentially. The middle zone contains high levels of both type II

Table 1.
Characteristics of Adult Tissue

	Articular Cartilage	Ligament		Tendon
		Cruciate	Collateral	
Cell type	Chondrocyte	Fibroblast/ Fibrochondrocyte	Fibroblast	Fibroblast
Cell density	Low	Low	Low	Low
	$(1.4–1.7 \times 10^4 \text{ cell/mm}^3)$[1]	$(2.7 \mu g \text{ DNA/mg dry wt})$[5]	$(2.6 \mu g \text{ DNA/mg dry wt})$[5]	$(1.5 \mu g \text{ DNA/mg dry wt})$[5]
Mitotic activity	Low	Low	Low	Low
Predominant collagen type	II	I	I	I
Vascularity	No	Low	Low	Low
Substrate for phenotype preservation	3D Hydrogel (viz,agarose)	2D Culture plates	2D Plates	2D Plates
Repair capacity	Low	Low	High	Low

collagen and proteoglycans, and the cells are spherical. The collagen fibrils form a coiled fibrous meshwork. The deep zone chondrocytes have a spherical morphology, and appear in columnar formation. The collagen fibrils are perpendicular to, and insert into, the calcified cartilage layer that overlies the subchondral bone *(3)*. The mechanical performance of articular cartilage, its main physiological function, is dependent on the architecture of its extracellular matrix and the ability of the constituent chondrocytes to maintain the matrix. Tissue-engineering approaches are challenged to restore this complex hierarchical structure and the continuity of the regenerated cartilage with adjacent tissue.

In monolayer culture, the normally spherical chondrocytes become polygonal, and, with continued passaging, appear quite fibroblastic. Their morphology, however, is readily maintained in three-dimensional (3-D) culture systems utilizing agarose or collagen gels *(4)*, and, most recently, it has been found, type II collagen sponges *(5)*. Alternatively, one can maintain morphology by placing the cells in a rotating flask, where there is little opportunity for the cells to adhere to the surface. Chondrocytes with retained morphology have been shown to produce type II collagen and proteoglycan in vitro *(4)*. These same cells, however, assume a fibroblast shape and produce type I collagen when passaged in monolayer culture, indicating an altered phenotype. When the cells are returned to a 3-D conformation, they revert to the chondrocytic phenotype and synthetic pathways. This ability to recover chondrocyte phenotype is limited, and it is suggested that chondrocytes in monolayer undergo no more than four passages in order to assure retention of the chondrocytic characteristics.

Tendons and ligaments, in contrast to articular cartilage, are mainly composed of uniaxially aligned type I collagen, and are generally vascularized, although the midsubstance of the tissues display few vessels. Tendon cells (tenocytes) and ligament cells are fibroblast-like cells that have shown some variability from dermal fibroblasts in certain characteristics (e.g., their integrins) *(6–8)*. These cells are easily grown, as are most fibroblasts, and may undergo multiple passaging without significant changes in their capabilities. Histologic evaluation of ligaments and tendons have demonstrated that the cells in the proximity of the insertion into bone display a different morphology. The cells at this transition zone are fibrocartilage cells and display a polygonal shape, but continue to produce type I collagen. Even though ligaments and tendons have relatively simple architectures and are comprised of cells that readily proliferate in vitro, they too offer challenges for tissue engineering. Templates for regeneration in vivo must allow for a distribution of the fibroblast-like cells, so that the synthesized matrix assumes a near-normal density and orientation, as well as composition. Material engineered in vitro must be implanted in such a way that subsequent remodeling results in mechanical coupling with the host tissues.

As would be expected, cells retrieved from embryonic, neonatal, and adult samples will display different mitogenic and biosynthetic characteristics. Immature articular cartilage is composed of very actively proliferating cells, and often it is a vascularized tissue. This is also true of ligament and tendon. The mature tissue is much more structured and displays little variability of cell morphology, except at the superficial zone of cartilage, where the cells are more fibroblast-like, and at tendon and ligament insertion sites at the interface with muscle and bone, where there are fibrocartilagenous cells.

By selecting the appropriate cell source and culture conditions, a high enough yield of cells can result from proliferation in vitro to allow for direct transplantation into a treatment site, or for seeding of a structured matrix in vitro for subsequent implantation. Regulation of cell growth and biosynthetic activity can be achieved with cytokines or chemokines, incorporated into a cell-seeded matrix, or administered to the cells in vitro prior to direct transplantation or matrix seeding.

This chapter describes the isolation and in vitro culture of chondrocytes, tenocytes, and ligament cells. Also included is a description of a 3-D culture system for chondrocytes (agarose). This system can be used as a control to evaluate the continued expression of chondrocytic features in vitro at varied passages. The cell techniques are described so that the cell isolation is a general procedure utilized for all of these tissues; when the culture conditions may vary slightly in terms of medium used and plating efficiency, those descriptions are presented individually. Since phenotypic expression is of paramount importance with chondrocytes, descriptions of passaging are also presented.

2. Materials

2.1. Tissue Dissection

2.1.1. Instruments and Equipment

1. Surgical blade handles, sizes 3 and 4.
2. Scalpel blades, surgical steel, sizes 20 and 10.
3. 4 Kelly clamps, straight.
4. 2 pair of toothed tissue forceps.
5. 2 pair of Rochester Oschner straight clamps, sizes 8 and 61/4.
6. 2 pair of cartilage forceps.
7. 1 stainless steel tray with cover.
8. Animal shaver.

2.1.2. Solutions

1. Dulbecco's phosphate buffered saline (PBS) (Gibco-BRL, Grand Island, NY) containing 100 U/mL penicillin and 100 µg/mL streptomycin (1% P/S) (Gibco-BRL). The PBS solution should be kept on ice.

2.1.3. Disposable Items

1. Sterile gloves.
2. Sterile drapes and sterile gauze.
3. Sterile gowns, caps, masks, and booties.
4. Betadine (povidone-iodine) solution or swabs.

2.2. Cell Isolation

2.2.1. Instruments and Equipment

1. #3 surgical blade handles.
2. #10 blades.
3. Cartilage forceps.
4. Small stainless steel spatula.
5. 2–3 100-mm Petri dishes.
6. 250-mL spinner flask.
7. Nylon mesh 70-µ pore size strainer (Becton Dickinson Labware, Franklin Lakes, NJ).

2.2.2. Solutions

1. PBS supplemented with 100 U of penicillin and 100 pg of streptomycin (1% P/S).
2. Collagenase (type I, 212 U/mL, Worthington Biochemical, NJ): Dissolve in 50–100 mL cell culture medium containing 1% P/S, for a final collagenase concentration of 0.15%. Solution is filtered through a 0.45-µm sterile filtration unit (Corning, Corning, NY).
3. Primary culture medium (*see* **Subheadings 2.3. and 2.4.**) (HyClone Laboratory).
4. Trypan blue stock solution (0.4% in normal saline): Dilute 1:10 at time of use (Gibco Laboratories Life Technologies, Grand Island, NY).

2.3. Chondrocyte Culture

1. Primary culture medium: Dulbecco's modified Eagle's medium/HAM's F-12 nutrient mixture [1:1] (DMEM/F-12) containing 2 mM L-glutamine, 10% heat-inactivated fetal bovine serum (FBS) (HyClone Laboratory, Logan, UT), 1% P/S, and 250 µg/mL fungizone (Gibco-BRL).
2. Passaging solutions.
   a. Trypsin–EDTA solution (Gibco-BRL): 0.05% trypsin in 0.53 mM EDTA-4Na.
   b. Primary culture medium.
   c. Trypan blue stock solution (as above).

2.4. Tenocyte and Fibroblast Culture

1. Primary culture medium: alpha minimal essential medium (αMEM) or DMEM containing 10% FBS, 1% P/S, and 250 µg/mL fungizone.
2. Passaging solutions.
   a. Trypsin–EDTA solution (Gibco-BRL).
   b. Primary culture medium.
   c. Trypan blue stock solution (as above).

2.5. Agarose 3-D Implantation

1. Seaplaque GTG agarose (FMC BioProducts, Rockland, ME) (low melting temperature).
2. Bio Rad agarose (Bio-Rad Laboratories, Hercules, CA) (high melting temperature).

3. Methods
3.1. Articular Dissection of Cartilage and Ligament

Whether retrieving the cartilage at necropsy (rabbit, dog) or appropriating a joint specimen from the local abattoir (bovine samples), the procedure should be performed in a sterile or clean environment.

1. The area around the knee is shaved, cleaned using betadine scrub solution, and then draped.
2. A straight midline longitudinal incision is made with a #20 blade, extending from the distal fourth of the femur to the tibial tuberosity.
3. The skin and subcutaneous fascia are retracted to expose the patellar tendon.
4. The patellar tendon is excised by a parapatellar incision that runs the length of the tendon from the tibia to the sartorious muscle.
5. The tendon is then bisected at approx one-eighth to one-quarter in. from the insertion sites. The fat pad and the patella are dissected from the tendon.
6. A deeper incision into the joint capsule exposes the femoral condyles and the tibial plateau.
7. The anterior cruciate ligament and the lateral and medial collateral ligaments are transected.
8. The joint can now be fully flexed in order to dissect the cartilage. Using a (#10 or #15) surgical blade, the cartilage is peeled from the condyles and the tibial plateau, as well as the trochlear portion of the femur. Approximately 1–2- (human), 0.5–0.8- (dog), or 0.2–0.3-mm-(rabbit) thick slices, with diameters of 2–5 mm, are obtained. Using forceps, the cartilage pieces are placed in sterile, cold PBS.

3.2. Dissection of the Achilles Tendon

1. The area is prepped and shaved, with the animal positioned either on its side or in a prone position.
2. An incision is made from the distal third of the tibia to the calcaneus. This exposes the tendon from the musculotendinous junction to the plantaris fascia of the calcaneus.
3. The skin and soft tissue are released and retracted.
4. The tendon is transsected approx one-eigth to one-quarter in. from the insertion.
5. The specimen is then placed in a sterile container in PBS on ice.

The patellar tendon may also be excised, as described in **Subheading 3.1.** for retrieval of cartilage.

3.3. Cell Isolation

The retrieved specimens should be used within 2–4 h of acquisition if possible and should be kept at 4°C until use.

1. Place the specimens in 10 mL of PBS in a 100-mL sterile Petri dish.
2. Add 2 mL of PBS to a second Petri dish to keep the specimens moist while dissecting.
3. Take each sample and dice to approx 2–3 by 1–2 mm in the second Petri dish.
4. Filter the collagenase solution using a 0.45 µ sterile filtration unit and place the filtered solution into the spinner flask.
5. Place the spatula under the floating cartilage pieces, retrieve the dissected samples on the spatula with the aid of a pair of forceps, and place them into a spinner flask. Repeat this procedure as needed.
6. Place the spinner flask on a stirring platform in a 37°C incubator for 4 h to overnight. When most of the sample has been digested (approx 95%), pour the suspension through the nylon mesh strainer in order to remove any matrix debris and undigested material.
7. Centrifuge the filtrate at 500g for 10 min in sterile 50 mL centrifuge tubes.
8. Discard the supernatant, wash the cell pellet with fresh medium, and resuspend 3× to remove any residual enzyme.
9. Resuspend the cell pellet in 1 mL of DMEM/F-12.
10. Perform a cell count using a hemocytometer and check for viability with the trypan blue exclusion test. Use a 1:1 dilution of the cell suspension and diluted trypan blue. Place a drop on the hemocytometer and count the total number of cells and the number of cells that have turned blue. To calculate the percent viability, subtract the number of blue cells from the total, then divide by the total number of cells and multiply by 100.
11. The cells are plated in 25-cm^2 tissue-culture flasks in 4–5 mL of culture medium at a concentration of 10^6 cells/mL for chondrocytes and 10^5 cells/mL for tenocytes and ligament cells. Flasks are placed in a 37°C incubator, 5% CO_2 atmosphere and 95% humidity.
12. The culture medium is changed every 2 d. The cells should adhere well to the plate and usually take 4 d to a week to become confluent.

3.4. Passaging of Cells

1. At confluence, remove the cell medium and add an equivalent amount of trypsin–EDTA solution.
2. Return flasks to the incubator for approx 15 min.
3. If available, check the cells under an inverted microscope to *see* if they have lifted off the surface.
4. Aspirate the solution and place into centrifuge tubes.
5. Centrifuge for 10–15 min at 400g.
6. Remove the supernatant and disperse the pellet in primary culture medium.

7. Centrifuge again and remove the supernatant. Repeat 2× in order to remove all of the trypsin.
8. Resuspend the pellet in 1 mL of medium and take a 50-µL aliquot for a cell count and estimate of cell viability.
9. Dilute the cell suspension appropriately and place in 75-cm^2 culture flasks according to the previously described cell densities. At each passage, one-quarter to one-half of the cells may be frozen for future use.

3.5. Chondrocytes in 3-D Agarose

Primary chondrocytes assume a polygonal shape, and, with each passage, establish a more fibroblastic morphology in monolayer. If cultured in agarose or in spinner flasks, these cells will continue to have the typical chondrocyte spherical morphology. The Bio-Rad agarose is used to provide a platform for the cell-incorporated Seaplaque agarose. This bilayer facilitates intact retrieval of the gel for histology.

1. Weigh out 1 gm of Bio-Rad agarose and add to 50 mL ddH$_2$O in a sterile bottle, and autoclave.
2. Transfer 1.5 mL of the Bio-Rad agarose into each well of a 6-well plate and allow to cool for at least 1 h or overnight.
3. Dissolve 1 gm of Seaplaque agarose in 25 mL of ddH$_2$O in a sterile bottle, and autoclave. Let cool for 15–20 min.
4. Add 10^6 cells in a 1:1 dilution of medium and Seaplaque agarose, using warm sterile pipets (37°C).
5. Layer 2.5 mL of the Seaplaque agarose–cell solution onto the Bio-Rad gel, using warm sterile pipets. Allow to set for 10–15 min.
6. Add 2.5 mL of primary culture medium on the top of each well.
7. Replace medium every 2–3 d.

4. Notes

1. If cells do not adhere to the culture plates within 24 h, the likelihood of an infection is manifest. Cells retrieved from animal sources are particularly susceptible to contamination. This may occur during or after the dissection of the tissue from the source. Therefore, sterile conditions should be maintained at all times.
2. Chondrocytes plated at a low density tend to become fibroblastic, and proliferate more slowly than chondrocytes plated at a density of between 10^5 and 10^6 cells/mm^2. The section is observed with neonatal or immature chondrocytes, which there is a high proliferative activity and a retention of the polygonal shape through approximately two passages.
3. When culturing tenocytes and ligament cells, they should be at low density, in order to decrease the number of passages required because these cells proliferate rapidly.
4. When preparing an agarose culture for chondrocytes, warm pipets should always be used for the Seaplaque agarose, in order to prevent premature gelation. The

culture medium in which the cells are to be diluted in order to incorporate them into the Seaplaque should also be warmed to 37°C, in order to distribute the cells evenly throughout the Seaplaque.
5. The use of fungizone is optional, but should definitely be utilized if there is any indication of fungal infections.

References

1. Sledge, C. B. (1993) Biology of the joint, in *Textbook of Rheumatology* (Kelley, W. N., Harris, E. D., Jr., Ruddy, S., Sledge, C. B., eds.), Saunders, Philadelphia **1**, pp. 1–21.
2. Meachim, G. and Stockwell, R. A. (1979) The matrix, in *Adult Articular Cartilage* (Freeman, M. A. R., ed.), Pitman, England) pp. 1–68.
3. Stockwell, R. A. and Meachim, G. (1979) The chondrocytes, in *Adult Articular Cartilage* (Freeman, M. A. R., ed.), Pitman, England) pp. 69–144.
4. Buschmann, M. D., Gluzband, Y. A., Grodzinsky, A. J., Kimura, J. H., and Hunziker, E. B. (1992) Chondrocytes in agarose culture synthesize a mechanically functional extracellular matrix. *J. Orthop. Res.* **10**, 745–758.
5. Nehrer, S., Breinan, H. H., Ramappa, A., Young, G., Shortkroff, S., Louie, L., et al. (1997) Matrix collagen type and pore size influence behavior of seeded canine chondrocytes. *Biomaterials,* **18**, 769–776.
6. Schreck, P. J., Kitabayashi, L. R., Amiel, D., Akeson, W. H., and Woods, V. L., Jr. (1995) Integrin display increases in the wounded rabbit medial collateral ligament but not the wounded anterior cruciate ligament. *J. Orthop. Res.* **13**, 174–183.
7. Amiel, D., Frank, C., Harwood, F., Fronek, J., and Akeson, W. (1984) Tendons and ligaments: a morphological and biochemical comparison. *J. Orthop. Res.* **1**, 257–265
8. Lyon, R. M., Akeson, W. H., Amiel, D., Kitabayashi, L. R., and Woo, S. L.-Y. (1991) Ultrastructural differences between the cells of the medical collateral and the anterior cruciate ligaments. *Clin. Orthop.* **272**, 279–286.

16

Culture and Identification of Autologous Human Articular Chondrocytes for Implantation

Ross Tubo and Francois Binette

1. Introduction

The disability and pain that result from damage to articular cartilage within the knee joint has stimulated the development of several approaches to facilitate the restoration of joint function *(1–9)*. Recently, cultured autologous chondrocytes, isolated from an individual's own cartilage, have been expanded in vitro, and then implanted into the damaged site for repair of damaged knee cartilage *(10)*. This remarkable process has been characterized by the modulation of gene expression during proliferation expansion and subsequent redifferentiation of cultured chondrocytes in vitro *(11)* and in vivo *(12)*. Since the unique biomechanical properties of hyaline articular cartilage have been shown to be intimately linked with the biochemistry of the tissue (*see* Buckwalter and Mow ref. *13* for review), we have developed an in vitro system to verify that proliferatively expanded chondrocytes retain their ability to redifferentiate, or re-express their hyaline articular cartilage phenotype. Although the methods described herein were developed for specific application to chondrocytes, the principles for evaluation of biochemical and molecular biological properties of tissue-engineered materials, in vitro, may be applied to the development of any functional, high quality, tissue engineered implant.

2. Materials

1. Cartilage tissue–biopsy transport medium: Normal articular cartilage specimens are transported in serum-free Dulbecco's modified Eagle's medium (DMEM, Life Technologies, Grand Island, NY, #11965–084), supplemented with 0.050 mg/mL of gentamycin (Life Technologies, #15750–011) to prevent contamination *(14)*.

Transport tubes contained 40 mL of 0.2 μ filter-sterilized cartilage-transport medium and is stable for 4 mo at 4°C.

2. Cartilage digestion–chondrocyte isolation media: Chondrocytes are freed from cartilage tissue by sequential enzymatic digestion using collagenase *(14)*, followed by a mixture of enzymes, as suggested by Watt and Dudhia *(15)*. Specifically, the first cartilage digestion is performed in DMEM containing 0.1% (w/v) collagenase (CLS-2, Worthington, Freehold, NJ, #LS004176). All subsequent digestions are performed in DMEM supplemented with 0.25% (w/v) collagenase and 0.1% (w/v) trypsin (Life Technologies, #27250). Trypsin is neutralized by adding fetal bovine serum (FBS, HyClone, Logan, Utah, #A-1115–L) to a final concentration of 10% (v/v).

3. Chondrocyte growth medium: Chondrocyte growth medium is prepared by supplementing DMEM with 10% (v/v) FBS, and with 0.050 mg/mL gentamycin. The medium is stored no longer than 1 wk prior to use at 4°C.

4. Trypsin–versene solution for passage of monolayer chondrocytes: Equal volumes of 0.1% (w/v) trypsin, in DMEM, and 0.04% EDTA, in Ca^{2+}- and Mg^{2+}-free PBS (pH 7.4) (Versene solution, Life Technologies, #15040–066), are combined to make trypsin–versene (TE) solution. The final concentrations of trypsin and EDTA in TE is 0.05 and 0.02%, respectively.

5. Chondrocyte freezing medium: Dimethyl sulfoxide (DMSO, Fluka, Ronkonkona, NY, #41641) and FBS are mixed to final concentrations of 20% (v/v) and 80% (v/v), respectively. The final concentration of DMSO and FBS for freezing is 10% (v/v) DMSO, and 50% (v/v) FBS. Concentrated chondrocytes in DMEM supplemented with 10% FBS are mixed 1:1 with DMSO/FBS.

6. Chondrocyte redifferentiation system: The following reagents are required for the evaluation of chondrocyte redifferentiation in alginate as previously described *(16)*:

   a. Potassium alginate (Improved Kelmar, Kelco, Rahway, NJ) solution is prepared as a 1.2% solution in 0.15 M sodium chloride/25 mM HEPES buffer at pH 7.0.
   b. Alginate polymerization solution consisted of a 102 mM $CaCl_2$ solution.
   c. DMEM supplemented with 10% FBS.
   d. Alginate depolymerization buffer consisted of a 0.15 M sodium chloride/25 mM HEPES buffer, pH 7.0, and 55 mM sodium citrate solution.
   e. RNA isolation reagents are obtained from the Rneasy™ kit (Qiagen, Chatsworth, CA, #74104).
   f. In vitro transcription reagents are obtained from the Maxiscript™ kit (Ambion, Austin, TX, #1314).
   g. RNase protection assay reagents are obtained from the Hybspeed™ RPA kit (Ambion, #1412).
   h. Radioactive RNA fragments are detected by phosphorimager (Fujifilm, Stamford, CT, model BAS-1500).
   i. Specific RNA probes for aggrecan *(17)*, type I collagen *(18)* and type II collagen *(19)* are prepared from published sequences.

j. Commercially available antibodies, recognizing type I collagen (Biodesign, Kennebunk, ME, T40202R), type II collagen (Biodesign, T40311R), and chondroitin SO_4 (Seikagaku, Rockville, MD, MO-225) are purchased for immunohistochemical analysis.

k. Incorporation and detection of 5 bromo-2'-deoxy-uridine (BrdU) is performed with the BrdU labeling and detection kit I (Boehringer Mannheim, Indianapolis, IN, #1296 736).

3. Methods
3.1. Day-One Biopsy Processing (10,14)

1. Disinfect and set up the biosafety cabinet (BSC). The following equipment is placed within the biosafety cabinet (BSC), or laminar flow hood, as required, for processing the cartilage biopsy: a balance (Mettler, Hightstown, NJ, Model AE240), one pair of fine tip forceps, one pair of blunt tip forceps, disposable scalpel, sharps pouch, racks for 50-mL or 250-mL tubes, 50-mL and 250-mL centrifuge tubes, disposable serological pipets (5-mL, 10-mL, and 25-mL), aspirating pipet, hemocytometer and coverslip, and vacuum hose.
2. Media for initial cartilage digestion: 20 mL of DMEM and 1.1 mL of 1.0% (w/v) collagenase are removed from the refrigerator/freezer and warmed to 37°C.
3. A Petri dish (100-mm) for weighing the cartilage biopsy, weighing dish is placed on the balance.
4. The thawed collagenase and the warmed bottle of DMEM are transported into the bsc.
5. Cartilage digestion solution is prepared by mixing 1 mL of 1.0% (w/v) collagenase with 9 mL of DMEM in a 50-mL tube. The final concentration of collagenase is 0.1% (w/v).
6. The biopsy transport medium is decanted from the biopsy tube and the cartilage biopsy is subsequently transferred into a sterile Petri dish (not the weighing dish) using sterile forceps, or by tapping the biopsy tube.
7. The biopsy (*see* **Note 1**) is rinsed twice with 10 mL of serum free DMEM in a Petri dish, by carefully swirling the rinse solution inside the dish until the entire biopsy has been rinsed. The rinse solution is then aspirated.
8. The biopsy is held with forceps and trimmed with disposable scalpel to remove any spurious bone that may be associated with the cartilage biopsy.
9. The trimmed cartilage is transferred to the weighing dish, and weighed.
10. The biopsy is finely minced within the weighing dish, using a sterile disposable scalpel, until all pieces are less than 0.5 mm in size. The mincing is performed in the presence of 1 mL of the 0.1% collagenase solution (**Subheading 3.1., step 5**) to prevent cartilage sticking to the Petri dish.
11. Using a wetted 10-mL wide-mouth pipet, the minced biopsy is transferred into a 50-mL centrifuge tube, if the biopsy is less than or equal to 300 mg. If the biopsy is greater than 300 mg, it is transferred to a 225-mL tube, referred to hereafter as the digestion vessel.

12. The mincing dish is rinsed with 5 mL of 0.1% collagenase solution and transferred to the digestion vessel. Rinsing is repeated until all pieces in the dish are transferred to the digestion vessel. The final volume of collagenase used is 10 or 20 mL, in a 50- or 225-mL tube, respectively.
13. The digestion vessel is then transferred to an incubator operating at 37°C for overnight incubation (18–20 h) (*see* **Note 2**).

3.2. Day-Two Biopsy Processing

1. The following items are placed in the BSC, as required, for d-2 biopsy processing: racks for 50- or 250-mL tubes, 50- and 250-mL centrifuge tubes, low-speed centrifuge, disposable serological pipets (5, 10, and 25 mL), aspirating pipet, hemocytometer and coverslip, vacuum hose, and tissue culture flasks (T-25, T-75, and T-150).
2. Aliquots of DMEM, supplemented with 10% FBS (60 mL), DMEM (20 mL), 0.25% (w/v) trypsin (10 mL), and 2.6 mL of 1.0% (w/v) collagenase, are prewarmed (37°C).
3. The digestion vessel containing cartilage incubated overnight is removed from the incubator and brought to the BSC. After the biopsy pieces settle, the supernatant is transferred into a sterile 50-mL centrifuge tube, using a prewetted 10-mL serological pipet. The supernatant is treated as described in **Subheading 3.2, step 5**.
4. Undigested cartilage pieces within the digestion vessel are subjected to further digestion, using 0.25% collagenase–0.1% trypsin. This digestion solution is prepared by adding 3.5 mL of DMEM, 2.5 mL of 1.0% (w/v) collagenase, and 4 mL of 0.25% trypsin to the digestion vessel. The pieces are incubated for 2.5–3 h at 37°C.
5. The volume of the supernatant from **Subheading 3.2., step 3** is brought up to either 20 mL for a 50-mL tube, or 30 mL for a 225-mL tube. Using a wetted 10-mL pipet, the cell suspension is dispersed 5–6 times, or until all cell aggregates are broken up. Cells are counted by hemocytometer.
6. The total number of cells recovered is calculated using the following formula:

$$\text{Total cells} = \text{Cells/mL} \times \text{Vol solution}$$

7. The remaining cell suspension is centrifuged at 1000 RPM for 5.0 min to concentrate the cells.
8. The total number of flasks needed is determined using the following formula, where N = total number of cells:

$$\text{No. of flasks (T-150)} = N/5.0 \times 10^5$$

9. The centrifuged cell pellet is then resuspended in an appropriate volume of DME/FBS (5 mL/5.0×10^5 cells). Cells are then inoculated into the flask(s) with 5 mL of the cell suspension and placed in an incubator at 37°C.

3.3. Processing Cells from Second Digestion

1. The BSC is cleared and cleaned as described in **Subheading 3.2., step 1**.
2. Following 2.5–3 h, the digestion vessel is removed from the incubator.

3. The second digestion is processed as in **Subheading 3.2., step 3**, and **Subheading 3.2., steps 5–9**.
4. Following completion of the second digestion, the remaining pieces of cartilage are added directly to a tissue culture flask for explant outgrowth. If original biopsy weight is less than or equal to 300 mg, inoculate into a T-75. If the original biopsy weight is greater than 300 mg, inoculate into a T-150. Cells are cultured at 37°C.

3.4. Chondrocyte Culture

3.4.1. Chondrocyte Expansion in Monolayer Culture (see **Note 3; ref. 11**)

1. Chondrocytes released from the digestion of human articular cartilage are proliferatively expanded in tissue culture flasks containing DMEM (Gibco-BRL), supplemented with 10% FBS (Hyclone).
2. Cultures are incubated at 37°C in a humidified 10% CO_2 environment.
3. Cultures are fed with fresh media every 2–3 d.
4. Cells are passaged at 80–90% confluence, using TE.
5. Recovered cells are diluted 1:10 at subculture into DMEM supplemented with 10% FBS, and seeded for further expansion in monolayer, or centrifuged at 700g for 5 min, prior to initiation of alginate suspension culture for differentiation assessment.

3.4.2. Chondrocyte Culture in Alginate Suspension (see **Note 4; ref. 16**)

1. Monolayer-cultured, dedifferentiated chondrocytes are pelleted by centrifugation (**Subheading 3.4., step 1.e.**), and then resuspended in the alginate solution at one million cells/mL.
2. Cells, in unpolymerized alginate solution, are loaded into a 10-cc syringe equipped with a 22-gage needle and added by dropwise expulsion into a 0.1 M calcium chloride solution, and allowed to cure for 10 min at room temperature. Upon contact with calcium chloride, the alginate in the solution polymerizes and entraps the cultured cells in suspension.
3. The calcium chloride solution is discarded and the beads are washed twice in 0.15 M sodium chloride, transferred to fresh DMEM supplemented with 10% FBS, and cultured at 37°C.
4. Alginate cultures are fed every 2–3 d.
5. Five-mL aliquots of alginate beads, corresponding to approx 5×10^6 cells, are collected at timed intervals between 4 d to 5 mo (140 d). The beads are washed twice with 0.15 M NaCl, and the chondrocytes are released by solubilizing the beads in an alginate depolymerization buffer.
6. The cells are pelleted by low speed centrifugation at 1000g for 5 min, and either quickly frozen in liquid nitrogen and kept at –80°C until RNase protection assay (**Subheading 3.5., step 1**), or fixed on a microscope slide for immunohistochemistry (**Subheading 3.5., step 2**).

3.5. Evaluation of Differentiation Potential

1. RNase Protection Assay (RPA) (*see* **Note 5**).
   a. Total Cellular RNA is prepared from frozen cell pellets (**Subheading 3.4., step 2.f.**) containing approx 5×10^6 cells, using the Rneasy kit according to the manufacturer's instructions. Typically, a yield of 5–15 µg of total RNA is obtained. The purified RNA preparation is either frozen in water or kept as an ethanol suspension at –80°C. All reagents are free of RNase.
   b. In vitro transcription of antisense RNA probes. The Ambion Maxiscript kit is used according to the manufacturer's instruction. cDNA fragments for human aggrecan *(14)* and type I *(15)* and type II *(16)* collagens are used as template for in vitro transcriptions. A plasmid carrying a fragment of the 18S rRNA (Ambion, # 7339) is also used as an internal control for normalization of the RNA samples.
   c. The antisense RNA probes, along with standard mol wt markers, are run on 7 M urea, 4% polyacrylamide 1X TBE gels, and visualized by autoradiography. The bands are localized on the gel and the ones corresponding to the full-length transcripts are cut and allowed to diffuse passively into the probe elution buffer (0.5 M NH$_4$OAC, 0.2% SDS, 1 mM EDTA) for 2 h at 37°C.
   d. The activity of the probes is quantified by scintillation counting.
   e. The (RPA) are performed using the Ambion Hybspeed RPA kit. Briefly, 1 µg of total RNA is co-precipitated overnight with an aliquot of the purified probes (25,000 cpm per probe) by the addition of 0.1 volume of 5 M NH$_4$OAC and 2.5 vol of ethanol.
   f. After a 10 min centrifugation, the RNA pellet is washed with 70% ethanol, dried, and solubilized in 10 µL of hybridization buffer.
   g. Probes and sample RNA are allowed to hybridize for 10 min at 68°C, and 100 µL of a 1:100 dilution of RNaseA/RNaseT1 mix in digestion buffer is added. The digestion is allowed to proceed for 30 min at 37°C before being terminated with the addition of inactivation/precipitation buffer.
   h. Samples are centrifuged for 10 min and the RNA pellets are redissolved in 10 µL of sample buffer.
   i. Protected RNA fragments are resolved by electrophoresis on an 8 M urea containing 4% polyacrylamide gel, and visualized by autoradiography.
2. Immunohistochemistry on cytospin.
   a. Approximately 50,000–100,000 cells are attached to the slide by centrifugation at 800g for 5 min, using a cytospin centrifuge (**Subheading 3.4., step 2.f.**). Several slides are prepared for each time point (2, 4, and 6 wk). The slides are allowed to air dry at room temperature and are fixed with 4% paraformaldehyde for 5 min.
   b. Primary antibodies to type I collagen, type II collagen, and chondroitin SO$_4$ are diluted 1:50 in phosphate buffered saline (PBS), and separately administered to cells centrifuged and fixed on microscope slides. Primary antibody reactions are carried out in a humid atmosphere at 37°C for 1 h. Normal rabbit IgG is used as a negative control in these experiments.

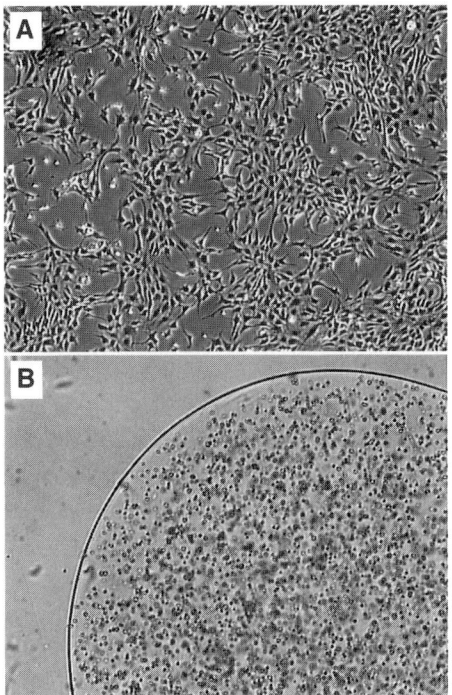

Fig. 1. Articular chondrocyte morphology in culture. Phase contrast micrographs of human articular chondrocytes cultured in (**A**) monolayer, or (**B**) alginate suspension.

    c. The slides are then washed three times in PBS, followed by incubation with a 1:200 dilution in PBS of rhodamine-conjugated goat antirabbit IgG as a secondary antibody, in the same conditions described for the primary antibodies.
    d. Hoechst dye at 1 µg/mL is included in some experiments with the secondary antibodies for nuclear staining.
    e. The slides are washed 3× in PBS, mounted, and examined with a fluorescent microscope.

3.6. Results

1. Minimal growth requirements: Adult articular chondrocytes are sufficiently expanded using DMEM supplemented with 10% FBS. The chondrocytes are observed to dedifferentiate under these conditions. Dedifferentiation has been shown to be reversible by transferring dedifferentiated cells into a suspension culture system *(11,16)*.
2. Morphology: Freshly isolated human articular chondrocytes have a rounded morphology that converts to an attached-fibroblastic morphology (**Fig. 1A**) when placed

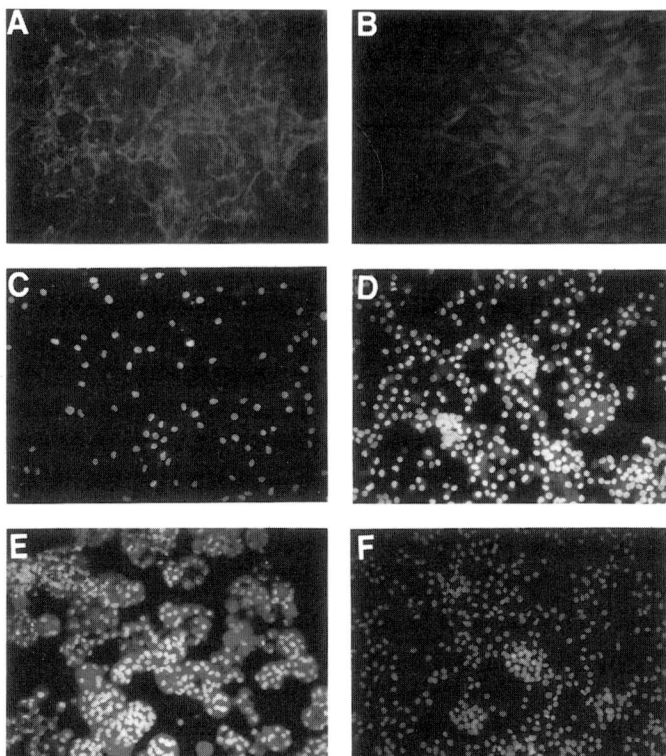

Fig. 2. Immunolocalization of type I and type II collagen in adult articular chondrocytes. Monolayer chondrocytes are probed with antibodies specific for type I collagen (**A**) or type II collagen (**B**). Chondrocytes released from alginate suspension culture at 2 wk (**C**), 4 wk (**D**), and 6 wk (**E**) are probed for type II collagen. Normal rabbit IgGs are used as a negative control primary antibody on 6-wk alginate cultures (**F**). Hoechst dye is used to detect the cell nucleus (**B–F**).

in monolayer culture. When monolayer-grown chondrocytes are placed in semisolid media, like alginate (**Fig. 1B**), the morphology is rounded, as in whole tissue.
3. Expression of antigens: During monolayer culture, chondrocytes turned off type II collagen and began to express type I collagen, which is consistent with dedifferentiation and a fibroblastic expression pattern. Using a semisolid media-culturing system, we are able to verify that chondrocytes expanded in vitro, expressing markers consistent with dedifferentiation, are capable of redifferentiating to express markers consistent with articular cartilage (*see* **Note 6; Fig. 2**).

4. Notes

1. Cartilage is harvested from the femoral condyle or tibial plateau of fresh cadavers (18–60 yr old). Cartilage specimens are transported in serum-free DMEM,

and stored at 4°C within 48 h of harvest. Cells are isolated by enzymatic digestion within 48 h of arrival at our laboratory.

2. The use of 0.1% collagenase overnight at 4°C, or 0.25% collagenase for 2–3 h at 37°C, is sufficient for maximal chondrocyte release from cartilage tissue. Increased chondrocyte recovery is not observed with the use of hyaluronidase *(15)* or DNase *(10)*.

3. Freshly isolated chondrocytes have been observed to require 2–3 d for attachment to the culture substratum. During early culture, microscopic inspection revealed the presence of a grainy substance, which is visually consistent with microbial contamination. However, this material is observed to dissipate when the culture media is replenished, and is judged to be residual pericellular cartilage matrix.

4. Articular chondrocytes have been shown to reversibly alter their gene expression profile in response to their three dimensional environment *(11)*. Once freed from their cartilage matrix and cultured in monolayer, chondrocytes decreased their expression of type II collagen and large aggregating proteoglycans (the key components of hyaline articular cartilage matrix), and began to proliferate rapidly and express a fibroblastic phenotype. Re-expression of hyaline articular cartilage markers has been demonstrated both in vitro *(11,12)* and in vivo *(12)*.

5. RNase protection assays are performed on total mRNA, isolated from monolayer cultured chondrocytes and alginate cultured chondrocytes harvested at timed intervals (**Fig. 3**). Total mRNA is probed with specific reverse transcribed sequences of the following genes: aggrecan, type II collagen, type I collagen, and 18S rRNA. RNA integrity and yield is controlled for by using 18S rRNA cDNA, as an internal standard. Chondrocytes grown in monolayer cultures expressed type I collagen mRNA. Following anchorage independent culture in alginate, the chondrocytes switched their collagen gene expression to predominantly type II collagen mRNA. All normal adult articular chondrocyte strains examined demonstrated switching on of type II collagen mRNA by 1–2 wk in suspension culture (data not shown).

6. Analysis of the ultimate functional potential of cultured cells is critical for providing high quality tissue engineered products for implantation. The approach presented here, for growing healthy cells and confirming their ability to function (differentiate), could be applied to any ex vivo cell expansion and implantation system.

References

1. Bert, J. M. and Maschka, K. (1989) The arthroscopic treatment of unicompartmental gonarthrosis: a five-year follow-up study of abrasion arthroplasty plus arthroscopic debridement alone. *Arthroscopy* **5**, 25–32.
2. Childers, J. C. and Ellwood, S. C. (1978) Partial chondrectomy and subchondral bone drilling for chondromalacia. *Clin. Orthop.* 114–120.
3. Coutts, R. D., Woo, S., Amiel, D., von Schroeder, H. P., and Kwan, M. K., (1992) Rib periochondral autografts in full-thickness articular cartilage defects in rabbits. *Clinical Orthop.* **275**, 263–273.

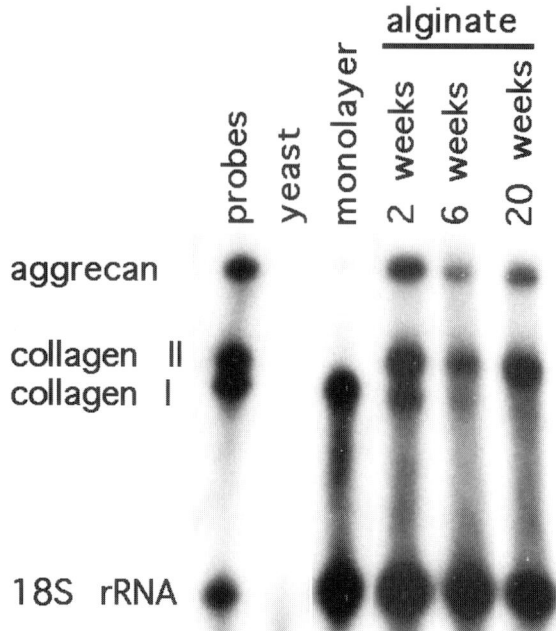

Fig. 3. RNase protection analysis of articular chondrocytes in alginate suspension. Total RNA from human articular chondrocytes growing in monolayer or in alginate suspension are hybridized with antisense RNA probes for aggrecan, type I- and type II- collagen, and 18S ribosomal RNA. Following hybridization, mixtures are subjected to RNase digestion. Protected fragments, specific for the probes, are separated on a 7 M urea containing 4% polyacrylamide gel and visualized by autoradiography. (Lane 1, undigested probes; lane 2, yeast RNA; lane 3, monolayer chondrocytes; and lanes 4–6, alginate suspension cultures at 2, 6, and 20 wk, respectively).

4. Grande, D. A., Pitman, M. I., Peterson, L., Mensch, D., and Klein, M. (1989) The repair of experimentally produced defects in rabbit articular cartilage by autologous chondrocyte transplantation. *J. Orthop. Res.* **7**, 208–218.
5. Homminga, G. N., Bulstra, S., Boumeester, P. M., and Van Der Linden, A. J. (1990) Perichondral grafting for cartilage lesions of the knee. *J. Bone. Joint. Surg. Br.* **72**, 1003–1007.
6. Johnson, L. L. (1986) Arthroscopic abrasion arthroplasty historical and pathologic perspective: present status. *Arthroscopy* **2**, 54–69.
7. O'Driscoll, S. W. and Salter, R. B. (1984) The induction of neochondrogenesis in free intra-articular periosteal autograft under the influence of continuous passive motion. *J. Bone. Joint. Surg.* 1248–1257.
8. Rodrigo, J. J., Steadman, J. R., Silliman, J. F., and Fulstone, H. A. (1994) Improvement of full-thickness chondral defect healing in the human knee after debridement and microfracture using continuous passive motion. *Am. J. Knee Surg.* **7**, 109–115.

9. Wakitani, S., Goto, T., Pineda, S. J., Young, R. G., Mansour, J. M., Caplan, A. L., and Goldberg, V. M. (1994) Mesenchymal cell-based repair of large, full-thickness defects of articular cartilage. *J. Bone. Joint. Surg.* **74,** 579–592.
10. Brittberg, M., Lindahl, A., Nilsson, A., Ohlsson, C., Isaksson, O., and Peterson, L. (1994) Treatment of deep cartilage defects in the knee with autologous chondrocyte transplantation. *N. Eng. J. Med.* **331,** 889–895.
11. Benya, P. and Schaffer, J. D. (1982) Dedifferentiated chondrocytes reexpress the differentiated collagen phenotype when cultured in agarose gels. *Cell.* **30,** 215–224.
12. Shortkroff, S., Barone, L., Hsu, H. P., Wrenn, C., Gagne, T., Chi, T., et al. (1996) Healing of chondral and osteochondral defects in a canine model: the role of cultured chondrocytes in regeneration of articular cartilage. *Biomaterials* **17,** 147–154.
13. Buckwalter, J. A. and Mow, V. C. (1992) Cartilage repair in osteoarthritis, in *Osteoarthritis, Diagnosis, and Medical/Surgical Management,* vol. 2, (Saunders, W. B., ed.), Philadelphia, pp. 71–107.
14. Aulthouse, A. L., Beck, M., Griffey, E., Sanford, J., Arden, K., Machado, M. A., and Horton, W. A. (1989) Expression of the human chondrocyte phenotype in vitro. *In Vitro Cell. Dev. Biol.* **25,** 659–668.
15. Watt, F. M. and Dudhia, J. (1988) Prolonged expression of differentiated phenotype by chondrocytes cultured at low density on a composite substrate of collagen and agarose that restricts cell spreading. *Differentiation* **38,** 140–147.
16. Bonaventure, J., Kadhom, N., Cohen-Solal, L., Ng, K. H., Bourguignon, J., Lasselin, C., and Freisinger, P. (1994) Reexpression of cartilage-specific genes by dedifferentiated human articular chondrocytes cultured in alginate beads. *Exp. Cell Res.* **212,** 97–104.
17. Doege, K. L., Sasaki, M., Kimura, T., and Yamada, Y. (1991) Complete coding sequence and deduced primary structure of the human cartilage large aggregating proteoglycan, Aggrecan. *J. Biol. Chem.* **266,** 894–902.
18. Kuivaniemi, H., Tromp, G., Chu, M. L., and Prockop, D. J. (1988) Structure of a full-length cDNA clone for the prepro alpha2 (I) chain of human type I procollagen. Comparison with the chicken gene confirms usual patterns of gene conservation. *Biochem. J.* **252,** 633–640.
19. Baldwin, C. T., Reginato, A. M., Smith, C., Jiminez, S. A., and Prockop, D. J. (1989) Structure of cDNA clones coding for human type II procollagen. The alpha 1 (II) chain is more similar to the alpha 1 (II) chain than to other alpha chains of fibrillar collagens. *Biochem. J.* **262,** 521–528.

17

Organogenesis of Skeletal Muscle in Tissue Culture

Herman Vandenburgh, Janet Shansky, Michael Del Tatto, and Joseph Chromiak

1. Introduction

Skeletal muscle structure is regulated by many factors, including nutrition, hormones, electrical activity, and tension. The muscle cells are subjected to both passive and active mechanical forces at all stages of development, and these forces play important but poorly understood roles in regulating muscle organogenesis and growth. For example, during embryogenesis, the rapidly growing skeleton places large passive mechanical forces on the attached muscle tissue. These forces not only help to organize the proliferating mononucleated myoblasts into the oriented, multinucleated myofibers of a functional muscle, but also tightly couple the growth rate of muscle to that of bone. Postnatally, the actively contracting, innervated muscle fibers are subjected to different patterns of active and passive tensions that regulate longitudinal and cross-sectional myofiber growth. These mechanically induced organogenic processes have been difficult to study under normal tissue culture conditions, resulting in the development of numerous methods and specialized equipment to simulate the in vivo mechanical environment *(1–4)*. These techniques have led to the engineering of bioartificial muscles (organoids), which display many of the characteristics of in vivo muscle, including parallel arrays of postmitotic fibers organized into fascicle-like structures with tendon-like ends. They are contractile, express adult isoforms of contractile proteins, perform directed work, and can be maintained in culture for long periods. The in vivo-like characteristics and durability of these muscle organoids make them useful for long-term in vitro studies on mechanotransduction mechanisms, and on muscle atrophy induced by decreased tension *(5–8)*. They have also been used as an implantable cell based device for the systemic delivery of recombinant proteins, such as growth hormone from postmitotic muscle fibers *(9)*. In this report, we describe a simple method for generating muscle organoids from either primary embryonic avian or neonatal rodent myoblasts. No complex mechanical stretching apparatus is needed, the procedures use readily available inexpensive materials, and large numbers of uniform in vivo-like organoids can be formed.

2. Materials
2.1. Well Construction

1. Silicone rubber sheeting: vulcanized, nonreinforced, 0.01 in. thick (Dow Corning, Silastic™; or Silicone Speciality Fabricators, Pasco Robles, CA) (*see* **Note 1**).
2. Silicone rubber tubing: 3/16 in. id × 5/16 in. od, medical grade (Baxter Healthcare, Deerfield, IL).
3. General Electric (Waterford, NY) RTV silicone sealant: 4 oz tube (local hardware store).
4. Stainless steel mesh: No. 100 (Newark Wire Cloth, Newark, NJ).
5. Stainless steel grade 316 screws, washers, and hex nuts: washers size 4 (od 5/16 in., id 1/8 in. and od 7/16 in., id 1/8 in.); screws size 4-40 (3/4 in.); hex nuts size 4-40 (1/4 in.) (local hardware store).
6. Velcro™: nonself-adhesive (local sewing center).

2.2. Tissue Cultures, General

1. Collagen: type 1 rat tail (Collaborative Biomedical, Bedford, MA).
2. NUNC tissue culture trays: 22 cm square (USA Scientific Plastics, Ocala, FL).
3. Matrigel™ (Collaborative Biomedical).
4. Airbrush: Badger model 200 with air pump, filter on air line (local art supply store).
5. Falcon 6-well tissue culture plates (Becton Dickinson, Franklin Lakes, NJ).

2.3. Avian Muscle Tissue-Culture Medium

1. 85/10/5 Growth medium: basal Eagle's medium containing 10% horse serum, 5% chicken embryo extract, penicillin (100 U/mL), and amphotericin B (5 µg/mL). Unless otherwise noted, all tissue-culture reagents are purchased from Sigma, St. Louis, MO.

2.4. Mammalian Muscle Tissue-Culture Media

1. Growth medium (GM): Ham's F-10 containing 20% fetal bovine serum, 2.5 ng/mL bFGF, penicillin (100 U/mL), and streptomycin (50 U/mL).
2. Fusion Medium (FM): DMEM (high glucose) with 10% horse serum, and penicillin (100 U/mL).
3. Maintenance medium (MM): DMEM (high glucose) with 10% horse serum, 5% FBS, penicillin (100 U/mL).

2.5. Immunocytochemical Staining

1. Histochoice™ tissue fixative (Amresco, Solon, OH).
2. Anti-sarcomeric tropomyosin (Sigma).

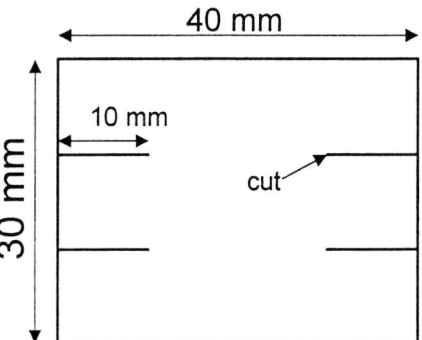

Fig. 1. Well pattern used to cut from silicone rubber sheeting. For multiple well preparation, 40 such patterns can be drawn on one cardboard sheet cut to the size of the silicone sheet (20 × 28 cm) and placed under the silicone rubber.

3. Vectastain™ ABC kit (Vector Labs, Burlingame, CA).
4. Sigmafast™ DAB tablet sets (Sigma).

3. Methods
3.1. Large Culture Well Construction (20 mm long × 10 mm wide × 10 mm deep)

1. A cardboard pattern sheet containing approx 40 well outlines (**Fig. 1**) is placed under three 20 × 28 cm silicone rubber sheets, and the well edges are cut with a scalpel. Three-dimensional wells are formed from the membrane pieces by folding the middle flap at each end up 90 degrees, and the two outside flaps up and in. The RTV silicone sealant is applied to the mating surfaces before folding with a 10-mL syringe containing a small pipet tip in place of a needle. After the adhesive has dried, the wells are cleaned with Pre-Zyme® (VHA, Irving, TX) and rinsed well with tissue culture grade water.
2. Pieces of stainless steel mesh 5 mm wide × 10 mm long are chemically cleaned by sonication in ethanol for 10 min, followed by 10 min sonication in acetone and 10 min sonication in 30% nitric acid. They are then washed with Pre-Zyme and rinsed 3–4× with tissue culture grade water. The pieces are bent 90 degrees in the middle, and glued to the bottom of the culture wells at each end with the RTV silicone sealant (**Fig. 2**). The screening provides a surface to which the cells can attach as they proliferate. Several hundred wells can be constructed in several days and the wells can be reused 2–3× before they begin to leak.

3.2. Notched Bracket Construction and Large Culture Well Attachment

1. The cleaned wells are secured to notched aluminum brackets (**Fig. 2**) using stainless steel screws placed through small holes cut in each end of the wells. A stain-

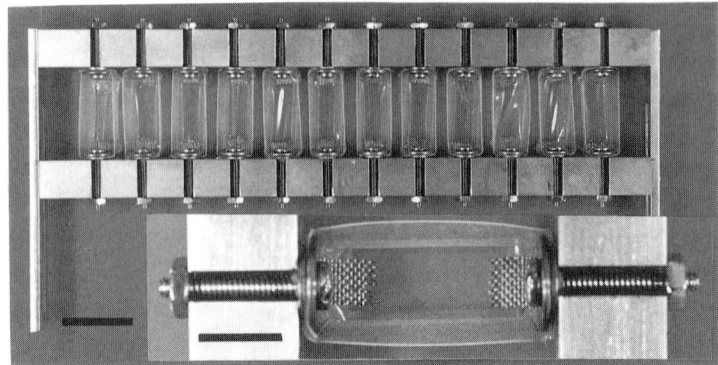

Fig. 2. Large culture wells formed from silicone rubber sheeting and secured to aluminum brackets. Wells measuring 20 mm long × 10 mm wide × 10 mm deep are glued from a precut thin silicone rubber sheet and formed into three-dimensional wells. Wells are secured to notched aluminum brackets using stainless steels screws, hex nuts, and washers. Each bracket holds 12 culture wells, and the spacing between the two notched aluminum bars is adjustable. The inset is an enlargement of one well, and shows the stainless steel screening at each end of the boat that provides a surface for cell attachment. Bars represent 10 and 25 mm in the inset and main figure, respectively. (Adapted with permission from ref. *12*.)

less steel washer, 7/16 in. od, is placed inside the well ends and a 5/16 in. od washer is placed outside the well ends (*see* **Notes 2** and **3**). Each bracket can hold up to 12 culture wells, and consists of two square bars (1 × 1 × 20 cm), with 12 3-mm-wide by 6-mm-deep notches on the top. The brackets are connected at their ends by slotted flat bars (10.5 × 1.2 cm), which allow the distance between notched bars to be adjusted. The wells are placed into the bracket notches and the screws tightened with hex nuts on the outside of the bracket bars to hold the wells in place. For optimal formation of organoids, the bar spacing is set at 30 mm; this stretches the wells approx 50% from their resting length of 20 mm.

2. After the wells are attached to the brackets and stretched, they are rinsed with tissue-culture-grade water, dried, and lightly sprayed 6–8 times with 0.1% (w/v) type 1 collagen in 1% (v/v) glacial acetic acid, using an airbrush sprayer. Thin coats are applied, so that the collagen does not bead, and are dried between coats with a heat lamp. The brackets are placed into large tissue-culture trays, and sterilized with ethylene oxide; they can also be autoclaved in glass Petri dishes (20 min maximum time) if desired, but the silicone rubber membrane should not contact the glass while it is hot (*see* **Note 4**).

3.3. Small Culture Well Construction

1. Silicone rubber tubing is cut into pieces measuring 29 mm in length. These pieces are cut longitudinally and a lengthwise section removed, leaving approximately two-thirds of the original tubing remaining.

2. Velcro™ pieces (3 × 4 mm from looped half) or stainless steel screens (3 × 6 mm, cleaned as described in **Subheading 3.1., step 2** and bent midway at a 90-degree angle) are glued with silicone sealant into the bottom of the tubing at each end, 1–2 mm from the ends.
3. When the sealant is dry, the tubing is glued into 6 well tissue culture dishes. A drop of silicone sealant is placed under each end of the tubing and two are glued side by side in each 35-mm well of the 6-well plate (**Fig. 3A**).
4. When the glue is dry, each well is rinsed 4× with distilled water. Tissue culture grade distilled water is then added to each well, and the plate is placed on a rotary shaker at 40 rpm for 30 min. Wells are drained, fresh tissue culture water is added to each well, and the plates are incubated at 37°C overnight.
5. Plates are drained, air dried, and sprayed with collagen, as described for large wells. They are sterilized with ethylene oxide.

3.4. Muscle Cell Tissue Cultures

1. On the day of culturing, the large wells are preincubated with Earle's balanced salt solution (EBSS) for several hours before cell plating (*see* **Note 5**).
2. For the avian muscle cells, plating density is $5-7.5 \times 10^6$/well in 1 mL of 85/10/5 growth medium. Cultures are fed daily with fresh 85/10/5. Avian myoblasts proliferate and fuse into multinucleated myofibers beginning 48 h after plating, align parallel to the direction of substratum tension, and become striated and contractile by 96–120 h.
3. Three–5 d after plating, the cell layer lifts off the bottom of the silicone rubber wells (while remaining attached to the screens at both ends), and the long edges of the cell layer roll in to form a muscle organoid similar to those formed using more complex stretching equipment (*2*; *see* **Note 6**). These organoids contain organized and contractile myofibers (**Fig. 3B**).
4. For mammalian muscle organoids, myoblasts are isolated from the forelimbs and hindlimbs of rat neonates following the method of Rando and Blau *(10)*. Primary cells are suspended in a chilled 1:6 solution of MATRIGEL: collagen (type 1, 1.6 mg/mL) prepared with GM (*see* **Note 7**). The cell suspension is kept on ice and pipeted into the wells using precooled pipet tips at a concentration of 4×10^6 cells/0.750 mL in the large wells and 2×10^6 cells/0.4 mL in the small culture wells. In the small culture wells with Velcro end supports, the cell:gel suspension is gently worked into the Velcro with the pipet tip, while pipeting the cells into the wells. The wells are placed in a 37°C incubator for 2–6 h to allow the Matrigel–collagen mixture to gel before carefully overlaying with 1 mL GM in the large wells, or 5 mL in the small wells contained within the 6-well plates. Cultures are maintained in GM for 3 d, FM for 3 d, and MM for up to 4 wk. The gel–cell mixture condenses and dehydrates during the first 2–3 d, pulling off the elastic substratum (**Fig. 3A**), and generating internal tensions to align the forming myofibers (**Fig. 3B, C**) (*see* **Note 8**).

3.5. Removal of Muscle Organoids from Cell-Culture Wells

1. The differentiated muscle organoids in the large culture wells can be removed from the brackets without tension release and transferred to other tissue-culture

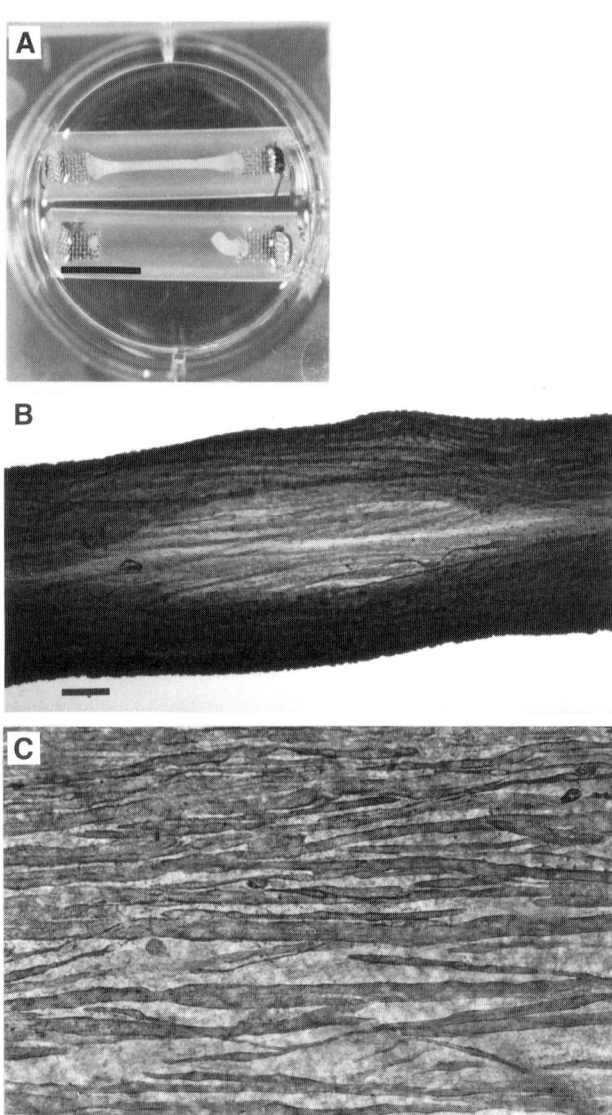

Fig. 3. Skeletal muscle organoids contain parallel arrays of myofibers expressing sarcomeric tropomyosin. Avian and mammalian myoblasts are grown and maintained in the small culture wells that are glued into 35-mm wells of 6-well plates. **(A)** Mammalian organoids, 48 h after plating. The upper organoid is well-attached to screening at the ends of the tubing; the lower organoid has detached from one end, resulting in the release of internal tension. **(B)** and **(C)** After 14–18 d in culture, muscle organoids are fixed and stained with an antibody against sarcomeric tropomyosin, followed by incubation with an avidin-biotinylated secondary antibody coupled to horseradish

chambers for long-term muscle growth/atrophy studies. This is accomplished by placing a stainless steel screen spacer, 1 cm-wide and the length of the culture well, into the wells just above the organoid. The spacer is bent into a shallow inverted U-shape, so that it does not touch the organoid. The wells can then be removed from the brackets by loosening the screws and slipping them out of the top of the bracket notches. Using this technique, muscle organoids can be removed from the brackets, and transferred to other growth chambers (e.g., Petri dishes or perfusion chambers).

2. The muscle organoids in the small culture wells have been used mainly for implantation into animals for growth and survival studies *(9)*. This is accomplished by carefully cutting the sides of the wells as narrow as possible with a sterile scalpel without disturbing the Velcro pieces at each end, to which the muscle organoid is attached. The resulting organoid is held under tension when implanted subcutaneously, and can survive for up to 12 wk in syngeneic animals *(9)*.

3.6. Immunocytochemical Whole-Mount Staining of Muscle Organoids

1. Prior to immunocytochemical staining, organoids are rinsed 3× quickly with PBS (10 mM PBS, pH 7.4, containing 138 mM NaCl and 2.7 mM KCl), fixed for 45–60 min in Histochoice, and rinsed over 30 min with four changes of PBS. At this step, organoids can be stored in PBS at 4°C, or the staining procedure can be continued.

2. Modified Dent's fixative *(11)*, with 1 part DMSO:4 parts 100% ethanol, is added to each organoid, and incubated for 3 h at room temperature, with hourly changes to fresh fixative. This is followed by a 4°C overnight incubation with fixative, and rinsing with PBS at room temperature with four changes over 60 min.

3. An antibody to sarcomeric tropomyosin is diluted 1:100 in 2% serum, 0.2% Tween in PBS (blocking buffer). Organoids are incubated overnight at 4°C with the primary antibody solution, followed by rinsing with PBS at room temperature with four changes over 2 h.

4. The secondary antibody is biotinylated antimouse IgG, and is diluted 1:200 in blocking buffer. Organoids are incubated overnight at 4°C with the secondary antibody solution, followed by rinsing with PBS at room temperature with four changes over 2 h.

5. Avidin/biotinylated horseradish peroxidase complex (A/B solution) is prepared by diluting reagents A and B from the Vectastain ABC kit 1:100 in blocking buffer. Organoids are incubated at room temperature with this solution for 2–3 h, followed by rinsing with PBS at room temperature with four changes over 2 h.

---

peroxidase, and development with diaminobenzidine to form a dark precipitate. **(B)** avian skeletal muscle organoid, whole-mount stained, showing aligned surface myofibers; **(C)** mammalian organoid formed from neonatal rat skeletal myoblasts, with myofiber alignment in the long axis of the organoid. Bars represent 10 mm, 200 µM, and 50 µM in (A), (B), and (C), respectively. (Adapted with permission from ref. *9*.)

6. Diaminobenzidine (DAB) peroxidase substrate solution (prepared according to manufacturer's protocol) is added to the organoids for 5 min at room temperature, forming a brown precipitate in muscle fibers containing sarcomeric tropomyosin. The substrate solution is removed and the organoids are rinsed and stored in distilled water.
7. Immunocytochemically stained organoids are removed from the culture wells, placed onto glass microscope slides, and cover-slipped with glycerol. Myofibers are visualized with a light microscope.

4. Notes

1. Silastic sheeting is translucent and more resilient than transparent silicone rubber sheeting from other manufacturers; therefore, the former is better suited for repetitive stimulation; the latter is better for morphological observations of the live cells.
2. Stainless steel screws and washers are cleaned before use in the same manner as described for the stainless steel mesh.
3. A small amount of RTV sealant can be placed under the washers if leakage around the screws is a problem. This is recommended only if necessary, since it makes reuse of the screws and washers difficult.
4. Sterile, collagen-coated wells can be stored at room temperature for at least 2–3 wk without adverse effects on cell attachment.
5. Small wells are not preincubated with EBSS, since it is essential that they are completely dry when the cell suspension is added, so that the viscous solution does not run off the ends of the tubing.
6. Organoids formed under the static tension methods described in this paper do not differentiate as well as those formed under dynamic stretching conditions *(2)*. For example, they do not form fascicles and the myofibers are not as densely packed.
7. The extracellular matrix gel required for mammalian, but not avian, organoid development may be because of the greater proliferative ability of connective-tissue-forming fibroblasts in the latter than the former cultures (unpublished observation).
8. Myoblast cell lines such as C2C12, which are devoid of fibroblasts, can also be formed into organoids, but require the Matrigen–collagen mixture used for primary mammalian cells.

Acknowledgment

Supported by NASA Grants NAGW-4674 and NAG2-914.

References

1. Vandenburgh, H. H. and Karlisch, P. (1989) Longitudinal growth of skeletal myotubes in vitro in a new horizontal mechanical cell stimulator. *In Vitro* **25,** 607–616.
2. Vandenburgh, H. H., Swasdison, S., and Karlisch, P. (1991) Computer aided mechanogenesis of skeletal muscle organs from single cells in vitro. *FASEB J.* **5,** 2860–2867.

3. Strohman, R. C., Byne, E., Spector, D., Obinata, T., Micou-Eastwood, J., and Maniotis, A. (1990) Myogenesis and histogenesis of skeletal muscle on flexible membranes in vitro. *In Vitro Cell. Dev. Biol.* **26,** 201–208.
4. Swasdison, S. and Mayne, R. (1992) Formation of highly organized skeletal muscle fibers *in vitro*: comparison with muscle development *in vivo*. *J. Cell. Sci.* **102,** 643–652.
5. Vandenburgh, H. H., Hatfaludy, S., and Shansky, J. (1989) Skeletal muscle growth is stimulated by intermittent stretch/relaxation in tissue culture. *Am. J. Physiol.* **256**(*Cell Physiol.* **25**), C674–C682.
6. Chromiak, J. A. and Vandenburgh, H. H. (1992) Glucocorticoid-induced skeletal muscle atrophy in vitro is attenuated by mechanical stimulation. *Am. J. Physiol. Cell. Physiol.* **262,** C1471–C1477.
7. Vandenburgh, H. H., Shansky, J., Karlisch, P., and Solerssi, R. L. (1993) Mechanical stimulation of skeletal muscle generates lipid-related second messengers by phospholipase activation. *J. Cell. Physiol.* **155,** 63–71.
8. Vandenburgh, H. H., Shansky, J., Solerssi, R., and Chromiak, J. (1995) Mechanical stimulation of skeletal muscle increases prostaglandin $F_{2\alpha}$ production, cyclooxygenase activity, and cell growth by a pertussis toxin sensitive mechanism. *J. Cell. Physiol.* **163,** 285–294.
9. Vandenburgh, H. H., Del Tatto, M., Shansky, J., LeMaire, J., Chang, A., and Payumo, F., et al. (1996) Tissue engineered skeletal muscle organoids for reversible gene therapy. *Human Gene Ther.* **7,** 2195–2200.
10. Rando, T. A. and Blau, H. M. (1994) Primary mouse myoblast purification, characterization, and transplantation for cell-mediated gene therapy. *J. Cell. Biol.* **125,** 1275–1287.
11. Dent, J. A., Polson, A. G., and Klymkowsky, M. W. (1989) A whole-mount immunocytochemical analysis of the expression of the intermediate filament protein vimentin in Xenopus. *Development* **105,** 61–74.
12. Shansky, J., DelTatto, M., Chromiak, J., and Vandenburgh, H. (1997) A simplified method for tissue engineering skeletal muscle organoids in vitro. *In Vitro Cell Dev. Biol.*, in press.

18

Hepatocyte Primary Cultures

Currently Used Systems and Their Applications for Studies of Hepatocyte Growth and Differentiation

George K. Michalopoulos

1. Introduction

Liver tissue has been used to study regeneration and carcinogenesis, as well as a whole array of liver-specific differentiated functions, including regulation and synthesis of blood proteins, coagulation factors, lipoproteins. The multiple homeostatic functions supported by liver mandate that liver is involved in studies affecting the regulation of components of particular circuits of homeostatic regulation. Given the complexity of studies in vivo, in which multiple interacting factors often defy a logical analysis of the results, establishment of primary cultures of hepatocytes was intended to provide a system in which all parameters for these studies could be controlled. One of the objectives leading to the creation of the early primary cultures was the establishment of conditions that would lead to the controlled growth and clonal expansion of differentiated hepatocytes. This "holy grail" of hepatocyte cultures has not been yet fully achieved. It has, however, become the guiding vision that led to numerous studies that yielded very useful knowledge on hepatic differentiation, regeneration, and carcinogenesis. In the more than 20 yr since the first primary cultures of hepatocytes *(1)*, and with literally thousands of publications using hepatocyte cultures, primary hepatocyte cultures have fulfilled most of the original intentions of their design. It was discovered by the use of hepatocyte cultures that epidermal growth factor (EGF) *(2)* and acidic fibroblast growth factor (aFGF) *(3)* are mitogens for hepatocytes. In addition, DNA synthesis in primary cultures of hepatocytes was used as the end point that led to the identification of a new growth factor, hepatocyte growth factor (HGF) *(4–5)*. Similar techniques were used to identify the mitoinhibitory effect of TGF-β1 and other cytokines, including IL-6 and TNF-α, on hepatocyte growth and differentiation. The tremendous amount of information generated by use of the hepatocyte cultures has led to some standardization of the employed techniques. At the same time, however, it led to many well established complex culture systems that were designed to overcome the limitations of the standard methodology. The purpose of this chapter is to discuss the flexibility that can be allowed in using the standard techniques, and the role of different variations in these techniques in the final outcome, in terms of quality and function of the hepatocyte cultures.

2. Hepatocyte Isolation

Perfusion of rat liver with collagenase has now become the standard tool for isolation of hepatocytes. Despite original and very complex schemes and buffer solutions involved in this process, the simple two-step calcium (Ca)–collagenase approach first described by Seglen *(6)*, as a modification of the original technique by Berry and Friend *(7)*, has become the standard method for the isolation for these cells. The methodologic details are described in **Subheading 6**. Typically, a simple Ca-free buffer is perfused through the inferior vena cava and enters the liver through the hepatic veins. In reversing the standard circulation route, portal vein is used as the outflow tract. The purpose of the Ca-free buffer is to remove most of the hepatic Ca, thus forcing a partial dissociation of the hepatocytes caused by break down of desmosomes into hemi–desmosomes. The second buffer typically contains Ca plus collagenase. Several types of collagenase have been used by different studies. The lack of standardization of nomenclature of these collagenases has made it difficult to compare the results of different brand names and methods of collagenase preparation. Typically, most of the commercial brands of collagenase described as working with collagenase perfusion of liver fulfill this objective. In view of the variability of the different brand products and the rather crude preparation methods involved in preparing collagenases for tissue dissociation, most laboratories prefer to do testing of a batch of collagenase from a specific supplier, which, after being purchased in large amounts, is used in a reproducible manner for multiple isolations of hepatocytes.

Variations of this approach have been tried on liver tissue from other organisms. The mouse technique is essentially the same as that for the rat. Adaptations of the collagenase technique have been published for human liver by several groups *(8–9)*. Typically, veins that are accessible on the cut surface of the hepatic tissue, or complete, anatomically preserved vascular channels, are used to perfuse human liver in a two-stage collagenase technique. The isolated cells are separated into hepatocytes and nonparenchymal cells by simple low-gravity centrifugation, as described in **Subheading 6**. Although the collagenase perfusion technique is overall highly reliable for the rat, the most important determinant of the outcome of the perfusion of human liver tissue is the degree of preservation of the original hepatic tissue. Usually, there is considerable variability in the results obtained when perfusing human liver. Hepatocytes isolated from perfusing rodent liver, or liver of other animals sacrificed specifically for that objective, typically have a very high viability. The viability of hepatocytes isolated from perfusion of human liver varies substantially, based on preservation damage inflicted on the original organ. An excellent recent review for production of mass numbers of porcine hepatocytes was also published recently *(10)*.

3. Standard Types of Primary Cultures of Hepatocytes

In these cultures, hepatocytes are maintained in simple media as sparse or dense monolayers on rigid substrates, such as plain plastic or collagen–coated surfaces. Given the limited stability of these cultures, they tend to be used for experiments during the first week after cell inoculation.

3.1. Dense Monolayers of Hepatocyte Cultures

Monolayer cultures composed of hepatocytes in high density offer specific advantages for the study of the regulation of specific hepatocyte components. The large number of cells in a dense monolayer ensures that the object of study is of sufficient quantity to yield reproducible results. Typically, 100,000 rat hepatocytes inoculated per cm^2 of culture establish a complete monolayer under conditions of optimal attachment. These conditions usually include coating of the plastic surface with type I collagen and addition of 5% fetal bovine serum (FBS), although near-optimal attachment can be achieved in serum-free conditions. The advantage of serum as an attachment-enhancing factor is often mitigated by complications in interpretation of results. Serum itself contains many cytokines with specific effects that can interfere with the interpretation of results. Most experiments with serum-supplemented media often forget that addition of serum in the control cultures is not sufficient to ensure that the observed results are directly caused by the addition of the tested cytokine in the experimental cultures. The tested cytokine may affect the disposition of one of the serum components of the hepatocytes, and that may indirectly be the reason for the observed effects. Even under the best conditions of attachment (type I collagen coating, as well as addition of serum), the monolayers are never complete. There are gaps and patches of empty areas, in an otherwise topologically continuous monolayer of hepatocytes. The monolayers usually become established within 24 h. Hepatocytes tend to have dome-shaped morphology for the first 12 h after attachment. At the end of the first day, they spread and acquire a larger surface. This enables them to make contact with each other, re-establish bile canaliculi, and become a continuous monolayer of cells. This apparent expansion of the surface occupied by single hepatocytes led in the past to an erroneous interpretation of cell growth. It should be emphasized that all of these changes in primary cultures occur in the absence of DNA synthesis, and they are a result of phenotypic transformation of cells, changing their volume and shape as a result of the culture conditions. The monolayers remain intact for 4–7 d. This is the useful time for experimental studies. Degenerative changes ensue, typically within 7 d for rat hepatocytes. The intercellular connections become severed, the cells lose the cytoplasmic granules, many of them undergo apoptosis, and become detached. Those that remain attached lose their characteristic morphology. Contaminating nonparenchymal cells, which at the time of isolation do not comprise more than 1–2% of the total cell population, gradually increase in numbers, even in the absence of serum. Hepatocytes produce several growth factors in vivo and in vitro, and they facilitate growth of contaminating cells. Typically, such contaminating cells are cells of Ito, which transform in culture into myofibroblasts *(11)*. Endothelial cells and bile duct cells are also occasionally seen as contaminating cell populations. These contaminating nonparenchymal cells tend to overpopulate the hepatocytes within 2 wk in culture. Although the above changes are typical for rat hepatocytes, it should also be pointed out that the rate of appearance of these changes varies from species to species. Human hepatocyte monolayers, once established, tend to stay intact for several weeks.

3.2. Studies on Hepatocyte Growth Regulation in Sparse Cultures

Hepatocytes in dense monolayers offer advantages for biochemical research. On the other hand, very little growth can be stimulated in such dense monolayer cultures. Several studies in the past have shown that there is an inverse correlation between hepatocyte density and the capacity to enter into DNA synthesis *(12–13)*. These studies have also shown that several hepatocyte functions become enhanced in dense cultures. Typically, hepatocytes in sparse cultures (less than 10,000 cells/cm^2) are used to study hepatocyte growth. These cultures have been a successful tool in generating information relevant to liver-growth biology and regeneration. Hepatocyte density at 10,000 cells inoculated/cm^2 is the optimal to maximize the rate of entry into DNA synthesis. In cell densities less than 10,000, DNA synthesis becomes diminished for reasons not understood *(13)*. This may reflect a certain conditioning of the tissue-culture medium by the inoculated cells that is required to allow better preservation of the viable cells. Different varieties of minimal essential or well supplemented media have been used for these studies. Most of these studies include 10^{-7} M insulin as a standard additive. In the absence of insulin, DNA synthesis induced by growth factors is substantially diminished. Usually, growth factors are added at the first or second d in culture. DNA synthesis reaches a peak at 24–72 h after addition of growth factors, and can be easily monitored by either thymidine incorporation into DNA or tritiated thymidine autoradiography. Autoradiography is an important complement in studies involving DNA synthesis. Contaminating cells often enter into DNA synthesis in conditions under which hepatocytes do not, even in the absence of FBS *(11)*. This can lead to the wrong conclusion that the observed cell proliferation is caused by proliferation of hepatocytes. In addition, several past studies have shown that tritiated thymidine degradation products can enter into protein and lipid components and create the erroneous impression of DNA synthesis, based on results obtained by merely measuring tritiated thymidine incorporated into TCA precipitable material *(14–15)*.

A select few cytokines can stimulate substantial DNA synthesis in hepatocytes in sparse monolayer cultures. The amount of DNA synthesis stimulated by these few cytokines typically exceeds five-fold over background. These growth factors are HGF *(4–5)*, EGF *(2)* and TGF-α *(16)*. These three polypeptides are the strongest mitogens for hepatocytes in primary culture, and are referred to as "primary mitogens". Weaker mitogenic effects have been observed for acidic fibroblast growth factor *(3)*. Several small molecules, such as norepinephrine *(17)*, estrogens *(18)*, and angiotensin II *(19)*, among others, have been shown to enhance the mitogenic effect of the primary mitogens. Because of this property, these compounds have been referred as "co-mitogens" or "secondary mitogens." TGF-β1 at sufficiently high concentrations can completely inhibit the mitogenic effect of these growth factors. TGF-β1 also induces apoptosis in hepatocyte cultures, especially when added in the absence of mitogenic factors in the medium *(20)*.

The DNA synthesis stimulated by the primary mitogens lasts for 3–4 d. After that time, degenerative changes ensue that resemble those seen in dense monolayers. Hepatocytes lose their characteristic appearance of granular cytoplasm and bright round nucleus. The cells become clear and have a fibroblastic morphology. Many of these cells die by apoptosis, and the overall number of cells decreases rapidly, as other nonparenchymal cells eventually start growing in small patches after d 10–11 in culture. It should be emphasized that, whereas the contaminating nonparenchymal cells grow at late culture, the DNA synthesis observed at the early cultures (first 5–6 d) is composed almost entirely of hepatocytes. This has been shown repeatedly by autoradiography. The reasons leading to hepatocyte death after a period of rapid growth, or in dense, non-replicating, monolayer cultures, are not clear. Several approaches have been developed that overcome these effects, as described below.

3.3. Gene Expression Changes in Standard Hepatocyte Cultures

Isolated hepatocytes undergo profound changes in gene expression patterns when placed in primary culture. Primarily, these are loss of differentiated functions and enhanced expression of cell-cycle-related genes. In contrast to quiescent hepatocytes in vivo, hepatocytes in culture express cell-cycle-related genes, such as jun, fos, myc, ras, and so on *(21)*. This is more evident in sparse cultures than dense monolayers. Detailed experimental studies have documented that the collagenase step of the liver-perfusion technique is the time when these changes appear to be initiated. Thus, hepatocytes entering in culture are already in a state resembling the G_1 phase of the cell cycle. The multiplicity of gene expression changes, especially the overall decline of the expression of hepatic differentiation functions, cannot, however, be ascribed solely to an apparent change in cell cycle status. Hepatocytes in regenerating liver maintain most of their specific functions, even though they progress through the S-phase and enter into mitosis. The gene expression changes in cultured hepatocytes are profound and effect practically all aspects of hepatocyte differentiation. These include albumin expression, coagulation proteins, cytochrome P450, lipoprotein expression, and, in general, any of the functions that are deemed as characteristic of the hepatocytic phenotype. Typically, the loss of these functions is also associated with a decrease and eventual elimination of the mRNA populations involved in the translation and production of the specific proteins. The decrease in mRNA is most often the result of decreased gene expression, although, in some instances, mRNA stability also appears to be a factor. Gene expression is maintained longer in dense monolayer cultures than in sparse monolayers. The reasons for these changes are not clear. Several remedies are used that partially correct the observed deficiencies. These remedies offer clues to potential mechanisms that lead to such profound effects.

4. Hepatocyte Cultures with Enhanced Phenotypic Stability

Cell hepatocyte culture systems have been described that overcome the previously mentioned changes in gene expression. These are as follows:

4.1. Systems Supplemented with Complex Matrix

Hepatocyte functions are maintained for a longer period of time when hepatocytes are cultured on top of collagen gels *(22)*. Cytochrome P450, albumin expression, and a variety of other cultures including induction of tyrosine aminotransferase (TAT), have been studied in these systems. The electron microscopic analysis of the cells showed that their morphological characteristics were also stable for prolonged periods of time. This was the first application using collagen gels for primary cultures of epithelial cells, and it was subsequently applied to many other epithelial cell-culture systems, including squamous cells, mammary epithelial cells, and so on *(23)*. The technique was further enhanced when hepatocytes were maintained as a sandwiched monolayer between two layers of collagen gels *(24)*. These cultures are characterized by even more enhanced phenotypic stability and maintenance of differentiated functions. The monolayers are relatively stable for prolonged periods of time, although they do show a progressive decline in some differentiated functions, notably the expression of albumin, expression of cytochrome P450, and so on. DNA synthesis can be stimulated in sandwiched hepatocytes by primary hepatocyte mitogens (HGF, EGF, and TGF-α). The levels of DNA synthesis stimulated are below those seen in cultures with the same cell density grown on rigid or collagen coated substrates *(25)*. Hepatocytes do not invade the collagen gels; they stay within the monolayer sandwich. To the contrary, when hepatocytes are mixed within the collagen gel and maintained in the gel matrix, in the hands of most investigators the viability is substantially diminished and the cells rapidly die.

The studies with the type I collagen gels documented that matrix is essential in maintaining hepatocyte differentiation. This paradigm was carried one step further by illustrating that several hepatocyte-specific functions could be maintained on hepatocytes cultured over patches of hepatic biomatrix extracted from rodent liver and applied in a discontinuous manner over a plastic surface. The cells over the patches of biomatrix maintained differentiation much longer than the cells on the plastic surface areas in the same dish. The components of the hepatic biomatrix associated with this effect have not been identified.

Phenotypic differentiation and gene expression patterns are even better enhanced when hepatocytes are inoculated over or under a layer of matrix extracted from EHS mouse sarcoma tumor *(26)*. This biomatrix, commonly known as Matrigel™, is now used extensively in many culture systems. Matrigel has a high concentration of laminin and glycosaminoglycans. It also includes cytokines such as TGF-β1. Hepatocytes maintained differentiated functions for prolonged periods of time when cultured under or in the midst of Matrigel matrix *(40)*. Hepatocytes tend to remain spheroidal and do not form extensive monolayers. The morphologic effects are quite pronounced, and the stability is superior to that of the other systems. Notable, however, is the inability to stimulate DNA synthesis by any of the standard mitogens in hepatocytes maintained in Matrigel *(27)*. It was originally thought that this is because of the contaminating presence of TGF-β1. However, several commercial preparations containing Matrigel with substantially depleted concentrations of TGF-β1 continue to have the same effects. The components of Matrigel that inhibit hepatocyte proliferation are not clear, and it is quite likely that the effect is caused by the overall composition of the matrix, rather to any other specific components. Many of the matrix signals are transmitted via complex interactions between matrix components and specific cellular receptors on the plasma membrane, called integrins. Despite the extensive literature or the effects of matrix on hepatocyte differentiation in primary culture, and an equally extensive literature on the expression of specific integrins in hepatocytes (notably the α5β1), very little is currently understood about the role of these integrins and the specific components of the matrices as mediators of phenotypic stability. In these stable cultures, the rate of mRNA transcription is dramatically enhanced and many differentiated functions are maintained, including induction of members of the cytochrome P450 family by phenobarbital, synthesis of a variety of proteins, and so on.

4.2. Cultures Supplemented with 2% DMSO

This system, described by Isom and coworkers *(28)*, allows formation of hepatocyte monolayers with phenotypic stability and maintenance of gene-expression patterns in chemically defined media supplemented with 2% DMSO. The optimal effect is seen when the chemically defined medium described by Isom et al. is used. The effect of 2% DMSO, however, on phenotypic stability of hepatocytes is noted, even when hepatocytes are cultured in simpler media, such as MEM *(29)*. The phenotypic stability is quite prolonged, and morphologic stability of up to 1 yr has been recorded. The components of the chemically defined medium include EGF and insulin. Very little DNA synthesis is seen in 2% DMSO, even with the presence of EGF. When DMSO is removed while EGF is retained, there is an abrupt rise in DNA synthesis that is well synchronized and lasts for 2–4 d. The phenotypic stability of the monolayer can be re-established when 2% DMSO is added back to the cultures *(29)*. Conversely, when EGF is removed in the presence of DMSO, there is an enhanced rate of cellular apoptosis. When EGF is reintroduced into the system, DNA synthesis can be stimulated, even in the presence of 2% DMSO *(30)*. The mechanism regulating this apparent homeostasis of the monolayer in vitro are not clear. The model is very useful, however, for studies of hepatocyte proliferation and differentiation.

4.3. Spheroids

In contrast to the phenotypic stabilization imparted by complex matrix systems, in other cultures, phenotypic stability is promoted when the cultured dishes are specifically treated to prevent attachments of the inoculated hepatocytes. Under these conditions, hepatocytes form aggregates that vary in size. Typically, these aggregates are maintained in rotating or gyrating cultures. Because of their shape, the term spheroids has been used to describe them. The viability of the spheroids is inversely proportional to their size. Large spheroids have areas of central necrosis, but small spheroids maintain viability throughout the entire spheroid. The formation of spheroids and their stability appears to be dependent on the medium used. Limited DNA synthesis is seen in general, although some DNA synthesis can be stimulated by primary hepatocyte mitogens. Hepatocyte-specific functions, such as albumin and cytochrome P450, have are well maintained in these systems *(31)*. The spheroids lose their three-dimensional architecture and transform into monolayers when they are transferred from nonadherent into adherent substrates, such as with collagen-coated plastic dishes. The spheroids rapidly spread and become transformed into patches of monolayers, which then undergo phenotypic differentiation. Although the mechanisms that promote phenotypic stability in spheroids are not entirely clear, it is assumed that the association of hepatocytes with each other results in synthesis of new matrix that is hepatocyte specific. Thus, hepatocytes are surrounded by matrix that is not derived from other biological systems but produced by hepatocytes themselves. Although the spheroids are composed primarily of hepatocytes, other cells that often contaminate hepatocyte cultures might also a play a role in the synthesis of the matrix that maintains the cohesiveness of the spheroids.

4.4. Mixed Cultures of Hepatocytes with Other Cells

Marked phenotypic stabilization of hepatocytes is also achieved when they are co-cultured with another cell type typically described as "liver epithelial cell" or "clear epithelial cell." These cells typically present as small patches of cells with clear cytoplasm, and are present from the time of the inoculation of suspensions obtained from hepatocyte collagenase perfusion. Their tendency to aggregate in patches and their clear cytoplasm easily distinguish them from primary hepatocytes. At the early days of hepatocyte cultures, however, many publications confused these cells with proliferating hepatocytes and many liver-cell lines were established based on these cells. Although the origin of these cells is not entirely agreed upon, the expression of vimentin and their overall appearance suggests that they are derived from the mesothelial capsule surrounding the liver. The most telling evidence of this association is the fact that these cells, in contrast to biliary epithelium (considered as alternative origin), are negative for GGT after their isolation, and they tend to be present as cell patches from the very beginning of the cultures.

Regardless of the origin of these liver epithelial cells, their addition to subconfluent hepatocyte monolayers confers long-term phenotypic stability on hepatocytes. The system was first introduced and best described by Guguen-Guillouzo et al. *(32)*. Hepatocytes are first inoculated in numbers sufficient to form subconfluent monolayer cultures (e.g., 50,000 cells/cm^2, in collagen-coated dishes in MEM media supplemented with 5% FBS). One day after the establishment of patches of hepatocyte monolayers, large numbers of clear epithelial cells are introduced, which attach and fill the gaps between the patches of hepatocytes. The combined cultures eventually become a completely confluent monolayer. There is extensive new matrix synthesis in the interface between the clear epithelial cells and the hepatocytes. The matrix synthesis is predominantly by the clear epithelial cells, suggesting that the hepatocytes may have an inductive role in this process *(33)*. Studies by Gebhardt have also shown that hepatocytes located immediately proximal to the clear epithelial cells become positive for glutamine synthetase, an enzyme only seen in vivo in the single layer of cells surrounding the central veins of the hepatic lobules *(34)*. Other co-culture systems have also been tried, such as growth of hepatocytes over feeder layers such as 3T3 cells, and so on. Particularly innovative was the technique of stimulating growth of hepatocytes from livers of rats treated with initiating carcinogens. These cells can be stimulated to grow spontaneously over feeder layers of 3T3 cells, only in the presence of the tumor promoter phenobarbital *(35)*. Enhanced phenotypic stability is also achieved when hepatocytes are grown over monolayers of skin fibroblasts *(36)*. This stability is less than that seen with co-cultures with clear epithelial cells.

It appears that, rather than a specific single component, it is the overall matrix synthesized by the nonhepatocyte components of the co-culture systems that is responsible for the enhancement of phenotypic stability. Conditioned media from co-cultures do not confer stability in hepatocyte monolayers. This shows that the co-culture effect is not the result of soluble factors synthesized by the nonhepatocyte components, but rather because of the matrix that these cells produce.

5. Hepatocyte Cultures with Clonal Growth

As noted above, proliferating hepatocytes in sparse monolayers and in media containing primary hepatocyte mitogens usually enter into degenerative changes and die by apoptosis within 4–7 d after stimulation of DNA synthesis. Typically, the DNA synthesis is self-limited and stops after about two or three rounds of cell proliferation. There is rarely an increase in hepatocyte numbers, and no sustained clonal growth can be demonstrated. Some of these limitations were recently overcome in cultures composed of chemically defined media containing high concentrations of nicotinamide, dexamethasone, transferrin, insulin, HGF, EGF, and so on. These cultures, originally described by Mitaka and Pitot as colonies of small cells surrounded by typical hepatocyte monolayers *(37)*, become even further enhanced in media supplemented by arginine and glucose *(38)*. The latter supplementation resulted in the formulation of a medium called hepatocyte growth medium (HGM), in which the entire hepatocyte population is transformed into clonally derived confluent colonies of small hepatocytes in the presence of HGF and EGF or TGF-α *(38)*. The cells grow as dense monolayers, maintaining very few morphological or biochemical characteristics related to the original hepatocytes. They do maintain, however, hepatocyte-associated transcription factors, such as HNF1, HNF3, and HNF4. The expression of C/EBP-α is decreased, but there is enhanced expression of the proliferation associated transcription factors AP1 and NFkB. The clonally growing hepatocytes best resemble the hepatoblasts seen in fetal liver or in models of carcinogenesis. Despite the phenotypic departure from mature hepatocytes, these cells revert to the fully mature hepatocyte phenotype upon addition of Matrigel, even when DNA synthesis is inhibited. The phenotypic reversal is complete and includes induction of cytochrome P450 and reestablishment of the morphology of the hepatocytes (including expression of glycogen, rough endoplasmic reticulum, and bile canaliculi. Transfection of the growing hepatoblasts with genetic constructs containing lac–Z demonstrates that single cells can give rise to clones. When the growing proliferating hepatocytes are maintained in type I collagen gels, and in the presence of HGF as the sole mitogen, they tend to acquire a ductal morphology resembling the ductular hepatocytes seen in rodent carcinogenesis or in human liver disease. Currently, this system using HGM medium is the only one in which normal hepatocytes can be maintained for prolonged periods in sustained clonal growth. It should be mentioned, however, that sustained clonal growth can also be seen in hepatocyte cultures derived from transgenic mice expressing TGF-α in the liver under the influence of the albumin promoter. Hepatocytes from these mice grow in the absence of any exogenously added growth factors and easily lead to establishment of differentiated hepatocyte cell lines.

6. Methods for Isolation and Maintenance of Rat Hepatocyte Cultures

6.1. Hepatocyte Isolation by Collagenase Perfusion

6.1.1 General Setup

The buffer used for liver perfusion is identical to the one originally described by Seglen et al. *(6)* It consists of 83 g of NaCl (.142 M), .5 g KCl (.007 M), 24 g HEPES powder (.01 M), and 1.9 g NaOH (.005 M) in 10 L of buffer. It is adjusted to a pH of 7.4 and sterilized by filtration. The sterile vessel containing the perfusion buffer medium is kept in a water bath at 37°C. A sterilized plastic tube exiting the vessel is threaded through a peristaltic pump, which is calibrated to deliver a perfusion rate of 12 cc/min. The plastic tube is filled with media and connected to a tuberculin syringe to which an 18-gage needle is attached. The plastic tube is long enough to allow multiple loops to be put back into the water bath prior to the site in which the tuberculin syringe is attached. The peristaltic pump acts as a heat sink and rapidly decreases the temperature of the buffer, after it passes through the peristaltic pump. Thus, the tube needs to be long enough so that loops of it can be placed back into the water bath to ensure that the temperature of the buffer to be delivered into the liver is indeed 37°C. This small detail is important, and it is the cause of most of the variation from one liver perfusion to the next, especially in situations ion which the room temperature (and thus the pump temperature) varies substantially through the year because of heating, air conditioning, and so on. No extra oxygenation is necessary other than the air normally dissolved in the system.

6.1.2. Rat Surgery

Rats are anesthetized by sodium pentobarbital (Nembutal, Abbott, N. Chicago, IL) (intraperitoneal injection of 0.1 cc/100 g rat weight). When the rats are deeply anesthetized, an operation is carried out to insert a catheter into the inferior vena cava. Following a 3-cm midabdominal incision starting immediately below the xiphoid, the 18-gage needle connected to the 1-cc tuberculin syringe is inserted into the vena cava at a point approx 0.25 cm above the exit point of the right renal vein. The metallic needles can easily tear through the vessel. Soft plastic trocar catheters inserted into vessel holes generated by ophthalmic scissors are sometimes preferred. The catheter or needle should be inserted by 2 cm inside the vessel. Prior to the insertion of the catheter, silk suture is inserted below the vena cava. After the catheter is inserted, it is fastened securely by tying the silk suture around it and around the vessel, to keep the catheter in place. Once the needle or catheter is secure, the peristaltic pump is turned on, the portal vein is sectioned (to create the outflow tract), and the chest cavity of the rat is opened. The inferior vena cava is ligated by a hemostatic clamp in the chest cavity, at a point as close to the heart as possible. This ensures that the clamping point is above the point of insertion of the hepatic veins in the inferior vena cava. When the surgical procedure is complete, the perfusion buffer enters through the inferior vena cava, proceeds upward, and meets the clamp above the hepatic veins. This forces it to reverse course and enter into the liver, where it eventually exits through the portal vein. The liver blanches and loses its dark color, to become pale brown. It should be noted that this procedure establishes a circulation pattern that is the reverse of the normal circulation in the liver animal. This is done because the inferior vena cava is a longer vessel than the portal vein, and thus easier to catheterize. The portal vein splits shortly after it forms and does not have sufficient length for surgical manipulation during the catheterization.

6.1.3. Liver Perfusion

Typically, 200–250 cc of perfusion buffer at 12 mL/minute are used in each of the two steps of the perfusion for a 200 gram rat. The first step uses the

Ca-free buffer described above. When this is complete, the tube is inserted into an adjacent vessel containing an equal quantity of the same buffer supplemented with calcium chloride ($CaCl_2$ {$2H_2O$: 6.36 g/10 L: 0.004 M). Collagenase is also added as powder immediately prior to beginning of the second perfusion step (100 U of collagenase, or 70–100 mg/250 cc of buffer. The exact concentration needs to be determined with every batch of collagenase). At the end of the second step (approx 40 min after the perfusion was started), the liver becomes very soft, with the consistency of a very viscous liquid.

6.1.4 Hepatocyte Isolation

For more details on the physical and chemical parameters of this technique, the outstanding review by Seglen *(6)*, first authored in 1976, remains the gold standard. In a simple summary form, the procedure is described here. The liver is resected with sterile surgical tools and transferred into a sterile beaker. 20 mL of ice-cold perfusion buffer (Ca-free) is added. The liver is minced with sterile long loose scissors. The scissors should not be tight, since they may damage cells as they are released. The mechanical mincing by the loose scissors releases the hepatocytes (singlets or doublets). After the liver is minced for 1–2 min, the suspension of the cells is filtered by pouring over a beaker covered by a Nitex filter of 100 µ in pore diameter (Polyamide Nylon mesh filter, Tetco, Briarcliff, NY). Hepatocytes and other cells enter through the pores, but undigested tissue is retained. The hepatocyte suspension is kept on ice throughout the whole process. The suspension is centrifuged in (typically) 50-mL sterile plastic conical tubes in very low gravity conditions ($50g$). If not centrifuged, hepatocytes settle to the bottom of the tube within about 10 min in unit gravity, because of their large size and weight compared to the other cells. The hepatocyte pellet is collected on the bottom of the conical tube, but the supernatant (containing the much smaller nonparenchymal cells, e.g., endothelial cells, Ito cells, bile duct cells, and cells from the mesothelial capsule) is decanted. This process is repeated three times altogether. The final cell pellet predominantly contains hepatocytes (95–97%, as originally described). The supernatant of the first centrifugation is retained, if it is desirable to isolate nonparenchymal cells. This supernatant can be centrifuged to higher g forces (500–$100g$) to isolate a pellet of the nonparenchymal cells as a mixture.

The hepatocyte pellet contains approx 70 million rat hepatocytes/cc of packed pellet (at $50g$ centrifugation). This is the most commonly used approach to count hepatocytes, since the isolated cells are present mostly as cell doublets and triplets, and rarely as single cells, thus defying the use of automated procedures such as cell sorting. Alternatively, a hemocytometer can also be used. Mixing of a drop of hepatocyte suspension with 0.4% trypan blue solution (Sigma, St. Louis, MO) and simple examination under the microscope allows evaluation of the viability of preparation (nonviable cells become blue within seconds). The pellet of the hepatocytes is kept on ice until ready to mix with the cell-culture medium.

6.1.5. Preparation and Maintenance of Cultures

The multiplicity of systems described above clearly suggests that the methodologies used after this point vary, depending on the system of culture. For standard dense monolayers (approx 100,000 cells/cm^2), or for sparse cultures used to study DNA synthesis (typically 10,000 cells/cm^2), the cells are suspended by use of a wide-bore pipet into the medium. Since hepatocytes are large, they are easily subjected to mechanical damage, and steps involving pipeting through small apertures or mixing with mechanical stirring bars, and so on should be at all times avoided. The culture plates can be plain or specifically coated plastic. The methodology for coating plastic with dry rat tail type I collagen, or for making collagen gels, has been described in the past *(39)*. Insulin at 10^{-7} *M* should be used in all media, unless specifically contraindicated for reasons of experimental design. A very important feature is the depth of the medium over the hepatocytes. It should not exceed 1–2 mm. Hepatocytes are very sensitive to oxygen deprivation, and die very quickly if the overlying medium is thicker than that. Typically, 1 cc of medium should be used with a standard 35-mm diameter cell culture plate. The small amount of medium forced by these considerations requires that the medium is changed frequently. This should be done every day for dense monolayers, and every 2–3 d for sparse cultures.

References

1. Bissell, D. M., Hammaker, L. E., and Meyer, U. A. (1973) Parenchymal cells from adult rat liver in nonproliferating monolayer culture. I. Functional studies. *J. Cell. Biol.* **59**, 722–734.
2. McGowan, J. A., Strain, A. J., and Bucher, N. L. R. (1981) DNA synthesis in primary cultures of adult rat hepatocytes in a defined medium, effects of epidermal growth factor, insulin, glucagon, and cyclic-AMP. *J. Cell. Physiol.* **180**, 353–363.
3. Kan, M., Huan, J., Mansson, P., Yasumitsu, H., Carr, B., and McKeehan, W. (1989) Heparin-binding growth factor type 1 (acidic fibroblast growth factor): a potential biphasic autocrine and paracrine regulator of hepatocyte regeneration. *Proc. Nat. Acad. Sci. USA* **86**, 7432–7436.
4. Michalopoulos, G., Houck, K. A., Dolan, M. L., and Luetteke, N. C. (1984) Control of hepatocyte replication by two serum factors. *Cancer Res.* **44**, 4414–4419.
5. Nakamura, T., Nawa, K., and Ichihara, A. (1984) Partial purification and characterization of hepatocyte growth factor from serum of hepatectomized rats. *Biochem. Biophys. Res. Commun.* **122**, 1450–1459.

6. Seglen, P. O. (1976) Preparation of isolated rat liver cells. *Methods Cell Biol.* **13,** 29–83.
7. Berry, M. N. and Friend, D. S. (1969) High-yield preparation of isolated rat liver parenchymal cells, a biochemical and fine structural study. *J. Cell. Biol.* **43,** 506–520.
8. Strom, S. C., Jirtle, R. L., Jones, R. S., and Rosenberg, M. R., Michalopoulos, G. (1982) Isolation, culture and transplantation of human hepatocytes. *J. Natl. Cancer Inst.* **68,** 771–775.
9. Guguen-Guillouzo, C., Campion, J. P., Brissot, P., Glaise, D., Launois, B., Bourel, M., and Guillouzo, A. (1982) High yield preparation of isolated human adult hepatocytes by enzymatic perfusion of the liver. *Cell Biol. Int. Rep.* **6,** 625–628.
10. Morsiani, E., Rozga, J., Scott, H. C., Kong, L. B., Lebow, L. T., McGrath, M. F., Moscioni, A. D., and Demetriou, A. A. (1994) Automated large-scale production of porcine hepatocytes for bioartificial liver support. *Transplantation Proc.* **26,** 3505–3506.
11. Maher, J. J., Bissell, D. M., Friedman, S. L., and Roll, F. J. (1988) Collagen measured in primary cultures of normal rat hepatocytes derives from lipocytes within the monolayer. *J. Clin. Invest.* **82,** 450–459.
12. Nakamura, T., Yoshimoto, K., Nakayama, Y., Tomita, Y., and Ichihara, A. (1983) Reciprocal modulation of growth and differentiated functions of mature rat hepatocytes in primary culture by cell–cell contact and cell membranes. *Proc. Natl. Acad. Sci. USA* **80,** 7229–7233.
13. Michalopoulos, G., Cianciulli, H. D., Novotny, A. R., Kligerman, A. D., Strom, S. C., and Jirtle, R. L. (1982) Liver regeneration studies with rat hepatocytes in primary culture. *Cancer. Res.* **42,** 4673–4682.
14. Schneider, W. C. and Greco, A. E. (1971) Incorporation of pyrimidine deoxyribonucleosides into liver lipids and other components. *Biochim. Biophys. Acta* **228,** 610–626.
15. Myers, D. K. and Ram, S. (1969) Incorporation of radioactive label from pyrimidines into liver lipids. *Can. J. Physiol. Pharmaco.* **47,** 731–733.
16. Mead, J. E. and Fausto, N. (1989) Transforming growth factor alpha may be a physiological regulator of liver regeneration by means of an autocrine mechanism. *Proc. Natl. Acad. Sci. USA* **86,** 1558–1562.
17. Cruise, J. L., Houck, K. A., and Michalopoulos, G. K. (1985) Induction of DNA synthesis in cultured rat hepatocytes through stimulation of alpha 1 adrenoreceptor by norepinephrine. *Science* **227,** 749–751.
18. Shi, Y. E. and Yager, J. D. (1989) Effects of the liver tumor promoter ethinyl estradiol on epidermal growth factor-induced DNA synthesis and epidermal growth factor receptor levels in cultured rat hepatocytes. *Cancer Res.* **49,** 3574–3580.
19. Houck, K. A. and Michalopoulos, G. K. (1989) Altered responses of regenerating hepatocytes to norepinephrine and transforming growth factor type beta. *J. Cell. Physiol.* **141,** 503–509.
20. Oberhammer, F., Bursch, W., Parzefall, W., Breit, P., Erber, E., Stadler, M., and Schulte-Hermann, R. (1991) Effect of transforming growth factor beta on cell death of cultured rat hepatocytes. *Cancer Res.* **51,** 2478–2485.

21. Etienne, P. L., Baffet, G., Desvergne, B., Boisnard-Rissel, M., Glaise, D., and Guguen-Guillouzo, C. (1988) Transient expression of c-fos and constant expression of c-myc in freshly isolated and cultured normal adult rat hepatocytes. *Oncogene Res.* **3,** 255–262.
22. Michalopoulos, G., Sattler, C. A., Sattler, G. L., and Pitot, H. C. (1976) Cytochrome P-450 induction by phenobarbital and 3-methylcholanthrene in primary cultures of hepatocytes. *Science* **193,** 907–909.
23. CITATION CLASSIC, (1991). *Current Contents,* **34,** 11.
24. Dunn, J. C., Yarmush, M. L., Koebe, H. G., and Tompkins, R. G. (1989) Hepatocyte function and extracellular matrix geometry, long-term culture in a sandwich configuration. *FASEB J.* 3 174–177.
25. Michalopoulos, G. K., Bowen, W., Nussler, A. K., Becich, M. J., and Howard, T. A. (1993) Comparative analysis of Mitogenic and Morphogenic effects of HGF and EGF on rat and human hepatocytes maintained in collagen gels. *J. Cell. Physiol.* **156,** 443–452.
26. Kleinman, H. K., McGarvey, M. L., Liotta, L. A., Robey, P. G., Tryggvason, K., and Martin, G. R. (1982) Isolation and characterization of type IV procollagen, laminin, and heparan sulfate proteoglycan from the EHS sarcoma. *Biochemistry* **21,** 6188–6193.
27. Bucher, N. L., Robinson, G. S., and Farmer, S. R. (1990) Effects of extracellular matrix on hepatocyte growth and gene expression, implications for hepatic regeneration and the repair of liver injury. *Semin. Liver Dis.* **10,** 11–19.
28. Isom, H. C., Secott, T., Georgoff, I., Woodworth, C., and Mummaw, J. (1985) Maintenance of differentiated rat hepatocytes in primary culture. *Proc .Natl. Acad. Sci. USA* **82,** 3252–3256.
29. Chan, K., Kost, D. P., and Michalopoulos G (1989) Multiple sequential periods of DNA synthesis and quiescence in primary hepatocyte cultures maintained on the DMSO-EGF on/off protocol. *J. Cell. Physiol.* **14,** 584–590.
30. Serra, R. and Isom, H. C. (1993) Stimulation of DNA synthesis and protooncogene expression in primary rat hepatocytes in long-term DMSO culture. *J. Cell. Physiol.* **154,** 543–553.
31. Yuasa, C., Tomita, Y., Shono, M., Ishimura, K., and Ichihara, A. (1993) Importance of cell aggregation for expression of liver functions and regeneration demonstrated with primary cultured hepatocytes. *J. Cell. Physiol.* **156,** 522–530.
32. Guguen-Guillouzo, C., Clement, B., Baffet, G., Beaumont, C., Morel-Chany, E., Glaise, D., and Guillouzo, A. (1983) Maintenance and reversibility of active albumin secretion by adult rat hepatocytes co-cultured with another liver epithelial cell type. *Exp. Cell. Res.* **143,** 47–54.
33. Baffet, G., Clement, B., Glaise, D., Guillouzo, A., and Guguen-Guillouzo, C. (1982) Hydrocortisone modulates the production of extracellular material and albumin in long-term cocultures of adult rat hepatocytes with other liver epithelial cells. *Biochem. Biophys. Res. Commun.* **109,** 507–512.
34. Schrode, W., Mecke, D., and Gebhardt, R. (1990) Induction of glutamine synthetase in periportal hepatocytes by cocultivation with a liver epithelial cell line. *Eur. J. Cell. Biol.* **53,** 35–41.

35. Chiao, C., Zhang, Y., Kaufman, D. G., and Kaufmann, W. K. (1995) Derivation of phenobarbital-responsive immortal rat hepatocytes. *Am. J. Pathol.* **146,** 1248–1259.
36. Michalopoulos, G., Russell, F., and Biles, C. (1979) Primary cultures of hepatocytes on human fibroblasts. *In Vitro* **15,** 796–806.
37. Mitaka, T., Mikami, M., Sattler, G. L., Pitot, H. C., and Mochizuki, Y. (1992) Small cell colonies appear in the primary culture of adult rat hepatocytes in the presence of nicotinamide and epidermal growth factor. *Hepatology* **16,** 440–447.
38. Block, G. D., Locker, J., Bowen, W. C., Petersen, B. E., Katyal, S., Strom, S. C., et al. (1996) Population expansion, clonal growth and specific differentiation patterns in primary cultures of hepatocytes induced by HGF/SF, EGF and TGF-α in a chemically defined (HGM) medium. *J. Cell Biol.* **132,** 1133–1149.
39. Strom, S. C. and Michalopoulos, G. (1982) Collagen as a substrate for cell growth and differentiation. *Methods Enzymol.* **82,** 544–555.
40. Bissell, D. M., Arenson, D. M., Maher, J. J., and Roll, F. J. (1987) Support of cultured hepatocytes by a laminin-rich gel. Evidence for a functionally significant subendothelial matrix in normal rat liver. *J. Clin. Invest.* **79,** 801–812.

19

Formation and Characterization of Hepatocyte Spheroids

Julie R. Friend, Florence J. Wu, Linda K. Hansen, Rory P. Remmel, and Wei-Shou Hu

1. Introduction

Several investigators have demonstrated that freshly harvested hepatocytes self-assemble into three-dimensional, compacted, freely suspended aggregates known as spheroids *(1–3)*. These aggregates have smooth, undulating surfaces and average approx 120 µm in diameter. Hepatocyte spheroids exhibit enhanced liver-specific activities and prolonged viability, compared to cells maintained as a monolayer *(4,5)*. Extensive cell–cell contacts, tight junctions, and microvilli-lined channels that resemble bile canaliculi have been observed between hepatocytes in spheroids *(6,7)*. Thus, these cells appear to mimic the morphology and ultrastructure of an in vivo liver lobule. Reorganization of hepatocytes into these three-dimensional structures is hypothesized to contribute to their enhanced liver-specific functions. Because of their enhanced function and tissue-like ultrastructure, hepatocyte spheroids show great promise for use in tissue-engineering applications and drug metabolism studies.

Spheroids can be formed both in static culture and in suspension culture. The spheroids formed by both methods appear to be functionally and structurally similar. To form spheroids in static culture, hepatocytes are plated on positively-charged polystyrene Petri dishes in hormonally defined medium. The hepatocytes initially spread to form a monolayer on the surface. Over time, the cells aggregate into multicellular islands, and, after 3 d of culture, hepatocytes will detach from the surface as spheroids. Approximately 35–40% of the hepatocytes initially plated will form into spheroids *(3)*.

Production of spheroids in static culture is not practical for generating large numbers of spheroids quickly. In order to produce spheroids in large quantity

Table 1
Media Composition

Component[a]	Units	Spheroid medium A	Spheroid medium B
L-glutamine (Gibco-BRL)	mg/mL	0.292	0.292
Bovine-porcine insulin (Eli Lilly, Indianapolis, IN)	U/mL	0.2	0.2
Epidermal growth factor (EGF)	ng/mL	50	5
Linoleic acid	ng/mL	50	50
Bovine serum albumin	µg/mL	500	500
Dexamethasone	nM		1
Glucagon	ng/mL		4
Transferrin	µg/mL		6.25
Liver growth factor	ng/mL		20
Selenium	ng/mL		6.25
$CuSO_4 \cdot 5H_2O$	µM	0.1	0.1
H_2SeO_3	nM	3	3
$ZnSO_4 \cdot 7H_2O$	pM	50	50
HEPES	mM	15	15
Penicillin (Celox, Hopkins, MN)	U/mL	100	100
Streptomycin (Celox)	µg/mL	100	100

[a]All components from Sigma unless otherwise noted.

for applications such as a bioartificial liver device, suspension culture is more suitable. Use of suspension culture for spheroid formation also allows for control and monitoring of environmental conditions, such as the pH, dissolved oxygen, or turbidity in the culture. Once put into suspension culture, hepatocytes begin to aggregate quickly, and compacted aggregates can be seen after 1 d in culture. By this method, approx 80% of the inoculated hepatocytes form into spheroids *(8)*. Forming spheroids in Primaria dishes is convenient and more cost-effective for small-scale laboratory investigations. For large-scale production of spheroids, suspension culture is much less labor-intensive and more efficient. Both methods will be described in this chapter.

2. Materials
2.1. Cell-Culture Medium Composition

The media used for spheroid formation in Petri dishes and spinners are slightly different in composition. These conditions have proven to give satisfactory results consistently. Further optimization may improve the spheroid formation process. For both media types, the basal medium is Williams' E (Gibco-BRL Laboratories, Grand Island, NY). **Table 1** lists the remaining components in the respective media. (*see* **Note 1**).

2.2. Other Culture Materials

1. 60 × 15 mm Primaria tissue-culture dishes (Falcon, Becton Dickinson, Lincoln Park, NJ) (*see* **Note 2**).
2. Spinner flasks, 250 mL (Wilbur Scientific, Boston, MA) (*see* **Note 3**).
3. Sigmacote (Sigma, St. Louis, MO).

2.3. Viability Staining

1. Fluorescein diacetate (FDA) (Sigma), 5 mg/mL stock solution in DMSO. Store at 4°C in dark.
2. Ethidium bromide (EB) (Sigma), 40 µg/mL stock solution in phosphate-buffered saline (PBS). Store at 4°C in dark.

2.4. Albumin Assay

1. Nunc Maxisorp plates (Nunc, Naperville, IL), 96-well.
2. 96-well plates (mixing plates) (Corning, Corning, NY).
3. Room temperature humidified chamber.
4. Polyclonal rabbit antirat albumin, peroxidase bound (Cappel, Durham, NC). To prepare stock solution, dissolve lyophilized antibody in dH_2O to a final concentration of approx 10 mg/mL. Aliquot in 10 µL units. Store at –20°C.
5. Bovine serum albumin (BSA) (Sigma).
6. Rat albumin (Sigma); 1 mg/mL stock solution in dH_2O. Aliquot in units of 150 µL and store at –20°C.
7. PBS-Tween-20 wash solution: 0.05% (w/v) Tween-20 (Bio-Rad, Melville, NY), 1 mM KH_2PO_4, 10 mM Na_2HPO_4, 20 mM NaCl, 2.5 mM KCl, pH 7.2. Store at 4°C. Warm to room temperature before use.
8. 2% (w/v) Tween-20 in PBS.
9. Developer: 0.05 M citric acid, 0.05 M sodium citrate, pH 4.2; add 0.5% ABTS (2,2'azino-di-[3-ethylbenzthiazoline-6-sulfonate]) (Boehringer Mannheim, Indianapolis, IN)

2.5. Ethoxyresorufin O-dealkylase (EROD) Assay for Cytochrome P450

1. β-naphthoflavone (BNF) (Aldrich, Milwaukee, WI), 100 mM stock solution in DMSO. Store in dark.
2. Dicumarol (Sigma), 10 mM stock solution in 0.5 N NaOH.
3. Probenecid (Sigma), 0.16 M stock solution in 0.5 N NaOH.
4. Induction medium: Cell culture medium described in **Subheading 2.1.**, containing 50 µM BNF.
5. Incubation medium: Williams' E medium without phenol red, 25 µM dicumarol (Sigma), 2 mM probenecid (Sigma); pH 7.2. Make fresh each time assay is performed.
6. Ethoxyresorufin (Molecular Probes, Eugene, OR), 1 mM stock solution in DMSO. Store in dark.

3. Methods

3.1. Spheroid Formation

The procedures described below are for rat hepatocytes (*see* **Note 4**). For either method, it is important to start with freshly harvested hepatocytes. For the harvesting protocol, please *see* Chapter 31, this volume.

3.1.1. Spheroid Formation in Static Culture

1. Inoculate freshly harvested hepatocytes at $3–8 \times 10^4/cm^2$ in 60×15 mm Primaria dishes in 5 mL of spheroid medium A. Culture cells at 37°C, 5% CO_2.
2. Change the medium by 24 h. At this time, hepatocytes should be attached and in a monolayer. They may just be beginning to aggregate (*see* **Note 5**). Medium may be changed by simply aspirating spent medium and gently adding fresh medium to the dish. Unless you are inducing cells for the EROD assay (*see* **Subheading 3.2.3.**), there is no need to change the medium again before spheroids form.
3. Three d after inoculation, if spheroids have detached from the surface, they may be harvested for use (*see* **Note 6**).

3.1.2. Spheroid Formation in Suspension Culture

1. Siliconize 250-mL spinner vessels by coating with Sigmacote, following manufacturer's instructions. Once siliconized, wash spinners thoroughly and autoclave at 121°C for 30 min to sterilize.
2. Add 100 mL spheroid medium B to spinner flask and put in incubator at 37°C, 5% CO_2 to allow pH and dissolved oxygen to equilibrate before cells are ready.
3. Adjust stirring speed of magnetic stirrer plate, so that the agitation speed in the spinner is about 90 rpm (*see* **Note 7**). The stir bar should rotate smoothly.
4. Inoculate hepatocytes at $3 \times 10^5/mL$ (*see* **Note 8**).
5. Change culture medium every 2–3 d. To change medium, record the mass of the spinner vessel. Allow spheroids to settle by gravity. Remove medium, being careful not to remove spheroids. Record the mass of the spinner flask. Add back fresh medium until the original mass of the spinner flask is attained (*see* **Note 9**). Spheroids may be kept in suspension culture for 3 wk or longer (*see* **Note 10**).

3.2. Spheroid Characterization

Once spheroids are formed, several assays are used to assess their viability functions. These include evaluating albumin synthesis, urea synthesis, lidocaine clearance, and *in situ* cytochrome P450 activity. Ureagenesis is measured using the urea nitrogen kit from Sigma. Other procedures are described below.

3.2.1. Determination of Viability

In order to examine the viability of spheroids, they are stained with fluorescein diacetate and ethidium bromide. Esterases in live cells cleave the fluores-

Hepatocyte Spheroids

cein diacetate to fluorescein. Thus, live cells appear green. Ethidium bromide is able to penetrate dead cells and stain their nuclei red. The staining protocol is described below.

1. Prepare the FDA/EB staining solution by adding 10 µL FDA stock solution to 5 mL EB stock solution (*see* **Note 11**).
2. Transfer an aliquot of spheroids to a centrifuge tube; allow them to settle by gravity.
3. Aspirate supernatant without removing spheroids.
4. Add 2 mL FDA/EB solution; let stand for 2 min.
5. Aspirate FDA/EB solution without removing spheroids. Wash 3× in PBS.
6. View on fluorescence microscope with green (fluorescein) filter set.

3.2.2. Albumin Assay

Albumin synthesis is measured using a competitive ELISA assay.
On the day before the assay:

1. Add 7.5 g BSA to 250 mL PBS-Tween to yield a 30 g/L solution. Gently mix and allow BSA to dissolve.
2. Filter PBS-Tween-BSA through 0.2-µm filter for sterility.
3. Prepare working antibody solution. Add 10 µL stock antibody solution to 100 mL PBS-Tween-BSA (*see* **Note 12**). Store overnight at 4°C to allow antibodies that cross-react with bovine albumin to bind.
4. To 120 mL PBS, add 24 µL rat albumin stock to yield a 100 ng/mL solution. Coat Nunc plates with this solution at 100 µL/well. Store overnight at 4°C.

The day of the assay:

1. Remove PBS-Tween from cold room and allow to warm to room temperature.
2. Prepare standard curve:
   Add 80 µL rat albumin standard to 7.92 mL PBS.
   Add 1 mL of this standard to a test tube.
   Add 500 µL PBS to seven other test tubes.
   Serially transfer 500 µL (use a fresh pipet tip for each tube) from tube 1 through tube 8.
   Add samples in duplicate to first row (row A) of first two mixing plates (140 µL/well).
3. Prepare mixing plates:
   To rows B–H, add 120 µL PBS-Tween-BSA.
   To row A, add 20 µL 0.5% Tween-20 in PBS (1 mL 2% Tween-PBS in 4 mL PBS).
4. Vortex all thawed samples to ensure sample homogeneity.
5. Add 140 µL sample (in duplicate) to each well of row A, using the edge columns for standards. (Thus, each plate should have a standard at either end and four samples in the wells between).
6. Using 12 channel pipetor, serially transfer 40 µL from row A through row H.
7. Add 50 µL working antibody solution to each well.

8. Incubate at 37°C for approx 2 h.
9. Remove Nunc plates from cold room.
10. Empty contents of Nunc plates into sink. Blot plates dry on a paper towel.
11. Fill each well of Nunc plates with ~200 µL PBS-Tween wash solution and let sit for 3 min.
12. Repeat **steps 10** and **11** twice more (three washes total) and end by repeating step 10.
13. Transfer 100 µL from each well of each mixing plate to its corresponding well in a Nunc plate.
14. Incubate in room-temperature humidified chamber for 2 h.
15. Begin to thaw developer, so that it is ready when needed.
16. Wash Nunc plates as described in **steps 10–12**.
17. To 130 mL developer, add 130 µL 30% H_2O_2.
18. Add 100 µL developer to each well.
19. Incubate in room-temperature humidified chamber for 1 h (*see* **Note 13**).
20. Using a plate spectrophotometer, read plates at 405λ–490λ.
21. Analyze the data using software that will allow four parameter or logit regression.

3.2.3. In situ *Microscope Observation of Cytochrome P450 Activity*

Cytochrome P450 activity is assessed by examining ethoxyresorufin O-dealkylation (EROD) with confocal microscopy. Ethoxyresorufin is a substrate for cytochrome P450IA1 when cells are induced with BNF *(9,10)*.

During spheroid culture, cells should be cultured in induction medium for at least 48 h before the assay is done, with medium changes daily.

Microscopic examination of EROD activity:

1. Remove aliquot of spheroids from culture; allow to settle by gravity.
2. Add incubation medium.
3. Incubate cells for 5 min at 37°C.
4. Add 20 µL ethoxyresorufin stock solution.
5. View on fluorescence microscope with red (rhodamine) filters within 15 min of the addition of ethoxyresorufin.

4. Notes

1. It is important that spheroids be formed in serum-free medium. Spheroid formation is inhibited in the presence of serum *(2)*.
2. Other size dishes may be used, but we have found that efficiency of spheroid formation is best in this size dish.
3. Spheroid formation is sensitive to the geometry of the flask. We use flasks 6 cm in diameter. They have one side arm, which is capped, and an aeration port filled with glass wool. The magnetic stir bar should be suspended. We use stir bars 4 cm in length, without any blades. The stir bar should hang approx 1 cm from the bottom of the flask. Too small a clearance will cause cell destruction. Too large a clearance will result in cells settling at the bottom of the flask.

4. Conditions for porcine hepatocyte spheroid formation are slightly different (*see* ref. *11*).
5. When inoculated onto a Primaria dish, hepatocytes attach and initially spread as a monolayer. After about 20–24 h, they slowly begin to form multicellular islands, and over the following 2 d form compacted spherical structures easily distinguished from loose aggregates of cells. Spheroids will detach from the surface after 3–4 d in culture.
6. If the spheroids remain in the Primaria dish in which they were formed past 3–4 d, they may reattach to the surface. If spheroids are transferred to another surface, it is important that this surface be nonadhesive. If spheroids are transferred to an adhesive surface, such as a collagen-coated surface, they will disassemble into a monolayer.
7. Proper agitation speed is crucial for spheroid formation. If the agitation speed is too fast, cell damage and death may occur. If it is too slow, cells will not be well suspended and may settle on the bottom of the flask. A homogenous environment is important in the early stages of spheroid formation.
8. Proper cell concentration is also important for spheroid formation. Oxygen is supplied by surface aeration only. If cell concentration is too high, the oxygen tension may not be sufficient, and cell damage may result.
9. This method for changing medium should result in an exchange of approx 70–80% of the spent medium. If a complete change of medium is desired, the contents of the flasks may be removed and gently centrifuged for 2–3 min at 34g. After centrifugation, remove the medium by aspiration, being careful not to disturb the spheroids. Resuspend the spheroid pellet in medium and add the cell suspension back to the flask.
10. Spheroids may be kept in suspension culture for weeks, but if you want to use them in some other application, you will most likely want to harvest them after 2–3 d. In suspension culture, the average size of the spheroids is very much dependent on when you harvest them. Over time, spheroids appear to collide with other spheroids to form larger aggregates. Thus, the earlier spheroids are harvested, the smaller their size. If spheroids become too large, they may develop necrotic centers.
11. For best results, this solution should be made fresh from stock solutions for each use.
12. You may need to adjust the concentration if new antibody stock is used. A reasonable goal is a maximum optical density reading after development of 0.5–0.6 to yield a reasonable sensitivity.
13. This concentration of H_2O_2 results in a maximal signal in 45–60 min, so it is important to read the plates within 1 h.

References

1. Landry, J., Bernier, D., Ouellet, C., Goyette, R., and Marceau N. (1985) Spheroidal aggregate culture of rat liver cells: Histotypic reorganization, biomatrix deposition, and maintenance of functional activities. *J. Cell Biol.* **101**, 914–923.
2. Koide, N., Sakaguchi, K., Koide, Y., Asano, K., Kawaguchi, M., Matsushima, H., et al. (1990) Formation of multicellular spheroids composed of adult rat hepato-

cytes in dishes with positively charged surfaces and under other nonadherent environments. *Exp. Cell Res.* **186,** 227–235.
3. Peshwa, M. V., Wu, F. J., Follstad, B. D., Cerra, F. B., and Hu, W. S. (1994) Kinetics of hepatocyte spheroid formation. *Biotechnol. Prog.* **10,** 460–466.
4. Koide, N., Shinji, T., Tanabe, T., Asano, K., Kawaguchi, M., Sakaguchi, K., et al. (1989) Continued high albumin production by multicellular spheroids of adult rat hepatocytes formed in the presence of liver-derived proteoglycans. *Biochem. Biophys. Res. Comm.* **161,** 385–391.
5. Tong, J. Z., Bernard, O., and Alvarez, F. (1990) Long-term culture of rat liver cell spheroids in hormonally defined media. *Exp. Cell Res.* **189,** 87–92.
6. Asano, K., Koide, N., and Tsuji, T. (1989) Ultrastructure of multicellular spheroids formed in the primary culture of adult rat hepatocytes. *Journal of Clinical Electron Microscopy* **22,** 243–252.
7. Peshwa, M. V., Wu, F. J., Sharp, H. L., Cerra, F. B., and Hu, W.-S. (1996) Mechanistics of formation and ultrastructural evaluation of hepatocyte spheroids. *In Vitro Cell. Dev. Biol.* **32A,** 197–203.
8. Wu, F. J., Friend, J. R., Hsiao, C.-C., Zilliox, M. J., Ko, W.-J., Cerra, F. B., and Hu, W.-S. (1996) Efficient assembly or primary rat hepatocyte spheroids for tissue engineering applications. *Biotechnol. Bioeng.* **50,** 404–415.
9. Burke, M. D., Thompson, S., Elcombe, C. R., Halpert, J., Haaparanta, T., and Mayer, R. T. (1985) Ethoxy-, pentoxy- and benzyloxyphenoxazones and homologues: a series of substrates to distinguish between different induced cytochromes P-450. *Biochem. Pharmacol.* **34,** 3337–3345.
10. Burke, M. D., Thompson, S., Weaver, R. J., Wolf, C. R., and Mayer, R. T. (1994) Cytochrome P450 specificities of alkoxyresorufin O-dealkylation in human and rat liver. *Biochem. Pharmacol.* **48,** 923–936.
11. Lazar, A., Peshwa, M. V., Wu, F. J., Chi, C. M., Cerra, F. B., and Hu, W. S. (1995) Formation of porcine hepatocyte spheroids for use in a bioartificial liver. *Cell Transplant* **4,** 259–268.

20

Isolation and Culture of Human Endothelial Cells

François Berthiaume

1. Introduction

Because the cell mass in a bioartificial tissue exceeds relatively small numbers, there is a requirement to incorporate a convective transport system that provides nutrients and removes waste products. This function is provided by the vascular tree in vivo, and it may be desirable that endothelial cells be used in the design of the vascular tree in the bioartificial tissue, because endothelial cells provide a nonthrombogenic surface inside normal blood vessels. Furthermore, there is an ongoing effort in the area of vascular graft development, and the development of these applications requires a reliable source of endothelial cells.

Human umbilical veins provide a suitable model of human endothelium, because they have been shown to respond to a vast array of chemical and mechanical stimuli. They have long been used as a source of human endothelial cells because of their ease of procurement and the ability to obtain pure cultures. In this chapter, we describe a methodology to isolate vascular endothelial cells from human umbilical veins, maintain them in culture, and characterize the expression of endothelial-specific markers. The procedures outlined below could be easily adapted to the isolation of endothelial cells from other vessels in the 2–5 mm diameter range.

2. Materials
2.1. Isolation of Endothelial Cells from Human Umbilical Veins

1. Umbilical cords: They can be readily obtained from a local hospital, since this tissue is normally discarded. A formal agreement with the hospital is required. Preferred length is ~25 cm per cord, and age should be less than 3 d old. The risk of bacterial contamination increases with older tissue samples. They should be placed in a sterile plastic bag and kept in the refrigerator until picked up.

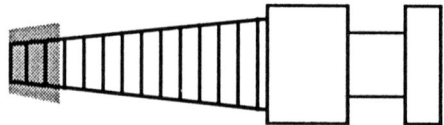

Fig. 1. Metal adapter with silicone tubing ring used to cannulate umbilical veins.

2. Collagenase solution (prepare less than 3 h prior to isolation):
   a. Dulbecco's modified phosphate buffered saline (DPBSS; Irvine Scientific, Santa Anna, CA).
   b. $CaCl_2$, 200 mM stock solution in water. Add 1% v/v to DPBSS. Filter through a 0.2-µm filter (do not autoclave).
   c. Collagenase powder (type A, Boehringer Mannheim, Indianapolis, IN) is dissolved in the calcium-supplemented DPBSS to a final concentration of 0.02% (w/v). Stir for 30 min at room temperature, centrifuge at 2000g for 20 min, and filter the supernatant through a 0.45-µm filter.
3. Hanks' balanced salt solution (HBSS; Ca^{2+}- and Mg^{2+}-free, Irvine Scientific).
4. Endothelial cell-culture medium:
   a. Medium 199 (Irvine Scientific).
   b. Defined fetal bovine serum (HyClone, Logan, UT). Low endotoxin (<0.2 ng/mL) lots are preferable. Add 20% (v/v) to Medium 199.
   c. L-glutamine, 200 mM in water (Sigma, St. Louis, MO). Add 1% v/v to medium.
   d. Penicillin–streptomycin: 5000 U/mL and 5 mg/mL in 0.9% NaCl, respectively (Sigma). Add 1% (v/v) to medium.
5. Surgical instruments, hardware:
   a. One sterile surgical blade.
   b. Hemostats (one per cord).
   c. Plastic cable ties. Cable-tie tensioning tool (Cole-Parmer, Niles, IL).
   d. Stainless steel adapters, metal, female (Baxter, Mac Gaw Park, IL, #BD3092).
   e. Silicone tubing (16 gage), cut in rings ~4-mm long, using a razor blade. Place one silicone tubing ring on the end of each adapter (**Fig. 1**).
   f. Two 500-mL beakers and two 500-mL Wheaton bottles.
   g. The surgical instruments are wrapped in aluminium foil, the glassware tops covered with foil, and all these items are autoclaved prior to use.
   h. Two 60-cc plastic disposable syringes (Becton Dickinson, Bedford, MA).
   i. Two pairs of sterile surgical gloves.

2.2. Serial Propagation of Human Endothelial Cells

1. Gelatin coating solution: 2% w/v sterile gelatin solution in DPBSS (Sigma). Dilute 10-fold in sterile DPBSS.
2. Endothelial cell-growth medium:
   a. Endothelial cell-culture medium, prepared as above.
   b. Add endothelial mitogen (Biomedical Technologies, Stoughton, MA) to a final concentration of 25 µg/mL.

c. Add heparin (Sigma) to a final concentration of 90 µg/mL.
3. Trypsin solution: trypsin powder (1:250, from porcine pancreas, Sigma) and ethylene diamine tetraacetic acid (EDTA, tetrasodium salt) are dissolved in HBSS, to a final concentration of 0.05% w/v and 0.02% w/v, respectively.

2.3. Characterization of Endothelial Cells

1. DiI-Ac-LDL is available in solution of 200 µg/mL (Biomedical Technologies). The vial should be centrifuged at 500g for 10 min before use. Dilute DiI-Ac-LDL in endothelial cell culture medium to a final concentration of 10 µg/mL.
2. Fixative: Mix equal volumes of 95% ethanol and acetone. Keep in refrigerator.
3. DPBSS + 0.1% bovine serum albumin (BSA).
4. Rabbit antihuman von Willebrand factor, IgG fraction of antiserum (Sigma). Dilute 1:200 in DPBSS with 0.1% BSA. IgG fraction of normal rabbit serum should be purchased as control (nonspecific antibody).
5. Affinity purified goat antirabbit IgG, fluorescein (FITC) conjugate (Sigma). Dilute 1:40 in DPBSS with 0.1% BSA.

3. Methods
3.1. Isolation of Endothelial Cells from Human Umbilical Veins

The method for isolating endothelial cells presented below is a modification of the procedures originally published by Jaffe et al. *(1)* and Gimbrone et al. *(2)*. Proper organization and the operator's care in following sterile handling techniques are critical for the success of the isolations.

3.1.1. Preparation and Cannulation of Umbilical Cords

1. Place all items in hood; a suggested layout is shown in **Fig. 2**. All items should be sprayed with 70% ethanol prior to placing in the laminar flow hood.
2. Place a 40 × 40 cm piece of foil in the hood and cover it with several layers of absorbing tissue. Spray generously with ethanol.
3. Remove cords from bags and clean with paper towels to remove excess blood on the outside. Lay them flat on the absorbing tissue in the hood. Nonsterile gloves can be used during this step.
4. All foil wrappings containing the surgical instruments are opened to allow access to the sterile contents without having to touch nonsterile surfaces. Similarly, foil covers on the beakers and bottles are loosened. One beaker is filled with the collagenase solution (50 mL per cord is needed) and another one with HBSS (100 mL per cord is needed).
5. Open a sterile glove package in the hood and lay flat on the absorbing tissue. Put the sterile gloves on and transfer the cords onto the paper wrapping that the gloves came in. Cut out approx 1 cm at each end of the cords. Insert one catheter in the vein at one end of each cord. The vein is the largest vessel and can easily be distinguished from the two arteries. Secure the catheter with two plastic cable ties. Fit each catheter with a one-way stopcock.

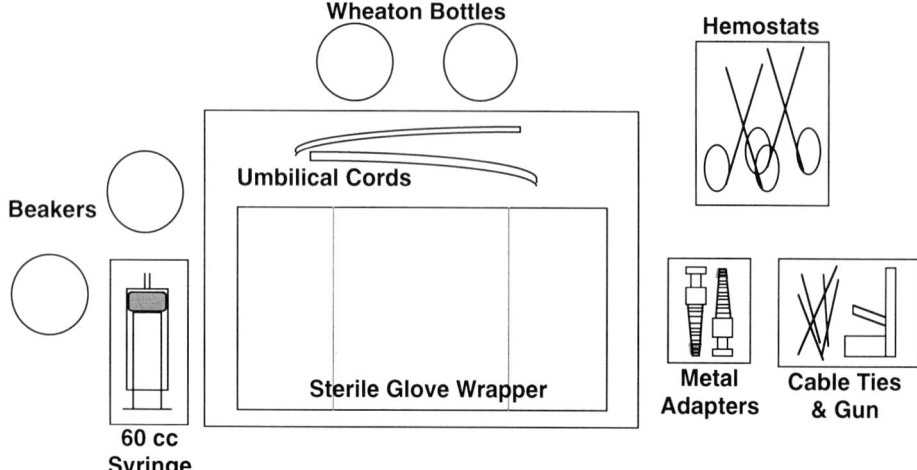

Fig. 2. Initial layout of instruments and glassware at beginning of isolation procedure. This layout minimizes the number and range of hand movements and thus reduces the risk of bacterial contamination.

3.1.2. Perfusion and Collagenase Digestion

1. Fill a 60-cc syringe with HBSS and perfuse each cord to remove blood and dilate the vein. The effluent is collected in one of the Wheaton bottles. One must proceed slowly at first because the vessel can be obstructed by clots. Excessive resistance to flow usually indicates that the catheter is not in the vein. Repeat the perfusion a second time with another 50–60 mL of fresh HBSS for each cord.
2. Fill the syringe with the collagenase solution and perfuse the cord until the last traces of the red-colored HBSS have come out. Clamp the end of the cord with a hemostat and continue filling the cord with the collagenase solution until slight distension. Leave in sterile hood for 30 min.

3.1.3. Harvesting of Endothelial Cells

1. Put on a new pair of sterile gloves and fill the 60-cc syringe with HBSS. Flush the cord contents with ~25 mL HBSS into a new, clean Wheaton bottle. Repeat the procedure for each cord.
2. Collect the bottle's contents into 50-mL centrifuge tubes. Pieces of tissue may remain at the bottom of the bottle, and care should be taken that they not be transferred to the tubes. Pellet at 650 g for 10 min.
3. Discard supernatants and resuspend the pellets in a total of 10 mL of endothelial cell-culture medium times the number of cords used in the isolation procedure.
4. Transfer the cell suspension to 75-mm tissue-culture flasks (one per cord). Place in a 5% CO_2 incubator overnight.
5. Rinse the flasks with DPBSS or HBSS to remove nonadherent blood cells. Add fresh endothelial cell culture medium. The resulting plating density is $1.5–3.5 \times 10^4$ endothelial cells/cm^2. Confluence is normally reached after 2–3 d.

3.2. Serial Propagation of Human Endothelial Cells

Subculture allows one to expand the number of cells; however, it is important to monitor certain key endothelial-specific characteristics, because the expression of differentiated function often decays with passage number. The suggested culture medium is based on the formulation of Thornton et al. *(3)*, which contains endothelial mitogen and heparin. Endothelial mitogen's main active ingredient is basic fibroblast growth factor, and heparin potentiates its effect on endothelial cells. Furthermore, heparin prevents the growth of a potential small number of contaminating smooth muscle cells. Commercial media and reagent sources, such as that sold by Clonetics (San Diego, CA), have a lower serum requirement and also provide excellent results for endothelial cell propagation. Interested readers should refer to the company's literature.

1. Surfaces (e.g., culture flasks, glass slides) are precoated with 0.2% gelatin in DPBSS for 1 h or longer at 37°C.
2. Endothelial cells at ~75% confluence are washed with HBSS, then trypsin solution is added (3–5 mL/75 cm^2).
3. Watch cells under the microscope, and, as soon as cells start to detach under gentle shaking, add endothelial cell culture medium.
4. Pellet the cells and resuspend in endothelial cell growth medium.
5. Surfaces are plated at 0.5–1 × 104 endothelial cells/cm^2. Confluency is obtained after 6–7 d. If further passaging is desired, cells should not be allowed to reach confluence and should be passaged when they reach ~75% confluence.

3.3. Characterization of Harvested Cells

The procedures outlined below are the most widely used to assess the expression endothelial differentiation characteristics in cultured cells. They can be used to evaluate the purity of the cells obtained from the isolation procedure, and to monitor the stability of the endothelial phenotype over several passages. For these procedures, it is best to prepare endothelial monolayers in multi-well plates (e.g., 96-well or 24-well) to save on reagents.

3.3.1. Staining for Scavenger Receptor

The scavenger receptor is mainly expressed in endothelial cells, macrophages, and microglial cells *(4–6)*. This receptor is responsible for the uptake of modified lipoproteins. Thus, incubation of live-cell monolayers with DiI-Ac-LDL (1,1'-dioctadecyl-3,3,3',3'-tetramethylindocarbocyanine perchlorate acetylated low-density lipoprotein) has been used as an indication of strain purity.

1. Incubate cells with DiI-Ac-LDL containing medium for 4 h at 37°C.
2. Examine by fluorescence microscopy using standard rhodamine optics.
3. Typical fluorescence localization is shown in **Fig. 3A**. Cells can also be removed from the substrate by trypsinization and analyzed by flow cytometry.

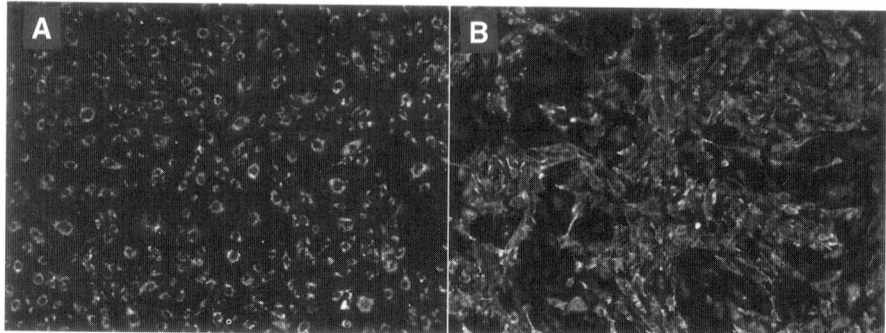

Fig. 3. Staining for endothelial-specific markers. **(A)** Internalized DiI-labeled acetylated LDL forms a punctate pattern in the cytoplasm around the nucleus. **(B)** Localization of von Willebrand factor by immunofluorescence staining.

3.3.2. Staining for von Willebrand Factor

Von Willebrand factor or factor VIII-related antigen is localized in the Weibel-Palade bodies inside the endothelial cell cytoplasm *(7)*. Von Willebrand factor is not found in other cells, except endocardium, platelets, and megakaryocytes.

1. Rinse the endothelial monolayer with DPBSS. Add fixative onto cells and leave at 4°C for 90 min. Rinse with DPBSS with 0.1% BSA.
2. Add the primary antibody and incubate for 60 min at room temperature. Some monolayers receive the specific anti-von Willebrand factor IgG; the others receive the control, nonspecific rabbit IgG.
3. Rinse the monolayer 3× (5 min each), add the fluorescein-labeled secondary antibody, and incubate for 90 min at room temperature.
4. Rinse the monolayer 3×, the last time in distilled water. Cells are ready to be examined by fluorescence microscopy.
5. Typical fluorescence localization is shown in **Fig. 3B**.

4. Notes

1. **Caution:** Gloves must be worn at all times while handling human tissues and cells, because they may carry transmissible pathogens.
2. If a large number of small, rounded cells on the surface of a primary endothelial cell culture is observed after removing blood cells the day after seeding, this is often an early sign of bacterial contamination. These are white blood cells adhering to the endothelial cells as a result of the induction of adhesion molecules by bacterial derived products. Such cultures must be discarded.
3. If the isolation yields fewer cells than expected and sparse seeding is obtained, the cells will never reach confluence, unless endothelial cell-growth medium con-

taining endothelial mitogen and heparin is used. Generally speaking, this medium should be used when culturing endothelial cells for more than 4 d.
4. After reaching confluence on plastic or glass surfaces, endothelial cells cultured for several additional days will often start to peel off. A more adherent substrate such as native collagen gel may be necessary for stable long-term culture.

References

1. Jaffe, E. A., Nachman, R. L., Becker, C. G., and Minick, C. R. (1973) Culture of human endothelial cells derived from umbilical veins. Identification by morphologic and immunologic criteria. *J. Clin. Invest.* **52,** 2745–2756.
2. Gimbrone, M. A., Jr., Cotran, R. S., and Folkman, J. (1974) Human vascular endothelial cells in culture. Growth and DNA synthesis. *J. Cell Biol.* **60,** 673–684.
3. Thornton, S. C., Mueller, S. N., and Levine, E. M. (1983) Human endothelial cells: use of heparin in cloning and long-term serial cultivation. *Science* **222,** 623–625.
4. Giulian, D. and Baker, T. J. (1986) Characterization of ameboid microglia isolated from developing mammalian brain. *J. Neurosci.* **6,** 2163–2178.
5. Voyta, J. C., Via, D. P., Butterfield, C. E., and Zetter, B. R. (1984) Identification and isolation of endothelial cells based on their increased uptake of acetylated-low density lipoprotein. *J. Cell Biol.* **99,** 2034–2040.
6. Pitas, R. E., Innerarity, T. L., Weinstein, J. N., and Mahley, R. W. (1981) Acetoacetylated lipoproteins used to distinguish fibroblasts from macrophages in vitro by fluorescence microscopy. *Arteriosclerosis* **1,** 177–185.
7. Aznar-Salatti, J., Bastida, E., Buchanan, M. R., Castillo, R., Ordinas, A., and Escolar, G. (1990) Differential localization of von Willebrand factor, fibronectin and 13-HODE in human endothelial cell cultures. *Histochemistry* **93,** 507–511.

21

Isolation and Culture of Microvascular Endothelial Cells

Lisa Richard, Paula Velasco, and Michael Detmar

1. Introduction

The cultivation of endothelial cells from large vessels, predominantly from human umbilical veins *(1,2)*, has become a routine procedure in many laboratories and has contributed to the development of modern vascular biology. However, there is convincing evidence that microvascular endothelial cells display a number of important functional differences, compared to large vessel-derived endothelial cells *(3)*, in particular, with regard to their growth factor response *(4,5)* and their regulation of adhesion molecule expression *(6–8)*. Since endothelial cells involved in the pathogenesis of tumor angiogenesis, wound healing, and acute and chronic inflammation are predominantly of microvascular origin, techniques have been developed to isolate endothelial cells from small vessels, most frequently from the skin *(5,9–13)*. The culture of human dermal microvascular endothelial cells (HDMEC) has remained problematic because of difficulties in cell isolation, low cell yields, and short lifespans of the isolated cells. In particular, potential contamination of HDMEC cultures with fibroblasts required time-consuming density-gradient centrifugations *(5,12)* or mechanical removal of fibroblasts *(10)*, and remained problematic after several cell passages. We established a simplified protocol that allows the rapid and reliable immunomagnetic isolation of a well characterized, 100% pure population of HDMEC from neonatal human foreskins. This technique is based on the endothelial cell-specific induction of E-selectin by tumor necrosis factor-alpha (TNF-α) *(14)*, predominantly in postcapillary venule endothelial cells *(15)*, and selection of E-selectin-expressing cells by Dynabeads coupled with an anti-E-selectin monoclonal antibody. Dynabeads are magnetic beads that can be conjugated to specific antibodies, and, after binding to specifically recognized target cells, allow for the immunomagnetic isolation of labeled cells, using a magnetic particle concentrator. Dynabeads coated with Ulex europaeus agglutinin, UEA-1 *(16)*, or with antibodies against thrombomodulin *(17)*, or platelet-endothelial cell-adhesion molecule PECAM-1 (CD31) *(18)*, have been previously used to isolate endothelial cells. However, these markers may cross-react with other skin-cell populations and/or may prevent detachment of the dynabeads because of their persistent expression on the cell surface. Therefore, we chose the transient induction of E-selectin by TNF-α in vitro *(19)* as a specific selection marker. This technique yields 100% pure cultures of HDMEC that express endothelial cell-specific markers, proliferate in response to the specific angiogenesis factor, vascular endothelial growth factor (vascular permeability factor; VEGF/VPF), form capillary-like tubes when overlaid with a collagen type I gel, and undergo a minimum of 35 population doublings before reaching proliferative senescence.

2. Materials

1. Foreskin-collecting medium: The medium for collecting and transporting neonatal foreskins is composed of Hanks' balanced salt solution (HBSS, Gibco-BRL, Grand Island, NY, #11201-092) supplemented with penicillin (400 U/mL), streptomycin (400 μg/mL), and fungizone (1.0 μg/mL) (all from Gibco-BRL, #15240-062). The foreskin-collecting medium is transferred to sterile tubes in 10-mL aliquots and stored at 4°C for up to 4 wk.
2. Trypsin solution: Dissolve 250 mg of porcine trypsin (Sigma, St. Louis, MO, #T-8128) in 100 mL phosphate-buffered saline without Ca^{2+} or Mg^{2+} (Gibco-BRL, #21600-069), to yield a final trypsin concentration of 0.25%. After sterilization through a 0.2-μm filter, the trypsin solution is transferred to sterile containers in 5-mL aliquots and stored at −20°C for up to 1 yr.
3. Endothelial cell-growth medium: The following stock solutions are required:
    a. Hydrocortisone acetate (Sigma, #H-4126), 1 mg/mL stock. Dissolve 10 mg of crystalline hydrocortisone acetate in 10 mL ethanol, transfer to sterile tubes in 500-μL aliquots, and store at −20°C for up to 1 yr.
    b. N6,2'-O-dibutyryl-adenosine 3':5'-cyclic monophosphate (da-cAMP, Sigma, #D-0627), $5 \times 10^{-2} M$ stock. Dissolve 100 mg of crystalline da-cAMP in 4 mL PBS, sterilize through a 0.2-μm filter, transfer to sterile tubes in 500-μL aliquots, and store for up to 1 yr.
    Endothelial growth medium is prepared as follows: Mix 500 mL endothelial cell basal medium (EBM, Clonetics, Walkersville, MD, #CC-3121) with one 500-μL aliquot of hydrocortisone acetate stock (final concentration: 1 μg/mL), and with one 500-μL aliquot of da-cAMP stock (final concentration: $5 \times 10^{-5} M$). Add 20% heat-inactivated fetal calf serum (FCS, Gibco-BRL, #16000-044), 100 U/mL penicillin, and 100 μg/mL Streptomycin.
4. E-selectin-coated magnetic beads: Dynabeads M-450 sheep antimouse IgG ST (Dynal, Lake Success, NY, #410.02) are suspended well by vortexing. Transfer

500 µL beads (15 mg) under sterile conditions into a sterile 1.5-mL microcentrifuge tube and add 10 µg of mouse antihuman E-selectin monoclonal antibody (Genzyme, Cambridge, MA, #2138-01). Mix overnight at 4°C by end-over-end rotation. Insert tube into magnetic particle concentrator for 2 min and discard the supernatant. The E-selectin-coated magnetic beads adhere to the wall of the tube and are collected in PBS containing 1% FCS. Remove tube from magnetic particle concentrator, add 1 mL PBS/1% FCS, mix well, and insert tube again into magnetic particle concentrator. Repeat this washing step 4×, add 500 µL PBS/1% FCS and store E-selectin-coated beads at 4°C for up to 3 mo. Before use, wash beads twice with PBS/1% FCS. For prolonged storage, addition of 0.02% sodium azide is recommended. Sodium azide is removed prior to use by three washes with PBS/1% FCS.

5. Cell-preservative medium: Prepare 10% (v/v) dimethyl sulfoxide (DMSO; Sigma, #D-2650) in EBM medium containing 20% FCS, 1 µg/mL hydrocortisone, and 5×10^{-5} M da-cAMP.

6. Collagen type I-coated culture dishes: Dilute collagen type I solution (Vitrogen 100; Collagen, Palo Alto, CA, #PC 0701) to 50 µg/mL in sterile PBS without Ca^{2+} or Mg^{2+}, and add 2 mL to each 100-mm culture dish (1.5 mL to each 60-mm dish, 1 mL to each 35-mm dish, 0.5 mL to each well of a 24-well plate). Incubate the dishes at 4°C for 1 h. Aspirate the collagen solution, add complete endothelial cell-growth medium, and incubate at 37°C for up to 24 h.

3. Methods
3.1. Establishment of Primary Cultures

1. Prepare the following in a laminar flow hood: Two pairs of sterile forceps, one pair of scissors, surgical scalpel with a no. 21 blade, 20 mL of foreskin-collecting medium.
2. Transport the neonatal foreskins, obtained after routine circumcision, in foreskin-collecting medium, on ice, to the laboratory.
3. Separate the inner and outer sheath of the foreskins using a scalpel, wash twice with foreskin-collecting medium, and incubate in a 100-mm culture dish, in 20 mL foreskin-collecting medium at 37°C for 2 h.
4. Cut foreskins with a scalpel blade into approx 2 × 2-mm squares, wash twice in PBS, and incubate in trypsin solution, dermal side down, at 4°C overnight, or at 37°C for 90 min.
5. Aspirate the trypsin solution, wash foreskins twice with PBS, and add 5 mL of complete endothelial cell-growth medium. Carefully remove the epidermal sheet from the dermis and discard the epidermis. Release endothelial cells and other dermal cells into the medium by scraping the upper side of the dermis with a blunt scalpel blade (no. 21) until the skin appears pink and slimy. The medium appears milky at this point.
6. Discard the remnants of dermal tissue and transfer the cell suspension through a sterile 100-µm nylon mesh into a sterile 50-mL centrifuge tube. Rinse the dish again with 5 mL endothelial growth medium, transfer the cell suspension through the nylon mesh, count the cells, and centrifuge at 300g for 5 min.

7. Resuspend the cells in complete endothelial growth medium, plate the cells at a seeding density of 2×10^4 cells/cm^2 on collagen type I-coated dishes, and incubate at 37°C at 5% CO_2 in a humidified atmosphere.
8. Four h after plating the cells, wash the dishes thoroughly with PBS to remove all nonadherent cells, add endothelial growth medium, and incubate again at 37°C. Thereafter, change the culture medium every 2–3 d until the cultures become 70–80% confluent. At this stage, cultures consist predominantly of a mixture of microvascular endothelial cells, fibroblasts, and epidermal keratinocytes.

3.2. Establishment of Pure Microvascular Endothelial Cell Cultures

1. When primary cultures have reached a stage of 70–80% confluency: Add 100 ng/mL recombinant human TNF-α (several suppliers; TNF-α for our experiments was provided by BASF-Knoll, Mannheim, Germany) directly into the culture medium for 6 h, to selectively induce the expression of E-selectin on dermal microvascular endothelial cells.
2. After 6 h: Add 5 μL of E-selectin-coupled magnetic beads/mL culture medium, and incubate at room temperature with gentle shaking for 5 min.
3. Wash cultures thoroughly four times with PBS without Ca^{2+} or Mg^{2+}, add 1 mL trypsin solution, incubate at 37°C for 5–10 min until the cells are dispersed as single cells, pipet the cell suspension several times up and down, and transfer the cell suspension to a sterile microcentrifuge tube.
4. Insert the microcentrifuge tube into a magnetic particle concentrator for 2 min, and discard the supernatant containing all E-selectin negative cells, including dermal fibroblasts and epidermal keratinocytes. Wash twice with PBS, add endothelial growth medium, and plate dermal microvascular endothelial cells at 5×10^3 cells/cm^2 on collagen type I-coated dishes. Change the culture medium every 2–3 d.
5. At 70–80% confluency: Repeat **steps 1–4**.
6. When subculturing dermal microvascular endothelial cells, split the cells when 90% confluent, at a ratio of 1:5. Seed the cells on collagen type I-coated dishes.

3.3. Results

Primary cultures at 70–80% confluency consist of a mixture of dermal microvascular endothelial cells, dermal fibroblasts, and epidermal keratinocytes. After immunomagnetic purification of primary cultures with anti-E-selectin dynabeads, first-passage cultures usually consist of >99% HDMEC. The vast majority of magnetic beads is shed from the cell surface within 2 h after subculture. Some magnetic beads are taken up by HDMEC; however, the incorporation of magnetic beads did not influence the growth behavior of HDMEC.

After immunomagnetic purification of first-passage cultures, second-passage cultures regularly consist of 100% HDMEC, as evaluated by their characteristic morphology (**Fig. 1**) and their staining with an anti-CD31 antibody. Staining for von Willebrand factor revealed the intracytoplasmic granular stain-

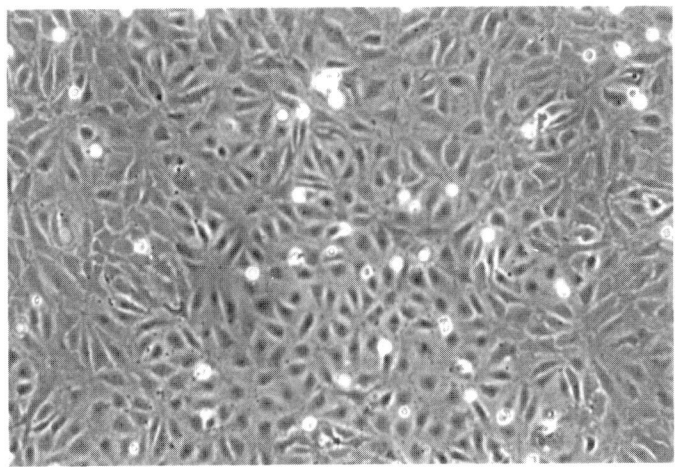

Fig. 1. HDMEC, isolated by Dynabeads coupled with an anti-E-selectin monoclonal antibody (MAb), passage 5. Monolayer culture with characteristic cobblestone morphology.

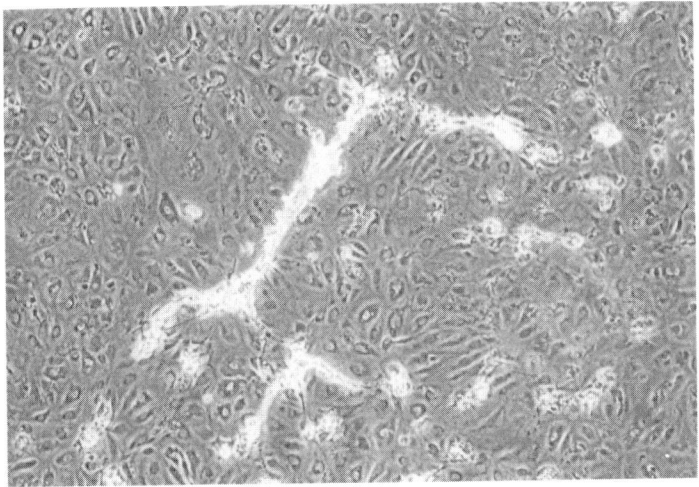

Fig. 2. Passage 6 HDMECs that have been maintained for 6 mo without subculture: Monolayer culture with few tube-like structures and no evidence of contaminating fibroblasts or other cell populations.

ing pattern characteristic for endothelial cells. HDMEC also showed a rapid uptake of acetylated LDL. Even after 6 mo without passage, HDMEC cultures are characterized by the persistence of typical monolayers with some degree of capillary-like tube formation (**Fig. 2**). When confluent HDMEC cultures are

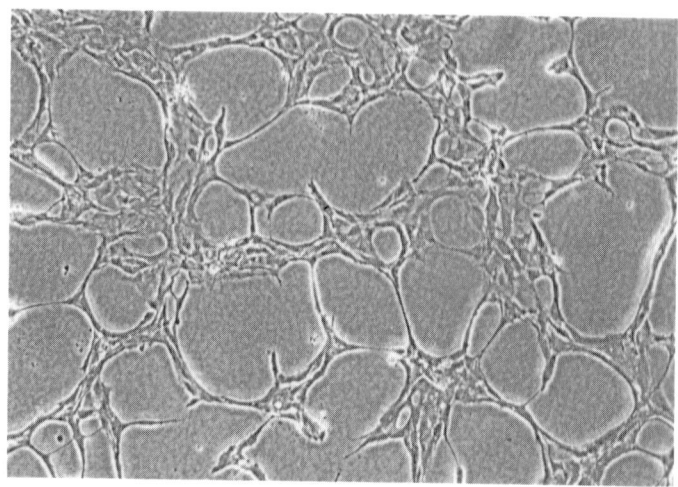

Fig. 3. Spontaneous formation of tube-like structures 16 h after overlay of a collagen type I gel onto confluent HDMEC, passage 5.

overlaid with a collagen type I gel (500 µg/mL in endothelial growth medium), they form capillary-like tubes within 4–6 h (**Fig. 3**). The expression of the VEGF/VPF receptors, KDR *(20)* and Flt-1 *(21)*, has been widely considered as a specific feature of endothelial cells. HDMEC isolated by the described immunomagnetic protocol express both high-affinity receptors for VEGF/VPF, KDR and Flt-1, as demonstrated by Scatchard analysis of ^{125}I-VEGF binding and by immunohistochemistry, using specific monoclonal antibodies against KDR and Flt-1. Accordingly, treatment with VEGF/VPF potently induced the proliferation of HDMEC in concentrations as low as 0.3 ng/mL. The mitogenic effect of VEGF was observed in HDMEC from passages 2–12. HDMEC isolated by the immunomagnetic protocol, using anti-E-selectin coupled Dynabeads, are highly sensitive to the induction of cellular adhesion molecules by cytokines such as IL-1-β and TNF-α. Treatment of HDMEC with TNF-α, in passages 2–8, potently induced the expression of intercellular adhesion molecule-1 (ICAM-1; CD54), vascular cell adhesion molecule-1 (VCAM-1; CD106), and endothelial–leukocyte adhesion molecule-1 (E-selectin, ELAM-1; CD62E).

4. Notes

1. Tissue source and collection: The sources for dermal microvascular endothelial cell cultures are human neonatal foreskins obtained from routine circumcisions. Foreskins are easily obtained, and have a markedly higher density of dermal microvessels than adult human skin. Although it is possible to isolate and culture

HDMEC from adult skin, the endothelial cell yield is considerably lower, and the life span of HDMEC from adult skin is shorter, compared to HDMEC obtained from neonatal foreskins. Foreskin-collecting medium is usually kept in a refrigerator near the surgical area, and specimens are collected over a period of 4 h, during which they are stored at 4°C. A storage period of 4 h does not affect the viability of HDMEC during later cultivation.
2. Sterilization of foreskins: It is important that the surgeon applies appropriate antiseptic protocols prior to circumcisions. In our experience, a 2-h incubation in collecting medium, containing an increased concentration of antibiotics, has helped to keep the rate of bacterial or fungal infections at a minimum.
3. Trypsinization: An overnight incubation at 4°C is usually sufficient to allow for a separation of the epidermis from the dermis. If the epidermis still remains attached, the trypsin solution should be replaced with fresh trypsin solution, and the specimens should be incubated at 37°C for an additional 30 min. It is our experience that the success of the protocol depends on the correct technique to mechanically isolate dermal endothelial cells by scraping the upper side of the dermis with a blunt scalpel blade, once the epidermal sheet has been removed. Alternatively, after trypsin digestion, collagenase (5 mg/mL, Sigma C-7657) can be added to the specimens for 10–20 min to facilitate tissue disintegration. However, when the collagenase digestion is allowed to continue for too long, the dermal tissue disintegrates, rendering the efficient mechanical release of microvascular endothelial cells more difficult.
4. Removal of epidermis from dermis: After trypsin digestion of foreskins, the epidermis is mechanically removed, using two pairs of forceps. One forceps holds the epidermal sheet, and the other holds the lower dermis. The epidermal sheet is then carefully removed and discarded.
5. Endothelial cell-growth medium: Several authors have reported that the addition of up to 30% human serum to the growth medium was necessary to sustain HDMEC growth. In our experience, pure HDMEC cultures can be maintained and efficiently propagated with only the addition of FCS. This avoids the high costs and potential contamination risks associated with human serum. In addition, the use of prepartum maternal serum has been suggested to enhance the growth of HDMEC in vitro *(22)*. However, this has not been necessary in our experience, since cultivation in 20% FCS enables the cultivation of pure HDMECs, which express several endothelial cell markers, and are characterized by an extended life-span of more than 35 population doublings. Moreover, prepartum maternal serum may contain a number of additional growth factors, possibly including VEGF, complicating the study of isolated effects of growth factors and cytokines. Epidermal growth factor has been used and recommended by most authors as an additional supplement to the endothelial growth medium for HDMEC cultivation. This seems to be rather unnecessary, since HDMECs in vitro do not express EGF receptors, do not react to the mitogenic effects of EGF, do not show any detectable expression of EGF receptors on Western blots, and do not show tyrosine phosphorylation after EGF stimulation (M. Detmar et al., unpublished data).

6. Collagen coating of culture dishes: We routinely coat culture dishes with 50 µg/mL collagen type I (Vitrogen). However, coating of culture dishes with fibronectin (50 µg/mL) may increase the seeding efficiency of primary endothelial cell isolates. Aside from primary endothelial cell cultures, we have been unable to detect significant differences in the plating efficiency of HDMEC on fibronectin vs collagen type I in later passages. Instead of Vitrogen, some authors plate HDMEC on gelatin-coated dishes, without any obvious advantage or disadvantage over Vitrogen.

References

1. Jaffe, E. A., Nachmann, R. L., Becker, C. G., and Minick, C. R. (1973) Culture of human endothelial cells from umbilical veins. Identification by morphologic and immunologic criteria. *J. Clin. Invest.* **52,** 2745–2756.
2. Maruyama, Y. (1973) The human endothelial cell in tissue culture. *Z. Zellforschung* **60,** 69–79.
3. Kumar, A. U., West, D. C., and Ager, A. (1987) Heterogeneity in endothelial cells from large vessels and microvessels. *Differentiation* **36,** 57–70.
4. Detmar, M., Imcke, E., Ruszczak, Z., and Orfanos, C. E. (1990) Effects of tumor necrosis factor-alpha (TNF) on cultured microvascular endothelial cells derived from human dermis. *J. Invest. Dermatol.* **95,** 219S–222S.
5. Detmar, M., Tenorio, S., Hettmannsperger, U., Ruszczak, Z., and Orfanos, C. E. (1992) Cytokine regulation of proliferation and ICAM-1 expression of human dermal microvascular endothelial cells in vitro. *J. Invest. Dermatol.* **98,** 147–153.
6. Lee, K. H., Lawley, T. J., Xu, Y., and Swerlick, R. A. (1992) VCAM-1, Elam-1, and ICAM-1–independent adhesion of melanoma cells to cultured human dermal microvascular endothelial cells. *J. Immunol.* **98,** 79–85.
7. Swerlick, R. A., Lee, K. H., Li, L. J., Sepp, N. T., Caughman, S. W., and Lawley, T. J. (1992) Regulation of vascular cell adhesion molecule 1 on human dermal microvascular endothelial cells. *J. Immunol.* **149,** 698–705.
8. Swerlick, R. A., Lee, K. H., Wick, T. M., and Lawley, T. J. (1992) Human dermal microvascular endothelial but not human umbilical vein endothelial cells express CD36 in vivo and in vitro. *J. Immunol.* **148,** 78–83.
9. Folkman, J., Haudenschild, C. C., and Zetter, B. R. (1979) Long-term culture of capillary endothelial cells. *Proc. Natl. Acad. Sci. USA* **76,** 5217–5221.
10. Davison, P. M., Bensch, K., and Karasek, M. A. (1980) Isolation and growth of endothelial cells from the microvessels of the newborn human foreskin in cell culture. *J. Invest. Dermatol.* **75,** 316–321.
11. Davison, P. M., Bensch, K., and Karasek, M. A. (1983) Isolation and long-term serial cultivation of endothelial cells from the microvessels of the adult human dermis. *In Vitro* **19,** 937–945.
12. Kubota, Y., Kleinman, H. K., Martin, G. R., and Lawley, T. J. (1988) Role of laminin and basement membrane in the morphological differentiation of human endothelial cells into capillary-like structures. *J. Cell Biol.* **107,** 1589–1598.

13. Kraeling, B. M., Jimenez, S. A., Sorger, T., and Maul, G. G. (1994) Isolation and characterization of microvascular endothelial cells from the adult human dermis and from skin biopsies of patients with systemic sclerosis. *Lab. Invest.* **71,** 745–754.
14. Bevilacqua, M. P., Stengelin, S., Gimbrone, M. A., Jr., and Seed, B. (1989) Endothelial leukocyte adhesion molecule 1: An inducible receptor for neutrophils related to complement regulatory proteins and lectins. *Science* **243,** 1160–1165.
15. Groves, R. W., Allen, M. H., Barker, J. N. W. N., Haskard, D. O., and MacDonald, D. M. (1991) Endothelial leucocyte adhesion molecule-1 (ELAM-1) expression in cutaneous inflammation. *Br. J. Dermatol.* **124,** 117–123.
16. Jackson, C. J., Garbett, P. K., Nissen, B., and Schrieber, L. (1990) Binding of human endothelium to Ulex europaeus-1 coated dynabeads: application to the isolation of microvascular endothelium. *J. Cell Sci.* **96,** 257–262.
17. Drake, B. L. and Loke, Y. W. (1991) Isolation of endothelial cells from human first trimester decidua using immunomagnetic beads. *Human Reprod.* **6,** 1156–1159.
18. Hewett, P. W. and Murray, J. C. (1993) Immunomagnetic purification of human microvessel endothelial cells using Dynabeads coated with monoclonal antibodies to PECAM-1. *Eur. J. Cell Biol.* **62,** 451–454.
19. Sepp, N. T., Gille, J., Li, L. J., Caughman, S. W., Lawley, T. J., and Swerlick, R. A. (1994) A factor in human plasma permits persistent expression of E-selectin by human endothelial cells. *J. Invest. Dermatol.* **102,** 445–50.
20. Terman, B. I., Dougher, V. M., Carrion, M. E., Dimitrov, D., Armellino, D. C., Gospodarowicz, D., and Bohlen, P. (1992) Identification of the KDR tyrosine kinase as a receptor for vascular endothelial cell growth factor. *Biochem. Biophys. Res. Commun.* **187,** 1579–1586.
21. deVries, C., Escobedo, J., Ueno, H., Houck, K., Ferrara, N., and Williams, L. T. (1992) The fms-like tyrosine kinase, a receptor for vascular endothelial growth factor. *Science* **255,** 989–991.
22. Karasek, M. A. (1989) Microvascular endothelial cell culture. *J. Invest. Dermatol.* 33S–38S.

22

Initiation, Maintenance, and Quantification of Human Hematopoietic Cell Cultures

Paul C. Collins, Sanjay D. Patel, William M. Miller, and E. Terry Papoutsakis

1. Introduction

The culture of hematopoietic cells for cell and gene therapies is a rapidly growing area within the field of applied hematology and tissue engineering. As evidenced by recent clinical trials *(1)*, ex vivo expanded hematopoietic cells offer great promise for the reconstitution of in vivo hematopoiesis in immunocompromised patients who have undergone chemotherapy. Other potential applications for ex vivo expansion include production of cycling stem and progenitor cells for gene therapy, expansion of dendritic cells for immunotherapy, and production of red blood cells (RBCs) and platelets for transfusions *(2)*.

The heterogeneous cell population contained in a hematopoietic culture is always changing as a result of the delicate balance between proliferation of certain cell types, their differentiation into other cell types, and the death of various cell populations. Through the combination of different cytokines, media, and physicochemical parameters, hematopoietic cultures can be directed toward lineages of interest. Complete evaluation of the performance of hematopoietic cultures requires the use of long-term assays such as the 2-wk methylcellulose assay, to detect progenitor or colony-forming cells (CFC), and the very primitive long-term culture-initiating cell (LTC-IC) assay (7 wk). Flow cytometry can be utilized to quantify cells bearing antigens, such as CD34 (primitive progenitors), CD15 and CD11b (granulocyte and monocyte postprogenitors), and glycophorin A (maturing erythrocytes).

In this chapter, we describe methods for the initiation and maintenance of hematopoietic cultures, the quantification of a variety of cell types, and the setup of mixed suspension cultures in spinner flasks.

From: *Methods in Molecular Medicine, Vol. 18: Tissue Engineering Methods and Protocols*
Edited by: J. R. Morgan and M. L. Yarmush © Humana Press Inc., Totowa, NJ

2. Materials
2.1. Isolation
1. Ficoll (Sigma, St. Louis, MO).
2. Phosphate-buffered saline (PBS) (Sigma).

2.2. Culture Devices and Setup
1. Human long-term medium (HLTM).
   a. 300 mL McCoy's 5A medium (Sigma).
   b. 50 mL horse serum (HS, StemCell Technologies, Vancouver, BC).
   c. 50 mL fetal bovine serum (FBS, StemCell Technologies) (*see* **Note 1**).
   d. 4 mL 100X nonessential amino acid solution (Irvine Scientific, Irvine, CA).
   e. 4 mL 100 mM sodium pyruvate (Sigma).
   f. 4 mL 100X L-glutamine (Sigma).
   g. 4 mL 100X α-thioglycerol (Sigma) solution (10^{-4} M in McCoy's 5A medium).
   h. 4 mL 100X MEM vitamin solution (Irvine Scientific).
   i. 4 mL hydrocortisone (Sigma) solution (10^{-4} M in McCoy's 5A medium) (*see* **Note 2**).
   j. 2.5 mL 50X essential amino acid solution (Irvine Scientific).
   k. 0.4 mL gentamycin sulfate (Sigma).
2. XVIVO-20 serum-free medium (BioWhittaker, Walkersville, MD).
3. Hematopoietic cytokines: stem cell factor (SCF, Amgen, Thousand Oaks, CA), Interleukin 3 (IL-3, Novartis, East Hanover, NJ), Interleukin 6 (IL-6, Novartis), granulocyte-colony stimulating factor (G-CSF, Amgen), granulocyte/monocyte-stimulating factor (GM-CSF, Immunex, Seattle, WA), erythropoietin (Epo, Amgen).
4. PBS (Sigma).
5. Trypsin (Sigma).
6. 24-well tissue culture plate (Falcon, Lincoln Park, NJ).
7. 25 cm^2 tissue culture flask (Falcon).
8. Model 1967-100 spinner flask assembly (Bellco, Vineland, NJ). Do not siliconize the flask (*see* **Notes 3** and **4**). Replace the standard agitator assembly with the agitator assembly designed for use with Bellco's microcarrier spinner flasks (1965 series). This agitator assembly contains a flat blade impeller that improves cell suspension over the 1967 model's magnetic stir bar (*see* **Note 5**). Remove the magnetic stir bar and flat blade from the agitator shaft assembly and look inside the shaft housing. If there is a small silicone component inside the shaft housing, remove it (*see* **Note 6**). Adjust the agitator shaft so that the impeller assembly hangs freely above the flask bottom. The impeller assembly should rest 1–2 mm above the flask bottom.

2.3. Quantitative Analysis
2.3.1. Total Nucleated Cells
1. Cetrimide solution: Combine all components and filter into bottles.
   a. 90 g cetrimide (Hexadecyltrimethylammonium bromide) powder (Sigma).

b. 25 g NaCl (Sigma).
c. 1.1 g EDTA (Sigma).
d. 3 L deionized water (dH$_2$O).

2.3.2. Colony-Forming Cell Assay

2.3.2.1. COLONY ASSAY MEDIUM

Complete colony assay medium (methylcellulose) can be purchased from StemCell Technologies (*see* **Note 1**). Alternatively, the colony assay medium can be made from individual components.

1. 2X IMDM + 2-mercaptoethanol.
   a. Prepare IMDM solution with twice the required IMDM (Sigma) and sodium bicarbonate, as specified in the manufacturer's instructions.
   b. Add 25.4 µL of 2-mercaptoethanol (2-ME) solution (Fisher Scientific, Itasca, IL) to 1000 mL of 2X IMDM.
   c. Add 2 mL of gentamycin sulfate (Sigma) to the 1000 mL of 2X IMDM + 2 ME.
2. Methylcellulose stock solution.
   a. Autoclave 800 mL of ddH$_2$O with a 4-in. magnetic stir bar in a 2-L Erlenmeyer flask.
   b. Place the autoclaved flask on a heated stir plate and add 33.2 g of methylcellulose powder (Gibco, Grand Island, NY) while the water is still hot.
   c. Slowly increase the heat level on the stir plate until the methylcellulose solution begins to boil. The slurry has a tendency to boil over, so watch the boiling process carefully. Remove the flask from the stir plate if necessary.
   d. Remove the flask from the stir plate to stop boiling. Repeat the boiling process 2×.
   e. Allow the methylcellulose solution to cool to room temperature. Slowly add 800 mL of the cold 2X IMDM solution prepared in **step 1**. The resulting solution will retain some of the cloudiness of the pure methylcellulose solution.
   f. Place the methylcellulose–IMDM solution onto a stir plate inside a cold (4°C) room. Allow the methylcellulose to stir over a 2-d period. The methylcellulose should now appear optically clear.
   g. Aliquot the methylcellulose stock solution into sterile 50-mL tubes and freeze at –20°C until needed.
3. Colony Assay Medium (CAM).
   a. Place the following components into each of four 50-mL sterile centrifuge tubes: 3.75 mL IMDM; 3.35 mL BSA (StemCell Technologies); 15 mL FBS; 27.5 mL methylcellulose stock solution; 0.4 mL sterile ddH$_2$O.
   b. Shake the contents of the tube until well mixed. Centrifuge at 750g for 20 min. A pellet of undissolved methylcellulose powder will be at the bottom of the tube. Decant the CAM into a sterile 200-mL container. Add the following growth factors at the concentrations given for liquid culture: SCF, IL-3, IL-6, G-CSF, GM-CSF, and Epo.

2.3.3. LTC-IC Assay

1. 24-well plates.
2. 75-cm² T-flasks.
3. LTBMC medium. LTBMC is formed from a base of 1X IMDM with 10% FBS, 10% HS, 4 mM glutamine, 100 µg/mL penicillin, 10 U/mL streptomycin, and 5 µmol/L hydrocortisone.

2.3.4. Flow Cytometry

1. PBS.
2. PAB. The following reagents are required: 4 L PBS; 20 g BSA fraction V (Sigma); 4 g sodium azide (Sigma).
   a. Dissolve BSA and sodium azide in PBS under stirring and low heat (overnight, if needed).
   b. Filter and store at 4°C. Expires in 3 mo.
3. NH_4Cl lysing solution. The following reagents will be required: 1 L dH_2O; 8.26 g NH_4Cl (Sigma); 1.0 g potassium bicarbonate (Sigma); 0.0037 g tetra sodium EDTA (Sigma). Stir until all the salts are dissolved. Store at room temperature. Expires in 3 mo.
4. Propidium iodide. The following reagents are required: 1 mL ethyl alcohol; 0.002 g propidium iodide (Sigma); 9 mL PBS.
   a. Dissolve propidium iodide in ethyl alcohol.
   b. Dilute the solution with PBS, wrap the container in aluminum foil, and store at 4°C. Expires in 1 yr.
5. Fluorescent antibodies: FITC IgG (Becton-Dickinson, San Jose, CA), PE IgG (Becton-Dickinson), FITC CD45 (Becton-Dickinson), PE CD45 (Coulter, Hialeah, FL), FITC CD15 (Becton-Dickinson), PE CD11b (Becton-Dickinson).
6. Polystyrene round-bottom tubes, 12 × 75 mm (Becton-Dickinson Labware, Franklin Lakes, NJ).
7. 1% paraformaldehyde fixative solution. The following reagents are required: 21.4 g cacodylate acid sodium salt (Sigma); 2 L dH_2O; 20 g paraformaldehyde (Sigma); 15 g NaCl (Sigma).
   a. Dissolve cacodylate acid sodium salt in water.
   b. Adjust pH to 7.2 using 1 N NaOH.
   c. Add paraformaldehyde, stirring on a hot plate in a fume hood until dissolved.
   d. Add and dissolve NaCl.
   e. Filter sterilize and store at 4°C. Expires in 1 yr.

3. Methods

3.1. Isolation

Hematopoietic cells for culture are obtained from three major sources: bone marrow aspirates, umbilical cord blood, and mobilized peripheral blood. Depending on the source and the nature of the intended experiment, separation and isolation of particular hematopoietic populations may be necessary. The two most commonly utilized separation techniques for hematopoietic cells are mononuclear cell (MNC) separation from whole blood (Ficoll) and CD34$^+$ cell selection from mononuclear cells.

3.1.1. MNC Separation from Whole Blood

Of the three cell sources mentioned above, both bone marrow aspirates and umbilical cord blood samples require separation of the MNC fraction because of their high RBC content. Mobilized peripheral blood is more variable in its RBC content, but can often be used as obtained from the apheresis process (*see* **Note 7**). Ficoll is slightly more dense than water (1.077 g/mL). Cells will be separated based on relative density, with denser RBCs and mature neutrophils passing through the Ficoll, and MNC retained on the Ficoll–cell interface.

1. Place 15 mL of 37°C Ficoll in a sterile 50-mL polypropylene tube. Withdraw 5 mL of sample into a 25-mL pipet. Using the same pipette, remove an additional 20 mL of 37°C PBS, bringing the total volume in the pipet to 25 mL. Slowly pipet the cell suspension on top of the Ficoll (*see* **Note 8**).
2. Repeat **step 1** with as many tubes as necessary to separate the entire sample.
3. Place the tubes into a centrifuge, being careful not to mix the two layers. Run the centrifuge at 300*g* for 30 min.
4. When the tubes are removed, you should notice a red pellet at the bottom of each tube and an opaque layer of cells at the Ficoll–PBS interface. Carefully remove the interface layer from the tube with a 10-mL pipet, being careful not to remove much of the Ficoll.
5. Combine all the interface layers into one or two 50-mL tubes and centrifuge at 300*g* for 10 min to pellet the MNC (*see* **Note 9**).
6. Aspirate the supernatant off the MNC pellet and wash the cells with the medium that will be used in the experiment. Count the cells and resuspend to a density that will facilitate the setup of the experiment.

3.1.2. CD34$^+$ Cell Selection

CD34 is an antigen expressed on more primitive hematopoietic cells. Cells at the colony-forming cell stage, and those more primitive, express CD34. Colony-forming cells are capable of greater proliferation than more differentiated cells, and, as such, cultures rich in CD34$^+$ cells (selected populations range from ~70–95% CD34$^+$, depending on technique) will undergo a larger total cell expansion than those with a lower CD34$^+$ content (such as MNC cultures, which are typically 0.1–10% CD34$^+$, depending on the cell source). The relatively undifferentiated nature of CD34$^+$ cells also allows for flexibility in directing the culture into lineages of interest. Several methods exist for separation of MNCs into CD34$^+$ and CD34$^-$ populations. All methods rely upon the use of an antibody that recognizes the CD34 antigen and subsequent recovery of the cell–antibody complex. The MiniMACS (Miltenyi Biotech, Sunnyvale, CA) CD34 separation column provides excellent separation efficiencies *(3)* in 2-h time (*see* **Note 10**). In this procedure, the CD34 antibody is attached to a magnetic microsphere and the positive population is trapped within a magnetic field, but the negative population is washed free. Use the protocol supplied by the manufacturer (*see* **Note 11**).

3.2. Culture Devices and Setup

Tissue-culture-treated polystyrene well plates and T-flasks provide an excellent growth environment for hematopoietic cells. Hematopoietic cultures in these systems are similar to those of other animal cell cultures, except that hematopoietic cells are largely nonadherent. The homogeneous environment of a stirred system is excellent for experiments requiring frequent sampling and characterization. The utilization of stirred systems presents some challenges, and we therefore present a method for the setup and culture of hematopoietic cells in 100-mL spinner flasks.

3.2.1. Cytokine Choice

The cytokines listed in Materials (**Subheading 2.2.**) will yield an acceptable expansion of total cells and all types of progenitor cells. The cytokines listed can be classified into three groups: a group acting on primitive hematopoietic cells (SCF); a group (IL-3, IL-6, GM-CSF) acting on a wide array of progenitor cells; and a group (G-CSF, Epo) acting on more mature, lineage-restricted cells. This list of cytokines is by no means exhaustive. There are several additional early-acting (Flt3-ligand), multilineage (PIXY321), and lineage-restricted (thrombopoietin, macrophage-colony stimulating factor) cytokines that can also be used *(4,5)*. It may not be possible to obtain all the recommended cytokines because of cost or availability. In this case, it is generally acceptable to use a combination derived by choosing at least one cytokine from each of the three aforementioned groups. Consult literature references to determine optimal concentrations if their use is desired. Additionally, if the production of a specific lineage of cells is desired in culture, consult literature references to determine which cytokines should be included in or excluded from the medium.

3.2.2. Biocompatibility of Materials

Tissue-culture-treated polystyrene, commonly utilized in the construction of well plates and T-flasks, is biocompatible with hematopoietic cells. However, other materials commonly used for the construction of culture devices for animal cells may not be compatible with hematopoietic cells. Serum can partially protect hematopoietic cells from the negative effects of some construction materials. Material compatibility is therefore especially important if the culture is carried out in serum-free medium. Silicone, glass, and polycarbonate are a few of the materials that adversely affect hematopoietic culture performance, as identified in a recent publication *(6)*. When designing custom culture systems, the relative performance of hematopoietic cells on these materials should be examined.

3.2.3. Static Culture Initiation and Maintenance

1. Using the medium of choice (*see* **Note 12**), add the proper concentrations of cytokines, and suspend the cells to the desired inoculum density (*see* **Note 13**). Recommended cytokine concentrations are: SCF (50 ng/mL), IL-3 (5 ng/mL), IL-6 (10 ng/mL), G-CSF (150 U/mL), GM-CSF (200 U/mL), Epo (3 U/mL).
2. Pipet the proper volume of cell suspension (*see* **Note 14**) into a 24-well plate or 25-cm^2 tissue-culture flask (*see* **Note 15**).
3. Incubate the cultures in a fully humidified atmosphere of 5% CO_2 and air, or 5% CO_2, 5% O_2, and balance N_2 (*see* **Note 16**).
4. Feed the cultures periodically (*see* **Note 17**), as follows:
   For well plates:
   a. Add 1 mL of fresh medium and cytokines to each well. If the volume in a well has reached its maximum (2 mL), then the contents of the well can be split into two wells, with 1 mL fresh medium and cytokines added to each well.
   For T-flasks:
   a. Remove one-half of the cell suspension from the flask, and place in a sterile 15-mL polypropylene tube. Centrifuge this fraction for 10 min at 300g.
   b. Using a pipet, gently remove the supernatant, and replace it with fresh, prewarmed (37°C) medium and cytokines.
   c. Gently resuspend the cell pellet, and return the cell suspension to the original T-flask.
5. To perform the desired cell assays on a given day for T-flask cultures, simply remove the required cell suspension volume. To perform the assays on well-plate cultures, the contents of the entire well must be harvested in the following manner:
   a. Withdraw the entire cell suspension from the well using a pipet.
   b. Wash the well with 250 µL PBS.
   c. Add 250 µL of prewarmed trypsin (37°C) to the well, and incubate the well plate at 37°C for 10 min. This step (along with the next two) can be neglected if only the nonadherent cell population is to be examined, as is the case in T-flasks.
   d. Add 250 µL of HLTM to the well, and scrape the well bottom using a pipet (*see* **Note 18**).
   e. Withdraw the adherent cell fraction, and wash the well with 250 µL PBS.

3.2.4. Spinner Flask Initiation and Maintenance

3.2.4.1. SPINNER CLEANING AND PREPARATION

1. Aspirate out the previous contents of the spinner. If the spinner flask is new, proceed to step 2.

2. Fill the spinner flask with ddH$_2$O and autoclave the flask for 20 min. Do not bleach the flask (see **Note 19**).
3. Discard the sterilized water and flush the flask well with water. Remove the flask side-arm caps and agitator assembly for separate washing.
4. Fill the flask with water and a detergent suitable for culture glassware. Scrub the inner walls of the spinner flask and rinse well with water.
5. Remove the agitator stir bar and flat blade from the shaft assembly. Wash the bar and blade well. Scrub the shaft housing and rinse the inside portion of the shaft well. Wash the side-arm caps. Reassemble the agitator assembly and the spinner flask assembly.
6. Rinse the entire spinner flask assembly well with deionized H$_2$O and then fill with 10 mL of ddH$_2$O. Sterilize the spinner flask for 20 min in an autoclave.
7. Aspirate out the sterile water in the spinner flask and rinse the inside of the flask with 20 mL sterile PBS. Swirl the flask to contact as much of the spinner surface as possible with PBS. Aspirate out the PBS. The flask is now ready to use.

3.2.4.2. Spinner Flask Culture Initiation

1. For a serum-containing medium, the minimum acceptable seeding density for cord blood or peripheral blood mononuclear cultures is 2×10^5 cells/mL; serum-free cultures require an inoculum density of 3×10^5 cells/mL. For CD34$^+$ cultures, 5×10^4 cells/mL is the minimum density for either type of medium. We have initiated MNC cultures as high as 1.5×10^6 cells/mL with good results. Inoculum densities for CD34$^+$ cultures should not exceed 1.2×10^5 cells/mL (see **Note 20**). A good intermediate seeding density for both serum-containing and serum-free MNC cultures is 5×10^5 cells/mL and 7.5×10^4 cells/mL for CD34$^+$ cell cultures.
2. Inoculate the culture with 30–50 mL of the cell suspension.
3. Place the spinner flask on a magnetic stir plate at 30–40 RPM in a 5% CO$_2$ incubator (see **Notes 21** and **22**). Spinner cultures will grow well in either a 5 or 20% oxygen environment.

3.2.4.3. Spinner Maintenance

Depending of the purpose of the experiment being conducted in the spinner, the culture duration will vary. We have carried out cultures as long as 28 d, although 10–14 d are usually sufficient to observe large changes in many hematopoietic cell populations. Regardless of the length, most cultures will require medium exchange. Many protocols can be used successfully to maintain sufficient nutrient levels and low byproduct levels. The protocol presented here is the most commonly utilized in our spinner cultures, and serves the majority of cultures well.

1. On d 4, use a pipet to remove 50% of the culture volume and place in a sterile 50-mL polypropylene tube. Pellet the cells by centrifugation at 300g for 10 min. Place the spinner flask, with its remaining cells, back into the incubator during this time (see **Note 23**).

Human Hematopoietic Cell Cultures

2. Aspirate the medium supernatant off the cell pellet and replace with an equivalent amount of fresh, prewarmed (37°C) medium plus cytokines. Pipet the medium up and down to resuspend the cell pellet. Return the cells to the spinner flask.
3. Repeat the feeding procedure every other day until the end of the culture (*see* **Note 24**).

3.3. Quantitative Analysis

3.3.1. Total Nucleated Cells

The increase or decrease in the number of total nucleated cells present in the culture is a good measure of the effectiveness of the particular experimental condition in stimulating hematopoietic expansion. Nucleated cells are of primary interest, because they are capable of further proliferation or differentiation. Mature RBCs do not have a nucleus, and therefore are of less interest in hematopoietic cell culture. For this reason, cell lysis is performed and nuclei are counted. The slow growth kinetics of most hematopoietic cultures makes a once-a-day analysis adequate.

Use a Coulter Counter or similar apparatus to count the cells (*see* **Note 25**).

1. A total of 10 mL of 37°C cetrimide (*see* **Note 26**) and culture sample should be mixed in a coulter vial, i.e., 9.9 mL cetrimide and 100 µL of sample. Other dilution ratios can be used (*see* **Note 27**).
2. Set the gates on the Coulter counter to measure particles (cell nuclei) between 2.5 and 7.5 µ.
3. Perform a count on cell-free cetrimide to determine the background level within the size range of interest.
4. Count the nuclei in the prepared Coulter vial and subtract the blank value. Based on dilution ratios, calculate the nucleated cell density present in the culture.

3.3.2. Colony-Forming Cells (CFC)

Counting total nucleated cells only measures overall proliferation of the culture. The heterogeneous nature of hematopoietic cell culture requires the assay of specific cell types within the nucleated population. Colony-forming cells (CFC) give rise to the mature, differentiated cells of their lineage. The measurement of CFC therefore provides insight as to the ability of the experimental condition tested to expand cells of a particular hematopoietic lineage. Assays for CFC rely on immobilization of the cultured cells in a semisolid medium, until the differentiated progeny of the CFC accumulate to a degree that they can be distinguished under a microscope. The CFC assay should be performed along with the nucleated cell count.

1. Place 3 mL of colony assay medium (CAM) into a sterile 17 × 100-mm polypropylene tube. Allow the medium to warm to 37°C (*see* **Note 28**).

2. Calculate the volume of cells necessary to inoculate the colony assay medium at a suitable density (*see* **Note 29**), such as 10,000 cells/mL. Multiple seeding densities are recommended, until experience with the particular parameters allows for the optimal choice to be made.
3. Pipet the culture sample into the CAM, and vortex well.
4. The CAM should be removed from the tube with a positive-displacement repeat pipeter (*see* **Note 30**), and aliquoted into 35-mm Petri dishes (1 mL per dish). The 35-mm dishes should be placed inside a larger Petri dish (150 mm) containing 15 mL of sterile water, which helps control local evaporation.
5. Culture for 13–15 d in a 5% O_2, 5% CO_2, balance N_2 incubator at 37°C (*see* **Note 31**).
6. Examine the cultures under a dark-field stereo microscope at ×25–35 magnification. Colonies of granulocytes, which arise from the colony-forming unit-granulocyte (CFU-G), will appear as highly refractile, white colonies of greater than 50 cells. Colonies of monocytes/macrophages, which arise from the CFU-M, are also white colonies of greater than 50 cells. They are usually distinguishable from granulocyte colonies through examination of individual cells of the colony. Individual cells of a monocyte colony are generally larger and less refractile than individual members of a granulocyte colony. Monocyte colonies are often less tightly packed than granulocyte colonies. CFU-GM (granulocyte/monocyte) give rise to colonies that contain cells from both the G and *M* lineages. Colonies arising from the burst-forming unit erythroid (BFU-E) are pink to deep red in appearance and contain subunit clusters, which can be numerous (*see* **Note 32**).
7. From the number of each CFC type counted and the plating density, calculate the number of each CFC type present in the sample. For example, if the colony assay was inoculated at 5,000 cells/mL and 50 granulocyte colonies were counted after incubation, the frequency of occurrence of CFU-G in the sample is 1/100.

3.3.3. Long-term Culture-Initiating Cell (LTC-IC)

The LTC-IC assay detects cells close to the stem-cell compartment. Many variations of the LTC-IC assay exist *(7–9)*, but most are centered around the concept of culturing cells for a 5-wk time period, after which the cells are plated into a standard CFC assay. Only very primitive cells will not have terminally differentiated in 5 wk, and hence will be able to form colonies in the CFC assay. Thus, these colonies represent LTC-IC. Because of the time involved in carrying out the LTC-IC assay, it is not usually performed as frequently as the CFC assay. We outline a modification of the method described by Koller et al. *(10)*.

3.3.3.1. STROMA PREPARATION

Most LTC-IC assays involve the use of accessory cells (known as stroma) from the bone marrow. Stromal cells produce growth factors and extracellular matrix molecules that support hematopoietic cells. Bone marrow aspirates are the traditional source of stromal cells for this assay (*see* **Note 33**).

1. Ficoll the bone marrow aspirate using the method described above for MNC isolation.
2. Seed T-75 flasks with 1×10^6 MNC/mL in a total of 10 mL of LTBMC medium.
3. Incubate at 33°C for 2 wk in a 5% CO_2 incubator. Perform a 50% medium exchange with LTBMC medium once a week.
4. The stromal cells should have formed a confluent layer on the bottom of the T-flask. Remove the medium from the flask and wash with 10 mL of PBS. Remove the stromal layer by trypsinizing at 33°C for 10 min. Pellet the collected stromal cells and wash twice with LTBMC. Determine the nucleated density of stromal cells. Resuspend the stromal cells so that the density is 5×10^4 cells/mL.
5. Irradiate the stromal cells at 20 cGy (*see* **Note 34**).
6. Prepare 24-well plates for the assay. The outer ring of wells should be filled with 2 mL of sterile water per well to reduce medium evaporation over the 5-wk culture period.
7. Pipet 1 mL of stromal cell suspension into each of the eight center wells. Each well should therefore contain 5×10^4 stromal cells. Each LTC-IC assay will use three wells. Use as many well plates as needed to carry out the desired number of LTC-IC assays.
8. Allow the stromal cells to attach to the bottom of the well plate overnight. The stromal layers are then ready for use in the LTC-IC assay.

3.3.3.2. LTC-IC Initiation

1. Remove 3×10^6 cells from the culture of interest. Pellet the cells and remove the supernatant. Resuspend in LTBMC medium at a density of 1×10^6 cells/mL. Place 1 mL of the suspension onto one of the eight prepared stromal layers within the well plate. The assay should be done in triplicate (*see* **Note 35**).
2. Each week, carefully remove 1 mL of medium from the top of the culture without removing cells at the bottom of the well. Replace the discarded medium with fresh LTBMC medium.
3. The culture is carried out for a total of 5 wk. At the end of the culture, remove all nonadherent cells, using a pipet. Trypsinize the adherent layer and combine with the nonadherent cells. Rinse the well with 250 µL of LTBMC to quench the trypsin. Determine the nucleated cell density of the suspension.
4. Initiate standard CFC assays at seeding densities of 2.5×10^4 and 5×10^4 cells/mL.
5. Enumerate the number of colonies at d 13–15 of culture in the same fashion as for the CFC assay. Most colonies will either be CFU-M or CFU-G. BFU-E appear more rarely.
6. An LTC-IC is generally thought to differentiate into 4 CFC. The LTC-IC frequency of the culture is therefore: Number of CFC counted/4 · [Colony assay seeding density].

3.3.4. Flow Cytometry

Flow cytometric techniques can be used to identify and quantify a specific population of cells in a heterogeneous mixture of other cells, based on cell size, granularity, and surface antigen expression. Because the expression of surface antigens on hematopoietic cells changes as the cells differentiate, flow cytometry can also be used to monitor cell differentiation in the cultures.

3.3.4.1. Cell Staining

1. A minimum of four polystyrene tubes will be required (three tubes to adjust the flow cytometer settings, and one analysis tube). Add $1 \times 10^5 - 1 \times 10^6$ cells to each of the tubes.
2. Centrifuge the tubes for 3 min at 3500g and decant the supernatant.
3. Vortex the pellet, and wash with 1 mL PAB (centrifuge and decant supernatant).
4. Add the antibodies to each pellet as described below (*see* **Notes 36** and **37**), vortex the pellet, and incubate the tubes in a 4°C refrigerator for 15 min.
   Tube 1: FITC isotype control/PE isotype control (*see* **Note 38**)
   Tube 2: FITC CD45
   Tube 3: PE CD45
   Tube 4: FITC CD15/PE CD11b (*see* **Note 39**)
5. Add 3 mL NH_4Cl red cell lysing solution, and shake each tube for 1 min. Centrifuge the tubes and decant the supernatant, blotting each tube 3× on a paper towel to remove all of the lysing solution.
6. Wash the cell pellets twice with PAB.
7. Resuspend the cell pellet in 1 mL PAB for immediate use, or 1% paraformaldehyde for later use (*see* **Notes 40** and **41**).

3.3.4.2. Data Acquisition

1. Turn the flow cytometer on. Depending on the manufacturer, different setup procedures (voltage adjustments, fluorescent bead calibration) will be required. Follow the manufacturer's instructions for preparing the instrument for data acquisition.
2. Set the machine to acquire data continuously (without saving to disk), place tube 1 on and display a dot plot of forward scatter (FSC) vs side scatter (SSC). Adjust the linear forward and side scatter values until the cell population(s) are centered on the plot (*see* **Fig. 1** and **Note 42**). Adjust the threshold on forward scatter to eliminate any debris.
3. Still using tube 1, display fluorescence channel 1 (FL1) vs fluorescence channel 2 (FL2). The isotype control should be uniformly negative for both FL1 and FL2. Any positive events are the result of nonspecific staining. Adjust the FL1 and FL2 voltages until the cell population sits in the lower left portion of the plot (negative for FL1 and FL2; *see* **Fig. 2** and **Note 43**). Place a quadrant around this population.
4. Still acquiring the data continuously, place tube 2 on, and display FL1 vs FL2. FITC CD45 should only be positive in FL1, but because of slight spectral emission overlap between FITC and PE, some fluorescence is detected by FL2. Therefore, adjust the FL2–%FL1 compensation until the cell population lies in the lower right quadrant (*see* **Fig. 3** and **Note 44**).
5. Still acquiring the data continuously, place tube 3 on, and display FL1 vs FL2. PE CD45 should only be positive in FL2, but because of slight spectral emission overlap between PE and FITC, and PE and propidium iodide, some fluorescence is detected by FL1 and FL3. Therefore, adjust the FL1–%FL2 compensation until the cell population lies in the upper left quadrant (*see* **Fig. 3**).

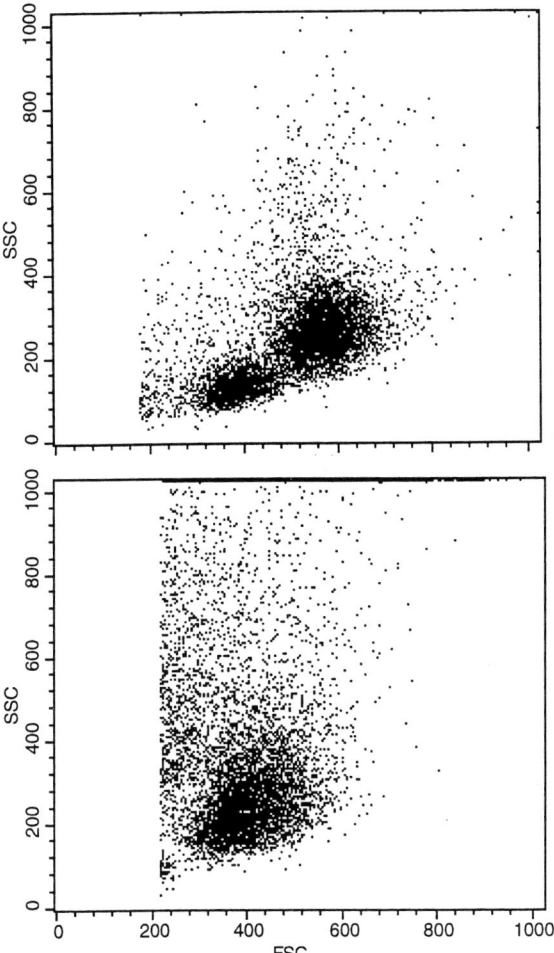

Fig. 1. Dot plots of forward scatter vs side scatter for uncultured peripheral blood cells (apheresis product) (top) and cultured cells (bottom). The lymphocytes (low forward and side scatter) and monocytes (intermediate forward scatter and low side scatter) are clearly separated. Note the absence of granulocytes (intermediate forward scatter and high side scatter) in the apheresis product.

6. Still using tube 3, display FL2 vs FL3. Adjust the FL3–%FL2 compensation until the cell population lies in the lower right quadrant.
7. Save the instrument settings for future use (*see* **Note 45**).
8. Add 10 μL of propidium iodide solution to tubes 1 and 4 (*see* **Note 46**).
9. The samples are now ready to be run. Place tube 1 on and acquire no less than 10,000 events. Save the data on a floppy disk. Repeat this step for each of the remaining tubes.

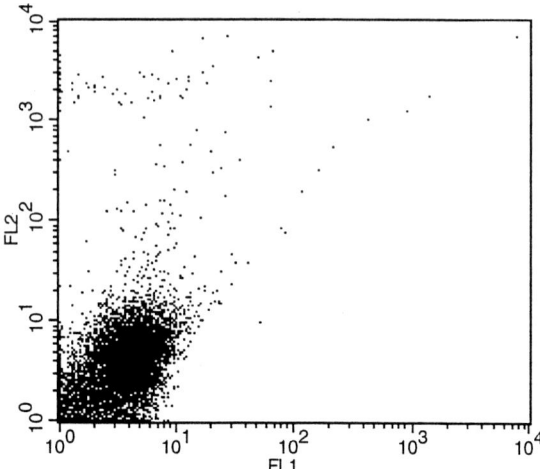

Fig. 2. Dot plot of peripheral blood cells stained with IgG control antibodies.

3.3.4.3. DATA ANALYSIS

1. Recall the saved file, and display a dot plot of side scatter vs FL3. Using the computer mouse, draw a region around the FL3 negative cells. These are the viable cells.
2. Display a plot of FL1 vs FL2, using only the viable cells determined in **step 1**. Using the quadrants established in tube 1, the percentage of viable cells that are $CD15^- CD11b^-$, $CD15^+CD11b^-$, $CD15^-CD11b^+$, and $CD15^+CD11b^+$ can now be determined. The amount of nonspecific staining in each quadrant (determined from tube 1) should be subtracted from each of these values to obtain more accurate results.

4. Notes

1. Both horse serum and fetal bovine serum may be obtained from alternate vendors such as HyClone (Logan, UT), Gibco, Sigma, and BioWhittaker. StemCell Technologies prescreens their serum, as well as serum components (such as BSA), for use in hematopoietic culture, which ensures good results. Serum lots from other vendors must be screened to determine if they are suitable for hematopoietic culture.
2. Hydrocortisone is only needed for the growth of adherent cell populations in bone marrow. If peripheral blood or cord blood is the source of the culture inoculum, hydrocortisone may be omitted from HLTM.
3. StemCell Technologies also sells an equivalent flask designed to culture hematopoietic cells. Siliconization is not necessary because the cells will not adhere to the wall. Monocytes will adhere to each other and form small clumps. The addition of the siliconization compound to the flask can only harm culture performance via introduction of toxic materials to the culture environment.
4. The 1965 model flasks designed for microcarrier cultures will also work well for hematopoietic cells, but require slightly more volume than the flat-bottom 1967 model.

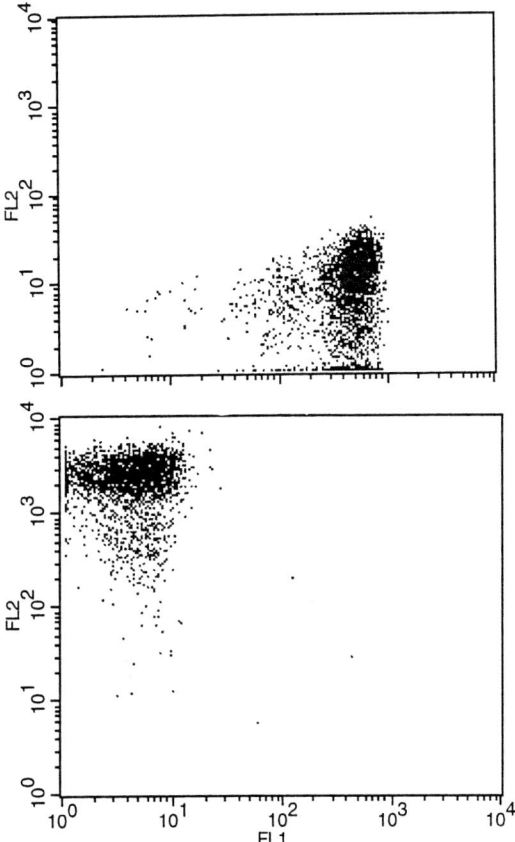

Fig. 3. Dot plot displays for compensation tubes #2 (top) and #3 (bottom). Display of FL2 vs FL3 for tube #3 is omitted.

5. The addition of the flat-blade impeller is critical to obtaining culture expansion. With just the magnetic stir bar, cells will clump under the agitator and die. Agitator flat-blade impeller to vessel diameter ratios of 0.6–0.8 work well.
6. The silicone component is intended to serve as a cushion between the glass shaft and the soft teflon flat-blade impeller. This component is difficult to clean, though, and can harbor leachables that negatively affect cell proliferation. The cell death associated with this component is much more pronounced in serum-free culture.
7. Bone marrow and umbilical cord blood can be cultured without MNC separation; however, the high RBC content can make visual observation of culture growth quite difficult. Peripheral blood samples should be low in red blood cell content because of the harvest procedure utilized (leukapheresis), which separates the MNC from the RBCs. However, variations in the leukapheresis protocol will alter the number of RBC in your sample.

8. It is critical that the cells be pipeted slowly onto the Ficoll layer. If the rate of pipeting is too rapid, cells will mix with the Ficoll and make separation impossible. When all of the cells are deposited on the Ficoll, you should note a clear demarcation between the Ficoll and blood.
9. Do not leave the cells in the Ficoll for any longer than necessary. Extended exposure of cells to the Ficoll can reduce viability.
10. For applications that require extremely high purity (>95%) of $CD34^+$ cells, we would recommend fluorescence-activated cell sorting (FACS). Additionally, the Mini-MACS system does not remove the antibody–magnetic bead complex from the cell. If this presents a concern, other methods exist that will remove the antibody *(3)*.
11. The protocol from Miltenyi is relatively straightforward and easy to use successfully. We caution that utilization of the supplied reagents beyond their expiration date will result in a low yield of $CD34^+$ cells. The protocol does require the use of degassed buffer for one step. This can be accomplished by retrofitting a bottle-top filter, so that the filter is replaced by a rubber cork. The buffer is poured into the bottle and the modified filter top is placed onto the bottle. Attach a vacuum hose to the filter nozzle and initiate vacuum. Gas bubbles will be drawn out of the buffer.
12. Either serum-containing or serum-free medium can be used. Serum-containing medium generally gives higher progenitor cell and total-cell expansion; however, if a more defined medium is desired (e.g., for clinical applications), acceptable expansion can be attained using XVIVO-20. Equilibrating the medium with the incubation atmosphere and temperature is recommended prior to suspending the cells.
13. The chosen inoculum density for MNC cultures should not exceed 500,000 cells/mL, because higher values will deplete key nutrients quickly, necessitating tedious feeding schedules. Additionally, to ensure reproducible cell expansion, the inoculum density should not be below 50,000 cells/mL. In general, lower-density cultures will exhibit a greater expansion of total cells and progenitor cells than higher-density cultures. However, if a large number of cells are required for specific assays (such as flow cytometry), high-density cultures are recommended. For $CD34^+$ cultures, an appropriate inoculum density is 20,000–50,000 cells/mL.
14. In the 24-well plate, inoculate 1 mL of the cell suspension into each of the eight center wells. The 16 outer wells should be filled with 1.5–2.0 mL sterile water to minimize media evaporation from the center wells. T-25 flasks should be inoculated with enough of the cell suspension to wet the surface (at least 5 mL), but not too much (more than 10 mL), in order to prevent oxygen transport limitations to the cells caused by excessive liquid height.
15. 6-well culture plates should be avoided, because they are prone to uneven evaporation patterns. Larger tissue-culture flasks can be used if a large culture volume is desired.
16. For low- and intermediate-density cultures, either atmosphere will allow acceptable total-cell expansions. Studies suggest that culturing at a reduced-oxygen tension may be beneficial for progenitor cell expansion *(11–13)*. For high cell density MNC cultures (>400,000 cells/mL), a 5% O_2 atmosphere should be avoided to preclude the possibility of creating a hypoxic microenvironment for the cells resting at the bottom of the cultures.

17. In general, the higher the seeding density of the culture, the more often feeding will be required. Typically, low-seeding-density cultures (50,000–100,000 cells/mL) will not need to be fed more than once over 15 d, medium-density cultures (100,000–400,000 cells/mL) should be fed once or twice, and high-density cultures (400,000–500,000 cells/mL) should be fed at least twice. $CD34^+$ cultures are usually fed once over 15 d. A good guideline for determining the proper time of feeding is depressed pH (<7.1, pale-orange to bright-yellow medium) or a high concentration of lactate (above 15 mM, indicating a feeding requirement). In all cases, it is important to remember to equilibrate the medium with the incubation atmosphere and temperature prior to feeding.
18. Generally, the adherent cell population will cover approx 30–40% of the well bottom. Scraping of the well bottom should focus on the perimeter, as most of the adherent fraction will be found there. Visual examination of each well using light microscopy will give a good idea as to when most of the adherent fraction has been removed.
19. This addition of water is not absolutely necessary, but highly recommended. In addition to killing cells in the flask, the hot water cleans the agitator assembly of any cell clumps that have adhered over the course of the prior culture. The removal of bleach from some flask components is extremely difficult, even with copious flushing, hence the use of steam sterilization as a decontamination step.
20. Occasionally, MNC cultures seeded as low as 1×10^5 cells/mL will proliferate, but 2×10^5 cells/mL is a more reliable density. Although low-seeding-density cultures may not exhibit expansion of total cells, they quite often will realize limited expansion of CFC. $CD34^+$ cell cultures carried out above 1.2×10^5 cells/mL will die within 6–8 d unless they are fed at least once a day beginning at d\ 4. Our seeding densities are designed for short-duration cultures (under 21 d). Longer-term spinner cultures may require seeding densities above 1×10^6 cells/mL *(14)*.
21. Agitation rates above 50 RPM will often negatively affect cell growth. Suspension is adequate at 30 RPM, and, at volumes of 30–50 mL, the agitator blade breaks the liquid surface, providing sufficient oxygen transport.
22. The standard animal-cell-culture magnetic stir plates sold by Bellco are intended for higher agitation rates. As such, the agitation motion at RPMs less than 100 is not smooth. We prefer the use of the Cellgro magnetic stir plate (Thermolyne), which is designed for agitation rates between 1 and 100 RPM.
23. If the serum-free medium XVIVO-20 is left outside of a CO_2 incubator too long, the pH will rise to > 8.0, and often will not return to normal (pH ~7.4) when placed back into a CO_2 incubator. We therefore suggest that cultures carried out in this medium not be left out of the incubator for more than 5–10 min.
24. This feeding protocol retains all cells throughout the culture. Cultures can be diluted as in fed-batch, or can be reduced in density by replacing medium without cell retention. It is our personal preference to retain the cells at constant volume.
25. A hemacytometer may be used to make whole cell counts. A difficulty in using this apparatus is that intact enucleated RBCs can be difficult to distinguish from

nucleated cells. Since the point of the nucleated cell count is to determine the density of nucleated cells, use of the hemacytometer can potentially result in inaccurate measurements, depending on the number of RBCs in the culture. We have not had much success in counting nuclei with the hemacytometer.

26. Prewarming cetrimide is recommended prior to cell counting. Depending on room temperature, cetrimide crystallization can occur. The crystal size is of the same order of magnitude as the nuclei size; therefore, crystallized cetrimide can artificially raise the measured cell density.

27. The choice of dilution ratio is important to the accuracy of the cell count. When the number of cells in the cetrimide is too high, clumping can occur. These clumps may be counted as a single nucleus, thereby reducing the measured cell density. When the number of cells is too low, the cell count may be of the same order of magnitude as the cetrimide blank. Total counts of 5,000–40,000 work well (*see* ref. *15*).

28. We have not noticed that unwarmed colony assay medium has an adverse effect on CFC viability; however, cool medium will not mix well and can increase variability between replicate plates.

29. Plating density is the most critical parameter for achieving accurate colony assays. The goal in choosing plating density is to space CFC far enough apart that their resultant colonies are easily distinguishable from each other. High plating densities result in overcrowding of the medium with CFC. This overcrowding increases the difficulty in determining distinct colonies. Low plating densities often do not detect less frequently occurring CFC, such as the CFU-Mix. A good number of total colonies to aim for is 100 per plate. Generally, we find that the CFC content of the culture rises for the first 4–6 d and then declines. We therefore adjust our densities in an attempt to maintain a constant number of colonies per plate each day. A plating density of 3,000 cells/mL works well until d 6 for mononuclear cultures. A good rule of thumb for d 7–14 is to multiply the day by 1,000 to get the plating density. $CD34^+$ cell cultures should be plated at 500 cells/mL for the first 5 d of culture, because of their high clonogenicity. The plating density should be slowly increased to 5,000 cells/mL until d 10. After d 10, the guidelines for MNC plating apply. If a growth factor combination other than the one suggested in the CFC protocol is used, these guidelines may not apply. For example, removal of SCF from the CAM will reduce the number of colonies detected at a given inoculum density.

30. The positive-displacement pipeter must be used to assure quantitative transfer into the Petri dishes. A standard pipeter with a standard pipet tip will leave CAM on the walls of the pipet. We use the Brinkmann (Westbury, NY) Eppendorf Repeater Pipetter (Brinkmann #22 26 000-6).

31. A standard CO_2 incubator can be used if a controlled O_2 incubator is not available. CFC detection is improved at low O_2 tensions *(11–13)*.

32. The description of the difference in granulocyte and monocyte/macrophage colonies applies to most cultured cells. However, the only accepted method to distinguish these cell types is to pluck the colonies from the plate and perform morphological analyses. For this reason, these two distinct colony types are often lumped into a single category, referred to as CFU-GM (G + *M* + GM), and are

indicative of total white CFC. This CFC designation is not to be confused with the true CFU-GM, which is a progenitor of both the CFU-G and CFU-M.

33. Murine stromal cell lines *(16,17)* and genetically engineered variations of these stromal cell lines *(18)* have also been used successfully for the LTC-IC assay.

34. Other methods for stopping stromal cell proliferation exist *(19)*, but irradiation is preferred. Irradiation prevents overgrowth of the stroma, while maintaining (and even enhancing) their ability to produce growth factors.

35. Accessory cells in the culture affect the outcome of the LTC-IC assay *(20)*. The assay is therefore inoculum-density-dependent. For this reason, the assay is carried out in triplicate at the one density of 1×10^6 cells/mL. The high seeding density is used to ensure that at least some LTC-IC are present in the sample of cells used in each assay well (LTC-IC exist at a very low frequency).

36. The amount of antibody to add depends on the source of the antibody. Generally, antibody solutions are added at 20 µL per 1×10^6 cells. Consult manufacturer or antibody data sheet to determine the appropriate amount.

37. Add the antibodies to the cell pellet, with the tubes on ice to minimize receptor internalization.

38. Because most nonspecific staining occurs via binding of the Fc region of the antibody to the cell, isotype controls should be of same class as other antibodies (generally IgG).

39. This antibody combination is used to identify developing granulocytes and monocytes. If quantification of other lineages is desired, different surface antigens must be investigated. A brief summary of surface antigens characteristic to specific hematopoietic lineages is given below:
    Monocytes: CD14, CD15, CD11b
    Granulocytes: CD16, CD15, CD11b
    Megakaryocytes: CD41a
    Erythrocytes: Glycophorin A, CD71
    Lymphocytes: CD3, CD19

40. Fixing in 1% paraformaldehye preserves the cells for up to 1 wk. However, identification of dead cells using propidium iodide is not possible after fixing the cells. Therefore, whenever possible, run cells on the flow cytometer live, because dead cell accumulation can become important in later stages of the culture.

41. If cells are not to be run immediately, wrap the tubes in aluminum foil and store at 4°C in the dark to minimize receptor internalization and maximize the fluorescence signal.

42. Uncultured cells are clearly separated into lymphocyte, granulocyte, and monocyte populations on dot plots of forward scatter vs side scatter. Peripheral blood apheresis products will not contain any granulocytes, since they will have been removed in the apheresis process. In contrast, cultured cells are seen as one diffuse population possessing a higher forward scatter (because they are larger than uncultured cells). Accordingly, forward and side scatter values will have to be changed to properly view this population. As a general rule, the cells to be analyzed (cultured or uncultured) can be located by looking for the population with

the highest forward-scatter. Anything with a lower forward scatter represents debris or mature red blood cells, which can be removed from view using the electronic threshold.

43. Often, a tail-like projection, which extends 45 degrees up and to the right, can be seen on this double-negative population. This tail is the result of two factors: dead cells that nonspecifically bind the antibodies, and autofluorescing monocytes. As a result, it may be difficult to move the entire population into the lower left quadrant without cutting off any of the dots (events). Focus only on the main cell population (excluding the tail) when adjusting the voltages, because much of the tail will disappear when the dead cells are removed in the data analysis. Also, make sure the axes for FL1, FL2, and FL3 are logarithmic.

44. Because of the presence of dead cells, sometimes two cell populations can be seen in these plots: a dense, compact cluster of cells, and a second, more diffuse smear of cells. The compact population is usually the viable cell population, and adjustment of compensation values should focus on this population. The dead cells will be gated out in the data analysis.

45. Generally, the flow cytometer settings for uncultured cells will remain constant between cell samples. Therefore, these settings can be recalled for future samples without needing to use tubes 1–3. Periodically, however, the flow cytometer setup procedure should be repeated to ensure the settings are still appropriate. This does not apply to cultured cells, whose settings often vary between samples. Additionally, the settings for cultured cells can change as the culture time progresses. Therefore, new settings should be determined for each day the culture is to be analyzed.

46. Add the propidium iodide no less than 1 min and no more than 1 h before the data are to be acquired. Propidium iodide cannot be used with fixed cells.

Acknowledgments

Supported by National Science Foundation Grant BES-9410751 and a predoctoral fellowship to P. C. Collins through the National Institutes of Health Predoctoral Biotechnology Training Grant (T32 GM08449).

References

1. Brugger, W., Heimfeld, S., Berenson, R. J., Mertelsmann, R., and Kanz, L. (1995) Reconstitution of hematopoiesis after high-dose chemotherapy by autologous progenitor cells generated ex vivo. *N. Engl. J. Med.* **333,** 283–287.
2. McAdams, T. A., Winter, J. N., Miller, W. M., and Papoutsakis, E. T. (1996) Hematopoietic cell culture therapies (Part II): clinical aspects and applications. *Trends Biotechnol.* **14,** 388–396.
3. de Wynter, E. A., Coutinho, L. H., Pei, X., Mars, J. C. W., Hows, J., Luft, T., and Testa, N. G. (1995) Comparison of purity and enrichment of $CD34^+$ cells from bone marrow, umbilical cord and peripheral blood (primed for apheresis) using five separation systems. *Stem Cells* **13,** 524–532.

4. McKenna, H. J., deVries, P., Brasel, K., Lyman, S. D., and Williams, D. E. (1995) Effect of *flt3* ligand on the *ex vivo* expansion of human $CD34^+$ hematopoietic progenitor cells. *Blood* **86,** 3413–3420.
5. Debili, N., Wendling, F., Katz, A., Guichard, J., Breton-Gorius, J., Hunt, P., and Vainchecker, W. (1995) The Mpl-ligand or thrombopoietin or megakaryocyte growth and differentiation factor has both direct proliferative and differentiative activities on human megakaryocyte progenitors. *Blood* **86,** 2516–2525.
6. Laluppa, J. A., McAdams, T. A., Papoutsakis, E. T., and Miller, W. M. (1997) Culture materials affect ex vivo expansion of hematopoietic progenitor cells. *J. Biomed. Mater. Res.* **36,** in press.
7. Quito, F. L., Beh, J., Bashayan, O., Basilico, C., Basch, R. S. (1996) Effects of fibroblast growth factor-4 (k-FGF) on long-term cultures of human bone marrow cells. *Blood* **87,** 1282–1291.
8. Gordon, M. Y. (1994) Plastic-adherent cells in human bone marrow generate long-term hematopoiesis in vitro. *Leukemia* **8,** 865–870.
9. Traycoff, C. M., Kosak, S. T., Grigsby, S., and Srour, E. F. (1995) Evaluation of ex vivo expansion potential of cord blood and bone marrow hematopoietic progenitor cells using cell tracking and limiting dilution analysis. *Blood* **85,** 2059–2068.
10. Koller, M. R., Manchel, I., Palsson, M. A., Maher, R. J., and Palsson, B. O. (1996) Different measures of *ex vivo* human hematopoietic culture performance are optimized under vastly different conditions. *Biotechnol. Bioeng.* **50,** 505–513.
11. Koller, M. R., Bender, J. G., Papoutsakis, E. T., and Miller, W. M. (1992) Beneficial effects of reduced oxygen tension and perfusion in long-term hematopoietic cultures. *Ann. NY Acad. Sci.* **665,** 105–116.
12. Koller, M. R., Bender, J. G., Miller, W. M., and Papoutsakis, E. T. (1992) Reduced oxygen tension increases hematopoiesis in long-term culture of human stem and progenitor cells from cord blood and bone marrow. *Exp. Hematol.* **20,** 264–270.
13. Koller, M. R., Bender, J. G., Papoutsakis, E. T., and Miller, W. M. (1992) Effects of synergistic cytokine combinations, low oxygen, and irradiated stroma on the expansion of human cord blood progenitors. *Blood* **80,** 403–411.
14. Zandstra, P. W., Eaves, C. J., and Piret, J. M. (1994) Expansions of hematopoietic progenitor cell populations in stirred suspension bioreactors of normal human bone marrow cells. *BioTechnology* **12,** 909–914.
15. Lin, A. A., Nguyen, T., and Miller, W. M. (1991) A Rapid Method for Counting Cell Nuclei Using a Particle Sizer/Counter. *Biotechnol. Techniques* **5,** 153–156.
16. Sutherland, H. J., Hogge, D. E., Cook, D., and Eaves, C. J. (1993) Alternative mechanisms with and without steel factor support primitive human hematopoiesis. *Blood* **81,** 1465–1470.
17. Croisille, L., Auffray, I., Katz, A., Izac, B., Vainchenker, W., and Coulombel, L. (1994) Hydrocortisone differentially affects the ability of murine stromal cells and human marrow-derived adherent cells to promote the differentiation of CD34++/CD38– long-term culture-initiating cells. *Blood* **84,** 4116–4124.

18. Hogge, D. E., Lansdorp, P. M., Reid, D., Gerhard, B., and Eaves, C. J. (1996) Enhanced detection, maintenance, and differentiation of primitive human hematopoietic cells in cultures containing murine fibroblasts engineered to produce human steel factor, interleukin-3, and granulocyte colony-stimulating factor. *Blood* **88**, 3765–3773.
19. Cicuttini, F. M., Martin, M., Salvaris, E., Ashman, L., Begley, C. G., Novotny, J., Maher, D., and Boyd, A. W. (1992) Support of human cord blood progenitor cells on human stromal cell lines transformed by SV40 large T antigen under the influence of an inducible (metallothionein) promoter. *Blood* **80**, 102–112.
20. Koller, M. R., Palsson, M. A., Manchel, I., and Palsson, B. O. (1995) Long-term culture-initiating cell expansion is dependent on frequent medium exchange combined with stromal and other accessory cell effects. *Blood* **86**, 1784–1793.

23

Methods to Isolate, Culture, and Study Osteoblasts

Mechteld V. Hillsley

1. Introduction

Primary cells are often desirable over clonal cell lines, because they are more likely to retain the presence and activity of certain enzymes and proteins that are often lost in clones. In addition, primary cells more closely resemble the cell in the actual animal than clones do. An often-cited drawback of primary cells is the heterogeneity inherent in them. Primary osteoblast cultures, for example, consist of preosteoblasts, osteoblasts, and osteocytes (if the culture is old enough), and possibly even a few fibroblasts. Such a mix of osteoblastic cells, however, could be deemed beneficial and more realistic. The most detrimental cell contaminant is the fibroblast. Fibroblasts, if present in the culture, will eventually take over. We have never had this problem with our cultures. Osteoblastic cultures isolated by the method described in this chapter have been shown to be osteoblastic in nature by exhibiting extensive alkaline phosphatase activity *(1–3)*, by forming nodules that stain positive for mineral by a Von Kossa stain *(1–5)*, and by showing a characteristic increase (three-fold) in intracellular cAMP in response to PTH stimulation (2.7×10^{-8} M for 15 min) *(1)*.

The following method is a procedure, utilized successfully in our lab, for harvesting rat calvarial osteoblasts. The culturing techniques employed are simple and cause little trauma to the cells. The main drawback is the long turn-around time from start of isolation to the time when enough cells have grown for use in an experiment (2–4 wk).

The technique employed is one based on a technique described by Ecarot-Charrier *(6)* and refined by Reich et al. *(1)*. It is a mechanical isolation procedure limiting the use of digestive enzymes, thereby limiting potential damage to the cells. The technique involves the selective migration of osteoblasts from calvarias onto glass chips. For reasons that are not totally clear, osteoblasts

From: *Methods in Molecular Medicine, Vol. 18: Tissue Engineering Methods and Protocols*
Edited by: J. R. Morgan and M. L. Yarmush © Humana Press Inc., Totowa, NJ

will migrate from dissected, periosteum free calvarias onto glass chips. Other cell types, such as fibroblasts, do not partake in this migration and are thereby kept separate from the osteoblasts. After 4–6 d of incubation with the glass chips, the calvarias are removed from the chips, and the cells that have migrated onto the chips are given time to grow and multiply. Only one enzymatic digestion is employed near the end of the isolation and culturing procedure, to remove the osteoblasts from the glass chips.

2. Materials

1. Sodium bicarbonate washing solution: Approx 2 g (the exact amount is not critical) of sodium bicarbonate is dissolved in 1 L of distilled water. This solution is made up just prior to use.
2. Basal medium: Medium 199 (HyClone Laboratories, Logan, UT) (*see* **Note 1**). Other media are made using this basal medium. Without antibiotics and other additives, it is only used for short time-periods and is not appropriate for cell growth.
3. Complete growth medium: Complete growth medium consists of basal medium supplemented with 10% fetal bovine serum (FBS) (HyClone) (*see* **Note 2**), 1% L-glutamine (200 mM), and 1% antibiotics (5000 U penicillin and 5 mg/mL streptomycin, both from Sigma, St. Louis, MO). The FBS must be heat-inactivated for 30 min at 57°C prior to use. All batches of serum should be tested for their ability to support osteoblast growth before use.
4. Buffers: Buffers used include calcium- and magnesium-free Hanks' balanced salt solution (HBSS) and Dulbecco's phosphate-buffered saline (DPBS).
5. Collagenase A:
   a. A 1 g/L collagenase stock solution is made by dissolving 0.5 g of collagenase A (Boehringer Mannheim, Indianapolis, IN) in 500 mL of DPBS, and adding 1% antibiotics (5000 U penicillin and 5 mg/mL streptomycin). The collagenase is nonsterile and contains quite a few particulates. The solution is thus sterilized by pressure-driven filtration (approx 20 psi) through a set of filters. The first filter is a 0.45-µm prefilter (Millipore Sterivex-HV), which is followed by the final 0.22-µm filter (Millipore Sterivex-GS, Bedford, MA). The 0.22-µm filter usually lasts for an entire batch. Often, however, the 0.45-µm prefilter gets clogged and must be replaced 1–2 times over 500 mL of collagenase solution. The sterile stock solution is aliquoted and frozen. Individual aliquots may be thawed and kept in the refrigerator for approx 1 mo (*see* **Note 3**). The stock solution should be sterile tested in combination with growth medium before use.
   b. The collagenase working solution of 0.2 mg/mL is made by diluting the stock solution fivefold with HBSS or any other calcium-free buffer. This should be done just prior to use.
6. Fibronectin: Human fibronectin (1 mg) (Biomedical Technologies, Stoughton, MA) (*see* **Note 4**) is reconstituted by adding 1 mL of a sterile 2 M urea, 0.05 M

Tris-HCL, pH 8.0, solution at room temperature, and letting it sit for at least 1.5 h. This solution is then diluted 20-fold with HBSS. Aliquots are stored in the freezer.
7. Metofane is the drug used to euthanize the rat pups prior to the cell-harvesting procedure. This, or other drug, is obtained through a veterinarian working with the institution's animal care and use committee (IACUC).

3. Methods
3.1. Preparation of Glass Chips

Glass chips are prepared in advance and may be stored in a sterile container indefinitely (*see* **Note 5**).

1. Cover slips (No. 1) are cleaned by placing them, preferably, in a slide rack to prevent sticking, in a sodium bicarbonate washing solution, which is then brought to a boil. When boiling commences, the heat is turned off. The cover slips should be rinsed thoroughly soon after boiling has stopped. Rinsing should include extensive flushing with distilled water, followed by a soak in clean distilled water (at least 15 min, and up to overnight), followed by another cycle of flushing rinses. The cover slips are then dried in a drier until completely dry.
2. Glass chips are made from the washed cover slips by wrapping the clean cover slips in several clean paper towels and crushing them by pounding the wrapped bundle with a hammer or other blunt object. Care must be taken not to overcrush the chips and thereby form only glass dust. **Caution:** Safety glasses and gloves should be worn during this step.
3. Glass chips in the 250–500-µm size range are separated from larger and smaller chips by sieving through appropriately sized sieves (*see* **Note 6**). This size range was chosen because smaller size chips will float, thereby becoming a problem in cell culturing; and larger chips tend to cover too large an area of the calvaria, thereby limiting the number of cells in contact with glass chips. Again, **Caution:** Safety glasses and gloves should be worn for this step.
4. Sieved glass chips are placed in small autoclavable containers (glass Petri dishes, for example) and steam autoclaved. They are now ready for use.

3.2. Harvesting of Rat Calvarias

Calvarias are harvested from 4- to 6-d old Sprague-Dawley rat pups. Younger rat pups are so small that the calvarias are tiny and paper thin, making them very difficult to handle and very easy to tear. Older rats provide fewer osteoblasts per calvaria, probably because the cells are dividing more slowly. The entire harvesting procedure (starting with **step 2**) is performed aseptically in a laminar flow hood. The instruments used are autoclaved before the procedure. Since generally a litter of rat pups will be handled at one sitting, the instruments are resterilized between pups by dipping in 70% ethanol after each use, and again dipping in 70% ethanol and flaming the remaining ethanol in a propane flame immediately before each use. The instruments used include sharp dissecting scissors, curved on the flat (for decapitation); a hemostat; blunt-tipped dissecting scissors (for cutting skin); delicate, sharp, spring-handled scissors (for cutting the calvarias); and four sets of forceps. All work involving the pups in the laminar flow hood is performed on or over autoclaved paper towels. These paper towels catch any spilled blood and waste scraps from the dissection. The time involved for this step, not counting setting up, is about 5 to 10 min per pup.

1. The rat pups are euthanized using Metofane. Liquid Metofane on a Q-tip is placed in a closed container with the pups. The pups breathe the vaporized metofane and die within 10–15 min.
   The following steps are performed sequentially on one rat pup at a time.
2. The rat pup is dipped in 70% ethanol (balance distilled water) to disinfect it, and is brought into the laminar flow hood. From this point on, all instruments used are dipped in 70% ethanol and flamed in a propane flame to burn off the ethanol.
3. Using the curved scissors, decapitate the rat so that the head falls onto the sterile paper towel, and dispose of the body. Use the hemostat to grasp the head by the nose (*see* **Note 7**). Then, using the blunt-tipped scissors, cut the skin open along the center back of the head to the forehead. Cut down toward the neck along the top of the head to form two squares of skin. Using blunt forceps, pull the skin off the skull (*see* **Note 8**). These forceps touch the outside of the skin and should not be used thereafter to touch anything cleaner than the animal skin.
4. The calvaria is now exposed and needs to be cut and removed. Using the delicate, sharp, spring-handled scissors, cut out the frontal and parietal bones in the same pattern as the skin was cut, but staying away from the edge of the skin. Using a different pair of forceps than those used in **step 3**, remove the pieces of calvaria (there should be two—one from each side of the skull) and place them in a Petri dish of warmed basal medium (*see* **Note 9**). The Petri dish is then placed in the incubator while the remaining pups are processed.

3.3. Stripping the Periosteum from the Calvaria and Covering It with Glass Chips

This step must be done immediately following the steps of **Subheading 3.2.** The cells are sitting in a medium that does not support growth (*see* **Note 10**). The following procedure should be done one dish at a time (i.e., one set of two calvarias at once). Again, between sets of calvarias, all instruments should be resterilized by dipping in 70% ethanol and flaming over a propane flame.

1. Over sterilized paper towels, using two sets of forceps, carefully scrape off the periosteal layer over the calvaria. This is a tedious process, and it is very easy to tear the calvaria. As pieces of periosteum are removed, they can be wiped off the forceps onto the paper towel.
2. The periosteum free calvarias are placed concave side up in clean Petri dishes (6-cm diameter) (*see* **Note 11**). A scoop of glass chips is carefully placed into each calvaria. The calvarias are then covered with warmed complete medium by

Osteoblasts

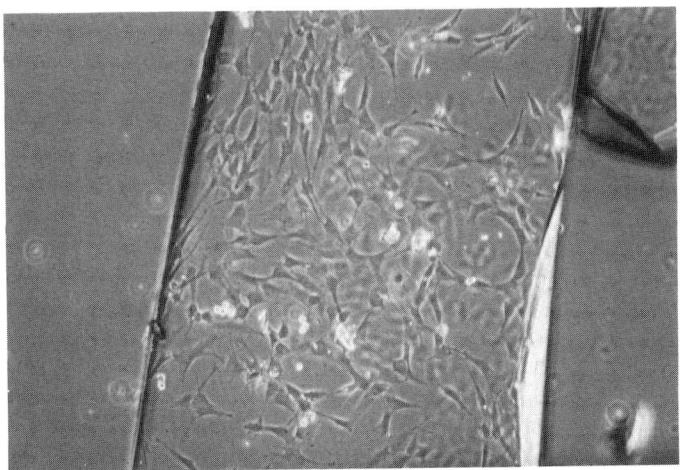

Fig. 1. Glass chip with osteoblasts that have migrated onto it. The chip has just been removed from the calvaria. ×100.

placing several drops on the calvarias with chips directly before filling the remainder of the dish. This is to prevent spilling the chips off the calvarias. The dish is then placed in the incubator for 4–6 d.

3.4. Removal of Glass Chips from the Calvarias

1. After 4–6 d of incubation with the glass chips, the calvarias are removed from the glass chips. This is done by carefully transferring each set of calvarias, with the glass chips, to a new tissue-culture dish (one set to one 6-cm diameter dish).
2. Add 4 mL of warmed complete growth media to each plate while trying to flush the chips off the calvarias. Repeat the flushing of the calvarias by repipeting the same media over the calvarias several times. Try to get all the chips dislodged in this manner, while taking care not to splash media onto the rim of the culture dish. When all the chips have been dislodged, remove the empty calvarias from the dish and discard. Place the Petri dish in the incubator.
3. At this point some of the glass chips are covered with osteoblasts (*see* **Fig. 1**). These osteoblasts will grow onto the tissue culture dish bottom. Over time there will be enough cells for an experiment. Cells should be fed every 3–4 d, more frequently as the density increases. Generally, cells are ready for use during the period 1–3 wk after removal of the chips from the calvarias.

3.5. Removing Osteoblasts from the Glass Chips and Seeding onto the Experimental Surface

When cells are ready to be used in an experiment, they generally need to be removed from the glass chips still in the Petri dish. This is accomplished by collagenase digestion (*see* **Note 12**).

1. Surface preparation: The surface onto which the cells will be plated generally needs to be pretreated with an attachment factor such as fibronectin. Glass microscope slides (75 mm × 38 mm) are cleaned following the same procedure as that used for cleaning the cover slips used in making glass chips. The cleaned slides are coated with 1 mL of 20 μg/mL fibronectin at least 1 h prior to seeding (*see* **Note 13**). Just prior to seeding, the slides with the fibronectin solution are rinsed using a buffer such as HBSS, DPBS, or basal medium (*see* **Note 14**).
2. Cells are first rinsed with a warmed calcium-free buffer such as HBSS to remove the serum and calcium, which will hinder collagenase activity and cell detachment.
3. The cells are incubated with 2–3 mL warmed collagenase working solution for 30 min (*see* **Note 15**). After this time, a solid tap to the Petri dish should dislodge a good number of the cells. The remainder can be dislodged using media irrigation. First, the collagenase is deactivated from further digestion by adding 4–6 mL complete growth media (approx 2 times as much growth media as initial collagenase solution). The cells are then dislodged from the Petri plate and from each other by vigorous irrigation. By irrigation is meant passing the media up and down a pipet and squirting it vigorously against the side or bottom of the Petri dish (*see* **Note 16**). If the cells are not adequately separated from each other, the seeded surface will result in large clumps of multilayered cells, rather than a smooth monolayer.
4. The cell suspension is next centrifuged for 10 min at 1000g. The cell pellet is then resuspended in the appropriate amount of complete media. It is critical that resuspension results in a suspension of individual cells (*see* **Note 17**). Dislodged cells are plated onto the desired surface in a small amount of media (1 mL cell suspension onto a 75 × 38 mm slide). The plating density varies and is dependent on the desired final density and the attachment efficiency. In my experience, seeding at 9×10^4 cells/cm^2 onto fibronectin-covered glass slides (75 × 38 mm) produced a confluent monolayer culture 3–4 d after seeding *(1,4)*. For other experiments, in which a much tighter monolayer was required and where the attachment efficiency was lower, cells were plated onto 4.2 cm^2 Millipore PCF membrane transwell inserts at a seeding density of 2.1×10^5 cells/cm^2 *(5)* (*see* **Note 18**).
5. The seeded cells are placed in the incubator and are allowed to settle and attach to the surface. Attachment occurs fairly quickly, so that the cells may be fed a larger amount of complete media from 2 to 24 h after seeding. This time to settle and attach in limited media is only necessary if there are attachment boundaries, such as keeping cells on the microscope slide instead of attaching to the entire Petri dish. If cells are desired over the entire contained area, no attachment time is required before adding additional growth media to the cells.

4. Notes

1. Medium 199 was used mainly in our lab. However, we have also used minimum essential medium (MEM) and other home-made base media successfully. Different media should be tested on the cells prior to use.

2. Osteoblasts isolated as described here grew well in FBS. They were less sensitive to serum quality than human umbilical vein endothelial cells (HUVEC). Our osteoblasts, however, died when bovine serum albumin (BSA) was used instead of FBS. BSA can be used in combination with FBS (1% BSA, 2% FBS) when necessary *(5)*; however, cells grow better and faster in 10% FBS.
3. Collagenase will digest itself, especially at warmer temperatures. Care should be taken to warm the refrigerated aliquot only when necessary, and to return the aliquot to the refrigerator as quickly as possible. Repeated freeze–thawing cycles of collagenase have been shown to be more damaging than storing the aliquot in the refrigerator.
4. Human fibronectin was used in our studies, because it happened to be the cheapest. Other sources of fibronectin have also worked as an attachment factor and could be used after testing.
5. Depending on the quality of the autoclave and steam used, repeated autoclaving may over time form deposits on the glass. Therefore, excessive autoclaving of glass chips should be avoided.
6. The size range of the chips is approximate. It was determined based on the sieves available. The size range may probably be shifted 20–50 μm in either direction, if needed, to accommodate sieve sizes.
7. When holding the pup head with the hemostat, be careful not to clamp too hard. This may result in cutting through the snout by the end of the procedure. Holding it too loose, however, will result in dropping the head and compromising sterility. Some people find it easier to rest the neck on the paper towel while performing the calvarial dissection.
8. It takes some force to pull the skin loose. It is best to wedge the skin between the two long sides of the forceps rather than by trying to use the tips of the forceps. Using only the tips will often result in pulling off small pieces of skin rather than the entire square.
9. Medium 199 is used because of convenience. Any buffer should also work, since the calvarias will only be there for a short period while the rest of the litter is processed.
10. Prolonged exposure of the calvaria to any buffer or medium before the removal of the periosteum softens the calvaria and makes it more difficult to remove all of the periosteal layer without ripping the calvaria itself.
11. The tissue-to-media ratio (or dish size) is important in the effectiveness of cell migration onto the chips. A 6-cm diameter Petri dish with 4 mL media works well for one set of calvarias (from a single rat pup). A larger dish (10-cm diameter) has also been used effectively for 2–3 sets of calvarias with 8–10 mL media. However, the smaller dishes seem to get slightly better results.
12. Trypsin digestion is commonly used in seeding cells. However, we have found in our work that collagenase is much less damaging to the cell surface than trypsin. For example, osteoblasts seeded using trypsin lost vitamin D_3 ($1,25[OH]_2D_3$) responsiveness (measured by osteocalcin activity in the media) for at least 8 d postseeding. Cells seeded using collagenase digestion, however, showed no such loss in vitamin D_3 responsiveness *(2)*.

13. Osteoblasts attach better to fibronectin-covered microscope slides than to uncoated slides. We have also seeded osteoblasts onto filters (Millipore PCF membrane transwell inserts). These filters require fibronectin pretreatment as well. We again used 1 mL of 20 µg/mL fibronectin per 4.2 cm^2 insert. Tissue culture plates do not require pretreatment.
14. When rinsing the slides, take care not to spill solution off the slide itself into the Petri dish. This will then provide a path for the cell suspension to follow, resulting in a fair number of the cells plating the Petri dish surface, rather than the slide surface.
15. The amount of collagenase used may need to be adjusted, depending on the age of the collagenase aliquot. Fresh collagenase should just cover the cell layer and may not require all of the 30 min. Older collagenase may require a larger quantity to be added to the cells and may also require up to 45 min of exposure. If more time or a large quantity is required, a new aliquot of collagenase should be opened.
16. When pipeting vigorously, take care not to place the pipet directly against the side or bottom of the Petri dish, since this will damage the cells. Instead, hold the pipet 1–2 cm away from the surface, and pipet, at high speed, a stream of fluid and cells against this surface. This impact will be enough to dislodge and separate cells, but the impact angle and force will not damage the cells significantly.
17. A seeding suspension consisting of clumps of cells will cause the cells to grow in isolated, multilayered clumps, rather than a monolayer covering the entire surface area.
18. Whether the osteoblasts grow into a monolayer or form large clumps of multilayered cells is very sensitive to the seeding density (as well as to the effectiveness of cell separation during the passing procedure, as discussed in **Note 17**). A seeding density that is too high will result in osteoblasts growing in clumps and forming nodules. A seeding density that is too low will also result in osteoblasts migrating toward each other and forming isolated clumps of cells. When a monolayer or smooth layer of cells is desired, the cells must be well dispersed in the medium and seeded at the correct seeding density. You may need to experiment with the seeding density to get the desired result.

References

1. Reich, K. M., Gay, C. V., and Frangos, J. A. (1990) Fluid shear stress as a mediator of osteoblast cyclic adenosine monophosphate production. *J. Cell Physiol.* **143**, 100–104.
2. Hillsley, M. V. (1990) The effects of fluid shear stress and 1,25-dihydroxy-vitamin D_3 on collagen and osteocalcin production by osteoblasts. MS Thesis. Pennsylvania State University.
3. Hillsley, M. V. and Frangos, J. A. (1997) Alkaline phosphatase in osteoblasts is down-regulated by pulsatile fluid flow. *Calc. Tissue Int.*, **60(1)**, 48–53.
4. Hillsley, M. V. and Frangos, J. A. (1994) Bone tissue engineering: the role of interstitial fluid flow. *Biotechnol. Bioeng.* **43**, 573–581 (review).

5. Hillsley, M. V. and Frangos, J. A. (1996) Osteoblast hydraulic conductivity is regulated by calcitonin and parathyroid hormone. *J. Bone Miner. Res.* **11,** 114–124.
6. Ecarot-Charrier, B., Glorieux, H., van der Rest, M., and Pereira, G. (1983) Osteoblasts isolated from mouse calvaria initiate matrix mineralization in culture. *J. Cell Biol.* **96,** 639–643.

24

Cryopreservation of Rat Hepatocytes in a Three-Dimensional Culture Configuration Using a Controlled-Rate Freezing Device

Michael J. Russo and Mehmet Toner

1. Introduction

From engineered tissues to transfected cell lines, the long term storage of living biologicals is desirable for a variety of medical, scientific, economic, and regulatory concerns, including transport, the expense of development, repeatability issues, and the point of use. Currently, the best option is cryogenic storage, placing the biomaterials in suspended animation at very low temperatures ($-196°C$), halting all chemical reactions, limiting genetic drift, and ensuring the maintenance of cell viability and function upon thawing *(1)*. Obtaining such an advantageous state, however, can be a difficult achievement. This problem becomes further complicated as we move toward next generation multicomponent products such as engineered skin and cartilage substitutes, composed of multiple cell types oriented in complicated three-dimensional geometries within an extracellular matrix scaffold *(2–4)*.

In order to survive freezing and warming protocols, cells must overcome extremes in solute concentration, volume, and viscosity, and even the formation of intracellular ice crystals *(5–7)*. These protocols must therefore be carefully designed in a cell specific fashion to minimize the damage that will be inflicted by these inhospitable conditions. There are many parameters that can be varied toward the optimization of a cryobiological protocol: the cooling rate, the type and concentration of cryoprotective agents (CPAs, additions made to the cytosol to aid the cell in overcoming freezing stresses; *see* **ref. *8***), the temperature at which extracellular ice is seeded, the temperature at which the cells are transferred to liquid nitrogen for storage, and the warming rate (*see* **Note 1**).

Here, we outline the construction of a controlled rate freezing device capable of executing multistep freeze–thaw protocols, with cooling rates of up to −100°C/min (for temperatures as low as −80°C) and warming rates in excess of +400°C/min. This design operates by discharging cooled nitrogen (N) gas (−160°C) into a channel underlying a heating device *(9)*. The sample platform is located on top of the heating unit, and the sample temperature is thereby determined by a precise balancing between the heat generated by the heating unit and the cooling power of the nitrogen gas. The control system employs a feedback system which can be programmed to follow a variety of cooling and warming protocols by dynamically determining the power output and thus the temperature to the heating unit.

Though this system is capable of freezing a variety of cultured and suspended cells, as well as tissue samples, the method described here utilizes the device for the freezing of cultured hepatocytes, sandwiched between two layers of collagen. From a tissue-engineering perspective, this is an interesting system because it involves the freezing of a multiphase three-dimensional unit composed of extracellular matrix (collagen), living cells (hepatocytes), and culture media, each with its own transport, mechanical, and freezing characteristics *(10)*. Medically, hepatocyte cryopreservation is necessary for ensuring a continuous supply of cells for use in bioartifical hepatic support devices, as well as for cell transplantation purposes *(11,12)*. Previously, most attempts at hepatocyte cryopreservation have been performed on cell suspensions, yielding low viability and function *(13,14)*. The collagen-sandwich culture system, by partially recreating the hepatocytes' *in situ* environment, results in long-term stable cultures, with hepatocytes expressing hepatospecific functions, such as albumin secretion and urea synthesis, for over 10 wk. By freezing hepatocytes in this culture configuration after they begin demonstrating phenotypic stability (d 7 of culture), we can demonstrate a stable, full recovery of albumin secretion following a 2–3 d postthaw period *(15)*.

2. Materials
2.1. Hepatocyte Isolation and Culture
2.1.1. Isolation of Hepatocytes from Rat Liver

The materials necessary for the isolation and purification of rat hepatocytes are covered in Chapter 33 of this volume.

2.1.2. Long Term Hepatocyte Culture

In addition to those materials for hepatocyte sandwich cultures listed in Chapter 33, this preservation method requires that the following additional materials be used for hepatocyte culture:

Cryopreservation of Rat Hepatocytes

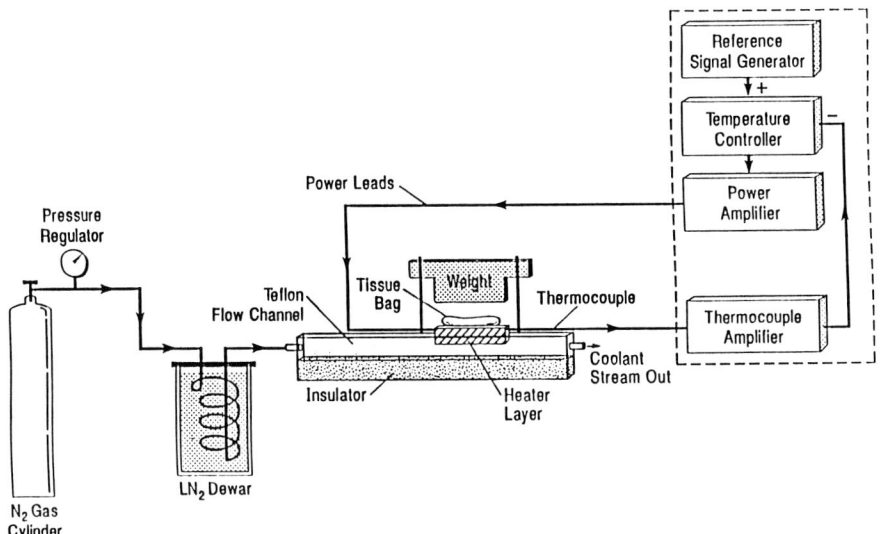

Fig. 1. Schematic overview of the controlled rate freezer unit *(9)*.

1. No. 1 cover glasses (24 × 60 mm, Baxter Healthcare, McGaw Park, IL).
2. Silicone rubber (GE Silicones, Waterford, NY).
3. Scalpel.

2.2. Hepatocyte Freezing

1. Freezing unit: A detailed description of the design and characterization of the freezing unit is given in **ref. 9**, and is shown in **Fig. 1**.
    a. Nitrogen gas tank with pressure regulator and flow meter.
    b. Liquid nitrogen.
    c. Liquid nitrogen dewar (3–5-in. diameter).
    d. Copper coil (3/8 in. OD; 1/8 in. ID; at least 10 turns of 3 in. diameter).
    e. Tygon tubing. (1/2 in. OD, 3/8 in. ID)
    f. Heating gun or hair dryer.
    g. Refrigerant Flow Channel (*see* **Fig. 2** for multilayered design).
        1. Bottom channel layer: Polypenco acetal polymer rectangular solid (Delrin, DuPont, Wilmington, DE), 33.7 × 9.5 ¥ 2.0 cm^3.
        2. Top channel layer: polypenco acetal polymer rectangular solid 33.7 × 9.5 × 0.5 cm^3.
        3. Tubing connectors (1-cm inlet diameter).
        4. Silicone RTV (GE Silicones).
        5. Cork, 40.0 × 12.0 × 2.0 cm^3
        6. Styrofoam, 60.0 × 30.0 × 1.0 cm^3.
    h. Heater layer (*see* **Fig. 3** for design).

TOP VIEW

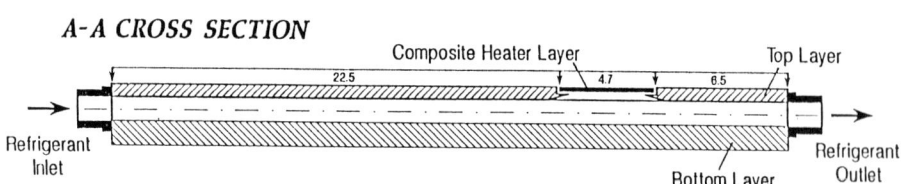

Fig. 2. Flow channel schematic, with heater layer location indicated. Dimensions are in mm *(9)*.

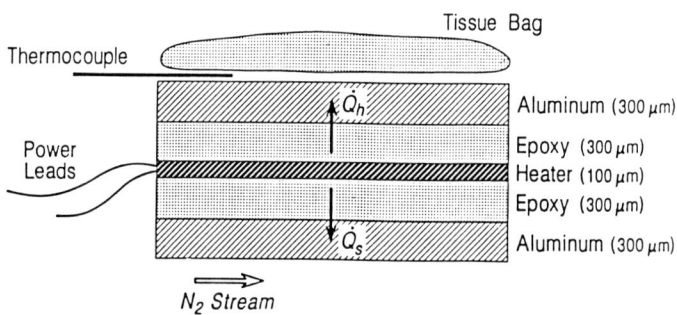

Fig. 3. Heater layer construction, with sample bag *(9)*.

1. Two aluminum plates, 7.5 × 4.7 cm² and 300-μm thick.
2. Kapton heater layer, 7.5 × 4.7 cm² and 100-μm thick (230 Ω, 10 W, HK-TBA, Minco, Minneapolis, MN).
3. Copper-constantan foil thermocouple (CO2-T, Omega, Stamford, CT).
4. Silicone RTV (GE Silicones).
i. Proportional integral controller (Interface Techniques, Cambridge, MA) and IBM™ compatible computer. The building and design of a controller is outside the scope of this method. Many commercially available control systems are designed to provide the temperature-reference signal and control, and the power amplifier necessary for the precision temperature control of the freezing unit.

Cryopreservation of Rat Hepatocytes 307

2. Hepatocyte sample preparation
   a. Coverslip with cultured hepatocytes as prepared in **Subheading 3.1.2**.
   b. Freezing solution: 2 M dimethylsulfoxide (99% pure; Sigma, St. Louis, MO) in Dulbecco's modified Eagle's medium (DMEM, Gibco-BRL, Gaithersburg, MD). Prepared fresh.
   c. Scotchpak pouch sealer and 50-μm thick, heat sealable polyester pouches (Kapak, Minneapolis, MN).
   d. Foil thermocouple (SA1-T, Omega).
   e. Stripchart recorder (Hewlett Packard, Temperature Module, model 17502A).

3. Method
3.1. Hepatocyte Isolation and Culture
3.1.1. Isolation of Hepatocytes from Rat Liver

Hepatocytes were isolated from 2–3-mo-old female Lewis rats by a modification of the method originally reported by Seglen *(16)*, and covered in detail in Chapter 33.

3.1.2. Long-term Hepatocyte Culture

This cell-culture technique, consisting of two layers of collagen, between which hepatocytes are sandwiched, is a modification of that covered in Chapter 33 of this volume. The primary difference is that here hepatocytes are cultured on coverslips which must then be cut out of the surrounding collagen gel.

1. Prepare the coverslips with an edge of silicone several mm high to prevent the slippage of the collagen sandwich. Since these will be included in the culture system, they must be sterilized prior to use.
2. Place two coverslips in a 100-mm plastic tissue-culture dish.
3. At least 60 min prior to cell seeding, coat the dishes containing the coverslips with 3 mL of type 1 collagen (1.11 mg/mL) and place them into a 37°C incubator for gelation. Care should be taken to ensure even coating of the coverslips, to prevent collagen from getting under the glass, and to keep the glasses away from the edge of the dish.
4. To each dish, gently add 9 mL of hepatocyte culture media containing 1×10^6 hepatocytes/mL. Distribute the solution while pipeting, and gently tap the dish to completely cover the collagen layer. Evenly distribute the cells by gentle shaking and tilting of the dish.
5. Incubate the dishes for 24 h at 37°C and 10% CO_2.
6. Aspirate the supernatant media and unattached cells and evenly apply another 3 mL of collagen over the cell layer. Incubate for 60 min at 37°C.
7. Add 9 mL of hepatocyte culture media and incubate for 24 h at 37°C and 10% CO_2.
8. Aspirate off the culture media and carefully cut out the coverslips from the collagen gel and transfer each to its own tissue-culture dish. Use only sterilized

instruments and avoid tilting the coverslip to prevent gel slippage. Add 9 mL of culture media and incubate.
9. Change culture media daily and collect samples for analysis as outlined in Chapter 33. Cells on d 7–9 of culture are to be used for freezing.

3.2. Hepatocyte Freezing

This subheading has two parts: the first dealing with the construction of the freezing unit and the second addressing hepatocyte freezing techniques.

3.2.1. Freezing Unit

This gives an overview for the construction of a controlled-rate freezing device for small samples.

1. Flow Channel. This work should be performed by someone experienced with the machining of plastics and resins. Machine the bottom of the flow channel from the Polypenco acetal polymer 33.7 × 9.5 × 2.0 cm^3-rectangular solid. Machine a 1.4-cm-deep entrance channel with a 1-cm-wide and 1-cm-long initial entrance, followed by a gradual 10-degree widening from 4 to 7 cm within a 21.5-cm length; maintain 7-cm width for next 8 cm; quickly narrow channel to the exit port diameter of 1 cm (**Fig. 2**; *see* **Note 2**).
2. Machine the top of the flow channel from a Polypenco acetal polymer 33.7 × 9.5 × 0.5-cm^3 rectangular solid. Machine 7.5 × 4.7-cm^2 opening, with a 0.5-cm bevel to hold the heater unit, as indicated in **Fig. 2**.
3. Seal top layer to bottom layer with silicone RTV (GE Silicones), carefully maintaining their proper orientation. Allow silicone to cure for 24 h, and check carefully for leaks.
4. Securely attach the tubing connectors at the entrance and exit ports with silicone RTV.
5. Heater Layer. As shown in **Fig. 3**, seal a single aluminum plate to the top and a single aluminum plate to the bottom of the Kapton heater layer, each with a 300-µm layer of RTV. Attach a copper-constantan thermocouple to the top aluminum layer, with heat conductive epoxy. Place the heater layer into the specially machined opening in the flow channel. Plug the thermocouple and power leads into the appropriate ports on the control system computer card.
6. Place the flow chamber onto the cork insulation and make a small styrofoam box to place over the setup. Make small holes in the styrofoam to allow for the inlet and outlet of nitrogen gas and for the thermocouple lead.
7. Coolant Source. Half fill the dewar with liquid nitrogen, and allow the temperature of the dewar to equilibrate until boiling stops. **Caution:** Always be very careful when handling liquid nitrogen to avoid severe burn injury. Wear gloves and protective clothing.
8. Attach tubing from the pressure regulator to the inlet end of the copper coil. Set the regulator at 5–10 psi and the flow meter at 100 cm^3/min. Open the nitrogen tank valve and allow nitrogen to run through the copper coil for several minutes.

Cryopreservation of Rat Hepatocytes

It is essential that this purge of the coil and tubing be done to avoid the accumulation of condensed water and, hence, ice, when the coil is placed into the dewar. Such an ice blockage will limit the performance of your cooling system, and could prove dangerous if pressure builds within the tubing or coil.

9. Place the copper coil into the dewar containing liquid nitrogen. **Caution:** the liquid nitrogen will splatter as it boils. After the nitrogen has stopped boiling, fill the remainder of the dewar with liquid nitrogen. Never turn off the nitrogen gas while the coil is in the dewar; always maintain a minimum flow rate of 100 cm^3/min.
10. Program the control system with the appropriate freezing protocol, starting at ambient temperature.
11. Attach tubing from the outlet of the copper coil to the refrigerant inlet port of the flow channel. Allow the flow channel to equilibrate with the cooled nitrogen gas (5 min), to reduce thermal gradients that will introduce large errors and limit the control systems ability to properly execute the freezing protocol.
12. The system is now ready to freeze the sample placed on its heating unit. During a freezing protocol maintain a high liquid nitrogen level in the dewar.

3.2.2. Freezing of Hepatocytes in Sandwich Culture (**Fig. 4**)

1. On d 7–9 of culture, aspirate the media from the dish and replace it with 9 mL DMEM containing 2 M DMSO. Allow these dishes to sit in ambient conditions for 1 h. After this incubation time, place a single coverslip and 3 mL of DMEM with 2 M DMSO into a polyester pouch and seal the pouch with the ScotchPac pouch sealer. Remove as much air as possible from the pouch and minimize its size.
2. Place a foil thermocouple (attached to a strip chart reader) to the top of the pouch for the monitoring of vertical temperature gradients.
3. Place the sample onto the heater unit and use a weighted piece of styrofoam to uniformly press on the pouch, thus ensuring maximum thermal contact.
4. Increase the flow meter to 5000 cm^3/min, and start the freezing protocol.
5. Freezing Protocol. A typical freezing protocol is shown in **Fig. 5**. The method outlined here is the protocol that best preserved hepatocytes under the conditions tested (15; see **Note 3**). Cool the pouch from 20° to –12°C at –10°C/min. Hold at –12°C for 3 min to allow for the pouch to reach thermal equilibration. Turn the heating unit off, allowing for a rapid drop in temperature and the spontaneous seeding of extracellular ice (at about –28°C). Upon the latent heat release indicated by an abrupt rise in temperature, turn the heater back on, and hold the temperature at –12°C for 15 min to permit adequate cell dehydration and complete formation of ice. Decrease the temperature from –12°C to –40°C at –10°C/min. These cells are now ready for storage at –196°C in liquid nitrogen.
6. Warming. Warming is best accomplished by quickly picking up the pouch with precooled forceps and immersing it in a 37°C sterile water bath.
7. Remove the coverslip from the pouch and transfer it to a tissue-culture dish. Add 9 mL of 37°C culture media and incubate for 30 min. Aspirate off media and replace it with fresh media to remove the DMSO, which has diffused out of the cells.

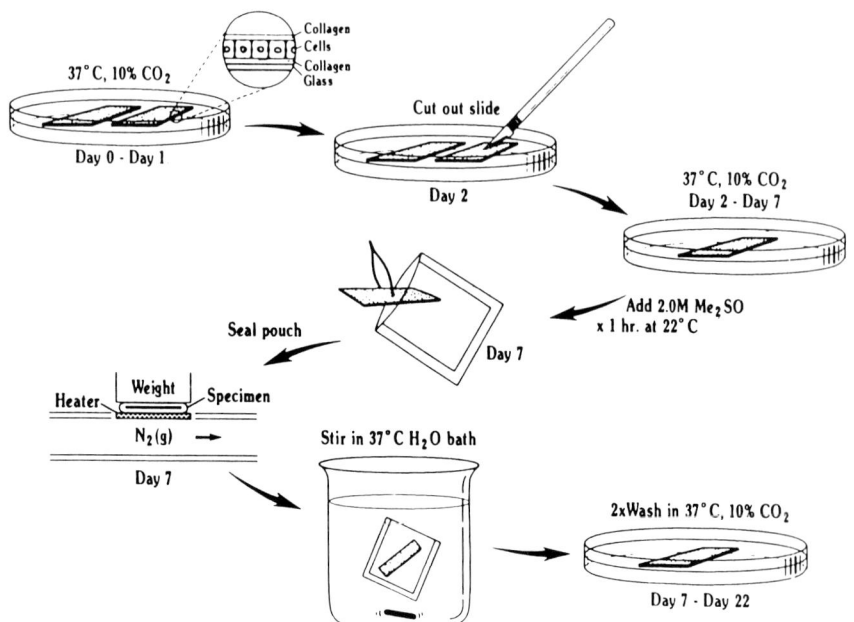

Fig. 4. Outline of hepatocyte cryopreservation method showing culture on coverslips, sample preparation, freezing, and thawing *(15)*.

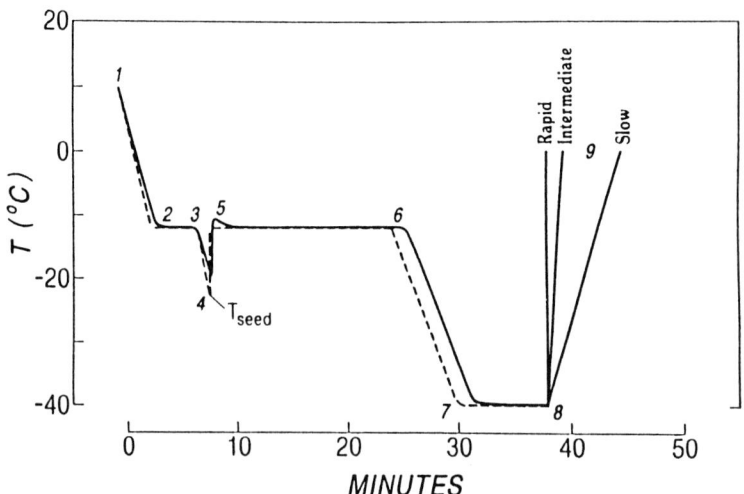

Fig. 5. Typical freeze thaw protocol with temperature as a function of time *(15)*. (1–2) Cooling at –10°C/min to –12°C. (2–3) Equilibration at –12°C. *(3)* Power off. (3–4) Temperature decrease until ice is seeded at 4. (4–5) Latent heat release and power back on. (5–6) Equilibrate at –12°C. (6–7) Cool at prescribed cooling rate to final temperature. (7–8) Hold at final temperature. (8–9) Thaw at various rates.

8. Lower the N flow to purge levels and remove the coil from the liquid nitrogen. Defrost the system with the hair dryer until the coil and the tubing are at room temperature. Turn off N gas flow.
9. Characterization of Cultured Hepatocytes. Cultures should be followed for 2 wk with media changes and samples taken daily. Methods to characterize the integrity and function of the postthaw hepatocytes are covered in Chapter 33.

4. Notes

1. The effect of freezing parameters can be elucidated in the context of the two hit hypothesis of cell damage. Upon the cooling of cells, ice will preferentially form outside of the cells in the extracellular solution. As the ice is formed in the extracellular solution, the remaining liquid fraction will become increasingly concentrated, inducing the cells to dehydrate against the imposed osmotic gradient. If the cooling rate is very rapid, a large amount of intracellular water will remain, as cells pass through low temperatures, resulting in lethal intracellular ice formation. If, however, the cooling rate is very slow, greater water efflux will occur, resulting in extreme dehydration and deleteriously high solute concentrations. One goal, therefore, is to find an appropriate intermediate cooling rate to maximize cell survival. The choosing of this cooling rate, however, is somewhat conflated by the ice-seeding temperature, the temperature at which cellular dehydration begins: Generally, the higher the seeding temperature, the greater the relative degree of cell dehydration caused by a higher level of water energetics. For most cell types, the optimal cooling protocol still results in a low yield. Therefore, improvements can be made by cryoprotective agents, which are added to increase intracellular viscosity, depress the formation of intracellular ice, and reduce the degree of water efflux. Similarly, because of damaging osmotic fluxes and ice recrystallization encountered during thawing, the warming protocol must likewise be optimized.
2. The flow channel is machined as described, because the gradual increase in channel width will reduce backflow and flow separations, thus increasing channel efficiency. This freezing apparatus can easily be scaled to efficiently accommodate other sample sizes. Similarly, other coverslip sizes can be used to reduce sample size and thermal mass, if necessary. Further, other gasses could be used as the coolant gas, though the channel may need a different design to accommodate the variation in flow. Helium, for example, can be used to achieve much higher cooling rates (-500 to $-1000°C/min$).
3. We explored several different cooling rates ($-0.25, -1.0, -5.0, -10.0, -16.0,$ and $-35.0°C/min$), final freezing temperatures ($-40°C$ and $-80°C$), and warming rates ($+5, +150,$ and $+400°C/min$), with varying results *(15)*. As evaluated by albumin secretion, cells performed best when cooled between -5 and $-10°C/min$. When cooled to $-40°C$, full postthaw recovery was independent of warming rate, but cells cooled to $-80°C$ required rapid thawing to attain even 75% functional recovery. The extracellular DMSO concentration can be varied to result in other intracellular DMSO concentrations following uptake during the incubation period. Here, 2 *M*

DMSO in DMEM results in 1.33 M intracellular DMSO after 1 h of incubation. Similarly, 0.5, 1.0, and 5.0 M extracellular DMSO results in 0.33, 0.66, and 3.33 M intracellular DMSO, respectively, after 1 h, with 3.33 M causing irreversible toxic damage *(17)*. Likewise, various other CPAs (ethylene glycol, propanediol, glycerol, and so on) can be used. Finally, since this freezer design allows easy access to the sample, one could seed extracellular ice by touching the bag with a spatula or forceps previously cooled in liquid nitrogen.

References

1. Coger, R. and Toner, M. (1995) Preservation techniques for biomaterials in *The Biomedical Engineering Handbook*. (Brönzino, J. D., ed.) CRC, Boca Raton, pp. 1557–1567.
2. Karlsson, J. O. M. and Toner, M. (1996) Long-term storage of tissues by cryopreservation: critical issues. *Biomaterials* **17,** 243–256.
3. Morgan, J. and Yarmush, M. (1997) Bioengineered skin substitutes. *Sci. Med.* **4,** 5–15.
4. Langer, R. and Vicanti, J. P. (1993) Tissue engineering. *Science* **260,** 920–926.
5. Mazur, P. (1984) Freezing of living cells: mechanisms and implications. *Am. J. Physiol.* **143,** C125–C142.
6. Karlsson, J. O. M., Cravalho, E. G., and Toner, M. (1993) Intracellular ice formation: causes and consequences. *Cryo-Lett.* **14,** 323–334.
7. Schwartz, G. J. and Diller, K. R. (1983) Osmotic response of individual cells during freezing. *Cryobiology* **20,** 61–77.
8. Arakawa, T., Carpenter, J. F., Lita, Y. A., and Crowe, J. H. (1990) The basis of toxicity of certain cryoprotectants: a hypothesis. *Cryobiology* **27,** 401–415.
9. Toner, M. and Borel-Rinkes, I. H. M. (1993) A controlled rate freezing device for cryopreservation of biological tissues. *Cryo-Lett.* **7,** 43–56.
10. Dunn, J. C. Y., Tompkins, R. G., and Yarmush, M. L. (1991) Long-term in vitro function of adult hepatocytes in a collagen sandwich configuration. *Biotechnol. Prog.* **7,** 237–245.
11. Yarmush, M. L., Dunn, J. C., and Tompkins, R. G. (1992) Assessment of artificial liver support technology. *Cell Transplant.* **1,** 323–341.
12. Sussman, N. L., Chang, M. G., Koussayer, T., He, D., Shong, T. ., Whisennand, H. H., and Kelly, J. H. (1992) Reversal of fulminant hepatic failure using an extracorporeal liver assist device. *Hepatology* **16,** 60–65.
13. Chesne, C. and Guillouzo, A. (1988) Cryopreservation of isolated rat hepatocytes: a critical evaluation of freezing and thawing conditions. *Cryobiology* **25,** 323–330.
14. Fuller, B. J., Morris, G. J., Nutt, L. H., and Attenburrow, V. D. (1980) Functional recovery of isolated rat hepatocytes upon thawing from −196°C. *Cryo-Lett.* **1,** 139–146.
15. Borel-Rinkes, I. H. M., Toner, M., Ezzell, R. M., Tompkins, R. G., and Yarmush, M. L. (1992) Long-term functional recovery of hepatocytes after cryopreservation in a three-dimensional culture configuration. *Cell Transplant.* **1,** 281–292.
16. Selgen, P. O. (1976) Preparation of isolated rat liver cells. *Methods Biol.* **13,** 29–83.
17. Borel-Rinkes, I. H. M., Toner, M., Ezzell, R. M., Tompkins, R. G., and Yarmush, M. L. (1992) Effects of dimethyl sulfoxide on cultured rat hepatocytes in a sandwich configuration. *Cryobiology* **29,** 443–453.

IV

CELL MATERIAL COMPOSITES

25

Microencapsulation of Enzymes, Cells, and Genetically Engineered Microorganisms

Thomas M. S. Chang

1. Introduction

Microencapsulation of biologically active material in the form of artificial cell was reported as early as 1964 *(1–4)*. However, it is only in the past 10 yr that many centers have extensively developed this *(5)*. More recently, we have concentrated on three areas of artificial cells for blood substitutes, enzyme therapy, and cell therapy. Space allows only a few examples to be given here.

HIV has stimulated extensive development in the past 10 yr. The early idea of crosslinked hemoglobin *(1,4)* has been developed as first generation blood substitute, now in phase III clinical trials by a number of groups *(6,7)*. Second-generation blood substitutes include the microencapsulation of hemoglobin in lipid vesicles. We are now developing a further generation of blood substitutes based on the use of nanotechnology to prepare 150-nm diameter, biodegradable, polymeric membrane nanocapsules containing hemoglobin and enzymes *(8,9)*.

Injection of enzyme artificial cells is effective in enzyme therapy for inborn errors of metabolism *(10)*, and for cancer therapy *(11)*. However, accumulation of the implanted artificial cells has been a problem. We have solved this problem recently by giving the enzyme artificial cells orally, as in the following example. Lesch-Nyhan disease is a very rare inborn error of metabolism, with accumulation of hypoxanthine. Being lipid-soluble, it can diffuse rapidly into the intestine. By giving the enzyme microcapsules orally to act in the intestine we can avoid the need for injection. Daily oral administration of microencapsulated xanthine oxidase resulted in the lowering of hypoxanthine in the plasma and cerebral spinal fluid in a patient with Lesch-Nyhan disease *(11,12)*.

Phenylketonuria (PKU) is the most common inborn error of metabolism in humans with enzyme defects resulting in the elevation of the amino acid phenylalanine. However, amino acids from the body do not diffuse rapidly into the intestine. We can now solve this problem, based on our new findings of extensive enterorecirculation of amino acids. We showed that there is an extensive enterorecirculation of amino acids between the body and the intestine. Large volumes of digestive juice containing enzymes and other proteins are secreted into the intestine. These are digested by tryptic enzymes in the intestine into amino acids. The amino acids formed in this way are reabsorbed into the body. We showed for the first time that amino acids formed from this source are higher in amount than those resulting from protein in ingested food. This means that when we give orally microcapsules containing a specific enzyme, it can remove the corresponding specific amino acid in the intestine and therefore prevent it from returning to the body. The net result is a depletion of this specific amino acid from the body. For example, microencapsulated phenylalanine ammonia lyase given orally once a day can selectively remove phenylalanine from the enterorecirculating amino acids. This explains our earlier observation of the effectiveness of this approach for lowering the elevated systemic phenylalanine in the PKU rats *(16)*, and more recently in the ENU2 phenylketonuric mice *(17)*. This approach can also be used to remove any of the other 26 amino acids in the body, with potential for treating other inborn errors of metabolisms and for amino acid dependent tumors. This is now being developed for clinical trials. This has resulted in renewed interest in the microencapsulation of enzymes.

We have also studied microcapsules containing multienzyme systems with cofactor recycling for multistep enzyme conversions *(18,19)*

The third major area of our present interest is based on the microencapsulation of cells and microorganisms. As early as 1965, this author successfully encapsulated cells and wrote *(20)*:

> ... microencapsulation of intact cells or tissue fragments ... the enclosed material might be protected from destruction and from participation in immunological processes, while the enclosing membrane would be permeable to small molecules of specific cellular product which could then enter the general extracellular compartment of the recipient. For instance, encapsulated endocrine cells might survive and maintain an effect supply of hormone. ... The situation would then be comparable to that of a graft placed in an immunologically favourable site." "There would be the further advantage that implantation could be accomplished by a simple injection procedure rather than by a surgical operation." "Microencapsulation of intact cells ... The erythrocytes were suspended in hemolysate rather than in the diamine solution; and a silicone oil [Dow Corning 200 fluid] was substituted for the stock organic liquid. The microencapsulation was then carried out... by the principle of interfacial polymerization for membranes of cross-linked proteins... A large number of human erythrocytes suspended in hemolysate within a microcapsule of about 500 μ diameter was prepared by the syringe [drop] method."

The author also described this in other publications *(2,4)*. However, it is only with the more recent interest in biotechnology that many groups around the world have extended this approach of cell encapsulation *(5,21)*. Some of our recent research interest is as follows.

We studied the use of encapsulated hepatocytes as a model for cell and gene therapy. Implantation of encapsulated hepatocytes increases the survival of fulminant hepatic-failure rats *(22)*. Encapsulated rat hepatocytes are not rejected after being implanted into mice *(23)*. Instead, there is an increase in viability after intraperitoneal implantation *(23)*. This is a result of the retention of hepatostimulating factors inside the microcapsules as they are secreted by the hepatocytes *(24)*. Gunn rat is the model for the Crigler-Najjar syndrome in humans caused by defects of the liver enzyme UDP-glucuronosyltransferase (UDPGT). Intraperitoneal implantation of artificial cells containing hepatocytes lowered the high systemic bilirubin levels *(25,26)*. Kinetic analysis shows that this is because of the hepatocytic UDPGT in artificial cells that conjugated bilirubin to the monoconjugated and diconjugated form for excretion in the urine as in normal animals *(27)*.

For encapsulation of high concentration of small cells we have to develop a two-step method for cell encapsulation in artificial cells. This is because we observe that some cells are exposed on the surface of artificial cells prepared by the standard method *(28)*. We have therefore developed a new method to prevent this problem *(28,29)*. This will be particularly useful for the encapsulation of hepatocytes and genetically engineered cells.

Another area is the study of artificial cells containing nonpathogenic genetically engineered microorganisms. However, they cannot be injected. We are studying the possibility of using oral microcapsules containing these cells. We started with basic research using *Escherichia coli* with *K. aerogenes* gene for urea removal *(30)*. For more than 30 yr investigators have been unable to find an oral treatment for uremia, mainly because of the inability to remove the large amount of urea. We have just found a new approach that combines artificial cells with gene expression genetic engineering technology *(30)*. Oral administration of a small amount of artificial cells containing genetically engineered *E. coli* DH5 cells once a day resulted in the decrease to normal of high urea level in uremic rats to normal *(31)*. This was maintained during the 21 d of treatment. Uremic rats survived during this period when treated with this approach. On the other hand, 50% of untreated control rats died *(31)*.

We have also studied the microencapsulation of erythropoietin (EPO)-secreting renal cells *(32)*, and we microencapsulated two other microorganisms, one that removed cholesterol *(33)*, another that converted substrates to L-DOPA *(34)*.

2. Materials
2.1. Microencapsulation of Enzymes
2.1.1. Cellulose Nitrate Membrane Microcapsules:

1. Hemoglobin solution containing enzymes: Fifteen grams of hemoglobin (bovine hemoglobin type 1, 2X crystallized, dialyzed, and lyophilized. Sigma, St. Louis, MO) was dissolved in 100 mL of distilled water and filtered (Whatman No. 42).
2. Water-saturated ether: Analytical grade ether is shaken with dH_2O in a separating funnel, then left standing for the two phases to separate, so that the water can be discarded.
3. Cellulose–nitrate solution: This solution is prepared by spreading 100 mL of USP collodion (Fisher Scientific, Montreal) in a evaporating disk in a well-ventilated hood overnight. This allowed the complete evaporation of its organic solvents, leaving a dry, thin sheet. The thin sheet of polymer is then cut into small pieces and dissolved in a 100-mL mixture containing 82.5 mL analytical grade ether and 17.5 mL analytical grade absolute alcohol.
4. Tween-20 (atlas powder) solution: The 50% (v/v) concentration solution is prepared by mixing equal volumes of Tween-20 and dH_2O, then adjusting the pH to 7.0. The 1% (v/v) concentration solution was prepared by mixing 1% of Tween-20 to the buffer solution used as the suspending media for the final microencapsulated enzyme system.

2.1.2. Polyamide Membrane Microcapsules

2.1.2.1. Two Solutions

1. The terepthaloyl organic solution is prepared just before use. 100 mg of terephthaloyl chloride (ICN K+K) is added to 30 mL organic solution (chloroform:cyclohexane, 1:4) kept in an ice bath. This was covered and stirred with a magnetic stirrer for 4 h and then filtered with Whatman no. 7 paper.
2. Diamine-polyethyleneimine solution is prepared just before use. 0.378 g $NaHCO_3$ and 0.464 g 1.6-hexadiamine (J. T. Baker) is dissolved in 5 mL dH_2O that contains the material to be encapsulated. Then the pH is adjusted to 9.0. Two mL 50% polyethyleneimine (ICN) is then added to the diamine solution and the pH readjusted to 9.0 and the final volume was made up to 10 mL with dH_2O.

2.1.2.2. 10 g/100 mL Hemoglobin Solution

Prepared as described above for cellulose nitrate microcapsules. Material to be encapsulated was dissolved in 5 mL of the hemoglobin solution, instead of H_2O as described above. The final pH is adjusted to 9.0.

2.2. Materials for the Encapsulation of Cells and Microorganisms
2.2.1. Materials for the Standard Method

1. Hepatocytes: Rat liver cells were isolated from 125–150 g male Wistar rats from Charles River Breeding, Montreal, QC, Canada.

2. Genetically engineered *E. coli*: From nonpathogenic bacteria (*E. coli* DH5).
3. Reagents for encapsulation: Low-viscosity sodium alginate (Keltone LV), mol wt 12,000–80,000, was obtained from Kelco, Clark, NJ. The reagents poly-L-lysine mol wt 15,000–30,000, type IV collagenase; and type I-S trypsin inhibitor were purchased from Sigma. HEPES (4-[2-hydroxyethyl]-1-piperazine ethane sulfonic acid) buffer was purchased from Boehringer Mannheim (Montreal). Analytical grade reagents included sodium chloride, sodium hydroxide, calcium chloride dihydrate, trisodium citrate dihydrate, and d-fructose. William's E medium (Gibco-BRL; Burlington, On), supplemented with streptomycin and penicillin (Gibco-BRL), was used for hepatocyte.

2.2.2. Two-Step Method for High Concentration of Cells or Microorganisms

Two differently sized droplet generators were used for microencapsulation.

1. The first droplet generator consisted of 2 co-axially arranged jets: the central jet consisted of a 26-gage stainless steel needle (Perfektum, Popper, New Hyde Park, NY), and a 16-gage surrounding air jet, through which the sample and air was passed, respectively. To prevent the extruding sample from occluding the outlet of the surrounding air jet, the tip of the sample jet was constructed so that the tip projected 0.5 mm beyond the end of the air jet.
2. The second droplet generator was a larger and a slightly modified variant of the first droplet generator described above. The second droplet generator was constructed with a 13-gage sample jet, and a 8-gage surrounding air jet. The ends of the jets were cut flush to each other. A 1.7 × 1.1-mm PTFE capillary tube (Pharmacia, Montreal, PQ) was inserted into the sample jet until it protruded approx 15 mm from the outlet of the sample jet. The end of the capillary tubing was tapered to facilitate shearing by the flow of passing air from the air jet. The capillary tubing was approx 3.2 m in length, and had the capacity of be filled with microspheres suspended in 2.5 mL of sodium alginate.

2.2.3. Materials for Preparation of Macroporous Microencapsules

1. Microorganism: *Pseudomonas pictorum* (ATCC #23328) was used because of its ability to degrade cholesterol. It was cultured first in nutrient broth (Difco) at 25°C, followed by harvesting and resuspension in a cholesterol medium. After culturing this suspension for 15 d at 25°C, it was used as an inoculum for biomass production. The culture was grown in bovine calf serum (Sigma) at 37°C for 36 h, and then harvested. This was used to prepare bacterial suspensions for immobilization. The concentration was about 0.4 mg of dry cell/mL.
2. Inoculum medium: Bovine calf serum was from Sigma. It was used in all experiments unless otherwise specified. Nutrient agar plates were prepared by dissolving 8 g of nutrient agar (Difco) in 100 mL of water. The solution was autoclaved for 15 min at 121°C, allowed to cool at 50°C, and poured into plastic Petri dishes (Fisher Scientific). The plates were stored at 4°C for up to 2 mo.

3. Cholesterol medium: The composition of the medium was: ammonium nitrate (0.1%); potassium phosphate (0.025%); magnesium sulfate (0.025%); ferric sulfate (0.0001%); yeast extract (0.5%); and cholesterol (0.1%), all dissolved in water, with pH adjusted to 7.0 and autoclaved for 15 min.

3. Methods
3.1. Microencapsulation of Enzymes
3.1.1. Cellulose Nitrate Membrane Microcapsules

Cellulose nitrate membrane microcapsules are prepared using an updated procedure based on earlier publications *(1–4,18)*.

1. Enzymes and other materials to be microencapsulated were dissolved or suspended in 2.5 mL of the hemoglobin solution. The final pH was adjusted to 8.5 with Tris buffer, and hemoglobin concentration adjusted to 10 g/dL.
2. 2.5 mL of this solution was added to a 150-mL glass beaker, and 25 mL of water saturated ether was added.
3. The mixture was immediately stirred with a Fisher Jambo magnetic stirrer at 1200 rpm (setting of 5) for 5 s.
4. While stirring was continued, 25 mL of a cellulose-nitrate solution was added. Stirring was continued for another 60 s.
5. The beaker was covered and allowed to stand unstirred at 4°C for 45 min.
6. The supernatant was decanted and 30 mL of n-butyl benzoate added. The mixture was stirred for 30 s at the same magnetic stirrer setting.
7. The beaker was allowed to stand uncovered and unstirred at 4°C for 30 min. Then the butyl benzoate was completely removed after centrifugation at 350g for 5 min.
8. Twenty-five mL of the Tween solution at 50% (v/v) concentration, pH 7.0 was added. Stirring was started at a setting of 10 for 30 s. 25 mL of water was added, and stirring continued at a setting of 5 for 30 s, then 200 mL of water added.
9. The supernatant was removed and the microcapsules were washed 3X with 200 mL of a 1% Tween-20 solution, pH 7.0. The microcapsules were then suspended in a suitable buffer. In properly prepared microcapsules, there should not be leakage of hemoglobin after the preparation.

3.1.2. Preparation of Polyamide Membrane Microcapsules by Interfacial Polymerization

Polyamide membrane microcapsules of 100 µm mean diameter were prepared using an updated method based on earlier methods *(1–4,18)*.

1. Enzyme is added to 2.5 mL of the hemoglobin solution with pH and concentrations adjusted as described in **Subheading 2.1.1., item 1.**
2. 2.5 mL of the diamine–polyethyleneimine solution was mixed for 10 s in a 150-mL beaker placed in an ice bath.

3. 25 mL of an 0.5% (v/v) Span 85 (Atlas Powder) organic solution (chloroform: cyclohexane, 1:4) was added and stirred in the Fisher Jambo magnetic stirrer at speed setting of 2.5 for 60 s.
4. 25 mL of the terephthaloyl–chloride solution prepared earlier was added, and the reaction was allowed to proceed for 3 min with the same stirring speed.
5. The supernatant was discarded and another 25 mL of the terephthaloyl–chloride solution was added.
6. The reaction was carried out with stirring for another 3 min. The supernatant was discarded.
7. Then 50 mL of the 0.5% Span 85 chloroform–cyclohexane solution was added and stirred for 30 s. The supernatant was discarded.
8. After this, the procedure of the use of Tween-20 as described for cellulose–nitrate microcapsules was used here for the transfer of the microcapsules into the buffer solution.

3.1.3. Lipid–Polymer Membrane Microcapsules that Retain Cofactors

Cofactors covalently linked to macromolecules like dextran or polyethyleneime can be retained within semipermeable microcapsules to be recycled enzymatically. However, linkage of cofactors to macromolecules increases steric hindrance and reduces their rate of reactions with enzymes. In biological cells like erythrocytes, free cofactors and multienzyme systems are all retained within the cells in free solution. Thus studies were carried out here to immobilize free cofactors inside microcapsules, with membranes impermeable to cofactors, but permeable to the initial substrates. This way, the free cofactor can function without steric hindrance in close proximity to the enzymes. Furthermore, all enzymes and cofactors inside the microcapsules are in free solution. Lipid–polyamide membrane microcapsules have been prepared *(4)*. These are permeable to lipophilic molecules, but with little or no permeability to hydrophilic molecules as small as K^+ and Na^+.

Lipid–polyamide microcapsules of 100 μm mean diameter, containing multienzyme systems, cofactors and α-ketoglutarate, were prepared *(35)*. The first part is similar to the procedure described earlier in this chapter under basic procedure for the preparation of polyamide microcapsules.

1. To 2.0 mL of the hemoglobin solution was added glutamic dehydrogenase, 12.5 mg (bovine liver, type III, 40 U/mg, Sigma); alcohol dehydrogenase, 6.25 mg (yeast, 330 U/mg, Sigma); Urease, 0.5 mg (51 U/mg Millipore); ADP, 1.18 mg; and either NAD^+ (0.52, 105, 2.11, or 21.13 mg) or NADH (21.13 mg) dissolved in 0.25 mL water. Finally, 0.25 mL α-ketoglutarate (56.5 mg), $MgCl_2$ (2.5 mg), KCl (0.93 mg) were added.
2. 2.5 mL of the hemoglobin–enzyme solution so prepared was added to 2.5 mL of the diamine–polyethyleneime solution.
3. The remaining steps were the same as described above, except that the Tween-20 steps were omitted here. Instead, after washing with the 0.5% Span 85 organic

solution, as described, the following steps were carried out to apply the lipids to the polyamide membranes.

4. The microcapsules were rinsed twice with 10 mL of a lipid–organic liquid. This was prepared earlier as follows: 1.4 g lecithin and 0.86 g cholesterol were added to 100 mL tetradecane and stirred for 4 h at room temperature. If a more permeable lipid membrane is required to allow urea to diffuse across, then the lipid compositions should be 0.43 g cholesterol and 0.7 g lecithin.
5. Then another 10 mL of the lipid–organic liquid was added and the suspension was slowly rotated for 1 h at 4°C on a multipurpose rotator.
6. After this, the supernatant was decanted and the lipid–polyamide membrane microcapsules were recovered and left in this form at 4°C, without being suspended in aqueous solution until it was added to the substrate solution just before the reaction.

The procedure takes practice and the microcapsules prepared must be tested for the absence of leakage of enzymes or cofactors before being used in experimental studies.

3.2. Microencapsulation of Cells and Microorganisms
3.2.1. Standard Method (Fig. 1)
3.2.1.1. PREPARATION OF RAT HEPATOCYTES

1. Each rat was anesthetized with sodium pentobarbital, and cannulated via the portal vein.
2. The thoracic vena cava was cut and the liver was perfused with a calcium-free perfusion buffer (142 mM NaCl, 6.7 mM KCl, 10 mM HEPES, pH 7.4) for 10 min at 40 mL/min.
3. Afterwards, the liver was perfused with a collagenase perfusion buffer (67 mM NaCl, 6.7 mM KCl, 100 mM HEPES, 5 mM CaCl$_2$, 0.05% collagenase, pH 7.5) for an additional 15 min at 25 mL/min.
4. The liver was then excised, placed in Williams' E medium supplemented with 100 µg/mL streptomycin and penicillin, and gently shaken to free loose liver cells from the liver tissue.
5. The cells were collected, filtered through a 74-µm nylon monofilament mesh (Cistron, Elmford, NY), and centrifuged to remove connective tissue debris, cell clumps, nonparenchymal cells, and damaged cells.
6. Isolated hepatocytes were prepared for encapsulation by first washing and suspending the cells with buffered saline (0.85% NaCl, 10 mM HEPES, 20 mM D-fructose, pH 7.4).
7. The cells were then mixed with a 4% stock solution of sodium alginate (4% sodium alginate, 0.45% NaCl), to make a cell suspension consisting of 20 × 10^6 cells/mL of 2% sodium alginate.

3.2.1.2. GENETICALLY ENGINEERED E. COLI DH5 CELLS
1. E. coli DH5 cells were grown in L.B. medium.
2. Log phase bacterial cells were harvested by centrifuging at 10,000g for 20 min at 4 0°C. The supernatant was discarded.

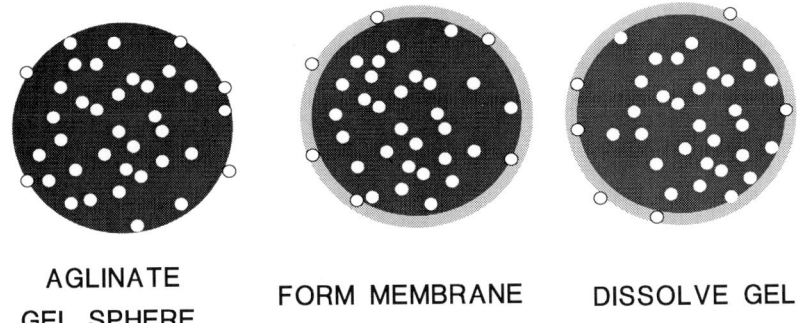

AGLINATE GEL SPHERE   FORM MEMBRANE   DISSOLVE GEL

Fig. 1. Standard method for bioencapsulation of cells and microorganisms. Step 1: formation of gelled alginate spheres containing cells. Some cells are present on the surface of the sphere. Step 2: formation of membrane around gelled alginate spheres. Those cells on the surface of the spheres would be entrapped into the membrane. Step 3: dissolution of content of spheres to form microcapsules with liquid contents. Membrane would be weakened by cells entrapped or exposed on the membrane. When used to encapsulate a high concentration of small cells or microorganisms, some cells may become entrapped in the membrane matrix. This can result in membrane weakness or perforation. Reproduced with permission from ref. *28*.

3. The cell mass was then washed and centrifuged at 10,000g for 10 min at 40°C for five times with sterile cold water to remove media components.
4. Bacterial cells were suspended in an autoclaved sodium alginate in ice-cold 0.90% sodium chloride solution.

3.2.1.3. ENCAPSULATION USING THE STANDARD METHOD

1. Hepatocytes or bacterial cells were suspended in an autoclaved sodium alginate in ice-cold 0.90% sodium chloride solution.
2. The viscous alginate suspension was pressed through a 23-gage stainless steel needle using a syringe pump (Compact infusion pump model-975 Harvard Apparatus, Mill, MA). Sterile compressed air, through a 16-gage coaxial stainless steel needle, was used to shear the droplets coming out of the tip of the 23-gage needle.
3. The droplets were allowed to gel for 15 min in a gently stirred, heat-sterilized and ice-cold solution of calcium chloride (1.4%, pH 7.2). Upon contact with the calcium chloride buffer, alginate gelation is immediate.
4. After gelation in the calcium chloride solution, alginate gel beads were reacted with polylysine (PLL), mol wt 16,100 (0.05% in HEPES buffer saline, pH 7.2) for 10 min. The positively charged PLL forms a complex of semipermeable membrane.
5. The beads were then washed with HEPES (pH 7.2) and coated with an alginate solution (0.1%) for 4 min.
6. The alginate-poly-L-lysine-alginate capsules so formed were then washed in a 3% citrate bath (3% in 1:1 HEPES-buffer saline, pH 7.2) to liquefy the gel in the microcapsules.

7. The APA microcapsules formed, which contains entrapped hepatocytes or genetically engineered bacteria *E. coli*, were stored at 40°C and used for experiments. The conditions were kept sterile during the process of microencapsulation.

3.2.2. Two-Step Method for High Concentration of Cells or Microorganisms (**Fig. 2**)

The standard method described above is not optimal for encapsulating high concentrations of cells or microorganisms. As shown schematically in **Fig. 1**, cells or microorganisms may be trapped in the membrane matrix. This can weaken the membrane. If cells are exposed to the surface, this may also resulted in loss of immunoisolation and rejection. As a result, we have developed a two-step method that prevented this problem *(28,29)*.

1. First, the hepatocytes suspended in sodium alginate were entrapped within solid calcium alginate microspheres. This was done by filling a 5.0-mL syringe (Becton Dickinson, Rutherford, NJ) with the cell suspension, and extruding the sample with a syringe infusion pump (Harvard Apparatus) through the sample jet of the first droplet generator. The droplets formed at the end of the sample jet were allowed to fall dropwise into a Pyrex dish (125.65 mm) containing 300 mL 100 mM CaCl$_2$ (100 mM CaCl$_2$, 10 mM HEPES, 20 mM D-fructose, pH 7.4). Every 5 min, the cells in the syringe were resuspended by gentle inversion of the syringe, to minimize the effect of cells sedimenting in the alginate solution. The air-flow and infusion rates through the droplet generator were 2.0–3.0 L/min and 0.28–0.39 mL/min respectively; and the clearance height between the end of the sample jet and the surface of the calcium solution was set at approx 20 cm. A strainer cup was fitted inside the dish to collect the droplets, and to facilitate the removal of the formed microspheres.
2. The microspheres were allowed to cure for approx 15 min, after which they were removed and temporarily stored in Hanks' balanced salt solution (Gibco-BRL) supplemented with 10%, 100 mM CaCl$_2$.
3. In the second step, 1.0 mL of formed microspheres were collected and washed three times with buffered saline (0.85% NaCl, 10 mM HEPES, 20 mM D-fructose, pH 7.4).
4. The final saline washing was aspirated, and 1.0 mL of 1.2–1.6% sodium alginate was added to the 1.0 mL of washed microspheres. The sodium alginate was prepared by diluting the 4% stock solution with buffered saline. With a 5-mL syringe, the length of the PTFE capillary tubing was filled with the sodium alginate and suspension of microspheres. The tapered end of the capillary tubing was inserted through the top of the sample jet of the second droplet generator until the tip of the tubing extended approx 15 mm beyond the end of the sample jet. The air flow and extrusion rates through the modified droplet generator were 7.0–9.0 L/min and 0.28–0.39 mL/min, respectively. The tip of the capillary tubing was set approx 20 cm above the surface of the calcium solution. With the 5-mL syringe still attached to the other end of the tubing, the microsphere suspension in the tubing was extruded with the Harvard infusion

Microencapsulation

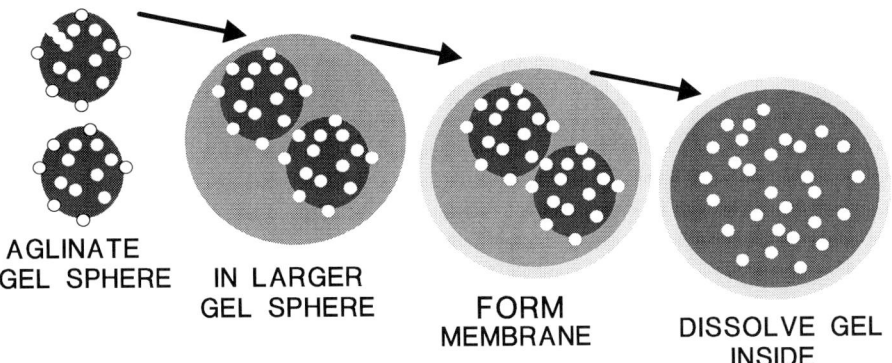

Fig. 2. Two-step method for bioencapsulation of a high concentration of smaller cells or microorganisms. Step 1: formation of small gelled alginate spheres containing cells. Step 2: to include the small gelled alginate spheres containing cells into larger gelled alginate spheres. This way, no cells are present on the surface of the larger gelled alginate spheres. Step 3: formation of membrane around the large gelled alginate spheres. Step 4: dissolution of content of spheres, including the smaller spheres inside, to form microcapsules with liquid contents. This helps to prevent entrapment of cells in the membrane matrix. Reproduced with permission from ref. *28*.

pump. Similarly, the drops formed at the end of the sample jet were allowed to fall dropwise into a Pyrex dish containing a strainer cup, and filled with 300 mL of 100 mM CaCl$_2$.

5. The spheres were allowed to cure in the calcium solution for approx 15 min, after which they were removed and washed with buffered saline.
6. The alginic acid matrix on the surface of the sphere was stabilized with poly-L-lysine by immersing 5 mL (settled volume) of macrospheres in 80 mL of 50 mg% poly-L-lysine (50 mg% poly-L-lysine, 0.85% NaCl, 10 mM HEPES, 20 mM D-fructose, pH 7.4) for 10 min.
7. The spheres were then drained, washed with buffered saline, and immersed into 200 mL of 0.2% sodium alginate (0.2% sodium alginate, 0.85% NaCl, 10 mM HEPES, 20 mM D-fructose, pH 7.4) for 10 min. to apply an external layer of alginate.
8. After 10 min, the spheres were collected and immersed in 200 mL 50 mM sodium citrate (50 mM sodium citrate, 0.47% NaCl, 20 mM D-fructose, pH 7.4) to solubilize the intracapsular calcium alginate. This may require up to 30 min, with frequent changes of the sodium citrate solution.

3.3.3. Macroporous Microcapsules

When using cells or microorganisms to act on macromolecules, the above methods cannot be used. Thus, in using microorganisms to act on cholesterol bound to lipoprotein, we have to encapsulate the microorganisms in macroporous microcapsules *(33)*.

1. A 2% agar (Difco) and 2% sodium alginate (Kelco) solution was autoclaved for 15 min and cooled to 45°C to 50°C.
2. *P. pictorum* suspended in 0.4 mL of 0.9% NaCl was added drop by drop to 3.6 mL of agar alginate solution at 45°C, while stirring vigorously.
3. Three mL of the mixture obtained was kept at all times at 45°C while it was being extruded through the syringe. The extruded drops were collected into cold (4°C) 2% calcium chloride, and allowed to harden. These agar–alginate beads were about 2 mm in diameter.
4. After 15 min, the supernatant was discarded and the beads were resuspended in 2% sodium citrate for 15 min.
5. Then they were washed and stored in 0.9% saline at 4°C.

When testing for immobilized bacterial activity, 1 mL of beads/microcapsules were placed in a sterile 50-mL Erlenmayer. 5 mL of serum were added and a foam plug was fitted. Samples were withdrawn at specified intervals. When empty beads or microcapsules were prepared, the bacterial suspension was replaced by saline, and all the other steps were kept the same.

4. Notes
4.1. Microencapsulation of Enzymes

1. Cellulose nitrate membrane microcapsules: Hemoglobin at a concentration of 10 g/dL is necessary for the successful preparation. Furthermore, this high concentration of protein stabilizes the enzymes during the preparation and also during reaction and storage *(36)*. When the material (e.g., NADH) to be encapsulated is sensitive to the enzymes present in hemoglobin, highly purified hemoglobin is used. This requires the use of purification using affinity chromatography on an NAD^+ sepharose column.
2. Crosslinking with glutaraldehyde: The long-term stability of microencapsulated enzyme activity can be greatly increased by crosslinking with glutaraldehyde *(36)*. This is done at the expense of reduced initial enzyme activity.
3. Oral administration: When using cellulose nitrate microcapsules containing enzymes for oral administration, the permeability of the membrane may need to be decreased. This is to prevent the entry of smaller tryptic enzymes. Permeability can be decreased by decreasing the proportion of alcohol used in dissolving the evaporated cellulose nitrate polymer.
4. Interfacial polymerization: Failure in preparing good microcapsules is frequently caused by the use of diamine or diacids that have been stored after they have been opened. A new unopened bottle will usually solve the problems. Unlike the cellulose nitrate microcapsules, in interfacial polymerization, the hemoglobin solution can be replaced by a 10% polyetheleneimine solution adjusted to pH 9.0. However, the microcapsules prepared without hemoglobin may not be as sturdy.

Crosslinking the microencapsulated enzymes with glutaraldehyde after the preparation of the enzyme microcapsules could also be carried out to increase the long term stability of the enclosed enzymes *(36)*, although this decreases the initial enzyme activity.

5. Multienzyme reaction: In multienzyme reaction requiring cofactor recycling, the cofactor can be crosslinked to dextran-70 and then encapsulated together with the enzymes. For example, NAD^+-N6-[N-(6-aminohexyl)-acetamide] was coupled to dextran T-70, polyethyleneimine, or albumin to form a water soluble NAD^+ derivative, and then encapsulated together with the multienzyme systems in the microcapsules *(18,19)*. This way both the cellulose nitrate microcapsules and polyamide microcapsules can be used. This allows for high permeation to substrates and products. However, linking cofactor to soluble macromolecules resulted in significant increases in steric hindrance and diffusion restrictions of the cofactor.

6. Retention of cofactors: Lipid–polyamide membrane microcapsules containing multienzyme systems, cofactors, and substrates can retain cofactors in the free form. Thus, analogous to the intracellular environments of red blood cells, free NADH or NADPH in solution inside the microcapsules is effectively recycled by the multistep enzyme systems that are also in solution. However, only lipophilic or very small hydrophilic molecules like urea can cross the membrane. For example, ammonia and urea equilibrate into the microcapsules to be converted into amino acids *(18,19)*. However, external alcohol instead of glucose had to be used as substrate for the conversion and recycling of NAD^+ to NADH. Some substrates, e.g., α-ketoglutarate, had to be encapsulated, since they cannot enter the lipid complexed microcapsules *(18,19)*.

4.2. Microencapsulation of Cells and Microorganisms

7. Standard method: Alginate concentration in the tested range, 1–2.25% (w/v), does not affect the bacterial cell viability or cell growth. Quality of microcapsules improves with increasing alginate concentration from 1–1.75% (w/v). The use of 2% (w/v) alginate resulted in perfectly spherical shape and sturdy microcapsules, with maximum number of encapsulated bacterial cells. An increase in liquid flow rate of the alginate–cell or bacterial suspension through the syringe pump, from 0.00264 to 0.0369 mL/min, resulted in increase in microcapsule diameter. The flow rate in the range of 0.00724–0.278 mL/min resulted in good spherical microcapsules.

At an air-flow rate of 2 L/min, the microcapsules had an average diameter of 500 ± 45 m diameter. At the air-flow rates above 3 L/min, microcapsules were irregular in shape.

These results indicate that alginate concentration, air-flow rate, and liquid-flow rate are critical for obtaining microcapsules of desired characteristics and permselectivity *(30)*. Using the following: 2% (w/v) alginate, 0.0724 mL/min liquid-flow rate, 2 L/min air-flow rate, which has enabled the development of suitable microencapsulation process for cells or microorganisms.

Microcapsules prepared this way are permeable to albumin, but impermeable to molecules with higher mol wt *(37)*. Thus, hepatostimulating factors *(24)* and globulin *(37)* cannot cross the membrane of the standard microcapsules.

8. Two-step method: This method prevents the entrapment of small cells in the membrane matrix. Microcapsules prepared this way when implanted are much more stable, with decrease in rejection *(28,29)*
9. Macroporous microcapsules: Temperature is a very critical parameter in the immobilization of *P. pictorum (33)*. A low temperature produces gelation of the polymer in the syringe or conduits. A high temperature prevents gelation, but increases the mortality rate of *P. pictorum*. Exposing *P. pictorum* to 55°C for 10 min or more can completely inhibit enzymatic activity. However, up to 20 min exposure to 45°C does not significantly inhibit cholesterol activity. Open pore agar beads stored at 4°C did not show any sign of deterioration. The beads retain their enzymatic activity even after 9 mo of storage.

Acknowledgments

This research has been carried out with support to TMSC from the Medical Research Council of Canada in both operating grants and the career investigator award. It is also supported by the Quebec MESST Virage Award of Centre of Excellence in Biotechnology to TMSC.

References

1. Chang, T. M. S. (1964) Semipermeable microcapsules. *Science* **146**, 524,525.
2. Chang, T. M. S., MacIntosh, F. C., and Mason, S. G. (1966) Semipermeable aqueous microcapsules: I. Preparation and properties. *Can. J. Physiol. Pharmacol.* **44**, 115–128.
3. Chang, T. M. S., MacIntosh, F. C., and Mason, S. G. (1971) Encapsulated hydrophilic compositions and methods of making them. Canadian Patent 873, 815, 1971.
4. Chang, T. M. S. (1972) Artificial Cells, Monograph, Charles C Thomas, Springfield, IL.
5. Chang, T. M. S. (1995) Artificial cells with emphasis on bioencapsulation in biotechnology. *Biotechnol. Annu. Rev.* **2**, 267–295.
6. Chang, T. M. S. (1997) Recent and future developments in modified hemoglobin and microencapsulated hemoglobin as red blood cell substiutes. *Artif. Cells Blood Substitutes Immobilization Biotechnol.* **25**, 1–24.
7. Chang, T. M. S. (1997) *Blood Substitutes: Principles, Methods, Products and Clinical Trials.* Karger/Landes, Austin, TX.
8. Chang, T. M. S. and Yu, W. P. (1996) Biodegradable polymer membrane containing hemoglobin for blood substitutes. U. S. A. Patent approved in 1996.
9. Yu, W. P. and Chang, T. M. S. (1996) Submicron polymer membrane hemoglobin nanocapsules as potential blood substitutes: preparation and characterization. *Artif. Cells Blood Substitutes Immobilization Biotechnol.* **24**, 169–184.
10. Chang, T. M. S. and Poznansky, M. J. (1968) Semipermeable microcapsules containing catalase for enzyme replacement in acatalsaemic mice. *Nature* **218**, 242–245.

11. Chang, T. M. S. (1971) The in vivo effects of semipermeable microcapsules containing L-asparaginase on 6°C3HED lymphosarcoma. Nature **229**, 117,118.
12. Chang, T. M. S. (1989) Preparation and characterization of xanthine oxidase immobilized by microencapsulation in artificial cells for the removal of hypoxanthine. *J. Biomater. Artif. Cells Artif. Organs* **17**, 611–616.
13. Palmour, R. M., Goodyer, P., Reade, T., and Chang, T. M. S. (1989) Microencapsulated xanthine oxidase as experimental therapy in Lesch-Nyhan Disease. *Lancet* **2**, 687,688.
14. Chang, T. M. S., Bourget, L., and Lister, C. (1992) Orgal administration of microcapsules for removal of amino acids, US Patent No 5,147,641.
15. Chang, T. M. S., Bourget, L., and Lister, C. (1995) A new theory of enterorecirculation of amino acids and its use for depleting unwanted amino acids using oral enzyme-artificial cells, as in removing phenylalanine in phenylketonuria. *Artif. Cells Blood Substitutes Immobilization Biotechnol.* **25**, 1–23.
16. Bourget, L. and Chang, T. M. S. (1986) Phenylalanine ammonia-lyase immobilized in microcapsules for the depletion of phenylalanine in plasma in phenylketonuric rat model. *Biochim. Biophys. Acta* **883**, 432–438.
17. Safos, S. and Chang, T. M. S. (1995) Enzyme replacement therapy in ENU2 phenylketonuric mice using oral microencapsulated phenylalanine ammonia-lyase: a preliminary report. *Artif. Cells Blood Substitutes Immobilization Biotechnol.* **25**, 681–692.
18. Chang, T. M. S. (1985) Artificial cells with regenerating multienzyme systems. *Methods Enzymol.* **112**, 195–203.
19. Gu, K. F., Chang, T. M. S. (1990) Production of essential L-branched-chained amino acids, in bioreactors containing artificial cells immobilized multienzyme systems and dextran-NAD⁺. *Appl. Biochem. Biotechnol.* **26**, 263–269.
20. Chang, T. M. S. (1965) Semipermeable aqueous microcapsules. PhD Thesis. McGill University.
21. Lim, F. and Sun, A. M. (1980) Microencapsulated islets as bioartificial endocrine pancreas. *Science* **210**, 908–909
22. Wong, H. and Chang, T. M. S. (1986) Bioartificial liver: implanted artificial cells microencapsulated living hepatocytes increases survival of liver failure rats. *Int. J. Artif. Organs* **9**, 335,336.
23. Wong, H. and Chang, T. M. S. (1988) The viability and regeneration of artificial cell microencapsulated rat hepatocyte xenograft transplants in mice. *J. Biomater. Artif. Cells Artif. Organs* **16**, 731–740.
24. Kashani, S. and Chang, T. M. S. (1991) Effects of hepatic stimulatory factor released from free or microencapsulated hepatocytes on galactosamine induced fulminant hepatic failure animal model. *J. Biomater. Artif. Cells Immobilization Biotechnol.* **19**, 579–598.
25. Bruni, S. and Chang, T. M. S. (1989) Hepatocytes immobilized by microencapsulation in artificial cells: effects on hyperbilirubinemia in Gunn Rats. *J. Biomater. Artif. Cells Artif. Organs* **17**, 403–12.
26. Bruni, S. and Chang, T. M. S. (1991) Encapsulated hepatocytes for controlling hyper-bilirubinemia in Gunn Rats. *Int. J. Artif. Organs* **14**, 239–241.

27. Bruni, S. and Chang, T. M. S. (1995) Kinetics of UDP-glucuronosyl-transferase in bilirubin conjugation by encnapsulated hepatocytes for transplantation into Gunn rats *J. Artif. Organs* **19**, 449–457.
28. Wong, H. and Chang, T. M. S. (1991) A novel two step procedure for immobilizing living cells in microcapsules for improving xenograft survival. *J. Biomater. Artif. Cells Immobilization Biotechnol.* **19**, 687–698.
29. Chang, T. M. S. and Wong, H. (1992) A novel method for cell encapsulation in artificial cells. USA Patent No. 5,084,350.
30. Prakash, S. and Chang, T. M. S. (1995) Kinetic studies of microecnapsulated genetically engineered *E. coli* cells containing *K. aerogenes* gene for urea and ammonia removal. *J. Biotechnol. Bioeng.* **46**, 621–626.
31. Prakash, S. and Chang, T. M. S. (1996) Microencapsulated genetically engineered live *E. coli* DH5 cells administered orally to maintain normal plasma urea level in uremic rats. *Nature Med.* **2**, 883–887.
32. Koo, J. and Chang, T. M. S. (1993) Secretion of erythropoietin from microencapsulated rat kidney cells: preliminary results. *Int. J. Artif. Organs* **16**, 557–560.
33. Garofalo, F. and Chang, T. M. S. (1991) Effects of mass transfer and reaction kinetics on serum cholesterol depletion rates of free and immobilized *Pseudomonas pictorum. Appl. Biochem. Biotechnol.* **27**, 75–91.
34. Lyold-George, I. and Chang, T. M. S. (1995) Characterization of free and alginate-polylysine-alginate microencapsulated *Erwinia herbicola* for the conversion of ammonia, pyruvate and phenol into L-tyrosine and L-DOPA. *J. Bioeng. Biotechnol.* **48**, 706–714.
35. Yu, Y. T. and Chang, T. M. S. (1981) Lipid-polymer membrane artificial cells containing multienzyme systems, cofactors and substrates for the removal of ammonia and urea. *Trans. Am. Soc. Artif. Intern. Organs* **27**, 535–538.
36. Chang, T. M. S. (1971) Stabilization of enzyme by microencapsulation with a concentrated solution o or by crosslinking with glutaraldehyde. *Biochem. Biophys. Res. Com.* **44**, 1531–1533.
37. Coromili, V. and Chang, T. M. S. (1993) Polydisperse dextran as a diffusing test solute to study the membrane permeability of alginate polylysine microcapsules. *J. Biomater. Artif. Cells Immobilization Biotechnol.* **21**, 323–335.

26

Methods for Microencapsulation with HEMA–MMA

Shahab Lahooti and Michael V. Sefton

1. Introduction

Encapsulation of cells in a membrane prior to implantation holds potential for controlling the adverse immune response that may be generated against the transplanted cells, by physically isolating the cells from the host's immune system. If successful, encapsulation eliminates or minimizes the adverse effects of immunosuppressive therapy and permits the use of xenogeneic cells. Ideally, the capsule membrane holds permselective properties, so that the passage of nutrients, growth factors, and the therapeutic product secreted by the cells occur readily across the membrane, but mediators of the immune system do not penetrate the membrane (**Fig. 1**). The major types of immunoisolation devices include intravascular arteriovenous shunts, diffusion chambers of tubular or planar geometry, and microcapsules *(1–6)*.

Our current standard microencapsulation technique utilizes a stationary coaxial needle assembly submerged in a flowing stream of hexadecane or dodecane (**Fig. 2;** *7*). Cells suspended in the appropriate tissue-culture medium (and extracellular matrix components) are pumped through the inner needle, while the polymer solution is pumped through the outer needle. This leads to the formation of droplets of polymer solution surrounding a core of cell suspension at the tip of the coaxial needles. The droplets are sheared from the tip of the needle by the flowing hexadecane (**Fig. 3**). Subsequently, upon travelling through a hexadecane layer, the droplets enter a phosphate-buffered saline (PBS) solution in a receiving dish. The hexadecane in the receiving dish is recirculated to the needle assembly by a pump. In the PBS, the polymer solvent is extracted to precipitate the polymer membrane, forming spherical microcapsules containing a core of cell suspension. The capsules are maintained in suspension in the PBS precipitation bath by a magnetic stirrer. The PBS solution contains 100–150

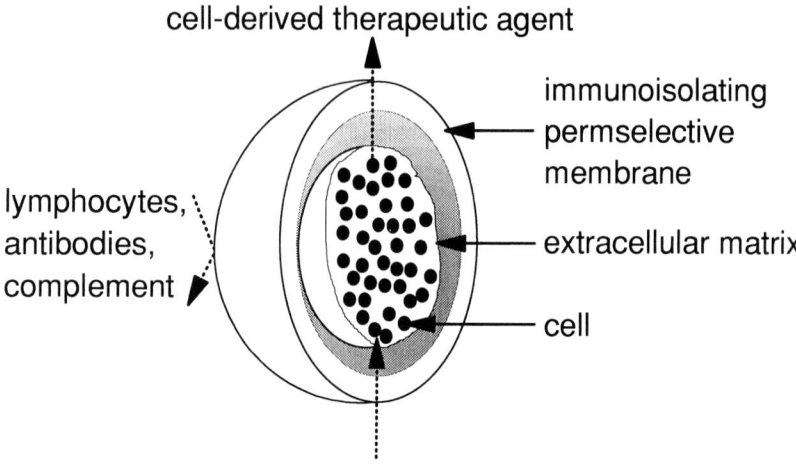

Fig. 1. Schematic drawing of a model microcapsule. Cells are encapsulated with or without an extracellular matrix in a permselective membrane, which provides isolation from the mediators of the immune system, but allows the passage of nutrients, growth factors, and the therapeutic product secreted by the cells.

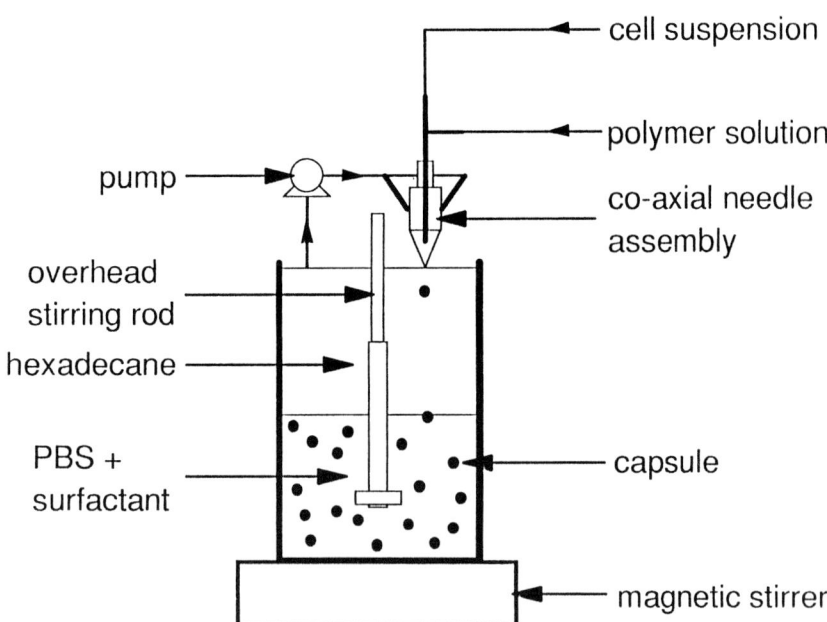

Fig. 2. Schematic drawing of the stationary-needle assembly microencapsulation setup producing small-diameter (~300–600 μm) capsules. Droplets of polymer solution surrounding a core of cell suspension formed at the tip of coaxial needles are sheared off by flowing hexadecane into a PBS bath, where the capsule membrane is formed by interfacial precipitation. A peristaltic pump recirculates the hexadecane and a magnetic overhead stirrer maintains the capsules in suspension.

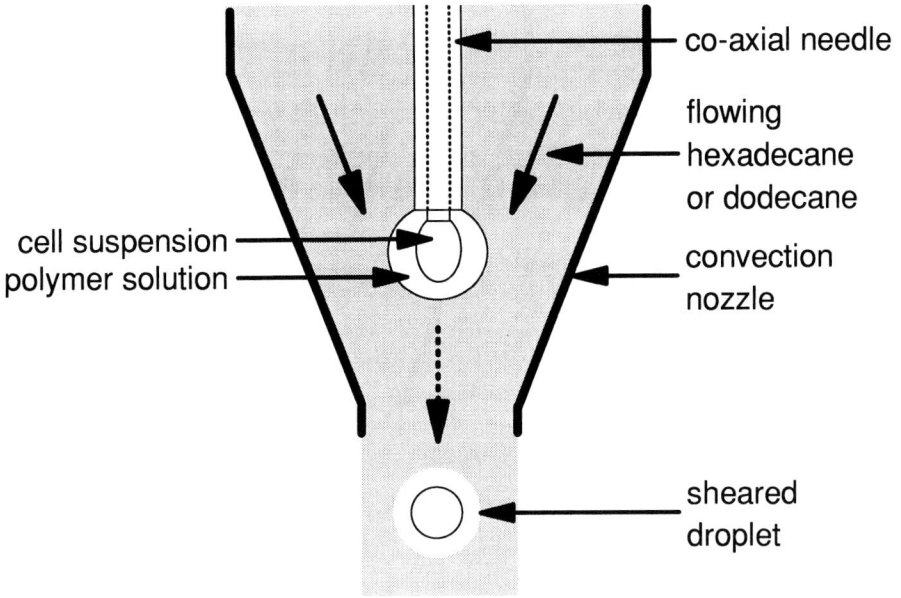

Fig. 3. Schematic drawing of the release of droplets from the stationary co-axial needle assembly. Cells suspended in the appropriate tissue-culture medium (and extracellular matrix component) are pumped through the inner needle; the polymer solution is pumped through the outer needle of a co-axial needle assembly. The co-axial needles are submerged in a flowing stream of hexadecane or dodecane. As a result, droplets of polymer solution surrounding a core of cell suspension are sheared from the tip of the needle assembly as they are formed. The rate of capsule production and the size of the capsules may be controlled by the flow rate of the hexadecane or dodecane and by the vertical position of the needle assembly within the nozzle.

ppm of a nonionic surfactant, Pluronic® L101 (BASF, Parsippany, NJ), which reduces the interfacial tension at the hexadecane–PBS interface, and thus facilitates the passage of the capsules through the interface. After washing the capsules in a fresh PBS solution, they are incubated in a Petri dish containing the appropriate tissue-culture medium for the cells at 37°C and in 95% air/5% CO_2.

In a previous setup, droplets were sheared from the coaxial needles by oscillating the needle in the vertical plane by a motor and cam assembly at a hexadecane–air interface (**Fig. 4**; *8*). This simpler setup resulted in formation of microcapsules with an od of ~700–900 μm and a wall thickness of 50–150 μm *(8)*. In the present setup, the higher shearing force produced by the flow of hexadecane leads to smaller diameter capsules. Microcapsules ranging in diameter from 300 μm to 600 μm may be produced by adjusting the hexadecane flow

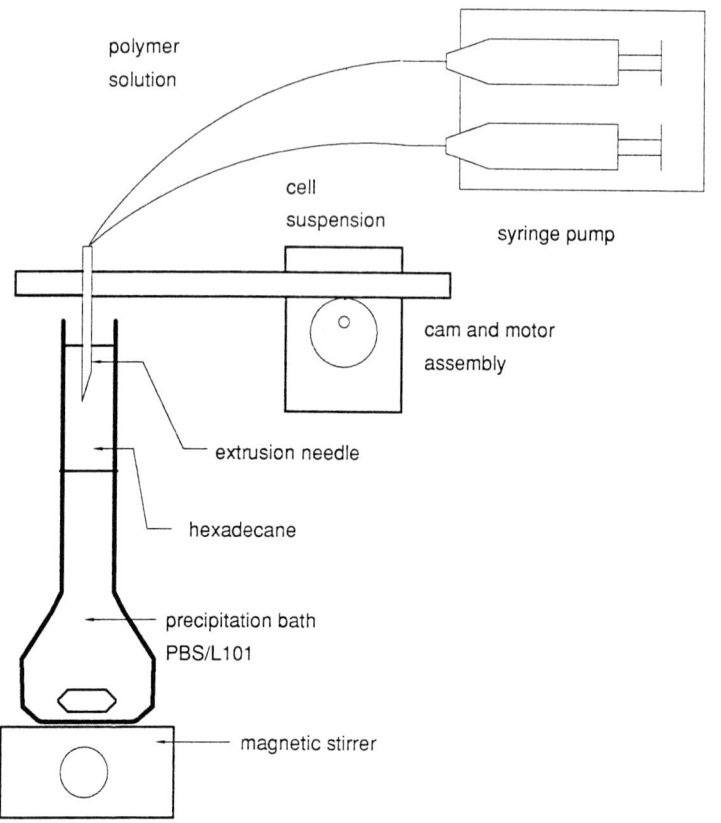

Fig. 4. Schematic drawing of the oscillating needle assembly microencapsulation setup producing large-diameter (~700–900 μm) capsules. Droplets of polymer solution surrounding a core of cell suspension are sheared from the tip of the coaxial needle assembly by the vertical oscillation of the needles at a hexadecane–air interface. The sheared droplets fall into a PBS bath, where the capsule membrane is formed by interfacial precipitation. The vertical oscillation of the co-axial needles is achieved by a cam and motor assembly. The capsules are maintained in suspension by a magnetic stirrer. Reproduced with permission from ref. *13*.

rate; the higher the flow rate: the smaller the capsules *(9)*. The polymer membrane has a thickness ranging from 20 μm to 90 μm (**Fig. 5**).

The polymer utilized for the formation of our capsules is a thermoplastic polyacrylate copolymer, namely, hydroxyethylmethacrylate-methylmethacrylate (HEMA–MMA, ~75 mol % HEMA), prepared by solution polymerization after careful monomer purification to reduce the crosslinker content *(5,10)*. This copolymer is hydrophilic, with a 25–30% (w/w) water uptake *(5,11)* con-

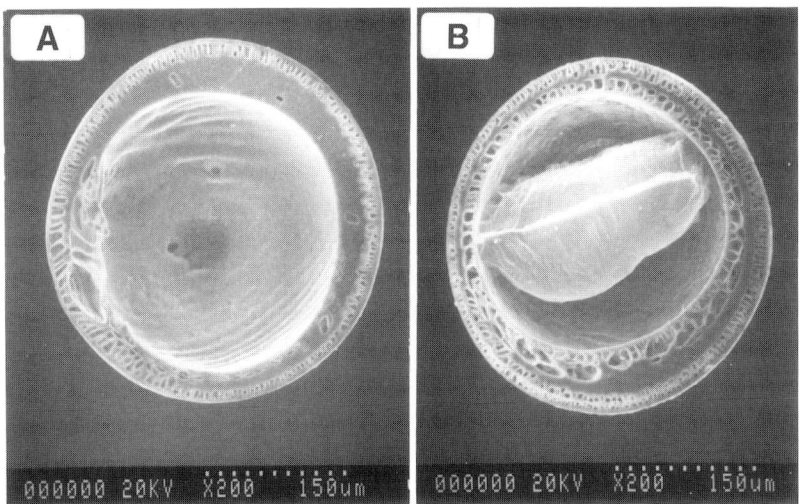

Fig. 5. Scanning electron micrographs of cut-open small diameter microcapsules prepared by the stationary needle assembly setup: **(A)** capsule produced with a 15% (w/v) Ficoll 400 core solution, **(B)** capsule produced with a 2-mg/mL collagen solution, in which the collagen gel can be seen in the core of the capsule.

sistent with the poly(HEMA) content, but has mechanical strength, toughness, and elasticity imparted by the poly(MMA) component. The water insolubility of the HEMA–MMA copolymer provides stability in the aqueous physiological environment, but necessitates the use of an organic solvent to prepare the polymer solution. Polyethylene glycol 200 (PEG 200) is used as the solvent, because cells tolerate PEG 200 without loss of viability, provided the direct contact between the cells and the solvent is limited, as is the case here; alternatively, triethylene glycol (TEG) may be used. HEMA–MMA is dissolved in PEG 200 at a concentration of 9–10% (w/v). The capsule structure can be modified by changing the ratio of the polymer solution to cells during co-extrusion. Also, the capsule membrane permeability and mol wt cut-off can be modulated by the polymer solution concentration or composition *(12)*, or the composition of the precipitation bath *(13)*.

Normally, the cell suspension contains 10–20% of Ficoll® 400 (Pharmacia, Uppsala, Sweden), a neutral, highly branched, hydrophilic polymer of sucrose. Ficoll 400 increases the viscosity of the core solution. As a result, less mixing of the cell suspension with the polymer solution occurs, a well defined core is produced, and cell viability is improved. Addition of a viscous extracellular matrix (ECM) solution, e.g., collagen, Matrigel® (Collaborative Research,

Bedford, MA), agarose, or chitosan to the cell suspension can also serve the same purpose. Collagen provides a matrix for attachment and better distribution of the cells within the core of the capsules. Matrigel, a reconstituted basement membrane, is prepared from a urea extract of Engelbreth-Holm-Swarm (EHS) tumors. Similar to collagen, Matrigel provides an extracellular matrix within the core of the capsules to which the cells attach and grow and maximize their expression of differentiated functions. Collagen and Matrigel are liquid at low temperatures (4°C), but are a gel at 37°C (gelling temperature ~22–35°C). Thus, it is necessary to cool the syringe and tube delivering the cell suspension to the needle assembly in order to prevent the solidification of collagen or Matrigel prior to droplet formation. Also, because of the lower cell-suspension flow rate in the stationary nozzle setup, the temperature surrounding the coaxial needle assembly is lowered by using a cooled flowing stream. In this case dodecane is used in place of hexadecane because of its lower melting point (–9.6°C vs 18.2°C). Also, the precipitation bath is warmed to about 30°C to promote gelation of the core collagen or Matrigel solution, as soon as the capsules are formed.

In contrast, in coencapsulation with agarose, the temperature of the precipitation bath is lowered to below 15°C, and the core syringe is maintained at room temperature. Typically, ultra-low gelling agarose, which is a liquid at 37°C, but forms a gel below 17°C, is used. Because of contact with the cooled precipitation bath, dodecane is used in coencapsulation with agarose. Agarose acts as an immobilization matrix that provides better distribution of the cells within the core, so that nutrient diffusion limitations are minimized. Other immobilization matrices may also be used. For example, in coencapsulation with chitosan, the deacylated form of chitin which is the main constituent of shrimp and crab exoskeletons, the neutral pH of the precipitation bath promotes precipitation of chitosan, thus, temperature adjustments of the setup are not required.

A variety of cell types, including the Chinese hamster ovary (CHO) fibroblast *(14)*, rat pheochromocytoma, PC12 *(15)*, human hepatoma, HepG2 *(16,17)*, murine fibroblast, L929 *(18)* cell lines, primary rat hepatocytes *(19)*, and rat islet tissue *(20)*, have been encapsulated with HEMA–MMA microcapsules. These studies have shown that the cells survive the encapsulation procedure, despite the exposure to shear forces and organic solvents/nonsolvents, and grow or function afterwards in vitro for periods from 2 to at least 6 wk. Also, mouse fibroblasts (2A-50), genetically engineered to secrete human growth hormone and β-glucuronidase *(21)*, have been microencapsulated to evaluate the use of encapsulated cells for gene therapy. The performance of encapsulated cells is generally evaluated by a colorimetric assay testing the metabolic activity of cells according to their ability to convert MTT to formazan *(14,22)*, by direct cell counts and trypan blue exclusion to test proliferation and viability *(14)*, by scanning

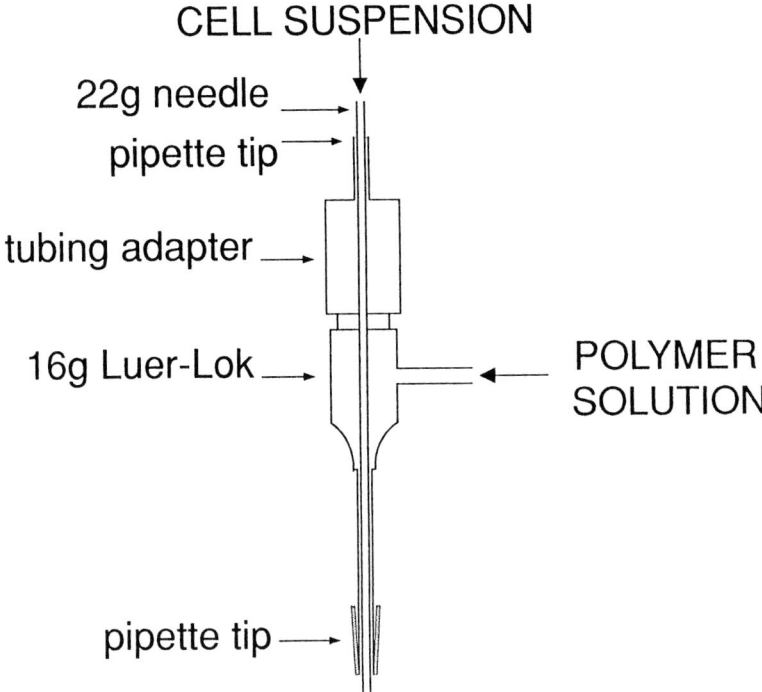

Fig. 6. Schematic of the needle assembly used in the oscillating-needle assembly microencapsulation setup. The coaxial needle assembly consists of an outer 16-gage Luer-Lok needle to which is added a side arm for the polymer solution, and an inner 22-gage needle that the cell suspension is pumped through. A plastic pipet tip is placed on the bottom of the 16-gage needle to facilitate the shearing off of droplets. Reproduced with permission from ref. *8*.

electron microscopy and histological sections for morphological assessment *(16)*, and by enzyme-linked immunosorbent assay techniques for protein secretion *(17)*.

2. Materials

1. Needle assembly: The coextrusion needle for the setup in which the assembly undergoes vertical oscillations consists of a 16-gage Luer-Lok needle that has been modified with a side-arm that acts as the inlet for the polymer solution (**Fig. 6**). A plastic pipet tip (Fisher Scientific, Don Mills, Ontario, Canada) is placed on the bottom of the 16-gage needle to facilitate the shearing off of droplets. The inner needle is a 22-gage needle fitted through a tubing adapter (male Luer-Lok tip to tubing, Becton Dickinson, Franklin Lakes, NJ) that is locked into the Luer-Lok mechanism. The inner needle extends through the outer needle, in a centered

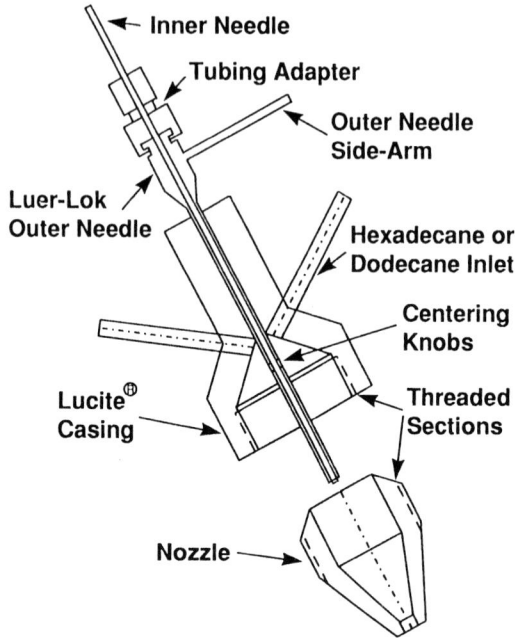

Fig. 7. Details of the needle assembly used in the stationary-needle assembly microencapsulation setup. The coaxial needle assembly consists of an outer 16-gage needle and an inner 22-gage needle. The core cell suspension is pumped through the inner needle and the polymer solution is pumped through the outer needle via a side arm. The needles are enclosed in a Lucite casing that forms a nozzle, providing a uniform flow of hexadecane around the needle tip. The needles have been tapered to facilitate droplet formation and shearing off. Also, the inner needle is equipped with three knobs in its midsection that provide centering within the outer needle without impeding the flow of the polymer solution. Adapted from ref. *21*.

position, and protrudes ~100 μm past the end of the pipet tip. The pipet tip and the 22-gage needle are blunt cut.

In the current setup, in which the coaxial needle is stationary within a flowing stream of hexadecane, modifications have been made to the needle assembly (**Fig. 7**). The pipet tip at the bottom of the 16-gage needle has been removed. Instead, the needle itself has been machined so that it is tapered on the inside and outside, thus facilitating the shearing off of droplets. The inner needle has also been tapered on the outside to facilitate better flow of the polymer solution around the core solution. Also, the inner needle is equipped with three knobs in its midsection that facilitate its centering within the outer needle without impeding the flow of the polymer solution. In addition, the needle assembly is enclosed in a Lucite® casing, which forms a convection nozzle that creates a uniform flow of hexadecane around the needle tip. The Lucite casing also has two inlets for the hexadecane flow. Thus, the hexadecane flows through the inlets into the casing, around the coaxial needle, and exits, along with the sheared droplet, through the convection nozzle, into the receiving dish.

2. Syringes: B-D, 1-cc, regular Luer-Lok tip syringes (Becton Dickinson) are used for both the polymer and core solutions. If desired, a Multifit, 2-cc, glass syringe with Luer-Lok tip (Becton Dickinson) that has a micro stir bar (7 mm L × 2 mm od, Cole-Parmer Instrument Co, Vernon Hills, IL) inside its barrel may be used to maintain the cells in suspension and prevent them from settling to the side of the syringe. The microstir bar is stirred by a microelectromagnetic stirrer (Cole-Parmer) that is suspended above the syringe. The stirring of the core solution, however, is not performed in coencapsulations with collagen or Matrigel, since the syringe is immersed in an ice-water bath. The core and polymer syringes are fitted with 22-gage and 18-gage blunt needles (Monoject®, Sherwood Medical, St. Louis, MO), respectively.

3. Pumps: The pump for hexadecane or dodecane is a Masterflex® L/S system with a standard drive (MR-7553-70) and a Quick Load pump head (Cole-Parmer). Also, a pulse dampener with a dead volume of 190 mL (MR-07596-20, Cole-Parmer) is used to attain a steady flow of hexadecane around the needle assembly. Separate syringe pumps are used for the core and polymer syringes (model A-99, Razel Scientific Instruments, Stamford, CT).

4. Tubings: The syringe containing the core solution is connected to the needle assembly with polyethylene tubing (PE-50, 0.58 mm id × 0.965 mm od, Clay Adams, Becton Dickinson, Parsippany, NJ). The syringe containing the polymer solution is connected to the needle assembly with Silastic® tubing (0.04 in id × 0.085 in od, Dow-Corning, Midland, MI). For hexadecane flow, the needle assembly, pump, and receiving dish are connected with Tygon® fuel and lubricant tubing (size 16, Masterflex, Cole-Parmer).

5. Receiving dish and stirrer: For the oscillating needle assembly setup, the receiving dish is a 250-mL volumetric flask (Pyrex, Corning, Corning, NY). A 8 mm od × 32 mm L (Fisher Scientific) stir bar is used to maintain the capsules in suspension. In the setup in which a flowing stream of hexadecane is used, a receiving dish with a larger opening is utilized. In this case, the receiving dish (155 mm L × 95 mm id) is custom-made from glass. An overhead stirring rod is used to prevent damaging the small capsules against the bottom of the dish. The stirring rod is composed of a metal rod that is placed in a bearing with a Teflon-coated magnetic stirring bar of 40 mm L × 10 mm od at the end.

6. HEMA–MMA solution: The appropriate mass of polymer is dissolved in PEG 200 (BDH, Poole, England) to make a 9–10% (w/v; the volume of the solvent rather than the volume of the final solution is considered) polymer solution. Complete dissolution of the powder requires about 3 d, and a transparent solution is produced.

7. Precipitation bath: Pluronic® L101 (BASF) is added to precooled (4°C) PBS (Dulbecco's calcium/magnesium-free, tissue-culture medium preparation, University of Toronto) to produce a 1500 ppm (10X) surfactant solution. The bottle

is maintained at 4°C overnight (and thereafter), in order for the L101 to go into solution. This stock solution is diluted 1:9 with PBS, prior to encapsulation.

8. Core solutions: The cells used for encapsulation are suspended in one of the following solutions to form the core mixture. In all cases, the type of tissue-culture medium used is dictated by the cells.

   a. Ficoll solution: Ficoll 400 is dissolved in an appropriate volume of tissue-culture medium by vigorous mixing, and the solution is sterile-filtered through a 0.2-μm filter (Millipore, Bedford, MA). Subsequently, the appropriate amounts of serum and antibiotics are added to produce a 10–20% (w/v) Ficoll solution. In addition, HEPES buffer (Gibco-BRL, Grand Island, NY) may be added to maintain the neutral pH of the medium.

   b. Collagen solution: A Vitrogen 100® (Bovine dermal type I collagen, Collagen Corp., Fremont, CA) solution is prepared according to the manufacturer's instructions. Eight mL of chilled 3 mg/mL Vitrogen 100 is mixed with 1 mL of 10X solution of buffered cell-culture medium containing phenol red. Then, 1 mL of 0.1 M NaOH (endotoxin-free, Sigma, St. Louis, MO) is added, and the solution is mixed. The pH of the solution is adjusted to 7.4 ± 0.2 by dropwise addition of 0.1 M HCl or 0.1 M NaOH. The solution is mixed with an appropriate amount of tissue-culture medium to produce a 1–2 mg/mL collagen solution. The collagen solution is maintained at 4°C and used within hours for encapsulation.

   c. Matrigel solution: Matrigel is mixed in a 1:1 (v/v) ratio with tissue-culture medium at a temperature of 4°C.

   d. Agarose solution: A 2% (w/v) agarose solution is prepared according to the manufacturer's instructions. SeaPrep® agarose (FMC BioProducts, Rockland, Maine), an ultra-low gelling temperature (<17°C) agarose, is weighed and added to a beaker containing stirring PBS (Dulbecco's calcium/magnesium free) at room temperature. The beaker is 2–4 times the volume of the solution. The beaker and solution are weighed and the beaker is covered with plastic wrap, which has a small hole that allows for ventilation. The beaker is transferred to a heater/stirrer unit and the solution is brought to a boil. The gentle boiling and stirring are continued until all of the agarose powder is dissolved (10 min). Sufficient hot PBS is added to the beaker to replace that which is lost, and the solution is mixed thoroughly. The agarose solution is maintained at 37°C and is used for encapsulation within 24 h. For encapsulation, the solution is mixed at a 1:1 (v/v) ratio with tissue-culture medium to obtain a 1% (w/v) agarose concentration.

9. Temperature control components: For coencapsulation with collagen or Matrigel in the stationary needle assembly setup, additional apparatus is required to cool the cell suspension to prevent gelation prior to formation of droplets at the tip of the needle assembly (**Fig. 8**). An ice-water bath is used for the core mixture syringe and the polyethylene tubing delivering the cell suspension to the needle assembly. The polyethylene tubing is also inserted in a water-jacket tube (silicone tubing, size 17, with size 16 side-arm, Masterflex, Cole-Parmer), which is

Microencapsulation with HEMA–MMA

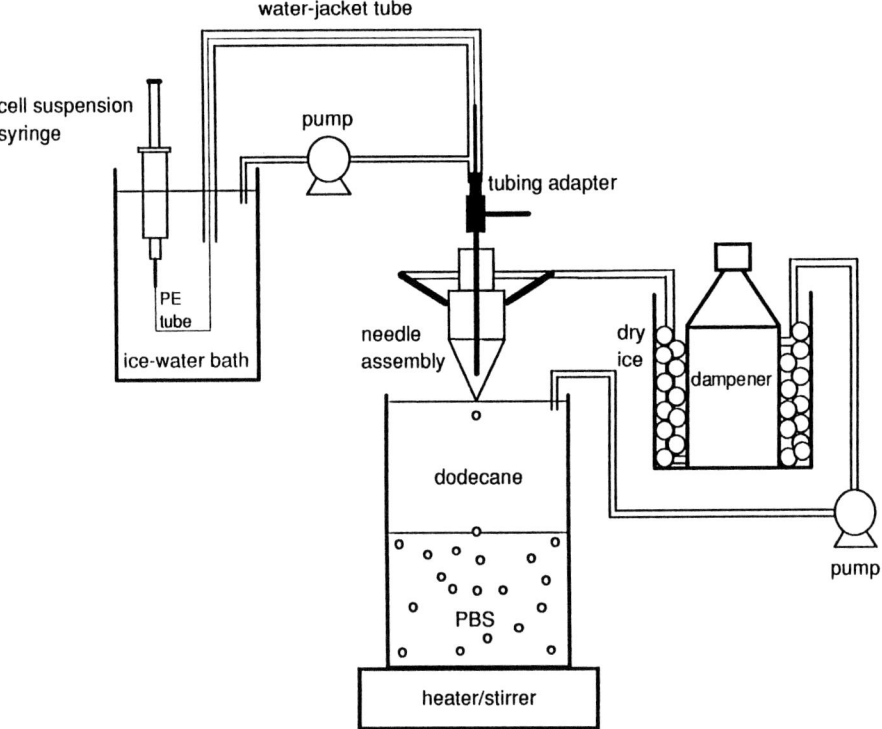

Fig. 8. Modifications to the stationary needle assembly microencapsulation setup for coencapsulation with collagen or Matrigel. Collagen and Matrigel solutions have a gelling temperature of ~22–35°C; thus, they must be maintained at a low temperature prior to droplet formation. The syringe containing the collagen or Matrigel solution is immersed in an ice-water bath. The polyethylene tubing delivering the cell suspension to the needle assembly is cooled by insertion in a water-jacket tubing, where cold water is recirculated by a pump. The flowing stream of dodecane is cooled by immersion of the dampener of the peristaltic pump used to recirculate the dodecane in dry ice. Also, once the capsules are formed, the precipitation bath is warmed to facilitate the gelling of the core solution.

fed by a pump (same model as the pump for dodecane). Also, the stream of dodecane surrounding the coaxial needle is cooled by immersing the dampener for the peristaltic pump in dry ice. In order to increase the exposure time of a given volume of circulating dodecane to the dry ice, a 500-mL bottle, which has inlet and outlet ports is used in place of the smaller 190 mL dampener. To facilitate the gelling of the collagen–Matrigel solutions in the core of the capsules, a magnetic stirrer/heater is used to warm the precipitation bath to approx 30°C. In the oscillating needle assembly setup, the flow of cell suspension is higher (*see*

Subheading 3.3.), therefore, immersion of the core syringe in ice-water bath is sufficient to prevent gelation prior to droplet formation. For coencapsulation with agarose, the receiving dish is placed in a cool water bath to reduce the temperature of the precipitation bath to below 15°C.

3. Methods
3.1. Stationary Needle Assembly, Small Diameter Microcapsules

1. To ensure sterility, all operations are performed in a sterile flow hood. Glassware and the Silastic tubing are steam-sterilized; the polyethylene and Tygon tubing are gas-sterilized. All chemicals are dispensed inside the flow hood.
2. The needle assembly is cleaned by aspirating water and then 95% ethanol through the core needle. The polymer needle is cleaned by aspiration of 95% ethanol only; ethanol is a solvent for HEMA–MMA. The needles are subsequently air-dried. The Lucite casing and nozzle are also cleaned with water and 95% ethanol, but overexposure to ethanol is avoided. To ensure sterility of the needle assembly, it is placed in an environment saturated with alcohol vapor 1 h prior to encapsulation. Prior to cleaning, the needle assembly is examined under a stereoscopic microscope to ensure the appropriate protrusion (~100 μm) and centering of the core needle with respect to the outer polymer needle, and the appropriate vertical position of the coaxial needles within the convection nozzle (4 mm from the tip of a nozzle that has a 20-degree angle and a 2.2 mm opening).
3. The magnetic stirrer, peristaltic pump, and syringe pumps are positioned in the flow hood. The syringe pumps are typically placed on stands.
4. The receiving dish is filled with 555 mL of PBS containing 150 ppm L101. The dish is placed on the stirrer and the overhead stirring rod is positioned in a centered, vertical position, so that the stirring rod is 2–3 cm from the bottom of the dish. The dish is subsequently filled completely with 500 mL of hexadecane or dodecane.
5. The polyethylene, Silastic, and Tygon tubings are connected to the needle assembly. The needle assembly is positioned above the receiving dish.
6. The other end of the Tygon tubing is connected to the dampener, and the dampener is connected to the receiving dish with a tube that is threaded through the peristaltic pump.
7. The pump is used to fill the Tygon tubing, dampener, and the needle assembly with hexadecane.
8. The cells are prepared and suspended in one of the core solutions mentioned above, typically at a concentration of 5×10^6 cells/mL. The core syringe is filled with the cell suspension, ensuring that air bubbles are removed. Subsequently, the syringe is fitted with the appropriate blunt-ended needle, connected to the polyethylene tubing, and placed within the syringe pump. The microstirrer is positioned above the syringe and is set to the lowest stirring speed.
9. Similarly, the polymer syringe is filled and installed in the syringe pump.
10. The peristaltic pump is started and the overhead stirrer is operated at approx 300 rpm. The speed of stirring should be such that the capsules are maintained in suspension

without creating undue mixing between the top hexadecane layer and the bottom PBS layer. Prior to stirring, the correct position of the nozzle over the receiving dish is checked by the location that the hexadecane jet impacts the hexadecane–PBS interface. It is desirable for the droplets to land at the hexadecane–PBS interface, at a position between the stirring rod and the wall of the receiving dish. The momentum of the droplets, the low interfacial tension at the interface, and the vortex created by the overhead stirrer facilitate the entry of the droplets into the precipitation bath.

11. To commence the encapsulation process, the syringe pumps are turned on. The core mixture and polymer solution are advanced manually to the tip of the needle assembly while the syringe pumps are on. First, the polymer is advanced to the tip of the needle assembly, and then the core solution is advanced; manual operation of the syringe pumps is stopped once the solutions have reached the tip of the needles. Typically, the initial contact between the polymer solution and the core solution produces a polymer precipitate at the tip of the needle. Thus, it is advisable to set the initial hexadecane flow rate slightly higher, in order to dislodge the precipitate from the tip of the needles. Once the initial disturbance is relieved, and good core and polymer droplet formation is achieved, the hexadecane flow rate is reduced to the operational level. The appropriate hexadecane flow rate may be checked by a measurement of the rate of capsule production.

12. Typical flow rates for the cell suspension, polymer solution, and hexadecane are 0.0055 mL/min, 0.0125 mL/min, and 118 mL/min, respectively. These flow rates result in the production of ~60 capsules/min with a 20-degree nozzle angle, 2.2 mm nozzle opening, and a 4-mm distance between the tip of the nozzle and the coaxial needles. These capsules have a final diameter of ~400 µm and a membrane thickness of ~50 µm. However, these flow rates may be adjusted to alter the size of the capsules and the membrane thickness.

13. In order to stop encapsulation, the core pump is turned off, then the polymer pump, and then the peristaltic pump. The needle assembly is removed from the receiving dish, all tubings are disconnected, and the hexadecane layer is removed for reuse.

14. The capsules are subsequently transferred to a spinner flask that contains fresh PBS (without Pluronic L101) and maintained in suspension for an additional 30 min. Once the stirring rod is removed from the receiving dish, the capsules will accumulate at the bottom of the dish. They should be transferred immediately to the spinner flask with a transfer pipet.

15. Subsequently, the capsules are washed in a Petri dish with tissue-culture medium, to remove any remaining traces of hexadecane or the precipitation bath. They are then transferred to a Petri dish containing the appropriate tissue-culture medium and maintained at 37°C in 95% air/5% CO_2. Alternatively, in order to prevent the capsules from adhering to one another, the capsules may be kept in suspension by a magnetic stirrer in the incubator. Grossly defective capsules are removed under a stereoscopic microscope.

3.2. Oscillating Needle Assembly, Large-Diameter Microcapsules

The encapsulation process for the stationary needle assembly setup follows the same principal steps as outlined above, with the following modifications:

1. The inner needle is inserted manually through the Luer-Lok end of the assembly into the outer needle, and its centering is adjusted. The plastic tip is inserted onto its sleeve on the outer needle, and the protrusion (~100 µm) of the inner needle from the plastic tip is adjusted with the aid of a stereoscopic microscope.
2. The 250-mL volumetric flask is filled with the precipitation bath, except for the top 7 cm in the neck, which is filled with a hexadecane overlayer.
3. After attaching the polyethylene and Silastic tubing, the needle assembly is installed on the arm of the cam and motor assembly, above the opening of the volumetric flask.
4. Once the syringe pumps and the magnetic stirrer (low setting) are started, the coextrusion needle is oscillated in the vertical plane by the motor and cam assembly at a frequency of 30/min, with one droplet released each stroke. The height of the assembly is adjusted so that the droplet formed and the tip of the needle assembly are completely immersed in the hexadecane layer. Because the needles are not enclosed, any initial precipitate can be easily dislodged from the tip of the needles.
5. The cell suspension is pumped to the needle tip at a flow rate of ~0.02 mL/min, and the ratio of polymer solution to cell flow rate varies from 1.5:1 to 2.67:1, depending on the cell suspension composition (viscosity).

3.3. Coencapsulation with Extracellular/Immobilization Matrices

1. For coencapsulation with agarose, the receiving dish is placed in a water bath to maintain its temperature below 17°C. The core syringe is maintained at room temperature. Also, dodecane is used, rather than hexadecane.
2. For coencapsulation with collagen or Matrigel, cooling of the core solution is required. In the oscillating needle assembly setup, after attaching the polyethylene tubing to the core syringe, the syringe and a portion of the polyethylene tubing are immersed in an ice-water bath.
3. In the stationary needle setup, the core flow rate is less than the flow rate in the oscillating needle setup (0.0055 mL/min vs 0.02 mL/min). Therefore, additional provisions are needed to cool the core suspension for coencapsulation with collagen or Matrigel (**Fig. 8**):
   a. After attaching the polyethylene, Silastic, and Tygon tubings to the needle assembly, the polyethylene tubing is inserted completely inside another tubing, so that the end of the polyethylene tubing that is not attached to the needle assembly protrudes to the outside. This outer tubing acts as a cold water-jacket tubing around the polyethylene tubing. The water-jacket tubing is press-fitted onto the tubing adapter of the needle assembly, thus sealing the cold water within the tube. The water-jacket tubing is transparent, to allow for the visual monitoring of the advancement of the core solution in the polyethylene tube.

b. The end of the polyethylene tubing that protrudes from the water-jacket tubing is attached to the core syringe. Subsequently, the filled core syringe, the portion of the polyethylene tubing that protrudes from the water-jacket tubing, and the tip of the water-jacket tubing are immersed in an ice-water bath.
c. The water-jacket tubing is equipped with a sidearm, near the needle assembly tubing adapter, that is also connected to the ice-water bath by a pump. Once the core syringe is immersed in the water bath, this pump is started, to circulate the cold water at a flow rate of ~100 mL/min, around the polyethylene tubing and the portion of the core needle within the water-jacket tubing.
d. Once the peristaltic pump is started to circulate the dodecane (hexadecane cannot be used here), the 500-mL bottle used as dampener is immersed in dry ice. Simultaneously, the heater of the magnetic stirrer is set to a low setting. By allowing the system to reach a steady state for approx 1 h, the temperature of the dodecane within the nozzle may be maintained below 10°C; the precipitation bath is at ~30°C.
e. When the peristaltic pump is stopped at the end of encapsulation, the dampener is removed from the dry ice in order to prevent freezing of the dodecane. After removing the needle assembly from the receiving dish, the water-jacket tube is emptied prior to disconnection of all tubings from the needle assembly.

4. Notes

1. Some of the modifications made to the needle assembly in our current standard microencapsulation setup may also be applied to the setup that utilizes oscillating coaxial needles: A metallic, tapered outer needle, and a tapered inner needle that has centering knobs, may be used.
2. The vertical positioning of the coaxial needles within the nozzle affects the shear rate that the droplets experience, and thus the rate of production and size of the capsules. The closer the needles are to the tip of the nozzle, the smaller the capsules and the higher the production rate. Care must be exercised in attaching the tubings to the needle assembly, in order to avoid altering the protrusion of the core needle and vertical position of the needle assembly (**Fig. 7**). A key feature is keeping the needles centered relative to each other. Also, nozzles with varying angles may be used to alter the shear force experienced by the droplets. The typical nozzle angle is 20 degrees.
3. For a thorough cleaning of the coaxial needles, the core needle is removed from the assembly through the Luer-Lok tubing adapter, and a mild soap solution is aspirated through the needles. Any particulates that may have adhered to the 2inner wall of the core needle are dislodged with a 24-gage needle that is ~6 in. in length. The needles are then rinsed with water and ethanol, and air-dried.
4. The polyethylene and Silastic tubings must be of adequate length to provide flexibility in maneuvering the needle assembly during the setup.
5. The interfacial tension of a 0.01% aqueous solution of Pluronic L101 vs mineral oil (Nujol) is 3.5 mJ/m^2 (BASF). Pluronic L101 attains its highest surface activity at its cloud point. The cloud point for a 0.01% aqueous solution is ~18°C. The

precipitation bath may be maintained at the L101 cloud point by immersing the receiving dish in a cool-water bath; however, the overlying hexadecane layer may freeze if the PBS precipitation bath is excessively cooled.

6. To prevent the presence of air bubbles in the hexadecane portion of the needle assembly, only the dampener is filled initially with hexadecane from the receiving dish. Then, the pump is run in the reverse direction to fill the needle assembly casing with hexadecane directly from the receiving dish.
7. If the initial polymer precipitate is not dislodged by the hexadecane flow rate, intervention with a bent needle that is inserted inside the nozzle is required. However, contact with the coaxial needles should be avoided.
8. Immediately after precipitation, the polymer tends to be sticky. Any prolonged contact between the capsules at this point will inevitably lead to agglomeration, which is extremely difficult to reverse. The tendency of the capsules to adhere to one another continues for days after precipitation, although with a reduced affinity. Capsules incubated with serum-free medium in a Petri dish gently attach to the dish and hence do not agglomerate. In the first week postencapsulation, the capsules experience a 20–30% reduction in overall size, become more elastic and tougher, and also, the initial white, opaque appearance of the capsule membrane becomes more transparent.
9. The tubes used for preparation of the collagen solution should be maintained on ice, in order to prevent premature precipitation of the collagen. Any precipitates that are formed should be removed, because they may disrupt the flow of the core solution. These precipitates have a tendency to rise to the top of the collagen solution, and may be aspirated easily.
10. Depending on the ambient temperature and the concentration of the extracellular matrix component used, it may be sufficient to immerse the dampener for the dodecane pump in an ice-water bath, rather than in dry ice. Because of the heat exchange between the dodecane overlayer and the PBS precipitation bath in the receiving dish, the setting on the heater may need to be adjusted accordingly.

Acknowledgments

Financial support was provided by the Natural Sciences and Engineering Research Council and the Medical Research Council of Canada. The technical assistance of Vlad Horvath in designing and fabricating the needle assembly and the protocol is greatly appreciated.

References

1. Lanza, R. P., Kühtreiber, W. M., and Chick, W. L. (1995) Encapsulation technologies. *Tissue Eng.* **1,** 181–196.
2. Colton, C. K. (1995) Implantable biohybrid artificial organs. *Cell Transplantation* **4,** 415–436.
3. Lanza, R. P., Hayes, J. L., and Chick, W. L. (1996) Encapsulated cell technology. *Nature Biotechnol.* **14,** 1107–1111.

4. Babensee, J. E. and Sefton, M. V. (1996) Protein delivery by microencapsulated cells, in *Controlled Drug Delivery: Challenges and Strategies* (Park, K., ed.), American Chemical Society, Washington, DC, pp. 311–332.
5. Sefton, M. V. and Stevenson, W. T. K. (1993) Microencapsulation of live animal cells using polyacrylates, *Adv. Poly. Sci.* **107**, 143–197.
6. Uludag, H., Kharlip, L., and Sefton, M. V. (1993) Protein delivery by microencapsulated cells. *Adv. Drug Del. Rev.* **10**, 115–130.
7. Uludag, H., Horvath, V., Black, J. P., and Sefton, M. V. (1994) Viability and protein secretion from human hepatoma (HepG2) cells encapsulated in 400 μm polyacrylate microcapsules by submerged nozzle-liquid jet extrusion, *Biotechnol. Bioeng.* **44**, 1199–1204.
8. Sefton, M. V., Uludag, H., Babensee, J., Roberts, T., Horvath, V., and De Boni, U. (1994) Microencapsulation of cells in a thermoplastic copolymer (hydroxyethyl meythacrylate-methyl methacrylate). *Methods Neurosci.* **21**, 371–386.
9. Hwang, J. R. and Sefton, M. V. (1995) Effect of microcapsule diameter on the permeability to horseradish peroxidase of individual HEMA-MMA microcapsules. *Journal of Controlled Release* **33**, 273–283.
10. Stevenson, W. T. K., Evangelista, R. A., Broughton, R. L., and Sefton, M. V. (1987) Preparation and characterization of thermoplastic polymers from hydroxyalkyl methacrylates. *Journal of Applied Polymer Science* **34**, 65–83.
11. Stevenson, W. T. K. and Sefton, M. V. (1988) The equilibrium water content of some thermoplastic hydroxyalkyl methacrylate polymers. *Jurnal of Applied Polymer Science* **36**, 1541–1553.
12. Hwang, J. R. and Sefton, M. V. (1995) The effects of polymer concentration and a pore formin agent (PVP) on HEMA-MMA microcapsule structure and permeability. *J. Membrane Sci.* **108**, 257–268.
13. Crooks, C. A., Douglas, J. A., Broughton, R. L., and Sefton, M. V. (1990) Microencapsulation of mammalian cells in a HEMA-MMA copolymer: effects on capsule morphology and permeability. *J. Biomed. Mater. Res.* **24**, 1241–1262.
14. Uludag, H. and Sefton, M. V. (1993) Metabolic activity and proliferation of CHO cells in hydroxyethylmethacrylate-methylmethacrylate (HEMA-MMA) microcapsules. *Cell Transplantation* **2**, 175–182.
15. Roberts, T., De Boni, U., and Sefton, M. V. (1996) Dopamine secretion by PC12 cells microencapsulated in a hydroxyethyl methacrylate-methyl methacrylate copolymer. *Biomaterials* **17**, 267–276.
16. Babensee, J. E., De Boni, U., and Sefton, M. V. (1992) Morphological assessment of hepatoma cells (HepG2) microencapsulated in a HEMA-MMA copolymer with and without Matrigel. *J. Biomed. Mater. Res.* **26**, 1401–1408.
17. Uludag, H. and Sefton, M. V. (1993) Microencapsulated human hepatoma cells: in vtro growth and protein release. *J. Biomed. Mater. Res.* **27**, 1213–1224.
18. Ung, D. Y.-P. (1993) The effect of cytokines on microencapsulated mammalian model cells, Bachelor of Applied Science Thesis, University of Toronto.
19. Wells, G. D. M., Fisher, M. M., and Sefton, M. V. (1993) Microencapsulation of viable hepatocytes in HEMA-MMA microcapsules: a preliminary study. *Biomaterials* **14**, 615–620.

20. Sefton, M. V. and Kharlip, L. (1994) Insulin release from rat pancreatic islets microencapsulated in a HEMA-MMA polyacrylate, in *Pancreatic Islet Transplantation, Vol. III, Immunoisolation of Pancreatic Islets* (Lanza, R. P. and Chick, W. L., eds.), R. G. Landes, Austin, TX, pp. 107–117.
21. Tse, M., Uludag, H., Sefton, M. V., and Chang, P. L. (1996) Secretion of recombinant proteins from hydroxyethyl methacrylate-methyl methacrylate capsules, *Biotechnol. Bioeng.* **51,** 271–280.
22. Uludag, H. and Sefton, M. V. (1990) A colorimetric assay for cellular activity in microcapsules. *Biomaterials* **11,** 708–712.
23. Rodriguez, F. R. (1994) Engineering of a microencapsulation system, Master of Applied Science Thesis, University of Toronto.

27

Micropatterning Cells in Tissue Engineering

Sangeeta N. Bhatia, Martin L. Yarmush, and Mehmet Toner

1. Introduction

Recent advances in tissue engineering have been facilitated by the ability to control the environment of cells in vitro. Modulation of cell–extracellular matrix, cell-substrate, and cell–cytokine interactions have proven to be useful in organotypic cultures of many kinds. In some cases, such as skin, bone marrow, and liver, the recovery of physiologic tissue function is enhanced by co-cultivation of two or more cell types together. The in vitro dependence of tissue function on cell–cell interactions is reminiscent of cellular cues during embryogenesis, or adult interactions between parenchymal cells and supporting stroma.

In vitro attempts to study co-cultures of two cell types have typically assessed the influence of nonparenchymal cell populations on parenchymal cells by variations in cell seeding density or addition of excised tissue or confluent coverslips to existing cultures. Alternatively, physical separation of cell cultures through use of conditioned media *(1,2)* or porous filter inserts *(3)* has been utilized. In addition, dynamic cell–cell interaction has been studied in monolayers of a primary cell type in the presence of a shearing fluid containing a secondary cell type *(4,5)*. One limitation of these co-culture systems is the inability to vary local cell seeding density independently of the cell number. Micropatterning technology, or the ability to spatially control cell placement at the single-cell level, allows the precise manipulation of cell–cell interactions of interest.

Microfabrication techniques have been widely utilized for the spatial control of cells in culture *(6–15)*. Many strategies have employed variations in charge *(16,17)*, hydrophilicity *(11,18)*, and topology *(11,19,20)* to mediate selective adhesion of one cell type by differential serum-protein adsorption, or variations in surface free energy. In addition, specific use of biomolecules alone *(21,22)*, or in conjunction with aminosilanes *(10,13)* or self-assembled monolayers *(23)*, have been used to micropattern cells. A variety of cell types have been examined with micropatterning techniques, such as neuroblastoma cells *(18)*, BHK epithelial cells *(11)*, hepatocytes *(23–25)*, and myocytes *(9)*, with spatial resolution on the micron scale. These studies examined a wide array of physiologic functions, such as neuronal growth cone guidance, effects of cell shape on function, and electrical coupling through gap junctions. However, these methods have not been adapted to the simultaneous co-cultivation of more than one cell type.

In this chapter, we describe a versatile method for generating two-dimensional, anisotropic, model surfaces capable of organizing a single cell type or two different cell types in discrete spatial locations. We have chosen a primary rat hepatocyte/3T3 fibroblast cell system because of its potential significance in both basic science and technology development, and based on widely reported interactions observed in this co-culture model *(1,2,26,27)*. We have used photolithography, existing strategies for surface modification of glass substrates with aminosilanes linked to biomolecules, and manipulation of serum content of cell culture media to pattern cells. The first cell type, hepatocytes, attach and spread on an immobilized collagen I pattern; the second cell type, 3T3 fibroblasts, undergoes nonspecific, serum-mediated attachment to the remaining unmodified areas. This co-culture technique allows the manipulation of the initial cellular microenvironment without variation of adhered cell number, thereby allowing measurement of the influence of local variations in cell–cell interaction on bulk tissue function.

2. Materials

2.1. Microfabrication of Substrates

1. Borosilicate substrates: 2 in. diameter × 0.02 in. thickness float glass wafers are custom made (Erie Scientific, Portsmouth, NH) to fit a standard 60-mm diameter (P-60) tissue-culture dish.
2. Chrome masks: 5 in. × 5 in. chrome masks are fabricated by a high precision photolithographic process from artwork. Artwork is converted by computer-aided design packages (i.e., KIK, Berkeley, CA) to machine language (GDS) to generate an optical pattern (pattern generator) on a chrome-evaporated–photoresist coated glass or quartz plate (Advanced Reproductions, N. Andover, MA). Alternatively, some university microfabrication facilities complement their wafer microfabrication with mask-making facilities (Microsystems Technology Laboratory, Massachusetts Institute of Technology, Cambridge, MA). Another inexpensive option for relatively large feature size (>100 µm) is reduction of artwork

onto microfiche using standard library services. Use 2 microfiche masks on top of one another to prevent imaging of microfiche defects. These microfiche masks are fairly fragile and are primarily useful for exploratory studies.
3. Cleaning solution (piranha solution): 3:1 mixture of H_2SO_4:30% H_2O, prepared at time of experiment. **Caution:** Extreme caution must be observed during chemical handling because of the corrosive nature of the chemicals in use. Pyrex containers must be utilized and chemical removal should be performed via aspiration and dilution of acid mixture.
4. Photoresist: Theoretically, any photoresist can be utilized that will adhere sufficiently to untreated borosilicate, remain intact during subsequent processing, and yet be removed with relative ease after surface modification. We use a positive photoresist, OCG 825-835 centistokes (Olin-Ciba-Geigy, West Paterson, NJ).
5. Developer: Use developer appropriate for selected photoresist. We use OCG 934 1:1 MIF (Olin-Ciba-Geigy).

2.2. Surface Modification of Substrates

1. 3-[2-(aminoethyl)amino] propyl trimethoxysilane (AS): AS solution is 2% AS (Aldrich, Milwaukee, WI) in water, prepared immediately prior to use. AS should be stored in a nitrogen box, or bottle should be gassed with nitrogen prior to closure and storage.
2. Phosphate buffered saline (PBS): Can be purchased at 10-fold concentration (Biofluids, Rockville, MD).
3. Glutaraldehyde: Stock solution is 25% glutaraldehyde in water (Fisher Chemical, Fair Lawn, NJ). Glutaraldehyde solution is prepared by diluting stock solution in PBS to a final concentration of 2.5% v/v, pH 7.4.
4. Collagen solution: Stock collagen I solution is prepared from rat tail tendons to a concentration of approx 1 mg/mL in 1 mM HCl (described in Chapter 20 of this volume). Collagen I solution is prepared by 1:1 dilution of stock solution (final concentration is approx 500 µg/mL) with water, pH 5.0, prepared immediately prior to use. Alternatively, commercial preparations of collagen (or other desired adhesive proteins) can be substituted for the collagen solution described here.
5. Acetone: General laboratory grade (Baxter Diagnostics, Deerfield, IL).
6. Ethanol: 70% in dH_2O.

2.3. Cell Culture

2.3.1. Fibroblast Culture

1. Fibroblast feeder layer: 3T3-J2 mouse fibroblast cell line (originally provided by H. Green, Department of Physiology and Biophysics, Harvard Medical School, Boston, MA).
2. Fibroblast medium (FM): Dulbecco's modified Eagle's medium (DMEM, high glucose, L-glutamine, 110 mg/L sodium pyruvate, Gibco-BRL, Grand Island, NY), bovine calf serum 10% (BCS, HyClone, Logan, UT), penicillin–streptomycin (Boehringer, Indianapolis, IN) 100 IU/mL–100 µg/mL.

3. Mitomycin C (Boehringer, Indianopolis, IN): 10 µg/mL in fibroblast medium.
4. Trypsin: 0.1% trypsin (ICN Biomedicals) in PBS.
5. Versene: 0.1% EDTA (Boehringer) in PBS.

2.3.2. Hepatocyte Isolation and Culture

1. Isolation of hepatocytes from rat liver: Isolation of rat hepatocytes by collagenase perfusion is described in detail in Chapter 20 of this volume. Alternatively, isolated, purified rat hepatocytes may be obtained from another laboratory.
2. Serum-free hepatocyte culture medium (SFM):
   a. DMEM with high glucose, without sodium pyruvate, without sodium bicarbonate (Gibco-BRL, #12100).
   b. Glucagon (Lilly, Indianapolis, IN). Add to final concentration of 7 ng/mL.
   c. Insulin (Squibb, Princeton, NJ). Add to final concentration of 0.5 U/mL.
   d. Epidermal growth factor (Becton Dickinson, Bedford, MA). Add to a final concentration of 20 ng/mL.
   e. Hydrocortisone (Upjohn, Kalamazoo, MI). Add to a final concentrataion of 7.5 µg/mL.
   f. Penicillin–streptomycin, respectively, 5000 U/mL and 5 mg/mL, in 0.9% NaCl, (Sigma, St. Louis, MO). Add 1% v/v to medium.
   g. L-proline: Add to final concentration of 40 µg/mL (Sigma) *(16)*.
   h. Filter-sterilize medium through a 0.2-µm pore membrane.

2.4. Cell Culture on Modified Substrates

1. Culture medium (CM): Identical to SFM, except for exclusion of L-proline and inclusion of 10% fetal bovine serum (FBS, JRH Bioscience, Lenexa, KS).
2. Bovine serum albumin (BSA): BSA solution of 0.05% w/w in water (Sigma). Sterilize by filtration through a 0.45-µm filter, and store at 4°C.
3. Sterilized deionized water: Prepare two autoclavable containers of 500 mL of deionized water. Autoclave to sterilize.
4. Sterilized vessels: Prepare two glass beakers by covering with autoclave wrap. Autoclave.

3. Methods

Microfabrication techniques are used to modify glass substrates with biomolecules. These modified substrates are utilized to pattern a single cell type, or to micropattern two cell types to control their level of interaction. **Figure 1** schematically depicts the overall process for one representative pattern, although this procedure can be utilized for a variety of spatial configurations.

3.1. Microfabrication of Substrates

We produced experimental substrates utilizing standard microfabrication techniques at Microsystems Technology Lab, MIT, Cambridge, MA. This microfabrication facility, like many others of its kind, is dedicated primarily to

Micropatterning Cells

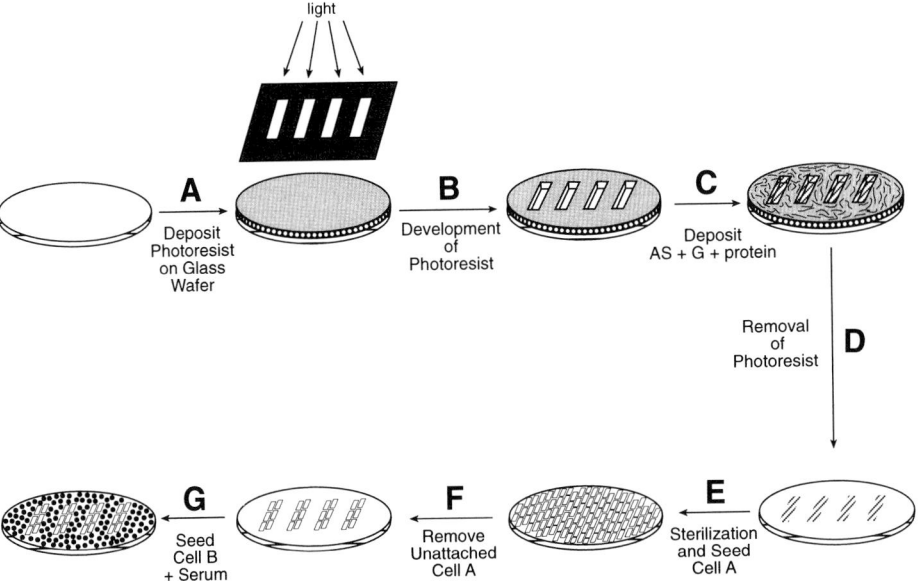

Fig. 1. Schematic of process for generating micropatterned cultures of a single cell type or two distinct cell types.

silicon wafer processing for electronic applications. As a result, the presence of potential contaminants, such as class III–V compounds, precious metals, alkali, and transition metals, are restricted. Therefore, the process described here utilizes separate facilities for surface fabrication and surface modification with biomolecules. In general, almost any integrated circuits manufacturing facilities could be used to perform these procedures.

1. Borosilicate wafers are cleaned by placement into wafer carriers. Place into a Pyrex vat, pour piranha solution over wafers, and wait 10 min. Rinse wafers 3X in a dump-to-resistivity tank, and nitrogen dry using a spin-dryer. In cases in which a spin-dryer is not available, manual drying is done by using a N_2 gas stream. This gas stream can either be accessed through a house N_2 gun, or a portable N_2 gas tank connected to tubing.
2. Dehydrate wafers promote adheszion of photoresist by baking for 60 min at 200°C.
3. Mount wafers on vacuum of a spin-coater chuck and coat with positive photoresist to a uniform layer of approx 1 μm. In our laboratory, this is accomplished as follows: Dispense photoresist at 500 rpm for 2 s, spread photoresist at 750 rpm for 6 s, and spin at 4000 rpm for 30 s (step A, **Fig. 1**).
4. Soft-bake for 30 min at 90°C to drive out excess solvent and anneal any stress in the film.

5. In order to create a latent image of the desired pattern in the resist layer, expose wafers to ultraviolet light in a defined pattern. We expose coated substrates to 365 nm UV light in a Bottom Side Mask Aligner (Karl Suss, Munich, Germany) through a 5 in. × 5 in. patterned chrome mask under vacuum-enhanced contact for 3 s, at a dose of 10W/cm^2.
6. To produce the final three-dimensional relief image, immersion develop the exposed photoresist in the appropriate developer. Complete removal of photoresist in exposed areas is critical to subsequent surface modification. Furthermore, residual photoresist is not easily detected on transparent (borosilicate) wafers; therefore, care should be taken to optimize this step of the protocol. Presence of residual photoresist can be assessed by inspection under light interference or fluorescent microscopy. We develop by immersion and agitation in a bath of developer for 70 s. Surfaces should then be rinsed three times under running deionized water and cascade rinsed (if possible) for 2 min (step B, **Fig. 1**).
7. Postbake patterned disks for 30 min at 120°C to drive off residual solvent and promote film adhesion.
8. Wafers can be stored in closed containers (preferably within a nitrogen box) at room temperature for at least 1 mo.
9. Finally, 24 h prior to surface modification, substrates can be exposed to oxygen plasma in order to dry-etch (remove) a small layer of photoresist. This ensures complete removal of photoresist from exposed borosilicate; however, if substrates are well developed and pattern dimensions are larger than 10 μm, this step may be skipped. We use a parallel-plate, plasma day etcher at a base vacuum of 50 mTorr in an O_2 atmosphere, and pressure of 100 mTorr at a power of 100W for 2–4 min, which corresponds to an etch rate of approx 0.1 μm/min.

3.2. Surface Modification of Substrates

There are a number of techniques for immobilizing adhesive proteins on solid substrates (*see* ref. *28*). Here we describe a technique modified from experimental methods developed by Lom et al. *(13)* and Britland et al. *(10)* (step C, **Fig. 1**). This silane-based coupling has advantages over thiol-based techniques *(19)* for patterning more than one cell type, because of the availability of the unmodified glass for attachment of a second cell type. In contrast, in some thiol-based techniques, outlying regions are modified with a nonadhesive compound (polyethylene glycol [PEG]), preventing attachment of a second cell type.

1. Patterned substrates are handled with wafer tweezers (Fluoroware, Chaska, MN). Pay special attention to the orientation of the patterned surface of the wafer; transparent substrates can be easily inverted without any obvious differences in appearance. Wafers are rinsed by immersion in deionized dH$_2$O in a glass 100-mm Petri dish. Repeat.
2. Silane immobilization of AS is performed by immersion of samples in AS solution for 30 s at room temperature, followed by two rinses in deionized water.

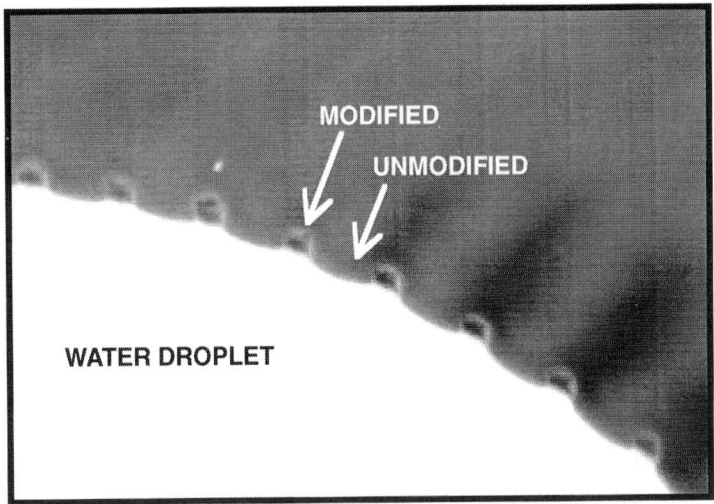

Fig. 2. Differential hydrophilicity of patterned glass surface.

3. Dry wafers with a stream of nitrogen gas to avoid drying artifacts.
4. Bake wafers in a closed container for 10 min at 120°C. Note: Temperature greater than 150°C will cause hardening of many photoresists and cause difficulty during removal.
5. Next, soak disks in a covered container of 2.5% v/v solution of glutaraldehyde in PBS (pH 7.4) for 1 h at 25°C, followed by two rinses in fresh PBS. Visually inspect wafers every 15 min to evaluate the integrity of the photoresist. In some instances, the glutaraldehyde solution can cause peeling of the photoresist and the process may need to be abbreviated.
6. To immobilize collagen, immerse wafers in 4 mL of collagen solution in a P-60 Petri dish for 30 min at 37°C.
7. To remove photoresist and expose underlying unmodified glass, float each wafer in acetone in a glass container and sonicate the container in a bath sonicator for 1–15 min. The duration of sonication is empirically determined by observation of the first wafer in each batch. Examine wafers for complete removal of photoresist (previously pink wafers will appear clear). Treat all wafers in a experimental batch identically to ensure comparability of immobilized protein layers on all substrates. We use 10 mL of acetone in a 100-mm glass Petri dish (step D, **Fig. 1**).
8. Rinse wafers twice by immersion into deionized water. As shown by Lom et al. *(13)* and others, modified areas should display differential wetting upon removal of substrate from water, thereby indicating successful patterned-surface modification (**Fig. 2**).
9. Wafers can be stored dry, in a covered container, at 4°C for at least 2 wk. We store wafers on a piece of filter paper (to absorb residual water and prevent sticking) in

60-mm Petri dishes. Storage of wafers in solution (i.e., PBS or ethanol) results in transfer of patterned protein to unmodified areas, presumably via desorption from modified areas and adsorption to unmodified areas; therefore, if the immobilized protein will tolerate unhydrated storage conditions, wafers should be stored dry. If immobilized protein requires hydration to maintain its bioactivity, storage time must be empirically determined—in our case, less than 48 h.

3.3. Cell Culture

Co-cultivation of two cell types has been proven to be useful in organotypic cultures of many kinds: skin, vasculature, bone marrow, as well as liver. Hepatocytes have been co-cultured with many secondary cell types, ranging from isolated biliary ductal epithilia *(29)* to cell lines such as murine 3T3 fibroblasts *(30)*. These co-cultures have been reported to sustain hepatocyte morphology and viability, form functional homotypic gap junctions, and produce elevated levels of various liver-specific markers, including albumin synthesis and various detoxification functions *(29,30)*. We chose co-culture of primary rat hepatocytes with 3T3-J2 fibroblasts, because of their ease of culture and availability, although genetically modified fibroblasts or alternative cell types could also be incorporated into this methodology.

3.3.1. Fibroblast Culture

1. Grow 3T3-J2 cells to preconfluence in 150-cm^2 flasks or other suitable dishes, 10% CO_2, balance air.
2. Passage cells by washing with 8 mL of versene, and incubating with 2 mL of trypsin solution and 2 mL of versene, for ~5 min at 37°C. Resuspend cells in 25 mL media to neutralize trypsin. Approximately 10% of the cells may be inoculated into a fresh tissue-culture flask containing a total of 50 mL of media. Cells may be passaged no more than 12 times.
3. A single 150-cm^2 flask will yield 4–8 × 10^6 cells at confluence. Time the passage and number of flasks so that a sufficient cell number will be available at preconfluence on the day of the experiment—in our case, 1 d after hepatocyte isolation.

3.3.2. Hepatocyte Isolation and Culture

1. Isolation of hepatocytes from rat liver: Isolation of rat hepatocytes by collagenase perfusion is described in detail in Chapter 20 of this volume. Hepatocytes are isolated from 2- to 3-mo-old adult female Lewis rats weighing 180–200 g, by a modified procedure of Seglen (1976). Routinely, 200–300 million cells can be isolated, with viability between 85 and 95%, as judged by trypan blue exclusion. Nonparenchymal cells, as judged by their size (<10 µm in diameter) and morphology (nonpolygonal or stellate), are typically less than 1%.
2. Assess cell concentration of stock suspension by hemocytometer and record.
3. Store hepatocytes on ice prior to use. In our hands, rat hepatocytes are viable and retain the capacity to attach and spread for at least 6 h on ice.

4. Dilute cell stock suspension in SFM to appropriate concentration (see **Note 4**). In our experiments, this concentration is $1–2 \times 10^6$ hepatocytes/mL. Prepare sufficient volume to perform multiple seedings for each experimental condition. In our experiments, we prepare 6–8 mL of cell suspension for each dish, which corresponds to 3–4 sequential seedings of 2 mL on each dish.

3.4. Cell Culture on Modified Substrates

Coordination of surface preparation, culture of primary cells, and culture of a cell line requires precise timing. Since isolated cells, such as hepatocytes, must be seeded within a few hours of isolation, surface preparation required on the day of cell seeding must be completed prior to receipt of cells from isolation. Similarly, the fibroblast population should have been expanded to sufficient cell numbers for trypsinization the following day.

1. Soak premodified wafer in 70% ethanol for sterilization. Use wafer tweezers to place wafer in P-60 with 5 mL of ethanol solution for at least 1 h, but not more than 24 h, at room temperature in a sterile hood (step E, **Fig. 1**).
2. In order to remove residual ethanol from surface of wafer, first pour autoclaved water into sterilized beaker. Flame sterilize wafer tweezers, allow to cool, and remove wafer from 70% ethanol under sterile conditions. Immerse wafer in water and agitate gently for approx 10 s, being sure to preserve orientation of the wafer.
3. In order to deter nonspecific attachment of hepatocytes on unmodified areas, the sterilized wafers will be coated with bovine serum albumin (BSA), a large, globular, negatively-charged protein thought to deter cell adhesion by nonspecific charge interactions for many cell types; however, the usefulness of this coating is cell-type dependent, and may vary *(14)*. For rat hepatocytes, BSA coating reduces nonspecific cell attachment to glass from 30% to negligible levels *(25)*. Place sterilized wafers in sterile P-60 dishes, add 4 mL of BSA solution to each dish, and place in incubator for 45 min at 37°C.
4. To remove residual BSA solution, use sterile tweezers to remove wafers from dishes under sterile conditions, and immerse in autoclaved water in sterile beaker, gently agitating for 10 s. Note: At this point, if one needs to extend the duration of the sample preparation period, substrates may be left under sterile conditions in dry Petri dishes.
5. Substrates are then rinsed with SFM to remove residual water. Add 3 mL of media to each dish, gently swirl solution, and aspirate. Repeat.
6. Seed hepatocytes by placing 2 mL of hepatocyte solution (in SFM) on each wafer. The underside of the wafers should still contain fluid from the SFM rinse, if not, add 0.2 mL of SFM to the dish to wet the underside of the wafer, to prevent wicking of cell suspension underneath the wafer. Agitate solution to disperse cell suspension, and place in incubator for 1–1.5 h at 37°C, 10% CO_2, balance air (step E, **Fig. 1**). Wafers should be periodically agitated (i.e., every 15 min) to promote maximal cell attachment.
7. At this point, pattern features should be visible because of selective cell adhesion. Typically, to ensure 100% confluence on patterned areas, hepatocyte seed-

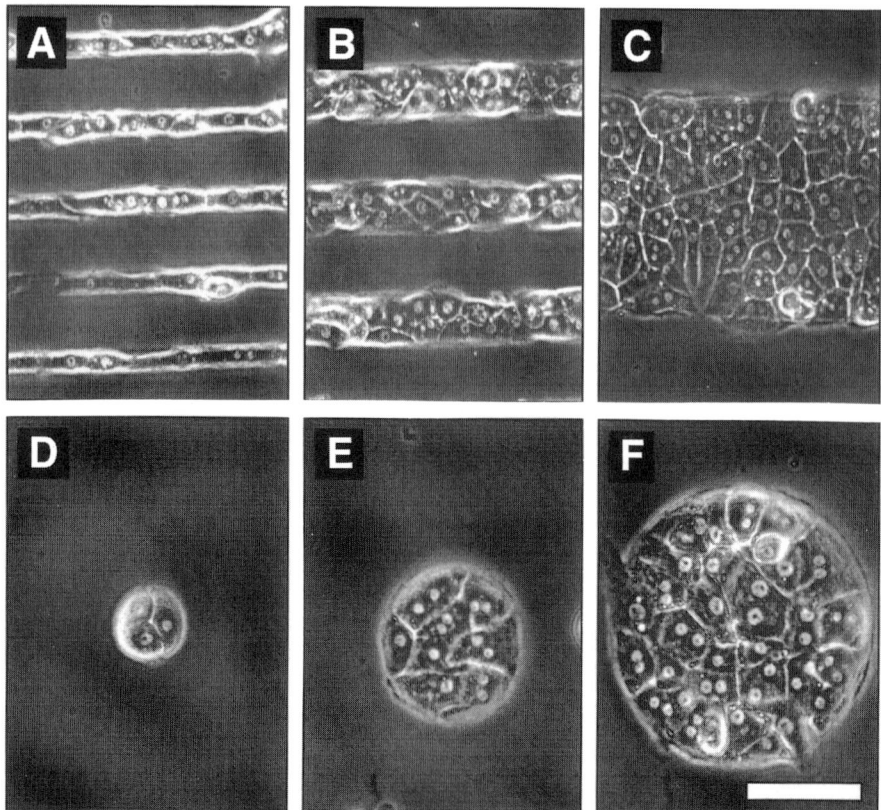

Fig. 3. Phase-contrast micrographs of micropatterned hepatocytes. Hepatocytes attached to linear strips of width (**A**) 20 μm, (**B**) 50 μm, and (**C**) 200 μm, and circular patterns of diameter (**D**) 50 μm, (**E**) 100 μm, and (**F**) 250 μm.

ing is repeated 2–3X. This is especially critical in cases of low cell viability, since dead cells render some modified sites inaccessible to live cells. Surfaces should be rinsed twice by pipeting and then aspirating 4 mL of serum-free media, reseeded with hepatocytes for 1.5 h, and rinsed again. Repeat as necessary (step F, **Fig. 1**). After the final rinse, incubate substrates in 2 mL of CM.
8. Hepatocytes are then allowed to spread over the remaining modified sites. Rat hepatocytes spread over greater than 10 h *(32)*; therefore, incubate patterned hepatocytes overnight. Cells (hepatocytes, endothelial cells) on small pattern features will confirm to the edges of the pattern (**Fig. 3**; *33,34*). Experimental studies can be performed using these micropatterned hepatocytes; otherwise, proceed with addition of second cell type.
9. The day after hepatocyte seeding, trypsinize the appropriate number of flasks of 3T3-J2s, centrifuge cells at 1000 rpm for 5 min, and resuspend in a small volume

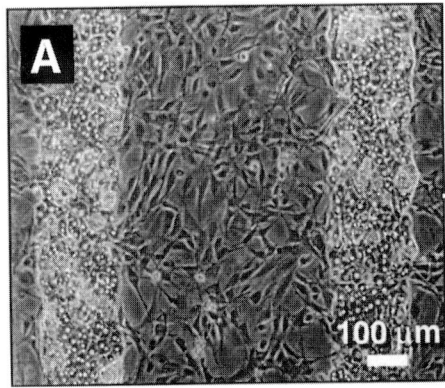 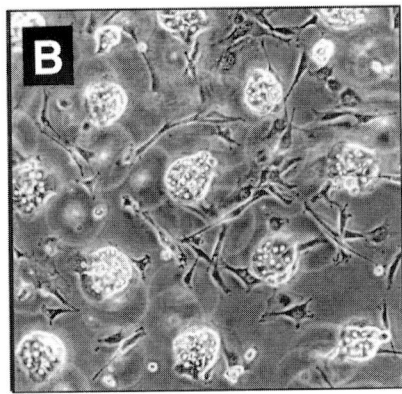

Fig. 4. Phase-contrast micrograph of micropatterned co-cultures with varying spatial configurations. (**A**) 200-μm strips of hepatocytes with 500-μm lane of intervening fibroblasts. (**B**) 100-μm islands of hepatocytes with 250-μm center-to-center spacing and intervening fibroblasts.

of fibroblast media (FM). Count with a hemocytometer and plate appropriate number of fibroblasts in 3 mL of FM per dish. We typically utilize 750,000 fibroblasts per dish; however, in some cases it is necessary to plate growth-arrested fibroblasts in greater numbers (step G, **Fig. 1**). (Growth-arrest fibroblasts by incubating each 150-cm^2 fibroblast flask with 15 mL of mitomycin C solution for 2 h, rinse with media, and trypsinize as usual. Incubate for at least 6 h to allow fibroblast spreading, aspirate media, and replace with 2 mL of CM.
10. Resultant micropatterned co-cultures of two distinct patterns are shown in **Fig. 4**. Proceed with experimental investigation. Pattern preservation is dependent on cell types, as well as pattern dimensions; for example, our patterns are stable on the order of weeks for collagen-modified areas of larger than a few hundred microns.

4. Notes

1. Microfabrication process: Facility regulations will largely determine the process utilized for preparation of patterned surfaces. Here, we have detailed the most versatile technique, whereby all biological materials are utilized subsequently in a separate facility. Others have pretreated wafers with organic materials in their own facilities, to allow for greater versatility *(13)*. This approach should also be considered where applicable. Facility specifications regarding dedicated equipment will also affect problems that may be encountered via contaminants, and so on (*see* **Note 6**).

   In addition, the overall process will vary with the characteristics of the desired protein. In our case, collagen I retained its bioactivity for hepatocyte attachment and spreading, despite treatment with acetone, ethanol, and dehydration. Other proteins may require modifications of this protocol to retain their bioactivity.

2. Surface modification: The method described here was intended to be robust. In many instances, users may be able to eliminate certain elements of this process. Exposure to plasma oxygen may be unnecessary, if patterns are well-developed and not contaminated during storage. Similarly, baking of wafers after AS modification may be unnecessary in some protocols. In some cases, covalent binding of protein to glass may be eliminated entirely by simply adsorbing protein to exposed glass. For example, collagen and some other proteins are known to irreversibly adsorb to certain substrates; however, it should be noted that reliance on adsorption alone is likely to result in different immobilized protein conformations than covalently bound proteins.
3. Cell-culture issues: Cell-culture issues involved with micropatterned cultures include cell-seeding protocols for hepatocytes, cell seeding protocols for fibroblasts, media formulation for co-cultures, and control of cell population. Our seeding protocol for rat hepatocytes ($2-4 \times 10^6$ per dish, with agitation every 15 min for 1–1.5 h, followed by 2 repeated seedings, which yields approx 250,000 attached hepatocytes) may be modified, depending on cell source, duration of viability in suspension, and cell aggregation at high densities. It is important to note that the photoresist pattern determines available wafer surface area for hepatocyte attachment. Since hepatocytes are seeding in excess, the number of attached hepatocytes is ultimately determined by the mask design. Fibroblast seeding protocols include specifications for cell number, passage restrictions, and timing of seeding (in our case, 750,000 per dish fibroblasts are seeded, before the twelfth passage, in a single seeding, and allowed to attach for 6 h in fibroblast media). Fibroblasts attach to unmodified glass, and, therefore, in contrast to hepatocyte number, fibroblast number is dictated by the number of viable cells seeded per substrate. Finally, selection of media for two cell types must be considered. Because of the rapid mitotic rate of fibroblasts, our media formulation contains high levels of hydrocortisone as a growth inhibitor, each type of culture will require some degree of media formulation.
4. Probing tissue function of micropatterned cultures: Cultures of two different cell types often require unique analysis. For example, selective markers of cell function (in our case, albumin or urea synthesis as markers of liver-specific function of hepatocytes), methods to quantitate relative growth of both cell populations (in our case, growth-arresting fibroblasts allowed measurement of total DNA as a vehicle for tracking both cell populations), spatial tracking of both cell populations (in our case, either immunofluorescent staining of cytokeratin, an intermediate filament in hepatocytes as compared to F-actin, a cytoskeletal polymer present in both cell type, or labeling of each cell population with a different long-lived intracellular fluorescent probe, such as Molecular Probes [Eugene, OR] Cell Tracker dyes).
5. Strategies for achieving selective cell adhesion of different cell types: For patterning of two different cell types, the general approach is to pattern an adhesive substrate on a background that is both nonadhesive to one cell type, and adhesive to a second cell type. The sequence of seeding two distinct cell populations can be

critical to this process. For example, hepatocytes must be patterned prior to fibroblast seeding; hepatocytes would attach and spread uniformly on the surface of a fibroblast patterned wafer. Other variables that can influence selective cell adhesion include: the kinetics of attachment (we allow 1.5 h hepatocyte attachment), the influence of ligand density (we use saturated levels), the affinity of different ligands (we use collagen I), the influence of charged coatings (BSA), and denaturation of adsorbed proteins to remove immobilized protein (8 M urea [Britland]).

6. Troubleshooting: Common problems include minimal cell adhesion to substrate, uniform cell adhesion to substrate, difficulty with lift-off of photoresist, and peeling of photoresist during processing. Surface wetting (**Fig. 2**) is often an important clue in targeting the source of difficulty, since a lack of pattern at that point indicates a problem upstream of sterilization and cell seeding.

   a. Lack of cell adhesion to the substrate can indicate many problems. The most common, however, is underdevelopment of the exposed photoresist, resulting in a lack of exposed borosilicate for protein immobilization. This can be alleviated by increasing development time, or increasing exposure to oxygen plasma prior to surface modification. This effect is often exacerbated with small (<5 µm) pattern dimensions.

   Alternatively, contaminants that coat the glass and prevent protein immobilization, or commonly utilized undercoatings for promoting photoresist adhesion, could produce this effect.

   b. Uniform cell adhesion: A lack of discernible pattern because of exuberant cell adhesion is frequently caused by a defect in the original photoresist coating. Photoresist can degrade and crack over time, allowing exposure of all areas of the substrate to surface modification. Alternatively, excessive exposure to oxygen plasma will strip the photoresist from the surface of the wafer completely.

   c. Peeling of photo during surface modification. The integrity of photoresist coatings varies with solvents. In some instances, photoresist will peel away from the wafer surface prematurely. This indicates either insufficient adhesion of the photoresist to glass (often because of an insufficient dehydration bake prior to photoresist coating) or insufficient baking after development to harden the photoresist.

   d. Difficulty with photoresist removal: Extended sonication in acetone for photoresist removal often indicates exposure of photoresist to elevated temperatures.

Acknowledgments

This work was supported in part by the Shriners Hospital for Children and the American Association for University Women (SNB). The authors would like to thank Octavio Hurtado and the Microsystems Technology Laboratory, Massachusetts Institute of Technology, Cambridge, MA for help with microfabrication.

References

1. Shimaoka, S., Nakamura, T., and Ichihara, A. (1987) Stimulation of growth of primary cultured adult rat hepatocytes without growth factors by coculture with nonparenchymal liver cells. *Exp. Cell Res.* **172**, 228–242.

2. Goulet, F., Normand, C., and Morin, O. (1988) Cellular interactions promote tissue-specific function, biomatrix deposition and junctional communication of primary cultured hepatocytes. *Hepatology* **8**, 1010–1018.
3. Morin, O., Goulet, F., and Normand, C., (1988) Liver sinusoidal endothelial cells: isolation, purification, characterization, and interaction with hepatocytes. *Revisiones Sobre Biologia* **15**, 1–46.
4. Lawrence, M. B., McIntire, L. V., and Eskin, S. G. (1987) Effect of flow on polymorphonuclear leukocyte/endothelial cell adhesion. *Blood* **70**, 1284–1290.
5. Lawrence, M. B., Smith, C. W., Eskin, S. G., and McIntire, L. V. (1990) Effect on venous shear stress on CD18-mediated neutrophil adhesion to cultured endohelium. *Blood* **75**, 227–237.
6. Hammarback, J. A. and Letourneau, P. C. (1986) Neurite extension across regions of low cell-substratum adhesivity: implications for the guidepost hypothesis of axonal pathfinding. *Devel. Biol.* **117**, 655–671.
7. Matsuda, T., Inoue, K., and Sugawara, T. Development of micropatterning technology for cultured cells. ASAIO Transactions, 36: M559-M562 (1990).
8. Corey, J. M., Wheeler, B. C., and Brewer, G. J. (1991) Compliance of hippocampal neurons to patterned substrate networks. *J. Neurosci. Res.* **30**, 300–307.
9. Rohr, S., Schölly, D. M., and Kléber, A. G. (1991) Patterned growth of neonatal rat heart cells in culture: morphological and electrophysiological characterization. *Circ. Res.* **68**, 114–130.
10. Britland, S., Perez-Arnaud, E., Clark, P., McGinn, B., Connolly, P., and Moores, G. (1992) Micropatterning proteins and synthetic peptides on solid supports: a novel application for microelectronics fabrication technology. *Biotechnol. Prog.* **8**, 155–160.
11. Stenger, D. A., Georger, J. H., Dulcey, C. S., Hickman, J. J., Rudolph, A. S., Nielsen, T. B., McCort, S. M., and Calvert, J. M. (1992) Coplanar molecular assemblies of amino- and pefluorinated alkylsilanes: characterization and gemoetric definition of mammalian cell adhesion and growth. *J. Am. Chem. Soc.* **114**, 8345–8442.
12. Clark, P., Britland, S., and Connolly, P. (1993) Growth cone guidance and neuron morphology on micropatterned laminin. *J. Cell Sci.* **105**, 203–212.
13. Lom, B., Healy, K. E., and Hockberger, P. E. (1993) A versatile technique for patterning biomolecules onto glass coverslips. *J. Neurosci. Methods* **50**, 385–397.
14. Ranieri, J. P., Bellamkonda, R., Jacob, J., Vargo, T. G., Gardella, J. A., and Aebischer, P. (1993) Selective neuronal cell attachment to a covalently patterned monoamine on fluorinated ethylene propylene films. *J. Biomed. Mater. Res.* **27**, 917–925.
15. den Braber, E. T., de Ruijter, J. E., Smits, H. T. J., Ginsel, L. A., von Recum, A. F., and Jansen, J. A. (1995) Effect of parallel surface microgrooves and surface energy on cell growth. *J. Biomed. Mater. Res.* **29**, 511–518.
16. Lee, J., Morgan, J. R., Tomkins, R. G., and Yarmush, M. L. (1993) Proline-mediated enhancement of hepatocyte function in a collagen gel sandwich culture configuration. *FASEB J.* **7**, 586–591.
17. Soekarno, A., Lom, B., and Hockberger, P. E. (1993) Pathfinding by neuroblastoma cells in culture is directed by preferential adhesion to positively charged surfaces. *NeuroImage* **1(2)**, 129–144.

18. Matsuda, T., Sugawara, T., and Inoue, K. (1992) Two-dimensional cell manipulation technology. *ASAIO J.* **38,** M243–M247.
19. Singhvi, R., Stephanopoulos, G., Wang, D. I. C. (1994a) Review: effects of substratum morphology on cell physiology. *Biotechnol. Bioeng.* **43,** 764–771.
20. Oakley, C. and Brunette, D. M. (1995) Topographic compensation: Guidance and directed locomotion of fibroblasts on grooved micromachined substrata in the absence of microtubules. *Cell Motil. Cytoskeleton* **31,** 45–58.
21. Gundersen, R. W. (1987) Response of sensory neurites and growth cones to patterned substrata of laminin and fibronectin in vitro. *Dev. Biol.* **121,** 423–431.
22. Hammarback, J. A., McCarthy, J. B., Palm, S. L., Furcht, L. T., and Letourneau, P. C. (1988) Growth cone guidance by substrate-bound laminin pathways is correlated with neuron-to-pathway adhesivity. *Dev. Biol.* **126,** 29–39.
23. Singhvi, R., Kumar, A., Lopez, G. P., Stephanopoulos, G. N., Wang, D. I. C., Whitesides, G. M., Ingber, D. E. (1994b) Engineering cell shape and function. *Science* **264,** 696–698.
24. Miyamoto, S., Ohashi, A., Kimura, J., Tobe, S., and Akaike, T. (1993) A novel approach for toxicity sensing using hepatocytes on a collagen-patterned plate. *Sensors Actuators B* **13–14,** 196–199.
25. Bhatia, S. N., Toner, M., Tompkins, R. G., and Yarmush, M. L. (1994) Selective adhesion of hepatocytes on patterned surfaces. *Ann. NY Acad. Sci.* **745,** 187–209.
26. Langenbach, R., Malick, L., Tompa, A., Kuszynski, C., Freed, H., and Huberman, E. (1979) Maintenance of adult rat hepatocytes on c3H/10T1/2 cells. *Cancer Res.* **39,** 3509–3514.
27. Kuri-Harcuch, W. and Mendoza-Figueroa, T. (1989) Cultivation of adult rat hepatocytes on 3T3 cells: expression of various liver differentiated functions. *Differentiation* **41,** 148–157.
28. Drumheller, P. D. and Hubbell, J. A. (1995) Surface immobilization of adhesion ligands for investigations of cell-substrate interactions, in *Biomedical Engineering Handbook* (Bronzino, J. D., ed.), CRC, pp. 1583–1596.
29. Guguen-Guillouzo, C., Clement, B., Baffet, G., Beaumont, C., Morel-Chany, E., Glaise, D., and Guillouzo, A. (1983) Maintenance and reversibility of active albumin secretion by adult rat hepatocytes co-cultured with another liver epithelial cell type. *Exp. Cell Res.* **143,** 47–54.
30. Donato, M. T., Gmez-Lechn, M. J., and Castell, J. V. (1990) Drug metabolizing enzymes in rat hepatocytes co-cultured with cell lines. *In Vitro Cell Dev. Biol.* **26,** 1057–1062.
31. Seglen,P. O. (1976) Preparation of isolated rat liver cells. *Methods Biol.* **13,** 29–83.
32. Rotem, A., Toner, M., Bhatia, S., Foy, B. D., Tompkins, R. G., and Yarmush, M. L. (1994) Oxygen is a factor determining in vitro tissue assembly: effects on attachment and spreading of hepatocytes. *Biotechnol. Bioeng.* **43,** 654–660.
33. Bhatia, S. N., Yarmush, M. L., and Toner, M. (1997) Controlling homotypic vs heterotypic interactions by micropatterning: co-cultures of hepatocytes and 3T3 fibroblasts. *J. Biomed. Mater. Res.* **34,** 189–199.
34. Chen, C., Mrksich, M., Huang, S., Whitesides, G. M., and Ingber, D. E. (1997) Geometric control of cell life and death. *Science* **276,** 1425–1428.

28

Methods for the Serum-Free Culture of Keratinocytes and Transplantation of Collagen–GAG-Based Skin Substitutes

Steven T. Boyce

1. Introduction

Objectives for dermal-epidermal skin substitutes for treatment of acute and chronic wounds include, but are not limited to: increased availability; stimulation of wound healing by transplantation of parenchymal cells; regulation of wound healing responses; and, predictable composition and efficacy to reduce mortality and morbidity. Particularly, the importance of including a cellular component has been demonstrated in experimental grafts of cells and biopolymers *(1–7)*. Transplanted cells may include normal, nontransformed populations isolated for primary culture, or genetically modified cells to deliver specific gene products of therapeutic interest *(8–10)*. Nontransformed cells may include autologous, allogeneic, or chimeric populations within composite grafts. A variety of approaches are directed toward repair of skin wounds by restoration of the functional anatomy and physiology of skin.

An hypothesis shared by several laboratories presumes that duplication of native anatomy and physiology of skin will provide a level of efficacy comparable to split-thickness skin graft. This common rationale has generated models that differ mostly in their selection of biopolymer for delivery of skin cells. Acellular dermis has been populated with keratinocytes and fibroblasts in preclinical models *(11,12)*. Biosynthetic analogs of skin have combined cultured skin cells with polylactic/polyglycolic (PLA/PGA) fabric *(13,14)*, polyethylene oxide/polybutylene terephthalate (PEO/PBT) *(3,15)*, collagen gels *(16–18)*, and collagen–glycosaminoglycan (GAG) sponges *(7,19,20)*. This chapter describes specific techniques for preparation and grafting to surgical wounds in athymic mice of cultured skin substitutes (CSS) from collagen–GAG substrates

From: *Methods in Molecular Medicine, Vol. 18: Tissue Engineering Methods and Protocols*
Edited by: J. R. Morgan and M. L. Yarmush © Humana Press Inc., Totowa, NJ

populated with normal human keratinocytes, melanocytes, and fibroblasts grown in serum-free or low-serum conditions.

Principles for CSS based on collagen–GAG substrates that duplicate native skin include separation of epidermal and dermal cells into respective histologic compartments, reformation of epidermal barrier and basement membrane before grafting, regeneration of skin pigmentation, and restoration of a vascular plexus. CSS before grafting possess these properties, except for cellular components of the vasculature. However, addition of endothelial cells to analogs of cornea *(6)* has been reported to stimulate more rapid and complete morphogenesis of the corneal epithelium. Conversely, absence of vascular cells and precursors confers limitations to clinical efficacy of CSS, including increased time to healing, and graft loss from microbial contamination and nutrient deprivation *(21–24)*. Furthermore, any functional phenotypes of CSS that develop in vitro are not stable, and eventually deteriorate unless they are transplanted and engraft to a viable wound.

Anatomic and physiologic deficiencies of CSS impose additional considerations to their clinical use. Therefore, clinical protocols to compensate for deficiencies of CSS must be designed to optimize their efficacy. Conformity with local protocols for surgical and nursing standards may expedite compliance of the staff with modalities of wound care. With the model of CSS described here, its surgical application in a single procedure permits its administration according to prevailing standards of burn treatment *(25–27)*. These standards have greatly influenced the development and modification of preclinical protocols described below. But, ultimately, experimental CSS must provide comparable efficacy to the prevailing standards of wound care, which depend on objective assessment of outcome *(28–31)*. After demonstration of efficacy, factors of cost-effectiveness must also be considered *(32)*. If all these requirements are satisfied, then reduction of mortality and morbidity with CSS can be realized. In the following section, materials and methods are described for culture of human skin cells, preparation of CSS from cells and collagen–GAG substrates, and closure of surgical wounds in athymic mice for preclinical studies with CSS.

2. Materials

Below are summarized the solutions and other materials used for isolation and culture of human epidermal keratinocytes and melanocytes, and dermal fibroblasts; preparation of CSS; and grafting of CSS to athymic mice. These materials provide consistent preparation and engraftment of CSS. Unless stated otherwise, all organic reagents were obtained from Sigma (St. Louis, MO); and inorganic reagents were obtained from Fisher Chemical (Fairlawn, NJ).

2.1. Solutions

1. Isotonic HEPES buffered saline (HBS).
2. 5% v/v Dettol (Reckitt and Colemen, Hull, UK).
3. 2.4 U/mL Dispase II (Boehringer Mannheim, Indianapolis, IN) in HBS.
4. 625 U/mL collagenase (Worthington Biochemicals, Freehold, NJ) in MCDB 153 to activate the enzyme with calcium and magnesium ions, plus 5% v/v bovine pituitary extract (BPE) to neutralize trypsin activities.
5. 0.025% trypsin–0.01% ethylenediaminetetraacetic acid (EDTA) w/v.
6. 10% v/v fetal bovine serum (FBS, Gibco-BRL; Grand Island, NY) in keratinocyte growth medium to neutralize trypsin.
7. Keratinocyte growth medium.
8. Fibroblast growth medium.
9. Melanocyte growth medium.
10. CSS maturation medium, as described below.
11. Basal medium for culture of human epidermal keratinocytes and melanocytes is MCDB 153 *(33–35)*, plus increased concentrations of six amino acids (histidine, isoleucine, methionine, phenylalanine, tryptophan, tyrosine), as described by Pittelkow *(23)*. Human fibroblasts are propagated in Dulbecco's modified Eagle's medium (DMEM, Gibco-BRL). Formulations of these media are presented in **Table 1**. These basal media are supplemented as described below to support selective cultures of each respective cell type.
12. Media supplements for selective cultures of skins cells: Supplements to basal nutrient media are added to stimulate mitosis of respective cells. By understanding the mitogenic responses of keratinocytes and melanocytes, basal medium MCDB 153 can be supplemented differently to prepare selective cultures of each cell type. Colony-forming efficiency of cultured keratinocytes in serum-free or biochemically defined medium may be increased by addition of lethally irradiated 3T3 feeder cells *(36,37)*, if strictly defined conditions are not required. DMEM is supplemented with 5% FBS, 10 ng/mL epidermal growth factor, 5 µg/mL insulin, and 0.5 µg/mL hydrocortisone. Media supplements for selective growth of these three cell types are summarized in **Table 2**.
13. HEPES buffered saline (HBS): HBS is formulated as a 10-fold concentrated stock, filter-sterilized, and stored refrigerated at 4°C. The concentrated stock is prepared from 71.49 g HEPES buffer (Research Organics; Cleveland, OH), 10 mL phenol red Stock 5 for MCDB 153 (*see* **Table 1**); 18.02 g glucose, 2.236 g potassium chloride, 76.97 g sodium chloride, and 2.68 g disodium phosphate heptahydrate. Adjust pH of the concentrate to 7.4 before filtration. HBS working solution is prepared by diluting exactly 100 mL of the concentrated stock to 1 L, adjusting the pH to 7.4, and filtering aseptically into a sterile bottle *(34)*.
14. Media for cryopreservation of selective cell cultures: Selective cultures of keratinocytes, melanocytes, or fibroblasts are cryopreserved in 70% v/v of their respective supplemented media, 20% v/v FBS, and 10% v/v dimethyl sulfoxide (DMSO).
15. Collagen–GAG substrates: Collagen–GAG substrates for transplantation of cultured cells are prepared, as described elsewhere *(20)*, to generate a thin (<0.5

Table 1
Basal Media Composition [M/L]

	MCDB 153	DME
Amino acids		
Alanine	1.0e–04	
Arginine	1.0e–03	4.0e–04
Asparagine	1.0e–04	
Aspartic acid	3.0e–05	
Cystine	2.0e–04	
Cysteine	2.40e–04	
Glutamic acid	1.0e–04	
Glutamine	6.0e–03	4.0e–03
Glycine	1.0e–04	4.0e–04
Histidine[a]	2.5e–04	2.0e–04
Isoleucine[a]	7.80e–04	8.0e–04
Leucine	5.0e–04	8.0e–04
Lysine	1.0e–04	8.0e–04
Methionine[a]	9.2e–05	2.0e–04
Phenylalanine[a]	9.2e–05	4.0e–04
Proline	3.0e–04	
Serine	6.0e–04	4.0e–04
Threonine	1.0e–04	8.0e–04
Tryptophan[a]	4.6e–05	8.0e–05
Tyrosine[a]	7.7e–05	5.0e–04
Valine	3.0e–04	8.0e–04
Vitamins and coenzymes		
Biotin	6.0e–08	
Folic acid	1.8e–06	9.1e–06
Lipoic acid	1.0e–06	
Niacinamide	3.0e–07	3.3e–05
D-calcium pantothenate	1.0e–06	8.4e–06
Pyridoxine	3.0e–07	
Pyridoxal	2.0e–05	
Riboflavin	1.0e–07	1.1e–06
Thiamine	1.0e–06	1.2e–05
Vitamin B12	3.0e–07	
Organic compounds		
Adenine	1.8e–04	
Choline chloride	1.0e–04	2.9e–05
Glucose	6.0e–03	5.6e–03
Inositol	4.0e–05	
Myoinositol	1.0e–04	

(continued)

Table 1
Basal Media Composition [M/L]

	MCDB 153	DME
Putrescine	1.0e–06	
Sodium acetate	3.7e–03	
Sodium pyruvate	5.0e–04	1.0e–03
Thymidine	3.0e–06	
Bulk Ions		
Calcium chloride	3.0e–05	1.8e–03
Potassium chloride	1.5e–03	5.3e–03
Magnesium chloride	6.0e–04	
Magnesium sulfate	8.2e–04	
Sodium chloride	1.3e–01	1.1e–01
Sodium phosphate (dibasic)	2.0e–03	
Sodium phosphate (monob)	9.1e–04	
Trace elements		
Cupric sulfate	1.10e–08	
Ferrous nitrate	2.5e–07	
Ferrous sulfate	5.0e–06	
Manganese sulfate	1.0e–09	
Molybdenum	1.0e–09	
Nickel chloride	5.0e–10	
Selenium	3.0e–08	
Silicon	5.0e–07	
Tin chloride	5.0e–10	
Vanadium	5.0e–09	
Zinc sulfate	5.0e–07	
Buffers and indicators		
HEPES	2.8e–02	
NaOH	1.0e–02	
$NaHCO_3$	1.4e–02	2.2e–02
Phenol red	3.3e–06	3.8e–05

^{a}Concentrations of six amino acids are increased as described by Pittelkow and Scott (23).

mm), symmetric sponge that is laminated on one side with a microporous film of the same biopolymer mixture. Dry, acellular substrates are packaged, sterilized by exposure to 2.5 mega-Rad gamma irradiation, and stored at room temperature until they are populated with cultured cells.

Frames to contain cell suspensions over the surface of collagen–GAG substrates after inoculation are cut from type-316 stainless steel to fit 150-mm diameter Petri dishes. Inoculation frames are square and measure exactly 97 mm o.d., and 85 mm i.d. The bottom surface of the frame is beveled from the outside edge

Table 2
Supplements for Selective Culture of Cells from Human Skin

	Keratinocytes	Melanocytes	Fibroblasts
Basal medium	MCDB 153[a]	MCDB 153[a]	DMEM
Epidermal growth factor	1 ng/mL	0	10 ng/mL
Insulin	5.0 µg/mL	5.0 µg/mL	5.0 µg/mL
Hydrocortisone	0.5 µg/mL	0	0.5 µg/mL
Bovine pituitary extract	0.5% v/v	0	0
Fetal bovine serum	0	4% v/v	5% v/v
Penicillin-streptomycin-fungizone	1% v/v	1% v/v	1% v/v
Transferrin	0	0.5 µg/mL	0
Basic fibroblast growth factor	0	0.6 ng/mL	0
α-Melanocyte stimulating hormone	0	17 ng/mL	0
Endothelin-1	0	2.5 ng/mL	0
Vitamin E (α-tocopherol)	0	1.0 µg/mL	0

[a]Plus increased concentrations of hydrophobic amino acids as described in **Table 1**.

to the inside edge, to generate a blunt wedge at the bottom of the frame to increase weight distribution where the frame contacts the substrate, and to decrease leakage of cells. Inoculation frames used in the author's laboratory were fabricated by custom order in a local machine shop. Frames are sterilized by steam autoclaving.

16. Maturation medium for cultured skin substitutes: To stimulate development of epithelial barrier in skin substitutes before grafting, it was observed that the keratinocyte growth medium was not satisfactory. Poor epithelial organization was observed for at least two major reasons: absence of essential fatty acids that are required for barrier formation, and an insufficient calcium concentration. Replacement of calcium alone was not sufficient to permit formation of epidermal barrier in vitro. However, studies to supplement MCDB 153 with lipids *(38)* stimulated partial formation of barrier lipids and structures analogous to stratum corneum of epidermis. This maturation medium is formulated from lipid prestocks (oleic acid, palmitic acid, arachidonic acid; vitamin E) carried in the medium with BSA, carnitine, and increased serine. Specifications of this formulation are summarized in **Table 3**.

Supports to lift skin substitutes to the air–liquid interface are used to fit a 150-mm diameter Petri dish, and consist of three parts. Square, seamless frames measuring 97 mm od and 87 mm id, and circular wire mesh measuring 12.5 cm in diameter, both consist of type-316 stainless steel. Both of these devices were fabricated in a local machine shop. The wire mesh and lifting frames are sterilized by steam autoclaving. The mesh is overlaid with a cotton filter pad (Schleicher and Schuell;

Table 3
Formulation of Lipid-Supplemented Maturation Medium for Skin Substitutes

	Pre-stock		50X Stock		Final conc.
Compound	mg/mL	[M]	μL/mL	[M]	[M]
Linoleic acid	21.05	7.5×10^{-2}	10	7.5×10^{-4}	1.5×10^{-5}
Palmitic acid	32.05	1.25×10^{-1}	10	1.25×10^{-3}	2.5×10^{-5}
Oleic acid	35.30	1.25×10^{-1}	10	1.25×10^{-3}	2.5×10^{-5}
Archidonic acid	10.65	3.5×10^{-2}	10	3.5×10^{-4}	7.0×10^{-6}
α-Tocopherol-Ac	47.40	1.1×10^{-1}	10	1.1×10^{-3}	2.2×10^{-5}
BSA (FFA free)	27.60	4.2×10^{-4}	950	4.0×10^{-4}	8.0×10^{-6}
	Pre-stock		100X Stock		Final conc.
Compound	mg/mL	[M]	μL/mL	[M]	[M]
Serine	105	$1 \times 10^{-9} M$	10	1.0	1.0×10^{-2}
Carnitine	0.161	$1 \times 10^{-9} M$	10	1.0×10^{-3}	1.0×10^{-5}

Keene, NH) that is cut to a 12-cm diameter, and acts as a wick for uniform distribution of medium under air-exposed skin substitutes.

17. Culture vessels: Selective cell cultures of human keratinocytes or melanocytes are grown in 75-cm² polystyrene flasks (Corning, Corning, NY). Fibroblasts are grown in 150 cm² flasks (Corning). Cultured skin substitutes are inoculated with cells and incubated in 150-mm diameter tissue-culture Petri dishes (Lux-Nunc, Gaithersburg, MD). All cells are cryopreserved in 2-mL screw-capped cryogenic vials (Corning). Assorted sterile tubes and pipets are from Corning, or Falcon (Fisher Scientific, Itasca, IL).
18. Solution for irrigation of wounds grafted with cultured skin substitutes: Solutions for topical irrigation of grafted wounds consisting of nutrients, mitogens, and antimicrobials have been shown to improve engraftment of avascular skin substitutes *(39,40)*. For preclinical studies the irrigation solution consists of:
    a. Basal nutrient medium MCDB 153 (*see* **Table 1**), supplemented with.
    b. Insulin at 5 μg/mL.
    c. Hydrocortisone at 0.5 μg/mL.
    d. Ciprofloxacin (Cipro™, Miles, West Haven, CT) at 20 μg/mL.
    e. Nystatin at 100 U/mL (Gibco-BRL).

2.2. Animal Surgery

1. Anesthesia: Animals are anesthetized with Avertin, which consists of 25 mg/mL tribromoethanol dissolved in 1.25% tertiary-amyl alcohol *(41)*. Although this compound provides stable anesthesia for the 30–60 min surgical procedure, any anesthetic with pharmacologic actions appropriate to the procedure may be substituted.

2. Resusitation: Resusitation solution after surgery consists of 3 mg/mL cephtazidime in saline.
3. Surgical supplies: Sterile supplies of several varieties are used to graft skin substitutes to wounds in athymic mice, including:
    a. Betadine™ swabs (Purdue Frederick Company; Norwalk, CT).
    b. 70% isopropyl alcohol swabs (Kendall Healthcare, Mansfield, MA).
    c. Skin Skribe™ surgical markers (Hospital Marketing Services, Naugatuck, CT).
    d. 4 × 4 in., 16-ply cotton gauze sponges (Johnson & Johnson Medical, Arlington, TX).
    e. Ethilon™ 6-0 sutures (Ethicon, Sommerville, NJ).
    f. N-Terface™ (Winfield Laboratories, Dallas, TX).
    g. Xeroform™ gauze (Sherwood Medical, St. Louis, MO).
    h. Vaseline™ gauze (Sherwood Medical).
    i. Steri-Strip™ compound benzoin tincture ampules (3M Surgical Medical, St. Paul, MN).
    j. OpSite™ (Smith & Nephew Medical, Hull, UK).
    k. Coban™ bandage (3M, Surgical Medical, St. Paul, MN).
4. Instruments for surgery:
    a. Dumont™ fine forceps, #3C pointed-tipped, and #2A flat-tipped (Ted Pella, Redding, CA).
    b. Iris scissors, needle driver, bandage scissors, and mosquito hemostat (Miltex Instrument, Lake Success, NY).
5. Aseptic care and housing of athymic mice:
    a. Autoclavable, filter-topped cages, bedding, and chow.
    b. Water containing 0.2 mg/mL benedryl (as diphenhydramine, Elkins-Sinn, Cherry Hill, NJ), and 1% v/v Septra™ (as trimethoprim and sulfamethoxizole, Burroughs-Wellcome, Research Triangle Park, NC).
    c. Heating pads (non-sterile).
    d. Laminar flow cage isolator (Forma Scientific, Marietta, OH).

3. Methods
3.1. Individual Cell Cultures in Selective Media
3.1.1. Primary Cultures of Skin Cells

Keratinocytes, melanocytes, and fibroblasts may be isolated from a single skin biopsy by serial disaggregation using enzymatic and mechanical techniques. Split-thickness skin is preferable to full-thickness skin, because penetration of enzymes is more rapid. All procedures are performed aseptically, and all materials are sterile. Dissecting instruments may be repeatedly sterilized during the procedures by dipping in 95% ethanol, followed by burning the ethanol in a gas flame and cooling the instruments before contacting the tissue.

1. Wash tissue in HBS, and trim away any subcutaneous tissue.
2. Disinfect skin by submersion for 15–30 s in 5% v/v Dettol solution, followed by three washes in HBS.

3. Cut skin into strips 2–3 mm wide, and transfer to Petri dishes containing several layers of cotton gauze saturated with 2.4 U/mL Dispase II in HBS. Incubate skin dermis side against gauze in Dispase II for 30–120 min, until epidermis readily separates from the dermis with forceps. Do not allow skin strips to float. After epidermis has loosened from dermis, transfer the skin strips to a fresh Petri dish containing HBS.
4. Mechanically separate epidermis from dermis, and place epidermis into a fresh Petri dish containing HBS. After all the epidermal sheets have been collected, seal the Petri dish with Parafilm™ and incubate for 3 h at 4°C. Because HBS contains no calcium, this incubation depletes calcium and loosens desmosomal junctions between keratinocytes.
5. Mince dermis into fine (0.5–1 mm^2) pieces, and transfer into a 50-mL centifuge tube containing 625 U/mL collagenase in culture medium, plus 5% v/v BPE, equilibrated to 37°C, and 5% CO_2 (collagenase activity is dependent on divalent cations, i.e., calcium and magnesium). Agitate the tissue pieces with a plugged Pasteur pipet and incubate at 37°C and 5% CO_2 for 30–90 min. Agitate the tissue pieces for 1–2 min at 20–30 min intervals during the digestion. After the tissue has digested to minimal fragments, centrifuge the suspension at 250g for 5 min at 4°C (excessive digestion will reduce the yield of viable cells). Aspirate the collagenase and resuspend the pellet in selective medium for fibroblasts, as described in **Table 2**. Repeat centrifugation and resuspension, and inoculate dermal cells and fragments into 150-cm^2 flasks containing 10 mL selective medium for fibroblasts, at an approximate ratio of 1 cm^2 dermis:50 cm^2 of culture surface. After 16–24 h, add 15 mL of medium per flask. Refresh medium at 2-d intervals.
6. Retrieve epidermal strips from 4°C incubation and collect HBS from dish. Mince epidermal strips into fine pieces, as was done with dermis. Transfer tissue fragments to a 50-mL centrifuge tube, add 1–2 mL of 0.025% trypsin–0.01% EDTA per cm^2 of epidermal tissue, and agitate 4 min using a plugged Pasteur pipet. Let tissue fragments settle, collect the cell suspension, and transfer it to selective medium for keratinocytes with 10% FBS on ice. Resuspend the epidermal fragments in HBS, agitate for 4 min, collect the cell suspension, and transfer to keratinocyte medium plus 10% FBS. Repeat resuspension, agitation, and collection of cell suspension into keratinocyte medium plus 10% FBS. Resuspend the epidermal fragments in a small volume of keratinocyte medium containing 10% FBS in a separate tube. Centrifuge the suspension (s) and epidermal pieces at 250g for 5 min at 4°C, and aspirate the supernatants. Resuspend the epidermal cells and tissue fragments into a small volume of selective medium for keratinocytes. Count the epidermal cell suspension. To grow keratinocytes, inoculate into 75-cm^2 flasks at 2–3 × 10^4 cells/cm^2 containing 15 mL of selective medium for keratinocytes, equilibrated to 37°C and 5% CO_2. To grow melanocytes, inoculate into 75-cm^2 flasks at 1 × 10^5 cells/cm^2 containing 15 mL of selective medium for melanocytes *(42)* (*see* **Table 2**), equilibrated to 37°C and 5% CO_2. Refresh keratinocyte medium at 2-d intervals. Refresh melanocyte medium three times per wk.

3.1.2. Cryopreservation of Cells by Controlled Rate Freezing

Before cells become confluent in primary culture, great efficiency is gained if they are cryopreserved for later use. A representative procedure for cryopreservation of keratinocytes is described below, but it may be modified for other cell types by substitution of respective media for cryoprotection.

1. Prepare cryopreservation medium by combining 7 parts selective medium for keratinocytes, plus 2 parts FBS, and 1 part dimethylsulfoxide. Sterilize by filtration through a 0.22-µm filter.
2. Harvest keratinocytes by washing cells with HBS, aspiration of HBS, and addition of 1 mL/75-cm^2 flask of cold 0.025% trypsin–0.01% EDTA. Incubate for 3–4 min at 37°C, or 4–5 min at room temperature. Gently tap the flask against the bench and examine in a phase-contrast contrast microscope. After greater than 90% of the cells have detached, collect the cells by serial washing of multiple flasks with a single 10 mL aliquot of HBS, and transfer the cell suspension to selective medium for keratinocytes, plus 10% FBS on ice. Repeat the HBS wash and combine suspensions. Centrifuge cells at 250g, aspirate the supernatant, resuspend in a small volume of selective medium for keratinocytes, and count cells. Repeat centrifugation and aspiration of supernatant, and resuspend keratinocytes in cryopreservation medium for keratinocytes at 0.5–3.0 × 10^6 cells/mL. Dispense into cryogenic vials, and freeze at a controlled rate (i.e., 1°C/min) in a CyroMed microprocessor-controlled freezer (Forma Scientific, Marietta, OH) with a program for suspensions of cultured cells, to retain best viability. If a controlled-rate freezer is not available, place vials into a sealed block of styrofoam with at least 1 cm covering all sides of all vials, place into a –70°C freezer overnight, and transfer to liquid nitrogen the next day. Store cells indefinitely in liquid nitrogen.

3.1.3. Recovery of Cryopreserved Cells into Culture

Recover keratinocytes by thawing rapidly in 70% isopropyl alcohol, making sure that no alcohol leaks into the vials as they warm. To assure the seal on the vial, swab the vial with 70% isopropyl alcohol, allow it to dry, release pressure from the vial by loosening and resealing the cap, and place the vial into 70% isopropyl alcohol in a 37°C water bath. As soon as keratinocytes are thawed, inoculate them at 0.3–1.0 × 10^4 cells/cm^2 into 75-cm^2 flasks containing 15 mL of selective medium for keratinocytes equilibrated to 37°C and 5% CO_2. Refresh medium after 12–24 h, and at 2-d intervals thereafter. Fibroblasts and melanocytes are handled similarly, but in their respective media. Incubate until sufficient cells are available to prepare cultured skin substitutes. Representative photomicrographs of selective cultures of keratinocytes, melanocytes, and fibroblasts, in log-phase growth after recovery from cryopreservation, are shown in **Fig. 1**.

3.2. Preparation of Cultured Skin Substitutes

A schematic diagram for preparation of cultured skin substitutes is presented in **Fig. 2**. Selective cultures of skin cells are harvested on successive days, and

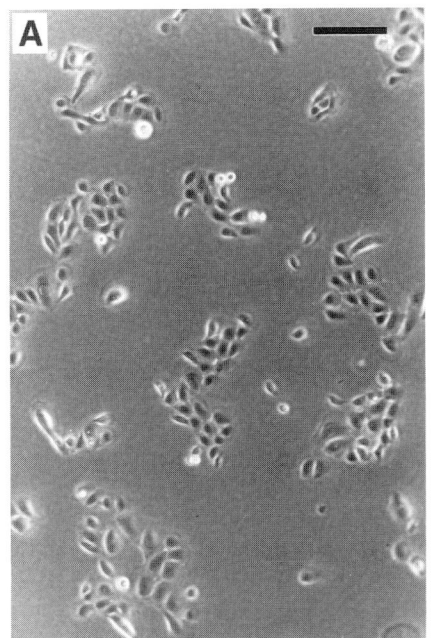
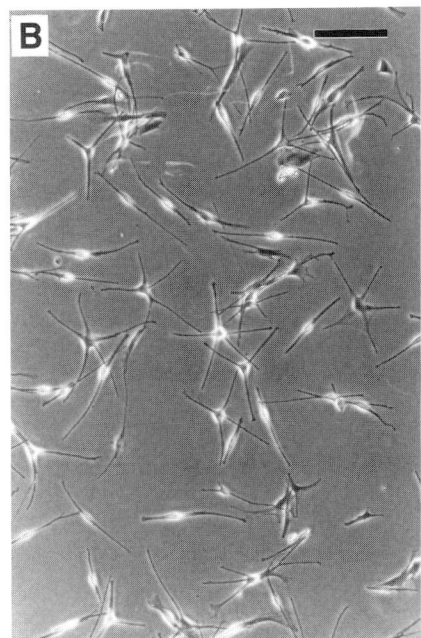
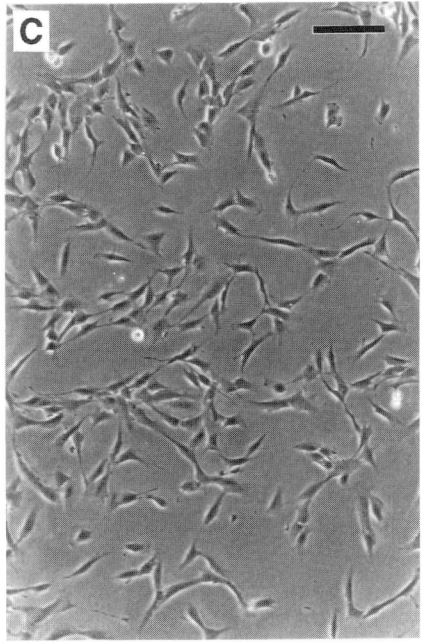

Fig. 1. Photomicrographs of normal human skin cells in log-phase, selective culture. **(A)** epidermal keratinocytes; **(B)** epidermal melanocytes; and, **(C)** dermal fibroblasts. Scale bar, 0.1 mm.

inoculated onto collagen-glycosaminoglycan substrates in submerged culture. Epithelium matures into an analog of epidermis by air-exposed incubation in lipid-supplemented medium. Epidermal melanocytes may be added to the skin substitute by co-inoculation with keratinocytes at a ratio of 1 melanocyte/30 keratinocytes.

3.2.1. Rehydration of Dry Collagen–GAG Substrates

Rehydrate dry collagen–GAG substrates (10 × 10 cm) by removal from sterile packaging, and place into 150 mm tissue-culture Petri dishes containing 75 mL HBS. Incubate 30 min and invert substrate in dish. Aspirate and refresh HBS, gently express air from the substrate, and incubate 30 min. Repeat procedure using selective medium for fibroblasts. Aspirate medium from dish, orient the substrate with porous side up, and install inoculation frame (*see* **Subheading 2.1.5.**) over substrate. Add 20 mL of selective fibroblast medium only to outside of frame, and equilibrate in cell culture incubator at 37°C and 5% CO_2.

3.2.2. Harvest and Inoculation of Cultured Fibroblasts

Harvest fibroblasts in log-phase growth, as described for keratinocytes in **Subheading 3.1.2., step 2**, except reduce the incubation time of fibroblasts in 0.025% trypsin–0.01% EDTA to 2–3 min. Collect cells in 10 mL HBS, transfer to a 50-mL centrifuge tube containing selective medium for fibroblasts, centrifuge at 250*g*, aspirate supernatant medium, resuspend fibroblasts in a small volume of their selective medium, and count cells. Adjust cell density to 4 × 10^6 cells/mL. Evenly inoculate 10 mL of the fibroblast suspension over the area within the inoculation frame (5 × 10^5 cells/cm^2). After 24 h, remove the inoculation frame, aspirate medium and unattached cells, and transfer the substrate with fibroblasts to a new 150-mm Petri dish. Wash the substrate with attached fibroblasts in 35 mL selective medium for keratinocytes, and carefully invert the dermal substitute, to orient it with the microporous side up, for inoculation of keratinocytes. Wash dermal substitute with 35 mL of selective medium for keratinocytes, aspirate medium, and install an inoculation frame, as before. Add 20 mL selective medium for keratinocytes only to the outside of the frame, and equilibrate in cell-culture incubator at 37°C and 5% CO_2.

3.2.3. Harvest and Inoculation of Cultured Keratinocytes

Harvest keratinocytes as described in **Subheading 3.1.2., step 2**, and, after centrifugation, resuspend the cell pellet in a small volume of selective medium for keratinocytes. Count the keratinocytes, and adjust the density to 8 × 10^6 cells/mL. Remove dermal substitute with inoculation frame from the incubator, and evenly inoculate 10 mL of the keratinocyte suspension over the area within the inoculation frame (1 × 10^6 cells/cm^2). Inoculation of keratinocytes onto dermal substitutes is defined as day 0 of incubation of cultured skin substitutes.

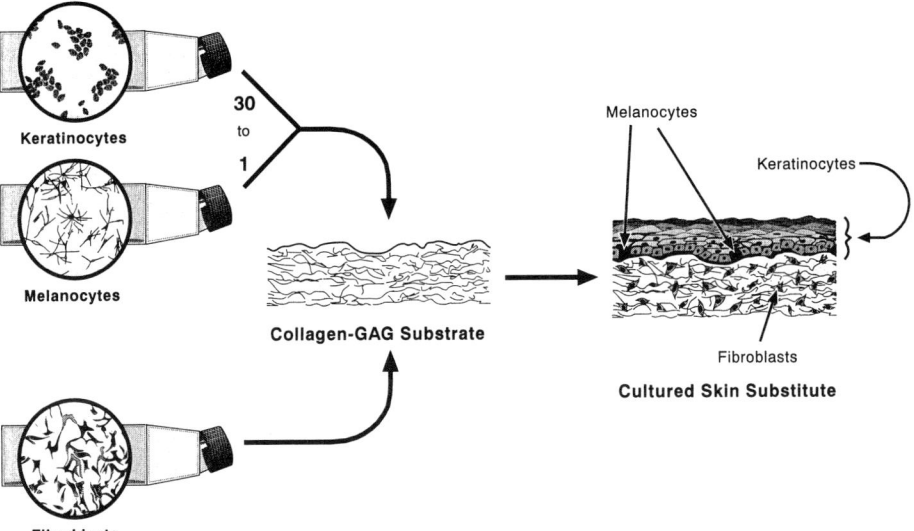

Fig. 2. Schematic diagram of preparation of cultured skin substitutes. Selective cultures of dermal fibroblasts are inoculated into the porous reticulations of a collagen-glycosaminoglycan (GAG) substrate. Selective cultures of epidermal keratinocytes are prepared to include or exclude selective cultures of epidermal melanocytes. If added, melanocytes are mixed with keratinocyte suspensions at a ratio of 1 melanocyte to 30 keratinocytes before inoculation onto the micro porous surface of a collagen–GAG substrate. The cell-biopolymer composite is incubated at the air-liquid interface to promote epithelial stratification and barrier formation.

3.2.4. Addition of Melanocytes to Cultured Skin Substitutes

Add melanocytes to skin substitutes by simultaneous harvest of selective cultures of melanocytes. After each cell type is counted, and before inoculation of keratinocytes, add melanocytes into the keratinocyte suspension at a ratio of 1 melanocyte:30 keratinocytes. Inoculate dermal substitutes, as described in **Subheading 3.2.3**.

3.2.5. Incubation Day 1

After 24 h (incubation d 1), add 10 mL selective medium for keratinocytes to the inside of the inoculation frame only.

3.2.6. Incubation Day 2

On incubation d 2, remove the inoculation frame and wash with maturation medium containing the lipid supplement described in **Table 3**. Increase the calcium concentration in the medium to 0.5 mM, as described in **Table 4**, and discontinue BPE. Incubate the cultured skin substitute submerged in medium.

3.2.7. Incubation Day 3

On incubation d 3, prepare a new 150-mm Petri dish containing a lifting frame, mesh and cotton support (*see* **Subheading 2.1., item 6**). Add 60 mL of maturation medium to the dish according to the formulation and schedule in **Table 4**. Equilibrate the assembly in a cell-culture incubator at 37°C and 5% CO_2. Aspirate medium from the dish containing the skin substitute, and trim away edges of the collagen–GAG substrate that were outside of the inoculation frame (no cells), and discard. Remove the Petri dish with lifting frame from the incubator, and transfer the cultured skin substitute to the top of the cotton pad on the stainless steel mesh. Transfer the skin substitute by laying a piece of N-terface of correct size (8.5 × 8.5 cm) on the epithelial surface. Lift one corner of the skin substitute and N-terface with a flat-tipped forceps, and hold an adjacent corner with another identical forceps. Lift the skin substitute, transfer it to the dish containing the lifting frame, and lay it on the cotton pad saturated with medium. To reposition the skin substitute, lift it from one site and replace it in another site, but do not drag or pull it. Return the lifted skin substitute to the incubator.

3.2.8. Incubation Day 4

On culture d 4, aspirate medium from the dish containing the skin substitute, and add 45 mL maturation medium for d 4, as described in **Table 4**. Incubate the skin substitute in this condition until use. Refresh maturation medium daily.

3.2.9. Preparation of Cultured Skin Substitutes for Surgery

On the day before animal surgery (ca incubation d 14), prepare cultured skin substitutes for grafting. Transfer skin substitutes to new 150-mm Petri dishes containing 10 mL maturation medium. With a straight-edge razor blade (Weck, Weck, NC) held in a hemostat and a fine-tipped forceps, cut skin substitutes into grafts exactly 2 × 2 cm. Transfer the individual grafts to the dish containing the lifting frame, and return to the incubator until the day of surgery. A photomicrograph of the histology of a representative skin substitute is presented in **Fig. 3**.

3.3. Animal Surgery

All care and use of animals is approved by the Institutional Animal Care and Use Committee of the University of Cincinnati. All procedures are performed with aseptic technique, and all materials are sterile. Surgical procedures are performed in a biological safety cabinet, and animals are housed in filter-topped cages in a laminar flow cage isolator.

Table 4
Schedule and Media for Maturation of Cultured Skin Substitutes

Factor	Day 1	Day 2	Day 3	Day 4 and after
BPE (% v/v)	0.5	0	0	0
Lipid supplement[a]	No	Yes	Yes	Yes
EGF (ng/mL)	1	1	1	0
Calcium (mM)	0.2	0.5	1.0	1.5
Air-exposure	No	No	Yes	Yes

[a]Lipid Supplement is formulated as described in **Table 3**. All media are MCDB 153 as described in **Table 1** supplemented with 5 µg/mL insulin and 0.5 µg/mL hydrocortisone.

3.3.1. Preparation of a Sterile Field for Surgery

Prepare materials for surgery by packaging of multiple units of each item into sterile specimen jars. Cover the inside of the biological safety cabinet with a sterile drape. Place electric heating pads set on "low" beneath the operating field, and beneath a cage to be used for recovery, to prevent hypothermia during and after surgery.

3.3.2. Anesthesia

Anesthetize an athymic mouse by intraperitoneal injection of 350–500 µL of avertin, as described in **Subheading 2.2.1**. Verify complete anesthesia by negative response to toe pinch.

3.3.3. Cleansing of the Surgical Site

Cleanse skin on (right) flank of animal by serial swabbing with Betadine, followed by 70% isopropanol. Demarcate a 2 × 2-cm area on the flank with a square template by marking the corners and midpoints of the sides of the template.

3.3.4. Preparation of a Full-Thickness Skin Wound

Carefully prepare a full-thickness skin wound to preserve the panniculus carnosus. With fine forceps and iris scissors, lift a corner of the demarcated site and incise through the epidermis and into the dermis. Continue the incision around the entire wound perimeter by placement of the scissors tip into the dermis, application of gentle forward and upward pressure, and closing the scissors to advance the incision. Apply gentle countertraction to the edges of the incision to promote its advancement. Periodically open the incision and verify visually that the panniculus carnosus remains intact. After the incision is complete, carefully lift one corner of the murine skin with a fine forceps, and dissect the panniculus carnosus from the dermis with the tip and edge of the scissors. An example of the completed dissection is presented in **Fig. 4A**.

Fig. 3. Histology of a cultured skin substitute. Keratinocytes (HK) are restricted to the outer surface of the dermal substitute (C-GAG-HF) to form separate compartments within the skin substitute. Total thickness is less than 0.5 mm. Scale bar, 0.1 mm.

3.3.5. Suturing and Dressing of Cultured Skin Substitutes onto Wounds

Place the cultured skin substitute with a backing of N-terface onto the wound. Secure the graft and dressing to the wound margin with stent-type suturing at eight points, as shown in **Fig. 4B**. Pack the graft with several layers of cotton gauze, and tie opposing sutures together, as shown in **Fig. 4C**. Apply benzoin adhesive around perimeter of wound, and cover the packing with OpSite to form a compartment over the graft that retains fluid, but is vapor permeable.

3.3.6. Resusitation

Resusitate the animals by intraperitoneal injection of 1.0 cc saline containing 3 mg of cephtazidime.

3.3.7. Bandaging of Dressings

Bandage the dressed graft with Coban, and prepare a small (3–5 mm) opening to inject irrigation solution. Inject 1.5 cc of irrigation solution (*see* **Subheading 2.1.8.**) through OpSite into gauze. Maintain mice on the heating pad until they recover from anesthesia. After animals recover from anesthesia, return them to their respective cages for housing. Keep cages on heating pads set on "low" until all dressings are removed. Supply water prepared with benedryl and Septra as described in **Subheading 2.2.5.2.**

3.4. Dressing and Care Protocol

3.4.1. Postoperative Days 1–14

Inject 1.0 cc/animal/d of irrigation solution into gauze packing over the wound.

3.4.2. Post-operative Day 14

Remove dressings and sutures and collect data (i.e., photographs, area tracings, noninvasive instruments) *(43,44)*. Replace dressings with N-terface contacting the graft, followed by Xeroform, cotton gauze, and Coban bandage until d 21.

3.4.3. Post-operative Day 21

Remove dressings, collect data and redress with N-Terface, cotton gauze, and Coban until d 28. Representative healing of wounds at d 21 after surgery is shown in **Fig. 5**. Healed skin from skin substitutes depleted of melanocytes is shown in **Fig. 5A**. Confirmation of engraftment of human keratinocytes is shown in **Fig. 5B** by immuno histochemical staining for HLA-ABC antigens *(2,40,45)*. Skin substitutes with added melanocytes produce uniformly black skin, as represented in **Fig. 5C**. Localization of melanocytes within the basal layer of keratinocytes is verified by immunohistochemical staining for the melanocyte marker, Mel-5 *(46)*.

3.4.4. Postoperative Day 28

Remove all dressings and swab healed skin with vaseline gauze. Collect data as needed.

4. Notes

4.1. Media Preparation

1. Storage of media: To increase reproducibility, prepare sufficient media and supplements for 2–3 mo, aliquot, and freeze at –20°C or –70°C. This simple practice also decreases labor intensity of medium preparation.
2. Reconstitution and supplementation of media: Reconstitute and supplement media for use not less than weekly. Even serum-supplemented media deteriorate at 4°C. Optimum performance of serum-free media is obtained if all components are as fresh as possible. Serum-free media are more subject to reduced performance if any component (s) lose (s) activity.
3. Use of media:
   a. Warm media to 37°C before use, minimize warm time.
   b. Keep labile reagents (especially trypsin) on ice during use.
   c. Avoid exposure of culture media to light, especially fluorescent light that degrades certain vitamins (i.e., riboflavin, B_2).
   d. Replace unused media chemicals no less than annually.

4.2. Cell Culture

4. Safety: Maintain universal precautions for laboratory staff. Human cells can carry pathogens (i.e., human immunodeficiency virus, hepatitis, cyto megalo virus). Certain sources (e.g., surgical discard) of human tissues may not require pathogen testing. Other sources (e.g., tissue banks accredited by the American Association of Tissue Banks) perform tests for human pathogens, and require negative

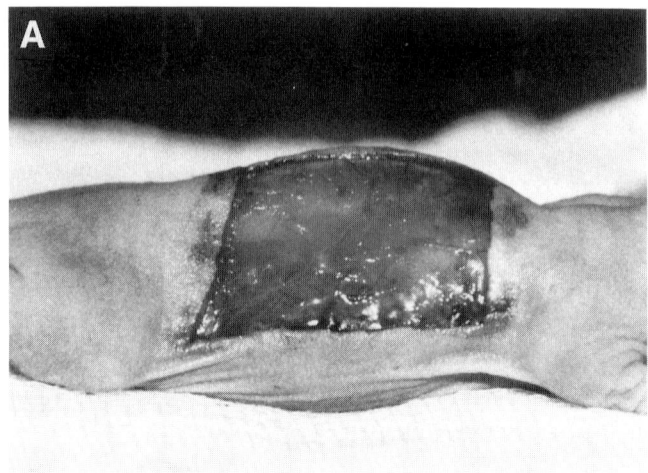

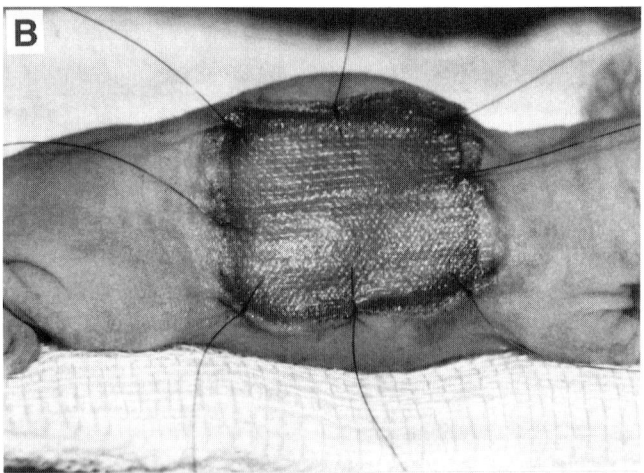

Fig. 4. Photographs of steps in the procedure to graft human cultured skin substitutes to athymic mice. (**A**) Murine skin removed with preservation of the panniculus carnosus. (**B**) Cultured skin substitute covered with N-Terface™ dressing is secured to the wound margin with stent-type sutures. *(continued on opposite page)*

   tests before release of tissue. However, negative tests for pathogens do not provide absolute assurance that pathogens are not present. Therefore, appropriate safety factors should be required to protect staff who handle human-derived materials. All human tissues and materials (media, culture vessels) that contact human cells and tissues should be treated as biohazards, and be treated lethally (e.g., bleach, incineration) for disposal.
5. Tissue: Isolate cells from tissue of optimum viability. Discarded tissue from elective surgery (i.e., neonatal circumcision, reduction mammoplasty) may be

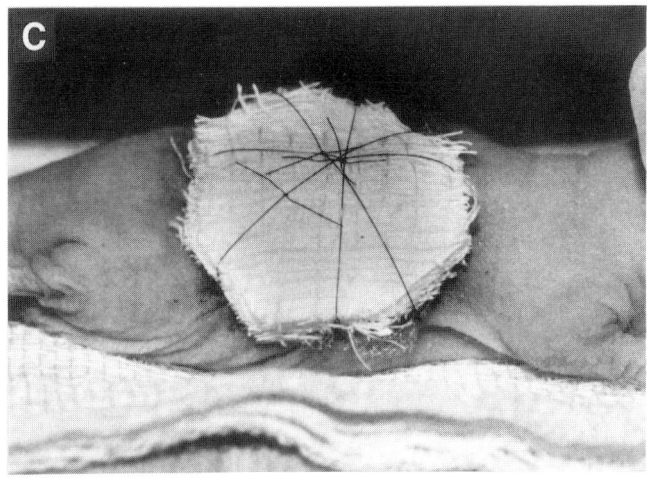

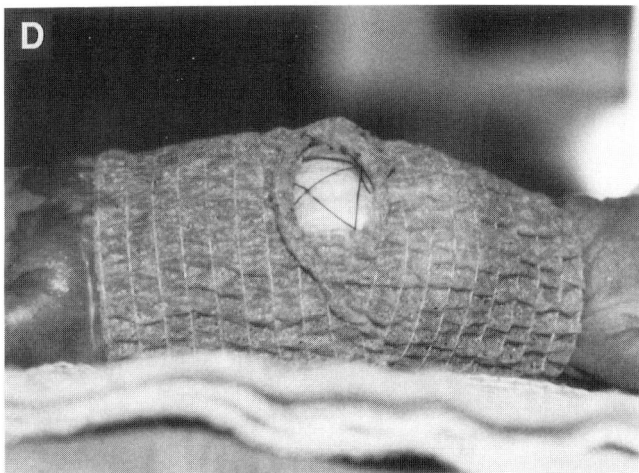

Fig. 4. (C) Grafted wound is packed with gauze and stent sutures are tied over to secure the packing. (D) Dressed wound is covered with OpSite™ to retain moisture in the wound, and is bandaged with Coban™. Irrigation solutions are injected through a small aperture in the Coban™, through the OpSite™ and into the gauze to modulate the healing wound.

obtained according to guidelines and policies of the local institutional review board. Cadaveric donors who are young (18–30 yr), otherwise healthy individuals are usually sources of high proportions of proliferative cells. Samples of split-thickness skin are best. Minimize ischemic time between acquisition of tissue and isolation of cells to preserve optimum cellular viability.

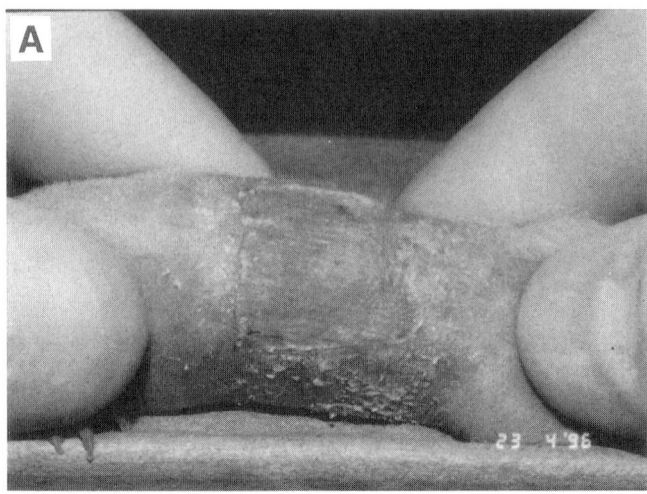

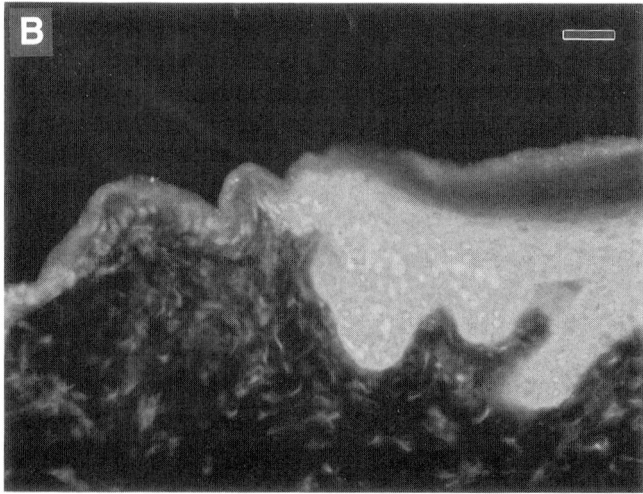

Fig. 5. Photographs of nonpigmented and pigmented human skin after engraftment of cultured skin substitutes. **(A)** By 21 d after grafting, the epithelium is fully functional and dry. **(B)** Persistence of human cells is verified by immuno histochemical staining of HLA-ABC antigens in healed epidermis (right of panel). Staining stops abruptly at the wound margin (center of panel), and murine epidermis (left of panel) is negative for human cell markers. **(C)** Cultured skin substitutes with added melanocytes generate black skin by 21 d after grafting. Localization of melanocytes to the basal layer of keratinocytes (arrowheads) is confirmed by immuno histochemical staining with the melanocyte-specific marker, Mel-5. The outer surface of the stratum corneum is identified with asterisks (*). Scale bar, 0.1 mm.

Transplantation of Skin Substitutes

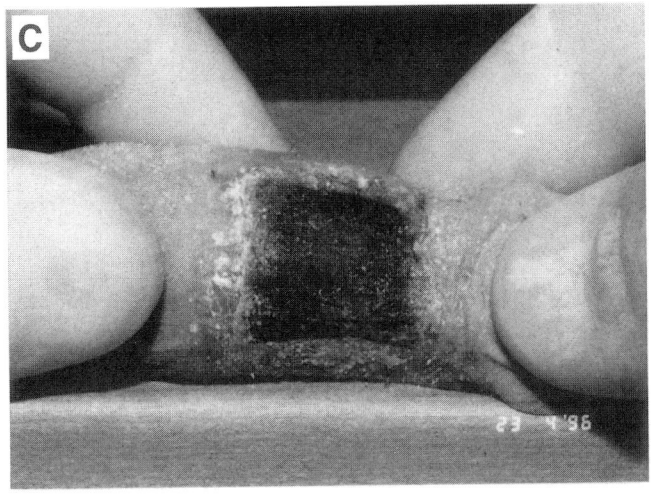

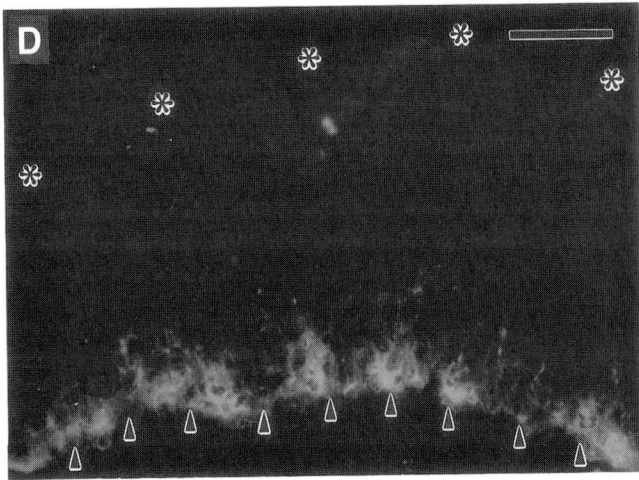

Fig. 5C and D.

6. Cryopreservation: Cryopreserve cells after primary culture in liquid nitrogen for long-term storage. If sufficiently large cell populations are propagated in primary culture, they may be distributed into aliquots and used for months or years. This simple step will assure continuous quality of results, and provide a performance standard for cellular responses to experimental conditions in preclinical studies.
7. Cell viability: Avoid repeated subculture before assay or surgery. All non-transformed cells have finite life spans, and cellular responses decline with time in culture.

4.3. CSS Preparation

8. Inoculate cells from log-phase cultures. Avoid confluence and density inhibition of keratinocyte cultures, especially. Density inhibition is not fully reversible, and, if keratinocytes become arrested, the epithelium of the skin substitute will be compromised.
9. Determine kinetics of optimum cellular viability and differentiation in vitro to predict the best surgical result. If skin substitutes are grafted too early, then stratification and barrier may not be optimal; if too late, then potential for cellular proliferation may be reduced.
10. Handle skin substitutes with a flexible, inert, nonadherent backing (i.e., N-terface polypropylene dressing) to reduce folding and increase mechanical strength during processing in vitro and surgical procedures.

4.4. Animal Surgery

11. Use a single gender of animals to reduce physiologic variations, and gender–gender interactions. Female athymic mice are very tolerant of handling and surgery.
12. Use animals 2–4 mo of age (young adults) to provide greatest vigor after surgery, and an observation period of more than 1 yr. Use animals from the same breeding lot for each experiment to avoid artifacts caused by age or other factors.
13. Be very careful to preserve the integrity and blood flow of the panniculus carnosus to optimize the vascularity of the woundbed for grafts. Although the panniculus carnosus is not required for engraftment of skin substitutes *(11)*, its presence increases the probability for engraftment.
14. Place chow cakes in the bedding of the cage for first 4 wk after surgery.
15. Check animals no less than daily while in dressings.

4.5. Conclusion

Preparation and grafting of cultured skin substitutes is a complex process, with high requirements for stringency to be successful consistently. However, the high stringency and special considerations for their routine use result from anatomic and physiologic deficiencies of the cultured grafts. Therefore, it may be expected that the stringency of preparative and grafting procedures for skin substitutes will decrease as graft composition becomes more homologous with native skin. Accomplishment of this goal will best serve all of the biomedical applications of cultured skin substitutes.

Acknowledgments

The author's work is supported by PHS grants GM50509 from the National Institutes of Health, and FD-R-672 from the United States Food and Drug Administration, and by grants from the Shriners Hospitals for Children.

References

1. Boyce, S. T., Glafkides, M. C., Foreman, T. J., and Hansbrough, J. F. (1988) Reduced wound contraction after grafting of full-thickness wounds with a collagen and chon-

droitin-6-sulfate (GAG) dermal skin substitute and coverage with Biobrane. *J. Burn Care Rehabil.* **94**, 364–370.
2. Boyce, S. T., Foreman, T. J., English, K. B., Stayner, N., Cooper, M. L., Sakabu, S., and Hansbrough, J. F. (1991) Skin wound closure in athymic mice with cultured human cells, biopolymers, and growth factors. *Surgery* **110**, 866–876.
3. Beumer, G. J., van Blitterswijk, C. A., Bakker, D., and Ponec, M. (1993) Cell-seeding and vitro biocompatibility evaluation of polymeric matrices of PEO/PBT copolymers and PLLA. *Biomaterials* **14(8)**, 598–604.
4. Doillon, C. J., Silver, F. H., and Berg, R. A. (1987) Fibroblast growth on a porous sponge containing hyaluronic acid and fibronectin. *Biomaterials* **8**, 195–200.
5. Krejci, N. C., Cuono, C. B., Langdon, R. C., and McGuire, J. (1991) In vitro reconstitution of skin: fibroblasts facilitate keratinocyte growth and differentiation on acellular reticular dermis. *J. Invest. Dermatol.* **97(5)**, 843–848.
6. Zieske, J. D., Mason, V. S., Wasson, M. E., Meunier, S. F., Nolte, C. J., Fukai, N., Olsen, B. R., and Parenteau, N. L. (1994) Basement membrane assembly and differentiation of cultured corneal cells: importance of culture environment and endothelial cell interaction. *Exp. Cell Res.* **214(2)**, 621–633.
7. Boyce, S. T. and Hansbrough, J. F. (1988) Biologic attachment, growth, and differentiation of cultured human epidermal keratinocytes on a graftable collagen and chondroitin-6-sulfate substrate. *Surgery* **103**, 421–431.
8. Morgan, J. R., Barrandon, Y., Green, H., and Mulligan, R. C. (1987) Expression of an exogenous growth hormone gene in transplantable human epidermal cells. *Science* **237**, 1476–1479.
9. Krueger, G. G., Morgan, J. R., Jorgensen, C. M., Schmidt, L., Li, H. L., Li, L. T., Boyce, S. T., Wiley, H. S., Kaplan, J., and Peterson, M. J. (1994) Genetically modified skin to treat disease: potentials and limitations. *J. Invest. Dermatol.* **103(5)**, 76s–84s.
10. Eming, S. A., Lee, J., Snow, R. G., Tompkins, R. G., Yarmush, M. L., and Morgan, J. R. (1995) Genetically modified himan epidermis overexpressing *PDGF-A* directs the development of a cellular and vascular connective tissue stroma when transplanted to athymic mice-Implications for the use of genetically modified keratinocytes to modulate dermal degeneration. *J. Invest. Dermatol.* **105**, 756–763.
11. Medalie, D. A., Eming, S. A., Tompkins, R. G., Yarmush, M. L., and Krueger, G. G. (1996) Evaluation of human skin reconstituted from composite grafts of cultured keratinocytes and human acellular dermis transplanted to athymic mice. *J. Invest. Dermatol.* **107(1)**, 121–127.
12. Griffey, E. S., Hueneke, M., Sukkar, S., Wainwright, D., and Livesey, S. A. (1995) Production of a human in vitro reconstituted skin and grafting to a nude mouse model. *Wound Rep. Reg.* **3(1)**, 92 (Abstract).
13. Hansbrough, J. F., Cooper, M. L., Cohen, R., Spielvogel, R. L., Greenleaf, G., Bartel, R. L., and Naughton, G. (1992) Evaluation of a biodegradable matrix containing cultured human fibroblasts as a dermal replacement beneath meshed skin grafts on athymic mice. *Surgery* **111(4)**, 438–446.
14. Cooper, M. L., Hansbrough, J. F., Spielvogel, R. L., Cohen, R., Bartel, R. L., and Naughton, G. (1991) In vivo optimization of a living dermal substitute employing

cultured human fibroblasts on a biodegradable polyglycolic or polygalactin mesh. *Biomaterials* **12**, 243–248.
15. Beumer, G. J., van Blitterswijk, C. A., and Ponec, M. (1994) Biocompatibility of a biodegradable matrix used as a skin substitute: an in vivo evaluation. *J. Biomed. Mater. Res.* **28(5),** 545–552.
16. Bell, E., Ehrlich, H. P., Buttle, D. J., and Nakatsji, T. (1981) A living tissue formed in vitro and accepted as a full thickness skin equilvalent. *Science* **211,** 1042–1054.
17. Bell, E., Ehrlich, H. P., Sher, S., Merrill, C., Sarber, R., and Hull, B. (1981) Development and use of a living skin equivalent. *Plast. Reconstr. Surg.* **67,** 386–392.
18. Germain, L., Michel, M., and Auger, F. A. (1993) Anchored skin equivalent cultured in vitro: a new tool for percutaneous absorption studies. *In Vitro Cell Dev. Biol.* **29A,** 834–837.
19. Hansbrough, J. F. and Boyce, S. T. (1984) What criteria should be used for designing artifical skin replacement and how well do the current materials meet these criteria. *J. Trauma* **24,** 31s–35s.
20. Boyce, S. T., Christianson, D. J., and Hansbrough, J. F. (1988) Structure of a collagen–GAG dermal skin substitute optimized for cultured human epidermal keratinocytes. *J. Biomed. Mater. Res.* **22,** 939–957.
21. Gallico III, G. G., O'Connor, N. E., Compton, C. C., Kehinde, O., and Green, H. (1984) Permanent coverage of large burn wounds with autologous cultured human epithelium. *New Engl. J. Med.* **311,** 448–451.
22. Cuono, C., Langdon, R., Birchall, N., Barttelbort, S., and McGuire, J. (1987) Composite autologous-allogeneic skin replacement: development and clinical application. Plast. Reconstr. Surg., 80, 626–635.
23. Pittelkow, M. R., and Scott, R. E. (1986) New techniques for the in vitro culture of human skin keratinocytes and perspectives on thier use for grafting of patients with extensive burns. *Mayo Clin. Proc.* **61,** 771–777.
24. Hansbrough, J. F., Boyce, S. T., Cooper, M. L., and Foreman, T. J. (1989) Burn wound closure with cultured autologous keratinocytes and fibroblasts attached to a collagen-glycosaminoglycan substrate. *JAMA* **262,** 2125–2130.
25. Tompkins, R. G., and Burke, J. F. (1996) Alternative wound coverings, in *Total Burn Care* (Herndon, D. N., ed.), W. B. Saunders, London, pp. 164–172.
26. Warden, G. D., Saffle, J. R., and Kravitz, M. (1982) A two-stage technique for excision and grafting of burn wounds. *J. Trauma* **22,** 98–103.
27. Tanner, J. C., Vandeput, J., and Olley, J. F. (1964) The mesh skin autograft. *Plast. Reconstr. Surg.* **34,** 287–292.
28. Boyce, S. T., Goretsky, M. J., Greenhalgh, D. G., Kagan, R. J., Rieman, M. T., and Warden, G. D. (1995) Comparative assessment of cultured skin substitutes and native skin autograft for treatment of full-thickness burns. *Ann. Surg.* **222(6),** 743–752.
29. Heimbach, D., Luterman, A., Burke, J. F., Cram, A., Herndon, D., Hunt, J., Jordon, M., McManus, W., Solem, L., Warden, G., and Zawacki, B. (1988) Artifical dermis for major burns; a multi-center randomized clinical trial. *Ann. Surg.* **208,** 313–320.
30. American Medical Association. (1993) The skin, in *Guides to Evaluation of Permanent Impairment* (Engelberg, A. L., ed.), American Medical Association, Chicago, IL, pp. 277–289.

31. Herndon, D. N. and Rutan, R. L. (1992) Comparison of cultured epidermal autograft and massive excision with serial autografting plus homograft overlay. *J. Burn Care Rehabil.* **13**, 154–157.
32. Gilpin, D. A., Barrow, R. E., Rutan, R. L., Broemeling, L., and Herndon, D. N. (1994) Recombinant human growth hormone accelerates wound healing in children with large cutaneous burns. *Ann. Surg.* **220(1)**, 19–24.
33. Boyce, S. T. and Ham, R. G. (1983) Calcium-regulated differentiation of normal human epidermal keratinocytes in chemically defined clonal culture and serum-free serial culture. *J. Invest. Dermatol.* **81, suppl 1**, 33s–40s.
34. Boyce, S. T. and Ham, R. G. (1985) Cultivation, frozen storage, and clonal growth of normal human epidermal keratinocytes in serum-free media. *J. Tiss. Cult. Meth.* **9**, 83–93.
35. Boyce, S. T. and Ham, R. G. (1986) Normal human epidermal keratinocytes, in *In Vitro Models for Cancer Research: Carcinomas of the Mammary Gland, Uterus, and Skin*, vol. 3 (Webber, M. M. and Sekeley, L. I., eds.), CRC, Boca Raton, FL, pp. 245–274.
36. Rheinwald, J. G. and Green, H. (1975) Formation of a keratinizing epithelium in culture by a cloned cell line derived from a teratoma. *Cell* **6**, 317–330.
37. Rheinwald, J. G. and Green, H. (1975) Serial cultivation of strains of human epidermal keratinocytes: the formation of keratinizing colonies from single cells. *Cell* **6**, 331–343.
38. Boyce, S. T. and Williams, M. L. (1993) Lipid supplemented medium induces lamellar bodies and precursors of barrier lipids in cultured analogues of human skin. *J. Invest. Dermatol.* **101**, 180–184.
39. Boyce, S. T., Medrano, E. E., Abdel-Malek, Z. A., Supp, A. P., Dodick, J. M., Nordlund, J. J., and Warden, G. D. (1993) Pigmentation and inhibition of wound contraction by cultured skin substitutes with adult melanocytes after transplantation to athymic mice. *J. Invest. Dermatol.* **100**, 360–365.
40. Boyce, S. T., Supp, A. P., Harriger, M. D., Greenhalgh, D. G., and Warden, G. D. (1995) Topical nutrients promote engraftment and inhibit wound contraction of cultured skin substitutes in athymic mice. *J. Invest. Dermatol.* **104(3)**, 345–349.
41. Cunliffe-Beamer, T. L. (1983) Biomethodology and surgical techniques, in *The Mouse in Biomedical Research*, vol. III (Foster, H. L., Small, J. D., and Fox, J. G., eds.), Academic, New York, NY, pp. 417–418.
42. Swope, V. B., Medrano, E. E., Smalara, D., and Abdel-Malek, Z. A. (1995) Long-term proliferation of human melanocytes is supported by the physiologic mitogens α-melanotropin, endothelin-1, and basic fibroblast growth factor. *Exp. Cell Res.* **217**, 453–459.
43. Boyce, S. T., Supp, A. P., Harriger, M. D., Pickens, W. L., Wickett, R. R., and Hoath, S. B. (1996) Surface electrical capacitance as a non-invasive index of epidermal barrier in cultured skin substitutes in athymic mice. *J. Invest. Dermatol.* **107(1)**, 82–87.
44. Serup, J., and Jemec, G. B. E. (1995) *Non-Invasive Methods and the Skin*, CRC, Boca Raton, FL.
45. Briggaman, R. A. (1985) Human skin grafts-nude mouse model: techniques and application, in *Methods in Skin Research* (Skerrow, D. and Skerrow, C. J., eds.), Wiley, New York, pp. 251–276.
46. Vijayasarahdi, S. and Houghton, A. N. (1991) Purification of an autoantigenic 75-kDa human melanosomal glycoprotein. *Intl. J. Cancer* **47(2)**, 298–303.

29

Use of Skin Equivalent Technology in a Wound Healing Model

Michael A. Vaccariello, Ashkan Javaherian, Nancy Parenteau, and Jonathan A. Garlick

1. Introduction

Re-epithelialization is defined as the reconstitution of cells into an organized, stratified squamous epithelium that permanently covers a wound defect and restores function *(1)*. Following wounding, keratinocytes are activated to undergo a series of phenotypic changes that have been well-characterized in vivo *(2–4)*. However, in vitro studies of re-epithelialization have often been limited by their inability to simulate the in vivo tissue. Wound models using skin explants *(5–8)* or submerged keratinocyte cultures *(9,10)* demonstrate only partial differentiation and hyperproliferative growth. These systems have been useful for studying keratinoctye migration *(11)*, but are limited in studying other aspects of re-epithelialization.

The development of novel models to study wound healing and re-epithelialization relies on the ability to engineer a tissue that mimics its in vivo counterpart. The skin equivalent (SE) is an in vitro tissue that consists of a stratified squamous epithelium grown on a collagen matrix populated with dermal fibroblasts (**Fig. 1**). The tissue-like character of this construct generates a three-dimensional, organotypic culture, demonstrating epithelial differentiation, morphology, and proliferation rates similar to that found in skin *(12,13)*. In this chapter, we describe techniques for construction of a model for wound healing in vitro that has been generated by adapting SEs *(14)*. Using this system, we have found that the chronology of events during re-epithelialization is similar to that reported in skin during wound healing *(15)*. Migration is initiated by keratinocytes at the wound margin, which form an epithelial tongue as progeny of proliferating cells move laterally to cover the wound defect (arrows, **Fig.**

From: *Methods in Molecular Medicine, Vol. 18: Tissue Engineering Methods and Protocols*
Edited by: J. R. Morgan and M. L. Yarmush © Humana Press Inc., Totowa, NJ

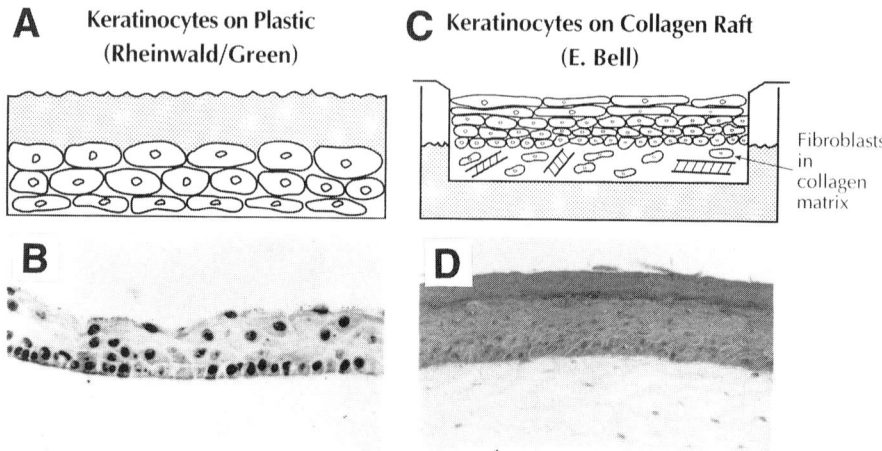

Fig. 1. Comparison of cultivation in submerged keratinocyte culture versus skin equivalent culture.(**A**) keratinocytes are cultured while submerged in medium on supporting feeder layers of mitotically inactive, metabolically active 3T3 fibroblasts. The resultant cultures are partially stratified, yet do not undergo full biochemical and morphologic differentiation (**B**). Alternatively, keratinocytes can be cultured on a collagen matrix containing fibroblasts while being fed at the air–liquid interface (**C**). Cultivation at this air–liquid interface results in skin equivalents demonstrating complete morphologic differentiation and cultures that more closely simulate the in vivo tissue (**D**).

2A). Proliferative activity continues to be high in the wound center as the epithelium stratifies to normal thickness (**Fig. 2B**). Using this model, the effect of growth factors, such as TGF-β1 *(16)*, and expression of matrix metalloproteinases *(17)* have been characterized during re-epithelialization of skin and oral mucosal keratinocytes. These applications demonstrate the potential of this model in studying cell phenotypes characteristic of the switch from a normal to a regenerative tissue during wound healing.

Since high-quality SEs provide the basis for the generation of the wound healing model, this chapter will first describe materials and methods needed in SE construction. Following this, the specific adaptation of these cultures to study wound healing will be detailed. It is hoped that by describing a model that recapitulates events occurring during re-epithelialization of wounds in vivo, further study of the nature and fate of keratinocytes mobilized during re-epithelialization may be facilitated.

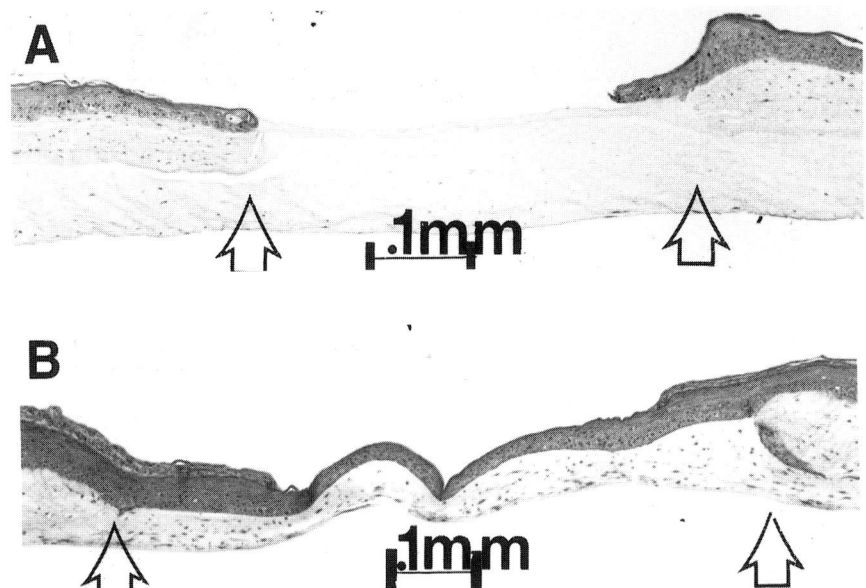

Fig. 2. Histologic appearance of skin equivalents during re-epithelialization. Skin equivalent cultures were wounded 3 d after being raised to the air–liquid interface. Eight hours after wounding, an epithelial tongue is seen to extend onto the central wound surface (**A**). By 48 h after wounding, the wound is completely covered by a stratified epithelium (**B**). Arrows mark the wound margins.

2. Materials
2.1. Submerged Cultures
2.1.1. Submerged Keratinocyte Culture Media (KCM)

1. Source of human keratinocytes: newborn foreskins (average size: 1–2 cm^2).
2. Keratinocyte tissue-culture medium:
    a. Dulbecco's modified Eagle's medium (DMEM):Ham's F12 medium (Gibco-BRL, Gaithersburg, MD) (3:1).
    b. Fetal bovine serum (FBS): 5% (HyClone, Logan, UT).
        1. Penicillin–streptomycin (Sigma, St. Louis, MO): Dissolve 2.42 g penicillin and 4.0 g streptomycin in 400 mL 2X dH$_2$O to make a 100X stock, filter-sterilize, aliquot and store at –20°C.
        2. HEPES (Sigma): Dissolve 47.24 g in 250 mL 2X dH$_2$O to make a 100X stock. Filter-sterilize, aliquot, and store at –20°C.
        3. Adenine (ICN): Dissolve 0.972 g in 2.4 mL of 4 N NaOH; bring vol to 400 mL with 2X dH$_2$O to make a 100X stock (18 mM), filter-sterilize, aliquot, and store at –20°C.

4. Cholera toxin (ICN, Costa Mesa, CA): Dissolve 1 mg in 1 mL 2X dH_2O, add this to 90 mL DMEM with 10 mL FBS to make a 1000X stock (1.2×10^{-7} M), filter-sterilize, aliquot, and store at $-20°C$.
5. Epidermal growth factor (#GF-010-9, Austral Biologicals, San Ramon, CA): Dissolve 10 µg/mL in 0.1% bovine serum albumin (BSA) to make a 1000X stock, filter-sterilize, aliquot, and store at $-20°C$.
6. Hydrocortisone (Sigma): Dissolve 0.0538 g in 200 mL 2X dH_2O to make a 500X stock (7.4×10^{-4} M), filter-sterilize, aliquot, and store at $-20°C$.
7. Insulin (Sigma): Dissolve 50 mg in 10 mL of 0.005 N HCl to make a 1000X stock (5 mg/mL), filter-sterilize, aliquot, and store at $-20°C$.

2.1.2. Fibroblasts for Submerged Keratinocyte Culture (3T3 Feeder Layers)

1. Irradiated 3T3-J2 cells are used as feeder cells.
2. 3T3 media:
   a. DMEM (Gibco-BRL).
   b. Bovine calf serum (BCS) 10% (HyClone).
   c. Penicillin–streptomycin (Sigma): as in **Subheading 2.1.1., step 2.b.1**.
   d. HEPES (Sigma): as in **Subheading 2.1.1., step 2.b.2**.

2.2. Skin Equivalent Cultures

2.2.1. Fibroblasts for Skin Equivalent Cultures

1. Source of human fibroblasts: newborn foreskin (average size: 1–2 cm^2).
2. Isolation of foreskin fibroblasts:
   a. Collagenase buffer: 130 mM NaCl, 10 mM Ca acetate, 20 mM HEPES. Solution adjusted to pH 7.2 and filter-sterilized.
   b. Dispase: Make 10X stock by dissolving 5 g in 200 mL ddH_2O, filter-sterlilize, and store $-20°C$.
   c. Collagenase (Worthington Biochemical, Freehold, NJ): 3 mg/mL in collagenase buffer.
   d. Trypsin: Prepared as a 1% stock by mixing 10X PBS (200 mL), 2 g glucose, 2 g trypsin, 0.2 g penicillin, and 0.2 g streptomycin. These components are dissolved in the PBS, then brought to 2 L with ddH_2O, filter-sterilized. Aliquot and store at $-20°C$.
   e. EDTA: Prepared as a 0.2% stock by dissolving 2 g in 1 L PBS and adjusting the pH to 7.45 with 4N NaOH. Autoclave and store at room temperature.
   f. PBS: Prepared as a 10X stock by dissolving 1.6 kg NaCL, 165 g $Na2HPO_4$, 40 g KH_2PO_4 and 40 g KCl in 20 L ddH_2O.
3. Fibroblast culture medium (FCM):
   a. DMEM (Gibco-BRL).
   b. Fetal bovine serum 10% (HyClone).
   c. Penicillin–streptomycin (Sigma): as in **Subheading 2.1.1., step 2.b.1**.
   d. HEPES (Sigma): as in **Subheading 2.1.1., step 2.b.2**.

Skin Equivalent Wound Healing Model

2.2.2. Collagen Matrix

1. A confluent culture of human foreskin fibroblasts (HFF).
2. 6-well deep tissue-culture tray containing special resin tissue-culture inserts with a 3-µm porous polycarbonate membrane (Organo Genesis, Canton, MA).
3. Sterile bovine tendon or rat tail acid-extracted collagen (1.0–1.3 mg/mL) in 0.05% acetic acid.
4. 10X minimum essential medium with Earle's salts (#12-684F, BioWhittaker, Walkersville, MD).
5. Newborn calf serum (HyClone).
6. L-glutamine (200 mM) (#17-605E, BioWhittaker).
7. Sodium bicarbonate (71.2 mg/mL).
8. Fibroblast culture medium.

2.2.3. Skin Equivalent Culture Media

1. DME base modified (#56430-10L, JRH Biosciences, Lenexa, KS).
2. Ham's F12 (Gibco-BRL).
3. L-glutamine (BioWhittaker) 200 mM is a 50X stock.
4. Hydrocortisone (Sigma).
5. ITT (Sigma) 500X stock:
   a. Bovine insulin: Dissolve in 0.0001 N HCl to a final concentration of 5 µg/mL.
   b. Human transferrin: Dissolve in ddH$_2$O to a final concentration of 5 µg/mL.
   c. Triiodothyronine: dissolve in acidified ethanol, dilute with double-distilled dH$_2$O to a final concentration of 20 pM.
6. EOP (Sigma) 500X stock:
   a. Ethanolamine: Reconstitute with ddH$_2$O to 10^{-4} M.
   b. O-phosphorylethanolamine: Dilute with ddH$_2$O to 10^{-4} M.
7. Adenine (Sigma): Dissolve 0.18 mM (500X stock) in acidified water warmed in a 37°C water bath.
8. Selenious acid (Aldrich, Milwaukee, WI): Dissolve 5.3×10^{-8} M (500X stock) with ddH$_2$O.
9. Calcium chloride: Dissolve 132.5 µg/mL (500X stock) in ddH$_2$O.
10. Progesterone: 2 nM solution is a 500X stock.
11. Serum (HyClone):
    a. Chelated newborn calf serum (cNBCS).
    b. Newborn calf serum (NBCS).

2.3. Wounding of Skin Equivalents

1. Sterile bovine tendon or rat tail acid-extracted collagen (1.0–1.3 mg/mL) in 0.05% acetic acid.
2. Sterile scalpel (gage 22), long forceps, and dental mirror.
3. Sterile 100 mm tissue-culture dish.

3. Methods
3.1. Submerged Cultures
3.1.1. Submerged Keratinocyte Cultures

Preparation of primary cultures of keratinocytes from foreskins can be followed, as previously described by Eming and Morgan *(18)*.

3.1.2. Fibroblasts for Submerged Keratinocyte Culture (Feeder Layers)

The clonal growth of keratinocytes requires co-cultivation with a metabolically active, nonproliferating feeder layer of 3T3-J2 fibroblasts. Grow 3T3-J2 fibroblasts in DME*M* containing 10% bovine calf serum, HEPES, and penicillin–streptomycin. When cells are 90% confluent, they are trypsinized, pelleted with centrifugation at 750g for 5 min, and irradiated by a gamma source of 2,000 Ci (Cs-137, 100% = 1,215 R/min) for 6.5 min. Irradiated cells are then plated at a density of 2×10^6/p100 in KC*M* before keratinocytes are added.

3.2. Culture of Skin Equivalents
3.2.3. Isolation of Dermal Fibroblasts

Fibroblasts are first isolated from human foreskins by using the connective tissue remnant after dispase separation of the epithelium used for submerged culture (*see* **Subheading 3.1.1.**). The connective tissue is then rinsed twice in PBS and placed in a 15-mL conical tube with 1 mL of collagenase in collagenase buffer. This mixture is incubated at 37°C for 30 min and is agitated every 5 min. Trypsin:EDTA mixed in a 1:1 ratio is added for 10 min at 37°C, at which time 1 mL of FC*M* is added to inactivate the trypsin, and cells are counted. These fibroblasts are grown so that they are densely confluent 1 d before the collagen matrix is to be cast. At this time, passage the cells at high density, so they will regrow to full confluence the next day, when they will be added to the collagen matrix. This extra passage ensures that a high fraction of fibroblasts are proliferating at the time of initiation of matrix construction (*see* **Note 2**).

3.2.2. Construction of the Collagen Matrix (**Fig. 3**, 1–5)

Mix the following components on ice, to generate an acellular and cellular layer for the SE. The goal is to create a thin layer of acellular collagen that will act as a substrate for the thicker layer of cellular collagen. This will prevent the cellular collagen from contracting completely from the insert.

1. Keeping all components on ice, mix the acellular matrix components in the order listed in **Table 1**. The color of the solution should be from straw-yellow to light pink, and any extreme variations in color may indicate a pH at which the collagen may not gel. If the final solution is bright yellow, slowly titrate sodium bicarbon-

Skin Equivalent Wound Healing Model

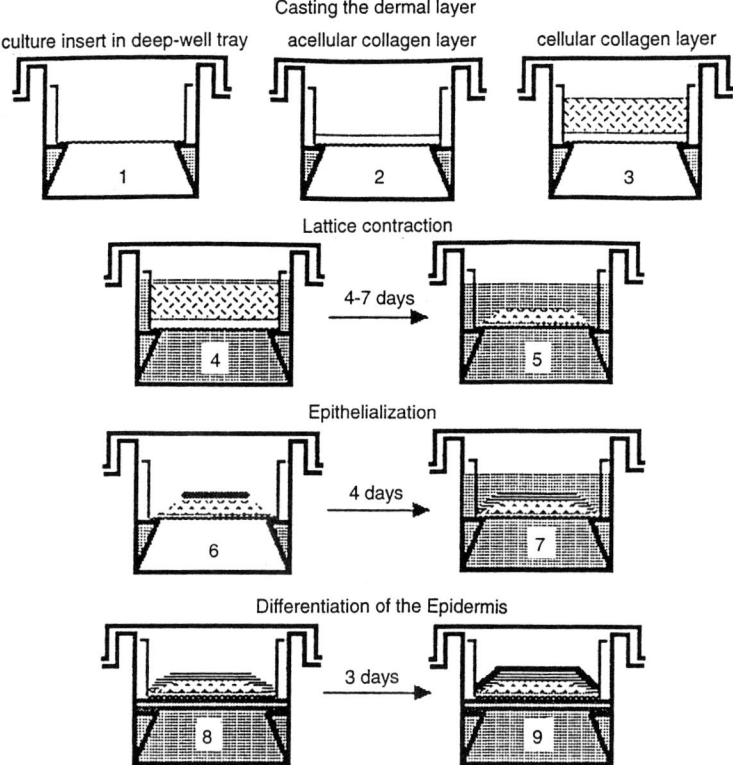

Fig. 3. Diagrammatic outline of skin equivalent protocol. 1. Culture insert in special deep well tray. 2. A 1-mL layer of acellular collagen is cast over the polycarbonate membrane of the insert. 3. A 3-mL cellular collagen layer is cast onto the acellular gel. 4, 5. The collagen gel contracts away from the sides of the insert, to form a contracted collagen lattice having a central raised area. 6. A suspension of epidermal keratinocytes is seeded onto the central area. 7. The epithelial sheet is allowed to develop submerged for 4 d. 8, 9 The developing culture is raised to the air–liquid interface and cultured for an additional 3 d until wounded.

ate drop by drop, until the appropriate color is noted. Add 1 mL to each insert, making sure the mixture coats the entire bottom of the insert. Once the gel has been poured, it should stand at room temperature, without being disturbed, until it polymerizes (10–15 min). As the gel polymerizes, the color of the matrix will change to a deeper pink color (**Fig. 3**, *2*).

2. While the acellular matrix layer is polymerizing, trypsinize and resuspend fibroblasts in FCM to a final concentration of 3.0×10^5 cells/mL. Resuspend these cells fully, since it will be harder to resuspend them properly once mixed with collagen.

Table 1
Components for the Collagen Matrix

	Acellular matrix for 6 mL (1 mL/insert)	Cellular matrix for 18 mL (3 mL/insert)
10X DMEM	0.59 mL	1.65 mL
L-Glutamine	0.05 mL	0.15 mL
Fetal Bovine serum	0.6 mL	1.85 mL
Sodium Bicarbonate	0.17 mL	0.52 mL
Collagen	4.6 mL	14 mL
Fibroblasts	—	4.5×10^5 cells in 1.5 mL fibroblast media

3. For the cellular matrix, again keep all components on ice, and mix in the order indicated in **Table 1**. Again, if the color needs to be adjusted, carefully titrate in a small amount of sodium bicarbonate (the acellular collagen layers can thus serve as a useful test of gelation conditions). The fibroblasts should be added last, to ensure that the mix has been neutralized by the addition of collagen, so that the cells are not damaged by an alkaline pH. Mix well and add 3 mL to the insert and allow it to gel at room temperature without disturbing (30–45 min) (**Fig. 3**, *3*). When the gels are pink and firm, they are covered with 12 mL of FC*M* (~10 mL in the well and 2 mL on top of the insert) and incubated for 4–7 d, until gel contraction is stable and complete. During this interval, there is a 50-fold decrease in the volume of the matrix, and considerable vertical shrinkage of the matrix is seen. A raised, mesa-like area is seen in the center of the matrix, and it is here that keratinocytes will be seeded (**Fig. 3**, *3–5*).

3.2.3. Growth of Keratinocytes on the Skin Equivalents

Keratinocytes are added to the collagen matrix as follows:

1. Use keratinocyte cultures when they are no more than 50% confluent. This is to minimize the number of differentiated cells plated on the matrix (*see* **Notes 1** and **5**). Thoroughly remove the 3T3 feeder cells from the culture by incubation in PBS–EDTA (10:1 ratio) for 2 min, and then remove 3T3s by gentle rinsing. Take care not to leave the culture too long in PBS–EDTA, because small keratinocyte colonies may detach. Rinse with PBS until the 3T3s have been removed.

2. Trypsinize the keratinocyte colonies with Trypsin–EDTA (10:1 ratio) for 5–10 min and add cells to a 15-mL tube, so that 5×10^5 keratinocytes will be plated in each insert. Centrifuge the cells at 750*g* for 5 min.

3. Remove all fibroblast media from the inserts containing the contracted collagen matrices.

4. Resuspend the keratinocytes so that they can be plated in a small volume, so that each insert will receive 50 µL of suspension containing 5×10^5 keratinocytes.

Table 2
Media Formulation for growth of SEs[a]

	Epidermalization I	Epidermalization II
DME	725	725
F12	240	240
L-glutamine	20	20
Hydrocortisone	2	2
ITT	2	2
EOP	2	2
Adenine	2	2
Selenium	2	2
$CaCl_2$	2	2
Progesterone	2	2
Serum	1 (cNBCS)	1 (NBCS)

[a]All volumes are in mL.

This can be done by eliminating all residual supernatant above the pellet. Use a sterile plastic 1-mL pipet with the appropriate volume of epidermalization I media to lift the pellet and transfer it to a sterile Eppendorf tube. Then use a 200- or 1000-µL pipetman to gently resuspend the cell pellet in the Eppendorf tube. The cell suspension can then be placed in the central, raised, mesa-like portion of the contracted collagen gel with a 200-µL pipetman (**Fig. 3, 6**).

5. Do not touch the plate for 1.5–2 h, while the keratinocytes adhere. At this point, add 12 mL of epidermalization I media (~10 mL in the well and 2 mL on top of the keratinocytes), and incubate cultures at 37°C, 7% CO_2 (**Fig. 3, 7**).
6. Cultures are fed for 7 d in the following way (media formulations seen in **Table 2**):
   a. Epidermalization I media: Submerged in 12 mL, first 2 d.
   b. Epidermalization II media: Submerged in 12 mL, next 2 d.
   c. Epidermalization II media: Grown at the air–liquid interface by feeding with 7 mL for 3 d (**Fig. 3, 8** and **9**).

The optimization of SE growth will vary with the strain of keratinocyte and fibroblast used (*see* **Note 2**).

3.3. Wounding of Skin Equivalents

SEs are wounded 7 d after the keratinocytes are seeded onto the collagen matrix. The optimal time of wounding may need to be determined (*see* **Note 3**). Six days before cultures are to be wounded, an additional collagen matrix is fabricated according to **Subheading 3.2.2**. This will be used as the substrate onto which the wounded SE will be transferred.

1. Aspirate all media from the SE, remove the insert from the 6-well plate, and place it upside down onto a sterile 100 mm dish.

2. Use the scalpel to cut out the entire polycarbonate membrane around the periphery of the insert. Place the cut-out SE onto a 100 mm dish, so that it rests on its polycarbonate membrane.
3. Trim the culture with the scalpel several mm from its raised, mesa-like region. This removes parts of the SE without keratinocytes (since keratinocytes are initially only seeded on the center mesa of the matrix) and facilitates removal of the culture from the insert membrane for transfer.
4. Place the scalpel's edge directly in the center of the culture and rock the blade back and forth, in order to create an incision 1.2 cm in length. This incision completely penetrates the epidermis, collagen matrix, and membrane (**Fig. 4A**). The culture should not be completely cut in half, and the two sides are to be kept attached.
5. At this point, place 20–30 µL of unpolymerized collagen onto the second collagen matrix that the culture is to be transferred to (*see* **Note 4**). This is to promote adherence of the two matrices, and to prevent the keratinocytes from migrating between the two matrices.
6. Use the forceps to gently lift the edge of the culture, so that it separates from the polycarbonate membrane. At this time, bring the dental mirror, which will serve as a spatula, close to the culture and drag the culture onto the mirror with a long forceps, leaving the membrane behind. The transfer will be facilitated if the mirror is slightly moistened with media.
7. Unfold any wrinkles in the culture by moving it back and forth on the mirror, using the forceps. Once the culture is smooth, pull one side of the culture slightly over the edge of the mirror; this is the site where the culture will be transferred onto the new matrix.
8. Bring the mirror directly over the second collagen matrix and lower it inside the insert, so that the edge of the mirror and the culture are in contact with the matrix (*see* **Note 7**). Slide the culture onto the collagen, using the forceps, as the mirror is pulled away slowly, leaving the culture on the collagen matrix (**Fig. 4B**).
9. Using the forceps, tease apart the incision in order to create an elliptical space between the two halves. This elliptical space will be 2–3 mm at its greatest width. It is important that the transferred culture is kept completely free of any folds or wrinkles.
10. Maintain the culture at air–liquid interface by adding 7 mL of epidermilization II media to the outer well (*see* **Note 6**). Incubate at 37°C with 7% CO_2, and change the media every 2 d until the end of the experiment (**Fig. 4C,D**).

An example of a wounded culture 3 d after wounding is seen in **Fig. 5**.

3.4. Processing of Skin Equivalents

Skin equivalents are fragile if handled improperly and require special care during processing. The following are steps that should ensure that samples can be cryosectioned without artifactual damage:

Skin Equivalent Wound Healing Model

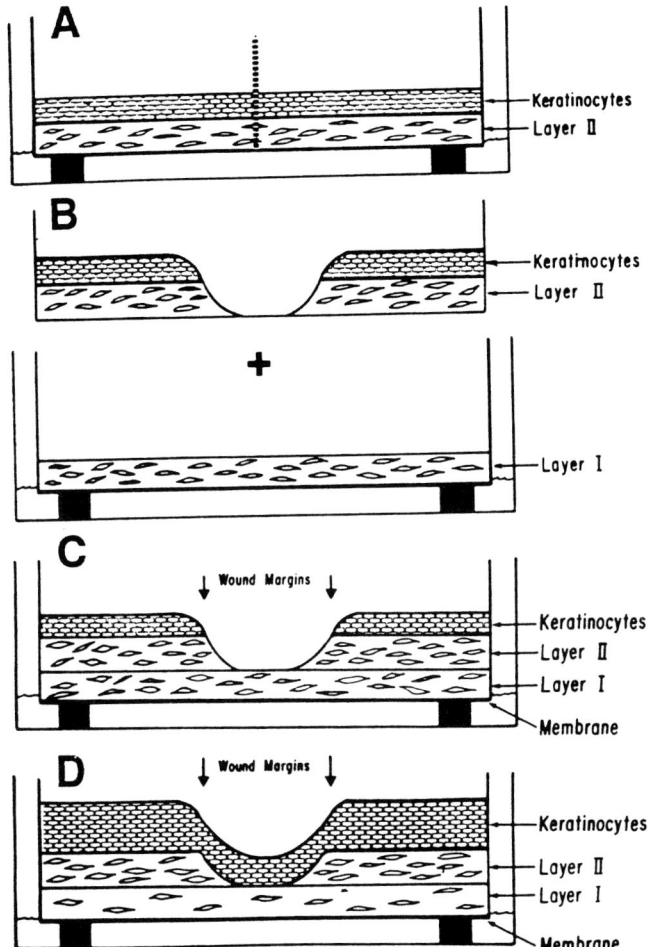

Fig. 4. Construction of composite organotypic coculture wounding model. (A) Schematic of stratified keratinocyte sheet growing on contracted collagen matrix containing fibroblasts (layer II). This skin equivalent has been cultured at the air–liquid interface for 3 d. A wound is formed by incising this culture (dotted line). (B) The wounded culture is then transferred onto a second contracted collagen matrix, which has been prepared 6 d in advance (layer I). (C) The resultant wounded skin equivalent consists of two layers of contracted matrix and one layer of epithelium. Wound margins are seen at the transition zone from layer I to layer II, and are noted with arrows. (D) Following re-epithelialization, stratified epithelium covers the wound bed. Reprinted with permission from **ref. 14**.

1. Cultures are rinsed twice in PBS, and SEs are separated from the plastic culture insert with a scalpel blade. Incision should include the polycarbonate membrane,

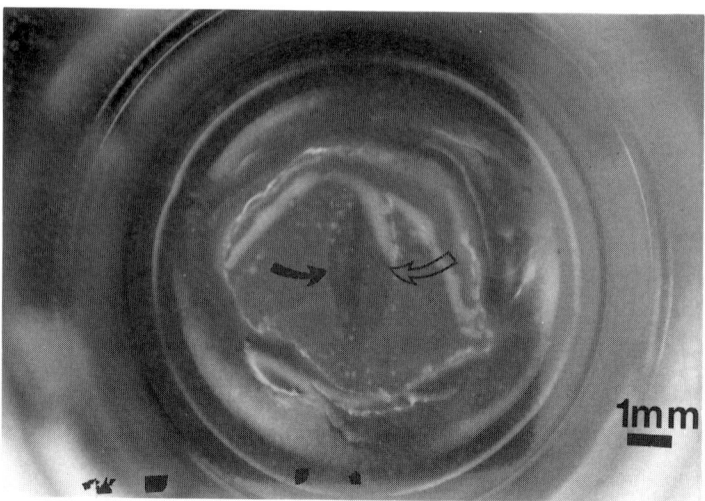

Fig. 5. Skin equivalent several days after wounding. A skin equivalent was wounded and transferred to a second collagen gel, as described. The original wound margins are marked by arrows.

because specimens will be easier to handle if still on the membrane. At this point, the wound can be sectioned to provide equal halves for formalin and frozen processing.
2. Immerse the SE for frozen section analysis in 1 M sucrose at room temperature for 1 h. This will remove some of the water from the collagen matrix and prevent fracture artifact upon freezing. After removal from sucrose, the SE should be noticeably firmer (should not flop over when standing on edge). As an alternative, SEs may be kept overnight at 4°C in 2 M sucrose, but incubation for any longer than this may result in the culture becoming somewhat brittle and difficult to cut.
3. A slow-freezing technique, in which the sample is frozen in liquid nitrogen vapor, is preferred, rather than snap-freezing in liquid nitrogen. This can be accomplished by using a three-quarter inch bottle cap as a template to make a mold with aluminum foil. Fill the mold with embedding compound (Tissue Freezing Medium [TBS]) to a thickness of one-half inch and place it on a test tube rack into a styrofoam box filled with liquid nitrogen just below the upper surface of the rack.
4. In order to embed the tissue specimen on edge, the embedding compound must be cooled to increase its viscosity. Place the aluminum mold on the rack so that it starts to chill in the liquid nitrogen vapor. Within 30 s, the embedding material turns white near its edges, and should be sufficiently viscous in the center to allow the culture to be stood on edge (The TBS embedding compound is more viscous than most, and is well-suited for this purpose). If necessary, hold the SE upright with a forceps. The SE will freeze within several minutes and specimens can be placed in a pillbox for storage at –70°C.

4. Notes

1. Optimization of SEs is essential in generating a responsive organotypic wounding model, and can be accomplished by assuring the presence of keratinocytes, which have a high growth potential in the SE. Most keratinocytes plated onto the collagen matrix will adhere well, but only cells capable of replication will grow after plating. Keratinocytes that have undergone terminal differentiation while in submerged culture will not generate a well-stratified SE, and will not re-epithelialize after wounding. It is therefore important to grow keratinocytes, so that a high growth fraction is present in the cultures at the time of passage to SEs. By growing keratinocytes as small colonies at high clonal density in submerged cultures on 3T3 feeder layers, terminal differentiation will be minimized and the fraction of replicating cells will be optimized. Alternatively, this can also be accomplished by growing keratinocytes in a low calcium media before passage to organotypic culture *(13)*.
2. We have found that optimal growth of SEs varies with different keratinocyte and fibroblast strains used. Keratinocyte strains should be tested to determine those providing the best morphologic differentiation of the SE. It is also apparent that the rate of re-epithelialization varies with the keratinocyte strain used. In addition, we have found great variability in the degree to which fibroblast strains support keratinocyte growth. This appears to be related to the degree to which fibroblasts are able to contract the collagen matrix. Fibroblast strains demonstrating more collagen gel contraction are better able to support keratinocyte growth, and may be used to determine fibroblast strains for optimal SE growth.
3. Another variable that needs to be determined for each keratinocyte strain is the time after airlift at which SEs should be wounded. We find that culture at the air–liquid interface for longer than 10 d results in an overly thickened stratum corneum, a flattened basal cell layer, and a failure to respond to growth stimuli. It is important that cultures not demonstrate this morphology at the time of wounding. In our hands, cultures grown at the air–liquid interface for 3–4 d provide the most responsive tissue for wounding. At this point in its maturation, the epithelium is well-stratified, demonstrates a thin stratum corneum and cuboidal basal cells, and shows indices of basal cell proliferation that approximate those in skin.
4. The small amount (20–30 µL) of nonpolymerized collagen added to the lower collagen matrix at the time of transfer of the wounded construct serves two purposes. First of all, it acts as a biological glue that promotes adherence of the two collagen layers. Secondly, it prevents the vertical migration of the epithelium between the two layers of collagen after transfer. Only when keratinocyte migration is limited to the wound surface do assays for keratinocyte phenotype during re-epithelialization become meaningful. For example, if keratinocytes migrated between the two layers of collagen, it would be impossible to measure proliferation of keratinocytes at the wound margin.
5. Several points regarding some subtleties of keratinocyte behavior in SEs are worth mentioning. The first concerns keratinocyte growth potential. Although SEs demonstrate a basal level of proliferation that is similar to skin, these cul-

tures have tremendous potential for proliferation, and are very responsive to external growth stimuli. Disruption of the cultures by wounding can result in a 10-fold increase in basal cell proliferation *(16)*, and addition of growth factors can modify proliferation, as well *(17)*.

Secondly, it should be kept in mind that, although keratinocytes grown in SEs share many features in common with skin keratinocytes, they do differ somewhat in phenotype. For example, integrin receptors not normally expressed in skin are constituitively expressed in keratinocytes. In addition, SEs are somewhat deficient in barrier function. The phenotype of keratinocytes grown in SEs have been compared to a newly re-epithelialized wound in vivo, in which morphologic differentiation is complete, but cells are still in a somewhat activated state.

6. An advantage of the SE wounding model is the ability to add soluble reagents directly to the cultures during re-epithelialization. This has been useful in following the effects of soluble mediators, such as TGF-β, on re-epithelialization, as well as on fibroblast and keratinocyte phenotype. Addition of reagents that assist in analysis of cell phenotype during re-epithelialization, such as a pulse of BrdU to determine proliferation indices, can also be added directly to media and incubated with cultures.

7. Perhaps the most difficult technical aspect of the model is the skill in transferring the wounded organotypic construct to the second collagen bed. An optimal SE construct has good tensile properties and should not tear during the transfer. The key lies in the manipulation of the dental mirror, which is used as a spatula to pick up and deliver the construct. This mirror has proved useful because of the size of its face and the angle of the bend in its handle. When the mirror is slightly moistened, the wounded construct can be pulled directly onto the mirror from its membrane. To deliver the construct, hold the face of the mirror at a 45-degree angle to the surface of the second matrix in its insert. First, allow the edge that is slightly overhanging from the mirror to touch the second matrix. Then, while holding the culture with a forceps at this area of contact, the mirror can be slowly removed from under the construct.

References

1. Clark, R. A. F. (1996) Wound repair: overview and general considerations, in *The Molecular and Cellular Biology of Wound Repair* (Clark, R. A. F., ed.) Plenum, New York, pp. 1–50.
2. Hertle, M. D., Kubler, M.-D., Leigh, I. M., and Watt, F. M. (1992) Aberrant integrin expression during epidermal wound healing and in psoriatic epidermis. *J. Clin. Invest.* **89**, 1892–1901.
3. Viziam, C. B., Maltotsy, A. G., and Mescon, H. (1964) Epithelialization of small wounds. *J. Invest. Dermatol.* **43**, 499–507.
4. Mansbridge, J. N. and Knapp, A. M. (1987) Changes in keratinocyte maturation during wound healing. *J. Invest. Dermatol.* **89**, 253–263.
5. Freeman, A. E., Eigel, H. J., Herman, B. J., and Kleinfeld, K. L. (1976) Growth and characterization of human skin epithelial cell cultures. *In Vitro* **2**, 352–358.

6. Stenn, K. S. (1978) The role of serum in the epithelial outgrowth of mouse skin explants. *Br. J. Dermatol.* **98,** 411–416.
7. Marks, S. and Nishikawa, T. (1973) Active epidermal movement in human skin in vitro. *Br. J. Dermatol.* **88,** 245–248.
8. Hintner, H., Fritsch, P. O., Foidart, J-M., Stingl, G., Schuler, G., and Katz, S. I. (1980) Expression of basement membrane zone antigens at the dermo-epibolic junction in organ cultures of human skin. *J. Invest. Dermatol.* **74,** 200–204.
9. Stenn, K. S., Madri, J. A., Tinghitella, T., and Terranova, V. P. (1983) Multiple mechanisms of dissociated epidermal cell spreading. *J. Cell Biol.* **96,** 63–67.
10. Stenn, K. S. and Milstone, L. M. (1984) Epidermal cell confluence and implications for a two-step mechanims of wound closure. *J. Invest. Dermatol.* **83,** 445–447.
11. Woodley, D. T., O'Keefe, E. J., and Prunieras M. (1985) Cutaneous wound healing: a model for cell-matrix interactions. *J. Am. Acad. Dermatol.* **12,** 420–433.
12. Bell, E., Ehrlich, H. P., Buttle, D. J., and Nakatsuji, T. (1981) Living tissue formed in vitro and accepted as skin-equivalent tissue of full thickness. *Science* **211,** 1052–1054.
13. Parenteau, N. (1994) Skin equivalents, in *The Keratinocyte Handbook*, vol. II) (Leigh, I. M. and Watt, F. W., eds.), Cambridge University Press, pp. 45–56.
14. Garlick, J. A. and Taichman, L. B. (1994) The fate of human keratinocytes during re-epithelialization in an organotypic culture model. *Lab. Invest.* **70,** 916–924.
15. Winter, G. D. (1972) Epidermal regeneration studied in the domestic pig, in *Epidermal Wound Healing* (Maibach, H. I. and Rovee, D. T., eds.), Year Book, Chicago, pp. 71–1112.
16. Garlick, J. A. and Taichman, L. B. (1994b) Effect of TGF-β1 on re-epithelialization of human keratinocytes in vitro: an organotypic model. *J. Invest. Dermatol.* **103,** 554–559.
17. Garlick, J. A., Parks, W. C., Welgus, H. G., and Taichman, L. T. (1996) Re-epithelialization of oral keratinocytes in vitro. *J. Dental Res.* **75,** 912–918.
18. Eming, S. A. and Morgan, J. R. (1996) Methods for the use of genetically modifed keratinocytes in gene therapy, in *Methods in Molecular Medicine, Gene Therapy Protocols* (Robbins, P., ed.), Humana, Totowa, NJ.

30

Preparation and Transplantation of a Composite Graft of Epidermal Keratinocytes on Acellular Dermis

Daniel A. Medalie and Jeffrey R. Morgan

1. Introduction

Loss of skin because of burns or ulcers is a major medical problem, and is the impetus for the development of skin substitutes and skin replacement technologies. Efforts in this area have focused on developing suitable substitutes for the epidermis or for the dermis, as well as ways to combine both technologies in a composite skin. In this chapter, we describe the construction of a skin substitute of cultured human keratinocytes on acellular human dermis.

Among the first efforts in this field was the development of a replacement for the epidermis by culturing keratinocytes, the primary cells of the epidermis (1). When grown to confluence, these cells form multicell layered epithelial sheets which can be detached from the culture dish, secured to gauze and grafted to skin defects (1). Several clinical trials have tested the efficacy of these epithelial sheets, as epidermal replacements for burns and ulcers (2,3). Despite scattered reports of the success of cultured epithelium (both autografts and allografts), their use is not standard care in the vast majority of burn centers (4,5). When compared to the success of split-thickness grafts, cultured epithelial sheets have deficiencies in several areas, including the initial graft take, adherence to the graft bed, susceptibility to infection, prevention of wound contracture and fragility of the graft sheets (6–8).

Other efforts have focused on the development of a substitute for the dermis in addition to the epidermis, and several types of dermal matrices have been developed. Addition of a dermal matrix to epithelial replacements adds the theoretical advantage of a thicker, more durable graft that more closely resembles the "gold standard" of split-thickness skin grafts. One of the first

From: *Methods in Molecular Medicine, Vol. 18: Tissue Engineering Methods and Protocols*
Edited by: J. R. Morgan and M. L. Yarmush © Humana Press Inc., Totowa, NJ

successes was a porous collagen and glycosaminoglycan (GAG) matrix *(9,10)* which was designed to be biodegradable and to serve as a template for the ingrowth of fibrovascular components. This collagen–GAG matrix has been modified by others so that cultured fibroblasts can be incorporated into the matrix, and cultured keratinocytes can be seeded onto the surface to form an epidermis *(11–14)*.

Another early success was the development of a skin equivalent based on a fibroblast contracted collagen lattice *(15)*. Fibroblasts, seeded in gelled collagen, contract the lattice into a dermal analog, and keratinocytes can be seeded onto this structure to make a skin equivalent *(15,16)*. Other skin substitutes being developed include fibroblast seeded meshes of biodegradable fibers of polyglycolic acid (PGA) and polyglactin (PGL), or nondegradable nylon fibers. The biodegradable matrices have been seeded with keratinocytes and transplanted to athymic mice, in which the fibers are absorbed in several weeks *(17,18)*.

We and others have had success with a dermal analog of acellular human dermis that is de-epithelialized and rendered completely acellular *(19–24)*. Acellular dermis is relatively nonimmunogenic and retains many of its structural elements after processing. The material is also durable, and can be lyophilized and stored at room temperature *(22,25–27)*. Composite grafts using the acellular dermis can be formed in vitro *(24,28–30)* and subsequently transplanted to athymic mice, generating a well-differentiated and fully pigmented epidermis with many of the same characteristics as normal skin *(19–21,23,31)*.

In this chapter, we describe methods for the preparation and storage of acellular human dermis, and the generation of composite tissue constructs by seeding cultured keratinocytes onto the acellular dermis. We then discuss how to successfully transplant these grafts to open wounds on athymic mice, and subsequently harvest grafts for analysis.

2. Materials
2.1. Cell Culture
2.1.1. Fibroblast Feeder Layer

1. Fibroblast feeder layer: 3T3-J2 mouse fibroblast cell line (originally provided by H. Green, Harvard Medical School, Boston, MA).
2. Fibroblast tissue-culture medium: Dulbecco's modified Eagle medium (DMEM, high glucose, L-glutamine, 110 mg/L sodium pyruvate, Gibco-BRL), bovine calf serum 10% (HyClone, Logan, UT), penicillin–streptomycin (Boehringer) 100 IU/mL–100 µg/mL.
3. Mitomycin C (Boehringer, Mannhiem, Indianapolis, IN): 15 µg/mL in serum-free DMEM.

2.1.2. Keratinocytes

1. Source for human keratinocytes: newborn foreskins (average size, 1–2 cm^2).
2. Keratinocyte tissue-culture medium (KCM):

a. DMEM/Ham's F12 medium (Gibco-BRL, Gaithersburg, MD) (3:1).
b. Fetal bovine serum 10% (JRH Bioscience).
c. Penicillin–streptomycin 100 IU/mL–100 µg/mL.
d. Adenine (6-aminopurine hydrochloride, Sigma, St. Louis, MO) make up fresh at time medium is prepared; prepare stock (50X) of 1.2 mg/mL in DMEM/F12 (3:1), adjust pH to 7.5 with 1 N NaOH; sterilize by filtration using a 0.45-µm filter; add 2 mL of stock to 100 mL of KCM; final concentration: 1.8×10^{-4} M.
e. Cholera toxin (Vibrio cholerae, type Inaba 569 B, Calbiochem, La Jolla, CA): Prepare concentrated stock of 10^{-5} M in dH$_2$O, store at 4°C; take 0.1 mL of concentrated stock and make up to 10 mL with DMEM (10% fetal calf serum); sterilize by filtration using a 0.45-µm filter, and aliquot in 1-mL portions (10^{-7} M), store at –20°C; add 0.1 to 100 mL of KCM; final concentration: 10^{-10} M.
f. Epidermal growth factor (mouse, Collaborative Biomedical, Bedford, MA): Resuspend lyophilized material in dH$_2$O to prepare stock of 10 µg/mL; sterilize by filtration using a 0.45-µm filter, and aliquot in 1-mL portions, store at –20°C; add 0.1 mL to 100 mL KCM with first medium change; final concentration: 10 ng/mL.
g. Hydrocortisone (chromatographic standard, Calbiochem): Prepare stock of 5 mg/mL in 95% ethanol, store at 4°C; take 0.4 mL of stock and make up to 10 mL with serum-free DMEM, sterilize by filtration using a 0.45-µm filter, and aliquot in 1-mL portions; add 0.2 mL to 100 mL of KCM; final concentration: 0.4 µg/mL.
h. Insulin (pork, 100 U/mL [3.8 mg/mL], Novo Nordisk, Danbury, CT): Add 0.13 mL of stock to 100 mL of KCM; final concentration: 5 µg/mL.
i. T/T3 stock (transferrin/triiodo-L-thyronine stock): T stock (transferrin, human, partially iron-saturated, Boehringer) 5 mg/mL in PBS; T3 stock (3,3',5-triiodo-L-thyronine, sodium salt, Sigma): Dissolve 13.6 mg in the minimum amount of 0.02 N NaOH, and make volume up to 100 mL with dH$_2$O, sterilize by filtration using a 0.45-µm filter, and store at –20°C (2×10^{-4} M); add 0.1 mL T3 stock to 9.9 mL of T stock, sterilize by filtration using a 0.45-µm filter, and aliquot in 1-mL portions, store at –20°C; add 0.1 mL of T/T3 stock to 100 mL of KCM; final concentration transferrin: 5 µg/mL, triiodo-L-thyronine: 2×10^{-9} M.
3. PBS (phosphate buffered saline): 138 mM NaCl, 2.7 M KCl, 8.1 mM Na$_2$HPO$_4$, 1.5 mM KH$_2$PO$_4$; sterilize by filtration using a 0.45-µm filter.
4. EDTA solution ([ethylenedinitriolo]tetraacetic acid disodium salt, Boehringer): 5 mM in PBS; sterilize by filtration using a 0.45-µm filter.
5. Trypsin solution (trypsin 1-300, ICN Biochemicals, Costa Mesa, CA): D-dextrose 0.1% (w/v), trypsin 0.1% (w/v) in PBS, pH 7.5; sterilize by filtration using a 0.45-µm filter, store at –20°C, avoid repeated thawing–freezing.
6. Trypsinizing flask (25 mL, Wheaton Scientific, Millville, NJ).

2.2. Harvesting of Human Cadaver Skin

1. Donor skin preparation: 70% isopropyl alcohol, Betadine surgical scrub and prep solution, topical mineral oil (sterile), 0.9% sodium chloride irrigation solution, O.R. scrub brush, disposable shaving razors.

2. Padget or Zimmer dermatome.
3. Dermatome blades.
4. Sterile drapes, sterile 4 x 4 sponges.
5. Sterile specimen containers.
6. Sterile gowns, gloves, mask, and protective goggles.
7. Sterile forceps and #15 sterile disposable scalpels (Feather Safety Razor Co., Japan).

2.3. Storage of Human Cadaver Skin

1. Sterile nylon gauze.
2. Sterile drapes.
3. Cryopreservation fluid (15% glycerol in Ringer's lactate).
4. Surgical instruments: sterile scissors, tissue forceps, sterile basins.
5. Sealable plastic or aluminum pouches.
6. Cryomed programmable freezer.
7. Liquid nitrogen.

2.4. Preparation of Acellular Human Dermis

1. Human cadaver dermis or freshly obtained skin (see **Note 1**).
2. Sterile PBS or DMEM (Gibco-BRL).
3. Antibiotics: Gentamycin (100 µg/mL), ciprofloxacin (10 µg/mL), amphoteracin B (2.5 µg/mL), penicillin–streptomycin (100 IU/mL–100 µg/mL).
4. Surgical instruments: Two pairs of fine forceps (tweezers) without teeth, #10 sterile disposable scalpels (Feather).
5. Sterile glass pipets.

2.5. Preparation of Composite Grafts in Vitro

1. 35-mm tissue-culture dishes.
2. Circular steel mesh screens (Fisher, Pittsburgh, PA).
3. Surgical instruments: two pairs of fine forceps (tweezers) without teeth.
4. Sterile glass pipets.

2.6. Grafting and Harvesting of Composite Grafts

1. Surgical instruments: Small, sharp scissors, sterile disposable scalpels (Feather), two pairs of fine tissue forceps (tweezers) without teeth, small needle driver, sterile glass pipets, 1-cc tuberculin syringes, razor blades.
2. 6-0 nylon sutures (Ethicon, Somerville, NJ), with a small (p-1) cutting needle.
3. Skin prep: 70% Isopropyl alcohol, Betadine surgical prep solution, tincture of benzoin skin adhesive, Neosporyn or Bacitracin antibiotic ointment.
4. Skin dressing: Telfa no-stick gauze (Kendall, Mansfield, MA), Tegaderm polyurethane occlusive dressing (3M) (5 x 6.5-cm size), 3M flexible Sports Band-Aid (must be at least 7/8 to 1 in. wide), 0.5-in. wide waterproof adhesive tape (Johnson and Johnson).
5. Protractor or template that can be used to draw squares on the back of the mouse.

6. Anesthesia (2,2,2-tribromoethanol, Aldrich, Milwaukee, WI): Prepare concentrated stock of 1.6 g/mL in 2 methyl-2 butanol, and store at 4°C. Before use, prepare a working solution by diluting 12 µL (concentrated stock)/mL 0.9% NaCl at 40°C. Sterilize by filtration using a 0.45-µm filter.
7. 35-mm tissue-culture dishes.
8. Formaldehyde (Sigma 37%): Dilute stock 1:10 (v/v) in PBS.
9. Dry ice (block form).
10. Camera for photo documentation (with macro lens).
11. Transparency film for tracing the grafts.
12. Athymic nude mice, NIH Swiss nu, 6–8 wk, average weight 20–30 g (outbred, TAC:N;NIFS-nuDF, Taconic, Germantown, NY).
13. Dental wax (Byte ryte, Mizzy Inc., Cherry Hill, NJ).

3. Methods
3.1. Harvest of Human Cadaver Skin

Cadaver skin with an average thickness of 0.015 in. can be obtained fresh or from any local skin bank. If it is obtained from a skin bank, it should have been cryopreserved according to the bank's protocols, using glycerol and controlled rate freezing and should be CMV-, hepatitis B- and HIV-negative. Described below is the method used for harvesting skin suitable for use in clinical and laboratory situations. (*see* **Notes 1 and 2**).

1. Prospective donor sites are completely shaved of all body hair with disposable razors, and then the area to be harvested is washed with regular tap water.
2. The person doing the prep and harvest does a full surgical scrub and then gowns in sterile attire, including mask, cap, and goggles. Prepping is performed in a circular fashion, starting in the center and working out.
3. First the Betadine surgical scrub is applied to the skin and the skin is scrubbed gently for several minutes with an OR scrub brush.
4. The area is washed with sterile saline and then painted three times, using 4 x 4 sterile gauze sponges in a circular motion with the Betadine skin-prep solution.
5. The Betadine is allowed to dry for 5 min, and then the skin is washed with 4 x 4 gauze soaked in 70% isopropyl alcohol prep solution.
6. The area to be harvested is now draped and all unprepped areas of the body are covered with sterile sheets or towels.
7. The person who prepped the area now regowns and regloves, or a second person harvests the grafts.
8. A Padget or Zimmer dermatome is used for the skin harvest. Lock a fresh dermatome blade into the apparatus, and then screw the maximum-width guard tightly down onto the blade.
9. To test the calibration of the dermatome, set the skin harvest thickness at 0.015 in. and slide a #15 scalpel blade between the dermatome blade and guard. The cutting portion of the scalpel blade should just fit between the two. This indicates that the calibration setting of 0.015 in. is correct.

10. Place the bevel of the dermatome on the donor site with the right hand. The angle of the dermatome to the skin is approx 45 degrees. The left hand rests firmly on the skin behind the instrument and provides firm countertraction. Depress the control button or lever and apply a steady and firm forward and downward pressure to begin the skin removal. To complete the cut, tilt the dermatome gradually, with the bevel upwards, and release the control.
11. Remove the skin from the dermatome with sterile forceps and place in a sterile container with a small quantity of saline. These containers can be stored at 4°C for 48 h or placed directly into a deep freezer.
12. In a laminar hood, pour cryopreservation medium into specimen containers and transfer the cadaver skin into the containers. These are then returned to the refrigerator for 2 h.
13. The skin is then removed from the containers and spread flat onto glycerolized sterile mesh gauze and placed into sealable plastic or aluminum pouches.
14. The skin is now ready to be cryopreserved and can be frozen in the Cryomed freezer, according to local skin banking protocols. We recommend a freezing rate of 1°C/min (*see* **Notes 2**)

3.2. Preparation of Acellular Human Dermis

Several authors have recommended ways to prepare de-epidermalized, acellular dermis *(22,27)*. We have followed a similar protocol.

1. The cryopreserved skin is warmed and then subjected to three rapid freeze–thaw cycles in liquid nitrogen to devitalize the cells, washed three times in sterile PBS, and then incubated at 37°C for 1 wk in sterile PBS with antibiotics (gentamycin at 100 µg/mL, ciprofloxacin at 10 µg/mL, amphoteracin B at 2.5 µg/mL, penicillin–streptomycin at 100 IU/mL–100 µg/mL).
2. At the end of 1 wk, the epidermis can be gently stripped or scraped from the dermis with forceps in a laminar flow hood under strict sterile conditions. Epidermis that remains adherent will detach spontaneously in the following weeks.
3. The dermis is maintained in antibiotic solution at 4°C for 4 more wk to remove any remaining cells. Again, sterile technique at all times is essential (*see* **Note 2**)
4. Prior to use, the acellular dermis should be washed three times with DMEM to remove residual antibiotics. (*see* **Note 3**)

3.3. Culture of Human Keratinocytes

The description of how to culture human keratinocytes using a fibroblast feeder layer is beyond the scope of this chapter, but is described in excellent detail elsewhere *(1)*. Included in **Subheading 2.1.** of this chapter are the specific culture media that we have used and found successful for cell culture.

3.4. Preparation of Composite Skin Grafts

1. The dermis is cut into 1.25–1.5 cm^2 pieces in a sterile fashion using forceps and a #10 scalpel blade, and then each piece is placed into a 35-mm tissue-culture dish,

papillary side up. The papillary side of the dermis can be distinguished from the reticular side by a rougher feel and duller sheen (*see* **Fig. 1**). To facilitate adherence of the dermis to the plastic, crosshatch scratches should be made on the bottom of the dish with a glass pipet, and then the dermis is allowed to incubate in the dish for 1 h at 37°C.
2. Keratinocytes that have been grown in culture are seeded onto the dermis by combining 500,000 freshly trypsinized keratinocytes in 0.25 mL of keratinocyte culture medium, and then applying the mixture to the surface of each piece of dermis. If possible, try to use the surface tension of the culture medium to maintain the cells on the dermis, rather than allowing the medium to run over the edge of the graft (*see* **Note 4**).
3. After 2 h of incubation at 37°C to allow preliminary cell attachment, an additional 1.75 mL of medium is gently added to the dish. Culture medium is changed every 3 d.
4. The composite grafts are maintained submerged in culture for a minimum of 5 d. During this time, a thin layer of keratinocytes (3–5-cells thick) will cover the papillary surface of the acellular dermis. The composite grafts can then be transplanted onto mice. Unlike grafts of sheets of keratinocytes, the composite grafts can be maintained submerged in culture for up to 2 wk without apparent detriment (and even longer if raised to the air–liquid interface) (*see* **Note 4**).

3.5. Preparation of Composite Skin Grafts Grown at the Air–Liquid Interface

Composite grafts that are maintained in the submerged state form a very thin epithelium. To generate grafts with a thicker, more stratified epithelium, the composites can be prepared in the exact same way as described above, but after being submerged for 3 d, they should be raised to the air–liquid interface *(24)*.

1. We use circular steel mesh screens that are approximately the size of a quarter, and are bent at the edges to allow them to sit approx 3 mm off the bottom of a tissue-culture dish.
2. The graft should be gently lifted at two corners with two pairs of forceps, and draped over the screen. Smooth out any wrinkles in the graft, but avoid scraping the surface of the graft.
3. The culture medium (KCM) should then be added to the dish, until it is level with, but not submerging, the graft. In this way, the dermis is exposed to the medium, and the keratinocytes are exposed to air. The keratinocytes survive by diffusion of nutrients through the dermis, and will begin to form a thicker, more stratified epithelium (**Fig. 2**; *see* **Note 5**).
4. Grafts should be left at the air–liquid interface a minimum of 7 d, and can be left for up to 3 wk.

3.6. Grafting of Composite Skin Grafts

Several methods have been described for the transplantation of human skin or skin equivalents to open wounds on athymic mice. The method described in this subheading was found by our group to maximize graft survival. The entire surgical procedure must be performed on anesthetized animals, using sterile conditions. All surgical procedures are performed in a laminar flow hood. The grafting procedure, when mastered, requires about 30 min/mouse.

1. Male nude mice can be rendered passive with a very brief CO_2 or ether treatment and then definitively anesthetized with approx 0.7–0.8 mL of 2,2,2-tribromoethanol (intraperitoneal injection with 25-gage 3/4 in. needle, 30 µL working solution/g of mouse).
2. When the animal is completely unconscious, it is transferred to a sterile operating environment.
3. The dorsum of the mouse is washed first with 70% alcohol and then Betadine. The Betadine is allowed to dry, and then a 1.25–1.5 cm^2 square is marked out on the mouse's side (the square should be no larger than the graft to be transplanted). The medial cephalad–caudad mark should be made just to the right or left of the midline, to avoid grafting over the spine. The cephalad medial-to-lateral mark should be made just inferior to the shoulder blade of either frontlimb.
4. Sharp scissors are used to create a full-thickness (including panniculus carnosus) defect down to the fascia of the dorsal musculature, and then the entire piece of skin is removed by scissor dissection. Bleeding may be encountered when coming across the anterior and posterior fat pads, and can be controlled by 1–2 min of gentle pressure (*see* **Note 6**)
5. Prior to removing the composite graft from the culture dish, a glass pipet or needle should be used to circumferentially scrape away any keratinocytes on the dish directly adjacent to the dermal composite. During culture, keratinocytes will grow off the edges of the dermis and attach to the plastic tissue-culture dish. If not removed, these keratinocytes may act as a tether that pulls the epithelium from of the dermal graft when it is lifted from the dish.
6. To transfer the graft to the wound, a square piece of Telfa no-stick gauze, which is roughly the same size as the graft, is laid over the graft in the culture dish. Two forceps are used to grab an edge of graft, along with Telfa, and lift it from the dish. The graft adheres lightly to the Telfa and thus does not bunch up. The graft is then laid into place over the wound, and the Telfa is gently removed (*see* **Fig. 3A**).
7. Once on the wound, either the graft or the wound edges are trimmed to generate a precise fit of the graft into the defect (a starting graft slightly bigger than the defect is ideal).
8. A 6-0 nylon stitch is used to anchor the graft at each corner of the wound. It is important to include mouse skin, mouse underlying musculature, and the graft with each stitch.
9. Antibiotic ointment is applied to a square piece of Telfa, which is then applied to the graft. One more piece of Telfa is placed on top of the first piece (to act as a compressive dressing), and then both pieces are secured with several 6-0 nylon sutures (tacked to the surrounding mouse skin, not the graft).
10. Tincture of benzoin (which makes the mouse skin sticky) is applied to the mouse skin on all sides of the wound.

Epidermal Keratinocyte Composite Graft

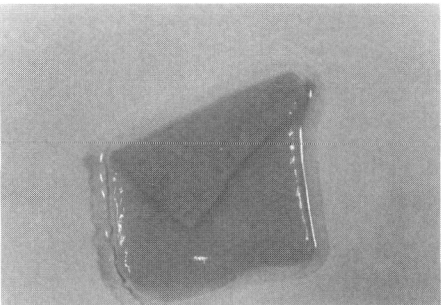

Fig. 1. Acellular human dermis that is oriented with the papillary side up. The top half of the dermis (arrow) has been flipped over to show the reticular side.

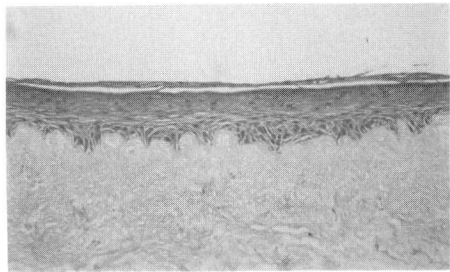

Fig. 2. Keratinocytes are seeded onto the dermis and allowed to grow submerged for 1 wk (data not shown). When subsequently raised to the air–liquid interface, they form a thicker, more stratified epithelium. Note the completely acellular dermal matrix.

11. A 5 by 6.5 centimeter piece of Tegaderm brand polyurethane occlusive dressing is cut in half and one half is applied over the Telfa dressing. The Tegaderm, which is adhesive, should form a tight seal for several millimeters around the wound and adhere tightly to the mouse skin (aided by the Benzoin) (see **Fig. 3B**).
12. At this time, approx 0.7 mL of keratinocyte culture medium is injected with a tuberculin syringe through the Tegaderm into the Telfa pads to keep the grafts moist for the first several days (see **Fig. 3B**).
13. Next, a 3-M flexible Sports Band-Aid is trimmed so that only the pad and a thin (5mm) adhesive strip around the pad remains. This Band-Aid is then placed over the Tegaderm on top of the wound. A 6-0 nylon suture is used to sew the Band-Aid to the surrounding mouse skin (Tegaderm included). This is a running "baseball" stitch that travels around the entire circumference of the Band-Aid (see **Fig. 3C**).
14. Finally, 1/2 in. waterproof tape is wrapped circumferentially around the mouse to cover the band-aid. Care must be taken not to wrap too closely to the mouse's hind legs, which might get caught in the tape as the mouse walks around after the surgery.

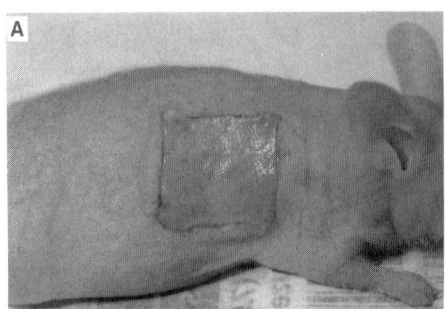
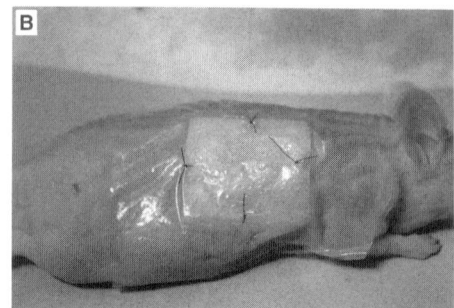
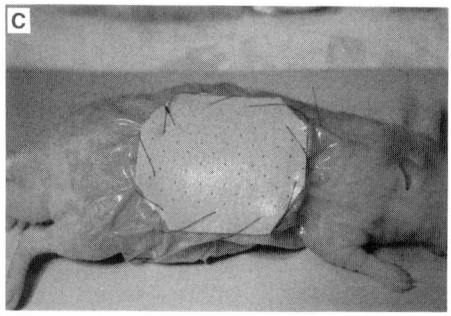

Fig 3. A typical graft after 1 wk of culture (**A**) is placed onto a full-thickness defect on the dorsum of an athymic mouse. The graft is dressed with Telfa and Tegaderm (**B**) and then a flexible band-aid (**C**).

15. If the whole procedure is performed correctly, the mouse retains its mobility and the dressing remains occlusive for at least 4–7 d (enough to allow epithelialization of the gaps between the mouse skin and the graft). The dressing should not be changed for at least 1 wk, and can, in fact, be left on indefinitely (*see* **Note 7**).

3.7. Graft Harvest

In as early as 7 d, the grafts integrate with the surrounding mouse skin. The epithelium undergoes further stratification and the dermis is infiltrated by mouse fibroblasts and capillaries *(19,21)*. By 6 wk, a fully differentiated, pigmented epidermis is generated that is quite durable and demonstrates a normal protective barrier, as evaluated by transepidermal water loss (**Fig. 4A**) *(20,23)*. The dermis is well populated by host fibroblasts and blood vessels, and passenger melanocytes have multiplied and repopulated the basal layers of the epidermis (**Fig. 4B**; *see* **Note 8**)

1. To harvest the graft, the animal should be sacrificed by CO_2 asphyxiation, rather than cervical dislocation (this prevents separation of the wound margins).

Epidermal Keratinocyte Composite Graft

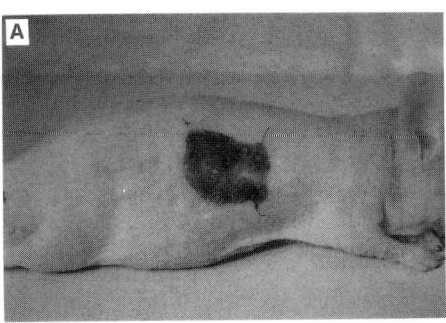
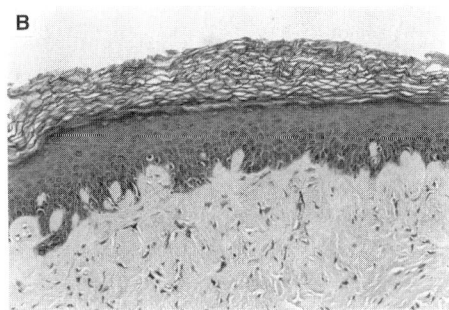

Fig. 4. By 6 wk, the graft is well incorporated into the surrounding mouse skin and has almost completely repigmented (**A**). Histology reveals a fully stratified epithelium with abundant keratin, an interdigitating dermal–epidermal junction, and a revascularized dermal matrix that is well populated with host fibroblasts (**B**).

2. The dressing is carefully removed so as not to dislodge the underlying graft epidermis. At this time, it is our practice to photograph and trace the graft while it is still on the mouse.
3. The entire graft is then harvested by making a full thickness incision down to, and through, dorsal thoracic musculature of the mouse, with sharp scissors. The harvest should include the muscle underlying the graft, but not the mouse ribs. Make sure to include a segment of normal mouse skin around the edges of the graft.
4. Depending on the experiment, the graft can be handled in several different ways. We like to divide the graft into two pieces, and save one-half for frozen sections and the other half for permanent histologic sections. To do this, spread the dissected graft on a piece of dental wax (Byte ryte, Mizzy) and pin the four corners to the wax. A razor blade can then be used to evenly divide the graft into two or more pieces.
5. One piece (still pinned to the wax) is then immediately placed in a solution of 10% formalin. The other piece is spread flat on a piece of aluminum foil, and then snap-frozen on a flat block of dry ice, labeled, and transferred to liquid nitrogen or a deep freezer. The formalin fixed pieces can be paraffin embedded, sectioned, and stained with hematoxylin and eosin to reveal the structure of the epidermis and dermis. The frozen pieces can be processed for cryosections and immunostained to reveal specific structures or proteins of the skin.

4. Notes

1. An alternative source of skin, which is readily available, is skin harvested in reduction mammoplasty, abdominoplasty, or panniculectomy procedures. Each of these procedures yields skin attached to underlying subcutaneous tissue. We do not recommend using full-thickness skin for the composite grafts; thus, this skin will have to be harvested from the subcutaneous tissue using a dermatome (Padget or Zimmer are

recommended), set at an average thickness of 0.015 in. It is much easier to harvest the skin while it is still attached. If the purpose of the harvested skin is only to generate an acellular dermal matrix, then normal cryopreservation methods, designed to protect the cells by controlled rate freezing may not not necessary. However, our laboratory used only skin from a skin bank that had been cryopreserved according to strict clinical use protocols, and thus we never experimented with the use of uncryopreserved skin. Freshly harvested skin can either be frozen according to the protocol outlined above or immediately subjected to the freeze–thaw cycles, and then placed into the antibiotic containing PBS. As with all human skin, it should be remembered that disease transmission is still a possibility, and full precautions should be taken.

2. It cannot be overemphasized how important sterile technique is at all stages of the preparation of the acellular dermis. Even the slightest contamination can blossom into fulminant bacterial overgrowth once the dermis is placed into an incubator.

3. When human skin is prepared as described above, the dermis is rendered completely acellular, but it retains key architectural elements, such as papillary projections, elastin fibers, and many of the basement membrane proteins *(21,24)*.

4. In the routine harvest and culture of keratinocytes from neonatal foreskins, passenger melanocytes have been observed to survive at about 10% of their normal skin number *(2)*. These melanocytes will adhere, along with the keratinocytes, to the supporting dermis, proliferate, and resume function *(19,21,31)*.

5. It should be noted that we were unable to find increased in vivo survival from raising the grafts to the air–liquid interface. Overall graft take for both graft types is excellent, and thus the added time delay of growing it at the interface is not necessary, unless the experiment depends upon varying thickness and maturity of the graft.

6. Many other researchers have described preserving the panniculus carnosus to aid in the survival of the transplanted skin graft. We do not find this to be necessary and have found that it serves to increase the length and difficulty of the procedure.

7. We do not use systemic antibiotics for the mice during or after the procedure. Good sterile technique results in few-to-no infections.

8. The cells used in our experiments were derived almost exclusively from darkly pigmented donor foreskins. This was done in order to more easily study the survival and subsequent function of passenger melanocytes. We have observed that melanocytes proliferate and resume function once transplanted to athymic mice. The composite grafts develop foci of dark pigment after several weeks in vivo, and these foci increase in size and number as the graft matures, generating a fully pigmented skin equivalent *(19,21,31)*.

References

1. Green, H., Kehinde, O., and Thomas, J. (1979) Growth of cultured human epidermal cells into multiple epithelia suitable for grafting. *Proc. Natl. Acad. Sci. USA* **76,** 5665–5668.
2. Gallico, G. G., O'Connor, N. E., Compton, C. C., Kehinde, O., and Green, H. (1984) Permanent coverage of large burn wounds with autologous cultured human epithelium. *N. Engl. J. Med.* **311,** 448–451.

3. Hefton, J. M., M. R. Madden, J. L. Finkelstein, and Shires, G. T. (1983) Grafting of burn patients with allografts of cultured epidermal cells. *Lancet* **2**, 428–430.
4. De Luca, M., Albanese, E., Bondanza, S., Megna, M., Ugozzoli, L., Molina, F., et al. (1989) Multicentre experience in the treatment of burns with autologous and allogenic cultured epithelium, fresh or preserved in a frozen state. *Burns.* **15**, 303–309.
5. Madden, M. R., Finkelstein, J. L., Staiano-Coico, L., Goodwin, C. W., Shires, G. T., Nolan, E. E., and Hefton, J. M. (1986) Grafting of cultured allogeneic epidermis on second- and third-degree burn wounds on 26 patients. *J Trauma* **26**, 955–962.
6. Compton, C. C., Hickerson, W., Nadire, K., and Press, W. (1993) Acceleration of skin regeneration from cultured epithelial autografts by transplantation to homograft dermis. *J. Invest. Dermatol.* **14**, 653–662.
7. Cooper, M. L., Andree, C., Hansbrough, J. F., Zapata, S. R., and Spielvogel, R. L. (1993) Direct comparison of a cultured composite skin substitute containing human keratinocytes and fibroblasts to an epidermal sheet graft containing human keratinocytes on athymic mice. *J. Invest. Dermatol.* **101**, 811–819.
7a. Cuono, C., Langdon, R., and McGuire, J. (1986) Use of cultured epidermal autografts and dermal allografts as skin replacement after burn injury. *Lancet.* **1**, 1123,1124.
8. Kangesu, T., Navsaria, H. A., Manek, S., Fryer, P. R., Leigh, I. M., and Green, C. J. (1993) Kerato-dermal grafts: the importance of dermis for the in vivo growth of cultured keratinocytes. *Br. J. Plast. Surg.* **46**, 401–409.
9. Yannas, I. V. and Burke, J. F. (1980) Design of an artificial skin. I. Basic design principles. *J. Biomed. Mater. Res.* **14**, 65–81.
10. Yannas, I. V., Burke, J. F., Gordon, P. L., Huang, C., and Rubenstein, R. H. (1980) Design of an artificial skin. II. Control of chemical composition. *J. Biomed. Mater. Res.* **14**, 65–81.
11. Boyce, S. T. and Hansbrough, J. F. (1988) Biologic attachment, growth, and differentiation of cultured human epidermal keratinocytes on a graftable collagen and chondroitin-6-sulfate substrate. *Surgery* **103**, 421–431.
12. Boyce, S. T., Christianson D. J., and Hansbrough, J. F. (1988) Structure of a collagen-GAG dermal skin substitute optimized for cultured human epidermal keratinocytes. *J Biomed Mater Res.* **22**, 939–957.
13. Hansbrough, J. F., Boyce, S. T., Cooper, M. L., and Foreman, T. J. (1989) Burn wound closure with cultured autologous keratinocytes and fibroblasts attached to a collagen-glycosaminoglycan substrate. *JAMA* **262**, 2125–2130.
14. Tinois, E., Tiollier, J., Gaucherand, M., Dumas, H., Tardy, M., and Thivolet, J. (1991) In vitro and posttransplantation differentiation of human keratinocytes grown on the human type IV collagen film of a bilayered dermal substitute. *Exp. Cell. Res.* **193**, 310–319.
15. Bell, E., Ehrlich, H., Buttle, D. J., and Nakatsuji, T. (1981) A Living tissue formed in vitro and accepted as skin-equivalent tissue of full thickness. *Science* **211**, 1052–1054.
16. Parenteau, N. L., Nolte, C. M., Bilbo, P., Rosenberg, M., Wilkins, L. M., Johnson, E. W., et al.(1991) Epidermis generated in vitro: practical considerations and applications. *J. Cell Biochem.* **45**, 245–251.

17. Hansbrough, J., Morgan, J., Greenleaf, G., Underwood, J. (1994) Development of a temporary living skin replacement composed of human neonatal fibroblasts cultured in Biobrane, a synthetic dressing material. *Surgery* **115**, 633–636.
18. Hansbrough, J. F., Morgan, J., Greenleaf, G., Parikh, M., Nolte, C., and Wilkins, L. (1994) Evaluation of Graftskin composite grafts on full-thickness wounds on athymic mice. *J. Burn Care Rehab.* **15**, 346–353.
19. Medalie, D., Eming, S. E., Tompkins, R. G., Yarmush, M. L., Kreuger, G. G., and Morgan, J. R. (1996) Evaluation of human skin reconstituted from composite grafts of cultured keratinocytes and human acellular dermis transplanted to athymic mice. *J. Invest. Dermatol.* **106**, 121–127.
20. Medalie, D. A., Eming, S. E., Collins, M. E., Tompkins, R. G., Yarmush, M. L., and Morgan, J. R. (1997) Differences in dermal analogs influence subsequent pigmentation, epidermal differentiation, basement membrane and rete ridge formation of transplanted composite skin grafts. *Transplantation*, **64**, 454–465.
21. Medalie, D. A., Tompkins, R. G., and Morgan, J. R. (1996) Evaluation of acellular human dermis as a dermal analog in a composite skin graft. *ASAIO J.* **42**, M455–M462.
22. Krejci, N. C., Cuono, C. B., Langdon, R. C., and McGuire, J. (1991) In vitro reconstitution of skin: fibroblasts facilitate keratinocyte growth and differentiation on acellular reticular dermis. *J. Invest. Dermatol.* **97**, 843–848.
22a. Langdon, R. C., Cuono, C. B., Birchall, N., Madri, J. A., Kuklinska, E., McGuire, J., and Moellman, G. E. (1988) Reconstitution of structure and cell function in human skin grafts derived from cryopreserved allogeneic dermis and autologous cultured keratinocytes. *J. Invest. Dermatol.* **91**, 478–485.
23. Choate, K. A., Medalie, D. A., Morgan, J. R., and Khavari, P. A. (1996) Corrective gene transfer in the human skin disorder lamellar ichthyosis. *Nature Med.* **2**, 1263–1267.
24. Prunieras, M., Regnier, M., and Woodley, D. (1983) Methods for cultivation of keratinocytes with an air-liquid interface. (Review). *J. Invest. Dermatol.* **81**, 285–289.
25. Kraut, J. D., Eckhardt, A. J., Patton, M. L., Antoniades, K., Haith, L. R. J., and Shotwell, B. S. (1995) Combined simultanous application of cultured epithelial autografts and alloderm®. *WOUNDS* **7**, 137–142.
26. Livesey, S. A., Herndon, D. N., Hollyoak, M. A., Atkinson, Y. H., and Nag, A. (1995) Transplanted acellular allograft dermal matrix. *Transplantation* **60**, 1–9.
26a. McKay, I., Woodward, B., Wood, K., Navsaria, H. A., Hoekstra, H., and Green, C. (1994) Reconstruction of human skin from glycerol-preserved allodermis and cultured keratinocyte sheets. *Burns* **20**, S19–S22.
27. Matouskova, E., Vogtova, D., and Konigova, R. (1993) A recombined skin composed of human keratinocytes cultured on cell-free pig dermis. *Burns* **19**, 118–123.
28. Guo, M. and Grinnell, F. (1989) Basement membrane and human epidermal differentiation in vitro. *J. Invest. Dermatol.* **93**, 372–378.
29. Heenen, M., Graef, C. D., Parent, D., Dobbeleer, G. D., and Galand, P. (1992) Renewal and differentiation of keratinocytes cultured on dead de-epidermalized dermis. *Cell Proliferation* **25**, 311–319.

30. Shakespeare, V. A. and Shakespeare, P. G. (1987) Growth of cultured human keratinocytes on fibrous dermal collagen: a scanning electron microscope study. *Burns Including Thermal Injury.* **13,** 343–348.
31. Higounenc, I., Demarchez, M., Regnier, M., Schmidt, R., Ponec, M., and Shroot, B. (1994) Improvement of epidermal differentiation and barrier function in reconstructed human skin after grafting onto athymic mice. *Arch. Dermatol. Res.* **286,** 107–114.
32. Staiano-Coico, L., Hefton, J. M., Amadeo, C., Pagan, Charry, I., Madden, M. R., and Cardon, C. (1990) Growth of melanocytes in human epidermal cell cultures. *J Trauma* **30,** 1037–1042.

31

Development of a Bioartificial Liver Device

Linda K. Hansen, Julie R. Friend, Rory Remmel, Frank B. Cerra, and Wei-Shou Hu

1. Introduction

Liver disease continues to be a challenge clinically, with 30,000 patients dying each year from liver failure *(1)*. Although liver transplantation can successfully treat many patients undergoing liver failure, the scarcity of donor organs severely limits this treatment's application. For this reason, many investigators are pursuing alternatives to total organ transplantation, from living donors to cell transplantation. One additional approach is the development of a hybrid, bioartificial liver as an extracorporeal device for the temporary treatment of acute liver failure. This approach has demonstrated early success, and may provide an important clinical treatment in the near future. In addition, a bioartificial liver reactor is useful for prolonged in vitro studies of hepatocyte function. This chapter will provide information on the design and use of such a reactor for in vitro applications.

The functions of the liver are numerous and varied, involving many complex biochemical reactions within hepatocytes that cannot be easily mimicked in an acellular environment. The development of a device to support and/or mimic normal liver function must therefore include living hepatocytes in an environment that maintains viability and promotes differentiated function. In addition, immunoprotection must be maintained for in vivo applications, and mass transfer limitations must also be addressed. Several models have been designed to accomplish this task, which utilize immobilized living hepatocytes in a hollow-fiber reactor. Many reviews have been published describing different devices developed in several laboratories *(2–5)*; this chapter will focus on the method of producing the Minnesota Bioartificial Liver (BAL). In this BAL, hepatocytes are suspended in a type I collagen solution and placed inside hol-

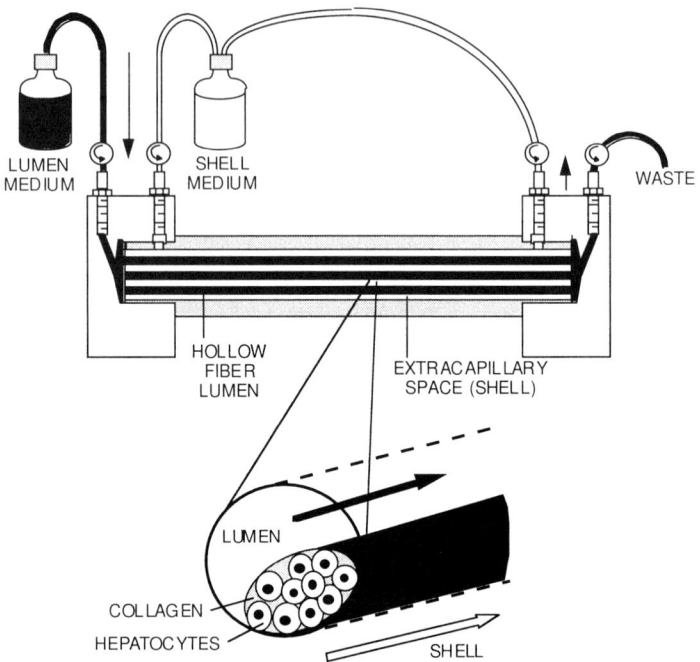

Fig. 1. Schematic Diagram of Minnesota BAL. Two perfusion streams are established within the hollow-fiber cartridge. Medium is perfused through the extracapillary compartment, or shell space ("shell medium"). During initial reactor operation, this medium may be recirculated. Collagen/hepatocyte suspension is injected into the intracapillary space. On collagen/cell contraction, a second perfusion stream is established (hollow-fiber lumen). Medium supplying hepatocytes with nutrients and growth factors ("lumen medium") is perfused through this lumenal space, allowing nutrient flow directly adjacent to immobilized cells. End caps keep the flow path separate. Shell medium is replaced with test media for detoxification during in vitro testing, or blood for in vivo applications (without recirculation). Inset demonstrates contracted collagen, entrapped hepatocytes, and the intercapillary lumen within a single hollow fiber.

low fibers. Incubation at 37°C promotes contraction of the collagen–cell suspension, creating a lumen within the hollow fiber. Media is perfused through the inner lumen, while media, or patient's blood for in vivo applications, can be perfused through the space outside the hollow fibers (**Fig. 1**). This design differs from others that place the hepatocytes outside the hollow fibers.

Because of the hybrid nature of the BAL, involving both living cells and a mechanical device, there are several steps involved in its assembly. The Method subheading will thus be divided into four subsections, each describing a separate step in the process (reactor setup, cell isolation, reactor loading, and reac-

Bioartificial Liver Device

tor operation). This information should provide sufficient instruction for producing an in vitro bioreactor for the study of living hepatocytes.

2. Materials

2.1. Reactor

Hollow fiber cartridge was obtained from Amicon, Danvers, MA, Model H1P100.

2.2. Media

1. Shell medium: Williams' Medium E (Gibco-BRL, Grand Island, NY), 2.2 mg/mL $NaHCO_3$, 50 µg/mL gentimycin, 0.2 U/mL insulin. Store at 4°C.
2. Lumen medium: Williams' Medium E (Gibco-BRL), 2.2 mg/mL $NaHCO_3$, 50 µg/mL gentimycin, 0.2 U/mL insulin, 500 µg/mL linoleic acid, 4 ng/mL glucagon, 20 ng/mL liver growth factor, 6.25 µg/mL transferrin, 5 ng/mL epidermal growth factor (EGF), 1 µM dexamethasone, 6.25 ng/mL selenium. Store at 4°C.
3. 4X Williams' E medium for collagen solution: (For 250 mL) one 1-L package of powdered Williams' E medium (Gibco-BRL), 2.2 g $NaHCO_3$, 50 mg gentamycin, 2 mg/mL linoleic acid, 16 ng/mL glucagon, 80 ng/mL liver growth factor (Sigma, St. Louis, MD), 25 µg/mL transferrin, 20 ng/mL EGF, 4 µM dexamethasone, 25 ng/mL selenium. Store at 4°C.

2.3. Cell Harvest and Preparation

1. Perfusion buffer I (Per I): 10X stock solution: 1.43 M NaCl, 67 mM KCl, 100 mM HEPES, pH 7.5; store at 4°C.
2. Perfusion buffer II (Per II): 1X stock solution: 1% bovine serum albumin, 67 mM NaCl, 6.7 mM KCl, 4.8 mM $CaCl_2$, 10 mM HEPES, adjust pH to 7.6; store at 4°C.
3. Collagen solution: Type I collagen solution (Vitrogen, Collagen, Santa Clara, CA) mixed 3:1 with 4X Williams' E medium.

3. Methods

3.1. Reactor and Media Preparation

The design of the hollow fiber reactor is shown in **Fig. 1**. The reactor consists of polysulfone hollow fibers sealed within a cartridge (Amicon). End caps are also available through Amicon, with ports allowing perfusion of media through both the inner hollow fiber lumen and the shell space outside the fibers. In addition, a loading cap was manufactured (University of Minnesota machine shop), which, like the end cap, slips over the end of the bioreactor cartridge, but has only one in-port into the fibers connected on the outside to tubing with a stopcock.

The reactor is hooked up to reservoirs of media with standard tubing and pumps. Additional monitors can be installed in line, if desired. For example, shell medium pH can be maintained using a pH electrode and microprocessor-controlled CO_2 switch *(6)*. In addition, the cartridges can be maintained in an automated cell culture device, such as the Maximizer, Jr. (Cellex Biosciences, Coon Rapids, MN). However, a cruder, more economical setup that maintains medium circulation, pH, and 37°C temperature is satisfactory.

3.2. Rat Hepatocyte Harvest

Several procedures have been published for the isolation of rat hepatocytes, most commonly derived from that of Seglen *(7)*, using a two-step *in situ* perfusion of the liver to loosen cells from the surrounding connective tissue. Similar procedures have also been published for hepatocyte harvest from other species *(8)*, which will not be discussed here. 4–6-wk-old Sprague-Dawley rats, weighing 200–250 g, are used.

3.2.1. Preoperative Procedure

1. Prepare the first perfusate buffer, Per I:
   a. Dilute 10X Per I stock up to 300 mL with dH_2O, to make 1X.
   b. Add 0.285 g EGTA.
   c. Adjust pH to 7.4 with 1 M NaOH.
   b. Sterile-filter, and place in 37°C water bath.
2. Prepare the second perfusate buffer, Per II:
   a. Add 0.1 g collagenase D to 200 mL Per II 1X stock.
   d. Sterile-filter, and place in 37°C water bath.
3. Aseptically prepare operating surface.
4. Run each perfusate through tubing of a multichannel pump to bleed lines free of air.
5. Anesthetize rat with ip injection of Nembutal stock (50 mg/mL) with 0.1 mL/100 g body weight, then shave and sterilize rat belly.

3.2.2. Operative Procedure

1. Make a transverse incision through the skin approx 1 cm below the xyphoid process, and make a similar incision through the abdominal wall.
2. Eviscerate the rat onto sterile gauze, exposing the inferior vena cava (IVC) and portal vein.
3. Inject 0.6 mL heparin into the IVC, using a 1-mL syringe, and allow heparin to circulate for 1 min.
4. Place two sutures loosely around the portal vein.
5. Insert an 18-gage iv catheter into the portal vein, and secure with suture ligatures.
6. Carefully remove the iv catheter needle, and allow the portal blood to back up to the iv catheter hub.
7. Connect the perfusion line to the catheter hub, turn on pump to start the Per I perfusion (25 mL/min), and cut the IVC.
8. Continue Per I perfusion until gone, or until the effluent from the IVC is clear.
9. Switch to Per II perfusion and reduce pump speed to 20 mL/min (*see* **Note 1**).
10. When Per II is gone, cut the liver free and place into a 60-mm Petri dish with 10–15 mL cold Williams' E media, cover, and take to sterile hood.

3.2.3. Cell Isolation

1. Using forceps and scissors, incise the liver capsule on all lobes, and gently agitate liver in the media to free the cells.
2. Gently pipet the cell suspension onto nylon gauze, and funnel into 50-mL centrifuge tube on ice, to remove clumps. Rinse gauze with additional cold Williams' E medium, using a second tube, if necessary. Cells should remain on ice throughout the remainder of the procedure (*see* **Note 2**).
3. Spin tubes at 50 g for 2 min at 4°C.
4. Resuspend pellet in cold Williams' E medium, and wash two more times, combining pellets from all tubes during last resuspension (*see* **Note 3**).
5. Dilute a small volume of the cell suspension 1:100 with trypan blue dye. Determine cell concentration and viability, using a hemocytometer (*see* **Note 4**).

3.3. Loading Reactor

3.3.1. Reactor Rinsing

The Amicon hollow-fiber reactor is stored in phosphate buffered saline (PBS), and thus requires no rinsing. Other hollow-fiber reactor models, however, may require rinsing in order to remove any potential residue from the hollow fibers that can harm the cells. This is done as follows:

1. In a sterile hood, vertically place the reactors, with end caps in place, in a ringstand with the in-ports at the bottom.
2. Hook up the out port from shell flow to the in-port of lumen flow, to create a single loop.
3. Turn on the pump at 1 L/min while checking for leaks.
4. When all PBS (e.g., 4 L) has been pumped through the system, continue to blow air through system for about 1 min, to remove PBS.

3.3.2. Preparation of Cell Suspension

1. Mix together a 3:1 solution of type I collagen (Vitrogen) and 4X Williams' E medium. The Amicon Model H1P100 hollow-fiber bioreactor can hold about 10 mL.
2. Use 1 N NaOH to bring up the pH gradually. Add a few drops of NaOH to the collagen–4X solution, and mix. Keep adding NaOH dropwise until color of 4X Williams' E mixture turns red, indicating pH ~7.2–7.4.
3. Pellet cells by spinning at 50g for 1 min at 4°C, to form a soft pellet. Remove supernatant and resuspend cells in the collagen–Williams' E solution, to yield a cell concentration of about 3–5 × 10^6 cells/mL. Since one Amicon reactor holds about 10 mL, approx 50 × 10^6 cells should be pelleted and brought up in 10 mL collagen–Williams' E solution for each reactor.

3.3.3. Reactor Loading

1. Attach a loading cap onto one end of the reactor, leaving the other end open. Reactor should be in a vertical position in a ringstand, with the loading cap positioned at the bottom end.

2. Load the collagen–cell mixture into a 35-cc syringe, using a 14-gage needle.
3. Remove the needle and attach the syringe to the lumen in-port of the reactor.
4. Slowly inject collagen–cell mixture until it fills the fibers from bottom to top. Close the stopcock of the loading cap.
5. Remove the bioreactor from the ringstand, remove the loading cap, and put end-caps in place.

3.4. Reactor Operation

1. Incubate reactor at 37°C.
2. About 10 min following incubation, put cartridge in line with medium, and begin perfusion of recirculated shell medium at about 30 mL/min.
3. Change shell perfusate after first 24 h, and every other day thereafter.
4. 24 h after shell perfusion began, during which time collagen gel should have contracted, begin perfusion of lumen medium at about 9 mL/hr.
5. Cell function and viability can be assessed over time by numerous methods. (Refer to **Note 5** and indicated references for further details.)

4. Notes

1. Watch the liver while the buffers perfuse through it. One should see an immediate blanching of the liver as the blood is replaced with perfusion buffer. If this does not occur, or only occurs in part of the liver, there is probably a bubble or other obstruction in the system. This will lead to lower yield and viability.
2. Hepatocytes will lose viability very quickly if they remain in suspension for more than a few hours, particularly if the solution approaches room temperature. Therefore, care must be taken to work quickly and keep the cell suspension on ice.
3. Collagenase perfusion yields a heterogeneous cell population. Although hepatocytes comprise the largest population of cell in the liver, there will also be other cells types, including lipocytes, Kuppfer cells, and others. The low-speed spins ($50g$, 2 min) preferentially separate the hepatocytes from other cell types, as well as dead cells because of their greater density, but an additional purification step, such as Percoll gradient centrifugation *(9)*, may be added if a homogeneous hepatocyte population is desired.
4. Most rat harvests should yield around $5 \times 10^8 - 1 \times 10^9$ cells. Ideally, viability should be 90% or greater. It is not recommended to use cells from a harvest that yielded an initial viability below 80%.
5. Hepatocytes are responsible for many metabolic, synthetic, and biotransformation functions in vivo. Quantitative analysis of these activities in the bioreactor will indicate the level of differentiated function maintained by the entrapped hepatocytes. Clearly, it is impossible to assay each of the many functions of the liver, but there are several quantitative assays that have been successfully employed to measure a variety of hepatocyte functions.

One of the simplest assays to employ is the measurement of albumin secretion. Media drawn from the lumen and/or shell space out-port can be analyzed for albumin protein concentration using enzyme-linked immunosorbent assay

(ELISA) *(10)*. Although this is a common assay, it is a measurement only of the synthesis of one protein that may be regulated differently than other functions that may be of more interest, depending on the application. Furthermore, albumin may also be released into the medium from dead, lysed cells, leading to an overestimation of *de novo* albumin synthesis and secretion.

Substrate metabolism by specific hepatic enzymes provides another useful parameter of hepatocyte function. Cytochrome P450 enzymes are involved in the phase I metabolic response, and this can be assessed by measuring lidocaine metabolism *(11)*. In addition, glucuronidation and sulfation (phase II) can be assessed by 4-methylumbelliferone conjugation by UDP-glucuronosyltransferase and sulfotransferase. Addition of either of these compounds to the shell medium prior to input, and analysis of the compound and its metabolites in the shell medium output, gives a good measure of the level of biotransformation activity *(10,12)*. Because of the diversity of hepatocyte functions and the potential for these functions to be regulated differently, the performance of a bioartificial liver device is best assessed using a variety of tests, as described here, to ascertain a spectrum of functions.

4.1. General Design Considerations

Several different hepatocyte bioreactor designs have been developed *(13)*. The reactor described in this chapter was designed to achieve high cultivation density and reduced mass transfer limitations, which are crucial in scaling up the reactor for clinical applications. Many reactor designs place the hepatocytes outside the hollow fibers (shell space), and perfuse medium only through the hollow fiber lumen. However, the interfiber distances and geometry are variable, resulting in mass transfer limitations in those areas farthest from the oxygen and nutrient source. Furthermore, in designs utilizing hepatocyte attachment to microcarriers, the space taken up by the microcarriers is essentially empty space. Placement of cells within the hollow fibers, and perfusing medium through the lumen created by collagen contraction, allows a much closer apposition between cells and medium, and can allow for a more efficient compaction of cells. In addition, the design described here creates space for a second perfusion stream, which, in clinical applications, can be utilized for patient blood, while still maintaining a nutrient stream in direct contact with the cells to which growth factors and other soluble mediators may be added.

Immobilization of hepatocytes within the collagen gel allows for self-association of hepatocytes. It is clear that cell–cell interaction and formation of three-dimensional structure that mimics that seen in vivo can enhance cellular function. Immobilization of hepatocytes onto a solid surface, such as a microcarrier, may inhibit these cell–cell interactions and formation of in vivo-like structures; suspension and contraction in a collagen gel can promote such interactions (*see* **Subheading 4.2.**). Additional insoluble matrix molecules may also be incorporated into the gel, which may enhance cell function and/or viability *(14)*.

4.2. Spheroids

Primary hepatocytes exhibit loss of differentiated function when removed from their native environment and placed into traditional tissue-culture conditions. Many studies have demonstrated that certain extracellular matrix substrates, and other conditions that mimic the native environment, can promote restoration of differentiated function. Under certain supportive conditions, hepatocytes will self-aggregate into multicellular structures, or spheroids, that express highly differentiated function. Cells within these spheroids possess cell polarity and cell–cell interactions very similar to those seen in vivo. Use of spheroids within the bioreactor may provide enhanced reactor function *(15,16)*. This topic is pursued further in Chapter 19.

References

1. National Center for Health Statistics (1993) *Vital Statistics of the United States*.
2. Yarmush, M. L., Dunn, J. C., and Tompkins, R. G. (1992) Assessment of artificial liver support technology. *Cell Transplantation* **1**, 323–341.
3. Kasai, S. K., Sawa, M., and Mito, M. (1994) Is the biological artificial liver clinically applicabale? A historic review of biological artificial liver support systems. *Artif. Organs* **18**, 348–354.
4. Dixit, V. (1994) Development of a bioartificial liver using isolated hepatocytes. *Artif. Organs* **18**, 371–384.
5. Jauregui, H. O., Chowdhury, N. R., and Chowdhur, J. R. (1996) Use of mammalian cells for artificial liver support. *Cell Transplantation* **5**, 353–367.
6. Shatford, R. A., Nyberg, S. L., Meier, S. J., White, J. G., Payne, W. D., Hu, W.-S., and Cerra, F. B. (1992) Hepatocyte function in a hollow fiber bioreactor: A potential bioartificial liver. *J. Surg. Res.* **53**, 549–557.
7. Seglen, P. O. (1976) Preparation of isolated rat liver cells. *Methods Cell Biol.* **13**, 29–83.
8. Sielaff, T., Hu, M. Y., Rao, S., Groehler, K., Olson, D., Mann, H. J., et al. (1995) A technique for porcine hepatocyte harvest and description of differentiated metabolic functions in static culture. *Transplantation* **59**, 1459–1463.
9. Kreamer, B. L., Staecker, J. L., Sawada, N., Sattler, G. L., Hsia, M. T. S., and Pitot, H. C. (1986) Use of a low-speed, iso-density Percoll centrifugation method to increase the viability of isolated rat hepatocyte preparations, *In Vitro Cell. Dev. Biol.* **22**, 201–211.
10. Nyberg, S. L., Shatford, R. A., Peshwa, M. V., White, J. G., Cerra, F. B., and Hu, W.-S. (1993) Evaluation of a hepatocyte entrapment hollow fiber bioreactor: a potential bioartificial liver. *Biotech. Bioeng.* **41**: 194–203.
11. Schroeder, T. J., Gremse, D. A., Mansour, M. E., Theuerling, A. W., Brunson, M. E., Ryckman, F. C., et al. (1989) Lidocaine metabolism as an index of liver function in hepatic transplant donors and recipients. *Transplantation Proc.* **21**, 2299–2301.

12. Nyberg, S. L., Mann, H. J., Remmel, R. P., Hu, W. S., and Cerra, F. B. (1993) Pharmacokinetic analysis verfies P450 function during in vitro and in vivo application of a bioartificial liver. *ASAIO J.* **39,** M252-M256.
13. Rozga, J., Williams, F., Ro, M.-S., Enuzil, D. F., Giorgio, T. D., Backfisch, G., Moscioni, A. D., Hakim, R., and Demetriou, A. A. (1993) Development of a bioartificial liver: properties and function of a hollow-fiber module inoculated with liver cells. *Hepatology* **17,** 258–265.
14. Hu, M. Y., Sielaff, T. D., and Cerra, F. B. (1994) Enhancement of cytochrome P450 function of collagen-entrapped hepatocytes by the addition of liver extracellular matrix components. *Transpl. Proc.* **26,** 3293.
15. Wu, F. J., Peshwa, M. V., Cerra, F. B., and Hu, W.-S. (1995) Entrapment of hepatocyte spheroids in a hollow fiber bioreactor as a potential bioartificial liver. *Cell Transplantation* **1,** 29–40.
16. Wu, F., Friend, J. R., Hsiao, C. C., Zilliox, M. J., Ko, W. J., Cerra, F. B., and Hu, W. S. (1996) Efficient assembly of rat hepatocyte spheroids for tissue engineering applications. *Biotechnol. Bioeng.* **50,** 404–415.

32

Methods for the Implantation of Liver Cells

Stephen S. Kim, Hirofumi Utsunomiya, and Joseph P. Vacanti

1. Introduction
1.1. Historical Perspectives

There have been many major advances in the field of liver transplantation in the past 30 yr. Orthotopic liver transplantation is currently the only established successful treatment for end-stage liver disease, with over 3000 liver transplantations being performed each year in more than 120 liver transplantation centers in the United States *(1)*. There are several major challenges, however, that impede the widespread practice and applicability of organ transplantation. These include the critical shortage of donor organs, the high cost and technical difficulty of the procedures, and the intensive postoperative care involved, including those associated with life-long immunosuppression. Among these obstacles, the critical scarcity of donor organs, especially in the pediatric population, is perhaps the most significant. Each year, end-stage liver disease accounts for 26,000 deaths in the United States *(2)*. Although the supply of donor organs has increased only slightly over the last 5 yr, the number of patients on the waiting list and the number of patients who die each year while on the waiting list have continued to grow at a disproportionate rate *(1)*.

These shortcomings have stimulated investigation into selective cell transplantation and the emergence of the field of tissue engineering as an alternative approach to the treatment of end-stage liver disease *(3–5)*. Tissue engineering is an interdisciplinary field that applies the principles of engineering and the life sciences toward the development of biological substitutes that restore, maintain, or improve tissue function *(4,5)*. Conceptually, the transplantation of only the essential tissue elements—in the case of the liver, the hepatocyte—has many advantages. These include:

From: *Methods in Molecular Medicine, Vol. 18: Tissue Engineering Methods and Protocols*
Edited by: J. R. Morgan and M. L. Yarmush © Humana Press Inc., Totowa, NJ

1. The potential for the alleviation of the donor organ shortage by utilizing cells from a small amount of donor tissue, and expanding them in vitro to create a potentially limitless supply.
2. The decrease in the risk and expense associated with major surgical procedures and protracted hospitalizations.
3. The potential of using autologous cells for transplantation, obviating the need for immunosuppression.
4. The capacity for liver-directed gene therapy to treat inborn errors of metabolism caused by single gene defects *(6)*.

Since the development of crucial techniques for the high-yield isolation of viable hepatocytes *(7,8)*, there has been a tremendous amount of interest and scientific investigation into hepatocyte transplantation. Hepatocyte transplantation has been performed in nearly every organ system, including the liver *(9–11)*, portal venous system *(12–15)*, spleen *(14,16–18)*, peritoneal cavity *(12–14,19)*, small bowel mesentery *(20–22)*, omentum *(20)*, lung *(23,24)*, pancreas *(25,26)*, renal capsule *(27,28)*, and subcutaneous tissue *(20,21,29,30)*, in a variety of animal models, including the mouse, rat, rabbit, pig, dog, and monkey. A number of studies have examined the use of hepatocyte transplantation in models of acute liver failure *(13,14)* and metabolic deficiency states *(12,15,31,32)*. Hepatocytes have been injected as cell suspensions or surgically implanted after being microencapsulated *(32,33)*, attached to microcarrier beads *(36–38)*, or seeded on polymer matrices *(20,34,35)*. Despite this vast spectrum of investigations, the optimal method and site of hepatocyte implantation has yet to be determined.

1.2. The Properties of the Optimal Method and Site of Implantation

There are many important properties that should be incorporated into the development of an optimal method of implantation of liver cells, based on established characteristics of the hepatocytes and the theoretical functions that the cells must fulfill after implantation. These include the following:

1. Hepatocytes are anchorage-dependent, and require an insoluble extracellular matrix for survival, organization, proliferation, and function. The extracellular matrices not only provide a surface for cell adherence, but they have profound influences on modulating cell shape and gene expression related to cell growth and liver-specific function *(39)*.
2. A large number of hepatocytes must be delivered and engrafted. An estimated 10–20% of the liver mass needs to be implanted to successfully replace liver function *(40)*.
3. Hepatocytes are highly metabolically active, and require rapid access to oxygen and nutrient supply.
4. The liver has a tremendous regenerative capacity in vivo. Liver cells are highly responsive to hepatotrophic stimulation *(41–44)*.

1.3. Overview of Methods for the Implantation of Liver Cells

Hepatocyte transplantation has been investigated in nearly every organ system in the body. For the purposes of this chapter, the methods that have demonstrated the most promising results in the rat model will be addressed:

1. Direct injection method:
   a. Into the portal venous system.
   b. Into the spleen.
2. Implantation associated with an extracellular matrix:
   a. Attached to microcarriers.
   b. After microencapsulation.
   c. Seeded on three-dimensional, biodegradable polymer scaffolds.

1.4. Advantages and Disadvantages of Different Methods of Implantation

1.4.1. Injection of Hepatocyte Suspension into Portal Venous System

The major advantages of hepatocyte implantation into the portal venous system, and, subsequently, into the host liver, include:

1. The utilization of the intact host hepatic extracellular matrix and hierarchical architecture of the host liver for cellular reorganization.
2. The interaction with other nonparenchymal liver cells and exposure to hepatotrophic factors released locally, and present in the portal system.
3. The immediate access to the blood supply for oxygen and nutrient delivery and waste removal.

The main disadvantages associated with direct injection of hepatocyte suspension into the portal venous system include:

1. The complications related to portal hypertension, portal vein thrombosis, systemic embolization, and host hepatic embolization and infarction.
2. The great difficulty of locating and differentiating the transplanted cells from the host cells after implantation.
3. The limited mass of cells that can be transplanted.

1.4.2. Injection of Hepatocyte Suspension into the Spleen

The major advantages of direct injection of hepatocytes into the spleen include:

1. The immediate access to the blood supply for oxygen and nutrient delivery and waste removal
2. The simplicity of the procedure, the ease of identifying transplanted hepatocytes in the spleen, and the relatively low risk of complications associated with splenic injection and temporary occlusion of the splenic vessels during injection *(45)*

3. The translocation of a large fraction of hepatocytes into the liver after splenic injection, if performed without concomitant occlusion of the splenic vessels *(10,46,47)*.

The main disadvantages associated with direct injection of hepatocytes into the spleen include:

1. The limited mass of cells that can be implanted into the spleen.
2. The translocation of hepatocytes into the portal system and the host liver after splenic injection, with the associated risks of portal hypertension, systemic embolization, and hepatic embolization.
3. The question of applicability in larger animal and human models.

1.4.3. Implantation of Hepatocytes Attached to Collagen-Coated Microcarriers into the Peritoneum

The major advantages of microcarrier-attached hepatocyte implantation in the peritoneum include:

1. The easy anatomic access and the minimally invasive procedure required.
2. The provision of an extracellular matrix to enhance cell attachment and function.
3. The large surface area available for the potential delivery of a large mass of hepatocytes.

The main disadvantages of microcarrier-attached hepatocyte implantation include:

1. The absence of a hierarchical tissue organization and structure of transplanted hepatocytes.
2. The lack of immediate access to a vascular supply.
3. The lack of evidence for long-term cell survival and function in vivo.

1.4.4. Implantation of Microencapsulated Hepatocytes into the Peritoneum

The major advantages of microencapsulated hepatocyte implantation in the peritoneum include:

1. The easy anatomic access, the minimally invasive procedure required, and the large surface area available for the potential delivery of a large mass of hepatocytes.
2. The provision of an extracellular matrix to enhance cell attachment and function.
3. The potential for immunoisolation of the microencapsulated hepatocytes, precluding the need for immunosuppression.

The main disadvantages of microencapsulated hepatocyte implantation include:

1. The absence of a hierarchical tissue organization and structure of transplanted hepatocytes.

2. The lack of immediate access to a vascular supply.
3. The lack of evidence for long-term cell survival and function in vivo.

1.4.5. Implantation of Hepatocytes Seeded on Biodegradable Polymer Scaffolds

The major advantages of hepatocyte transplantation on biodegradable polymer scaffolds include:

1. The three dimensional scaffold structure and the extracellular matrix components that can be incorporated into the polymer may enhance cell attachment, function, and tissue reorganization.
2. The use of a biodegradable polymer scaffold obviates concerns regarding foreign materials and long-term biocompatibility.
3. The potential to customize the scaffold size, design, and intrinsic properties for optimization of cell attachment, survival, and long-term function.

The main disadvantages of hepatocyte transplantation on biodegradable polymer scaffolds include:

1. The lack of immediate access to a vascular supply.
2. The lack of evidence for long-term cell survival and function in vivo.

2. Materials

2.1. Materials Universally Used in the Various Methods of Hepatocyte Implantation

1. Surgical instruments: scapel, hemostats, tissue forceps, metzenbaum scissors, sterile cotton applicators (Ethicon, Somerville, NJ).
2. Methoxyflurane (Pittman-Moore, Mundelein, IL).
3. 70% isopropyl alcohol prep pads (Baxter Health Care, Deerfield, IL).
4. Clinidine solution, povidone iodine (Clinipad, Guilford, CT).
5. Williams' E media (Sigma, St. Louis, MO) supplemented with 1 g sodium pyruvate (Sigma) and 5 mL penicillin G sodium (10,000 U/mL)/streptomycin sulfate (10,000 µg/mL) and L-glutamine (2.92 mg/mL) (Gibco-BRL, Grand Island, NY) per 500 mL of media.
6. 3-0 Prolene, polypropylene monofilament suture with a cutting needle (Ethicon).
7. Clippers to shave rat fur.

2.2. Direct Injection into the Portal Venous System

1. 26-gage needle (Becton Dickinson, Rutherford, NJ).
2. 1-cc sterile syringe (Becton Dickinson).

2.3. Direct Injection into the Spleen

1. 26-gage needle (Becton Dickinson).
2. 1-cc sterile syringe (Becton Dickinson).

3. Small vascular clamp (Biomedical Research Instruments, Rockville, MD).
4. 4-0 silk suture (Ethicon).

2.4. Implantation of Hepatocytes Attached to Collagen-Coated Microcarriers

1. Cytodex 3 type 1 collagen-coated dextran microcarriers (Pharmacia, Piscataway, NJ).
2. Phosphate buffered saline (PBS) without calcium and magnesium (Sigma).
3. Dulbecco's modified Eagle's medium (DMEM; Gibco-BRL) with 10% fetal calf serum (FCS; Gibco-BRL).
4. 10-mL sterile syringe (Becton Dickinson).
5. 13-gage needle (Becton Dickinson).

2.5. Implantation of Microencapsulated Hepatocytes

1. Microencapsulation material: a mixture of 2% (viscosity = 266 cps) sodium alginate (Kelco Gel LV, Kelco, San Diego, CA) and 1.7 mmol/L Matrigel (Collaborative Research Inc., Bedford, MA) (*see* **Note 1**).
2. Poly-L-lysine (Sigma).
3. 13-gage needle (Becton Dickinson).
4. 20-mL sterile syringe (Becton Dickinson).

2.6. Implantation of Hepatocytes on Biodegradable Polymer Scaffolds

1. Poly-L-lactic acid (PLLA) (Boehringer Ingelheim, Germany).
2. Chloroform (EM Science, Gibbstown, NJ).
3. Sodium chloride particles sieved to a size of 250–500 µm.
4. Teflon cylinders (Cole Palmer, Chicago, IL).
5. 1% aqueous poly-vinyl alcohol (PVA) (Aldrich, Milwaukee, WI) solution.
6. Ethylene oxide (H.W. Anderson, Chapel Hill, NC) warm gas sterilization.
7. 5-0 Prolene, polypropylene monofilament suture (Ethicon).
8. 8-0 Monofilament nylon suture (Ethicon).

3. Methods

3.1. Preparation of Isolated Rat Liver Cells

The most commonly used methods for the isolation and preparation of rat liver cell suspensions are based on an enzymatic digestion procedure previously described by Berry and Friend *(7)* and Seglen *(8)*. The concentration of hepatocyte suspensions can be varied, depending on the need.

3.2. Injection of the Hepatocyte Suspension into Portal Venous System

1. The hepatocyte pellet obtained after the isolation procedure is resuspended in Williams' E media to make a 2×10^7 cells/mL suspension.

2. The recipient rat is anesthetized using methoxyflurane inhalational anesthesia, and is weighed.
3. After the abdomen is shaved, it is sterilely prepped with 70% isopropyl alcohol and clinidine solution.
4. A midline incision is made, and the small bowel is eviscerated to the animal's left side, exposing the portal vein.
5. Using a 1-cc syringe, 1 mL of the hepatocyte suspension is injected into the portal vein using a 26-gage needle slowly over 1–2 min (*see* **Note 2**).
6. After the injection is completed, the needle is withdrawn and hemostasis obtained by applying pressure with a sterile cotton applicator for several minutes (*see* **Note 3**).
7. The abdominal contents are returned to their anatomic positions and the abdomen is closed in two layers, using 3-0 prolene suture.
8. The rat is monitored postoperatively for any signs of complications.

3.3. Injection of Hepatocyte Suspension into the Spleen

1. The hepatocyte pellet obtained after the isolation procedure is resuspended in Williams' E media, to make a 2×10^7 cells/mL suspension.
2. The recipient rat is anesthetized using methoxyflurane inhalational anesthesia, and is weighed.
3. After the abdomen is shaved, it is sterilely prepped with 70% isopropyl alcohol and clinidine solution.
4. A midline incision is made and the small bowel is eviscerated to the animal's right side, exposing the spleen and the splenic hilum.
5. A small nontraumatic vascular clamp is placed on the splenic hilum to occlude the splenic vessels within the hilum one-half min before to 5 min after the hepatocyte injection *(45)*.
6. Using a 1-cc syringe, 1 mL of the hepatocyte suspension is injected into the inferior pole of the spleen using a 26-gage needle, slowly 1–2 min (*see* **Notes 4**).
7. After the injection is completed, the needle is withdrawn and hemostasis obtained by ligating the injection site with a 4-0 silk suture.
8. The abdominal contents are returned to their anatomic positions and the abdomen is closed in two layers, using 3-0 Prolene suture.
9. The rat is monitored postoperatively for any signs of complications.

3.4. Implantation of Microcarrier-Attached Hepatocytes in the Peritoneum

1. Microcarrier preparation and hepatocyte attachment *(36,37)*: Cytodex 3 collagen-coated dextran microcarriers are hydrated in 125 mL of PBS and incubated at 37°C for 90 min. The PBS is removed and the microcarriers are washed with 50 mL of DMEM/10% FCS. Additional DMEM/10% FCS is added, to make a total volume of 125 mL. The microcarrier suspension is then transferred into 175-mL tissue-culture flasks and incubated (5% CO_2/humidified air) at 37°C for 60 min. The hepatocyte pellet obtained after the isolation procedure is resuspended in a small volume of DMEM/10% FCS, and added to the microcarrier suspension in the

tissue-culture flasks. This cell–microcarrier suspension is incubated (5% CO_2/humidified air) at 37°C for at least 2.5 h, to allow the cells to attach to the microcarriers. The microcarrier-attached cells are then washed with PBS and resuspended in DMEM/10% FCS to make a 2×10^6 cells/mL suspension.
2. The recipient rat is anesthetized using methoxyflurane inhalational anesthesia, and is weighed.
3. After the abdomen is shaved, it is sterilely prepped with 70% isopropyl alcohol and clinidine solution.
4. A small midline incision is made into the peritoneum.
5. Using a 10-mL syringe, 5 mL of the microcarrier–hepatocyte suspension is injected into the peritoneal cavity, using a 13-gage needle.
6. After the injection is completed, the abdomen is closed in two layers, using 3-0 Prolene suture.
7. The rat is monitored postoperatively for any signs of complications.

3.5. Implantation of Microencapsulated Hepatocytes in the Peritoneum

1. Microencapsulation of hepatocytes *(32,33)*: The hepatocyte pellet obtained after the isolation procedure is resuspended at room temperature in a mixture of 2% sodium alginate and 1.7 mmol/L Matrigel. A droplet-generating apparatus is used to form microdroplets of the mixture with a diameter of approx 300–700 µm. The alginate microdroplets are incubated at room temperature in 28 umol/L poly-L-lysine for 15 min, to form an outer envelop of polylysine. The polylysine-coated alginate microdroplets are further reacted with a 0.2% sodium alginate solution, to complete the formation of the alginate-polylysine-alginate (APLA) membrane. The microencapsulated hepatocyte suspension contains approx 1×10^7 cells/mL.
2. The recipient rat is anesthetized using methoxyflurane inhalational anesthesia, and is weighed.
3. After the abdomen is shaved, it is sterilely prepped with 70% isopropyl alcohol and clinidine solution.
4. A small midline incision is made into the peritoneum.
5. Five mL of the microencapsulated hepatocyte suspension is suspended in 10 mL of media.
6. Using a 20-cc syringe, the 15-mL microencapsulated hepatocyte suspension is injected into the peritoneal cavity, using a 13-gage needle.
7. After the injection is completed, the abdomen is closed in two layers, using 3-0 Prolene suture.
8. The rat is monitored postoperatively for any signs of complications.

3.6. Implantation of Hepatocytes on Biodegradable Polymer Scaffolds

1. Polymer scaffold preparation (*see* **Note 5**): Chloroform is used to dissolve the PLLA polymer to yield a 5% solution. This solution is loaded into Teflon cylinders packed with sodium chloride particles (250–500 µm). After the chloroform

Implantation of Liver Cells

has evaporated, the polymer scaffolds, shaped like disks (18 mm × 1 mm), are placed in distilled water for 48 h to remove the salt particles. The water is changed three times daily. After the salt-leaching process, the polymer scaffolds are highly porous (95%), with pore sizes ranging from 250 to 500 μm. The polymer disks are then immersed in a 1% aqueous PVA solution for 16 h, to coat the surfaces with PVA. The scaffolds are subsequently dried over 24 h, then sterilized with ethylene oxide.

2. One to 2 wk prior to hepatocyte implantation on polymer scaffolds, the recipient animals undergo an end-to-side portacaval shunt. After the animal is anesthetized with methoxyflurane inhalational anesthesia, the abdomen is shaved and sterilely prepped with 70% alcohol and clinidine solution. A midline incision is made and the portal vein and infrahepatic inferior vena cava (IVC) are identified. The portal vein is clamped and divided. The one end of the portal vein closest to the liver is ligated, and the other end closest to the bowel is anastomosed end-to-side to the IVC, using 8-0 monofilament nylon suture. The abdomen is closed in two layers, using 3-0 Prolene suture. The animal is monitored postoperatively for any signs of complications (*see* **Note 6**).

3. The hepatocyte pellet obtained after the isolation procedure is resuspended in Williams' E media to make a 5×10^7 cells/mL suspension. Four hundred μL of the hepatocyte suspension containing 2×10^7 cells are seeded on the polymer scaffold prior to implantation.

4. The recipient rat is anesthetized using methoxyflurane inhalational anesthesia, and is weighed.

5. After the abdomen is shaved, it is sterilely prepped with 70% isopropyl alcohol and clinidine solution.

6. A midline incision is made and the small bowel is eviscerated.

7. The cell–polymer constructs are implanted between the mesenteric leaves, and secured in place using 5-0 Prolene suture.

8. The abdominal contents are returned to their anatomic positions and the abdomen is closed in two layers, using 3-0 Prolene suture.

9. The rat is monitored postoperatively for any signs of complications.

4. Notes

1. 1.7 mmol/L bovine dermal collagen (Vitrogen 100, Collagen, Palo Alto, CA) may be substituted for the Matrigel; however, there is evidence that Matrigel may provide a better attachment substratum that may enhance the function of the microencapsulated hepatocytes *(33)*.

2. There is a wide variation in volumes and concentrations of hepatocyte suspensions infused into the portal vein, in volumes ranging from 0.05 to 4 mL, and concentrations between 2.5×10^6 to 3×10^7 cells. It is important to keep in mind that the risk of complications from portal venous injection is probably higher with increasing volume or concentration of suspensions infused.

3. Other methods used to obtain hemostasis include the application of Surgicel (Ethicon) and Gelfoam (Upjohn, Kalamazoo, MI).

4. There is a wide variation in the concentrations of hepatocyte suspensions injected into the spleen, ranging between 4×10^6 to 1×10^8 cells. The volume injected is usually ≤ 1 mL.
5. Biodegradable polymer scaffolds can be fabricated from polyglycolic acid (PGA) (Davis and Geck, Danbury, CT) and copolymers of polylactic and polyglycolic acids (PLGA) (Medisorb, Cincinnati, OH) in addition to PLLA. The size and shape, the pore size and porosity, and the chemical composition of the scaffolds can be manipulated, depending on the requirements. Different extracellular matrices can be incorporated or coated on the polymer scaffolds to enhance cell attachment, function, and reorganization.
6. Hepatotrophic stimulation is important for the engraftment and survival of heterotopically transplanted hepatocytes *(43,44)*. Since hepatotrophic factors are known to exist in the portal blood *(41,42)*, the creation of a portacaval shunt (PCS) may allow these factors to bypass the liver and enter the systemic circulation in higher concentrations. Partial hepatectomy has also been shown to have hepatotrophic effects, and can be performed using the techniques previously described *(48)*.

References

1. 1996 Annual Report of the U. S. Scientific Registry for Transplant Recipients and Organ Procurement and Transplantation Network—Transplant Data: 1988–1995. UNOS, Richmond, VA, and the Division of Transplantation, Bureau of Health Resources Development, Health Resources and Services Administration, U. S. Department of Health and Human Services, Rockville, MD.
2. American Liver Foundation (1996) Fact Sheet: Hepatitis, Liver and Gallbladder Diseases in the United States. Cedar Grove, NJ, American Liver Foundation.
3. Russell, P. S. (1985) Selective transplantation: an emerging concept. *Ann. Surg.* **201,** 255–262.
4. Skalak, R. and Fox, C. F., eds. (1988) *Tissue Engineering*, Riss, New York.
5. Langer, R. and Vacanti, J. P. (1993) Tissue engineering. *Science* **260,** 920–926.
6. Raper, S. E. and Wilson, J. M. (1993) Cell transplantation in liver-directed gene therapy. *Cell Transplant.* **2,** 381- 400.
7. Berry, M. N. and Friend, D. S. (1969) High yield preparation of isolated rat liver parenchymal cells. *J. Cell Biol.* **43,** 506–520.
8. Seglen, P. O. (1976) Preparation of isolated rat liver cells. *Methods Cell Biol.* **13,** 29–83.
9. Zhang, H., Miescher-Clemens, E., Drugas, G., Lee, S. M., and Colombani, P. (1992) *J. Pediatr. Surg.* **27,** 312–316.
10. Ponder, K. P., Gupta, S., Leland, F., Darlington, G., Finegold, M., DeMayo, J., et al. (1991) Mouse hepatocytes migrate to liver parenchyma and function indefinitely after intrasplenic transplantation. *Proc. Natl. Acad. Sci. USA* **88,** 1217–1221.
11. Gupta, S., Aragona, E., Vemuru, R. P., Bhargava, K. K., Burk, R. D., and Chowdhury, J. R. (1991) Permanent engraftment and function of hepatocytes delivered to the liver: implications for gene therapy and liver regeneration. *Hepatology* **14,** 144–149.
12. Matas, A. J., Sutherland, D. E. R., Steffes, M. W., Mauer, S. M., Lowe, A., Simmons, R. L., and Najarian, J. S. (1976) Hepatocellular transplantation for metabolic deficiencies: decrease of plasma bilirubin in gunn rats. *Science* **192,** 892–894.

13. Sutherland, D. E. R., Numata, M., Matas, A. J., Simmons, R. L., and Najarian, J. S. (1977) Hepatocellular transplantation in acute liver failure. *Surgery* **82**, 124–132.
14. Sommer, B. G., Sutherland, D. E. R., Matas, A. J., Simmons, R. L., and Najarian, J. S. (1979) Hepatocellular transplantation for treatment of D-galactosamine-induced acute liver failure in rats. *Transplantation Proc.* **11**, 578–584.
15. Holzman, M. D., Rozga, J., Neuzil, D. F., Griffin, D., Moscioni, A. D., and Demetriou, A. A. (1993) Selective intraportal hepatocyte transplantation in analbuminemic and gunn rats. *Transplantation* **55**, 1213–1219.
16. Mito, M., Kusano, M., Onishi, T., Saito, T., and Ebata, H. (1978) Hepatocellular transplantation: morphological study on hepatocytes transplanted into rat spleen. *Gastroenterol. Jpn.* **13**, 480–490.
17. Kusano, M. and Mito, M. (1982) Observations on the fine structure of long-survived isolated hepatocytes inoculated into rat spleen. *Gastroenterology* **82**, 616–628.
18. Jiang, B., Sawa, M., Yamamoto, T., and Kasai, S. (1997) Enhancement of proliferation of intrasplenically transplanted hepatocytes in cirrhotic rats by hepatic stimulatory substance. *Transplantation* **63**, 131–135.
19. Makowka, L., Rotstein, L. E., Falk, R. E., Falk, J. A., Zuk, R., Langer, B., Blendis, L. M., and Phillips, M. J. (1981) Allogeneic and xenogeneic hepatocyte transplantation. *Transplantation Proc.* **13**, 855–859.
20. Vacanti, J. P., Morse, M. A., Saltzman, W. M., Domb, A. J., Perez-Atayde, A., and Langer, R. (1988) Selective cell transplantation using bioabsorbable artificial polymers as matrices. *J. Pediatr. Surg.* **23**, 3–9.
21. Uyama, S., Kaufmann, P. M., Takeda, T., and Vacanti, J. P. (1993) Delivery of whole liver-equivalent hepatocyte mass using polymer devices and hepatotrophic stimulation. *Transplantation* **55**, 932–935.
22. Johnson, L. B., Aiken, J., Mooney, D., Schoo, B. L., Griffith-Cima, L., Langer, R., and Vacanti, J. P. (1994) The mesentery as a laminated vascular bed for hepatocyte transplantation. *Cell Transplantation* **3**, 273–281.
23. Selden, C., Gupta, S., Johnstone, R., and Hodgson, H. J. F. (1984) The pulmonary vascular bed as a site for implantation of isolated liver cells in inbred rats. *Transplantation* **38**, 81–83.
24. Sandbichler, P., Then, P., Vogel, W., Erhart, R., Dietze, O., Philadelphy, H., et al. (1992) Hepatocellular transplantation into the lung for temporary support of acute liver failure in the rat. *Gastroenterology* **102**, 605–609.
25. Jaffe, V., Darby, H., Selden, C., and Hodgson, H. J. F. (1988) The growth of transplanted liver cells within the pancreas. *Transplantation* **45**, 497,498.
26. Vroemen, J. P. A. M., Buurman, W. A., van der Linden, C. J., Visser, R., Heirwegh, K. P. M., and Kootstra, G. (1988) Transplantation of isolated hepatocytes into the pancreas. *Eur. Surg. Res.* **20**, 1–11.
27. Ricordi, C., Lacy, P. E., Callery, M. P., Park, P. W., and Flye, M. W. (1989) Trophic factors from pancreatic islets in combined hepatocyte-islet allografts enhance hepatocellular survival. *Surgery* **105**, 218–223.
28. Ricorde, C., Callery, M. P., Lacy, P. E., and Flye, M. W. (1989) Pancreatic islets enhance hepatocellular survival in combined hepatocyte-islet-cell transplantation. *Transplantation Proc.* **21**, 2689–2690.

29. Jirtle, R. L., Biles, C., and Michalopoulos, G. (1980) Morphologic and histochemical analysis of hepatocytes transplanted into syngeneir hosts. *Am. J. Pathol.* **101**, 115–126.
30. Jirtle, R. L. and Michalopoulos, G. (1982) Effects of partial hepatectomy on transplanted hepatocytes. *Cancer Res.* **42**, 3000–3004.
31. Takeda, T., Kim, T. H., Lee, S. K., Langer, R., and Vacanti, J. P. (1995) Hepatocyte transplantation in biodegradable polymer scaffolds using the dalmatian dog model of hyperuricosuria. *Transplantation Proc.* **27**, 635–636.
32. Dixit, V., Darvasi, R., Arthur, M., Brezina, M., Lewin, K., and Gitnick, G. (1990) Restoration of liver function in gunn rats without immunosuppression using transplanted microencapsulated hepatocytes. *Hepatology* **12**, 1342–1349.
33. Dixit, V., Arthur, M., Reinhardt. R., and Gitnick, G. (1992) Improved function of microencapsulated hepatocytes in a hybrid bioartificial liver support system. *Artif. Organs* **16**, 336–341.
34. Mooney, D. J., Kaufmann, P. M., Sano, K., McNamara, K. M., Vacanti, J. P., and Langer, R. (1994) Transplantation of hepatocytes using porous, biodegradable sponges. *Transplantation Proc.* **26**, 3425–3436.
35. Mooney, D. J., Park, S., Kaufmann, P. M., Sano, K., McNamara, K., Vacanti, J. P., and Langer, R. (1995) Biodegradable sponges for hepatocyte transplantation. *J. Biomed. Mater. Res.* **29**, 959–965.
36. Demetriou, A. A., Whiting, J. F., Feldman, D., Levenson, S. M., Chowdhury, N. R., Moscioni, A. D., Kram, M., and Chowdhury, J. R. (1986) Replacement of liver function in rats by transplantation of microcarrier-attached hepatocytes. *Science* **233**, 1190–1192.
37. Demetriou, A. A., Whiting, J., Levenson, S. M., Chowdhury, N. R., Schechner, R., Michalski, S., Feldman, D., and Chowdhury, J. R. (1986) New method of hepatocyte transplantation and extracorporeal liver support. *Ann. Surg.* **204**, 259–271.
38. Demetriou, A. A., Reisner, A., Sanchez, J., Levenson, S. M., Moscioni, A. D., and Chowdhury, J. R. (1988) Transplantation of microcarrier-attached hepatocytes into 90% partially hepatectomized rats. *Hepatology* **8**, 1006–1009.
39. Mooney, D., Hansen, L., Vacanti, J. P., Langer, R., Farmer, S., and Ingber, D. (1992) Switching from differentiation to growth in hepatocytes: control by extracellular matrix. *J. Cell Physiol.* **151**, 497–504.
40. Asonuma K., Gilbert, J. C., Stein, J. E., Takeda, T., and Vacanti, J. P. (1992) Quantitation of transplanted hepatic mass necessary to cure the gunn rat model of hyperbilirubinemia. *J. Pediatr. Surg.* **27**, 298–301.
41. Marchioro, T. L., Porter, K. A., Brown, B. I., Otte, J. B., and Starzl, T. E. (1967) The Effect of Partial Portacaval Transposition on the Canine Liver. *Surgery* **61**, 723–732.
42. Starzl, T. E., Francavilla, A., Halgrimson, C. G., Francavilla, F. R., Porter, K. A., Brown, T. H., and Putnam, C. W. (1973) The origin, hormonal nature, and action of hepatotrophic substances in portal venous blood. *Surg. Gynecol. Obstet.* **137**, 179–199.
43. Sano, K., Cusick, R. A., Lee, H., Pollok, J. M., Kaufmann, P. M., Uyama, S., Mooney, D., Langer, R., and Vacanti, J. P. (1996) Regenerative signals for heterotopic hepatocyte transplantation. *Transplantation Proc.* **28**, 1859,1860.

44. Kaufmann, P. M., Sano, K., Uyama, S., Takeda, T., and Vacanti, J. P. (1994) Heterotopic hepatocyte transplantation: assessing the impact of hepatotrophic stimulation. *Transplantation Proc.* **26,** 2240,2241.
45. Nieto, J. A., Escandon, J., Betancor, C., Ramos, J., Canton, T., and Cuervas-Mons, V. (1989) Evidence that temporary complete occlusion of splenic vessels prevents massive embolization and sudden death associated with intrasplenic hepatocellular transplantation. *Transplantation* **47,** 449,450.
46. Gupta, S, Vemuru, R. P., Lee, C. D., Yerneni, P. R., Aragona, E., and Burk, R. D. (1994) Hepatocytes exhibit superior transgene expression after transplantation into liver and spleen compared with peritoneal cavity or dorsal fat pad: implications for hepatic gene therapy. *Hum. Gene Ther.* **5,** 959–967.
47. Rajvanshi, P., Kerr, A., Bhargava, K. K., Burk, R. D., and Gupta, S. (1996) Efficacy and safety of repeated hepatocyte transplantation for significant liver repopulation in rodents. *Gastroenterology* **111,** 1092–1102.
48. Higgins, G. M. and Anderson, R. M. (1931) Experimental pathology of the liver: I. Restoration of the liver of the white rat following partial surgical removal. *Arch. Pathol.* **12,** 186–202.

33

Isolation and Long-Term Maintenance of Adult Rat Hepatocytes in Culture

François Berthiaume, Ronald G. Tompkins, and Martin L. Yarmush

1. Introduction

Long-term and stable hepatocyte culture systems have a wide variety of uses, both in basic science and in the development of hepatocyte-based applications. In most cases, long-term cultures of hepatocytes are superior to traditional cultures in collagen-coated dishes, which only transiently express a low level of liver-specific function during the first wk in culture *(1,2)*. The collagen sandwich provides a system capable of maintaining long term and stable function of hepatocytes with which to study liver physiology *(3,4)*. This system is now used along with several other long-term hepatocyte culture techniques that have been developed since the mid 1970s. These other methods include the use of special extracellular matrix (ECM) materials, such as an extract from the Engelbreth-Holm-Swarm sarcoma grown in mice *[5]* *(*under the commercial appellations of Matrigel and Biomatrix, Biomedical Technologies, Stoughton, MA), co-culture with mesenchymal, endothelial, or epithelial cells *(6–8)*, special culture media (e.g., dimethyl sulfoxide supplementation or arginine-free formulas), and culture at high seeding densities. In the context of studying liver physiology and morphogenesis of the liver plate, the sandwich culture system appears to be particularly well-suited, since it exhibits in vivo-like ECM geometry, has relatively flexible medium requirements, and individual cell morphology and structure can be easily visualized. One disadvantage, however, is that, in current practice, the ECM layer on top of the cells may present a transport barrier that can slow down the exchange of nutrients, products, and chemical signals with the bulk of the medium. This could be a major disadvantage for kinetic studies requiring addition and removal of factors over a time-scale of less than 1 h (e.g., amino acid transport studies). This problem could be overcome by developing techniques for layering very thin films on top of the cells. Current applications that have used sandwiched rat hepatocyte cultures include: the investigation of hepatic tissue physiology and toxicology *(9–11)*, the development and optimization of hepatocyte and liver preservation techniques *(12,13)*, and the development of clinical applications (e.g., bioartificial liver) *(14–16)*. The following is a detailed protocol to isolate hepatocytes from rat livers and place them in long-term culture, using the collagen sandwich technique.

2. Materials

All unspecified chemicals were from Sigma (St. Louis, MO).

2.1. Isolation of Hepatocytes from Rat Liver

1. Stock solutions and media (can be made in large quantities and well in advance):
   a. *N*-(2-hydroxyethyl)piperazine-*N'*-2-ethanesulfonic acid (HEPES) stock solution, 1000 mM.
   b. Calcium chloride solution, 11 mM.
   c. 10X HBSS: Dissolve 4.0 g KCl, 0.6 g KH_2PO_4, 80.0 g NaCl, 3.5 g $NaHCO_3$, 0.9 g $Na_2HPO_4 \cdot 7H_2O$, and 10.0 g D-glucose in 975 mL of distilled water. Adjust the pH to 7.4. Sterilize by filtration through a 0.2 μm pore membrane.
   d. 10X Dulbecco's modified Eagle's medium (DMEM): dissolve powder with high glucose, without sodium pyruvate, without sodium bicarbonate (Gibco-BRL, Grand Island, NY, #12100) in one-tenth the recommended amount of water. Filter-sterilize through a 0.2-μm porous membrane.
2. Solutions prepared the day of isolation:
   a. Krebs-Ringer buffered saline (KRB): 154 mM NaCl, 27 mM KCl, 5.5 mM D-glucose, and 5.4 mM $NaHCO_3$, 20 mM HEPES. Adjust pH to 7.4.
   b. Ethylenediamine tetraacetic acid solution (EDTA): 1 mM EDTA in KRB; stir at low heat until dissolved.
   c. Collagenase solution (prepare less than 3 h prior to isolation): Dissolve 0.05% w/v collagenase powder (type IV, Sigma) in KRB and add 5% v/v $CaCl_2$, solution. Stir for 30 min at room temperature, filter through a 0.45-μm-pore membrane, and then a 0.2-μm membrane. Keep at 37°C until use.
   d. Percoll (Pharmacia, Milwaukee, WI): Mix 43.2 mL manufacturer's stock with 4.8 mL of 10X DMEM.
3. Rats (Charles River, Wilmington, MA): Female Lewis; preferred size is 180–220 g.
4. Surgical instruments, hardware:
   a. Perfusion apparatus (**Fig. 1**).
   b. Autoclaved/sterile surgical instruments (surgical blade, scissors, toothed tweezers, fine curved forceps).
   c. 6-0 silk sutures.
   d. 16-gage 2 in Angiocath™ iv catheter/needle unit (Becton Dickinson Vascular Access, Sandy, Utah).

Adult Rat Hepatocytes in Culture

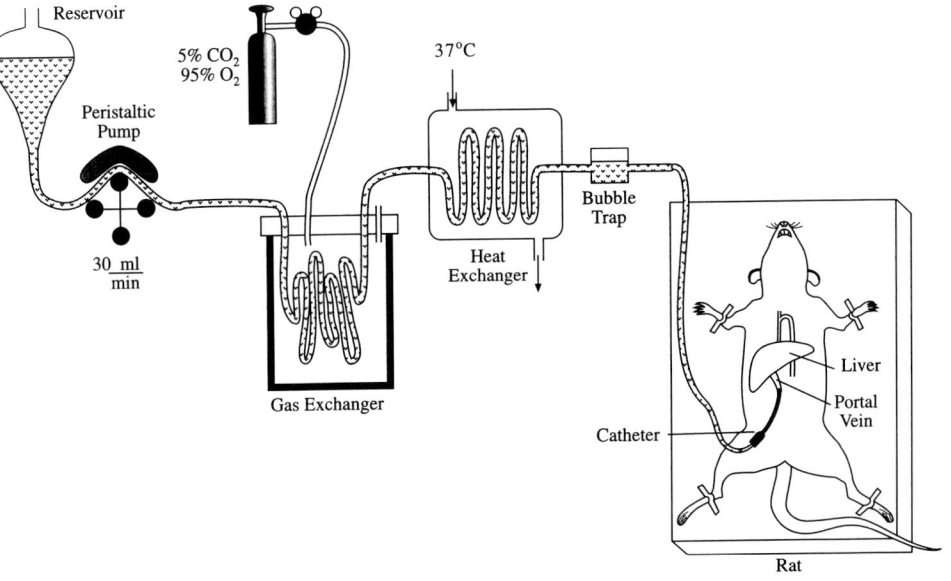

Fig. 1. Collagenase perfusion system for hepatocyte isolation.

    e. Surgical prep solution (Denison Pharmaceuticals, Pawtucket, RI).
5. Collagen: 1.1% rat tail tendon collagen in 1 mM HCl. It is prepared as follows *(17)*: Tendons are dissected from rat tails and stirred in 200 mL of 3% v/v acetic acid overnight at 4°C. The solution is filtered through layers of cheesecloth and centrifuged at 12,000g for 2 h. The supernatant is precipitated with 40 mL 30% w/v sodium chloride added in a dropwise fashion. The pellet is collected by centrifugation at 4000g for 30 min. The pellet is dissolved in 50 mL of 0.6% (v/v) acetic acid, and the solution dialyzed against 500 mL 1 mM HCl 5X. For sterilization, 0.15 mL chloroform is added to the solution which is then stirred for 2 d loosely capped to allow evaporation of the chloroform. The yield is approx 100 mg/rat tail.
6. Hepatocyte culture medium:
    a. DMEM with high glucose, without sodium pyruvate, without sodium bicarbonate (Gibco-BRL).
    b. Defined fetal bovine serum (JRH Biosciences, Lenexa, KS). Add 10% v/v to DMEM.
    c. Glucagon (Lilly, Indianapolis, IN): Add to DMEM to a final concentration of 7 ng/mL.
    d. Insulin (Squibb, Princeton, NJ): Add to DMEM to a final concentration of 0.5 U/mL.
    e. Epidermal growth factor (Becton Dickinson, Bedford, MA): Add to DMEM to a final concentration of 20 ng/mL.
    f. Hydrocortisone (Upjohn, Kalamazoo, MI): Add to DMEM to a final concentration of 7.5 µg/mL.

g. Penicillin–streptomycin, respectively, 5000 U/mL and 5 mg/mL in 0.9% NaCl (Sigma): Add 1% (v/v) to medium.
h. Filter-sterilize medium through a 0.2-μm membrane.

7. Characterization of hepatocytes.
    a. Fixative: 4% paraformaldehyde in phosphate buffered saline (PBS), freshly prepared. For making 100 mL, first boil 50 mL of dH_2O on a hot plate and add 4 g paraformadehyde while stirring. Add 2–3 drops of 6 M NaOH to cause the paraformaldehyde to dissolve. Add 40 mL of dH_2O and 10 mL of 10X PBS (Biofluids, Rockville, MD), and allow to cool down on ice. Keep at 4°C until use.
    b. 0.1% Triton X-100 in 1X PBS.
    c. 0.1% bovine serum albumin (BSA) in 1X PBS.
    d. Rhodamine-phalloidin (Molecular Probes, Eugene, OR): Dissolve 300 U in 1.5 mL methanol and dilute 100-fold in 1X PBS.
    e. Biotin Blocking System X0590, Peroxidase Blocking Reagent S2001, Rabbit Primary Universal Peroxidase Kit K0684, Dako, Carpintera, CA. Prepare streptavidin reagent and substrate-chromogen reagent as described in manufacturer's protocol.
    f. Anti-rat albumin, rabbit IgG fraction (Cappel, Cochranville, PA): Dilute 25 μL product in 10 mL 1XPBS.

3. Methods
3.1. Isolation of Hepatocytes from Rat Liver

The method for isolating endothelial cells presented below is a modification of the procedures originally published by Seglen et al. *(18)*. Proper organization and the operator's care in following sterile handling techniques are critical for the success of the isolations. Furthermore, it is extremely important to handle collagenase-digested tissue and dissociated liver cells as gently as possible, because they are very sensitive to mechanical damage.

3.1.1. Perfusion and Collagenase Digestion

1. Fill the reservoir with 70% ethanol. Turn on the pump and set flow rate to 50 mL/min. Flush the system with 70% ethanol for 15 min (this requires a total amount of ethanol of 750 mL).
2. When the ethanol level is almost at the bottom of the reservoir, fill with dH_2O and let the system flush with water. Refill the reservoir with water and let the system flush again.
3. When the water level is almost at the bottom of the reservoir, add 100 mL KRB–EDTA to the reservoir. When the liquid level is almost at the bottom of the reservoir, add 400 mL KRB–EDTA to the reservoir and place the outlet of the system in the reservoir to allow recirculation. Turn on the gas.
4. Anesthetize rat with ether and place it on a perforated plate. Sterilize the ventral area using surgical prep solution. Open the abdominal cavity and push the intes-

tines to the side to expose the portal vein and inferior vena cava. Place a ligature around the portal vein proximal to the liver near the hilus and another one distally between the mesenteric veins. Tie ligatures loosely.

5. Insert the catheter into the portal vein and slide the cannula past the ligature proximal to the liver, without going beyond the portal bifurcation. When blood starts to flow slowly out of the catheter, tie the distal ligature. Connect the outlet of the perfusion system to the catheter. Immediately cut open the inferior vena cava and tie the second ligature.
6. Open the thoracic cage through the diaphragm, and cut open the inferior vena cava above the diaphragm.
7. When the fluid level in the reservoir is near the bottom, add the collagenase solution to the reservoir. Care should be taken to prevent bubbles reaching the liver. If a bubble is seen to go past the bubble trap, temporarily detach the perfusion system outlet from the catheter to let the bubble pass.
8. When all the collagenase has gone through the liver, disconnect the catheter from the perfusion system and place the liver in a 100-mm dish containing ~15 mL cold KRB. Place on ice and take to a laminar flow hood.
9. The perfusion system should be extensively flushed with dH_2O, and then stop the pump.

3.1.2 Purification of Liver Cell Suspension

All the following operations must be performed while keeping the cells at 0–4°C.

1. Hold the liver at the tip of the vascular tree with tweezers, and, using a surgical rake, break open the liver lobes. Using a combination of agitation and raking, disperse the brown-colored material around the liver's vascular tree (the latter is not digested by the collagenase).
2. Prewet a 250-μm mesh-size nylon filter (Small Parts, Miami Lakes, FL) with sterile KRB, and pass the cell suspension through the mesh by using a gentle back and forth swirling motion. Use additional sterile KRB to clear any remaining cells from the filter. Repeat procedure using a 62-μm mesh-size nylon filter (Small Parts).
3. Aliquot the cell suspension in 50-mL conical centrifuge tubes and centrifuge at 202g for 5 min. Resuspend all pellets in KRB to a final total volume of 50 mL. Centrifuge at 202g for 5 min. Resuspend all pellets in KRB to a final total volume of 50 mL and distribute in four 12.5-mL aliquots in 50-mL conical tubes.
4. Add 12 mL Percoll per tube. Cap tight and mix by flipping the tube up and down. Centrifuge at 60g for 5 min. Aspirate supernatants, which may contain many cells, and resuspend the pellet in KRB to a final volume of 50 mL.
5. Centrifuge cell suspension at 202g for 5 min and resuspend in KRB to a final volume of 30 mL. Keep on ice until use.
6. For cell counting, take a 100-μL aliquot of cell suspension and mix to 800 μL of 1X PBS and 100 μL of 0.4% trypan blue solution (Sigma). Yields are typically in

the range of 150–250 × 10^6 viable cells, and the fraction of viable cells is ~80–90%.

3.2. Long-term Culture of Freshly Isolated Rat Hepatocytes

The following procedure is for cultures in standard 60-mm dishes. Other size dishes can be used as long as the seeding density is maintained at around 10^5 cells/dish.

1. Place dishes on a small tray that fits in the incubator.
2. Mix 9 parts of of 0.1% rat type I collagen in 1 mM HCl with 1 part 10X DMEM at 0–4°C. Precoat 60-mm culture dishes (tissue-culture-treated) by pouring 1 mL of the reconstituted collagen solution in each dish, spreading it on the entire surface of the dish before gellation occurs. Incubate dishes at 37°C and 10% CO_2/90% air atmosphere to allow the collagen to gel and the pH to equilibrate to 7.4.
3. Wet each collagen-coated dish with 2 mL of hepatocyte culture medium. Aspirate medium.
4. Resuspend the cells to a final concentration of 10^6 cells/mL, using hepatocyte culture medium. Pipet 2 mL of this suspension per 60 mm dish. This yields cultures that are approximately half confluent. The hepatocyte suspension settles very quickly, and thus it is recommended to load the pipet with enough suspension to seed only three dishes at a time. The suspension should be swirled before loading the pipet again, in order to ensure reproducibility in the seeding.
5. Just prior to placing the dishes in the incubator, use a quick back and forth motion to spread the cells evenly over the entire collagen surface in the dishes. **Caution:** Do not use a circular motion, since this would have the opposite effect.
6. Place overnight in an incubator at 37°C in a 10% CO_2 atmosphere. The next day, aspirate the medium and place 1 mL/dish of reconstituted collagen on top of the cells. Place 30 min in the incubator to gel the collagen, and add 2 mL of fresh hepatocyte culture medium.
7. Cultures are fed daily.

3.3. Characterization of Cultured Hepatocytes

The procedures outlined below are designed to assess the expression of differentiated, liver-specific characteristics in cultured hepatocytes. Because freshly isolated cells may require as much as 1 wk of culture before differentiation, these assays are best done at 1 wk postseeding, or later.

3.3.1. Staining for Intracellular Albumin

Albumin is exclusively synthesized in liver by hepatocytes. Thus, positive albumin staining is unequivoqual evidence for differentiated hepatocytes. Lack of staining indicates lack of expression of differentiated function or a contaminating liver nonparenchymal cell. The following protocol was developed by S. Bhatia *(19)*.

1. Rinse cultures with PBS and fix with 4% paraformaldehyde at 0–4°C for 30 min. Rinse with PBS again, and add 0.1% Triton X-100 for 10 min.

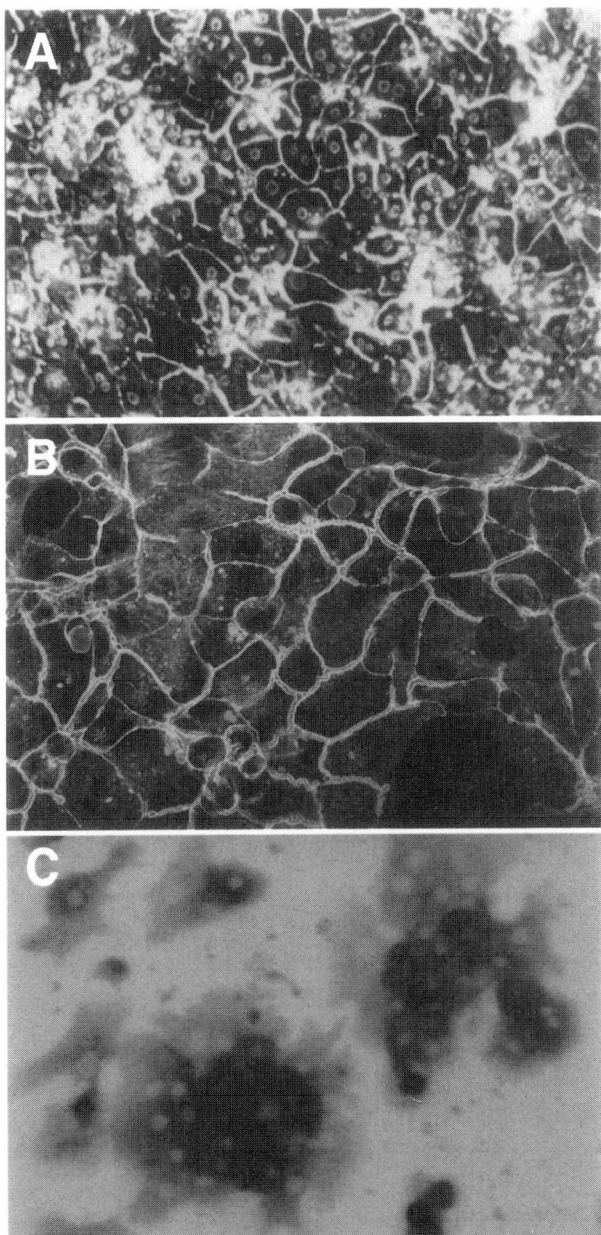

Fig. 2. Morphological appearance and hepatocellular-specific characteristics. (A) Hepatocyte monolayer as seen under the phase microscope. (B) Staining for actin microfilaments using rhodamine–phalloidin. (C) Intracellular albumin staining by immunofluorescence techniques.

2. Rinse with PBS, and incubate with avidin and biotin blocking reagents for 20 min each.
3. Rinse with PBS, and treat with peroxidase blocking reagent. Remove excess blocking reagent without rinsing.
4. Incubate with antirat albumin antibody solution for 20 min.
5. Rinse with PBS. Treat with antirabbit IgG peroxidase staining kit.
6. Observe by transmission microscopy. Typical morphology is shown in **Fig. 2**.

3.3.2. Staining for Actin Microfilaments

Differentiated hepatocytes have been shown to form intercellular bile canaliculi, which are surrounded by a belt of actin microfilaments *(10)*. Actin microfilaments are not normally found elsewhere in the cytoplasm and do not form stress fibers, except in poorly differentiated cultures, for example, on a single collagen gel *(20–21)*. Thus, actin staining, being extremely simple and specific, is a useful way to assess the state of differentiation of hepatocytes in cultures.

1. Rinse cultures with PBS and fix with 4% paraformaldehyde at 0–4°C for 30 min. Rinse with PBS again, and add 0.1% Triton X-100 for 10 min.
2. Rinse with PBS 3X, and incubate with rhodamine-phalloidin for 30 min at 37°C.
3. Rinse with PBS 3X, and observe by fluorescence microscopy, using a standard rhodamine filter set. Typical morphology is shown in **Fig. 2**.

4. Notes

1. The expression of differentiated functions and markers by sandwiched hepatocytes may vary, depending on the chemical composition of the substrate. For example, intercellular gap junctions have not been detected in type I collagen sandwich cultures, but are expressed when using Matrigel *(22)*.
2. Hepatocytes have relatively high metabolic and oxygen-consumption rates. For this reason, the oxygen tension near the cells is usually considerably lower than at the medium–gas phase interface. Thus, when seeding $2 \yen 10^6$ cells/60 mm dish, it is not recommended to add more than 1 mm thickness of medium on top of the cells, to ensure adequate diffusion from the gas phase to the cells. Furthermore, the oxygen tension near the cells is a function of seeding density, a factor that may have to be taken into account when comparing cell function at different seeding densities. A more complete discussion of this issue can be found in **ref. *23***.

Acknowledgments

This work was supported in part by the Shriners Hospitals for Children. The authors would like to thank Erika Swinnich for the preparation of **Fig. 1.**, and Ulysses Balis for **Fig. 2C**.

References

1. Clayton, D. F. and Darnell, J. E., Jr. (1983) Changes in liver-specific compared to common gene transcription during primary culture of mouse hepatocytes. *Mol. Cell Biol.* **5,** 2623–2632.

2. Clayton, D. F., Harrelson, A. L., and Darnell, J. E., Jr. (1985) Dependence of liver-specific transcription on tissue organization. *Mol. Cell Biol.* **5,** 2623–2632.
3. Dunn, J. C. Y., Yarmush, M. L., Koebe, H. G., and Tompkins, R. G. (1989) Hepatocyte function and extracellular matrix geometry: long-term culture in a sandwich configuration. *FASEB J.* **3,** 174–177.
4. Berthiaume, F., Moghe, P. V., Toner, M., and Yarmush, M. L. (1996) Effect of extracellular matrix topology on cell structure, function, and physiological responsiveness: hepatocytes cultured in a sandwich configuration. *FASEB J.* **10,** 1471–1484.
5. Bissell, D. M., Arenson, D. M., Maher, J. J., and Roll, F. J. (1987) Support of cultured hepatocytes by a laminin-rich gel. Evidence of a functionally significant subendothelial matrix in normal rat liver. *J. Clin. Invest.* **79,** 801–812.
6. Kuri-Harcuch, W. and Mendoza-Figueroa, T. (1989) Cultivation of adult rat hepatocytes on 3T3 cells: expression of various liver differentiated functions. *Differentiation* **41,** 148–157.
7. Guillouzo, A., Delers, F., Clément, B., Bernard, N., and Engler, R. (1984) Long term production of acute-phase proteins by adult rat hepatocytes co-cultured with another cell type in serum-free medium. *Biochem. Biophys. Res. Commun.* **120,** 311–317.
8. Conner, J., Vallet-Collom, I., Daveau, M., Delers, F., Hiron, M., Lebreton, J.-P., and Guillouzo, A. (1990) Acute-phase-response induction in rat hepatocytes co-cultured with rat liver epithelial cells. *Biochem. J.* **266,** 683–688.
9. Bader, A., Rinkes, I. H. B., Closs, E. I., Ryan, C. M., Toner, M., Cunningham, J. M., Tompkins, R. G., and Yarmush, M. L. (1992) A stable long-term hepatocyte culture system for studies of physiologic processes: cytokine stimulation of the acute phase response in rat and human hepatocytes. *Biotechnol. Prog.* **8,** 219–225.
10. LeCluyse, E. L., Audus, K. L., and Hochman, J. H. (1994) Formation of extensive canalicular networks by rat hepatocytes cultured in collagen-sandwich configuration. *Am. J. Physiol.* **266,** C1764–C1774.
11. Rotem, A., Matthew, H. W. T., Hsiao, P. H., Toner, M., Tompkins, R. G., and Yarmush, M. L. (1995) The activity of cytochrome P450 IA1 in stable cultured rat hepatocytes. *Toxicol. In Vitro* **9,** 139–149.
12. Borel-Rinkes, I. H. M., Toner, M., Sheehan, S. J., Tompkins, R. J., and Yarmush, M. L. (1992) Long-term functional recovery of hepatocytes after cryopreservation in a three-dimensional culture configuration. *Cell Transplantation* **1,** 281–292.
13. Stefanovich, P., Toner, M., Ezzell, R. M., Sheehan, S. J., Tompkins, R. G., and Yarmush, M. L. (1995) Effects of hypothermia on the function, membrane integrity, and cytoskeletal structure of hepatocytes. *Cryobiology* **23,** 389–403.
14. Bader, A., Knop, E., Böker, K., Frühauf, N., Schüttler, W., Oldhafer, K., et al. (1995) A novel bioreactor design for in vitro reconstruction of in vivo liver characteristics. *Artif. Organs* **19,** 368–374.
15. Taguchi, K., Matsushita, M., Takahashi, M., and Uchino, J. (1996) Development of a bioartificial liver with sandwiched-cultured hepatocytes between two collagen gels. *Artif. Organs* **20,** 178–185.
16. Koike, M., Matsushita, M., Taguchi, K., and Uchino, J. (1996) Function of culturing monolayer hepatocytes by collagen gel coating and coculture with nonparenchymal cells. *Artif. Organs* **20,** 186–192.

17. Elsdale, T. and Bard, J. (1972) Collagen substrata for for studies on cell behavior. *J. Cell. Biol.* **54,** 626–637.
18. Seglen, P. O. (1976) Preparation of isolated rat liver cells. *Methods Biol.* **13,** 29–83.
19. Bhatia S. (1997) Controlling cell-cell interactions in hepatic tissue engineering using microfabrication. PhD Thesis, Harvard–MIT.
20. Dunn, J. C. Y., Tompkins, R. G., and Yarmush, M. L. (1991) Long-term in vitro function of adult hepatocytes in a collagen sandwich configuration. *Biotechnol. Prog.* **7,** 237–245.
21. Ezzell, R. M., Toner, M., Hendricks, K., Dunn, J. C. Y., Tompkins, R. G., and Yarmush, M. L. (1993) Effect of collagen gel configuration on the cytoskeleton in cultured rat hepatocytes. *Exp. Cell Res.* **208,** 442–452.
22. Moghe, P. V., Berthiaume, F., Ezzell, R. M., Toner, M., Tompkins, R. G., and Yarmush, M. L. (1996) Role of extracellular matrix composition and configuration in maintenance of hepatocyte polarity and function. *Biomaterials* **17,** 373–385.
23. Yarmush, M. L., Toner, M., Dunn, J. C. Y., Rote, A., Hubel, and Tompkins, R. G. (1992) Hepatic tissue engineering. Development of critical technologies. *Ann. NY Acad. Sci.* **665,** 238–252.

34

Design and Fabrication of a Small Caliber Hybrid Arterial Bioprosthesis

Gilbert J. L'Italien and William M. Abbott

1. Introduction

Despite two generations of vascular prosthetic research, the optimal arterial surgical replacement for aortocoronary, femoropopliteal, and femorodistal reconstruction remains autogenous saphenous vein. It has become apparent from numerous investigations that the main reasons for the advantageous clinical performance of autogenous saphenous vein are its antithrombogenic endothelial surface and its elastic properties which resemble those of the host artery. The thromboresistant surface of the autogenous vein assures satisfactory patency, and favorable biomechanical properties minimize the development of occlusive neointimal hyperplasia. Clearly, these factors must be given serious consideration in the design of an arterial prosthesis. The ideal arterial replacement will possess a thromboresistant surface, which can withstand arterial fluid shear rates, and biomechanical properties that are similar to those of the host artery. In this chapter, we describe the design and fabrication of a hybrid blood vessel substitute that is comprised of three components: a naturally compliant silicone rubber matrix; a basement membrane composed of 0.1% fibronectin; and a living, confluent human saphenous vein endothelial cell monolayer that can withstand arterial fluid shear rates and arterial cyclic strain. We will describe in detail the comparative biomechanical properties of the hybrid prosthesis, and we will characterize the attachment, growth, and differentiation properties of the vascular wall cells comprising the hybrid prosthesis.

2. Materials

2.1. Fabrication of Silicone Rubber Tube Matrix

2.1.1. Preparation of the Silicone Rubber Solution

1. Silastic medical grade elastomer (Solution A: MDX4-4210, Dow-Corning Medical, Arlington, TN).
2. Curing agent (Solution B: 1,1,1 trichloroethane, Fisher Scientific, Pittsburgh, PA).

2.1.2. Fabrication of Silicone Rubber Tubes

1. Glass rods (Fisher Scientific).
2. 1% gelatin (Difco, Detroit MI).
3. Harvard infusion pump (Model 907; Harvard Apparatus, Natick, MA).
4. 15-336-26 ultrasonic cleaning solution (Fisher Scientific).

2.1.3. Application of the Fibronectin Basement Membrane Prior to Cell Culture

1. 0.01% fibronectin (Collaborative Biomedical).

2.2. Cell Culture

2.2.1. Establishment of Vascular Wall Cell Cultures

1. 0.1% collagenase (CLS II, Worthington).
2. 0.5% human serum albumin (Fraction V, Sigma,, St. Louis, MO).
3. 0.1% soy bean trypsin inhibitor (Sigma).
4. M199 (Gibco-BRL, Grand Island, NY).
5. Hanks' salts (Sigma).
6. HEPES (Sigma): 25 mmol/L.
7. Sodium bicarbonate (Sigma): 0.2%.
8. Fetal calf serum (FCS; Hyclone defined): 20%.
9. Heparin (Sigma): 150 mg/mL.
10. L-glutamine (Gibco-BRL): 2 mmol/L.
11. Penicillin and streptomycin (Gibco-BRL): 100 U/mL, 100 mg/mL, repectively.
12. EC growth supplement (ECGS; Collaborative Research): 50 µg/mL.

2.2.2. Application of Cells onto the Silicone Rubber Tube Matrix

1. Blood-gas analyzer (Model 1306, Instrumentation Laboratories, Lexington, MA).
2. M199 supplemented with 20% FCS.
3. 50 µg/mL ECGS, and 150 mg/mL heparin.

2.2.3. Arterial Pulsatile Fluid Shear and Cyclic Strain Analysis

1. Bellows pump (Model 14250-007; Gorman Rupp, Bellville, OH).
2. Electromagnetic flowmeter (Statham SP2202, Statham Instruments, Oxnard, CA).
3. P23ID pressure transducer.
4. Strip chart recorder (Hewlett-Packard, Model 7702B).

5. Video-tracked motion analyzer (VMA; Motion Analysis, Santa Rosa, CA).

2.2.4. Determination of Cell Viability and Maintenance of Differentiated Products After Application of Physiological Pressure, Pulsatile Flow, and Wall Strain

1. Particle counter (Coulter).
2. Di-I-acetylated LDL (BTI, Stoughton, MA).
3. 4% paraformaldehyde (Fisher Scientific).

3. Methods
3.1. Fabrication of Silicone Rubber Tube Matrix
3.1.1. Preparation of the Silicone Rubber Solution

Silicone rubber solution is prepared from a Silastic medical grade elastomer and a curing agent. The elastomer consists of 30% Silastic medical grade elastomer and 70% of the solvent. The curing agent is mixed 1:2.5 with the solvent to generate solution B. The final solution is prepared by mixing solution A with solution B in a 10:1 ratio. The obtained solution is thoroughly mixed with a magnetic stirrer, poured into a 50-mL glass cylinder, degassed under vacuum for 30 min, and than allowed to sit for 24 h before use. The entire procedure is performed at room temperature in a chemical fume hood.

3.1.2. Fabrication of Silicone Rubber Tubes

Before immersion, glass rods are coated with 1% gelatin at room temperature in distilled water by dipping, to facilitate the separation of the Silastic tube from the rod. The gelatinized rods are dried in an oven at 70°C. for 10 min prior to use. A vertically mounted Harvard infusion pump is used to lower the glass rod into the Silastic solution at a constant rate (4 mm/s), thus achieving a uniform wall thickness. After dipping, the Silastic covered rods are placed horizontally in a drying oven a 70°C for 20 min. By varying the number of dips and the diameter of the mandril, a range of elastic moduli and calibers are achievable that approximate those observed in human blood vessels. The Silicone rubber tubes (5 mm id × 115 mm in length) are then cleaned ultrasonically for 10 min in a 1:20 dilution of 15-336-26 ultrasonic cleaning solution in distilled water, followed by two 10-min ultrasonic washes in distilled water only. Tubes are sterilized under UV light for 48 h in a laminar flow hood, then autoclaved. The resultant tube matrices are transparent and readily visible with conventional phase contrast and fluorescence microscopy. Tubes are cannulated with a Luer connector and fastened to the connector via 2.0 silk suture for delivery of biologic materials. Cannulated tubes are mounted to a specially designed tube holder (*see* **Fig. 1**) comprised of a transparent plastic box and tube

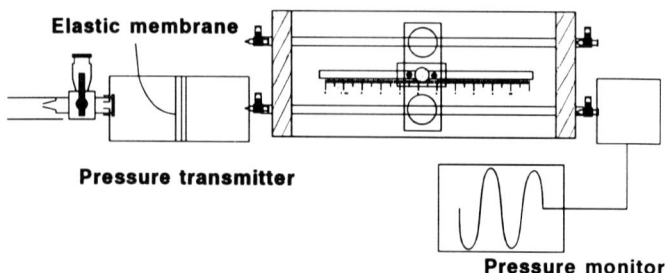

Fig. 1. Schematic diagram of the tube holder.

occluder. The box is designed to hold two tubes and to accommodate the axis of a rotation device (not shown). The rotation device is used to slowly rotate (10 revolutions/h) solutions and cell suspensions within the Silastic tubes, to achieve even coating or dispersion. The tube occluder is a rectangular, transparent segment of plastic mounted to a threaded rod, which permits both vertical and longitudinal movement. Thus, different positions along the length of the tube can be occluded for microscopic observation.

3.1.3. Application of the Fibronectin Basement Membrane Prior to Cell Culture:

Solutions of 0.01% fibronectin in sterile dH_2O are prepared just prior to use. Tubes are filled with the fibronectin solution, then closed (via three-way valve inserted into the connector piece) and rotated at 10 revolutions/h for 1 h at room temperature. The fibronectin was removed and replaced with complete medium, and the tubes are rotated for an additional hour at room temperature. Prior to cell seeding, tubes are emptied of complete medium, and cell suspensions are added at a density of 10^5 cells/cm^2 surface area.

3.2. Cell Culture

3.2.1. Establishment of Vascular Wall Cell Cultures

Human saphenous vein endothelial cells (HSVEC) are isolated by enzymatic treatment (0.1% collagenase, 0.5% Human serum albumin, and 0.1% soy bean trypsin inhibitor) of the lumenal surface of the vessels. Primary cultures are established, maintained, and passaged on gelatin (1%)-coated tissue-culture plastic. Cells are plated to the fibrinectin-coated silicone rubber tubes in passages 3–6. HSVEC are cultured in M199 containing Hanks' salts and supplemented with HEPES (25 mmol/L) to which are added sodium bicarbonate (0.2%), FCS (20%,), heparin (150 mg/mL,), L-glutamine (2 mmol/L), penicillin and streptomycin (100 U/mL and 100 mg/mL, respectively), and ECGS (50 µg/mL).

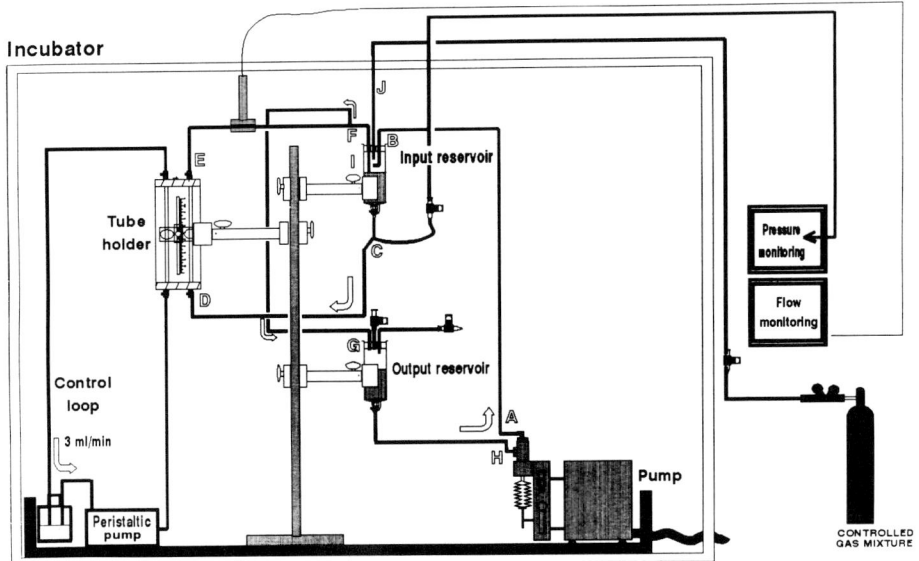

Fig. 2. Schematic diagram of the pumping system.

3.2.2. Application of Cells onto the Silicone Rubber Tube Matrix

Suspensions of known cell densities are applied to two parallel tubes and incubated at 37°C in an atmosphere of 95% air/5% CO_2, with rotation at 10 revolutions/h. Routinely, plating is at a density of 1×10^5 cells/cm^2. The total internal surface area of each tube, including connecting pieces, is 24 cm^2.

The porosity of the silicone rubber to gas exchange permits the pH of the culture medium to be maintained at neutrality, as determined visually from the indicator color of the culture medium. This can be corroborated from direct measurements of pH and pO_2 in the culture medium, using a blood-gas analyzer. Measurements can be made on aliquots of medium withdrawn from the tubes.

After 24–48 h rotational incubation, cells can be examined visually and photographed with an inverted microscope using phase contrast optics. Cells can be maintained on the tubes with M199 supplemented with 20% FCS, 50 µg/mL ECGS, and 150 mg/mL heparin.

3.2.3. Arterial Pulsatile Fluid Shear and Cyclic Strain Analysis

After the initial 24 h incubation period, tubes may be subjected to physiological conditions of combined pressure, flow, and wall strain, equivalent to that observed in peripheral arteries (pressure: 125/75 mmHg, with mean of 100 mmHg; pulsatile flow: 235/100 mL/min, with mean of 120 mL/min; frequency: 1 Hz; compliance : 5 + 1%/ mmHg 10^{-2}). To accomplish this, a specially designed flow and pressure delivery system is required. The pulsatile component of the system is a bellows pump driven by an adjustable eccentric cam that generates pulsatile flow, pressure, and diameter waveforms (*see* **Fig. 2**). Pump outflow is connected to a damping inflow reservoir, which serves to eliminate bubbles and to regulate pulse pressure by varying the air volume. The inlet line to the tube must be isodiametric with the hybrid bioprosthesis, and long enough to ensure that velocity profiles under pulsatile flow are parabolic and fully developed. The mean pressure of the system is adjusted by external pressurization of the inflow reservoir. Pulsatile flow rates can be measured using a cannulating electromagnetic flowmeter. Pressure can be measured using a pressure transducer coupled to a strip chart recorder.

Biomechanical properties (compliance, elastic modulus) can be measured using a video-tracked motion analyzer (VMA). The VMA uses a high-contrast imaging technique to measure the two-dimensional translational movement of an object. The image is converted to a gray-scale pixel file, and the motion of the object can be computed using the commercially available VMA software.

3.2.4. Determination of Cell Viability and Maintenance of Differentiated Products After Application of Physiological Pressure, Pulsatile Flow, and Wall Strain

In addition to the use of trypan blue exclusion, viability of cells, after culture on silicone rubber tubes in the presence of pulsatile flow, pressure, and strain, may be assessed by subsequent passage and growth determination. The presence of cell-specific proteins in these cultures may also be examined.

Specifically, after harvest of cells from tube cultures, aliquots of the cell suspension can be plated at known densities, on fibronectin-coated plastic culture wells. At 24 h and progressive time intervals following plating, the cells may be harvested and counted. Cell harvest from tubes is accomplished as follows: The medium is removed, tubes are rinsed in Hanks' solution, and cells are enzymatically removed from the lumenal surface with trypsin-EDTA, then collected for counting. The dissociation is stopped by the addition of complete medium or serum, and cells are pelleted by centrifugation. Pellets are resuspended in trypan blue in PBS. Cells may be examined and counted in a hemocytometer of particle counter. These determinations serve to demonstrate the maintenance of sterility and viability in the tube cultures. EC cultures can also be incubated in di-I-acetylated LDL, or fixed in fresh 4% paraformaldehyde and immunostained with antibody to factor VIII-related antigen.

3.3. Results of Analyses

3.3.1. Effects of Pulsatile Flow, Strain, and Pressure on the Adhesion and Orientation of HSVEC to Silicone Rubber Matrix

In our experience, 48-h application of physiological levels of pulsatile flow, pressure, and strain should not cause detachment of vascular wall cells from

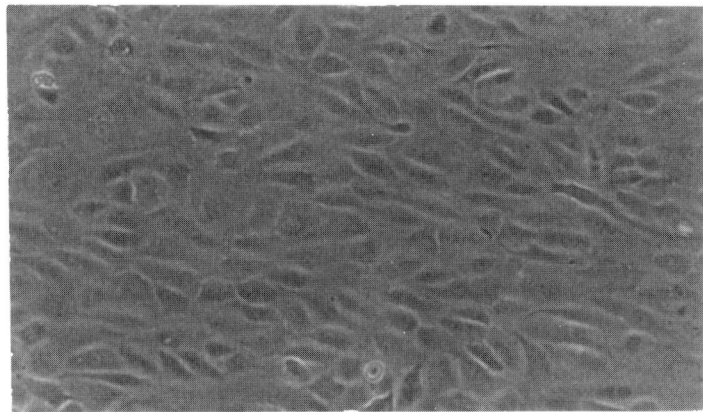

Fig. 3. Phase contrast micrographs of HSVEC of the hybrid bioprosthesis. After 48 hr of pulsatile flow and cyclic strain, dense monolayers remained. Cells consistently demonstrate alignment in the direction of flow.

the tube surface. Dense monolayers remained in the presence of 48 h of pulsatile flow, strain and pressure (**Fig. 3**). Cell alignment occurs consistently in the direction of flow, i.e., along the axis of the tube.

To quantify the actual numbers of cells retained on control tube surfaces, cells were enzymatically harvested from the tubes and total cell numbers determined. The mean percentage of cells recovered from the bioprosthesis (i.e., adhesion) was 98 ± 10%.

3.3.2. Cell Viability and Maintenance of Differentiated Phenotype on Silicone Rubber Tube Matrices

Viability, as assessed by trypan blue exclusion of cells recovered from the bioprosthesis, was 89 ± 2%. To further assess viability, cells isolated from tubes were replated at known densities onto fibronectin-coated wells of tissue-culture plastic, and grown for up to 12 d. Initial cell densities after 24 h incubation, as well as the final cell numbers, were determined. Although the absolute numbers varied between experiments, the final cell densities reached were similar for cells harvested from tubes as for cells grown on flat silicone rubber surfaces (**Fig. 4**).

Immunostaining with antibodies to factor VIII-related confirmed that cells isolated from the tubes retained differentiated characteristics.

3.3.3. Biomechanical Properties of the Hybrid Bioprosthesis

The stability of the elastic properties of the hybrid bioprosthesis was assessed over a period of 28 d of pulsatile flow and cyclic strain at a mean

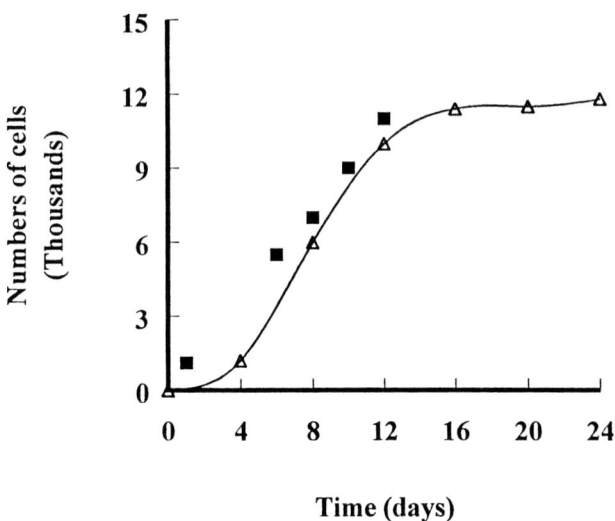

Fig. 4. Growth curves of HSVEC on flat surfaces of silicone rubber (SIL) coated with fibronectin and cells isolated from the hybrid bioprosthesis after 48 h of pulsatile flow and cyclic strain. Cells were plated at known densities onto the indicated surfaces, and harvested and counted at regular intervals over an approx 3-wk period. △, Flat; ■, hybrid bioprosthesis.

pressure of 100 mmHg and a mean flow rate of 150 mL/min (mean shear ~3dyns/cm^2). No change was observed in the circumferential compliance or diameter of the bioprosthesis during the 28 d period (**Table 1**).

We also compared the elastic modulus (E_{inc}) of the bioprosthesis to that of a human femoral artery (**Table 2**). At physiologic pressures (90–110 mmHg), the values E_{inc} were similar for both the artery and hybrid prosthesis.

A range of compliance values can be obtained for the bioprosthesis by varying the wall thickness (i.e., dip frequency) of the silicone rubber matrix. As shown in **Table 3**, there is a linear relationship between thickness:radius ratio and compliance (r = .988); the range of compliance values shown encompasses that of a variety of canine and human blood vessels.

4. Notes

1. Biomechanical computations: Values of dynamic compliance (independent of phase) were calculated from the systolic and diastolic diameter (D), and pressure (P) waves generated by the pumping system as follows:

$$C_{circ} = (D_{sys} - D_{dias}) / D_{dias} \times (P_{sys} - P_{dias}) \tag{1}$$

Compliance values were then averaged for 8–10 waves and reported as units of %dimensional change/mmHg × 10^{-2}.

Hybrid Arterial Bioprosthesis

Table 1
Results of 28-d Stability Experiment

Time (days)	Diameter (±SD)	Compliance (±SD)
0	6.14 ± .02	3.58 ± .33
7	6.28 ± .02	3.64 ± .37
14	6.24 ± .03	3.76 ± .36
21	6.19 ± .02	3.67 ± .26
28	6.21 ± .03	3.54 ± .27

Table 2
Comparison of Einc for the Hybrid Bioprosthesis and a Human Femoral Artery

Pressure (mmHg)	Hybrid bioprosthesis	Human femoral artery
90	13.3×10^6 dyn cm^{-2}	10.0×10^6 dyn cm^{-2}
100	13.0×10^6 dyn cm^{-2}	11.0×10^6 dyn cm^{-2}
110	12.1×10^6 dyn cm^{-2}	14.6×10^6 dyn cm^{-2}

Table 3
Comparative Compliance of Biologic Vessels and Matching Values for the Hybrid Bioprosthesis

Type of vessel	Vessel compliance	Hybrid prosthesis compliance	T/R
Canine carotid a.	13.9 + 0.8	14.1 + 1.2	.06
Canine femoral a.	11.5 + 0.9	11.5 + 0.6	.07
Human femoral a.	5.9 + 1.0	6.0 + 1.1	.09
Canine femoral v.	5.5 + 0.6	5.4 + 0.7	.10
Human saphenous v.	4.5 + 0.7	4.6 + 0.5	.11

An incremental elastic modulus was computed for both the circumferential and longitudinal displacements according to formulae described by Dobrin *(5)*. For the circumferential modulus:

$$E_{inc_{circ}} = \Delta T / (\Delta R / R_{dias}) \quad (2)$$

where $T = PR/h$

Here, T and R represent the difference between calculated values of T (stress) and R (radius) at systole and diastole, respectively, and h is wall thickness. The incremental longitudinal modulus was obtained as follows:

$$E_{inc_{long}} = \Delta T / \Delta L / L_{dias} \quad (3)$$

where $T = PR^2 / [R^2 - (R-h)^2] + F / [\pi R^2 - \pi(R-h)^2]$

Here, T is the sum of stresses caused by pressure (*P*) and traction force (*F*). Estimates of *F* were obtained from a series of in vitro experiments under pulsatile physiologic pressures, as follows: Length was measured in the pressurized vessel, which was subsequently coupled at one end to an isometric force transducer (Grass Instruments, F10, Quincy, MA). The mean longitudinal traction force was 30*g*. All values of *T* and E_{inc} are expressed as dyns/cm^2.

2. Cell Culture: The silicone rubber tubes must be thoroughly cleaned ultrasonically for 10 min in a 1:20 dilution of 15-336-26 ultrasonic cleaning solution (Fisher Scientific) in distilled water, followed by two 10-min ultrasonic washes in distilled water only. Exceedingly low levels of contaminant (e.g., the curing agent) will result in cytotoxicity and/or loss of adherence of cells to the fibronectin basement membrane. For the same reason, all tubing and reservoir stoppers must be comprised of silicone rubber.

Acknowledgment

We gratefully acknowledge the gift of silicone rubber solutions from Thomas W. Broadhagen and Linda M. Veresh of Dow Corning, Midland, MI, and also the suggestions they provided. In addition, we wish to thank Patricia M. Joseph, Pulmonary Unit, Massachusetts General Hospital, for her help with the use of the blood-gas instrumentation.

References

1. Sumpio, B. E. (1993) *Role of Hemodynamics in Vascular Endothelial Biology.* R. G. Landes, Austin, TX.
2. Levesque, M. J., Nerem, R. M., and Sprague, E. A. (1990) Vascular endothelial cell proliferation in culture and the influence of flow. *Biomaterials* **11**, 702–707.
3. Iba, T., Shin, T., Sonoda, T., Rosales, O., and Sumpio, B. E. (1991) Stimulation of endothelial secretion of tissue-type plasminogen activator by repetitive stretch. *J. Surg. Res.* **50**, 457–460.
4. Sterpetti, A. V., Cucina, A., Santoro, L., Cardillo, B., and Cavallaro, A. (1992) Modulation of arterial smooth muscle cell growth by haemodynamic forces. *Eur. J. Vasc. Surg.* **6**, 16–20.
5. Sutcliffe, M. C. and Davidson, J. M. (1990) Effect of static stretching on elastin production by porcine aortic smooth muscle cells. *Matrix* **10**, 148–153.
6. Berguer, R., Higgins, R. F., and Reddy, D. J. (1980) Intimal hyperplasia. An experimental study. *Arch. Surg.* **115**, 332–335.
7. Buck, R. C. (1983) Behavior of vascular smooth muscle cells during repeated stretching of the substratum in vitro. *Atherosclerosis* **46**, 217–223.
8. Diamond, S. L., Eskin, S. G., and McIntire, L. V. (1989) Fluid flow stimulates tissue plasminogen activator secretion by cultured human endothelial cells. *Science* **243**, 1483.
9. Diamond, S. L., Sharefkin, J. B., Dieffenbach, C., Frasier Scott, K., McIntire, L. V., and Eskin, S. G. (1990) Tissue plasminogen activator messenger RNA levels increase

in cultured human endothelial cells exposed to laminar shear stress. *J. Cell. Physiol.* **143**, 364–371.
10. Dobrin, P. B., Littooy, F. N., Golan, J., et al. (1988) Mechanical and histologic changes in canine vein grafts. *J. Surg. Res.* **44**, 259–265.
11. Hume, W. R. (1980) Proline and thymidine uptake in rabbit ear artery segments in vitro increased by chronic tangential load. *Hypertension* **2**, 738–743.
12. Leung, D. Y. M., Glagov, S., and Mathews, M. B. (1977) Elastin and collagen accumulation in rabbit ascending aorta and pulmonary trunk during postnatal growth. Correlation of cellular synthetic response with medial tension. *Circ. Res.* **41**, 316–323.
13. Sumpio, B. E., Banes, A. J., Link, G. W., and Johnson, G. J. (1988) Enhanced collagen production by smooth muscle cells during repetitive mechanical stretching. *Arch. Surg.* **123**, 1233–1236.
14. Sumpio, B. E. and Banes, A. J. (1988) Response of porcine aortic smooth muscle cells to cyclic tensional deformations in culture. *J. Surg. Res.* **44**, 696–701.
15. Sumpio, B. E., Banes, A. J., Levin, L. G., and Johnson, G. J. (1987) Mechanical stress stimulates aortic endothelial cells to proliferate. *J. Vasc. Surg.* **6**, 252–256.
16. Sumpio, B. E., Banes, A. J., Link, G. W., and Iba, T. (1990) Modulation of endothelial cell phenotype by cyclic stretch: inhibition of collagen production. *J. Surg. Res.* **48**, 415–420.
17. Sumpio, B. E. and Banes, A. J. (1988) Prostacyclin synthetic activity in cultured aortic endothelial cells undergoing cyclic mechanical deformation. *Surgery* **104**, 383–389.
18. Sumpio, B. E. and Widmann, M. D. (1990) Enhanced production of endothelium-derived contracting factor by endothelial cells subjected to pulsatile stretch. *Surgery* **108**, 277–282.
19. Towne, J. B., Quinn, K., Salles-Cunha, S., et al. (1982) Effect of increased arterial blood flow on localization and progression of atherosclerosis. *Arch. Surg.* **117**, 1469–1474.
20. Wolinsky, H. (1970) Response of the rat aortic media to hypertension. Morphological and chemical studies. *Circ. Res.* **26**, 507–522.
21. Zwolak, R. M., Adams, M. C., and Clowes, A. W. (1987) Kinetics of vein graft hyperplasia: Association with tangential stress. *J. Vasc. Surg.* **5**, 126–36.
22. Sharefkin, J. B., Diamond, S. L., Eskin, S. G., McIntire, L. V., and Dieffenbach, C. W. (1991) Fluid flow decreases preproendothelin mRNA levels and suppresses endothelin-1 peptide release in cultured human endothelial cells. *J. Vasc. Surg.* **14**, 1–9.
23. Carosi, J. A., Eskin, S. G., and McIntire, L. V. (1992) Cyclical strain effects on production of vasoactive materials in cultured endothelial cells. *J. Cell. Physiol.* **151**, 29–36.
24. Dewey, C. F. J., Bussolari, S. R., Gimbrone, M. A. Jr., and Davies, P. F. (1981) The dynamic response of vascular endothelial cells to fluid shear stress. *J. Biomech. Eng.* **103**, 177–185.
25. Eskin, S. G., Navarro, L. T., O'Bannon, W., and DeBakey, M. E. (1983) Behavior of endothelial cells cultured on Silastic and Dacron velour under flow conditions in vitro: implications for prelining vascular grafts with cells. *Artif. Organs* **7**, 31–37.
26. Frangos, J. A., Eskin, S. G., McIntire, L. V., and Ives, C. L. (1985) Flow effects on prostacyclin production by cultured human endothelial cells. *Science* **227**, 1477–1479.

27. Sterpetti, A. V., Cucina, A., D'Angelo, L. S., Cardillo, B., and Cavallaro, A. (1992) Response of arterial smooth muscle cells to laminar flow. *J. Cardiovasc. Surg.* **33,** 619–624.
28. Upchurch, G. R., Jr, Banes, A. J., Wagner, W. H., et al. (1989) Differences in secretion of prostacyclin by venous and arterial endothelial cells grown in vitro in a static versus a mechanically active environment. *J. Vasc. Surg.* **10,** 292–298.
29. Benbrahim, A., L'Italien, G. J., Milinazzo, B. B., Warnock, D. F., Dhara, S., B. S., Gertler, J. P., Orkin, R. W., and Abbott, W. M. (1994) A compliant tubular device to study the influences of wall strain and fluid shear stress on cells of the vascular wall. *J. Vasc. Surg.* **20,** 184–194.
30. Hasson, J. E., Wiebe, D. H., Sharefkin, J. B., and Abbott, W. M. (1986) Migration of adult human vascular endothelial cells: effect of extracellular matrix proteins. *Surgery* **100,** 384–91.
31. McGuire, P. G., Castellot, J. J., and Orkin, R. W. (1987) Size-dependent hyaluronate degradation by cultured cells. *J. Cell. Physiol.* **133,** 267–276.
32. Sambrook, J., Fritsch, E. F., and Maniatis, T. (1989) *Molecular Cloning, A Laboratory Manual,* 2nd ed. Cold Spring Harbor Laboratory, Cold Spring Harbor, NY.

35

Methods for the Immunoisolation and Transplantation of Pancreatic Cells

Anthony M. Sun

1. Introduction

Although the administration of insulin by injection is clearly a life-saving intervention for patients devoid of β-cells, this approach falls short of the remarkable titration of insulin delivery and consequent control of glucose levels achieved by normal, healthy individuals. In the absence of the physiological control of the plasma glucose concentrations, the daily injection of insulin has not been able to prevent the common complications of the disease, namely nephropathy, retinopathy, and neuropathy, as well as vascular complications. This has been confirmed in the recent diabetes control and complications trial which has demonstrated that intensive treatment of patients with insulin-dependent diabetes mellitus (IDDM), with tight glycemic control close to the control range, effectively delays the onset, and slows the progression, of the various diabetic complications *(1)*. Therefore, it becomes mandatory to develop methods, applicable early in the course of the disease, and in any type 1 diabetic patient, for obtaining perfect metabolic control without increasing the risk of severe hypoglycemia. Consequently, the transplantation of islet tissue, either as whole pancreas or as isolated islets, has been pursued, because these techniques can provide near-normal blood glucose control, and thus have the potential to prevent diabetic complications.

Because of the many problems associated with the whole-pancreas transplantation, grafts of isolated pancreatic islets represent the most promising approach to the restoration of the endocrine function of the pancreas in patients with IDDM. However, the problem of immune rejection remains. This has meant that the transplantation of insulin-producing tissue has largely been confined to diabetic recipients who already have a kidney transplant or who are receiving a kidney simultaneously with an islet transplant, because of the risks of long-term immunosuppression, which are considered to outweigh the potential benefits of transplantation in a newly diagnosed diabetic patient.

To tackle the problem of immune rejection, a number of studies have investigated techniques to decrease the immunogenicity of transplanted islets, and thus reduce for immunosuppression. These approaches, known as immunoalteration, are based on the hypothesis that islet immunogenicity is caused by passenger leukocytes *(2)* or dendritic cells *(3)*, and not by the endocrine cells. Manipulations designed to destroy the dendritic cells and decrease islet immunogenicity include: culturing the islets for prolonged periods at room temperature *(4)* or in 95% O2 *(5)*, exposing the islets to UV radiation *(6)*, cryopreservation *(7)* of islets, and pretreating the islets with antibody to Ia antigen plus complement *(8)*. Despite considerable progress in the use of immunoalteration in past years, this approach continues to have serious limitations. There is still no effective method of consistently protecting islet transplants from rejection by means of immunoalteration, with or without immunosuppression. No successful transplantation of immunoaltered islets into large animals has been reported, and indications are that such transplants would require immunosuppressive therapy in combination with donor islet pretreatment. Although depleting donor tissue of passenger leukocytes before transplantation may temporarily prevent rejection, it does not eliminate the possible vulnerability of the transplanted tissue to a recurrence of the original autoimmune attack.

Because of the difficulties with overcoming the problem of immunorejection and autoimmune rejection, the concept of immunoisolation has been advanced. This is achieved by enclosing the pancreatic islets by semipermeable and biocompatible membrane (bioartificial pancreas). The enclosed islets would act in the host as they had in the donor, provided the surrounding membrane was impermeable to higher mol wt antibodies, but permeable to oxygen, glucose, other substances, and/or the internally generated hormones. Thus, the encapsulated cells would respond to external substrate concentrations (e.g., blood glucose), and the required hormone (e.g., insulin) would be secreted into the systemic circulation. The clinical benefit of this approach is that diabetic patients would be provided with normal pancreatic islets that not only would be protected from immunorejection, but would secrete, in addition to insulin, other hormones, such as glucagon, somatostatin, pancreatic polypeptides, and possibly other islet proteins, in response to physiological demand. This approach has the potential not only to allow allogenic transplantation without immunosuppression, but also to allow the use of xenografts. In the form of a vascular implant, the islets can be distributed in a chamber surrounding the membrane, and the device is implanted as a shunt in the vascular system *(9,10)*.

The limitations of the vascularized devices are all associated with the creation of an arteriovenous shunt, i.e., requirement of major surgery, possible vascular thrombosis, and a potential for cardiac stress caused by a significant volume of shunted blood. Most important, an accidental breakage of such a unit may result in a serious hemorrhage.

Alternatively, the islets can be immunoisolated within diffusion chambers and placed intraperitoneally, subcutaneously, or in other sites. Numerous devices of this type have been evaluated during the past several decades *(11)*. These include disk-shaped diffusion chambers, Millipore cellulosic membranes, hollow-fiber diffusion chambers, and wider-bore tubular membrane chambers. Membrane materials used to make these devices include polyvinylchloride, polyamides (nylon), polypropylene, cellular nitrate, and cellulose triacetate.

The method of islet transplantation using such devices is frequently referred to as macroencapsulation. The common feature of this approach is that a great number of islets are placed within the chamber of the device. This severely limits the diffusion dynamics of the system. Consequently, the problem of cell death or dysfunction as a result of oxygen supply limitations, or accumulation of wastes or other agents, is likely to be severe. Chambers retrieved several weeks after implantation often contain a central necrotic core. Another problem is the frequent breakage of the diffusion chambers, which can cause not only loss of islet function, but also intraperitoneal inflammatory responses. The problems posed by the large volume of the implanted devices, as well as a possible fibrosis of the chamber surfaces, can further aggravate the situation. Typically, the use of these devices can result only in a short-term amelioration of diabetic hyperglycemia.

To overcome these problems, we developed semipermeable, alginate-polylysine-alginate (APA) biocompatible capsules to enclose individual islets (microencapsulation). The encapsulation of individual islets results in removing most problems associated with the macroencapsulation concept as discussed above. Since our original invention in 1980 *(12)*, many applications of this concept have been developed, with many centers now studying cell encapsulation. We have considerably improved the capsule construction since its conception in 1980. The biocompatibility of the capsules was improved by replacing the crude alginate by a more homogenous preparation of lower viscosity and high purity, and by replacing the outer polyethyleneimine layer by a second layer of low-viscosity alginate. The size of the capsules was reduced from the original diameter of 0.8 mm to the present 0.25–0.35 mm. This was made possible by construction of an electrostatic droplet generator *(13)*. The smaller capsules are stronger, with improved sphericity and surface smoothness. It has been demonstrated that the size of the microcapsules, their sphericity, surface smoothness, and, most important, the strength of the capsules, all have an important bearing on the life-span of the graft.

In our earlier experiments using rodents as both donors and recipients, we demonstrated that both allografts and xenografts of microencapsulated pancreatic islets were protected from immunorejection, and that in both streptozotocin-induced and spontaneously diabetic mouse and rat recipients, diabetes could be reversed for the life-span of the animals *(14–17)*. Recently, in an unprecedented preclinical study, we have demonstrated that intraperitoneal xenografts of microencapsulated porcine islets into naturally diabetic nonhuman primates result in normalization of diabetic hyperglycemia for more than 1 yr, without recourse to exogenous insulin or immunosuppression *(18)*. This is the first report ever on long-term discordant xenograft function resulting in physiological glycemic control without recourse to immunosuppression in a large-animal model.

The use of large animals and, eventually, humans in transplantation studies necessitates the development of a plentiful source of islet tissue. The shortage of human donor islets makes it obvious that if the goal of islet transplantation is to succeed, it is necessary to develop methods making it possible to transplant nonhuman islets. Currently, the xenograft of porcine islets remains the most favorable solution. The choice of porcine islets is justified, because it would represent an unlimited source of tissue, and because porcine insulin is very similar to human insulin in its structure, and was used for decades in the treatment of human diabetic patients. In addition, in the pig the regulation of insulin secretion by glucose and other nutrients is very similar to that in humans.

Porcine islet isolation is difficult, however, because of the islets' marked fragility, and because of the rapid dissociation of the pancreas into single cells during the isolation procedure *(19–24)*. In the pig pancreas, loose vascular channels are present within the islet, which makes it very easy for the islet to fragment during the isolation procedure. The method of porcine islet isolation developed in this laboratory *(25,26)* is highly reproducible in terms of both the quantity and quality of the isolated islets. In the procedure, porcine pancreata are perfused and digested with collagenase, and the islets are then purified on dextran density gradients. In order to avoid any damage to the islets, no mechanical devices nor any strenuous treatment is employed. A thorough separation of islets from the exocrine tissue is critical, since the dextran or Ficoll density purification has proven to be ineffective for those islets not completely free of acinar growth attachment.

The viability and physiological competence of the isolated islets were demonstrated in both in vitro and in vivo studies *(25,26)*. The rate of insulin secretion of microencapsulated islets in culture studies was shown to be very similar to that of free islets. Likewise, the kinetics of insulin release in the perifusion study of encapsulated and free islets followed a similar pattern.

The physiological competence of porcine islets isolated by our new method in vivo was unequivocally demonstrated in our preclinical study *(18)*, in which xenografts of microencapsulated porcine islets reversed diabetes in naturally diabetic monkeys for over a year, as mentioned earlier in this chapter.

2. Materials

2.1. Pancreatic Islets Isolation (as an Example of Endocrine Tissue Isolation)

Rat islets are isolated by the collagenase-digestion technique, and either handpicked or purified through discontinuous Ficoll gradients. Briefly, pancreatic tissue is dressed and perfused with Hanks' balanced salt solution. The minced tissue is then digested for 12–15 min at 37°C with collagenase (Sigma, St. Louis, MO)(12–15 mg/4 mL of Hanks' solution). Islets can be handpicked from digest with the aid of a dissecting microscope, or the tissue is mixed with 25% Ficoll in a centrifuge tube. Three Ficoll concentrations (23, 20, and 11%) are then layered above this suspension. Centrifugation is carried out for 10 min at 800g. The islets are harvested from the interface of the 23 and 20% Ficoll layers. The cells can either be microencapsulated immediately or cultured for 1–3 d prior to encapsulation.

2.2. Sodium Alginate Solutions

To prepare sodium alginate solutions, 3.0 g of sodium alginate (Kelco, San Diego, CA) is sprinkled or sifted slowly into 100 mL of distilled water, with stirring. Then 100 mL of 1.8% NaCl solution is added and well-mixed. The mixture is then centrifuged at 3000g for 1 h at 4°C. The supernatant is sterilized by filtering through a filtration unit (0.2 µm) and stored at 4°C. The 0.15% sodium alginate solution is prepared by diluting the 1.5% solution 10-fold with physiological saline.

2.3. Calcium Lactate Solutions

To prepare 2.72% calcium lactate solution (100 mM), 68.0 g of calcium lactate is dissolved in 2,500 mL (final volume) of distilled water.

2.4. Poly-L-lysine Solution (PLL)

Fifteen mg of PLL (Sigma) is dissolved in 30 mL of saline.

2.5. Sodium Citrate Solutions

Trisodium citric acid (3.23 g) is dissolved in 100 mL of distilled water, to which 100 mL of saline is added and mixed (55 mM sodium citrate solution).

2.6. Culture Medium (for Pancreatic Islets)

Medium RPMI 1640 is supplemented with calf serum (7.5%), penicillin (100 U/mL), and streptomycin (100 µg/mL).

3. Methods
3.1. Microencapsulation Techniques
3.1.1. Air-jet Technique

The air-jet technique represented the original procedure for cell microencapsulation, when it was first described in 1980 *(12)*. A syringe pump with a 10-mL syringe connected to a special jet is used for making sodium alginate islet droplets. For each preparation, 2000–3000 islets in 0.2 mL of saline is mixed gently with 1 mL of 1.5% sodium alginate, transferred to the 10-mL syringe, and connected to the air jet. The distance from the tip of the air jet 23-gage needle to the surface of the collecting fluid is set precisely at 4 cm. The syringe pump and air flow are then turned on to extrude sodium alginate droplets containing islets into 50 mL of the 1.1% $CaCl_2$ solution in a beaker. During extrusion, the islets are kept in suspension by gently rotating a small magnet inside the 10-mL syringe. After the extrusion process is completed, the spherical calcium alginate gel droplets are transferred to a 50-mL polystyrene test tube with a conical bottom, and allowed to settle before withdrawing the supernatant down to 5 mL, using a vacuum aspirator. The gel droplets are washed once with 30 mL of 55% $CaCl_2$, once with 0.28% $CaCl_2$, and then suspended in 25 mL of 0.1% CHES solution for 3 min. After aspirating the CHES solution, the capsules were washed with 1.1% $CaCl_2$ and suspended in 25 mL of 0.05% (w/v) poly-L-lysine for 6 min. After further washing with CHES, $CaCl_2$, and saline, the microcapsules are incubated in 0.15% sodium alginate for 4 min and washed with saline. The capsules are then suspended in 10 mL of 55 m*M* sodium citrate solution for 5 min. The final product is washed twice with saline and once with medium CMRL-1969, and then transferred to culture flasks for incubation at 37°C, until required for in vitro and in vivo studies.

The air-jet technique results in relatively large capsules of approx 0.8 mm in diameter.

The capsule construction has improved considerably over the past years. It has been demonstrated that the capsule size is an important factor in the kinetics of insulin release by encapsulated pancreatic islets, and in the access to nutrients and oxygen. Glucose tolerance in diabetic mice improved significantly after the mice received rat islets enclosed in capsules measuring 0.3 mm in diameter. The smaller capsules allow for:

1. Increased strength, with lesser chance of rupture.
2. Increased cell viability, because encapsulated cells have easier access to oxygen and nutrients.

3. Faster cell response to glucose fluctuations, since the dead space in the capsules is reduced.
4. Significant reduction in the volume of capsules needed for transplants, since the capsule volume is directly proportional to the cube of the capsule radius. Thus, if one million microencapsulated islets should be necessary to normalize hyperglycemia in a type 1 diabetic patient, the overall volume of the intraperitoneally implanted capsules (300 μ in diameter) would only amount to 14.1 mL, dropping to just 8.1 mL if the capsule diameter decreases to 250 μ.
5. Less susceptibility to cell overgrowth on capsular surfaces, because the smaller capsules have greater mobility.

3.1.2. Electrostatic Droplet Generator

The generation of smaller capsules was made possible by construction of an electrostatic droplet generator *(13)*. The size of the capsules was reduced from the original diameter of 0.8 mm to the present 0.25–0.35 mm. The smaller capsules are stronger, with improved average sphericity and surface smoothness. The size of the microcapsules, their sphericity, surface smoothness, and, most important, the strength of the capsules, all have an important bearing on the life-span of the graft. In an in vitro perifusion study, rat islets encapsulated in the new capsules showed a response to glucose challenges that was nearly comparable to that of free, unencapsulated islets. In vivo, following transplantation with the smaller capsules, there is much less contact irritation, which in turn leads to a considerably smaller probability of cell overgrowth on capsular surfaces. In the process of droplet generation, the electrostatic droplet generator uses alternate current frequency of 60 Hz, at the pulse length of 0.1 ms.

In the encapsulation procedure, pancreatic islets are suspended in 1.5% (w/v) purified sodium alginate at a concentration of approx 3–5 × 10^3 islets/mL. Spherical droplets are formed by an electrostatic field interaction, coupled with syringe pump extrusion (Razel A 99 syringe pump, pump speed 55–65, Stamford, CT) and are collected in 100 mM calcium lactate solution. To achieve this, the negative pole is attached to the loop, which is submerged in the calcium lactate solution, while the positive pole is attached to the needle. The tip of the needle is positioned about 1 cm from the surface of the calcium lactate solution. Calcium lactate solution temperature is 20°C (all solutions used in the encapsulation procedure are filter-sterilized). The gelled droplets are washed with 0.9% saline prior to suspension in 0.05% poly-L-lysine for 5 min. The droplets are again washed with 0.9% saline and suspended in 0.15% sodium alginate for 4 min. After another wash with 0.9% saline, the capsules are allowed to react with 55 mM sodium citrate for 4 min, and finally washed with 0.9% saline, and with culture medium. Most capsules contain one islet, and have a diameter of 0.25–0.35 mm. The entire microencapsulation procedure is illustrated in **Fig. 1A,B**.

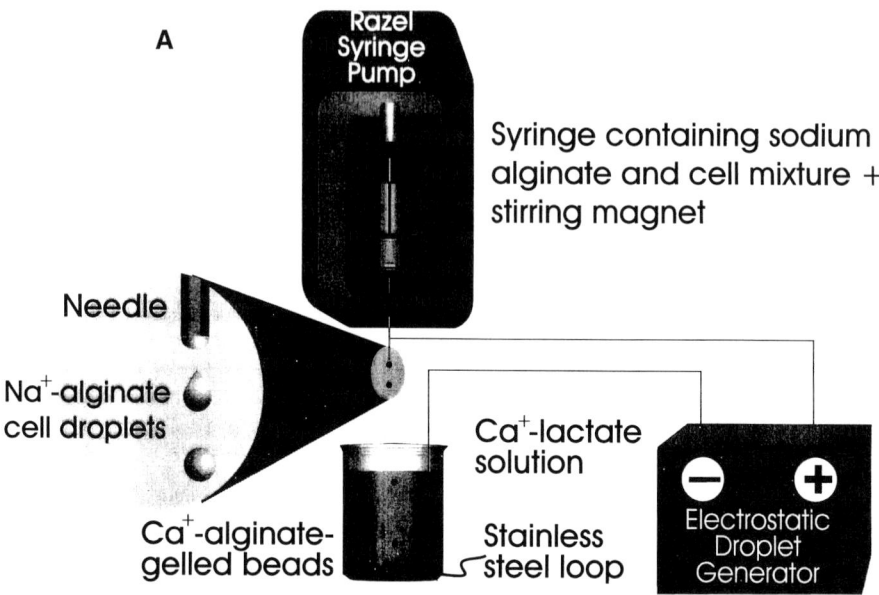
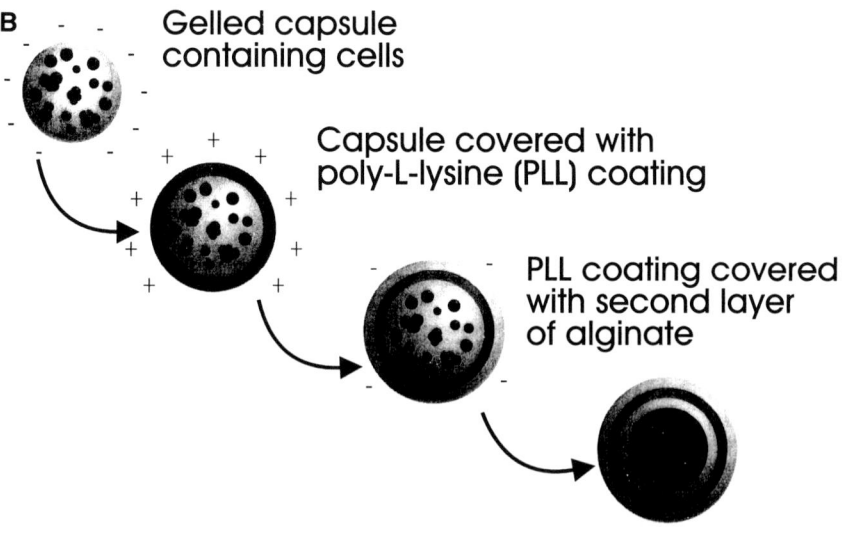

Fig. 1. Process of generation of alginate-polylysine-alginate microcapsule using the electrostatic droplet generator.

3.1.3. Capsule Improvement

In an effort to further improve the strength of the capsule, we studied the parameters affecting the capsules strength *(27)*. It can be postulated that the membrane strength is directly proportional to the thickness of the capsule. Consequently, we studied quantitatively different reaction conditions during the synthesis of the membrane.

1. Concentration of sodium alginate. The alginate concentration was found to affect the membrane thickness within the range of 1–2%. The osmotic pressure at this concentration is not harmful to cells or tissues. Similarly, the calcium concentration also has a bearing on the membrane thickness. The thickness is at its maximum when the concentration of calcium lactate used in the encapsulation procedure is 100 mM.
2. Reaction time of sodium alginate with calcium lactate. Another factor influencing the membrane strength is the reaction time of sodium alginate with calcium lactate, i.e., gelatination time. If the gelation time is shorter than 5 min, the membrane strength will suffer. The membrane strength increases with time; however, if the reaction time exceeds 1 h, instead of increasing, the membrane thickness slightly decreases. The washing time of calcium alginate beads with sodium chloride solution is also critical for the capsule strength. The membrane thickness decreases as the treatment time increases, especially during the first 15 min.
3. Parameters affecting the membrane strength and membrane thickness. The reaction of alginate beads with polylysine represents probably the most important step regarding the strength of the APA capsules. Four different parameters are essential in determining the resulting thickness of the capsular membrane:
    a. Membrane thickness increases with increasing time, during which alginate beads react with polylysine. It is therefore possible to control membrane thickness by regulating the reaction time.
    b. The membrane thickness increases with increasing polylysine concentration.
    c. The membrane strength increases with increasing mol wt of polylysine. (Polylysines with mol wt of 24,000 are now used for APA capsule preparation, because of the mol wt cutoffs of the resulting membrane—around 60,000 kDa.
    d. It can be demonstrated that the membrane thickness increases with a decrease of the pH of the polylysine solution.
4. Microencapsulation of cells and tissues other than pancreatic islets. It should be noted here that the capsules' chemical composition can be altered for encapsulation of cells and tissues other than pancreatic islets, thus suiting their special requirements. For instance, microencapsulation of hepatocytes (artificial liver) requires a very strong capsule, in order to avoid capsule rupture when encapsulating large numbers of cells needed for a proper function of the artificial liver. To achieve this, high sodium alginate concentrations have to be used and a long reaction time between alginate and polylysine is required.

In other applications, permeability of the APA membrane can be altered to suit the specific metabolic requirements of the encapsulated cells or tissues.

3.2. Transplantation of Microencapsulated Pancreatic Islets

3.2.1. Allotransplantation

Transplants (both allografts and xenografts) were performed in diabetic mice, rats, and monkeys *(14–18)*. In the procedure, microencapsulated islets are delivered into the peritoneal cavity of the recipient by injection, using an 18-gage catheter (Insyte, Becton Dickinson, Sandy, UT). Single allografts of APA-microencapsulated rat islets, when implanted intraperitoneally into rats with streptozotocin-induced diabetes (4.5×10^3 islets per rat), resulted in the reversion of diabetic hyperglycemia for 21 mo *(14)*. In addition to normalizing blood glucose concentrations, the encapsulated rat islet transplants reversed polyuria and polydypsia in the recipients. The recipients demonstrated a very rapid increase in body wt, and there was no evidence of cataracts. By contrast, untreated diabetic controls showed no significant weight increase during the same period, and developed cataracts within 2–3 mo. Capsules containing intact viable islets, as revealed by histological and insulin secretion studies, were recovered 156, 365, and 648 d postransplantation. The surfaces of most capsules were free of cell attachment and were physically intact, with the enclosed islets clearly visible. Scanning electron microscopy revealed essentially smooth interior and exterior capsular surfaces.

In the next stage, rat islet allografts microencapsulated in APA membranes were implanted intraperitoneally in spontaneously diabetic BB rats *(15)*. The BB rat represents a valuable experimental model of human type 1 diabetes. The course of development of the disease in this model resembles that of human type 1 diabetes, in that there is an abrupt onset of insulin-dependent, ketosis-prone diabetes and lymphocytic insulitis, with virtually complete destruction of the pancreatic β-cells. Therefore, the autoimmune etiology of this animal allows investigation of the effectiveness of pancreatic islets in an immunoisolated environment.

The encapsulated allografts reversed the diabetic state and maintained normoglycemia for up to 6 mo. Body wt and urine volumes were normal during this period, and no cataracts were detected in the transplant recipients. Thus, encapsulation of islets provided total protection from the host immune system, as well as from the autoimmune destruction, while eliminating the need for immunosuppression.

3.2.2. Xenotransplantation

Following the successful allotransplantation experiments, the first successful xenotransplants of microencapsulated islets were performed. Rat islets mircoencapsulated in APA.microcapsules were implanted intraperitoneally into BALB-c mice with streptozotocin-induced diabetes *(16)*. As a result, normoglycemia was restored in the recipient animals for up to 308 d.

In another xenotransplantation study, rat islets were implanted into spontaneously diabetic NOD mice, an ideal animal model of insulin-dependent diabetes *(17)*. Single intraperitoneal transplants of 800 islets per recipient resulted in an effective restoration of normoglycemia in all experimental animals for up to 230 d.

The development of a plentiful supply of pancreatic islets, as described earlier in this chapter, led to the initiation of a preclinical study in which microencapsulated porcine islets were xenotransplanted into nine spontaneously diabetic cynomologus monkeys *(18)*. After one, two, or three transplants of $3-7 \times 10^6$ islets per recipient, seven of the monkeys became insulin-independent for over 1 yr, with fasting blood glucose levels in the normoglycemic range. Glucose clearance rates in the transplant recipients were significantly higher than before the graft administration, and the insulin secretion during glucose tolerance tests was significantly higher, compared to pretransplant tests. Porcine C-peptide was detected in all transplant recipients throughout their period of normoglycemia, but none was found before the graft administration. Hemoglobin A_{1C} levels dropped significantly within 2 mo after transplantation. Ketones were detected in the urine of all recipients before the graft administration, but all experimental animals became ketone-free 2 wk after transplantation. Capsules recovered from two recipients 3 mo after the restoration of normoglycemia were found physically intact, with enclosed islets clearly visible. The capsules were free of cellular overgrowth. Examination of internal organs of two of the animals involved in our transplantation studies for the duration of 2 yr revealed no untoward effect of the extended presence of the microcapsules.

References

1. The Diabetes Control and Complications Trial Research Group (1993) The effect of intensive treatment of diabetes on the development and progression of long term complications in insulin—dependent diabetes mellitus. *N. Engl. J. Med.* **329**, 977–986.
2. Sollinger, H. W., Stratta, R. J., D'Alessando, A. M., Kalayoglu, M, Pirsch, J. D., and Belzer, F. O. (1988) Experience with simultaneous pancreas-kidney transplantation. *Ann. Surg.* **208**, 475–483.
3. Faustman, D., Hauptfeld, V., Davie, M., Lacy, P. E., and Shreffler, D. C. (1980) Murine pancreatic beta cells express H2K and H2D but not Ia antigens. *Exp. Med.* **151**, 1563–1568.
4. Ricordi, C., Kraus, C., and Lacy, P. E. (1988) Effect of low-temperature culture on the survival of intratesticular rat islet allografts. *Transplantation* **45**, 465–468.
5. Woehrle, M., Markmann, J. F., Silvers, W. K., Barker, C. F., and Naji, A. (1986) Transplantation of cultured pancreatic islets to BB rats. *Surgery* **100**, 334–340.
6. Kenyon, N. S., Strasser, S., and Alejandro, R. (1990) Ultraviolet light immunomodulation of canine islets for prolongation of allograft survival. *Diabetes* **39**, 305–311.

7. Cattral, M., Wornock, G., Evans, M., and Rajotte, R. (1991) Transplantation of purified single-donor cryopreserved canine islet allografts with cyclosporine. *Transplantation Proc.* **23,** 777,778.
8. Alejandro, R., Latif, Z., Noel, J., Shienvold, F. L., and Mintz, D. H. (1987) Effect of anti-Ia antibodies, culture and cyclosporin on prolongation of canine islet allograft survival. *Diabetes* **36,** 269–273.
9. Sun, A. M., Parisius, W, Healy, G. M., Vacek, I., and Macmorine, H. G. (1997) The use, in diabetic rats and monkeys, of artificial capillary units containing cultured islets of Langerhans (artificial endocrine pancreas). *Diabetes* **26,** 1136–1139.
10. Sullivan, S. J., Maki, T., Borland, K. M., Mahoney, M. D., Solomon, B. A., Muller, T. E., Monaco, A. P., and Chick, W. L. (1991) Biohybrid artificial pancreas: long-term implantation studies in diabetic pancreatectomized dogs. *Science* **252,** 718–721.
11. Lanza, R. P., Sullivan, S. J., and Chick, W. L. (1992) Islet transplantation with immunoisolation. *Diabetes* **41,** 1503–1510.
12. Lim, F. and Sun, A. M. (1980) Microencapsulated islets as bioartificial endocrine pancreas. *Science* **210,** 908–910.
13. Hommel, M., Sun, A. M., and Goosen, M. F. A. (1984) Droplet generation. Canadian Patent 458605.
14. Sun, A. M., O'Shea, G. M., and Garapetian, H. (1985) Artificial cells containing islets as bioartificial pancreas. *Prog. Artif. Organs* **1985,** 601.
15. Fan, M., Lum, Z., Fu, X., Levesque, L., and Sun, A. M. (1990) Long-term reversal of diabetes in BB rat by transplantation of microencapsulated pancreatic islets. *Diabetes* **39,** 519–522.
16. Lum, Z. P., Tai, I. T., and Sun, A. M. (1992) Xenografts of microencapsulated rat islets result in prolonged reversal of the diabetic state in streptozotocin-induced diabetic BALB-c mice. *Transplantation* **53,** 1180–1183.
17. Lum, Z. P., Krestow, M., Tai, I., Norton, J., Vacek, I., and Sun, A. M. (1991) Prolonged reversal of diabetic state in NOD mice by xenografts of microencapsulated rat islets. *Diabetes* **40,** 1511–1515.
18. Sun, Y., Ma, X., Zhou, D., Vacek, I., and Sun, A. M. (1996) Normalization of diabetes in spontaneously diabetic cynomologus monkeys by xenografts of microencapsulated porcine islets without immunosuppression. *J. Clin. Invest.* **98,** 1417–1422.
19. Ricordi, C., Finke, E. H., and Lacy, P. E. (1986) A method for the mass isolation of islets from the adult pig pancreas. *Diabetes* **35,** 649–653.
20. Crowther, N. J., Gotfredsen, C. F., Moody, A. J., and Green, I. C. (1989) Porcine islet isolation, cellular composition and secondary response. *Horm. Metabol. Res.* **21,** 590–595.
21. Marchetti, P., Zappello, A., Giannarell, R., Masiello, P., Masoni, A., Casanovi, E., et al. (1988) Isolation of islets of Langerhans from the adult pig pancreas. *Transplantation Proc.* **20,** 707,708.
22. Marchetti, P., Socci, C., Davalli, A. M., Staudacher, C., Pierangelo, B., Vertova, A., Sassi, I., Gavazzi, F., Pozza, G., and Di Carlo, V. (1991) Automated large-scale isolation, in vitro function and xenotransplantation of porcine islets of Langerhans. *Transplantation* **52,** 209.

23. Ricordi, C., et al. (1990) Isolation of the elusive pig islet. *Surgery* **107,** 688–694.
24. Basta, G., Falorni, A., Ostricioli, L., Brunetti, P., and Calafiore, R. (1955) Method for mass retrieval, morphologic, and functional characterization of adult porcine islets of Langerhans: a potential non-human pancreatic tissue resource for xenotransplantation in insulin-dependent diabetes mellitus. *J. Invest. Med.* **43,** 555.
25. Sun, Y., Ma, X., Zhou, D., Vacek, I., and Sun, A. M. (1993) Porcine pancreatic islets: isolation, microencapsulation and xenotransplantation. *Artif. Organs* **17,** 727–733.
26. Zhou, D., Yang, B., Sun, Y., Vacek, I., Sun, A. M. (1996) Effects of collagenase concentration on the puritry and viability of isolated porcine pancreatic islets for use in xenotransplantation studies. *Xenotransplantation* **3,** 11–17.
27. Ma, X., Vacek, I., and Sun, A. M. (1994) Generation of alginate-poly-L-lysine-alginate (APA) biomicrocapsules the relationship between the membrane strength and the reaction conditions. *Biomat. Artif. Cells Immobilization Technol.* **22,** 43–69.

36

Methods for the Study of Nerve Cell Migration and Patterning

Helen M. Buettner and Hsin-Chien Tai

1. Introduction

Recent advances in our ability to microfabricate tissue culture environments *(1)*, and to store and retrieve microscopy data in a digital format, have made it increasingly possible to study fundamental aspects of neuronal migration and response to environmental cues. Neuronal migration refers to the migration of the growth cone at the tip of an extending neurite, a process that guides the neurite to its destination during neural development and regeneration *(2)*. This chapter describes techniques for the observation of neuronal migration on both plain and patterned surfaces through the use of high-resolution, phase-contrast videomicroscopy. These techniques are particularly appropriate for analyzing the dynamics of single growth cone behavior in the presence of two-dimensional environmental microfeatures similar in scale to the growth cone dimensions (1–50 µm). Growth cone response to topographical features *(3)* or other three-dimensional environments are beyond the scope of this chapter. Substrate preparation procedures are derived from microlithography techniques first used in the microelectronics industry and adapted to cell culture systems *(4–7)*. Glass cover slips are used as the substrate support to provide the required optical clarity for high-resolution microscopy, and patterns are created using laminin or collagen.

The major procedures described in this chapter include:

1. Neuronal cell culture. This section focuses on one neuronal cell type, chick dorsal root ganglion (DRG), which represents an important model system for studies of neuronal migration, and complements the neuronal types described elsewhere in this volume. However, protocols for the culture of other types of neurons could be substituted here, and used with the procedures following this section. The

advantages of the chick DRG system include the fact that it is a relatively trouble-free source of primary cells, and produces growth cones that are somewhat larger than many other vertebrate neurons, corresponding to more easily distinguishable features.

2. Substrate preparation. Micropatterned cell culture surfaces are prepared using microlithography techniques. A photoresist pattern is first created on a glass cover slip support, which is then further processed to yield a final patterned, neuronal-growth-promoting, biological substrate. The use or services of a microfabrication facility is assumed for the photoresist patterning step of the protocol. The method described can be used to create either laminin or collagen patterns. It may be suitable for other neuronal growth substrates that have not been tested specifically with this protocol.

3. Videomicroscopy. Although an important feature of neuronal migration and patterning studies, videomicroscopy has become a relatively commonplace laboratory tool, and only the essential points of the technique are described here. The primary requirements are a microscope equipped with a good video camera, a personal computer with frame grabbing capabilities, and plenty of storage capacity for the image data.

2. Materials

2.1. Tissue Culture of Chick Dorsal Root Ganglion (DRG) Neurons

2.1.1. Dissection

1. Incubated fertile chicken eggs.
2. Ca^{2+}-free, Mg^{2+}-free Hanks' balanced salt solution (HBSS): Prepare stock from 1 part HBSS 10X liquid (containing no calcium chloride, magnesium chloride, magnesium sulfate, sodium bicarbonate, or phenol red (Gibco-BRL, Gaithersburg, MD), and 9 parts dH_2O. Filter sterilize with a 0.2 µm filter.
3. Ethanol: 70% in dH_2O.
4. Blunt end forceps, sterilized.
5. Dissecting forceps (Dumont No. 5, Fine Science Tools, Foster City, CA), sterilized.
6. 35-mm sterile plastic Petri dishes.

2.1.2. Cell Culture

1. Serum-free F12 culture medium (*see* **Note 1**):
   a. F12 medium: F12 nutrient mixture (146 mg/L L-glutamine, 1176 mg/L sodium bicarbonate, Gibco-BRL; *see* **Note 2**)
   b. Glucose-glutamine-pen–strep (GGPS): Mix D-glucose 150 mg/mL, L-glutamine 50 mM, penicillin–streptomycin 2500 µg/mL–2500 U/mL (all components from Sigma, St. Louis, MO) with F12 medium. Filter-sterilize using a 0.2-µm filter and aliquot into 4-mL portions. Stable for at least 3 mo at –20°C.
   c. Insulin–transferrin–selenium (ITS) stock: Dissolve ITS™ Premix lyophilized powder (25 mg insulin, 25 mg transferrin, 25 µg selenious acid, Collaborative Biomedical, Bedford, MA) in 5 mL F12 medium. Stable for 3 mo at –20°C (*see* **Note 3**).

Nerve Cell Migration and Patterning

    d. Putrescine stock: Dissolve 16.1 mg/mL putrescine powder (Sigma) in F12 medium. Dilute 1:100 in F12 medium to make 100 mM stock solution. Filter-sterilize with a 0.2-µm filter. Aliquot into 5-mL portions, and store at –20°C.

    e. Progesterone stock: Dissolve 0.63 mg/mL progesterone (Sigma) in 96% ethanol. Dilute 1:100 in F12 medium to make a 2×10^{-5} M stock solution. Filter-sterilize with a 0.2-µm filter. Aliquot into 5-mL portions, and store, tightly wrapped, with Parafilm at –20°C.

    f. Hormone stock: Combine 5 mL each of ITS, putrescine, and progesterone stock solutions. Aliquot into 300-µL portions. Stable for 3 mo at –20°C.

    g. Bovine pituitary extract (BPE): Trim bone, connective tissue, blood vessels, and membranes from 20–25 thawed bovine pituitaries (Pelfreeze Biologicals, Rogers, AK; *see* **Note 4**). Wash with distilled water until the rinse water is free of blood, then rinse with Ca^{2+}-free, Mg^{2+}-free HBSS, and drain. Add the pituitaries and 2.4 mL Ca^{2+}-free, Mg^{2+}-free HBSS/g pituitary to a blender and homogenize at low speed in 10 1-min pulses with 2–3 min rests on ice between pulses (*see* **Note 5**). Pour homogenate into a beaker. Rinse blender with additional HBSS and add to homogenate. Stir 90 min in 4°C cold room. Centrifuge 40 min at 10,000g and 4°C. Dialyze supernatant against phosphate buffered saline (PBS; Dulbecco's PBS, 0.2 g/L KCl, 0.2 g/L KH_2PO_4, 8.0 g/L NaCl, 1.15 g/L Na_2HPO_4, pH 7.4, Gibco-BRL) at 4°C, using Spectra/Por 4 dialysis tubing (mol wt cutoff 12,000–14,000, 16 mm diameter, 2 mL/cm, Fisher Scientific, Springfield, NJ). Dialyze twice, for at least 8 h each. Dialyze against F12 medium at 4°C, once, for at least 8 h (*see* **Note 6**). Centrifuge 1 h at 100,000g and 4°C. Filter twice through a 0.45-µm filter, then once through a 0.2-µm filter. Measure BPE concentration with a Bradford Assay (Bio-Rad, Cambridge, MA), using BSA as a standard. Aliquot into 0.15-mg portions. Stable for 3 mo at –4°C. Procedure from Tsao et al. *(8)*, as described by Baird and Raper *(9)*.

    h. Nerve growth factor (NGF): Dissolve 100 µg 7S NGF (mouse submaxillary gland, Collaborative Biomedical) in 5 mL sterile F12 medium. Aliquot into 100-µL portions. Stable for 1 mo at –20°C.

    i. F12 culture medium: Add 4 mL GGPS, 300 µL hormone stock, 0.1–0.2 mg/mL BPE, and 100 µL NGF to 95 mL of F12 medium to make 100 mL of serum-free F12 culture medium.

2. Tissue Culture Substrate (*see* **Subheading 2.2.**).

2.2. Substrate Preparation

2.2.1. Cleaning

1. Cover slips, 22 mm, No. 1 round (Fisher Scientific; *see* **Note 7**).
2. Alconox paste: Mix Alconox powder (Fisher Scientific) with a small amount of water to form a viscous slurry.
3. Gloves, powder-free.
4. Forceps.
5. Deionized, distilled water.
6. Ultrasonic cleaner (55 kHz, 70 W, Fisher Scientific).
7. Absolute ethanol.

2.2.2. Acid Etching

1. Filtered air.
2. 9:1 v/v sulfuric acid:hydrogen peroxide: Mix 9 parts sulfuric acid (98 wt%) to 1 part hydrogen peroxide (30 wt%).
3. Forceps, Teflon-coated.
4. Deionized, distilled water.
5. Petri dish, plastic, 60 mm.
6. Parafilm.
7. Desiccator.

2.2.3. Photoresist Patterning

1. Hexamethyldisilazane (HMDS; PCR, Gainesville, FL).
2. Microposit S1813 photoresist (Shipley, Marlboro, MA).
3. Microposit MF-319 developer (Shipley).
4. 10-mL sterile plastic syringes.
5. Acrodisc CR PFTE syringe filters (HPLC certified, 0.2 µm, Fisher Scientific).
6. Deionized water.
7. Filtered nitrogen gas.
8. Parafilm.
9. Aluminum foil.
10. Spin coater.
11. Contact printer: UV source 8 mW, 365 nm.
12. Desiccator.
13. Forceps.
14. Photomask.
15. Oven.

2.2.4. Protein Patterning

1. Use (a) or (b):
   a. Collagen: Rat tail type I (Collaborative Biomedical). As needed, dilute stock solution to 150 µg/mL in 0.02 M acetic acid, and filter-sterilize with a 0.2-µm filter.
   b. Laminin: EHS mouse tumor (Collaborative Biomedical). Aliquot stock solution into single use portions and store at –70°C. Stable for 9 mo. As needed, dilute one aliquot to 50 µg/mL in Ca^{2+}-free, Mg^{2+}-free HBSS.
2. Germicidal UV light source.
3. 1% aminopropyltriethoxysilane (AMPS): Combine 1 part 3-aminopropyltriethoxysilane (98%, Aldrich, Milwaukee, WI) with 99 parts dH_2O. Make fresh as needed.
4. Ultrasonic cleaner.
5. PBS: Dulbecco's PBS (0.2 g/L KCl, 0.2 g/L KH_2PO_4, 8.0 g/L NaCl, 1.15 g/L Na_2HPO_4, pH 7.4, Gibco-BRL).
6. Glutaraldehyde: Dilute 1 part 25% glutaraldehyde, aqueous, with 9 parts PBS to make 2.5% glutaraldehyde.

Nerve Cell Migration and Patterning

7. Acetone, electronic grade.
8. Bovine serum albumin (BSA): Bovine serum albumin, Fraction V (J. T. Baker, Phillipsburg, NJ) dissolved at 0.5 w/v% in PBS. Make fresh as needed.

2.3. Video Microscopy
2.3.1. Culture Chamber Preparation

1. Microscope slides.
2. Alconox paste.
3. Ethanol: 70% in dH_2O.
4. Lens paper.
5. Parafilm.
6. Metal washer, ≤15 mm id, ≥25 mm od.
7. Valap: Melt together a 1:1:1 mixture of Vaseline, lanolin, and paraffin. Cool to room temperature, and store covered until ready to use.
8. Paintbrush.
9. Windex solution: Dilute Windex window cleaner with dH_2O in a 1:1 ratio.
10. Q-tips.

2.3.2. Video Microscopy

1. Microscope equipped with a high-resolution video camera.
2. Time-lapse VCR.
3. Personal computer equipped with frame-grabbing capabilities.
4. Hard drive, ≥1 Gb.
5. Tape drive, same storage capacity as hard drive.
6. Microscope stage incubator with temperature controller.

2.3.3. Data Analysis

1. Image analysis software (e.g., *NIH Image*).

3. Methods
3.1. Tissue Culture of Chick DRG Neurons
3.1.1. Dissection

1. Remove egg from egg incubator and wipe the large end with 70% ethanol (*see* **Note 8**).
2. Crack the shell at the large end of the egg by tapping with blunt-ended forceps, and remove the shell to create an opening. Scoop out the embryo and place in a 35-mm Petri dish containing 2 mL Ca^{2+}-free, Mg^{2+}-free HBSS.
3. Remove head and internal organs with dissecting forceps, and transfer embryo to fresh Petri dish of Ca^{2+}-free, Mg^{2+}-free HBSS.
4. Under a dissecting microscope, use dissecting forceps to remove the skin from the back and around the legs.
5. Close the dissecting forceps, and insert into the area between the spinal column and the leg. Slowly open the forceps parallel to the spinal column, to tease tissue

apart. Repeat until leg is separated from body. Transfer the leg into a fresh Petri dish of Ca^{2+}-free, Mg^{2+}-free HBSS, and repeat for the other leg.
6. Remove exposed DRGs from the interior side of the top of the leg, and transfer into a fresh Petri dish of Ca^{2+}-free, Mg^{2+}-free HBSS (*see* **Note 9**).

3.1.1. Cell Culture

1. Place prepared substrate cover slips into separate 35-mm Petri dish, and pipet 750 μL F12 culture medium into each dish.
2. Cut DRG in half and place one-half DRG in the center of each substrate.
3. Incubate 4 h at 37°C and 5% CO_2, to allow explants to attach to the substrate.
4. Add 1.25 mL of medium to each dish, and incubate overnight (*see* **Note 10**).

3.2. Substrate Preparation

This subheading describes the preparation of micropatterned substrates for patterning nerve growth. It is divided into a series of procedures that can be performed with some space of time between each, if desired, keeping in mind the tendency of the photoresist to degrade within several weeks, following application in **Subheading 3.2.3**. To prepare a uniform substrate, the same procedure can be followed, omitting the photomask in **Subheading 3.2.3, step 4**. Many other procedures exist for preparing a uniform substrate, and are readily available in the general literature on neuronal growth and migration. However, this method provides a direct control for comparison with results obtained using the patterned substrates.

3.2.1. Cleaning

1. Wearing powder-free rubber gloves, rub each cover slip between fingers with Alconox paste for 30 s.
2. Rinse cover slips individually in hot tap water until the water runs off cleanly from the cover slip, then place in a 400-mL beaker of deionized, distilled water.
3. Sonicate the cover slips in the beaker for 10 min.
4. Transfer the cover slips individually to a fresh beaker of deionized, distilled water, and repeat step 3.
5. Repeat step 4.
6. Rinse with 2 L of running deionized, distilled water.
7. Transfer the cover slips individually into a beaker of absolute ethanol, and store covered with Parafilm until ready to acid etch.

3.2.2. Acid Etching

1. Remove cover slips individually from ethanol and blow dry with a stream of 0.22-μm filtered air.
2. Etch cover slips in 9:1 v/v H_2SO_4/H_2O_2 for 20 min.
3. Transfer the cover slips individually into a fresh beaker of deionized, distilled water.

Nerve Cell Migration and Patterning

4. Rinse with five changes of deionized, distilled water, then blow dry with a stream of 0.22-μm filtered air.
5. Seal cover slips in a Petri dish with Parafilm, and store in a desiccator until ready to apply photoresist.

3.2.3. Photoresist Patterning (see **Note 11**)

1. Spin coat cover slips individually with 5–7 drops of HMDS (an adhesion promoter) for 30 s at 3000 rpm. Apply HMDS with a 10-mL sterile plastic syringe, filtered through a 0.2-μm Acrodisk filter, while cover slip is spinning (see **Note 12**).
2. Spin coat cover slips individually with 5–7 drops of Microposit S1813 photoresist for 30 s at 3000 rpm. Apply photoresist with a 10-mL sterile plastic syringe, through a 0.2-μm Acrodisk filter, while cover slip is spinning.
3. Bake cover slips at 110°C for 5 min to cure the photoresist.
4. Using a contact printer, expose each cover slip to UV light through a photomask for 30 s (see **Note 13**).
5. Develop exposed photoresist in Microposit MF-319 developer for 30 s, making sure to keep track of the coated side of the cover slip (see **Note 14**).
6. Rinse cover slip in deionized water and dry with a stream of filtered nitrogen gas.
7. Place cover slips in Petri dishes, seal with Parafilm, wrap in aluminum foil, and store in a desiccator for up to 3 wk at room temperature, until ready to pattern with protein (see **Note 15**).

3.2.4. Protein Patterning (10)

1. Examine photoresist-patterned cover slips under a microscope for pattern regularity, discarding any with defective patterns.
2. Sterilize cover slips overnight (8–12 h) under the germicidal UV lamp of a tissue-culture hood.
3. Sonicate cover slips in 1% AMPS for 2 min.
4. Rinse cover slips in deionized, distilled water.
5. Transfer the cover slips into a cover slip holder, and bake at 110°C for 10 min.
6. Incubate photoresist-side-up for 1 h in 2.5% glutaraldehyde.
7. Wash 3× with fresh PBS.
8. Pipet 200 μL of 50 μg/mL laminin or 150 μg/mL collagen onto the cover slip, and incubate for 1 h.
9. Transfer the cover slips into the cover slip holder again, making sure that the photoresist side of the cover slips faces the same direction, and sonicate in acetone for 8 min.
10. Transfer each cover slip into its own 35-mm Petri dish containing PBS, and wash 3× with more PBS.
11. Incubate for 1 h in 0.5 mL of 0.5% BSA.
12. Rinse twice with Ca^{2+}-free, Mg^{2+}-free HBSS, and once with F12 culture medium. The cover slips are ready for use as cell culture substrates at this point.

3.3. Video Microscopy

The following method yields a sealed chamber with a volume of ≈30 μL for maintaining the neuronal culture during extended videomicroscopy observation. The sealed chamber prevents evaporation of the culture medium, is suitable even for microscopes with little clearance between condenser and objective, and can be adapted to any objective of interest by matching the cover slip thickness in **Subheading 2.2.1.** to the working distance of the objective. DRG cultures remain healthy, with no evidence of change or deterioration, for 6–18 h following the preparation.

3.3.1. Culture Chamber Preparation

1. Clean a microscope slide with Alconox paste, and rinse thoroughly with deionized, distilled water.
2. Spray 70% ethanol on both sides of the slide and wipe dry in a circular motion with crumpled lens paper, paying particular attention to the center of the slide.
3. Check that the slide looks clean with no streaks visible to the eye. Repeat steps 1 and 2, as necessary, until slide is clean.
4. Place four 3 × 11 mm Parafilm shims on the center of the microscope slide, so that their outer edges coincide with the edges of a 22-mm square, leaving gaps between the shims at the corners of the square.
5. Place the slide onto a heated metal washer, with the strips of Parafilm over the metal. Leave until the Parafilm melts in place onto the slide, approx 15 s.
6. Pipet 200 µL of F12 culture medium from the Petri dish into the area delimited by the Parafilm on the slide.
7. Remove one cover slip culture from its 35-mm Petri dish. Place one edge of the cover slip, explant side down, on an outside edge of the Parafilm. Allow the rest of the cover slip to fall gently onto the liquid, making sure that no air bubbles are trapped. This step should be performed smoothly, but rapidly, to ensure that the culture does not dry out.
8. Blot excess liquid from the microscope slide outside the Parafilm boundaries, and allow to evaporate dry before continuing to the next step. Briefly holding the slide up to a 60 W incandescent light bulb will speed the drying, but care must be taken not to overheat the culture.
9. Brush melted Valap across one edge of the cover slip in a single smooth stroke. Repeat on the opposite side. Rest the slide on a cool surface, such as the surface of the lab bench, to help set the Valap.
10. Very lightly press the cover slip with a blunt object and blot any liquid that escapes.
11. Brush Valap onto the other two edges.
12. Repeat step 10 and repair any leaks with Valap, until the cover slip is completely sealed to the microscope slide.
13. Using a Q-tip, clean the outside of the cover slip with a 1:1 solution of Windex and water, then wipe with ethanol and allow to evaporate dry. Repeat, if necessary, to remove any streaks from the surface. Do the same for the microscope slide on the opposite side.

3.3.2. Video Microscopy

1. Place the sealed culture chamber on the incubated microscope stage of an inverted microscope maintained at 37°C, and bring into focus, using a ×63 or ×100 oil immersion objective for high-resolution work, or other objective of choice.

2. Allow the culture to equilibrate with its environment for at least 10 min before recording microscopy data.
3. Connect the video camera on the microscope to both a time-lapse VCR and to the video input of a personal computer equipped with frame-grabbing capabilities (*see* **Note 16**).
4. Record migration behavior at desired time intervals of desired duration (*see* **Note 17**).

3.3.3. Data Analysis

It is much easier to obtain visual data than to analyze it. Although sophisticated techniques for analyzing dynamic neuronal migration characteristics are being developed *(12)*, they are not generally available. However, some basic measurements can be made with readily available software.

Using image analysis software, such as *NIH Image*, open each digital image in sequence and make the desired measurements. Several examples using NIH image include:

1. Neurite length. Using the ruler tool, measure the distance from the cell body to the neurite tip.
2. Neurite trajectory. Record the coordinates of the same neurite tip through a series of images.
3. Growth-cone size and shape. Using the polygon tool, outline the growth-cone, and record area and shape characteristics provided by the software.

4. Notes

1. Alternatively, the less completely defined, but much simpler, serum-containing formulation may be used. Serum-containing F12 culture medium: F12 nutrient mixture (146 mg/L L-glutamine, 1176 mg/L sodium bicarbonate), penicillin–streptomycin 5000 µg/mL–5000 U/mL, fetal bovine serum 10%, chick serum 10% (all components from Gibco-BRL), 7S NGF 50 ng/mL (*see* **Subheading 2.1.2.**, step 1h., Collaborative Biomedical).
2. Inclusion of the sodium bicarbonate provides a medium suitable for cell culture in a 5% CO_2 environment. Substitution of 4.766 g/L HEPES for the sodium bicarbonate yields a medium appropriate for incubation in a normal air environment.
3. Since the transferrin in the ITS Premix is derived from a human source, contact with it, or any solutions containing it, should be prevented by wearing gloves and protective clothing.
4. Thaw pituitaries in distilled water. They will comprise the reddish-brown and yellow tissue.
5. **Caution:** To avoid the risk of electrocution, place a plastic bag over the bottom of the blender pitcher during the rests on ice.
6. For efficient dialysis, divide the supernatant into four batches of 20–25 mL each, and simultaneously dialyze each batch against 1 L of the PBS or F12 medium.
7. Cover slip specifications are dictated by the following considerations:
    a. Round cover slips tend to pattern better than square cover slips in **Subheading 3.2.3.**

b. The diameter of the cover slips should be slightly less than the width of the microscope slide used in **Subheading 3.3.1**.

c. The thickness of No. 1 cover slips (≈150 μm) provides the appropriate working distance for high-resolution oil immersion objectives in **Subheading 3.3.2.**, although this must be determined for the specific objective in use *(11)*. This information can typically be found in the microscope manual, or obtained from the technical representative for a given microscope.

d. Fisher cover slips tend to produce the most uniform substrates in our own trials (unpublished observations).

8. Eggs are typically used at embryonic days 6–10 for neuronal migration studies. This represents a significant window in the 21-d development of the chick embryo; the qualitative aspects of the dissection procedure change significantly with each day of development, although the specifics remain the same. At day E6, features will be small and tissues very soft. At day E10, features will be much larger and tissues somewhat resistant.

9. Depending on the precise location of the separation, some DRGs may be recovered from the matching site on the body. There are 8 DRGs per leg, although less than 100% recovery is typical. The DRGs look like tiny balloons attached by individual strings to the leg. Another set of ganglia, the sympathetic chain ganglia, lie closer to the spine, and can be distinguished from the DRGs by the slightly smaller size of the sympathetic chain ganglia and their arrangement in a chain or series, rather than on individual stalks.

10. Adding all of the medium at the time of plating results in convection currents that tend to move the DRG off the cover slip before it becomes attached. This is not a problem if the substrate comprises the total area available for attachment. Incubation for 6–8 h provides the opportunity for neurites to initiate and extend past the dense portion of the neurite halo, so that individual growth cone migration can be observed.

11. This procedure is typically performed in a microfabrication facility, in which a spin coater, contact printer, oven, fume hood, and adequate ventilation are standard. Lom et al. *(5)* describe a way to perform this procedure in the laboratory, although this requires the construction of some nonstandard apparatus. Limited commercial services are also available (e.g., Bhatia et al., **ref. 7**).

12. Filtering the HMDS and photoresist is important for preventing contamination of the tissue-culture and for removing particulates. Resting the cover slip on a chuck in the spin coater that is slightly smaller than the cover slip (e.g., 16–18-mm diameter) will help prevent transfer of photoresist to the underside of the cover slip.

13. Exposure time may vary, depending on the configuration and condition of the equipment.

14. Cover slips should be developed individually or in a holder that maintains space between them, to ensure good contact between the developer and the surface. Swirling the cover slips in the developer also aids in this respect.

15. It may be possible to extend the life of the photoresist pattern by storing in a desiccator at 4°C.

16. Simultaneously recording the data on videotape provides a convenient record of the dynamic behavior for later demonstration purposes, and does not degrade the quality of the digital images saved to the computer. The alternative is to record only digital images online, and then to output those from the computer to the VCR at a later time.

17. The time-lapse interval and the duration of recording are determined by the specific behavior of interest, and frequency of data required to observe significant events. Typical choices for time-lapse recording of neuronal migration range from 1/30 to 1/300 of real time, or 1 frame every 1–10 s, as required to provide a smooth video. Digital images, because of their high storage requirements ($\approx$1 Mb/image), are often recorded at more conservative intervals. Very dynamic behavior may require one image every 5–10 s. Other behavior may be recorded much less frequently. The duration of recording may be minutes or hours.

References

1. Hoch, H. C., Jelinski, L. W., and Craighead, H. C., eds. (1996) *Nanofabrication and Biosystems: Integrating Materials Science, Engineering, and Biology* Cambridge University Press, New York, NY.
2. Alberts, B., Bray, D., Lewis, J., Raff, M., Roberts, K., and Watson, J. D. (1994) *Molecular Biology of the Cell*, 3rd ed., Garland, New York, NY.
3. Clark, P., Connolly, P., Curtis, A. S. G., Dow, J. A. T., and Wilkinson, C. D. W. (1991) Cell guidance by ultrafine topography in vitro. *J. Cell Sci.* **99,** 73–77.
4. Clark, P., Britland, S., and Connolly, P. (1993) Growth cone guidance and neuron morphology on micropatterned laminin surfaces. *J. Cell Sci.* **105,** 203–212.
5. Lom, B., Healy, K. H., and Hockberger, P. E. (1993) A versatile technique for patterning biomolecules onto glass coverslips. *J. Neurosci. Methods* **50,** 385–397.
6. Kleinfeld, D., Kahler, K. H., and Hockberger, P. E. (1988) Controlled outgrowth of dissociated neurons on patterned substrates. *J. Neurosci.* **8,** 4098–4120.
7. Bhatia, S. N., Toner, M., Tompkins, R. G., and Yarmush, M. L. (1994) Selective adhesion of hepatocytes on micropatterned surfaces. *Ann. NY Acad. Sci.* **745,** 187–209.
8. Tsao, M. C., Walthall, B. J., and Ham, R. G. (1982) Clonal growth of normal human epidermal keratinocytes in a defined medium. *J. Cell Physiol.* **110,** 219–229.
9. Baird, J. L. and Raper, J. A. (1995) A serine proteinase involved in contact mediated repulsion of retinal growth cones by DRG neurites. *J. Neurosci.* **15,** 6605–6618.
10. Bhatia, S. N., Toner, M., and Yarmush, M. L. Personal communication.
11. Smith, R. F. (1990) *Microscopy and Photomicroscopy: A Working Manual*, CRC, Boca Raton, FL.
12. Gwydir, S. S., Buettner, H. M., and Dunn, S. (1994) Non-rigid motion analysis and feature labeling of the growth cone. *Proc. IEEE Workshop on Biomedical Image Analysis*, Seattle, WA.

V

MEASUREMENT TECHNIQUES

37

Estimating Number and Volume of Islets Transplanted Within a Planar Immunobarrier Diffusion Chamber

Kazuhisa Suzuki, Clark K. Colton, Susan Bonner-Weir, Jennifer Hollister, and Gordon C. Weir

1. Introduction

Cell therapy involving the transplantation of cells or tissues with specific differentiated functions has potential in the treatment of human disease. However, the need for immunosuppressive drugs may lead to a variety of serious side effects. One approach to minimizing or eliminating systemic immunosuppression is immunoisolation, in which the transplanted tissue is enclosed in a semipermeable barrier or membrane in order to protect it from immune rejection, thereby creating what has been termed an implantable biohybrid artificial organ. Devices of this type are under study for the treatment of a wide variety of diseases, for example, diabetes.

In implanted devices, the implanted cells or tissues are separated from the body by an immunobarrier membrane. Cells can either be encapsulated at a high tissue-like density or dispersed in an extracellular matrix, such as agar, alginate, or chitosan. Ideally, the membrane prevents components of the cellular (and possibly humoral) immune responses from entering into the vicinity of the transplanted tissue, but permits passage of the secreted product, for example, insulin. At the same time, the transport properties of the graft, membrane, and surrounding tissue must permit sufficient access of nutrients, such as glucose and oxygen, and the removal of secreted metabolic waste products, such as lactic acid, carbon dioxide, and hydrogen ions. Transplanted cells must be supplied with nutrients by diffusion from the nearest blood supply, through surrounding host tissue, the immunoisolation membrane, and the graft tissue itself. Oxygen supply is thought to be the limiting factor in determining how much tissue can be supported in a specific configuration.

Transplanted insulin-secreting pancreatic islets contained within immunobarrier membrane devices may be partially protected from allo- or xenograft rejection and from autoimmune attack. Although there are some reports about the function and morphological appearance of islets contained within such devices *(1–11)*, little is known about such parameters as chamber volume, islet volume, islet number, islet necrosis, and fibrosis under these conditions. In order to better characterize the potential of this approach, it is important to quantify islet number and volume within membrane devices.

In this chapter, we describe a recently developed method to measure chamber volume, islet number, and islet volume *(12)*. This method can be used to determine how much islet tissue can be supported by a planar diffusion device. In previous studies with planar devices, sections have been taken perpendicularly to the plane of the membrane. This generates extraordinarily large numbers of sections and is impractical for analyzing chamber and islet parameters for an entire device. In this method, we use sections taken parallel to the nominal plane of the membrane, and a fraction of these are analyzed to estimate device parameters. This is the first rigorous method for determining how much islet volume must be contained within an implanted device to normalize glucose levels in an animal model of diabetes *(13)*.

2. Materials

2.1. Animals

1. Inbred male B6AF1 mice, 25–30 g (Taconic, Germantown, NY).
2. Streptozotocin, 180 mg/kg, ip (Sigma, St. Louis, MO).
3. Portable glucometer (One Touch II, Lifescan, Milpitas, CA).
4. Methoxyflurane (Metofane®, Pitman-Moore, Mundelein, IL), used like ether inhalant with gauze, until animal is asleep.
5. Veterinary tissue adhesive (Vetbond™, 3*M* Animal Care, St. Paul, MN).

2.2. Islet Isolation

1. Collagenase solution, 1 mg/mL, 2 mL/mouse (Boehringer Mannheim, Indianapolis, IN).
2. Histopaque, used pure (Histopague®-1077, Sigma).

2.3. Islet Encapsulation

1. Sodium alginate (Protan, Copenhagen, Denmark) solution, 16 g/L in HEPES-saline buffer.
2. $BaCl_2$ solution: 2.38 g/L HEPES, 7.01 g/L NaCl, 4.89 g/L Ba $Cl_2 \cdot 2H_2O$.
3. HEPES-saline buffer: (N – [2 hydroxyethyl] piperazine-N'-[2-ethane sulfonic acid]), 1.19 g/L, NaCl 8.48 g/L.

2.4. Planar Diffusion Chamber

Planar diffusion chamber (Baxter Healthcare, Round Lake, IL): Tissue is placed into a gap between two membrane laminates separated by a silicone rubber washer (id 0.70 cm, cross-sectional thickness 125 µm), which seals the gap and defines its thickness *(14,15)*. Each laminated membrane consists of three layers: an inner membrane of polytetrafluoroethylene (PTFE) with a hydrophilic surface (Biopore™, Millipore, Bedford, MA) with a nominal pore size of 0.45 µm; a PTFE membrane (W. L. Gore, Elkton, MD) with a nominal pore size of 5 µm and a thickness of about 15 µm; this, in turn, is laminated to a highly open outer meshwork about 125-µm thick of polyester fibers to provide support. The inner membrane excludes cells, but allows some penetration of all dissolved molecules. The second layer, with or without the third layer, promotes neovascularization at the interface with host tissue *(16)*. The membrane-tissue sandwich is held in place between a titanium ring housing and a titanium sealing ring, which is pressed into the housing by a hand press. When compressed by the titanium rings, the silicone rubber washer dimensions change. The id becomes about 0.66 cm and the washer (gap) thickness about 100 µm *(15)*. In an idealized device, in which the membranes maintain this geometry uniformly, the calculated contained volume would be 3.5 µL.

2.5. Sample Preparation

1. Tissue cassette (HistoPrep®, Fisher Scientific, Pittsburgh, PA).
2. Bouin's solution: 850 mL picric acid, 50 mL glacial acetic acid, and 150 mL 37% (v/v) formaldehyde in ddH$_2$O.
3. Warm liquid agar (Bacto-Agar, Difco, Detroit, MI): 7% (w/w) in ddH$_2$O.
4. Biopsy sponges (Fisher).
5. 10% buffered formalin (Fisher).

2.6. Quantification of Device Parameters

1. Photographs (×10 magnification, Kodak Technical Pan Film TP 135, ASA 50).
2. Electronic planimeter (SigmaScan™, Jandel Scientific, San Rafael, CA).

3. Methods
3.1. Animals, Islet Isolation, and Implantation

Mice are used as donors and recipients, although other animals can be used. Recipients are made diabetic with streptozotocin at least 1 wk prior to transplantation. Blood glucose levels are measured with a portable glucometer before transplantation. Mice are only used as recipients if their blood glucose levels are over 300 mg/dL, in order to test the efficacy of the implant in normalizing blood glucose concentration. Islets are isolated using a modification of the method of Gotoh et al. *(17)*, in which the pancreatic duct is distended with a collagenase-containing solution. After purification on a Histopaque gradient, islets with diameters between 75 and 250 μm are handpicked and encapsulated within alginate microspheres. Islets are suspended in a 0.3% (w/w) alginate solution, and $BaCl_2$ is used to crosslink the alginate to form the microspheres *(18,19)*. Encapsulation is helpful, because it prevents clumping of islets within the capsules. The method of analysis described in this chapter is applicable when islets are encapsulated by any method, unencapsulated, or suspended in a matrix. Approximately 500 encapsulated islets are loaded into a diffusion chamber. Recipient mice are anesthetized with the inhalant anesthesia methoxyflurane, and, following a midline incision, the membrane device is placed in the epididymal fat pad and fixed in place with veterinary tissue adhesive. Two wk after implantation, the device is removed from the anesthetized recipient and prepared for morphological evaluation.

3.2. Sample Preparation

The implanted device is removed from the recipient, placed in a tissue cassette, and fixed in Bouin's solution overnight. The device is then washed in water and placed in 10% buffered formalin for a further 24 h. Then the silicone spacer and titanium rings are gently removed. The membranes are held together with forceps and pushed back into the outer titanium ring. A drop of warm liquid agar is placed on a microscope slide, and the membranes with the metal ring are placed upside down into this pool of agar. Another drop of liquid agar is placed on the top membrane, and, on top of this, another microscope slide is placed to sandwich the membranes between the glass slides. Once the agar hardens at room temperature, the ring is removed, along with excess agar. The membranes, embedded in agar, are then placed between biopsy sponges, returned to the tissue cassette, and again placed into 10% buffered formalin, before being processed for paraffin embedding.

Serial sections (7 μm in thickness) are made in the plane of the membranes (*see* **Note 1**). The total number of cut sections is counted to determine the thickness of the device. The sections are then stained with hematoxylin.

3.3. Quantification of Device Parameters

In order to determine the total volume of the device chamber, V_C, the total volume of islets, V_I, and the total number of islets, M, photographs are taken of every evaluated section, using a light microscope (e.g., Olympus BH-2). Several photographs are required of each section to incorporate the entire area of that section. Prints (5 × 7 in.) are made into montages of each section. The device chamber area, A_C, which is the space surrounded by the membrane (s), of each evaluated section, and the profile area, A_I, of each individual islet, are measured using an electronic planimeter.

Estimating Islet Number and Volume

The device is assumed to be a shallow cylinder that is analyzed layer by layer. The sections are divided from top to bottom into N sequential zones. The top of each zone consists of an evaluated section, followed by a number of nonevaluated sections beneath it. The evaluated sections are selected for quality of section and stain at intervals of no more than nine sections (*see* **Note 2**). The data obtained from the evaluated section at the top of each zone are assumed to be representative of all of the nonevaluated sections in that zone, and are used to estimate the chamber volume, islet volume, and islet number of that zone. Because each section is 7-µm thick, the thickness of each zone is determined by the number of sections in that zone multiplied by 7-µm thickness.

3.3.1. Volume of Chamber

The volume of the device chamber, V_{Cn}, for a given zone, n, is calculated as the chamber area of the evaluated section, A_{Cn}, multiplied by the thickness of the zone, h_n,

$$V_{Cn} = A_{Cn} h_n \tag{1}$$

The total volume of the device chamber, V_C, for the entire device can be then obtained by summation over all of the N individual zones (*see* **Note 3**):

$$V_C = \sum_{n=1}^{N} V_{Cn} \tag{2}$$

3.3.2. Volume of Islets

The islet volume in the device can be obtained from similar equations, using the islet areas instead of device chamber areas. It is assumed that the islet can be represented as a cylinder in each zone. The islet volume in zone n is calculated as the total islet profile area in the evaluated section, A_{In}, multiplied by the thickness of the zone, h_n.

$$V_{In} = A_{In} h_n \tag{3}$$

The total islet volume in the device, V_I, is obtained by summation over the N individual zones (*see* **Note 4**):

$$V_I = \sum_{n=1}^{N} V_{In} \tag{4}$$

3.3.3. Number of Islets

For transplantation, islets with a diameter between 75 and 250 µm are handpicked using a calibrated eyepiece micrometer on a dissecting microscope. Because of this size, a portion of each islet is found on many more than one 7-µm section. The number of sections in which each islet is observed depends on the size of the islet and the thickness between evaluated sections. When an islet profile in an evaluated section is measured by planimetry, neither the size of the entire islet nor the plane of section is known. Furthermore, portions of many islets are observed in each evaluated section.

Different parts of an islet are included in different adjacent zones, the details of which depend on the diameter of the islet and the thickness of the zones. To estimate the number of islets, it is assumed that the islet profile area observed is actually a section through a cylindrical islet of equal diameter and height. The portion of the *m*th islet contained within zone *n*, q_{nm}, is the ratio of the volume of that islet within zone n to the volume of the entire islet, which is assumed to be a cylinder having a diameter (and height) equal to the equivalent diameter observed in the evaluated section:

$$q_{nm} = A_{Inm}h_n / A_{Inm}D_{nm} = h_n / D_{nm} \qquad (5)$$

where A_{Inm} is the profile area and D_{nm} is the equivalent diameter of the *m*th islet in zone *n*. Since

$$A_{Inm} = \pi D^2_{nm} / 4 \qquad (6)$$

q_{nm} can also be written as

$$q_{nm} = h_n / 2(A_{Inm} / \pi)^{1/2} \qquad (7)$$

The total cumulative portions, Q_n, of all islets contained in zone n is obtained by summing individual islet portions over all of the M_n islets observed in the evaluated section of zone *n*:

$$Q_n = \sum_{m=1}^{N_n} q_{nm} \qquad (8)$$

The total number of islets in the device is obtained by summing the portions of islets in all *N* zones (*see* **Notes 5** and **6**):

$$M = \sum_{n=1}^{N} Q_n \qquad (9)$$

4. Notes

1. To optimize the efficiency of this method, it is desirable to make sections that are as parallel to the original flat surface of the membrane as possible, even though the device, when embedded in paraffin, is usually not completely flat, but slightly askew. Some difficulties can be experienced with the sectioning. Sometimes the edge of a membrane is caught by the knife and pops out, making further analysis of that block impossible. Even though the method of enrobing the membranes in

agar and embedding the sample in paraffin usually works well, the membranes can still slip apart.
2. Obviously, as the number of zones and sections evaluated is increased, the more accurate is the estimation of device parameters, but the quantification process will take much longer, and the method becomes impractical at some point. By embedding and sectioning the device as flat as possible, the number of serial sections needed to go through all of the device volume is minimized. To estimate the optimal number of sections for quantification *(12)*, the device and islet volumes of one device have been measured by evaluating every other section throughout the serial sections (total of 13 evaluated sections and 13 zones). These results were compared with device and islet volumes obtained if only every fourth, sixth, eighth, or tenth section was evaluated. There was no change in values until samples were taken only every 10 sections. If every other section through only half of the device were used, the values were highly divergent. Therefore, it is recommended that every fourth to ninth section (each of which is selected because it is nicely sectioned and stained) be evaluated through the whole device.
3. In an illustrative example *(12)*, the total calculated volume of the chamber was 1.78 µL, roughly half of the 3.5 µL calculated for an idealized device with planar membranes. This difference is not surprising, because it is assumed that the two membranes remain flat and separated from each other by about 100 µm in the idealized device. Clearly, the membranes have some flexibility, and some bulging must occur when islets with diameters of 75–250 µm are placed into the chamber. With this flexibility, the height of the internal space can be expected to increase to 300 µm or more. In the areas between the islets, the membranes approach each other, which leads to a loss of volume of the internal space. This latter reduction more than counterbalances the increase from the bulging around the islets.
4. Although the islets handpicked for transplantation in the device are spherical or ovoid, considerable variation in shape is observed. To estimate islet volume and number, it is assumed that a circular islet profile is obtained in each zone as a slice through a cylindrical islet in that zone. In determining the chamber and islet volume of each zone, this assumption provides a reasonably close estimate, because each zone is relatively thin. If each zone is thin and all sections are evaluated for estimating the volume of device and islets, the error would be mathematically negligible. Therefore, this method of calculation should be suitable for an accurate estimate of chamber and islet volume.
5. Calculation of islet number is more complicated than total chamber or islet volume, because the volume of an islet within a zone is considered to be a portion of an islet, that is, the volume within a zone divided by the total volume of the cylindrical solid. This calculational procedure is not based on theoretical considerations, but rather is empirically derived, and used on the basis that it works reasonably well. To examine its validity, calculations have been carried out *(12)* by assuming that sections (each representing one zone) are cut through a spherical islet, and the calculations described here are applied to each section. The

result is that the portions sum to 1.38 islets. Thus, this method gives an estimate of islet number that is high by about 38%.
6. This method can be used to estimate not only the volume of the device, volume of islets, and islet number, but also the volume of fibrosis, necrosis, and other possible conditions that may decrease the function of a loaded device *(13)*.

References

1. Altman, J. J., Penfornis, A., Boillot, J., and Maletti, M. (1988) Bioartificial pancreas in autoimmune nonobese diabetic mice. *ASAIO Transact.* **34,** 247–249.
2. Brunetti, P., Basta, G., Faloerni, A., Calcinaro, F., Petropaolo, M., and Calafiore, R. (1991) Immunoprotection of pancreatic islet grafts within artificial microcapsules. *Int. J. Artif. Organs* **14,** 789–791.
3. Colton, C. K. and Avgoustiniatos, E. S. (1991) Bioengineering in development of the hybrid artificial pancreas. *ASME J. Biomech. Eng.* **113,** 152–170.
4. Lacy, P. E., Hegre, O. D., Gerasimidi-Vazeou, A., Gentile, F. T., and Dionne, K. E. (1991) Maintenance of normoglycemia in diabetic mice by SC xenografts of encapsulated islets. *Science* **254,** 1782–1784.
5. Brauker, J., Martinson, L. A., Loudovaris, T., Hill, R. S., Carr-Brendel, V., Hodgson, R., et al. (1992) Immunoisolation with large pore membranes: allografts are protected under conditions that result in destruction of xenografts. *Cell Transplantation* **1,** 164.
6. Hill, R. S., Young, S. K., Jacobs, S. A., Martinson, L. A., and Johnson, R. C. (1992) Membrane encapsulated islets implanted in epididymal fat pads correct diabetes in rats. *Cell Transplantation* **1,** 168.
7. Lanza, R. P., Borland, K. M., Staruk, J. E., Appel, M. C., Solomon, B. A., and Chick, W. L. (1992) Transplantation of encapsulated canine islets into spontaneously diabetic BB/Wor rats without immunosuppression. *Endocrinology* **131,** 637–642.
8. Lanza, R. P., Borland, K. M., Lodge, P., Carretta, M., Sullivan, S. J., Muller, T. E., et al. (1992) Treatment of severely diabetic pancreatectomized dogs using a diffusion-based hybrid pancreas. *Diabetes* **41,** 886–888.
9. Lanza, R. P., Beyer, A. M., Staruk, J. E., and Chick, W. L. (1993) Biohybrid artificial pancreas. *Transplantation* **56,** 1067–1072.
10. Scharp, D. W., Swanson, C. J., Olack, B. J., Latta, P. P., Hegre, O. D., Doherty, E. J., et al. (1994) Protection of encapsulated human islets implanted without immunosupression in patients with type I or type II diabetes and in nondiabetic control subjects. *Diabetes* **43,** 1167–1170.
11. Colton, C. K. (1995) Implantable biohybrid artificial organs. *Cell Transplantation* **4,** 415–436.
12. Suzuki, K., Bonner-Weir, S., Hollister, J., and Weir, G. C. (1996) A method for estimating number and mass of islets transplanted within a membrane device. *Cell Transplantation* **5,** 613–625.
13. Suzuki, K., Bonner-Weir, S., Hollister, J., Colton, C. K., and Weir, G. C. (1997) Number and volume of islets transplanted in immunobarrier devices. *Cell Transplantation*, in press.

14. Dudek, R. W., Lawrence, I. E., Jr., Hill, R. S., Johnson, R. C. (1991) Induction of islet cytodifferentiation by fetal mesenchyme in adult pancreatic ductal epithelium. *Diabetes* **40**, 1041–1048.
15. Brauker, J., Martinson, L. A., Young, S. K., and Johnson, R. C. (1996) Local inflammatory response around diffusion chambers containing xenografts. *Transplantation* **61**, 1671–1677.
16. Brauker, J. H., Carr-Brendel, V. E., Martinson, L. A., Crudele, J., Johnston, W. D., and Johnson, R. C. (1995) Neovascularization of synthetic membranes directed by membrane microarchitecture. *J. Biomed. Mater. Res.* **29**, 1517–1524.
17. Gotoh, M., Maki, T., Kiyoizumi, T., Satomi, S., and Monaco, A. P. (1985) An improved method for isolation of mouse pancreatic islets. *Transplantation* **40**, 437,438.
18. Zekorn, T., Entenmann, H., Horcher, A., Schnettler, R., Klock, G., Bretzel, R. G., Zimmermann, U., and Federlin, K. (1992) Barium-alginate beads for immunoisolated transplantation of islets of langerhans. *Transplantation Proc.* **24**, 937–939.
19. Schrezenmeir, J., Hering, B. J., Gero, L., Wiegnad-Dressler, J., Solhdju, M., Velten, F., et al. (1993) Long-term function of porcine islets and single cells embedded in barium-alginate matrix. *Horm. Metab. Res.* **25**, 204–209.

38

Quantitative Measurement of Cell–Cell Adhesion Under Flow Conditions

Carroll L. Ramos and Michael B. Lawrence

1. Introduction

Cell–cell and cell–matrix contacts play an important role in immunological processes and tissue organization. It has been shown that in many cases, cell adhesion is mediated by specialized adhesion receptors that are typically anchored in the cell membrane with mol wt ranging from 50 to 200 kDa. Binding of adhesion receptors anchored in the apposing cell membranes with their counterligands appears to be critical for many forms of cell–cell communication *(1)*.

Quantitative measurement of cell–cell and cell–substratum adhesion can be used to compare the contributions of distinct receptors to an adhesive interaction. For instance, by measuring specific cell adhesion quantitatively, the modulation of receptor avidity or affinity in response to biological stimuli can be tracked *(2)*. Other applications of adhesion force assays are in the study of adhesion receptor ligation on cell signaling processes and in the characterization of the strength of cell–biomaterials interactions. **Table 1** contains a list of some of the commonly used methods for assessing cell–cell adhesion, with selected references. In this protocol, we will focus on the use of parallel-plate flow chambers, highlighting the strengths and limitations of the assay relative to other methods in the Notes section.

Assessment of adhesion requires that a mechanical force be applied to the cells to separate them from other cells or the substrate. In almost all forms of cell–cell adhesion assays, fluid shear, or washing, is used to distinguish bound from unbound cells. Parallel-plate flow chambers generate controlled, highly reproducible washing conditions (detachment forces), allow direct visualiza-

Table 1
Cell–Cell and Cell–Substrate Adhesion Assays

Assay	Principle	Ref.
Stampler-Woodruff	Leukocyte suspension layered onto rotating tissue specimen (nonstatic)	13
Plate-binding assay	Leukocyte suspension allowed to settle in wells of an Elisa plate coated with adhesive substrate (static)	14
Smith-Hollers chamber	Cells injected into sealed chamber, allowed to settle, then chamber inverted (static)	15,16
Centrifugation assay	Cells in sealed ELISA plates are centrifuged at varying speeds	17,18
Cone and plate viscomter	Suspension of cells exposed to uniform shear	9
Couette viscometer	Suspension of cells exposed to uniform shear	19
Capillary tube	Suspension of cells exposed to varying shear forces; exposure of anchored cells to defined shear force; some designs employ a recirculating loop	20,21
Radial flow chamber	Anchored cells on adhesive substrate experience fluid shear force based on position from center of disk	22
Parallel-plate flow chamber	Anchored cells on adhesive substrate experience defined shear stress	23–25

tion of cell–substrate interactions, have relatively low cost, and can be adapted to study a number of different cell and receptor types.

2. Materials

1. Flow chamber apparatus
   a. Custom rectangular polycarbonate (Lexan) flow-chamber block machined from specifications (two ports on each long side, vacuum port on one end, two flow slots and one vacuum hole on face; Godwin Machine Works [Houston, TX]; specifications obtainable from authors); commercial flow chambers are available from sources such as Glycotech [Rockville, MD] or CytoDyne [San Diego, CA].
      1. Silicone rubber sheeting for hand-cut gaskets that are applied over the face of the flow chamber (4 × 8 × 0.020 in.; Technical Products [Decatur, GA], #500-5, gasket specifications obtainable from authors).
      2. Tygon tubing (1.6 mm id; Bio-Rad [Richmond, CA], #731–8215).
      3. Intramedic Luer stub adapters (15 gage; Becton Dickinson [Sparks, MD], #427560).

4. 60-, 10-, and 1-cc sterile, disposable syringes (Becton Dickinson [Franklin Lakes, NJ]).
5. Standard three-way stopcocks with male luer slip adapter.
6. Latex tubing (5/32 × 3/64 in.) cut into one-half-in. lengths used as connectors.
7. Vacuum grease (silicone) (Dow-Corning [Midland, MI]).

2. Inverted phase-contrast microscope (e.g., Nikon Diaphot 300 [Melville, NY]), with a long working-distance condenser and ×10, or ×20 objectives.
3. Video camera (e.g., Vicon [Melville, NY], #VC2410-24).
4. Video recorder (consumer models are adequate), video monitor, and video timer (e.g., Fora #VTG-33)
5. Syringe pump allowing infusion and withdrawal (Harvard Apparatus [South Natick, MA], #55-2226).
6. Vacuum apparatus (side-arm flask connected to laboratory vacuum).
7. Temperature-controlled water bath with 15 and 50 mL conical tube racks.
8. Polystyrene, bacteriological (nontissue-culture) Petri dishes (150 × 15 mm, Falcon 1058; 100 × 15 mm, Falcon 1029; Becton Dickinson [Lincoln Park, NJ]) for cell or protein immobilization.
9. Glass slides (70 × 40 mm; Corning [Corning, NY]), for cell or protein immobilization; custom sizes at thicknesses of 1 or 2 can be made to order by Corning Glassworks.
10. Disposable, sterile, polypropylene conical tubes (15 mL; 50 mL; Falcon, Becton Dickinson, #2097 and #2098).
11. Pyrex 10 × 15 × 2 in. rectangular dish (Corning).
12. Buffers and assay media.
    a. Hanks' balanced salt solution (HBSS; Gibco-BRL [Grand Island, NY], #24020-059) supplemented with 20 mM HEPES, pH 7.4, and 1% human serum albumin (HSA) (Sigma [St. Louis, MO], #A1653): suitable for leukocytes isolated from blood
    b. HBSS without calcium and magnesium can be obtained (Gibco-BRL #14170-021) and supplemented with 1–2 mM $CaCl_2$ or $MgCl_2$, depending on the nature of the adhesive interaction studied (e.g., addition of calcium alone for assaying selectin-dependent rolling interactions of leukocytes may be useful to avoid integrin-mediated interactions that are magnesium-dependent).
    c. RPMI-1640 (Gibco-BRL, #11875-093) supplemented with 20 mM HEPES, pH 7.4, and 1% HSA: suitable for cultured cell lines of myeloid or lymphoid origin, and some transfected cell lines.
    d. M199 (Gibco-BRL, #11150-026) supplemented with 20 mM HEPES, pH 7.4, and 1% HSA: suitable for assays using cultured endothelium as an adhesive substrate.
    e. 1X phosphate-buffered saline (PBS), pH 7.4.
    f. 10X PBS: 80 g NaCl, 2 g KCl, 11.5 g $Na_2HPO_4 \cdot 7H_2O$, 2 g KH_2PO_4 added to sufficient deionized, distilled water to make 1 L, adjust to pH 7.4 at room temperature.
    g. 50 mM Tris, pH 9.5, with 0.02% azide (adsorption buffer).

h. 50 mM Tris, pH 9.5, with 0.02% azide and 3% HSA (blocking buffer A); solution should be made fresh before use.

i. 1X PBS, pH 7.4, with 0.02% azide and 0.1% Tween-20 (Bio-Rad [Hercules, CA], #170–6531) (blocking buffer B).

3. Methods
3.1. Assembly of Custom Polycarbonate Flow-Chamber Apparatus

1. Using a disposable pipet tip, apply a thin film of vacuum grease to the face of the flow chamber block, being careful to avoid covering the slots or vacuum hole with grease. Reapply as necessary.
2. Place the cut silicone rubber gasket over the vacuum grease-coated face of the flow chamber. Position the gasket using a pipet tip to gently press downward on the gasket, so that the vacuum grease is evenly spread below. Excess grease that extends over the flow area can be removed by applying a small amount of liquid detergent and gently washing with gloved fingers under running water. During the first several usages, the gasket will compress and stretch a little, so it is best to make gap-thickness measurements several times. Once seated, several experiments may be performed before it is necessary to add more grease. Do not rub harshly or dry the flow block with wipes, since the polycarbonate is easily scratched. The gasket may be washed with mild detergent and water. Ethanol will cause the gasket to stretch and thus should not be used for rinsing.
3. Add 200 mL 10X PBS to 1800 mL deionized, distilled water, and dispense this solution (2 L) into the Pyrex rectangular dish. Lower the flow-chamber block, with the gasket side up, into the PBS. Using a disposable Pasteur pipet filled with PBS, dispense and fill several times over the flow slots, to remove air bubbles from the side-port channels.
4. Cut four lengths of Tygon tubing (two approx 12–15 in. and two 1/4 in.). Using pliers, remove the blunt needles from two Intramedic Luer stub adapters. Into one end of each piece of 12–15 in. Tygon tubing, insert a blunt needle. Over the needle and Tygon tubing, place a 1/2 in. piece of latex tubing, leaving a small amount of needle exposed. Insert the exposed needle into a 1/4 in. length of Tygon tubing, and place this end of the tubing system into the Luer ends of the three-way stopcock. Secure the fitting by pulling a portion of the latex tubing over the Luer outlet on the stopcock. Attach Intramedic luer stub adapters to a 60-cc and 10-cc syringe, and insert the syringes into the open (opposite) ends of the two Tygon tubing preparations.
5. Fill the tubing systems with PBS by drawing 5–10 mL PBS into each syringe while the stopcock end is placed into a container of buffer. Immerse the filled tubing systems into a PBS-filled Pyrex dish containing the flow chamber. Push the syringe plungers to dispel air from the tubing and stopcocks. While immersed, insert the ends of the stopcocks into each portal on one side of the flow chamber. Insert 1-cc syringes attached to stopcocks and filled with PBS into the two por-

tals opposite of the inlets and outlet manifolds of the flow chamber. Insure that air bubbles are removed from the tubing and syringes to facilitate an airtight system. The 1-cc syringes can be used to remove air bubbles, or when only small numbers of cells are being analyzed, to inject a bolus of cells into the chamber flow channel. In addition, the syringes allow loading of 1 mL antibody or stimulus solutions onto the adhesive substrate in the flow chamber.

3.2. Preparation of Adhesive Substrates

3.2.1. Immunoaffinity Purified Native Adhesion Receptor Proteins

1. Dilute the adhesion receptor or matrix protein in 50 mM Tris, pH 9.5, with 0.02% azide, to a concentration of 0.1–10 µg/mL. To ensure efficient adsorption, proteins purified from cell membrane preparations that are in nonionic detergents (e.g., octylglucoside) need to be diluted, so that the protein–detergent solution is below the critical micelle concentration.
2. Cut a 70 ¥ 40-mm slide from a 150 ¥ 15-mm polystyrene Petri dish, using a surgical scalpel. Latex gloves should be used to prevent coating the slides with oils from skin. We have found that nontreated polystyrene (Falcon 1058) is comparable to tissue-culture polystyrene for adsorption of protein and cells. Alternatively, a 70 ¥ 40-mm glass slide may be used. Wash the slide with a stream of deionized, distilled water, followed by rinsing with 100% ethanol. Repeat washing two additional times, and place the slides upright to air-dry. Scrubbing or drying with tissues (Kimwipes) should be avoided, to prevent scratching of the slide surface.
3. Using a laboratory marking pen, draw a 10-mm diameter circle on the underside of the slide. Mark a small dot in the center of the circle and place the slide markings side down in a 100 ¥ 15 Petri dish. Dispense 50–100 µL of the diluted protein preparation onto the slide in the center of the marked circle. To prevent evaporation and drying of the edges of the protein solution, place several 100-µL drops of Tris–azide or PBS on the slide surface. Place the lid on the Petri dish and incubate the slide for 1.5 h at room temperature, or overnight at 4°C.
4. Blocking of the plate to eliminate nonspecific interactions with plastic or glass: Dispense 500 µL of the freshly prepared blocking buffer A onto the protein solution on the plate. Using a Pasteur pipet connected to a vacuum source begin aspirating the solution on the plate while simultaneously dispensing 3–5 mL of blocking buffer A with a pipeter. **Caution:** The center area of the plate containing the adsorbed protein should never be allowed to dry or be exposed to air, since this may destroy the activity of the protein. Cover most of the plate with blocking buffer A. Incubate 1 h at room temperature, or overnight at 4°C. The plate can be stored for several days at 4°C.
5. Before the experiment, wash the plate in the same manner just described except using 2–3 mL of 0.1% Tween-20 solution (blocking buffer B), followed by rinsing with 2–3 mL PBS, pH 7.4.

3.2.2. Recombinant Adhesion Proteins

Recombinant Adhesion Proteins are Frequently fused with the Fc domain of human IgG or IgM, and thus introduce the potential for Fc–receptor interactions.
1. Adsorb the recombinant fusion protein and block the plate as described for purified native proteins.
2. Before the experiment, incubate the plate with 1 mL of 10 µg/mL polyclonal goat antihuman IgG (Fc) F (ab')$_2$ (Biodesign International [Kennebunk, ME], #W99334G) or human Ig*M* (Fc5µ) (#W99340G) for 15 min at room temperature.

3.2.3. Monolayers of Vascular Endothelial Cells (3)

1. Characterized normal human umbilical vein endothelial cells (HUVEC) can be obtained commercially from Clonetics [San Diego, CA], or others, as proliferating cells in T-25 flasks, #CC-2617 or cryopreserved cells, #CC2517. Culture media containing serum, antibiotics, and growth factors can also be obtained from Clonetics, #CC-3124. Alternatively, endothelial cells can be obtained from human umbilical veins, using collagenase digestion (an excellent protocol source: Freshney, R. I. (1994). Culture of specific cell types: endothelium, in *Culture of Animal Cells*, 3rd ed. Wiley-Liss, New York, pp. 333–334). Typically, primary or first passage endothelial cells are preferable for many types of studies because of the enhanced responsiveness to in vitro agonists relative to high-passage-number endothelial cells ($p > 4$).
2. Under aseptic conditions, collect cells from culture flasks, using 0.05% trypsin and 0.02% EDTA (Gibco-BRL, #25300-047) and wash into complete cell-culture media (with serum, antibiotics, and growth factors) to neutralize the trypsin. Resuspend cells in complete culture media and seed (0.5–1 × 10^6 cells/mL) onto sterile (autoclaved) 70 × 40-mm glass slides coated with fibronectin or gelatin *(3)* and placed in sterile Petri dishes. Place the slides in a 5% CO_2, 37°C incubator. Slides can be used for adhesion assays upon reaching a confluent monolayer in 1–3 d, depending on seeding density.

3.2.4. Platelet Monolayers (4)

1. 70 × 40-mm glass slides should be soaked in 70% nitric acid chromerge and washed thoroughly with deionized, distilled water. Dry the slides with ethanol and allow to air-dry. A number of matrix proteins or peptides can be absorbed to the clean glass surface. Plastic slides will work, as well, for binding proteins such as fibronectin or fibrinogen, to which platelets will adhere. Alternatively, some investigators dip the clean slide twice for 2 min in 4% 3-aminopropyl-triethoxysilane (Sigma, #A3648) in acetone *(4)*. Rinse the slides with acetone, followed by deionized, distilled water, and place in sterile Petri dishes. Place the slides into a 37°C incubator overnight to dry.
2. Collect human peripheral venous blood into acid citrate dextrose (ACD) solution (Sigma, #C3821) at a ratio of 1 vol ACD anticoagulant to 8 vol whole blood.

Centrifuge at 600g for 5 min at room temperature and transfer the upper platelet-rich plasma layer into a conical tube. Dilute to 2×10^8/mL in room temperature PBS, pH 7.4, with 0.3% bovine serum albumin (BSA), fraction V, essentially fatty-acid-free (Sigma, #A6003).

3. Dispense a sufficient volume of platelet suspension to cover the treated slides, and incubate for 30 min at room temperature to allow a platelet monolayer to form. The confluency of the monolayer can be controlled by the platelet seeding density and the addition of thrombin or ADP to promote platelet binding to the substratum.

3.3. Incorporation of Adhesive Substrates into the Flow-Chamber Apparatus

1. Submerse the Petri dish containing the slide with the adhesive substrate in the PBS-containing rectangular Pyrex dish (*see* **Subheading 3.1., step c**). If a cellular substrate (i.e., endothelium or platelets) is being tested, then it is important to add Ca^{2+} (1 mM) and Mg^{2+} (1 mM) to the buffer to maintain cell–substrate adhesion. Carefully invert the substrate slide while under buffer, and place over (substrate side down) the upper face of the submerged polycarbonate flow apparatus. Lower the slide onto the gasket of the flow chamber, so that the adhesive substrate is in the center of the rectangular flow chamber block. Without applying downward pressure on the slide, connect the tubing from the vacuum source into the end portal of the flow chamber. This is best accomplished by letting the slide rest on the flow chamber, without using your hands to stabilize it, and connecting the vacuum tubing. The system is airtight and properly sealed if the buffer in the connected syringes is not drawn out when the vacuum is applied. If buffer is drawn out, disconnect the vacuum tubing and reassess the position of the gasket, to determine whether it was displaced during incorporation of the substrate slide. Frequently, the gasket is slid to one side or the other during assembly. This is the most common problem that prevents sealing. If necessary, reapply the gasket to the face of the flow chamber, and avoid pressure on the gasket during placement of the slide on the chamber.
2. Remove the assembled flow chamber from the rectangular dish, and rinse the outer surface of the slide with a stream of deionized, distilled water to remove salts. Dried salts on the slide result in a poor microscopic image. Transport the entire flow-chamber system (syringes, tubing, and flow chamber) to the stage of the inverted phase contrast microscope. Place the flow chamber (slide down) on the stage over the objective opening, and secure it by placing stage clips over the tubing that exits from the polycarbonate block.
3. Remove the 10-cc syringe from its tubing, and place the tubing end into a 50-mL conical tube containing a suitable assay buffer (e.g., M199 supplemented with 10 mM HEPES, pH 7.4, and 1% HSA for endothelial monolayers, or HBSS supplemented with 10 mM HEPES, pH 7.4, and 1% HSA for adsorbed-protein receptors and

platelet monolayers). Eliminate air from the end of the tubing before placing into the assay media, to avoid drawing air into the flow-chamber system. Mount the 60-cc syringe (attached to the flow chamber by tubing) onto the syringe pump. Set the diameter according to the syringe chart provided by the manufacturer. Set the rate to mL/min and the flow direction to withdraw. Initiate flow at 1–3 mL/min (for a 60-cc Becton-Dickinson syringe) and allow assay media to be withdrawn through the flow chamber over the adhesive substrate for 2–3 min, to equilibrate the system. Assays can be performed at room temperature, or at 37°C, by placing the conical tubes containing assay media or cell suspensions in a temperature-controlled water bath.

4. Transfer the tubing in the assay media to a conical tube containing a cell suspension. Adhesion of leukocytes, such as neutrophils, leukemia cell lines, or transfected cells, can be studied over a range of cell concentrations, typically 0.5–5×10^6 cells/mL.

5. Select the desired microscope objective (e.g., ×10 or ×20) and adjust the position of the flow chamber, using the stage manipulators. If using adsorbed protein on plastic slides, focus on the center dot in the marked circle outlining the protein area (*see* **Subheading 2., step a.3**). Start the flow at 3–4 mL/min, and watch for cells to enter the field of view. Focus on the cells, so that they are phase bright and sharply spherical against the background.

3.4. Scanning for Adhesive Interactions and Recording of Data

1. To begin scanning for adhesive interactions, move the field of view to the edge of the marked circle, at the point where cells are beginning to enter the area of the adsorbed protein. In the case of endothelial or platelet monolayers, any starting point near the cell entry point can be selected, but it is best to analyze near the center of the channel, and midway along the length of the channel. When a bolus of cells enters the field at 3–4 mL/min, decrease the pump speed in stages, until interactions become detectable. Using the focus adjustment and the video monitor, begin scanning entire fields (width of the monitor screen) at 5–30 s intervals. The scanning sequence should include 15–20 fields of view, and can proceed down the center of the substrate (with the flow of cells), followed by a shift of view and scanning up the plate (against the flow of cells).

2. Adherence assay: The scanning process is repeated for a series of flow rates at the same time point and the number of adherent cells counted from video tape recordings of the scans.

3. Attachment rate assay (rate of cell accumulation on adhesive substrate at a given wall shear stress): The scanning process is repeated over time at the same flow rate (applied wall shear stress) and the number of adherent cells accumulated at each time point is quantitated from video tape recordings of the scans.

4. Detachment assay (resistance of cell adhesion to fluid shear stress): While injected cells (or cells infused through the inlet port) are settling onto the lower wall of the chamber (30–90 s), the tubing is transferred to assay media, in order to rinse away unbound cells (for example, 1–2 mL/min for 2–3 min). Flow rates

may be increased at 30–40 s intervals, followed by scanning at each point to measure numbers of cells remaining bound at each flow rate (applied wall shear stress).

5. Tests for specificity: Monoclonal antibodies against adsorbed adhesion molecules, as well as those expressed on leukocytes, endothelial monolayers, or platelet monolayers, can be utilized to assess whether flowing cells are specifically binding to adhesion molecules of interest. Cell suspensions to be perfused through the flow chamber can be treated with monoclonal antibody at 5–20 µg/mL (1:100–1:500 for ascites fluid), followed by direct introduction into the flow-chamber system without washing to remove unbound antibody. To treat adsorbed proteins or cell monolayers with monoclonal antibodies, antibody solutions (5–20 µg/mL) can be infused over the adhesive substrate for 1–2 min, followed by stopping of the flow and incubation for 5–15 min. Alternatively, 1-mL antibody solutions can be injected through the flow-chamber side port using the 1-cc syringe positioned near the top or cell-entry area of the flow chamber. To accomplish this, remove the withdrawal syringe from the pump and close off the tubing from the cell/buffer reservoir, using a hinged paper clip as a clamp. While slowly depressing the lower 1-cc side port syringe, remove the upper 1-cc syringe–stopcock combination from its flow chamber side port. Fill the syringe with antibody solution, and tap to dispel air. Depress the plunger to form a drop at the opening of the attached stopcock. Reattach the syringe to the flow chamber, using a drop-to-drop connection to avoid air bubbles. Slowly inject the contents of the syringe into the flow chamber.

4. Notes

Two general types of experiments are typically performed using parallel-plate flow chambers. The most broadly applicable is the detachment, or shear resistance assay, which can be used to evaluate cell–cell, cell–matrix (with any number or combination of integrin–ligand pairs), or cell–biomaterial interactions. In this format, either proteins or cells, such as endothelial, epithelial, or any anchorage-dependent cell, are immobilized or cultured on one wall of the chamber. The second format uses continuous flow to create constraints of force and contact time on cell adhesion, similar to that imparted by blood flow on cells of the vasculature. This second method has recently proved useful for estimating forces and bond lifetimes of adhesion receptors *(5–7)*.

1. Static–flow or detachment flow assay: Cells may be introduced into the flow chamber by injection through the side port (if only small numbers of cells are available) and allowed to settle onto the lower wall of the flow chamber. The sedimentation velocity for most mammalian cells is approx 1 µm/s, but can be estimated from Stokes' Law: $v_t = 2R^2 (\rho - \rho_s)/9\mu$, where ρ is the density of the cell, ρ_s is the density of the suspending media, μ is the viscosity of the suspending media, R is the radius of the spherical particle or cell, and v_t is the sedimentation velocity.

   Problems associated with the detachment assay include inadequate time allowed for sedimentation and variation between fields of view in the number of cells because of inadequate mixing of the injected cell suspension. Clumping of cells either in the stock vial or during sedimentation can significantly affect the spatial distribution of cells in the chamber. For interactions in which cells form relatively stable interactions, such as those through integrin receptors with extracellular matrix molecules such as fibronectin, it is helpful to begin with a relatively low flow (under 1 dyn/cm^2 wall shear stress), to remove nonadherent cells before beginning scans of the number of bound cells. To determine when the flow chamber's solution has been exchanged (for example, if a drug is being infused to modulate adhesiveness), the rinsing media can be doped with small beads 10-μ diameter (Duke Scientific [Palo Alto, CA], #9010) or erythrocytes as flow markers, to indicate when the new solution has completely displaced the original solution.

Since most flow chambers are transparent to allow microscopic visualization, the number of input cells can be directly counted and their fates determined. Among the critical variables that must be controlled are the magnitude of the flow and its duration at each level. Over the time-scale of seconds, cells can deform in response to the shear, and possibly increase their contact area and number of adhesive bonds. For example, with selectin-mediated leukocyte rolling, the rapid bond-formation rates appear to stabilize adhesions with increasing shear.

2. Attachment of cells under steady flow conditions: A second approach to quantifying cell–cell adhesion, using a flow chamber that has been widely applied to the study of blood cell adhesion to natural and artificial surfaces, is the attachment, or continuous-flow, assay. Adhesion of cells to surfaces in this protocol takes place under flow conditions, which imparts both constraints of force and contact time. Higher flow rates exert higher levels of stress on the initial bonds formed, and, concurrently, limits the contact time, because the fluid velocity is greater. Even if a bond creates a strong enough crossbridge to hold a cell under physiologic levels of shear, the window for the association of the ligand–receptor pair may be too brief for the docking event to take place. The time-scale for a bond-formation event, therefore, scales with the shear rate at approximately the reciprocal of the shear rate (1/G) *(8,9)*. Mathematical analyses exist for deriving the actual forces imparted by the flow *(6,10,11)*. The continuous flow minimizes nonspecific cellular interactions with the substrate, since contact times are short enough to remove the effect of matrix protein secretion from the mechanisms of adhesion.

3. Data presentation: Data from flow chamber assays is typically presented as number of cells bound per unit area of visualized substrate. Data from detachment assays, using a parallel plate flow chamber, is typically presented as a percent remaining bound as a function of increasing wall shear stress. A force or shear resistance titration can be generated that indicates the average strength of adhesion of a population of cells. In continuous-flow assays, the rate of accumulation as a function of time and area can be determined and reported as number of cells bound per unit area. Calculating a cell flux is possible by dividing the number bound by the time, creating an adhesion parameter of cells/area/time. Cell accu-

mulation linear with time can be a useful parameter, but attachment of cells may create nonlinear effects because of coverage of binding sites. Additionally, if flow rates are low enough, sedimentation will have some effect on the number of cells near the wall *(12)*.

4. Hydrodynamic forces and relevant equations for wall shear stress: The level of flow of the rinsing media is controlled by the syringe pump. The wall shear stress generated by the flow, which is laminar under most conditions (Reynolds number on the order of 1 to 2), is a description of the flow rate that allows comparison of the effect of fluid forces with other systems of differing dimensions and geometries. The Reynolds number (Re) is defined as $Dv\rho/\mu$, where D is the chamber gap, v is the mean velocity of the fluid, ρ is the density, and μ is the viscosity. Since the fluid shear stress varies with distance from the wall of the flow chamber (parabolic velocity profile), the typical flow rate specification is the wall shear stress. When Re is low (where viscous forces dominate interial forces), the shear stress correlates with the actual stresses imparted on the adherent cell by the fluid flow. Calculation of the magnitude of the stresses acting on an adherent cell depends in part on the geometry of the cell and its mechanical properties, which determine to what degree it deforms in response to the fluid shear stress. However, to a reasonable approximation in the case of comparative studies of a specific cell type, adherent cells experience nearly the same level of detachment force. Of course, a rounded cell will probably experience a different force than a highly spread one, and their contact areas may differ considerably, but over a population of cells, such effects may average out.

Both time of flow and the magnitude can be varied easily by a syringe pump, with wall shear stresses down to 0.1 dyn/cm^2 attainable (though the step motor in many syringe pumps may introduce a noticeable degree of periodicity in the flow, which can be compensated by using smaller syringes to allow the motor to turn faster). At a wall shear stress of 0.1 dyn/cm^2, the force acting on a cell can be estimated within an order of magnitude by multiplying the wall shear stress (since the cell is very close to the wall) by the surface area of the cell. Cells such as leukocytes, which form transient tethers through some types of adhesive receptors, and pivot around these stationary points *(6)*, exert a force on the bond estimated by the pivot arm and the fluid force on a sphere in a shear flow *(10)*. Solutions for the shear force generated by the flow, based on varying assumptions and simplification, have been reported for endothelial cells and platelet thrombi in flow chambers. Close to the wall, the parabolic velocity profile can be approximated by a linear-velocity gradient based on the wall shear rate, which is useful for distinguishing adherent from nonadherent cells that are tumbling along in the flow. The shear rate is multiplied by the viscosity of the perfusion media, to give the shear stress. The shear rate is the change in velocity with distance from the wall of the chamber or tube. The fluid shear stress is defined by: $\tau = \mu(dv/dr)$, where dv/dr is the shear rate (s^{-1}), μ is the viscosity (Poise), and τ is the shear stress (dyn/cm^2). The wall shear stress is given by: $\tau_{wall} = 3\mu Q/2ba^2$, where Q is the volumetric flow rate (cm^3/s), b is the channel width (cm), μ is the viscosity

(Poise), and a is the half channel height (cm). For example, at a wall shear rate of 100 s^{-1}, the fluid velocity is approx 100 µm/s at 1 µ from the wall. Although this is only an approximation of the velocity profile, which is actually described by a parabolic equation, it does provide a scaling factor. If the viscosity of the suspending media is increased, the shear stresses are increased proportionally. This strategy has recently been applied to investigate the force dependence of cell-adhesive crossbridges independent of the shear rate *(9)*.

References

1. Springer, T. A. (1995) Traffic signals on endothelium for lymphocyte recirculation and leukocyte emigration. *Annu. Rev. Physiol.* **57,** 827–872.
2. Xiao, Y. and Truskey, G. A. (1996) Effect of receptor-ligand affinity on the strength of endothelial cell adhesion. *Biophys. J.* **71,** 2869–2884.
3. Abbassi, O., Kishimoto, T. K., McIntire, L. V., Anderson, D. C., and Smith, C. W. (1993) E-selectin supports neutrophil rolling in vitro under conditions of flow. *J. Clin. Invest.* **92,** 2719–2730.
4. Buttrum, S. M., Hatton, R., and Nash, G. B. (1993) Selectin-mediated rolling of neutrophils on immobilized platelets. *Blood* **82,** 1165–1174.
5. Kaplanski, G., Farnarier, C., Tissot, O., Pierres, A., Benoliel, A. M., Alessi, M. C., Kaplanski, S., and Bongrand, P. (1993) Granulocyte–endothelium initial adhesion. Analysis of transient binding events mediated by E-selectin in a laminar shear flow. *Biophys. J.* **64,** 1922–1933.
6. Alon, R., Hammer, D. A., and Springer, T. A. (1995) Lifetime of the P-selectin–carbohydrate bond and its response to tensile force in hydrodynamic flow. *Nature (Lond.)* **374,** 539–542.
7. Lawrence, M. B., Kansas, G. S., Kunkel, E. J., and Ley, K. (1997) Threshold levels of fluid shear promote leukocyte adhesion through selectins (CD62L,P,E). *J. Cell Biol.* **136,** 717–727.
8. Capo, C., Garrouste, F., Benoliel, A., Bongrand, P., Ryter, A., and Bell, G. I. (1982) Concanavalin-A-mediated thymocyte agglutination: a model for a quantitative study of cell adhesion. *J. Cell Sci.* **56,** 21–48.
9. Taylor, A. D., Neelamegham, S., Hellums, J. D., Smith, C. W., and Simon, S. I. (1996) Molecular dynamics of the transition from L-selectin to β_2 integrin-dependent neutrophil adhesion under defined hydrodynamic shear. *Biophys. J.* **71,** 3488–3500.
10. Goldman, A. J., Cox, R. G., and Brenner, H. (1967) Slow viscous motion of a sphere parallel to a plane wall. II. Couette flow. *Chem. Eng. Sci.* **22,** 653–660.
11. Folie, B. J. and McIntire, L. V. (1989) Mathematical analysis of mural thrombogenesis. Concentration profiles of platelet-activating agents and effects of viscous shear flow. *Biophys. J.* **56,** 1121–1141.
12. Munn, L. L., Melder, R. J., and Jain, R. K. (1994) Analysis of cell flux in the parallel plate flow chamber: implications for cell capture studies. *Biophys. J.* **67,** 889–895.
13. Stamper, H. B., Jr. and Woodruff, J. J. (1976) Lymphocyte homing into lymph nodes: in vitro demonstration of the selective affinity of recirculating lymphocytes for high-endothelial venules. *J. Exp. Med.* **144,** 828–833.

14. Dustin, M. L. and Springer, T. A. (1989) T-cell receptor cross-linking transiently stimulates adhesiveness through LFA-1. *Nature (Lond.)* **341**, 41–46.
15. Smith, C. W., Rothlein, R., Hughes, B. J., Mariscalco, M. M., Schmalstieg, F. C., and Anderson, D. C. (1988) Recognition of an endothelial determinant for CD18-dependent neutrophil adherence and transendothelial migration. *J. Clin. Invest.* **82**, 1746–1756.
16. Smith, C. W., Kishimoto, T. K., Abbassi, O., Hughes, B., Rothlein, R., McIntire, L. V., Butcher, E. C., and Anderson, D. C. (1991) Chemotactic factors regulate lectin adhesion molecule 1 (LECAM-1)-dependent neutrophil adhesion to cytokine-stimulated endothelial cells in vitro. *J. Clin. Invest.* **87**, 609–618.
17. McClay, D. R., Wessel, G. M., and Marchase, R. B. (1981) Intercellular recognition: quantitation of initial binding events. *Proc. Natl. Acad. Sci. USA* **78**, 4975–4979.
18. Ward, M. D., Dembo, M., and Hammer, D. A. (1995) Kinetics of cell detachment: effect of ligand density. *Ann. Biomed. Eng.* **23**, 322–331.
19. Xia, Z. and Frojmovic, M. M. (1994) Aggregation efficiency of activated normal or fixed platelets in a simple shear field: effect of shear and fibrinogen occupancy. *Biophys. J.* **66**, 2190–2201.
20. Bell, D. N., Spain, S., and Goldsmith, H. L. (1989) Adenosine disphosphate-induced aggregation of human platelets in flow through tubes. I. Measurement of concentration and size of single platelets and aggregates. *Biophys. J.* **56**, 817–828.
21. Bargatze, R. F., Kurk, S., Butcher, E. C., and Jutila, M. A. (1994) Neutrophils roll on adherent neutrophils bound to cytokine-induced endothelial cells via L-selectin on the rolling cells. *J. Exp. Med.* **180**, 1785–1792.
22. Kuo, S. C. and Lauffenburger, D. A. (1993) Relationship between receptor/ligand binding affinity and adhesion strength. *Biophys. J.* **65**, 2191–2200.
23. Sung, L. A., Kabat, E. A., and Chien, S. (1985) Interaction energies in lectin-mediated erythrocyte aggregation. *J. Cell Biol.* **101**, 652–659.
24. Palecek, S. P., Loftus, J. C., Ginsberg, M. H., Lauffenburger, D. A., and Horwitz, A. F. (1997) Integrin–ligand binding properties govern cell migration speed through cell-substratum adhesiveness. *Nature (Lond.)* **385**, 537–540.
25. Truskey, G. A. and Pirone, J. S. (1990) The effect of fluid shear stress upon cell adhesion to fibronectin-treated surfaces. *J. Biomed. Mater. Res.* **24**, 1333–1353.

39

Quantitative Measurement of the Biological Response of Cartilage to Mechanical Deformation

R. Gregory Allen, Solomon R. Eisenberg, and Martha L. Gray

1. Introduction
1.1. General Introduction

Cartilage functionality is defined, in part, in terms of the ability of the extracellular matrix to support a mechanical load. It has been shown that such mechanical loading can influence the biological response of the chondrocytes that are embedded in the extracellular matrix. Cultured tissue explants have served as useful models for studying such chondrocyte-mediated responses to mechanical deformation. The explant paradigm facilitates the control of mechanical and biological variables that may influence cellular behavior. This chapter presents the means for assessing the biologic response of cartilage to controlled mechanical stimuli, and lays the foundation to further explore the response of cartilage to mechanical stimuli in the presence of other factors, such as cytokines and growth factors (e.g., IL-1β, TGF-β, IGF).

A requirement for observing and reporting the effects of mechanical loading on cartilage metabolism is the ability to quantify the deformation or stress state of the tissue. This becomes a key consideration in all aspects of experimental design, including tissue harvest, culture, mechanical intervention, and assessing metabolic activity. In constructing the experiment, one must carefully consider the geometry of the test sample and the macroscopic variables associated with the mechanical stimulus, because the choice of boundary conditions and method of loading determine the extent to which the internal mechanical environment may be conceptualized. The macroscopic mechanical parameters of relevance are load and displacement (load can be more completely defined with compressive, tensile, and tangential forces, and hydrostatic and osmotic pressures), which relate to continuum variables of stress and strain. These are dependent variables: The application of one specifies the other, and, therefore, necessitates methods for controlling one and measuring the other.

Ultimately, the mechanical behavior of the tissue in the vicinity of the cells must be considered, for therein lies the key to identifying the causal relationship between mechanical stimulus and biological response. Applied displacements, or forces, produce deformations of the extracellular matrix and the cells. For these deformations to occur, fluid must flow and redistribute. This flow is induced by hydrostatic pressure gradients. Consolidation of the matrix alters the fixed-charge density, thereby affecting the osmotic pressure of the interstitial fluid and altering pericellular chemical properties like pH and ionic strength. Time-dependent material deformations and pressure gradients are present during transient mechanical processes and during dynamically applied displacements or forces. These, in turn, induce fluid flows and effect physicochemical changes, like electrokinetic separation of charged metabolites. It has been demonstrated repeatedly that cells are receptive to static and dynamic mechanical perturbations of their local environment, as assessed by the subsequent modulation of metabolic activity. However, the precise nature of the transduction mechanisms have yet to be clearly elucidated. This tissue-level view suggests that the mechanisms might include cell deformation, fluid flow, pressure, electrical phenomenon, and physicochemical effects—and data exists supporting each of these mechanisms. The careful control of the macroscopic environment, combined with models estimating the tissue-level mechanical environment in which the cells reside, has allowed investigators to begin to experimentally evaluate these transduction mechanisms, and is the emphasis of this chapter.

In practice, the vicissitudes of experimentation become apparent, because our understanding of the mechanical environment is limited, in some cases, by constraints imposed by tissue-culture techniques, or by incomplete mechanical descriptions of the tissue deformation state. For example, in our laboratory, one well-established long-term culture protocol entails the static axial compression of radially unconfined, cylinder-shaped cartilage explants. The experimental system was designed to facilitate quantitative investigation of the effects of compression (matrix consolidation and fluid exudation) on cell metabolism. However, the interpretation of the results requires the understanding that matrix compaction is not the sole deformational result of axial compression. The radially unconfined nature of the explant gives the sample the freedom to bulge, creating tensile stresses caused by matrix extension in the radial direction. The application of time-varying loads or displacements introduces additional complications. In this dynamic environment, the tissue experiences spatially complex pressure gradients and fluid flows, the magnitude and distribution of which depends on frequency. Given that cells reside throughout the sample, and that the mechanics are complicated functions of space and frequency, the mechanical stimulus seen by each cell can be different. It is also important to keep in mind that many of the methods used to determine biological activity have limited spatial resolution. Most measurements based on radiotracer incorporation typically assess the bulk (spatially averaged) response, although autoradiography has the capability to localize regions of synthesis and assembly.

Laboratory procedures detailed here discuss the formulation of culture medium and the explantation and preparation of cartilage into disks. Note that from an engineering perspective, the geometrically friendly cylinder, in some cases, simplifies the normalization of loads and displacements into stresses and strains. Presented are methods describing the radially unconfined, axially applied static and dynamic compression of cultured cartilage disks. The biological response of cartilage, mediated by the resident chondrocytes, is monitored via biochemical and radioisotope measurements of synthesis and degradation of the extracellular constituents. Some common methods for quantifying biological responses to mechanical stimuli are described at the end of this chapter, and references are provided for direction to experimental procedures routinely employed by others.

1.2. Conceptual Framework

Cartilage consists primarily of an aqueous electrolyte and a sparse population of cells that reside in a plentiful extracellular matrix. There are two principal components of the matrix: collagen, which forms an isoelectric fibrillar network that is thought to be the primary determinant of cartilage tensile strength; and proteoglycans, which, with the tissue fluid, constitute a polyanionic gel that resists compression. The collagenous network contains the aggregating proteoglycans and other noncollagenous proteins through chemical binding reactions and steric hindrances. The negatively charged components of the proteoglycans, glycosaminoglycans, cause positive ion (or counter-ion, e.g., Ca^{2+}, H^+, Na^+) concentrations to be elevated over their respective concentrations in the culture medium. Similarly, concentrations of negatively charged ions (or co-ions, e.g., Cl^-, SO_4^{2-}) are depressed in order to maintain tissue electroneutrality.

Mechanical compression of cartilage disks alters the stress and deformation state within the extracellular matrix. Compression also causes changes in hydrostatic pressure, fluid content, and osmolality, and induces fluid flows and electric fields (streaming potentials). Chondrocytes may respond directly to applied stress, or to any of these compression-induced physical phenomena that occur in their environment. To facilitate experimental design, included is a brief discussion of the tissue-level mechanical environment during commonly used methods of loading.

The application of a static compression initiates a series of events that result in an equilibrium deformation. During the mechanical transient (creep or stress relaxation [*see* **Note 1**]) that follows compression, fluid convection and exudation ultimately leads to an equilibrium mechanical deformation. During this transient phase, counter ions are convectively separated from the fixed negative charges of the proteoglycans, generating electric fields collinear with the fluid flow. The reduction in volume (caused by fluid loss) alters the physicochemical environment by concentrating the negatively charged proteoglycans, which further partitions counter-ions into the tissue, increasing their concentrations (and decreasing pH), and simultaneously decreasing co-ion concentrations. Thus, from the perspective of the cell, many physical forces and flows occur during the transient phase that may trigger a cellular response. At equilibrium, the cell may be affected directly by the applied deformation, or by the concomitant changes in the physicochemical environment.

Dynamic compression is the periodic application of a load or displacement, usually superimposed on a steady-state static compression. In contrast to static compression, pressure gradients, fluid flows, and electric fields persist with the same frequency as the applied excitation. The mechanical environment during dynamic compression is spatially complicated and dependent upon boundary conditions, but modeling efforts have provided significant insight for certain limiting conditions *(1,2)*. Common to these models is the finding that fluid velocity and hydrostatic pressure in dynamically compressed, radially unconfined disks, exhibit both an axial and radial dependence, which vary with the frequency of the applied excitation. Given a relatively low-excitation frequency (e.g., 0.0001 Hz in a 3-mm diameter disk), the fluid velocity and pressure distributions are expected to be spatially quasiuniform. As frequency is increased, however, pressure gradients develop and fluid flows arise, with flows becoming increasingly restricted to the region adjacent to the radially unconfined surface of the disk. A cartilage disk mechanically loaded at a relatively high loading frequency (e.g., 0.1 Hz) would experience permeability-related fluid flow restrictions, elevating hydrostatic pressures in the bulk of the tissue. Fluid flows are expected at the radial surface (within 0.3 mm at 0.1 Hz for a 3-mm diameter sample), where the highest rates of fluid flow occur with correspondingly steep pressure gradients *(3)*.

In the discussion above, and in the methods described below, our focus is the direct application of mechanical loads or displacements. However, it is important to mention two alternative approaches used for assessing the biological effect of mechanical stimuli. These involve altering the hydrostatic or osmotic pressures of the bathing media. The mechanical effect of varying hydrostatic pressure, via a hydraulic actuator, is a spatially uniform change in pressure throughout the culture media and cartilage sample. Assuming that the solid and fluid phases are individually incompressible, tissue deformations, fluid flows, and all of the secondary mechanical events are absent. The effects of hydrostatic pressure on the biological response of cartilage have been studied, and the methods are described elsewhere *(4)*. In contrast, varying osmotic pressure can induce fluid flow, tissue deformation, and so on. This is achieved by altering the osmolality (osmotic pressure) of the bathing medium with a tissue impermeant solute, or by placing the tissue in dialysis tubing and adding a solute (e.g., polyethylene glycol) to the bathing solution outside the dialysis tube. When the osmotic pressure of the bathing solution is made hypertonic, fluid leaves the tissue, causing the tissue volume to decrease and the matrix to deform as though compressed. As with direct mechanical compression, osmotically induced fluid loss has the capacity to mechanically deform cells and effect physicochemical changes by altering the concentrations of polyelectrolytes and the density of fixed charges. A more complete treatment of this topic is detailed elsewhere *(5–8)*.

2. Materials
2.1. Culture Medium

25 mM HEPES-buffered Dulbecco's modified Eagle's medium (DMEM) (Gibco-BRL, Gaithersburg, MD, 12320-032), DMEM (JRH Biosciences, Lenecia, KS, 51447-78P), 10 mM nonessential amino acids (Sigma P-7145), 100 mM L-proline (Sigma, St. Louis, MO, P-0830), 200 mM L-glutamine (Sigma G-6392), 20 mg/mL L-ascorbate (Sigma A-4544), antibiotic/antimycotic (Sigma, A-7292), and heat-inactivated fetal calf serum (FCS) (**Note 2**; HyClone, Logan, UT).

2.2. Dissection

Autoclave or alcohol sterilize, as applicable: a vibrating sawblade; dissection tools, including scalpels, razor blades, forceps, hemostats, and weighing spatulas; several pairs of sterile gloves, a mask (if dissecting outside a biosafety cabinet) a topical antiseptic (e.g., Betadine); gauze; a drill press (configured to a rotational speed of ≤200 rpm) and coring bit (we use a custom built 3/8-inch ID bit); sterile specimen cups; and multiwell culture plates. Our rinsing solution consists of 1% by volume of antibiotic/antimycotic (Sigma, A-7292) in 500 mL of Hanks' balanced salt solution (Gibco-BRL, 14025-092). We use a Sledge microtome (American Optical, Buffalo, NY) to prepare explants into slices of uniform thickness. For preparing these slices into disks, assemble the following materials in a biosafety cabinet: forceps, rinsing solution, multiwell culture plates with ~0.5 mL/well complete media, dermal punch (Miltex Instrument, Lake Success, NY), and a cutting block. The function of the cutting block is to prevent damage to the punching tool; we use a block (approx 5 × 8 in. by 0.5-in. thick) of polysulfone, which is autoclavable.

2.3. Culture Techniques

Multiwell culture plates, forceps, media, sterile Pasteur pipets for exchanging media, and microcentrifuge tubes for media cryopreservation.

2.4. Mechanical Stimulus

Custom-built compression devices (*see* **Subheading 3.4.1.**).

2.5. Biochemical Techniques

2.5.1. Digestion

Papain (Sigma, P-3125), L-cysteine hydrochloride (Sigma, C1276), Na_2HPO_4, Na_2EDTA, 0.20-µm syringe filter (Gelman Sciences, Ann Arbor, MI, #4192), 10 mL syringe and a 25-gage needle, one 2-mL cryogenic vial per tissue sample (Corning), 70% ethanol sterile wipes, and a hot-water bath.

2.5.2. DMB Assay

2.5.2.1. Dye Solution

DMB chloride (Polysciences, Warrington, PA, #3610), NaCl (Mallinckrodt, Paris, KY, 7544), glycine (Sigma, G-7403), sodium azide (Fluka, Switzerland, #71290), 100% ethanol (reagent grade, Pharmco, Brookfield, CT), in HCL (Fisher Scientific, Hampton, NH, SA49-100, 50-mL centrifuge tube, flea magnetic stirrer, aluminum foil, mask, gloves, and demineralized water.

2.5.2.2. Glycosaminoglycan Standard Solution

Chondroitin sulfate (Sigma, C4384) (**Note 12**), demineralized water, and several cryopreservation vials.

2.5.2.3. Spectrophotometry

The spectrophotometer should be configured to measure absorbance at 525 nm. Materials should include several disposable pipet tips, 500-mL pump-action dispenser filled with DMB dye solution, and po lystyrene cuvets or 4.5 mL microplates.

2.5.3. DNA Assay

2.5.3.1. 10X TEN Buffer

Tris (Mallinckrodt H590), EDTA, and NaCl (Mallinckrodt 7544), demineralized water, HCl (~14 M HCl), and a 0.22-µm filter

2.5.3.2. Concentrated Dye Stock Solution

Hoechst 33258 (Polysciences, Warrington, PA, #09460), sterile demineralized water, and a foil-wrapped glass or brown plastic bottle.

2.5.3.3. Spectrofluorometry

Fluorometer, acryl cuvets (10 × 10 × 48 mm, Sarstedt, Newton, NC, 67-755), several disposable pipet tips, and a 500-mL pump-action dispenser filled with working dye solution.

2.5.4. Radioactive Labeling

2.5.4.1. 2 M Byanidine HCL Solution

[^{35}S]sulfate (Na$_2^{35}$SO$_4$, NEX-041, New England Nuclear, Boston, MA) and tritiated proline (1-[5-^{3}H]proline, NET-573, New England Nuclear).

2.5.3.2. Start Counting

Liquid scintillation counter, Omni vials and caps (Wheaton, Hillsboro, OR, 66021-23), Ecolume scintillation fluid (ICN, Costa Mesa, CA, HS 82470), guanidine hydrochloride (Aldrich, Milwaukee, WI, G11705), sodium acetate (Sigma S-8625), demineralized water, and 1 M HCl.

3. Methods

3.1. Culture Medium Formulation

Our complete media consists of low glucose DMEM buffered with 10 mM N-2-hydroxyethylpiperazine-N'-2-ethanesulfonic acid (HEPES), and fortified with 0.1 mM nonessential amino acids, 0.4 mM L-proline, 2 mM L-glutamine, 100 U/mL of penicillin, 0.10 mg/mL streptomycin, and 0.25 µg/mL amphotericin-B, 1% by volume of heat-inactivated fetal calf serum (FCS), and 0.50 µg/mL L-ascorbate. We prepare culture media in 500 mL batches and store it between 2 and 6°C. Complete media is comprised of:

1. 200 mL of 25 mM HEPES-buffered DMEM.
2. 293 mL of DMEM without HEPES.
3. 5 mL of 10 mM nonessential amino acids.
4. 2 mL of 100 mM L-proline.

Twenty minutes prior to changing the medium, add 1% by volume of each of the following substances, which have limited stability at culture conditions, and warm in a hot water bath at 37°C.

5. 200 mM L-glutamine.
6. 20 mg/mL L-ascorbate.
7. Antibiotic/antimycotic.

8. Heat-inactivated FCS.

Example: For a desired total volume of 10 mL of complete media, add 100 μL of each of the four supplemental substances to 9.6 mL of base media, preequilibrate to culture temperature, and deliver to explants.

3.2. Dissection

The ultimate goal is to prepare plane-parallel disks (i.e., cylindrical disks of uniform thickness). The first step is to harvest the cartilage so that it is possible to section the tissue using a microtome. To maintain a reasonable degree of sanitation, dissect as much of the tissue as possible within a biosafety cabinet. A vibrating saw has proven to be a useful device in harvesting tissue from larger joints (e.g., femoropatellar groove, humoral head). This tool, which is similar to a bone saw, is commercially available in hardware stores. Blades can be custom made to accommodate the larger joints of adult animals. For irrigating cartilage exposed during the explantation process, and for lubricating the cutting edges of power tools (e.g., bone-saw blade, drill bit), prepare a rinsing solution of HBSS supplemented with an antibiotic/antimycotic. Combine 500 mL HBSS and 5 mL antibiotic/antimycotic into a squirt bottle and store at 2–6°C. Use sterile specimen cups or multiwell culture plates for the bathing of tissue prior to sectioning. We use one 50-mL sterile specimen cup containing ~20 mL rinsing solution for each distal ulna joint. For tissue that requires coring, we use a 24-well culture plate containing ~2 mL/well rinsing solution.

3.2.1. Tissue Explantation Procedures

The synovial joint from which the cartilage is to be dissected should be delivered to the laboratory on ice, to preserve cellular viability. To prevent infection, the joint should be intact and encapsulated with tissue. If the dissection is not immediate, store the tissue between 2–6°C. All tissue used in our studies is bovine (calf and adult) in origin: 1–2 wk-old distal ulna or hind leg femoropatellar groove and humoral head. In the vernacular of the local abattoir, the terminology for the distal ulna, humoral head, and femoropatellar groove are front knee, shoulder, and hind quarter, respectively.

Begin by applying the topical antiseptic liberally to the external tissues of the joint (e.g., musculature); it is important that contact of the antiseptic with the cartilage be avoided.

3.2.1.1. Distal Ulna

Under quasisterile conditions aided by the biosafety cabinet, relieve the distal ulna of its surrounding musculature, and periosteal and perichondral connective tissues (**Fig. 1**). Isolate the ulna from the radius (the ulna is the smaller

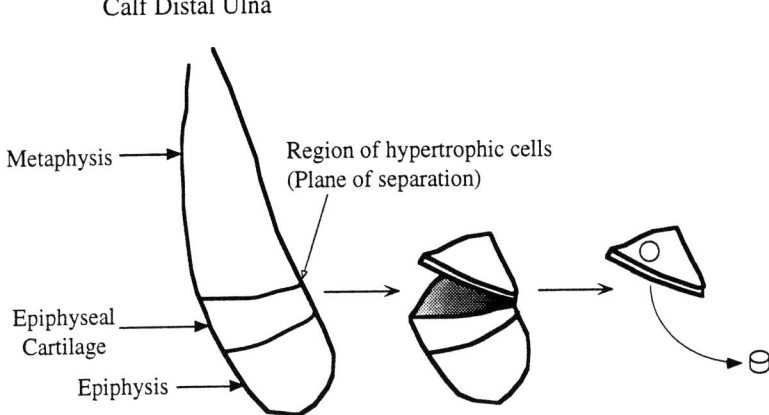

Fig. 1. Harvesting disks from the calf distal ulna.

of the two and has a large ~1-cm-thick epiphyseal cartilage region that is easily visualized). Separate the epiphysis from the metaphysis: bending the ulna until it snaps in half is a suitable means for accomplishing this. Separation occurs at the interface of the hypertrophic cells of the growth plate and calcified cartilage region of the metaphysis. Immerse the epiphyseal apparatus (which includes articular cartilage, the underlying secondary center of ossification, epiphyseal cartilage [a.k.a. progenitor cartilage], and growth plate) into the rinsing solution contained in the sterile specimen cup, and discard the remaining tissue.

3.2.1.2. FEMOROPATELLAR GROOVE AND HUMORAL HEAD

We dissect these joints on the quasisanitary benchtop. With a scalpel or razor blade, begin by removing as much connective tissue from the joint and diaphysis as possible, taking care to leave the joint capsule intact (which is the best possible means of thwarting an infection). Then, cut through the capsule and sever any ligaments, so that the cartilaginous surface of interest is completely exposed. Transect the diaphysis with the bone saw, so that the shaft can be clamped in a vise (*see* **Note 3**), securing the joint so that the articular cartilage faces up. With the joint immobilized, drill as many cores as possible. Configure the drill press to have a slow rotational speed (≤200 rpm), and align the joint so that the drill bit is perpendicular to the tissue surface (**Fig. 2**). While cutting, continuously lubricate the drill bit with rinsing solution. To free the cores from the underlying cancellous bone, transect the chondyle with the bone saw (severing the cores 10–15 mm beneath the surface), and expel the cores with a probe. Submerge the cartilaginous ends of the cores in rinsing solution (we use a 24-well culture plate with ~2 mL of rinsing solution in each well).

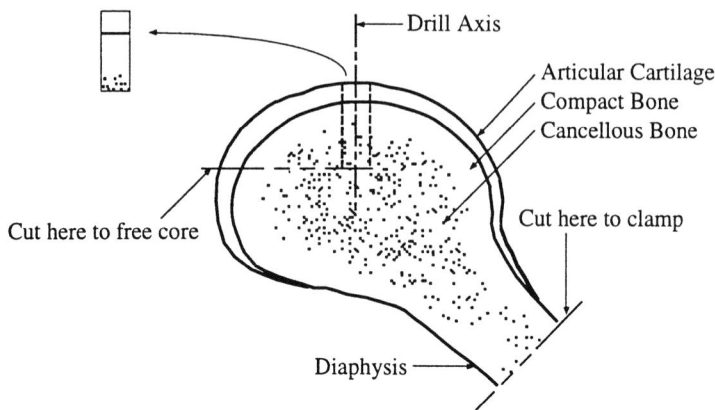

Fig. 2. Drilling cores from the humerus.

3.2.2. Preparation of Explant Disks

We prepare the explanted tissue into cylinder-shaped disks. The dimensions vary with the experiment, but typical diameters and thicknesses are 2–4 mm and 0.8–2.0 mm, respectively. Choosing the explant dimensions deserves some consideration. Accurate determination of biosynthetic activity may dictate large samples, because each specimen requires an adequate cell population. Alternatively, small samples may be required in order to obtain all samples from a single animal, thereby eliminating interanimal variations. Furthermore, the explant dimensions are limited by nutrient transport considerations. Our studies utilizing axially confining culture systems (i.e., radial nutrient transport) have found biosynthetic rates for 2- and 3-mm diameter samples to be equivalent, suggesting these dimensions are acceptable.

3.2.2.1. Distal Ulna

Clamp the epiphyseal bone into the sledge microtome, so that the epiphyseal cartilage is unconstrained and poised for sectioning. The initial ~800-µm-thick slice creates a flat surface, removes the growth plate, and generally leaves ~4 mm of epiphyseal cartilage for sectioning into plane-parallel slices. With forceps, place the slices into a pool of rinsing solution on the cutting block. Excise disks from each slice with a dermal punch, place into the culture plate, and begin incubation.

3.2.2.2. Femoropatellar Groove and Humoral Head

The drilled-out cores are generally comprised of 5 mm (young) or 0.6 mm (adult) of cartilage atop ~1 cm of subchondral bone. Clamp a core into the sledge microtome for sectioning. Face off the material with an initial section. Note that drilling perpendicularly into the chondyle minimizes the amount of material that must be discarded at this step. Also, it is possible to retain the articular surface, with careful attention being paid to making the core perpendicular to the surface. Section the remaining underlying cartilage into the desired thickness, place slices into pooled rinsing solution, excise disks from each slice with the punching tool, and place the disks into culture.

3.3. Culture Techniques

The environmental conditions of the incubator should be set to 37°C, 5% CO_2, and, to minimize media evaporation, at least 95% relative humidity. Culture the free-swelling cartilage explants in 0.5 mL complete media for at least 2 d before initiating any mechanical stimulation (*see* **Note 4**). Media should be exchanged on a regular basis; the frequency of media collection depends on the time-points required for assessing biological activity.

3.4. Mechanical Stimulus

In our laboratory, cartilage disks are cultured and mechanically loaded in one of two custom built compression devices. The simpler of the two, a static compression device, can be easily fabricated by any machine shop, and is relatively easy to use. As suggested by the name, this device does not allow dynamic compression (i.e., no continuous oscillatory loading), and typically controls either displacement or load. Alternatively, a mechanical spectrometer enables the application of static or dynamic displacement-based deformations, and simultaneously allows for the monitoring of the associated transient or dynamic loads. Both compression devices are designed to axially confine cartilage samples between two surfaces, the bottom of the compression chamber and a compression post located superiorly (*see* **Note 5**). To ensure tissue viability, the radial surfaces of the explants are unconstrained, thereby allowing the tissue to exchange waste products for nutrients with the culture medium. Uncompressed samples, which serve as untreated controls, are cultured in the same incubator as the mechanically challenged samples.

Discard disks that have obvious visible damage incurred during the explantation process, do not exhibit plane-parallel faces, or do not retain cylindrical shape. Note that all explanted plugs will swell by approx 10% of their original prepared thickness: Use a micrometer to measure sample thicknesses before mechanical stimulation.

3.4.1. Compression Devices

3.4.1.1. STATIC COMPRESSION CHAMBERS

The static compression chamber, depicted schematically in **Fig. 3**, is essentially a 24-well culture plate secured into an anodized aluminum frame (*see*

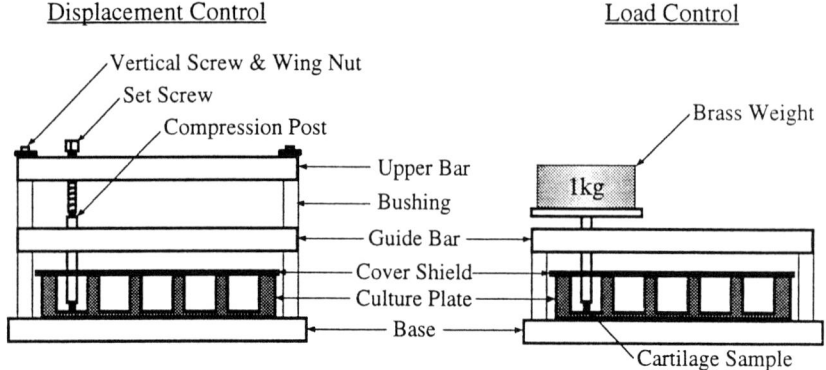

Fig. 3. Static compression devices.

Note 6). A quartz post transfers axial loads or displacements through a rigid, anodized aluminum guide bar and compresses the cartilage sample against the bottom of the culture well. In displacement control, the set screw is tightened down on the compression post so that the distance from the end of the screw to the bottom of the culture well equals the length of the compression post plus the desired cartilage thickness. Precision shims are used to prescribe the specimen thickness. In load control, mechanical stress is applied by placing weights (of known mass) on the compression posts (*see* **Note 7**). Media is exchanged by inserting a sterile Pasteur pipet under the Lexan cover shield. These studies can involve a large number of samples, since each device can accommodate 10 cartilage disks and a single incubator can contain several compression devices.

3.4.1.2. Mechanical Spectrometer

The mechanical spectrometer enables various kinds of intervention by controlling static or periodic displacements. The mechanical spectrometer generally allows automated control of displacement or load (in some instruments, the user has a choice of which to control; others are designed to allow control of only one). These instruments also provide a means of measuring both load and displacement. These machines can be custom built and are also available commercially (e.g., Dynastat, Instron, MTS, Vitrodyne). Tradeoffs in choosing instrumentation include the number of samples for which independent measurements are desired and the duration of an experiment (how long will an incubator environment be needed). The sample chambers resemble those of the static compression devices, but the details depend on the spectrometer. The methods will assume the cartilage sample is positioned between the bottom of a culture well and a platen.

3.4.2. Preparation for Displacement-Based Mechanical Stimulation

3.4.2.1. STATIC COMPRESSION CHAMBERS (DISPLACEMENT CONTROL)

1. Autoclave-sterilize all components of the compression device, forceps, weighing spatulas, and several Pasteur pipets (one per sample for extracting media, and a single pipet for adding media).
2. Assemble the entire compression device in the biosafety cabinet.
3. Place cartilage sample in the compression chamber and add culture media (ensure that the sample is completely immersed).
4. Impose the prescribed displacement. Specific details for applying the mechanical stimulus depend on the apparatus being used. The displacement should not be applied too rapidly, in order to minimize the risk of a devitalizing injurious compression *(9)* or sample expulsion. The macroscopic engineering strain to which the cartilage sample is confined is the quotient of the applied displacement and the uncompressed thickness.
5. Begin culture.

3.4.2.2. MECHANICAL SPECTROMETER

1. Autoclave-sterilize the compression chambers, forceps, and weighing spatulas.
2. Place samples into the compression chambers, lock them into position in the testing system, and add culture media so that samples are immersed.
3. To ensure that the compression posts contact the cartilage surface, gradually drive the compression posts down until the load cell instrumentation indicates contact (*see* **Note 8**).
4. Close the incubator door and allow the interior environment to reach culture conditions.
5. Begin mechanical stimulation (*see* **Note 9**).

3.4.3. Sample Protocols

3.4.3.1. COMPRESSION

The most straightforward procedure for inducing mechanical stress is the application of a compressive displacement. The dose-dependent effects of compression can be elucidated by modulating the degree of static compression: Commonly used compression levels are 0, 10, 20% (depicted in **Fig. 4**), and 50% (*see* **Note 10**). As previously mentioned, after the displacement is imposed, it takes some time for the load to reach a new steady state (stress relaxation). This time can be estimated (*see* **Note 1**) or measured by the mechanical spectrometer. Studies of the metabolic response to static compression must be considerably longer than this time. For 3-mm diameter samples, the stress relaxation time is typically on the order of 25 min. For load control, the rule of thumb is that the creep time is four-times longer than the stress relaxation time. The biological response to static compression (observed via radiolabeling periods ranging from 6 to 12 h) is presumed to be dominated by the cell and tissue deformation,

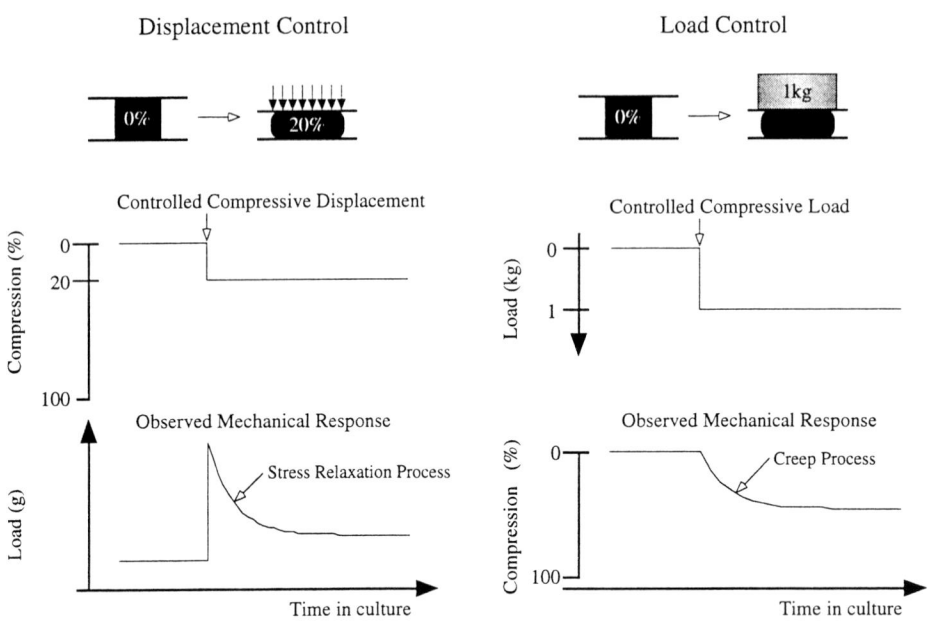

Fig. 4. Static compression.

and physicochemical changes resulting from compression. Data thus far have not revealed a role for the transient physical phenomena in eliciting the response. All subsequent methods will focus on displacement control. The reader can extend from these discussions how to perform load control studies.

3.4.3.2. COMPRESSION RELEASE

Contrived as an extension of the time-invariant static compression paradigm, the compression-release cycle (**Fig. 5**) entails a provisional deformation followed by a return to the original condition. This technique has been used to examine recovery (studied by adding radiolabels subsequent to the release stroke), and to examine a low-frequency dynamic mechanical environment developed by constructing a pulse-train of concatenated compression–release cycles.

3.4.3.3. OSCILLATORY COMPRESSION

The application of a continuous oscillatory displacement superimposed upon an offset compression (**Fig. 6**) induces additional stimulatory flow and deformational phenomena. Assuming that the biological response intrinsic to the offset compression has been characterized, the implications of dynamic compression can now be studied. By taking advantage of the complicated

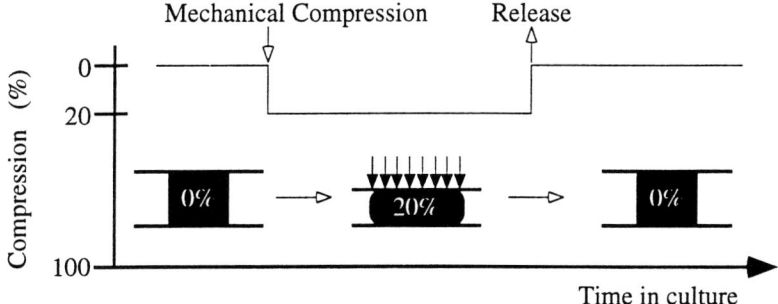

Fig. 5. Compression–release cycle.

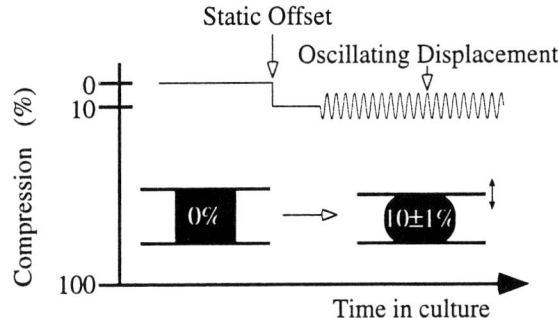

Fig. 6. Dynamic compression.

mechanical behavior of the tissue, modulating the frequency and magnitude of the dynamic compression creates regimes in which different physical stimuli are pronounced. For example, small-amplitude oscillations minimize the exchange of tissue fluid with the bathing medium; therefore, fluctuating physicochemical (fixed-charge density, osmolality) properties associated with tissue hydration can be dismissed as possible stimulatory factors. The static offset is required if a symmetric oscillation is to occur. Furthermore, the loads required and the ability of the tissue to reswell limit the amplitude of dynamic compression to 1–2%, values that are likely to be physiological.

3.5. Biochemical Techniques

This section begins with a procedure for the digestion of cartilage samples, a necessary step for further biochemical analysis. Included are protocols for the assessment of sulfated glycosaminoglycan and DNA content, and the introduction of radiolabels for monitoring biosynthesis. Measure sample wet and dry (lyophilized) wt prior to any biochemical analysis. Weight measurements aid in the interpretation of biochemical data.

3.5.1. Digestion of Cartilage Explants

It is necessary to solubilize cartilage samples for biochemical analysis. Papain is a sulfhydryl protease of wide specificity, which, when activated with 0.010 M cysteine, degrades most proteinaceous substrates (*see* **Note 11**). Sufficient digestion occurs after incubation (~16 h at 60°C) with enzyme:substrate ratios of ~1:50 (e.g., 0.125 mg papain per 6.25 mg tissue dry wt) *(10)*. Papain should be kept sterile at 2–6°C. Sterilization is necessary if DNA measurements are to be made on the digests.

3.5.1.1. 1L of Phosphate Buffer with EDTA (PBE)

1. Weigh 14.20 g Na_2HPO_4 (0.100 M).
2. Weigh 3.36 g Na_2EDTA (0.010 M).
3. Add to 1.0 L of demineralized water.
4. Adjust pH to 6.5 with 1 N NaOH.
5. Autoclave-sterilize.
6. Store at 2–6°C.

3.5.1.2. 100 mL Digestion Solution

Preparing excess digestion solution is not recommended, because its activity is unstable (solution is active for about a week, when stored at 2–6°C). This volume is sufficient for digesting 200 ~20 mg (dry wt) cartilage samples (0.5 mL/sample).

1. Weigh 176 mg cysteine.
2. Dissolve in small volume of PBE (e.g., 1–10% of final volume).
3. Sterilize through 0.22-μm syringe filter.
4. Bring total volume to 100 mL with PBE.
5. Add papain to 0.125 mg/mL (Sigma's suspension is 25 mg/mL, so pipet 0.5 mL). Allow approx 5 min for the stock papain solution to warm at room temperature, since the refrigerated suspension of papain is quite viscous. Extracting papain from the rubber-septum sealed container should be conducted in the aseptic biosafety cabinet: clean the septum with a sterile 70% ethanol wipe, withdraw the desired volume with a syringe, and rewipe the septum.

3.5.1.3. Digesting the Cartilage Explants

1. Measure and record the wet and dry (lyophilized) explant wt.
2. Deposit each cartilage sample within a 2-mL cryogenic vial.
3. Add 0.5 mL of digestion solution to each sample.
4. Incubate in a hot-water bath for 12–16 h at 60°C. Samples should be fully submerged in the digestion solution and devoid of adherent air bubbles.
5. If the samples are not completely dissolved, vortex and continue incubation.

3.5.2. DMB Assay for Glycosaminoglycan Content

The dimethylmethylene blue assay is a well-established method for the rapid determination of the total sulfated glycosaminoglycan (GAG) content of cartilage samples *(11)*. Although the assay cannot directly quantify synthesis or catabolism (i.e., the incremental change in the total amount of GAG), it is useful for measuring the amounts released to the media or retained within the explant.

3.5.2.1. ONE LITER OF DMB CHLORIDE DYE STOCK SOLUTION

This quantity will accommodate 500 samples when apportioned at 2.0 mL per sample. Scale the following formulation accordingly, if larger volumes are required. The required materials and their final concentrations are: DMB chloride (46.0 µM), NaCl (40.6 mM), glycine (40.5 mM), sodium azide, 3.1 mM 1 N HCl (~10 mM), demineralized water (~95% final volume).

1. Dissolve 0.016 g of DMB into 10.0 mL of 100% ethanol in a closed, flat-bottomed, 50-mL centrifuge tube. Add a flea magnetic stirrer, cap, wrap in aluminum foil (to block out light), and stir at room temperature until dissolved (4–16 h). Since DMB is moderately soluble in low-pH water, this procedure will enhance solubility in subsequent steps.
2. Weigh out 2.37 g NaCl and 3.04 g glycine and add both to 950 mL demineralized water.
3. Wearing a mask and gloves, weigh out 0.20g sodium azide and add to the solution of step 2 (**Caution:** Sodium azide is toxic and readily absorbable through skin).
4. Add the dissolved DMB to the solution. Rinse the 50-mL centrifuge tube with water to collect any precipitate. Trap the flea in the 50-mL centrifuge tube with a stir bar.
5. Titrate the solution to pH = 3.0 (±0.05) with 1 N HCl. Typically, ~10 mL of HCl are required. **Caution:** Wear a lab coat and safety glasses when handling HCl.
6. Bring the volume to 1 L with water.
7. Verify that the pH remained at 3.0 ± 0.05.
8. Check the absorbance at 525 nm and write the observed value on the bottle. The OD should be between 0.30 and 0.38 when read against an air reference in a plastic cuvet.
9. Filter the solution with large filter paper into a brown/amber storage bottle. The storage bottle must be cleaned prior to use, because residual precipitates may induce further precipitation. To minimize the precipitation of the dye, do not agitate the solution.
10. Note the date of the solution on the bottle. The dye remains stable for approx 3 mo.
11. Store the solution at room temperature, preferably in a dark cabinet.

3.5.2.2. 10 ML OF 1000 µG/ML CHONDROITIN SULFATE STANDARD STOCK SOLUTION

1. Weigh 10 mg chondroitin sulfate.
2. Add to 10 mL demineralized water.
3. Vortex and aliquot 0.5 mL into cryogenic vials.
4. Freeze until needed at or below −20°C.

3.5.2.3. ASSESSING GLYCOSAMINOGLYCAN CONTENT

1. Prepare samples in duplicate:
   a. For samples containing large amounts of GAG, prepare standard solutions of serial dilutions of chondroitin sulfate: 1000, 500, 250, 125, 62.5, and 0 µg/mL (a blank of demineralized water). For samples containing dilute glycosaminoglycan solutions, prepare five serial dilutions from 125 µg/mL, including a blank. Vortex and add 20 µL of each dilution to a cuvet or microplate.
   b. Vortex the digested cartilage samples and add a small aliquot indentical to that of the standard (e.g., 20 µL) of each sample to a cuvet or microplate.
2. Pump 2 mL of DMB solution to a cuvet and record the optical absorbance at 525 nm. Repeat process for the remaining cuvets. Alternatively, add 200 µL to microplate wells and measure the absorbance. DMB must be added immediately prior to measurement, because precipitation is likely.
3. The range of acceptable measurements span 0.3 (blank) to .8 (largest standard), and duplicates should be within 5%.

3.5.3. Hoescht 33258 Dye Assay for DNA Content

This biochemical technique provides an index for radiotracer incorporation potential, by measuring total DNA content *(12)*. For a given explant, the incorporation of radiolabeled metabolites is mediated by its resident cells. Since the quantity of DNA is related to the abundance of cells, knowledge of DNA content (i.e., cell population density) may enable the comparison of incorporation rates among samples with dissimilar cell populations. The following protocols describe the preparation of a concentrated buffer (10X TEN) and a stock dye solution (the combination of which furnishes a working dye solution).

3.5.3.1. 10X TEN BUFFER

For the preparation of 1.0 L of 10X Tris-EDTA-NaCl stock buffer (10X TEN), the required materials and their final concentrations are Tris (100 mM), EDTA (10 mM), and NaCl (1.0 M). We typically prepare 3.0 L.

1. Mix 12.10 g Tris-base, 3.72 g EDTA, and 58.44 g NaCl into 0.95 L sterilized, demineralized water. Allow 30 min for all the salts to completely dissolve.
2. Adjust pH to 7.40 with concentrated HCl (~14 M HCl).
3. Add water to bring the volume to 1.0 L.
4. Filter sterilize with 500 mL capacity 0.22-µm filter.
5. Store under sterile conditions at 4°C.

3.5.3.2. CONCENTRATED DYE STOCK SOLUTION

1. Add 10 mg of Hoechst 33258 to 10 mL of sterile demineralized water.
2. Store in foil-wrapped glass or brown plastic bottle at 4°C in a dark place. Stable for 6 mo. Note preparation date and expiration date on the bottle.

3.5.3.3. Working Dye Solution

1. Prepare a 1X TEN (10 mM Tris-HCl, 1 mM EDTA, 0.1M NaCl) solution by diluting 10X TEN 1:10. Take precautions to ensure the sterility of the stock 10X TEN.
2. Add dye stock solution to 1X TEN to a final dye concentration of 0.1 µg/mL. First, shake the dye stock solution, then combine 10 µL of dye solution and 100 mL of 1X TEN.

3.5.3.4. Assessing DNA content

1. Vortex the digested samples and dispense a small aliquot (~50 µL) in duplicate into cuvets.
2. Pump 2 mL Hoescht dye solution into each cuvet. This solution is stable.
3. Measure fluorescence and compare with standards.

3.5.4. Radioactive Labeling

The biological response of cartilage to mechanical stress is most commonly monitored by introducing radioactive precursors to the culture medium. Presumably, the rate of incorporation of radiolabeled metabolites into structural macromolecules is related to the rate of biosynthesis. Since sulfate is incorporated into glycosaminoglycans and amino acids into proteins, the uptake of [^{35}S]sulfate and tritiated amino acids from the culture medium allows for dual monitoring of newly synthesized glycosaminoglycans and proteins, respectively (*see* **Note 13**). The incorporation of both β-emitting radiolabels can be observed quantitatively through liquid scintillation counting, or spatially through autoradiography *(3,9)*. Radiolabel concentrations in the culture media should be on the order of 5–50 µCi/mL for both the [^{35}S]sulfate (Na$_2^{35}$SO$_4$) and tritiated amino acids (1-[5-^{3}H]proline). To achieve at least 3000 counts/min (which should be sufficiently greater than background) we typically use 20 µCi/mL [^{35}S]sulfate and 10 µCi/mL [^{3}H]proline for 8 h labeling of 2–4-mm diameter, 1–2-mm thick disks. These concentrations can be varied accordingly for substantially longer or shorter labeling times. Control studies should be performed to determine appropriate radiolabel concentrations.

3.5.4.1. 2 M Guanidine HCl Solution

1. 191.06 g guanidine hydrochloride.
2. 68.04 g sodium acetate.
3. Mix with 900 mL water.
4. Adjust pH to 6.8 with 1 M HCl.
5. Bring volume to 1 L.
6. Store at room temperature.

3.5.4.2. Scintillation Counting

1. Aliquot 40 µL of digested cartilage or media samples into vials in duplicate.
2. Aliquot 10 µL of the original radioactive media and 30 µL of water into vials in duplicate.

3. Add 460 μL of guanidine HCl and 2 mL of Ecolume to each vial.
4. Cap and vortex all vials.
5. Place vials in scintillation counter. Wait approx 1 h before measuring the chemiluminescence to allow any residual fluorescence to extinguish.
6. Adjust aliquot sizes to yield counts of at least 3000 counts/min.

4. Notes

1. Stress relaxation in cartilage is the transient physical process of matrix and fluid redistribution in response to applied displacements. It is monitored indirectly by observing the load (**Fig. 4**). Immediately after applying the displacement, the load is maximum. As fluid is redistributed, the load reduces, until leveling off at a new steady value needed to maintain the tissue at the imposed displacement. Similarly, creep is the transient response to applied loads. During stress relaxation and creep, fluid transport effects will be present. The time-course for stress relaxation, the process relevant to methods described here, as described by Armstrong's model of unconfined compression, is a function of the disk's radius, a, the equilibrium confined-compression modulus, H_A, and the hydraulic permeability, k *(1)*. The stress relaxation time constant, to first order approximations, is defined as: $\tau = a^2/\{H_A k\}$ (as a rule of thumb, the creep time is approximately 4X longer). Typical values for H_A and k in articular cartilage are 0.5 MPa and $3.0 \times 10^{-15} m^4/(Ns)$, respectively. For a 4-mm-diameter sample, for example, one would expect a characteristic stress-relaxation time of ~45 min; note that halving the radius would decrease this time fourfold.
2. To heat-inactivate serum, incubate in a hot-water bath at 56°C for 30 min. Swirl the serum at least twice during the incubation period to avoid gelation of serum proteins. Dispense 0.5 mL of sera into cryopreservation vials and freeze at –20°C. Thaw at room temperature when needed.
3. Our vise is essentially a stainless steel tube attached to a universal joint that is mounted in a custom-built acrylic tray. The shaft of the long bone is inserted into the tube and immobilized by tightening the three set screws. The universal joint allows for orienting the cartilage surface beneath the drill bit. The tray serves as a reservoir for collecting spent rinsing solution.
4. GAG synthesis after explantation is highly variable *(13–15)*; we have determined that GAG synthesis is stable after 2 d of culture (in free-swelling conditions) *(16)*. However, others have noted longer periods, albeit for adult ovine tissue *(17)*. Waiting for the stabilization of the GAG (and protein) synthesis is important for ascribing any observed biological variations in synthesis to mechanical stimulation.
5. Compression posts should be made of a material that is several orders of magnitude stiffer than cartilage (i.e., modulus of >100 MPa).
6. This particular configuration allows each cartilage sample to be cultured in its own well. A variety of other devices have been constructed by our lab and other labs. The interested reader is encouraged to contact the authors if they wish assistance in obtaining or designing a compression chamber.

7. For load-controlled static compression, individual weights can be fashioned from lead (or containers of lead shot) so that they can rest on top of the compression post. The stress applied to the tissue is the product of the mass (as measured on a balance) and 9.81 m/s^2 (acceleration because of gravity) divided by the surface area of the disk (e.g., to achieve 250 kPa—a stress that provides a nominal 20% compression—for a 3-mm-diameter disk, a mass of 180 g is required). The size of the weights can make the approach difficult, especially for larger diameter samples.
8. When the compression system establishes contact with the cartilage explant, an initial displacement is imposed, which the load cell instrumentation will indicate as an offset or tare load. The tare load should not exceed 5 g. The effect of the tare load upon the slope of load-displacement relationship (i.e., stiffness) of the cartilage is assumed to be negligible. Moreover, it is assumed that the tare load elicits a negligible initial compressive displacement, therefore a 0% macroscopic strain may be defined at the point of initial contact (this definition of strain assumes the reference length to be the swelled sample thickness as measured immediately prior to mechanical loading).
9. All load-displacement behavior is attributed solely to the cartilage, because the stiffness of the compression system (in the absence of cartilage) ranges from 200 to 400 kN/m; whereas a 4-mm-diameter by 2-mm-thick cartilage disk generally exhibits an equilibrium stiffness of approx 3.0 kN/m. The dynamic stiffness of cartilage increases dramatically with frequency in the range of 0.05 to 1.0 Hz; therefore, system compliance may introduce an artifact in determining the stiffness of cartilage.
10. Compression levels (i.e., macroscopic strain) of 0, 10, 20, and 50% for a tissue sample that was initially 800-µm thick correspond to final compressed thicknesses of 800, 720, 640, and 400 µm, respectively. In stating the macroscopic strain, it is important to also state the reference thickness (e.g., cut thickness, or swelled thickness, measured immediately prior to mechanical testing).
11. Other digestion procedures can be used: proteinase K, hydrolysis, and so on. We use papain, which is suitable for the most commonly used biochemical assays listed here.
12. Different chondroitin sulfate standards yield slightly different absorbances for the same concentration. Therefore, it is important to use the same chondroitin sulfate standard (preferably the same lot) if it is desirable to compare across experiments.
13. In order to interpret radiolabel incorporation as reflecting synthesis, it is important to establish that incorporation is linear with time, and varies appropriately as medium concentrations of cold label (nonradioactive precursor) are changed. Using common media formulations does not appear to pose problems for sulfate incorporation in any cartilage system. Others have shown that for the –addition of 0.4 mM nonradioactive proline to DMEM, the specific activity of proline incorporation appears to reflect the specific activity of the medium. Control experiments should be performed, and data in the literature should be consulted for other amino acids.

References

1. Armstrong, C., Lai, W., and Mow, V. (1984) An analysis of the unconfined compression of articular cartilage. *J. Biomech. Eng.* **106**, 165–173.
2. Kim, Y., Bonasser, L., and Grodzinsky, A. (1995) The role of cartilage streaming potential, fluid flow and pressure in the stimulation of chondrocyte biosynthesis during dynamic compression. *J. Biomech.* **28**, 1055–1066.
3. Kim, Y., Sah, R., Grodzinsky, A., Plaas, A., and Sandy, J. (1994) Mechanical regulation of cartilage biosynthetic behavior: physical stimuli. *Arch. Biochem. Biophys.* **311**, 1–12.
4. Hall, A., Urban, J., and GehI, K. (1989) The effects of hydrostatic pressure on matrix synthesis in articular cartilage. *J. Orthop. Res.* **9**, 1–10.
5. Maroudas, A. and Bannon, C. (1981) Measurement of swelling pressure in cartilage and comparison with the osmotic pressure of constituent proteoglycans. *Biorheology* **18**, 619–632.
6. Maroudas, A., Mizrahi, J., Ben Haim, E.. and Ziv, I. (1987) Swelling pressure in cartilage. *Adv. Microcirc.* **13**, 203–212.
7. Maroudas, A., Wachtel, E., Grushko, G., Katz, F., and Weinberg, P. (1991) The effect of osmotic and mechanical pressures on water partitioning in articular cartilage. *Biochim. Biophys. Acta* **1073**, 285–294.
8. Schneiderman, R., Keret, D., and Maroudas, A. (1986) Effects of mechanical and osmotic pressure on the rate of glycosaminoglycan synthesis in the human adult femoral head cartilage: an in vitro study. *J. Orthop. Res.* **4**, 393-408. I
9. Quinn, T. (1996) Articular cartilage: matrix assembly, mediation of chondrocyte metabolism, and response to compression. PhD, MIT.
10. Oegema, T., Carpenter, B., and Thompson, R. (1984) Fluorometric determination of DNA in cartilage of various species. *J. Orthop. Res.* **1**, 345–351.
11. Farndale, R., Buttle, D., and Barrett, A. (1986) Improved quantitation and discrimination of sulphated glycosaminoglycans by use of dimethylmethylene blue. *Biochim. Biophys. Acta* 173–177.
12. Kim, Y., Sah, R., Doong, J., and Grodzinsky, A. (1988) Fluorometric assay of DNA in cartilage explants using Hoechst 33258. *Anal. Biochem.* **174**, 168–176.
13. Hascall, V., Handley, C., McQuillan, D., Hascall, G., Robinson, H., and Lowther, D. (1983) The effect of serum on biosynthesis of proteoglycans by bovine articular cartilage in culture. *Arch. Biochem. Biophys.* **224**, 206–223.
14. Lane, J. and Brighton, C. (1974) In vitro rabbit articular cartilage organ model I. Morphology and glycosaminoglycan metabolism. *Arthritis Rheum.* **17**, 235.
15. McKenzie, L., Horsburgh, B., Ghosh, P., and Taylor, T. (1977) Organ culture of human articular cartilage: studies on sulphated glycosaminoglycan synthesis. *In Vitro* **13**, 423–428.
16. Gray, M. L., Pizzanelli, A. M., Grodzinsky, A. J., and Lee, R. C. (1988) Mechanical and physicochemical determinants of the chondrocyte biosynthetic response. *J. Orthop. Res.* **6**, 777–792.
17. Torzilli, P. A., Grigiene, R., Huang, C., Friedman, S. M., Doty, S. B., Boskey, A. L., and Lust, G. (1997) Characterization of the cartilage metabolic response to static and dynamic stress using a mechanical explant test system. *J. Biomech.* **30**, 1–9.

40

Measuring Receptor-Mediated Cell Adhesion Under Flow

Cell-Free Systems

Daniel A. Hammer and Debra K. Brunk

1. Introduction

Leukocytes must bind to vascular endothelium under conditions of flow to perform their appropriate physiological functions, which require trafficking into and out of the tissue space surrounding blood vessels. Trafficking into tissues is required of neutrophils during the acute inflammatory response, and during trafficking of lymphocytes into lymphoid tissue. Egress from blood lumen to tissue involves a series of adhesion-dependent steps, each of which involve different leukocyte adhesion receptors and counterreceptors on the endothelium. Transient adhesion, or rolling, of leukocytes over endothelial cells is a prerequisite to firm attachment and transendothelial migration *(1,2)*.

Many of the molecules mediating rolling are in the selectin adhesion molecule family, consisting of three selectins, L-, P-, and E-selectin. Selectins are selective lectins that bind to carbohydrates. A current research focus in cellular immunology is to determine which counterreceptors bind to these selectins. In static assays, it has been shown that all three selectins bind sialyl Lewis^x (sLe^x), a sialyated, fucosylated carbohydrate, which is widely distributed on both glycoprotein and glycolipid components of the neutrophil *(3,4)*, and which is constitutively expressed by human umbilical vein endothelial cells *(5–8)*. For P-selectin, the physiological ligand, P-selectin glycoprotein ligand-1, has been identified. This ligand bears sLe^x, but also contains other critical features that are required for recognition, including several tyrosines that can be, and are, functional when sulfated. Thus, the exact contribution of sLe^x to recognition is not clear. The ligands for the other selectins, E- and L-selectin, have not been conclusively demonstrated, and thus the importance of sulfated, fucosylated carbohydrates in their recognition is also not clear. Also, cells containing proteins bearing sLe^x do not always support rolling interactions.

Thus, although static assays may propose putative receptor–ligand pairs, they do not provide information on how sLe^x interacts with the selectins under dynamic conditions (i.e., can these molecules mediate rolling?). An additional confounding aspect of rolling is the role of cellular features, such as clustering of receptors on microvilli and cell deformability and intracellular signaling. It would be interesting to know if sLe^x–selectin interactions have the requisite properties of recognition to mediate rolling in the absence of cellular features. Indeed, Pierres and coworkers have suggested that the initial formation of bonds may be a passive process, which could be studied using inert substrates as carriers for ligands *(9)*. Therefore, we have developed a cell-free system to determine whether binding between sLe^x and E-selectin can mediate rolling under flow *(10,11)*. Such a clean system will allow us to identify receptor–ligand pairs that mediate rolling, and to better understand the functional properties of receptor recognition required for dynamic adhesion. This system can reproduce the dynamics of rolling adhesions seen with leukocytes over endothelium in vitro, both in average rolling velocity and in the fluctuations in velocity *(10,11)*. Such a system allows us to unambiguously determine which carbohydrate ligands support rolling on which selectin adhesion molecules, and provides a clean method for establishing the blocking interaction of antagonists targeted to these molecules. This chapter explains how this cell-free system is produced, characterized, and tested.

2. Materials
2.1. Chemical reagents

1. NeutrAvidin (Pierce, Rockford, IL).
2. Biotin (Sigma, St. Louis, MO).
3. FITC-biotin (Molecular Probes, Eugene, OR).
4. Biotinylated-(sLe^x)$_4$ (Syntesome, Munich, Germany; Glycotech, Rockville, MD).
5. Monoclonal antibody (MAb) to sLe^x (KM93, IgM, Kamiya Biomedical, Thousand Oaks, CA).
6. FITC-labeled mAb anti-IgM (Pharmingen, San Diego, CA).
7. Quantum 26 calibration microspheres (Flow Cytometry Standards, San Juan, PR).
8. E-selectin-IgG chimera (E-selectin-IgG), a gift from Brian Brandley (Rush Medical Center, Chicago, IL) *(12)*.
9. Silanated glass microscope slides (Sigma).
10. Flexiperm wells (Heraeus Instruments, South Plainfield, NJ).
11. Anti-E-selectin (68-5H11, IgG$_1$, Pharmingen).
12. FITC antimouse IgG, directed at the Fc portion of IgG (Sigma).

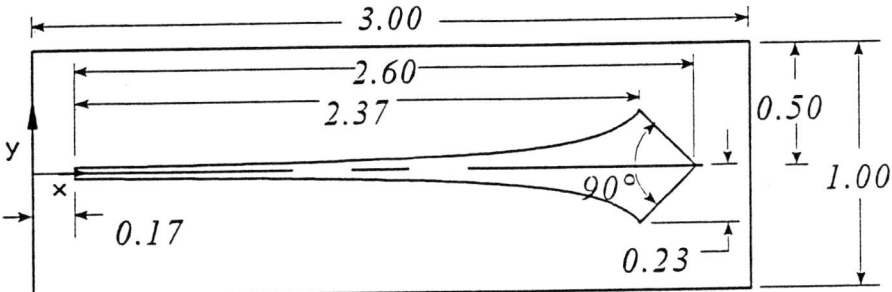

Fig. 1. Top view of flow channel, illustrating dimensions of the tapered channel. Units are in inches. Template is cut from reinforced Duralastic sheeting, 250 μm thick. The width of the channel is dependent on the axial placement along the length of the channel.

13. Human IgG$_1$ (Sigma).
14. Duralastic sheeting (Allied Biomedical, Goose Creek, SC).
15. Carbodiimide [1-(3-dimethylaminopropyl)-3-ethylcarbodiimide hydrochloride] (Aldrich, Milwaukee, WI).

2.2. Equipment

1. Photoscan (Nikon, Garden City, NJ).
2. Nikon Diaphot inverted microscope (Nikon).
3. 67S video camera (Dage-MTI, Michigan City, IN).
4. S-VHS video tape recorder (JVC, Elmwood Park, NJ).
5. Infusion–withdrawal syringe pump (Harvard Apparatus, South Natick, MA).
6. Computer for image analysis (Apple Macintosh, Cupertino CA, with Scion LG-3 frame grabber board, running NIH Image software).

2.3. Flow Chamber

The tapered channel design is based on the theory of Hele-Shaw flow between two parallel plates *(13)*. In the theory of this design, the wall shear stress, τ is given $\tau = (6\mu Q/h^2 w_1)(1 - z/L)$, where μ is viscosity, Q is flow rate, h is the channel height, w_1 is width, L is length, and z is the axial position. The plates are separated by 250-μm Duralastic sheeting, which when compressed results in a thickness of $h = 175$ μm (thus, $h = 175$ μm). The chamber is typically constructed in a machine shop from Lexan polymer. A design drawing for the Hele-Shaw template used for our *(10)* is shown in **Fig. 1**.

2.4. Apparatus

Experiments take place in a flow chamber on an inverted Nikon microscope. Two separate inlets lead to the chamber through a Y-connector, one from a syringe pump and the other from a hand-held syringe, from which beads and antibodies can be injected. The outlet of the chamber is equipped with a

Y-connector also, one exit leading to waste, the other to a buret for measuring flow rate. The events in the flow chamber are captured by a video camera and recorded on videotape for later analysis. Image analysis is performed by an Apple Macintosh computer with a video frame-grabber board and NIH Image software. Ample hard-disk storage is needed (several Gbytes recommended) for storing images.

3. Methods
3.1. Microsphere Preparation
3.1.1. Attach NeutrAvidin to Microspheres

NeutrAvidin was coupled to carboxylated polystyrene microspheres using carbodiimide chemistry *(14)*. The protocol was based on a protocol supplied by Polysciences (Warrington, PA). The detailed protocol is as follows:

1. Place 222.2 µL of 2.5% suspension of carboxylated microspheres (Polysciences, Warrington, PA) into a 1.5-mL lube tube.
2. Add 1 mL carbonate buffer (0.1 M, pH 9.6) to bead suspension, and vortex.
3. Centrifuge for 3 min at 6000 rpm (~2000g).
4. Remove supernatant carefully using a pipet, and discard supernatant.
5. Repeat steps 2–4.
6. Add 1 mL phosphate buffer (0.02 M, pH 4.5) and 2.5 µL Tween-20 (diluted 1:10 in phosphate buffer) to bead pellet and vortex.
7. Centrifuge for 3 min at 6000 rpm, remove supernatant, and discard.
8. Repeat steps 6–7 twice.
9. Add 0.3 mL phosphate buffer to bead pellet, and vortex.
10. Mix 0.35 mL of a 2% solution of carbodiimide [1-(3-dimethylaminopropyl)-3-ethylcarbodiimide hydrochloride] (Aldrich, WI). Prepare carbodiimide no sooner than 15 min prior to use.
11. Add 0.3 mL carbodiimide solution dropwise to bead suspension.
12. Mix for 3.5 h at room temperature with end-to-end mixing.
13. Centrifuge for 3 min and discard supernatant (hazardous waste).
14. Add 1 mL borate buffer (0.2 M, pH 8.5) to bead pellet, vortex, and centrifuge for 3 min at 6000 rpm (2000g). Discard supernatant.
15. Repeat step 14 twice.
16. Add 453 µL borate buffer and 3 µL Tween-20 (diluted 1:10 in borate buffer) to bead pellet and vortex.
17. Add 144 µL of 1 mg/mL NeutrAvidin to bead suspension.
18. Leave overnight at room temperature with end-to-end mixing.
19. Next day, centrifuge suspension for 10 min.
20. Remove supernatant and discard.
21. Resuspend pellet in 0.6 mL 0.1 M ethanolamine in borate buffer and incubate 30 min at room temperature with end-to-end mixing.

22. Centrifuge for 3 min at 6000 rpm (2000g), remove supernatant and discard.
23. Resuspend bead pellet in 0.6 mL PBS+ (PBS, pH 7.4, 1% BSA, 1 mM CaCl$_2$, 1 mM MgCl$_2$, sterile-filtered) and incubate for 1 h at room temperature with end-to-end mixing.
24. Wash twice with PBS+.
25. Resuspend bead pellet in 0.6 mL PBS+, vortex, and store in refrigerator at 4°C prior to use.

3.1.2. Attaching Biotinylated-Carbohydrates to Spheres.

Biotin-linked carbohydrates are commercially available from GlycoTech (Rockville, MD).

1. Remove 10^6 NeutrAvidin-coated microspheres for flow cytometry.
2. Mix the required concentration of biotin-sLe^x in PBS+. A concentration greater than 0.1 mg/mL will saturate the NeutraAvidin-coated microspheres.
3. Centrifuge NeutrAvidin-coated microspheres for 3 min, remove and discard supernatant.
4. Wash NeutrAvidin-coated microspheres once with PBS+.
5. Add 50 mL of biotin-sLe^x solution to microshere pellet, vortex, and incubate for 45 min, with occasional vortexing.
6. Centrifuge microspheres for 3 min; remove and discard supernatant.
7. Wash spheres once with PBS+.

3.1.3. Preparing Spheres for Flow Cytometry

After binding sLe^x to the spheres, the density of sLe^x can be determined with flow cytometry.

1. Prepare 30 μg/mL solution of anti-sLe^x antibody, such as KM93, IgM (Kamiya Biomedical, Thousand Oaks, CA) in MES+ buffer.
2. Wash biotin-sLe^x spheres twice in 50 mM MES+ (pH 5.5, containing 1% BSA, 1 mM CaCl$_2$, 1 mM MgCl$_2$) sterile filtered.
3. Add 50 μL of KM93 solution to biotin-sLe^x microspheres, and incubate for 30 min, with occasional vortexing.
4. Centifuge for 3 min, remove and discard supernatant.
5. Mix 60 μg/mL solution of FITC-labeled anti-IgM mAb (Pharmingen) in 50 mM MES+.
6. Add 50 mL of FITC-mAb to micropheres, and incubate 30 min in dark, with occasional vortexing.
7. Repeat **step 4**.
8. Wash once in MES+, once in PBS +, and once in PBS+ (-BSA).
9. Add 300 mL PBS+ (-BSA, +1% formaldehyde) to microspheres, and store in refrigerator until flow cytometry is performed.
10. Determine the fluorescence of these beads in a flow cytometer. Conversion of fluorescence peaks to site densities was done using Quantum 26 calibration microspheres (Flow Cytometry Standards) by comparing the fluorescence of the measured beads to that of the standards, with the additional information of the F:P ratio (the number of fluorescent probe/mol of protein), and an assumption about the number of secondary antibodies that binds per primary antibody (usually 1:1).

3.2. Selectin Substrate Preparation

1. Incubate slides in PBS overnight.
2. Incubate E-selectin-IgG on silanated glass microscope slides (Sigma) in Flexiperm wells (Heraeus Instruments, South Plainfield, NJ) at concentrations from 0.1 to 1.6 µg/mL for 2 h.
3. Wash slides twice with PBS.
4. Incubate in PBS+ (1% BSA, 1 mM CaCl$_2$, 1 mM MgCl$_2$) for 1 h to block nonspecific adhesion.

3.3. Determination of surface density of E-selectin.

In addition to **steps 1–3** in **Subheading 3.2.**, perform the following:

1. Add 20 µg/mL anti-E-selectin (68-5H11, IgG1, Pharmingen) in PBS+ to wells for 30 min.
2. Add 30 µg/mL Fc-specific FITC antimouse IgG (Sigma) in PBS+ to wells for 30 min.
3. Prepare a negative control using the same method, replacing the E-selectin-IgG with human IgG1.
4. Measure fluorescence in counts/s with a Photoscan (Nikon) attached to a Nikon Diaphot inverted microscope, and then convert to site densities using a calibration curve constructed from known concentrations of fluorescent IgG.

3.4. Flow Chamber Calibration

In this chamber, the width varies as a function of axial position down the length of the chamber.

1. Assemble the chamber with a silanated glass slide in PBS.
2. Attach a guide on the microscope stage, consisting of equidistant ticks.
3. Place a field-finder microscope slide (Fisher) into the chamber.
4. Use the manual control of stage position to relate the distance in the slide to distance on the guide.
5. At each axial position, use a reticle or photo mask to measure the width of the chamber, and compare the width measurements to that theoretically predicted by the design equation. This requires scrolling across the field of view several mutliples of the reticle dimension to span the entire width.
6. Place chamber on microscope stage and measure the height of the channel, using the fine verticle focus.
7. Perfuse PBS through the chamber, and measure the flow rate.
8. Suspend uniform polystyrene microspheres (10 µm in diameter) in PBS, and inject microsphere suspension into perfusion inlet.
9. Focus objective on bottom plate.
10. Record motion of microspheres using video camera and videotape recorder. Results of flow chamber calibration are shown in **Fig. 2**.

3.5. Adhesion Experiments

E-selectin-IgG-coated slides were placed in the well of the flow chamber, the chamber was assembled in PBS, then placed on the microscope stage. For a

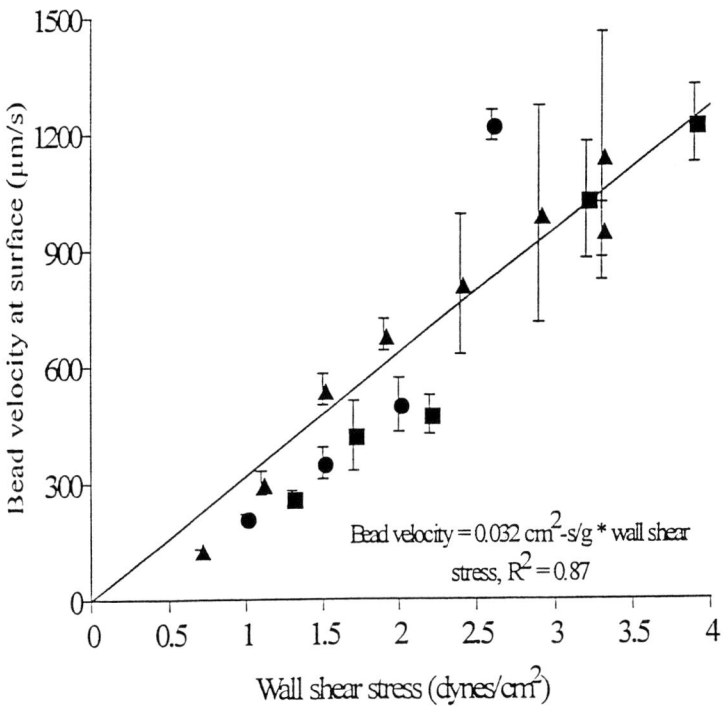

Fig. 2. Calibration of Hele-Shaw flow chamber, showing bead velocity near the plate surface as a function of wall shear stress. Wall shear stress was varied either by changing the flow rate, Q, by observing at different axial positions, z. Velocity is proportional to axial position, as predicted by theory. ■, $z = 3$ cm, varying Q; ●, $z = 5$, varying Q; ▲, $Q = 134$ μL/min, varying z.

channel height of 175 μm, the perfusion buffer or microsphere suspension flow rate must be 128 μL/min to obtain a range in wall shear stress from 0 to 4 dyn/cm² down the length of the channel. The chamber is perfused with PBS (+1 mM CaCl$_2$, 1 mM MgCl$_2$) for 15 min, then the microsphere suspension (5 × 10^5/mL) is introduced. Data is collected by stepping down the chamber from inlet to outlet in 0.5-cm steps, allowing about 1 min between steps. Microsphere interaction with the surface is recorded at a total magnification of ×200 for future analysis. All experiments are done at 23°C.

3.6. Antibody Blocking and Control Experiments

Antibody blocking and antibody control experiments are necessary to demonstrate specificity.

1. E-selectin-IgG-coated slides are incubated for 45 min in 20 μg/mL anti-E-selectin (BBA2, IgG$_2$, R&D Systems, Minneapolis, MN), which should block

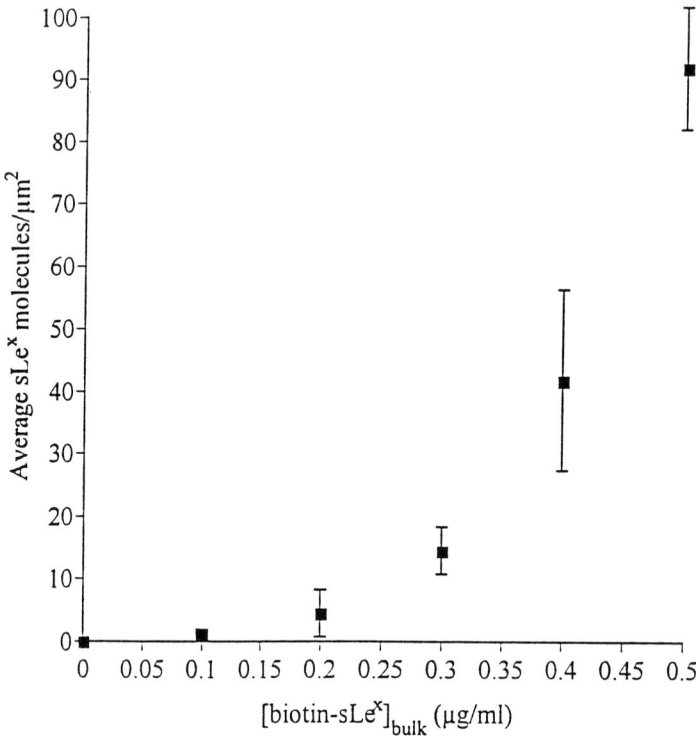

Fig. 3. Average sLe^x molecules per bead area as a function of solution concentration of biotin-sLe^x, as achieved by varying the ratio of biotin-sLe^x to biotin-Le^x in solution, keeping the total concentration of biotin-carbohydrate constant at 0.5 µg/mL.

binding; or with anti-vascular cell adhesion molecule-1 (VCAM-1) (51-10°C9, IgG$_1$, Pharmingen) or anti-P-selectin (AK4, IgG$_1$, Pharmingen), which are positive controls.
2. After antibody incubation, slide is inserted into the chamber.
3. Add 1 µg/mL of same antibody to the perfusion buffer and the microsphere solution.
4. Perform experiment as normal.

3.6. Data Analysis

1. To determine the particle flux, the number of transient interactions in a 0.03-mm^2 area were manually counted as a function of time. Transient attachment included all particles that interacted with the surface, i.e., rolling and stop–start interactions. Particles were considered firmly attached if they remained stationary for >10 s.
2. For rolling cells, velocity measurements are obtained from recorded data using a SCION LG-3 frame grabber in a Power Macintosh 7100 running NIH Image,

public domain image analysis software. Particle coordinates were obtained with 1 pixel accuracy (1 μm) by clicking the mouse at the center of a rolling microsphere. Average velocity was calculated by dividing the total displacement of a rolling particle by the observation time. Instantaneous velocity was determined by dividing the displacement of a rolling particle by the time between incremental captured frames.

4. Notes

1. NeutrAvidin is an altered form of avidin that has had all of its surface carbohydrate groups removed, thus removing any nonspecific binding, and lowering the isoelectric point to 7, compared to 9 for avidin. It is thus easier to work with, and displays less nonspecific binding induced by electrostatics.
2. Bead lot variability within single vendors is large, so stick with a lot that works. In addition to the Polysciences beads, we also found that beads from Bangs laboratories (Fishers, IN) worked well.
3. Make sure PBS+ is as fresh as possible.
4. One can reduce the amount of biotin-sLex on the bead surface by co-incubating it with a neutral biotinylated carbohydrate, such as biotin-Le^x in different ratios in which the total concentration is kept at a value >0.1 μg/mL. **Figure 3** illustrates changes in sLex-bead surface density one can achieve with this method.
5. The accuracy of our determination of sLe^x surface density was greatly increased by using an IgG$_3$ against sLe^x (provided by Anil Singhal, Biomembrane Institute, Seattle, WA).

References

1. von Andrian, U. H., Chambers, J. D., McEvoy, L. M., Bargatze, R. F., Arfors, K.-E., and Butcher, E. C. (1991) Two-step model of leukocyte-endothelial cell interaction in inflammation: distinct roles for LECAM-1 and the leukocyte β$_2$ integrins in vivo. *Proc. Natl. Acad. Sci. USA* **88,** 7538–7542.
2. Lawrence, M. B. and Springer, T. A. (1991) Leukocytes roll on a selectin at physiologic flow rates: distinction from and prerequisite for adhesion through integrins. *Cell* **65,** 859–873.
3. Fukuda, M., Spooncer, E., Oates, J. E., Dell, A., and Klod, J. C. (1984) Structure of sialylated fucosyl lactosaminoglycans isolated from human granulocytes. *J. Biol. Chem.* **259,** 10,925–10,935.
4. Symington, F. W., Hedges, R., and Hakomer, S. I. (1985) Glycolipid antigens of human polymorphonuclear neutrophils and the inducible HL-60 myeloid cell line. *J. Immunol.* **134,** 2498–2506.
5. Majuri, M.-L., Pinola, M., Niemela, R., Tiisala, S., Natunen, J., Renkonen, O., and Renkonen, R. (1994) 2,3-sialyl and 1,3-fucosyltransferase-dependent synthesis of sialyl Lewis x, an essential oligosaccharide present on L-selectin counterreceptors, in cultured endothelial cells. *Eur. J. Immunol.* **24,** 3205–3210.
6. Polley, M. J., Phillips, M. L., Wayner, E., Nudelman, E., Singhal, A. K., Hakomori, S. I., and Paulson, J. C. (1991) CD62 and endothelial cell-leukocyte adhesion molecule 1 (ELAM-1) recognize the same carbohydrate ligand, sialyl-Lewis x. *Proc. Natl. Acad. Sci. USA* **88,** 6224–6228.

7. Foxall, C., Watson, S. R., Dowbenko, D., Fennie, C., Lasky, L. A., Kiso, M., Hasegawa, A., Asa, D., and Brandley, B. K. (1992) The three members of the selectin receptor family recognize a common carbohydrate epitope, the sialyl Lewis^x oligosaccharide. *J. Cell Biol.* **117,** 895–902.
8. Berg, E. L., Magnani, J., Warnock, R. A., Robinson, M. K., and Butcher, E. C. (1992) Comparison of L-selectin and E-selectin ligand specificities: the L-selectin can bind the E-selectin ligands. *Biochem. Biophys. Res. Commun.* **184,** 1048–1055.
9. Pierres, A., Tissot, O., Malissen, B., and Bongrand, P. (1994) Dynamic adhesion of CD8-positive cells to antibody-coated surfaces: the initial step is independent of microfilaments and intracellular domains of cell binding molecules. *J. Cell Biol.* **125,** 945–953.
10. Brunk, D. K., Goetz, D. J., and Hammer, D. A. (1996) Sialyl Lewis^x/E-selectin-mediated rolling in a cell free system. *Biophys. J.* **71,** 2902–2907.
11. Brunk, D. K. and Hammer, D. A. (1997) Quantifying rolling adhesion with a cell-free assay: E-selectin and its carbohydrate ligands. *Biophys. J.* **72,** 2820–2833.
12. Watson, S. R., Imai, Y., Fennie, C., Geoffroy, J. S., Rosen, S. D., and Lasky, L. A. (1990) A homing receptor-IgG chimera as a probe for adhesive ligands of lymph node endothelial venules. *J. Cell Biol.* **110,** 2221–2229.
13. Usami, S., Chen, H.-H., Zhao, Y., Shien, S., and Skalak, R. (1993) Design and construction of a linear shear stress flow chamber. *Ann. Biomed. Eng.* **21,** 77–83.
14. Kuo, S. C. and Lauffenburger, D. A. (1993) Relationship between receptor/ligand binding affinity and adhesion strength. *Biophys. J.* **65,** 2191–2200.

41

In Vitro and In Vivo Quantification of Adhesion Between Leukocytes and Vascular Endothelium

Rakesh K. Jain, Lance L. Munn, Dai Fukumura and Robert J. Melder

1. Introduction

When a leukocyte enters a blood vessel, it may continue to move with flowing blood, collide with the vessel wall, adhere transiently or stably, and finally extravasate *(1)*. These interactions are governed by both local hydrodynamic and adhesive forces. The former are determined by the vessel diameter, fluid velocity, viscosity, and hematocrit, and the latter by the number, strength and kinetics of bond formation between adhesion molecules, and by surface area of contact *(1–6)*. Cellular deformability affects both types of forces *(7–9)*. Two families of cell adhesion molecules (CAMs) are involved in leukocyte rolling and stable adhesion. In general, the selectins (P, L, and E) mediate rolling, while the IgG superfamily members (ICAM-1 and VCAM-1) on endothelial cells, with their cognate receptors (β_2 and β_1 integrin receptors) on the leukocytes, mediate firm adhesion, with some overlap in these functions *(10–12)*. The expression of CAMs on the endothelial cells and leukocytes can be modulated by cytokines secreted by a variety of cells (e.g., cancer cells, fibroblasts, macrophages) *(13,14)*. Cellular deformability can be modulated by altering the cytoskeleton, membrane, or cytoplasm, with the cytoskeleton playing the dominant role *(7,15,16)*. In this chapter, we describe methods to quantitate cellular deformability in vitro, CAM expression in vitro, leukocyte–endothelial interaction (LEI) in vitro, and LEI in vivo.

2. Materials

2.1. Cellular Deformability In Vitro

The materials required for this procedure depend on the culture requirements of the cells to be used in the procedure, and on the experimental objective of the testing procedure. In general, aside from the equipment, the following materials are needed:

1. Culture medium, such as RPMI 1640 for lymphocytes (Mediatech, Fisher Scientific, Pittsburgh, PA).
2. Cell population.
3. Glass capillary stock (Sutter Instrument, Novato, CA).
4. Inverted microscope (Nikon Diaphot, Garden City, NY).
5. Camera (Panasonic, Secaucus, NJ).
6. Video cassette recorder (Panasonic).
7. Time code generator (WJ-810, Panasonic).
8. Video contrast enhancer (Model 605, Colorado Video, Boulder, CO).
9. Monitor (PM205A, Ikegami, Utsunomiya, Japan).
10. Pressure application system: damping chamber, three-way manifold, two water reservoir bottles, connecting tubes and a fast response solenoid valve (type 124, Bürket, Orange, CA).
11. Micropipet aspiration system: micropipet (4–7 µm od and 2–4 µm id), micromanipulator (M0202, Narishige, Greenvale, NY), temperature-controlled observation system (such as the Microwarm Plate, Schlueter Instrument, Boulder, CO).
12. Transducer (precision ±1.0%, Gould, Valley View, OH).
13. Digital display and chart recorder (or an A/D converter and computer).

2.2. Adhesion Molecule Expression In Vitro

1. Surface-dependent cells.
2. 2 × 4 microwell plate (Nunc, Labtek, Naperville, IL).
3. Matrix component (s) for coating surface of plate (e.g., Sigma, St. Louis, MO).
4. Paraformaldehyde: 2 mg/mL in PBS (Sigma).
5. Bovine serum albumin: 0.1% in PBS (Sigma, Hybritech).
6. Antibodies: primary, secondary, control.
7. Propidium iodide (Sigma).
8. Mounting buffer: glycerol:PBS (1:9) with 1 mg/mL paraphenylenediamine (Aldrich, Milwaukee, WI).

2.3. L–E Interactions In Vitro

1. Histopaque (Sigma).
2. L-phenylalanine methyl ester (E.I. du Pont de Nemours, Wilmington, DE).
3. Antibodies: CD19 (clone J4.119), CD16 (B73.1), CD15 (clone MMA).
4. Paramagnetic beads (Advanced Magnetics, Cambridge, MA).
5. Calcein acetoxymethyl ester (Molecular Probes, Portland, OR).
6. Endothelial cells (Clonetics, San Diego, CA).
7. EGM medium (Clonetics).
8. Fibronectin (Sigma).
9. Variable speed syringe pump (Harvard Apparatus, South Natick, MA).
10. Hanks' balanced salt solution (HBSS; Sigma).

11. Parallel-plate flow chamber (Machine Shop, Department of Radiation Oncology, MGH, Boston, MA).

2.4. In Vivo L–E Observation

1. Animals: mice (30–35 g body wt), rats (200–250 g body weight).
2. Anesthesia (ketamine and xylazine mixture): For mice, mix 1 mL of ketamine HCl (Ketalar 100 mg/mL, Parke-Davis, Morris Plains, NJ) and 100 µL of xylazine (Xyla-Ject 100 mg/mL, Phoenix Pharmaceutical, St. Joseph, MO), and make up to 10 mL with 0.9% sterile sodium chloride; for rats, mix 1 mL of ketamine and 100 µL of xylazine; final dose will be 10 mg ketamine/1 mg xylazine/100 g body wt (mice) and 9 mg ketamine/0.9 mg xylazine/100 g body wt (rats).
3. Cannulation tube: Cut polyethylene tubing (PE-10, Becton Dickinson, Sparks, MD) at 12 cm length, break off needle part of 30-gage needle (Becton Dickinson) by folding several times with needle holder, insert base end of this needle to PE-10 tube, and insert regular 30-gage needle to the other end of PE-10 tube.
4. Rhodamine 6G (Rho-6G, Molecular Probes): Mix 100 mg of Rho-6G powder with small amount of 95% EtOH to dissolve and make up to 100 mL with 0.9% sterile sodium chloride, sterilize by passing through 0.22-µm filter and aliquot in 1-mL portions to syringes; store at –20 to –70°C; final concentration is 0.1%.
5. Fluorescent latex beads: 0.8%; 1.0 µm (Polysciences, Warrington, PA).
6. Polyacrylate stage with rectangular hole (20 × 50 mm) covered with thin cover glass for mesentery.
7. Polyacrylate stage with microscope slide glass observation platform (4–6 mm height) for ear.
8. Symmetrical sandwich titanium chamber (Machine shop, Department of Radiation Oncology, MGH, Boston, MA), polyacrylate tube (id: 25 mm) and chamber holding stage (Machine shop) for dorsal skin chamber.
9. Circular glass cover slip 8 mm in diameter (Assistent, Germany), and polyacrylate stage with stereotactic apparatus for cranial window.

3. Methods

3.1. Cellular Deformability In Vitro

There are several methods to measure the deformability of cells in vitro, including filtration *(17,18)*, cell poker method *(19)*, and the micropipet aspiration technique *(15,20–22)*. We prefer the latter, because it permits quantitative estimation of intrinsic biophysical parameters, and thus allows interlaboratory comparison. In this method, a known negative pressure (suction) is applied on the cell membrane via a micropipet, and the resulting displacement of cell into the pipet is measured as a function of time (**Fig. 1A**). The resulting deformation data can itself be compared among different cell types, or the same cell type under different physiological conditions. If necessary, these data can be analyzed using an appropriate mathematical model to yield viscoelastic parameters *(22)*. Some

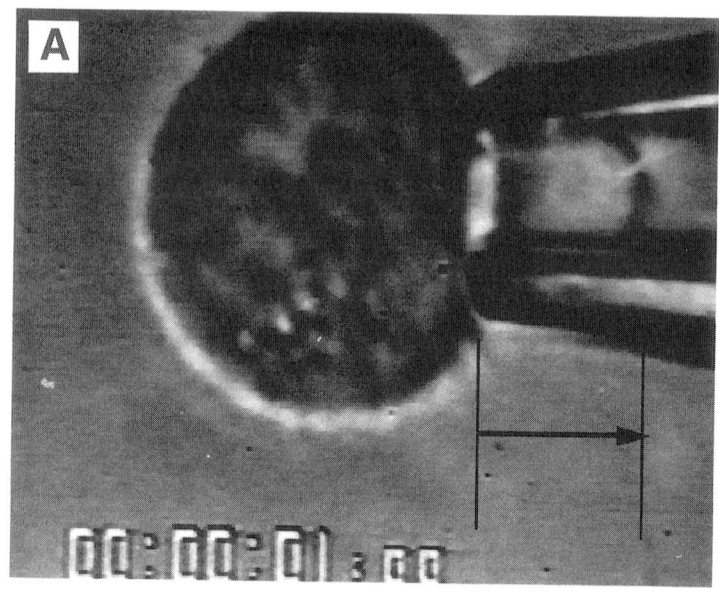
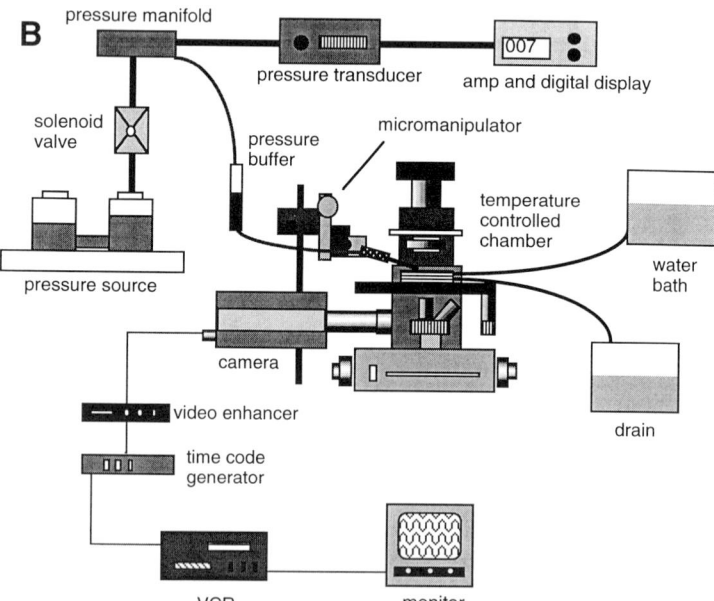

Fig. 1. Micropipet aspiration for cell deformability: **(A)** schematic, and **(B)** apparatus.

useful findings in leukocyte biology, using this technique, include demonstration that IL-2-mediated activation increases lymphocyte rigidity, and that thioglycolic acid reduces their rigidity *(15,22)*.

3.1.1. Equipment

An inverted microscope (Nikon Diaphot) equipped with Hoffman modulation optics is connected to a television camera (Panasonic), as shown in **Fig. 1B**. A video cassette recorder (IAG-6500, Panasonic), with a time code generator (WJ-810, Panasonic), video contrast enhancer (Model 605, Colorado Video), and a television monitor (PM205A, Ikegami), is used to record and display the images of the cells during manipulation and deformation. The pressure application system consists of a damping chamber, a three-way manifold, two water reservoir bottles, connecting tubes, and a fast response solenoid valve (type 124, Bürket). A damping chamber is used to diminish any pulse in pressure that may be associated with the opening of the solenoid. This chamber (approx 10 mL vol) is half-filled with saline, and its height above the microscope stage is adjusted to give a slight negative pressure (<0.5 cm of water) to a micropipet, enabling the investigator to pick up individual cells from the suspension.

The micropipet aspiration system consists of a micropipet and micromanipulator (M0202, Narishige, Greenvale, NY), and should include a temperature controlled observation system (such as the Microwarm Plate, Schlueter Instrument), although equally satisfactory devices may be devised and manufactured by the investigator to meet their individual needs. Clean glass micropipets, with 4–7 μm od and 2–4 μm id, can be made using a manual micropipet puller (Model 720, Kopf, Tujunga, CA) or an automated puller (Sutter). The pipets that are made with these devices must be selected according to the following criteria:

1. The opening must be uniform around its circumference, and without chips or cracks.
2. The opening diameter should be no larger than one-third the diameter of the cells to be tested.
3. The walls of the pipet near the opening should be nearly parallel.
4. The opening must be perpendicular to the axis of the pipet.

The selected pipets are generally filled with culture medium prior to mounting on the micromanipulator and connecting to the pressure system. A rapid pressure drop is applied by activating the solenoid valve, which simultaneously triggers the time code generator (Panasonic) through a relay circuit. In this way, the deformation history of the cells may be recorded, along with the time scale from the instant of the pressure drop. The pressure is monitored by a transducer (precision ±1.0%, Gould) connected to a digital display and chart recorder (or an A/D converter and computer).

3.1.2. Micropipet Aspiration Procedure

1. A suspension of cells, approx 10^3–10^4/mL, is placed into the observation dish on the microscope stage and permitted to settle for several minutes.

2. Individual cells in the suspension are selected and picked up with the tip of the micropipet with slight negative pressure and held within 15–20 μm of the chamber bottom.
3. A sustained pulse –25 mm water pressure is rapidly applied by triggering the solenoid valve for 2 s, and the movement of the cell membrane within the micropipet is recorded on video tape. Observations are generally limited to 2 s, in order to avoid the more complex mechanical system associated with the aspiration of whole cells into the pipet.
4. Consecutive video frames at 0.03 s intervals for each measurement are observed from playback of the videotape, and measurements of deformation are either taken directly from the video monitor or the images may be digitized for measurement using software packages on PCs (such as the NIH Image 1.59 program, available from the National Institutes of Health, Bethesda, MD).

3.1.3. Data Analysis

The cell diameter must be noted for each observation, as well as the internal and external pipet diameter. The edge of the cell membrane within the micropipet is observed in sequential frames after the application of negative pressure. The cellular deformation [$d(t)$] is defined as the difference between the length of the displaced membrane at time, t, and its original location at $t = 0$. Each displacement value must then be normalized as $d(t)/D_p$, where D_p = internal micropipet diameter, and the mean values for each test group of cells is plotted as a function of time after pressure initiation. The resulting curves reflect the deformation history of the cells and may be used to model the viscoelastic behavior during deformation *(15,21)*. Alternatively, the resistance factor (R_f), which provides a useful index of the cellular resistance to displacement in a microvascular network *(20–22)*, may be calculated as the product of the cell diameter and the normalized displacement at 1 s.

3.2. CAM Expression In Vitro

The most widely used method to measure expression of molecules on the surface of cells is flow cytometry. Although this is suitable for blood-borne cells or cells grown in suspension culture, it may not be appropriate for cells that are part of an organized tissue or that are grown as a monolayer. In addition to possible alteration in CAM expression and loss of some cells because of chemical/mechanical dissociation, required for flow cytometry, one loses the spatial information on CAM expression. For these reasons, we have recently developed a new method—target sampling fluorometry (TSF)—to measure CAM expression in intact monolayers of cells (*14*; **Fig. 2**) This method has yielded useful information about why lymphocytes bind heterogeneously to an endothelial monolayer.

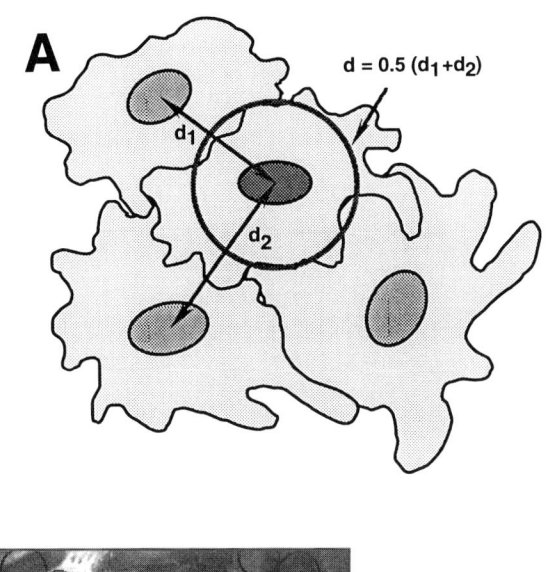

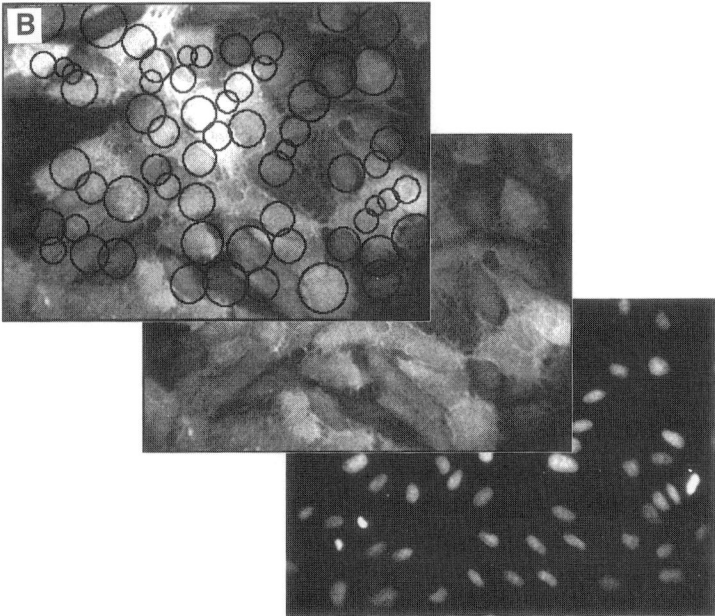

Fig. 2: Targeted sampling fluorometry: **(A)** ROI selection criteria, **(B)** schematic of principle and **(C)** sample histograms for VCAM-1 upregulation after TNF-α stimulation. Adapted with permission from **ref. *14***.

1. Maintain human umbilical vein endothelial cell (HUVEC) cultures (Clonetics) in EGM medium (Clonetics) supplemented with 10% fetal calf serum.

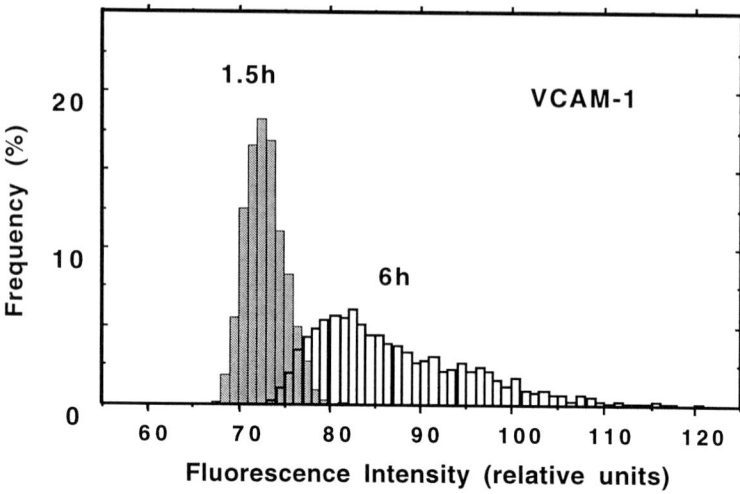

Fig. 2C.

2. Form endothelial cell monolayers in the wells of a 2 × 4 microwell plate (Nunc, Labtek). These slides consist of a standard glass microscope slide with 8 tissue-culture wells attached, to allow multiple-sample treatments on one slide. The wells are 0.9-cm square and can be removed after sample fixation for sample mounting. First, coat the glass surface with fibronectin (Sigma) at 6 mg/cm^2 for 30 min, wash with PBS, and then seed with a suspension of HUVECs. Use a low seeding density (2 × 10^4 cells/well) to produce an even monolayer that is confluent in 4–5 d.
3. Once confluent, activate the cells with the appropriate agent.
4. After activation, fix the monolayers with paraformaldehyde (2 mg/mL in PBS) for 15 min, wash four times with PBS, and then immunostain using a two-step procedure: First incubate with primary antibody directed against the adhesion molecule, and then with a goat F (ab')$_2$ antimouse IgG Ab conjugated with fluorescein (Tago, Burlingame, CA). All antibody incubations should be performed in the presence of 0.1% bovine serum albumin (BSA) to eliminate nonspecific protein binding. Primary antibodies used in this lab are: IgG controls (IgG$_{2a}$, κ and IgG$_1$, κ, Becton Dickinson), anti-E selectin (CD62E; clone H18/7, Becton Dickinson), anti-P selectin (CD62P; clone GA6, Becton Dickinson), anti-ICAM-1 (intercellular adhesion molecule-1, CD54; clone 84H10, AMAC, Westbrook, ME), and anti-VCAM-1 (vascular cell adhesion molecule-1, clone 1G11, AMAC). Perform all incubations at 37°C for 1 h. To determine the appropriate amounts of antibodies to use, it is necessary to first perform a titration to obtain the level that saturates the system. This must be done with both the primary and secondary antibodies.
5. After the second incubation, wash the layer 4×, and then incubate with 1 mg/mL propidium iodide to stain cell nuclei. After 5 min at 27°C, wash the layers four times with PBS, and mount for observation. The mounting buffer consists of

glycerol:PBS (1:9), with 1 mg/mL paraphenylenediamine (Aldrich) added to retard photobleaching of the samples *(23)*. All fixation, washing, and incubation solutions should contain the divalent cations Ca^{2+} and Mg^{2+}, to minimize cell detachment from the surface.

6. Gather the images for quantification using an epifluorescence microscope and a camera sensitive enough to image the fluorescence (e.g., a SIT or I-CCD camera). The linearity of the imaging system within the relevant intensity range must be verified with standard fluorescent samples *(24)*.
7. Digitize images with a frame grabber (DataTranslation, Marlboro, MA, or Scion, Frederick, MD) and store for later analysis. A ×10, 0.3 NA objective gives a 1.2 × 0.9-mm field of view consisting of approx 700 cells. Record three fields for each well, with two images taken per field: record one image in the 488/530 fluorescein channel, and the other in the 488/600 channel (for nuclei location via propidium iodide staining). No overlap of PI fluorescence should be detected in the fluorescein channel.
8. Record images of unstained HUVEC monolayers in one well of each slide in the 488/530 channel, to provide background fluorescence and lighting variation information.
9. The image analysis is performed using a macro within NIH Image 1.61 (available via anonymous FTP at zippy.nih.nimh.gov) that automates the process. This macro is available via FTP at simon.mgh.harvard.edu. The following processing and analysis steps are performed:

a. The images of cell nuclei stained with propidium iodide are first analyzed to locate the x, y coordinates of the cells. To identify individual cell nuclei, produce binary versions of the gray scale images by autothresholding each rectangular region of a 20 × 20 grid superimposed on the image. This procedure eliminates problems in the segmentation caused by shading variations, and results in an accurate binary representation of the HUVEC nuclei. The x, y coordinates of the centroid of each object in the binary image are recorded.
b. The background fluorescence images are averaged and applied to the immunostain image to correct for lighting variation over the surface. Pixel-by-pixel division using the averaged background image eliminates the shading effects and corrects for day to day lamp intensity variations.
c. The x, y coordinates of the cell nuclei are then used to place Regions of Interest (ROIs) over each cell in the corresponding processed gray scale images of adhesion molecule staining (**Fig. 2A**). The ROI diameters are adjusted in regions of higher cell density to minimize sampling overlap: The distances to the two nearest neighbors are determined and averaged to get the appropriate ROI diameter (**Fig. 2B**).
d. For each ROI, the mean gray level is recorded. These values are then presented in histogram form (**Fig. 2C**).

3.3. Leukocyte–Endothelial Adhesion In Vitro

Initial studies on L–E adhesion were done under static conditions. In these studies, leukocytes were allowed to adhere to endothelial monolayers for a

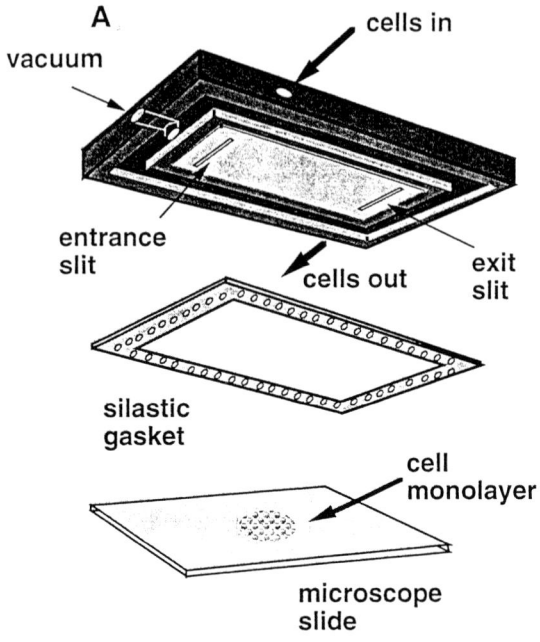
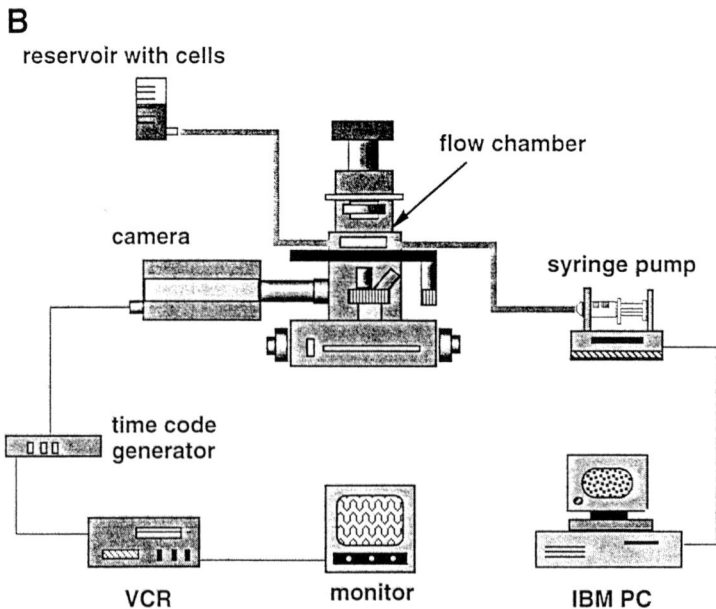

Fig. 3. Leukocyte-endothelial interactions in vitro: **(A)** schematic of parallel plate flow chamber, and **(B)** experimental setup. Adapted with permission from **ref. 35**.

fixed duration, and then nonadhered cells were washed off using a standard protocol *(25,26)*. To simulate adhesion under physiologically relevant flow conditions, a flow-chamber method, initially used for platelet adhesion studies *(27)*, was introduced *(28)*. In this method, cells are grown on the bottom plate of a parallel-plate flow cell (**Fig. 3**). The cell suspension is flowed through the chamber at a known flow rate. The fraction of cells rolling and adhering to the EC layer are measured using video-microscopy and image analysis. We have made three key modifications in this method: determined the duration of flow at each flow rate, so that the cell flux at the surface is equal at each flow rate; demonstrated that addition of RBCs to leukocytes is essential to mimic physiological flow conditions; and showed that keeping the flow vertical, instead of horizontal, eliminates effects of density differences that influence cell delivery to the surface *(29)*. Although these modifications are not widely used at the time of this writing, we recommend them.

3.3.1. Leukocyte Isolation

1. Isolate human T-cells from leukopheresis products of normal platelet donors by centrifugation over a Histopaque gradient (Sigma).
2. Deplete the lymphocyte layer of monocytes by incubation with PME (L-phenylalanine methyl ester, E.I. du Pont de Nemours) for 15 min *(30)*.
3. Deplete the remaining cells of the B-cell, NK cell and monocyte populations using antibodies directed against CD19 (clone J4.119, AMAC, Westbrook, ME), CD16 (B73.1, Becton-Dickenson) and CD15 (clone MMA, Becton-Dickenson) at 10 µg antibody/15 × 10^6 cells, followed by adsorption to a paramagnetic substrate for 30 min (Advanced Magnetics) and removal by a magnetic field. The lymphocytes isolated from this procedure are typically 72% CD4+ and 28% CD8+ cells by flow cytometric analysis.
4. If desired, label cells with the membrane-permeant form of calcein, calcein-AM (calcein acetoxymethyl ester, Molecular Probes) or similar dye. Calcein is well retained by the cells, has an extremely strong signal in the 488/530 channel, and causes no detectable change in the adhesion properties of these cells *(31,32)*. Load the cells with the dye immediately prior to the flow chamber studies by incubation at 24°C with 1 mM calcein-AM for 30 min.

3.3.2. Endothelial Monolayer Preparation

1. Maintain HUVEC cultures (Clonetics) in EGM medium (Clonetics) supplemented with 10% fetal calf serum.
2. Form endothelial cell monolayers in the wells of a 2 × 4 microwell plate (Nunc, Labtek). These slides consist of a standard glass microscope slide with 8 tissue-culture wells attached, to allow multiple-sample treatments on one slide. The wells are 0.9-cm square and can be removed after sample fixation for sample mounting. First, coat the glass surface with fibronectin (Sigma) at 6 mg/cm^2 for 30 min, wash with PBS, and then seed with a suspension of HUVECs. Use a low

seeding density (2×10^4 cells/well) to produce an even monolayer that is confluent in 4–5 d.
3. Once confluent, activate the cells with the appropriate agent.

3.3.3. Flow Chamber Experiments

Adhesion studies require a parallel plate flow chamber machined in-house (Department of Radiation Oncology, Massachusetts General Hospital, **Fig. 3A**). A silastic gasket forms a small gap between a plexiglass block and microscope slide, when the apparatus is assembled and held together with vacuum. Our chamber width is 1.25 cm, the length is 5.5 cm, and the gap size is 78 μm. The parallel plate flow chamber provides a controlled environment for determinations of the shear stress at which cells in suspension can bind to endothelial cell monolayers. It can be used in a number of different protocols, depending on the type of adhesion data required.

3.3.3.1. Detachment Assays

1. Begin perfusing flow chamber with the cell suspension at an intermediate flow rate, until the cell flux past the monolayer is steady.
2. Stop the flow and allow cells to settle onto the monolayer. The time of stoppage will depend on the cell concentration, chamber height, and cell type (sedimentation rate).
3. After allowing cells to bind, start flow at a low flow rate, and then step up the velocity at regular intervals, until there are few cells left adherent, or the monolayer has detached.
4. From the videotape, quantify the number of cells remaining at each shear stress level.

For a more complete description of this type of experiment, *see* **ref. *33***.

3.3.3.2. Single Flow Rate Capture Assays

The flow chamber can also be used to investigate how well cells adhere to the monolayer under dynamic conditions. The basic form of this assay is performed at a single flow rate. If all experiments are carried out at the same flow rate and sampling is performed at the same distance from the inlet to the chamber, then normalization of the data is simplified.

1. Start perfusion of cell suspension (10^6 cells/mL) at an intermediate flow rate. When the cell flux has stabilized, change the shear rate to the target value.
2. Allow adhesion to occur for 5–15 min.
3. While the suspension is still flowing, record five separate fields on videotape, moving horizontally across the chamber.
3. Quantify the number of cells per field that were stably adherent for 5 s or more, and, if desired, the number of cells that were rolling. Normalize to the cell flux passing through the chamber (flow rate multiplied by cell concentration).

3.3.3.3. Multiple Flow Rate Capture Assays

A more informative and, consequently, more complicated method for cell-capture assay involves the use of multiple flow rates *(12,34,35)*. By decreasing the flow rate of cell-containing media over the monolayer and assessing the number of cells bound at each wall shear stress, the relationship between shear force and binding efficiency can be determined. However, the rate of binding depends on the delivery of cells to the surface, as well as the intrinsic cell–surface interactions; thus, only if the cell flux to the surface is known can the resulting binding curves be interpreted correctly.

1. Start flow at a high wall shear stress (e.g., 4.1 dyn/cm^2), then decrease the flow rate in steps.
2. At each shear level, quantify the cumulative bound cell density at five fields sampled horizontally across the chamber. Monolayers in different runs should be sampled at approximately the same distance from the entrance slit, to avoid flux differences to the surface.
3. In order to facilitate normalization, the time interval at each flow step can be adjusted, so that the total number of cells delivered to the surface is constant for each shear-stress level *(35)*. The wait time, t_{wi}, between successive shear rates can be found by equalizing the number of cells passing at flow rate, i, according to:

$$G_i = Cx/2\,(u_{i-1}/u_i + 2vx/u_i h + 1) + Cu_t\,(1 + vx/u_i h)(t_{wi} - x/u_i) \qquad (1)$$

where G_i is the number of cells passing near the monolayer at step i, u_i and u_{i-1} are the current and previous flow velocities, respectively, and t_{wi} is the time spent at flow rate, i (including sampling time), C is the cell concentration, h is the characteristic distance of interaction, x is the distance from the entrance slit, v is the constant sedimentation velocity, and w is the chamber width *(35)*.

4. Calculate the capture efficiencies. If the time interval at each flow step was adjusted so that the total number of cells delivered to the surface was constant for each shear stress level, then normalization is straightforward. The capture efficiency is defined as the number of cells that bind (N_b) at a given shear stress divided by the total number of cells (G_i) that have passed near the surface ($E = N_b/G_i$). In order to extract the most information from the experiment, the cumulative binding curves can be fit to the equation $N_b = E_o'\,e^{-\kappa S}$, where S is the shear stress, and E_o' and κ are adjustable parameters. The value of E_o is then calculated from $E_o = E_o'\,[1 - e^{-\kappa \Delta S}]$, where ΔS is the difference in shear stress values for successive steps. Thus, $E = E_o'\,e^{-\kappa S}$, with κ giving the sensitivity of binding to shear stress, and E_o representing the theoretical efficiency at zero shear. Differences in the resulting parameters, κ and E_o, can be compared between various treatment groups with an appropriate statistical test (e.g., *t*-test or Mann-Whitney).

3.3.3.4. Flow Cell Measurements with Erythrocytes in the Medium

Additional complications occur if the suspending medium contains large numbers of cells or particles other than the binding species. For example, if blood or RBC suspensions are being used, sedimentation of the more dense RBCs to the surface of a horizontal flow chamber precludes delivery of the leukocytes to the monolayer. In this case, the system may be modified by constructing a vertical flow chamber, thus minimizing the influence of gravity on the flux at the surface. This has the added advantage of simplifying the normalization procedure, since **Eq. 1** only applies in a horizontal chamber. In the vertical case, normalization can be accomplished by simple estimates of the cell concentration and the velocity near the surface.

In non-Newtonian fluids such as blood, adhesion data should be expressed in terms of shear rate, to avoid complications of estimating the local shear force on a cell at the wall. Although the shear stresses at the wall can be estimated, the actual force on a cell rolling or adhered to the wall may be much higher than this force would indicate, and will be transient in time because of stochastic RBC collisions *(29,36)*. All data are therefore presented on the basis of equivalent shear rates, assuming parabolic flow.

The corresponding range of the maximum fluid velocities (assuming parabolic flow) was 1.0–4.1 mm/s, and the average fluid velocities ranged from 0.68 to 2.7 mm/s, respectively. The wall shear rates calculated as $S_w = 3/2 \cdot Q/2b^2w$ (with b the chamber half-height) were 53–210 s^{-1}.

3.4. L–E Adhesion In Vivo

The prerequisites for LEI studies in vivo are: availability of acute or chronic tissue preparations that permit monitoring of cells in vivo; use of optical labels that are nontoxic and specific to the cell population of interest; and availability of equipment and techniques that permit measurement of leukocyte flux, numbers of rolling and adherent cells, and flow parameters in blood vessels of choice. Acute preparations widely used include mesentery or cremaster muscle of rodents *(37)*. Commonly used chronic preparations include the rabbit ear chamber *(16,38–41)*, dorsal skin-fold chamber in rodents *(12,41)*, and cranial windows in mice and rats (*12,42,43*; **Fig. 4**). These tissue preparations can be used to monitor the normal tissue as a control or after treatment with appropriate chemical. The chronic windows permit transplantation of syngenic vs xenogenic tissues (normal or tumor) in immunocompetent vs immunodeficient hosts, respectively. The endogenous cell population is usually labeled with rhodamine 6G. The exogenously injected leukocytes can be labeled with a number of tracers (e.g., calcein), depending on the duration of experiment (short-term vs long-term). The physiological parameters (such as vessel diameter and RBC velocity to calculate the shear rate) are usually measured optoelectronically, using videomicroscopy and image analysis (**Fig. 4**). In recent years, the use of transgenic and knock-out mice has provided stunning insight into the LEI *(44)*.

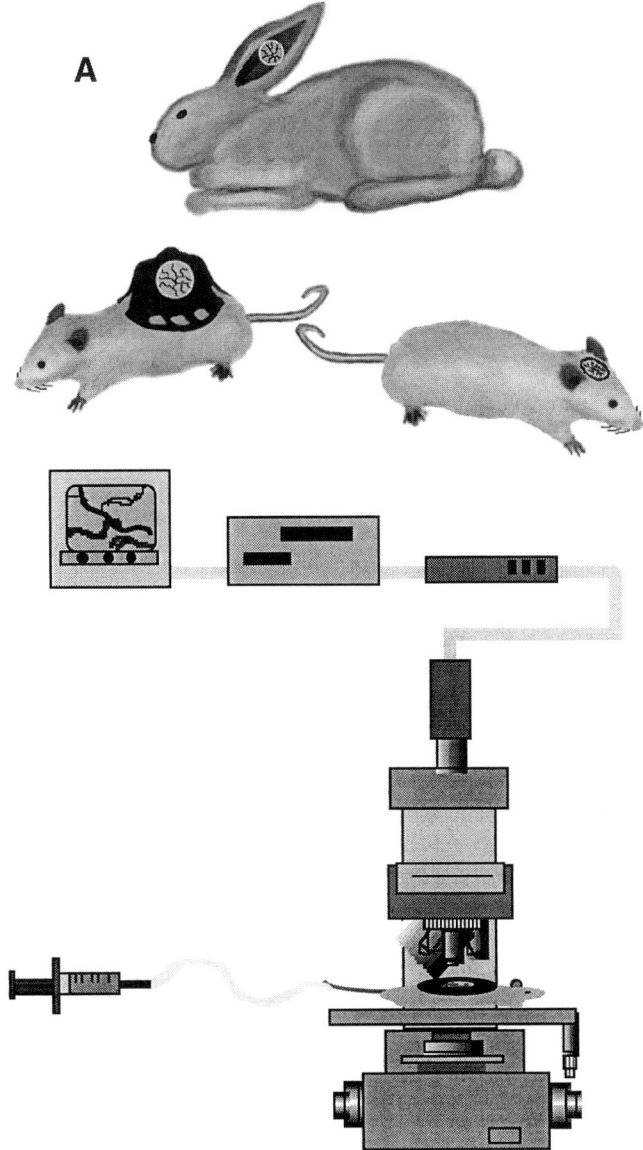

Fig. 4. Various chronic window preparations (**A**) and experimental setup (**B**). Parameters measured include vessel diameter, red blood cell velocity, leukocyte flux, rolling velocity, and adhesion density.

3.4.1. Preparation of Acute Tissue Observation: Mesentery

1. Fast the animals for 24 h before the observation.

2. Anesthetize the animals (sc injection of ketamine and xylazine mixture, 10 μL/g body wt for mice, 1 μL/g body wt for rats; final dose should be ketamine 10 mg/xylazine 1 mg per 100 g body wt for mice, and ketamine 9 mg/xylazine 0.9 mg per 100 g body wt for rats).
3. Place the catheter (30-gage needle and PE-10 tubing) into tail vein for fluorescence tracer injection; fix the cannulation site by a cyanoacrylate adhesive.
4. Shave abdominal skin by electric shaver. Clean the skin with antimicrobial Betadine solution. Open abdomen via midline incision.
5. Expose ileocecal portion, gently develop intestinal loop onto thin glass part of the polyacrylate stage using Q-tip immersed in saline, avoiding direct touch and tension to mesentery.
6. Gently straighten the intestine and fix by cotton immersed in saline, so that mesentery will be unfolded. Keep mesentery moist and warm by superfusion with warm saline (37°C). Animal body temperature is also controlled by means of heating pad.
7. Place the stage on inverted/orthotopic intravital fluorescence microscope (Axioplan; Zeiss, Oberkochen, Germany); observe mesenteric microcirculation using ×20 to ×40 objective lens.
8. Visualize transilluminated image by a charge-coupled device (CCD) video camera (AVC-D7, Sony, Tokyo, Japan) and record the image for 60 s using S-VHS video cassette recorder (SVO-9500MD, Sony).
9. Inject 20 μL of 0.1% Rho-6G for in vivo labeling of whole leukocyte population, visualize leukocytes by epi-illumination, using specific filter set for Rho-6G (excitation peak at 528 nm and emission peak at 550 nm) by intensified CCD video camera (C2400-08, Hamamatsu Photonics K.K., Hamamatsu, Japan) and record the image for 30 s.
10. In case of high-velocity vessel (>1 mm/s), inject 10 μL of 0.8% FITC-fluorescence beads (1.0 μm), visualize, and record FITC fluorescence image for 60 s, using ICCD camera with high-speed electronic shutter (C5909 Hamamatsu Photonics K.K.).
11. Repeat steps 7–9 at each observation time-point or other area.

3.4.2. Preparation of Acute Tissue Observation: Ear Microcirculation

1. Remove the hair of ear skin using hair-removing cream (8 min) 24 h before the observation.
2. Anesthetize the mice.
3. Cannulate the tail vein.
4. Drop saline on the microscope slide glass on the observation platform, place the ear gently onto the slide glass, drop saline at the edge of the ear, put thin cover glass on the top, and allow to disperse saline between the ear surface and the slide/cover glass to hold the ear to the slide and to enhance image quality.
5. Observe ear skin microcirculation (*see* **Subheading 3.4.1.**).

3.4.3. Preparation of Chronic Tissue Observation: Dorsal Skin-Fold Chamber

1. Anesthetize the mice (sc injection of ketamine and xylazine mixture 10 μL/g body wt; final dose should be ketamine 10 mg/xylazine 1 mg/100 g body wt).
2. Shave and deplete entire back of mouse using electric shaver and hair removing cream (8–12 min).
3. Assemble the sandwich titanium chamber: side 1, with three screws and nuts; side 2, close the window with circular cover slip and fix it by O-ring.
4. Suture on top and bottom of dorsal skin and pull up these sutures to make and extend double layer of the skin.
5. Suture at the top part of side 1 titanium chamber onto skin for temporal fixation.
6. Make two small holes in the double layer of skin, so that screws of the chamber can penetrate the skin.
7. Remove a circular area (15-mm diameter) of one layer of skin.
8. Remove membranous tissue carefully from the remaining layer as much as possible without tissue damage to get good optical quality.
9. Place side 2 titanium chamber to sandwich the extended double layer of the skin and fix the screws.
10. Remove the temporary sutures and suture the top part of the chamber to the skin.
11. Cover the remaining skin in the chamber with scotch tape, to protect from scratching.
12. Allow the mice to recover for 48 h after microsurgery and anesthesia.
13. Place a mouse in polyacrylate tube (id 25 mm), fix the chamber to the stage, and place the stage under the microscope.
14. Observe skin microcirculation (*see* **Subheading 3.4.1.**).

3.4.4. Preparation of Chronic Tissue Observation: Cranial Window

1. Anesthetize the animals (sc injection of ketamine and xylazine mixture 10 μL/g body wt for mice, 1 μL/g body wt for rats; final dose will be ketamine 10 mg/xylazine 1 mg/100 g body wt for mice, and ketamine 9 mg/xylazine 0.9 mg/100 g body wt for rats.
2. Fix the animal's head by a stereotactic apparatus and clean up the skin on top of the frontal and parietal regions of the skull with antimicrobial Betadine solution.
3. Make a longitudinal incision of the skin between the occiput and forehead, cut the skin in a circular manner on top of the skull, and scrape off the periosteum underneath to the temporal crests.
4. Draw a 6-mm circle over the frontal and parietal regions of the skull bilaterally, make a groove on the margin of the drawn circle using a high-speed air-turbine drill (CH4201S; Champion Dental Products, Placentia, CA) with a burr tip size (0.5 mm in diameter), separate the bone flap using a blunt microblade from the dura mater underneath, and keep the dura mater moist with physiological saline.
5. Make a nick close to the sagital sinus, pass iris microscissors through the nick, and cut the dura and arachnoid membranes completely from the surface of both hemispheres, avoiding any damage to the sagital sinus.

6. Seal the window with an 8-mm circular cover glass by adhering to the bone using a histocompatible cyanoacrylate glue.
7. Allow the animals to recover for 7 d after microsurgery and anesthesia.
8. Anesthetize the animals, cannulate the tail vein, put the animals on a polyacrylate plate, with the head fixed by plastic modeling compound, and put the plate under the microscope.
9. Observe the pial microcirculation (*see* **Subheading 3.4.1.**)

3.4.5. Data Analysis

1. Vessel diameter: In each observation, transilluminated tissue images should be recorded for 60 s and the video tapes will be analyzed off-line. The vessel diameter in μm (*D*) will be measured using an image-shearing device (digital video image shearing monitor, model 908; IPM, San Diego, CA) *(45)*.
2. The RBC velocity (V_{RBC}): V_{RBC} can be measured using the four-slit apparatus (Microflow System, model 208°C, video photometer version; IPM) equipped with a personal computer (IBM PS/2, 40SX; Computerland, Boston, MA) *(45)*. In case of high-velocity vessel (>1 mm/s), the fluorescent bead method can be used to estimate maximum blood velocity *(24)*. The distance a bead moves within a vessel in a given time interval (defined by the frequency of the camera shutter) can be determined by measuring the distance between two positions of the bead in the same track, using an image processing system (NIH Image Ver. 1.58; Macintosh IIfx, Apple Computer, Cupertino, CA). The maximum bead velocity can be calculated by dividing the maximum distance in more than 300 tracks during 1-min recording by the given time interval.
3. The mean blood flow rates of individual vessels (*Q*) and shear rate: *Q* can be calculated using *D* and the mean V_{RBC} (V_{mean}) as follows: $Q = \pi/4 \times V_{mean} \times D^2$; $V_{mean} = V_{RBC}/\alpha$ ($\alpha = 1.3$, for blood vessels < 10 μm; linear extrapolation $1.3 < \alpha < 1.6$ for blood vessels 10 and 15 μm; and $\alpha = 1.6$ for blood vessels > 15 μm) *(45)*. Shear rate will be calculated for each vessel as: Shear rate = $8 \times V_{mean}/D$ *(42)*.
4. Leukocyte–endothelial interactions (LEI): LEI can be measured by off line analysis of fluorescence images of leukocytes *(42)*. The numbers of rolling (N_r) and adhering (N_a) leukocytes are counted for 30 s along a 100-μm segment of a vessel. Rolling is defined as a short-term interaction and slow movement along vessel wall (<50% of V_{RBC}); adhesion is defined as stable binding at a certain position for more than a 30 s period. The total flux of cells for 30 s is also measured (N_t). The ratio of rolling cells to total flux (Rolling count) is calculated as follows: Rolling count (%) = $100 \times N_r/N_t$. The density of adhering leukocytes (Density) is calculated as follows: Density (cells/mm^2) = $10^6 \times N_a/(\pi \times D \times 100$ μm).

4. Notes

1. Cellular deformability:
   a. It is absolutely critical that the cells being tested do not adhere to the pipet tip during the testing procedure. Cells that rapidly adhere to the glass pipet, such as macrophages, can occlude the tip following the testing procedure, thus

preventing any additional measurements with the same pipet. This can be avoided by using a suspending medium that is free of calcium and magnesium, or by including EDTA in the testing medium.
   b. Nonviable cells and cell fragments can also be aspirated into the pipet tip, and block the fluid flow or ruin the optical clarity of the pipet tip. Thus, caution must be exercised when working with cell populations that have an overall low viability.
   c. The source of pressure in the hydraulic system must be free of oscillation. Although this is reduced with the in-line dampening chamber, it can be further suppressed by including a few drops of liquid soap into the flasks of water that provide the driving pressure for the system (thus reducing the surface tension of the fluid in these flasks).
2. Adhesion molecule expression:
   a. Even though the mounting medium contains a photobleach inhibitor, it is good practice to minimize sample illumination, so that photobleaching does not significantly affect the measurements.
   b. During the fixing and staining steps, washing should be performed as gently as possible (minimize fluid shear on the cells) to avoid damaging the adherent cells. We have found that the inclusion of divalent cations in the fixing and washing medium helps maintain an intact HUVEC monolayer.
   c. In some cases, the PI staining is nonuniform, with some nuclei staining more brightly than others. In this case, an additional PI application is necessary. It is also acceptable to include a detergent wash (after the cell fixation) to permeabilize the membranes, allowing easier access for the PI.
3. L–E interactions in vitro:
   a. After the addition of PME to deplete the monocyte population, an aggregate of DNA may form. It is necessary to remove this from the centrifuge tube and wash out as many of the trapped lymphocytes as possible with gentle agitation.
   b. Isolated T-cells should generally be stored at 4°C and used within 24 h, but have been shown to retain adhesive capabilities for as long as a week under these conditions.
   c. A VCR with jog/shuttle capabilities helps in the quantification of the video tapes. A dry-erase marker is also useful for making grids and for keeping track of bound –endothelial interactions in vivo:
   a. All surgical procedures should be done under aseptic conditions and all materials should be sterilized prior to use. If there is any sign of tissue injury, inflammation, or compromised hemodynamics, the tissue preparations should be eliminated from the observation.
   b. Ear setup is suitable for the baseline observation, since surgical preparation is not required. Mesentery set up has best optical quality and can be used for regional treatment by superfusion *(46)*.
   c. Chronic preparation can be used for long-term observation, regional treatment by continuous *(47)* or one-time superfusion *(42)*, growth factor incorporated gel assay *(48)*, and tumor study *(24,42,45,47)*. The procedure for the tumor implantation is as

follows: dorsal skin chamber; remove the coverslip and inoculate 2 μL of very dense tumor-cell suspension (approx 2×10^5 cells) or 1-mm-diameter fragment of tumor tissue onto the center of the chamber and close the chamber again using a cover slip; cranial window: before sealing the window with glass cover slip, inoculate a piece of tumor tissue (1 mm in diameter) at the center of the window.

References

1. Jain, R. K., Koenig, G. C., Dellian, M., Fukumura, D., Munn, L. L., and Melder, R. J. (1996) Leukocyte–endothelial adhesion and angiogenesis in tumors. *Cancer Metastasis Rev.* **15**, 195–204.
2. Bell, G. I. (1979) A theoretical model for adhesion between cells mediated by multivalent ligands. *Cell Biophys.* **1**, 133–147.
3. Dembo, M., Torney, D. C., Saxman, K., and Hammer, D. (1988) The reaction-limited kinetics of membrane-to-surface adhesion and detatchment. *Proc. R. Soc. Lond. B* **234**, 55–83.
4. Lawrence, M. B. and Springer, T. A. (1991) Leukocytes roll on a selectin at physiological flow rates: distinction from and prerequisite for adhesion through integrins. *Cell* **65**, 859–873.
5. Lipowsky, H. H., House, S. D., and Firrell, J. C. (1988) Leukocyte endothelium adhesion and microvascular hemodynamics, in *Vascular Endothelium in Health and Disease* (Chien, S., ed.), Plenum, New York, pp. 85–93.
6. Wattenbarger, M. R., Graves, D. J., and Lauffenburger, D. A. (1990) Specific adhesion of glycophorin liposomes to a lectin surface in shear flow. *Biophys. J.* **57**, 765–777.
7. Lipowsky, H. H., Riedel, D., and Shi, G. S. (1991) In vivo mechanical properties of leukocytes during adhesion to venular endothelium. *Biorheology* **28**, 53–64.
8. Sasaki, A., Jain, R. K., Maghazachi, A. A., Goldfarb, R. H., and Herberman, R. B. (1989) Low deformability of lymphokine-activated killer cells as a possible determinant of in vivo distribution. *Cancer Res.* **49**, 3742–3746.
9. Schmid-Schönbein, G. W. (1990) Leukocyte biophysics. *Cell. Biophys.* **17**, 107–135.
10. von Andrian, U. H., Hansell, P., Chambers, J. D., Berger, E. M., Torres-Filho, I., Butcher, E. C., and Arfors, K. E. (1992) L-selectin function is required for beta 2-integrin-mediated neutrophil adhesion at physiological shear rates in vivo. *Am. J. Physiol.* **263**, H1034–H1044.
11. von Andrian, U. H., Chambers, J. D., McEvoy, L. M., Bargatze, R. F., Arfors, K.-E., and Butcher, E. C. (1991) Two-step model of leukocyte-endothelial cell interaction in inflammation: distinct roles for LECAM-1 and the leukocyte β_2 integrins in vivo. *Proc. Natl. Acad. Sci. USA* **88**, 7538–7542.
12. Melder, R. J., Munn, L. L., Yamada, S., Ohkubo, C., and Jain, R. N. (1995) Selectin- and integrin-mediated T-lymphocyte rolling and arrest on TNF-α-activated endothelium: augmentation by erythorcytes. *Biophys. J.* **69**, 2131–2138.
13. Springer, T. A., Anderson, D. A., Rosenthal, A. S., and Rothlein, R., eds. (1988) *Leukocyte Adhesion Molecules: Structure, Function, and Regulation.* Springer-Verlag, New York.
14. Munn, L. L., Koenig, G. C., Jain, R. K., and Melder, R. J. (1995) Kinetics of adhesion molecule expression and spatial organization using targeted sampling fluorometry. *Biotechniques* **19**, 622–631.

15. Melder, R. J. and Jain, R. K. (1994) Reduction of rigidity in human activated natural killer cells by thioglycollate treatment. *J. Immunol. Methods* **175**, 69–77.
16. Sasaki, A., Melder, R. J., Whiteside, T. L., Herberman, R. B., and Jain, R. K. (1991) Preferential localization of human adherent lymphokine-activated killer (A-LAK) cells in tumor microcirculation. *J. Natl. Cancer Inst.* **83**, 433–437.
17. Betticher, D. C., Keller, H., Maly, F. E., and Reinhart, W. H. (1993) The effect of endotoxin and tumor necrosis factor on erythrocyte and leucocyte deformability in vitro. *Br. J. Haematol.* **83**, 130–137.
18. Welch, D. R., Lobl, T. J., Seftor, E. A., Wack, P. J., Aeed, P. A., Yohem, K. H., Seftor, R. E., and Hendrix, M. J. (1989) Use of the membrane invasion culture system (MICS) as a screen for anti-invasive agents. *Int. J. Cancer* **43**, 449–457.
19. Downey, G. P., Doherty, D. E. Schwab, B. D., Elson, E. L., Henson, P. M., and Worthen, G. S. (1990) Retention of leukocytes in capillaries: role of cell size and deformability. *J. Appl. Physiol.* **69**, 1767–1778.
20. Traykov, T. T. and Jain, R. K. (1987) Effect of glucose and galactose on red blood cell membrane deformability. *Int. J. Microcirc. Clin. Exp.* **6**, 35–44.
21. Melder, R. and Jain, R. (1992) Kinetics of interleukin 2 induced changes in rigidity of human natural killer cells. *Cell Biophys.* **20**, 161–176.
22. Sasaki, A., Jain, R. K., Maghazachi, A. A., Goldfarb, R. H., and Heberman, R. B. (1989) Low deformability of lymphokine-activated killer cells as a possible determinant of in vivo distribution. *Cancer Res.* **49**, 3742–3746.
23. Krenik, K. D., Kephart, G. M., Offord, K. P., Dunnette, S. L., and Gleich, G. J. (1989) Comparison of antifading agents used in immunofluorescence. *J. Immunol. Methods* **177**, 91–97.
24. Yuan, F., Salehi, H. A., Boucher, Y., Vasthare, U. S., Tuma, R. F., and Jain, R. K. (1994) Vascular permeability and microcirculation of gliomas and mammary carcinomas transplanted in rat and mouse cranial window. *Cancer Res.* **54**, 4564–4568.
25. Gallik, S., Usami, S., Jan, K.-M., and Chien, S. (1989) Shear stress-induced detachment of human polymorphonuclear leukocytes from endothelial cell monolayers. *Biorheology* **26**, 823–834.
26. Hochmuth, R. M., Mohandas, N., Spaeth, E. E., Williamson, J. R., Blackshear, P. L. and Johnson, D. W. (1972) Surface adhesion, deformation and detachment at low shear of red cells and white cells. *Trans. Am. Soc. Artif. Organs* **18**, 325–332.
27. Hubbell, J. A. and McIntire, L. V. (1986) Visualization and analysis of mural thrombogenesis on collegen, polyurethane and nylon. *Biomaterials* **7**, 354–363.
28. Lawrence, M. B., McIntire, L. V., and Eskin, S. G. (1987) Effect of flow on polymorphonuclear leukocyte/endothelial cell adhesion. *Blood* **70**, 1284–1290.
29. Munn, L. L., Melder, R. J., and Jain, R. K. (1996) Role of erythrocytes in leukocyte–endothelial interactions: mathematical model and experimental validation. *Biophys. J.* **71**, 466–478.
30. Leung, K. H. (1989) Human lymphokine-activated killer (LAK) cells I. Depletion of monocytes from peripheral blood mononuclear cells by L-phenylalanine methyl ester and optimization of LAK cell generation at high density. *Cancer Immunol. Immunother.* **30**, 247–254.

31. Weston, S. A. and Parish, C. R. (1990) New fluorescent dyes for lymphocyte migration studies: analysis by flow cytometry and fluorescence microscopy. *J. Immunol. Methods* **133,** 87–97.
32. Weston, S. A. and Parish, C. R. (1992) Calcein: a novel marker for lymphocytes which enter lymph nodes. Cytometry 13, 739–749.
33. Menter, D. G., Patton, J. T., Updyke, T. V., Kerbel, R. S., Maamer, M., McIntire, L. V., and Nicolson, G. L. (1992) Transglutaminase stabilizes melanoma adhesion under laminar flow. *Cell Biophys.* **18,** 123–143.
34. Melder, R. J., Koenig, G., Munn, L. L., and Jain, R. K. (1997) Adhesion of activated natural killer cells to TNFα-treated endothelium under physiological flow conditions. *Nat. Immun.* **15,** 154–163.
35. Munn, L. L., Melder, R. J., and Jain, R. K. (1994) Analysis of cell flux in the parallel plate flow chamber: implications for cell capture studies. *Biophys. J.* **67,** 889–895.
36. Schmid-Schönbein, G. W., Fung, Y.-C., and Zweifach, B. W. (1975) Vascular endothelium–leukocyte interaction. Sticking shear force in venules. *Circ. Res.* **36,** 173–184.
37. Fukumura, D., Yuan, F., Monsky, W., Chen, Y., and Jain, R. K. (1997) Effect of host microenvironment on the microcirculation of human colon adenocarcinoma. *Am. J. Pathol.* **150,** 679–688.
38. Ohkubo, C., Bigos, D., and Jain, R. K. (1991) IL-2 induced leukocyte adhesion to the normal and tumor microvascular endothelium in vivo and its inhibition by dextran sulfate: implications for vascular leak syndrome. *Cancer Res.* **51,** 1561–1563.
39. Nugent, L. J. and Jain, R. K. (1984) Extravascular diffusion in normal and neoplastic tissues. *Cancer Res.* **44,** 238–244.
40. Zawicki, D. F., Jain, R. K., Schmid-Schönbein, G. W., and Chien, S. (1981) Dynamics of neovascularization in normal tissues. *Microvasc. Res.* **21,** 37–47.
41. Dudar, T. E. and Jain, R. K. (1983) Microcirculatory flow changes during tissue growth. *Microvasc. Res.* **25,** 1–21.
42. Fukumura, D., Salehi, H. A., Witwer, B., Tuma, R. F., Melder, R. J., and Jain, R. K. (1995) Tumor necrosis factor alpha-induced leukocyte adhesion in normal and tumor vessels: effect of tumor type, transplantation site, and host strain. *Cancer Res.* **55,** 4824–4829.
43. Melder, R. J., Salehi, H. A., and Jain, R. K. (1995) Interaction of activated natuiral killer cells with normal and tumor vessels in cranial windows in mice. *Microvasc. Res.* **50,** 35–44.
44. Yamada, S., Mayadas, T. N., Uan, F., Wagner, D. D., Hynes, R. O., Melder, R. J., and Jain, R. K. (1995) Rolling in P-selectin-deficient mice is reduced but not eliminated in the dorsal skin. *Blood* **86,** 3487–3492.
45. Leunig, M. Yuan, F., Menger, M. D., Boucher, Y., Goetz, A. E., Messmer, K., and Jain, R. K. (1992) Angiogenesis, microvascular architechture, microhemodynamics, and interstitial fluid pressure during early growth of human adenocarcinoma LS174T in SCID mice. *Cancer Res.* **52,** 6553–6560.
46. Kurose, I., Fukumura, F., Miura, S., Suematsu, M., Sekizuka, E., Nagata, H., and Tsuchiya, M. (1993) Nitric oxide mediates vasoactive effects of endothelin-3 on rat mesenteric microvascular beds *in vivo. Angiology* **44,** 483–490.

47. Fukumura, F., Yuan, F., Endo, M., and Jain, R. K. (1997) Role of nitric oxide in tumor microcirculation: blood flow, vascular permeability, and leukocyte-endothelial interactions. *Am. J. Pathol.* **150,** 713–725.
48. Dellian, M., Witwer, B. P., Salehi, H. A., Yuan, F., and Jain, R. K. (1996) Quantitation and physiological characterization of angiogenic vessels in mice: effect of basic fibrooblast growth factor, vascular endothelial growth factor/vascular permeability factor, and host microenvironment. *Am. J. Pathol.* **149,** 59–72.

42

Quantitative Measurement of Shear-Stress Effects on Endothelial Cells

Maria Papadaki and Larry V. McIntire

1. Introduction

Over the past 20 yr, great strides have been made toward understanding the role of fluid hemodynamic forces in the vascular wall homeostasis at the molecular level. In vivo studies have demonstrated that blood vessels are adaptive to physiological changes in blood flow, with vessels tending to enlarge in areas of high flow and tending to reduce their lumen diameter in low-flow regimes *(1,2)*. Furthermore, altered hemodynamics have been implicated in the pathogenesis of many cardiovascular disorders, such as thrombosis, atherosclerosis, and vessel wall injury. Vascular endothelial cells serve as a barrier between perfused tissues and flowing blood, and they are believed to act as a sensor of the local biomechanical environment. The hemodynamic forces generated in the vasculature include frictional wall shear-stress, cyclic strain, and hydrostatic pressure *(3)*. For the purpose of this chapter, we will focus on methods for examining the link between fluid wall shear-stress and endothelial cell function. Advances in our understanding of the effects of shear-stress on endothelial cell function require that cell populations be exposed to controlled, well-defined, flow-induced shear-stress environments. Since in vivo studies have the inherent problem that they cannot quantitatively define the shearing forces or separate their effects from the other components of the hemodynamic system, in vitro flow studies using cultured cells are extensively used.

The objective of this chapter is to review the methodologies and the flow chambers used to expose cultured endothelial cells to fluid shear-stress. There are two major ways of introducing shear-stress in cultured cells: those that

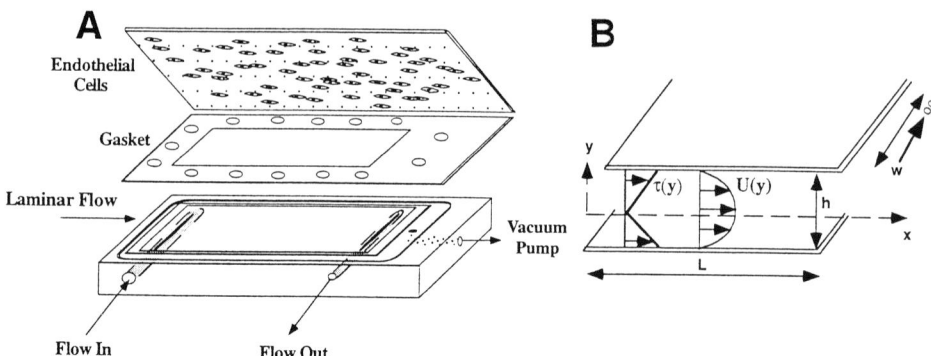

Fig. 1. **(A)** Rectangular parallel-plate flow chamber. The glass slide, the silicon gasket, and the polycarbonate plate are held together by a vacuum that is maintained at the periphery of the slide. **(B)** Velocity and shear-stress profile for flow between parallel-plates. The velocity profile is parabolic and shear-stress changes linearly with y. The wall shear-stress is constant (except very near the lateral edges); w is the channel width, h is the channel height, and L is the channel length.

produce uniform shear-stress throughout the fluid phase, and those in which shear-stress changes with the distance from the solid–liquid interphase *(4–6)*. Some of the commonly used instruments, together with the advantages and experimental problems associated with each one of them, are described in the following pages. In order to illustrate the fluid dynamics within those devices, several simple designs are considered in detail; devices with more complex geometries are briefly mentioned.

2. Methods
2.1. Parallel-Plate Flow Chamber

The parallel-plate flow chamber is the most widely used instrument for studying the effects of fluid shear-stress on endothelial cells (or any other kind of anchorage-dependent cells) *(6, 7)*. A schematic of the most common parallel-plate flow chamber geometry is shown in **Fig. 1**. This parallel-plate flow chamber consists of a rectangular polycarbonate plate, a silicon gasket, and a glass slide *(8)*. Endothelial cells are cultured on the glass slide, which can be coated with a variety of adhesive substrates. A vacuum forms a seal to provide a channel of parallel geometry, and to ensure uniform channel height, h. The polycarbonate plate has entrance and exit manifolds for the tissue-culture medium machined into the plastic block, as well as a vacuum-sealing ring. In other designs of the parallel-plate flow chamber the vacuum is replaced by evenly torqued screws. For short-term experiments tissue-culture medium is perfused through the chamber using a syringe pump; for long-term experiments, flow is produced by a constant hydrostatic pressure head, which is created by the vertical distance between two reservoirs *(8)*. Continuous pumping of the fluid medium from the lower to the upper reservoir is employed, so that there is an overflow of excess medium into the lower reservoir. The overflow serves to maintain a constant hydrostatic pressure head, to prevent air bubble entry into the flow chamber, and to provide good mass transfer for medium oxygenation. The wall shear-stress, τ_w, on the cell monolayer for a Newtonian fluid is calculated using the momentum balance:

$$\tau_w = \frac{6\mu Q}{wh^2} \qquad (1)$$

where Q is volumetric flow rate, μ is the viscosity of the fluid/tissue-culture medium, and w is the channel width.

The parallel-plate approximations are only valid if the height of the flow channel is much smaller than the width and length of the channel, L; $h/w \ll 1$, $h/L \ll 1$ *(5,6)*. In practice, a height:width ratio value of at least 1:50 is used. In order to ensure uniform velocity across the channel width, the reservoir diameter, from which the medium is discharged, should be large enough so that the pressure drop along the parallel-plates is larger than the pressure drop in the reservoirs and the kinetic energy of the inlet stream. The entrance length, L_e, must be small compared to the chamber length, in order for **Eq. 1** to be valid for nearly all the cell monolayer except the extreme lateral edges of the flow chamber. The entrance length is given by *(6,8)*:

$$L_e = 0.04\, h\, \mathrm{Re} = 0.04\, h\, \left(\frac{\tau_w h^2 \rho}{6\mu^2}\right) \qquad (2)$$

where Re is the Reynolds number; and ρ is the fluid density. For the low Reynolds numbers normally generated, the entrance length is approx 400× smaller than the length of the flow chamber.

The advantages of the parallel-plate flow chamber are:

1. It makes possible study of the effects of constant shear-stress on endothelial cells over a defined time-period.
2. The device is simple in design, assembly, and operation
3. The endothelial cells can be grown under flow conditions, and can be observed under a microscope, or visualized in real time, utilizing video microscopy
4. This type of chamber is preferable for studies on the metabolism of endothelial cells, in which a large number of cells needs to be harvested for measurement of mRNA
5. It permits continuous sampling of the incubation medium for secreted metabolites *(8)*.

The parallel-plate flow chamber, in its original design, is capable of producing well-defined wall shear-stress in the physiological range of 0.01–30 dyn/cm^2.

One disadvantage is that it is difficult to discriminate between effects caused by shear-stress and those caused by the hydrostatic pressure, since both of them increase in a manner proportional to the flow *(6,9)*. Furthermore, the parallel-plate flow chamber cannot effectively be used to study effects of shear-stress on suspended cells, because shear-stress varies over the cross section of the chamber, and sedimentation and aggregation of suspended cells can take place.

A variation of the rectangular parallel-plate flow chamber is the use of dual parallel flow paths employing a round tissue-culture dish as the substrate for endothelial cell growth *(10)*. Those chambers are commonly used to study the various receptor–ligand interactions in cell–cell and cell–substrate adhesion under defined hydrodynamic conditions. The area of the flow section in these round chambers is approx 16× smaller than the rectangular chamber, which makes them ideal for studies when cell numbers are limited. In addition, these round chambers are more convenient, since cells are cultured in plastic 35-mm tissue-culture dishes instead on glass slides.

Other investigators have developed designs in which the parallel-plate approximations are valid even at relatively high Reynolds numbers. Koslow et al. used a pair of parallel-plates with a divergent entrance and a convergent exit *(11)*. At the divergent portion of the chamber, flow was fluctuating, but at the central area, where cells were placed (on a cover slip), flow was laminar. The maximum shear rate studied was 1590/s (Re = 138). Viggers et al. used a parallel channel that combined bell-shaped entrance with baffles, which helped to establish a constant fluid velocity at the beginning of the laminar flow section *(12)*. Those same researchers used three-dimensional finite element numerical analysis to calculate wall shear-stress, and they found that the cell cover slip can examine shear-stresses over 200 dyn/cm^2, and that the velocity gradient is uniform across the central 80% of the cover slip.

Eskin et al. subjected bovine endothelial cells, seeded inside a square glass capillary tube, to flow, by connecting the tube into a closed loop consisting of two reservoirs and a pump *(13)*. The analytical solution for the velocity profile for pressure-driven low-Reynolds-number flow of a Newtonian fluid in a square tube is more complicated than the one for the parallel-plate flow chamber *(14)*. Similarly, Cooke et al. used six rectangular tissue-culture microcappilary tubes per experiment to examine endothelial cell growth, and to perform adhesion studies under well-defined flow conditions *(15)*.

2.2. Cone and Plate

The cone-and-plate device is a well-defined rheological system that was initially introduced for measuring fluid shear-stress effects on suspended cells *(5,6,16)*. More recently, this assembly has been used to study flow effects on anchorage-dependent cells. As shown in **Fig. 2**, the bottom plate is not moving and can accommodate several cover slips *(16)*. Shear-stress is produced on

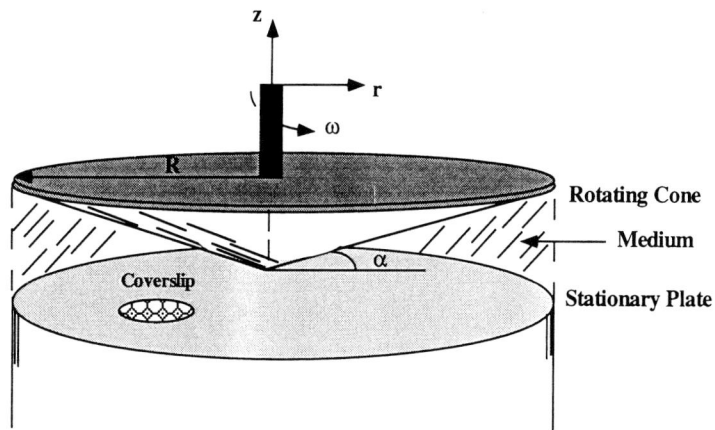

Fig. 2. Cone-and-plate device. Endothelial cells or any other kind of anchorage dependent cell are cultured on cover slips, which then are inserted on the stationary plate. For very small cone angles ($\alpha \ll 1$) and low-rotation speeds ω, shear-stress is uniform throughout the fluid phase between the cone and the plate. R is the cone radius.

cells seeded on the cover slips by rotation of the cone, which forces the fluid between the cone and the plate to flow. The cone is driven by a motor through magnetic coupling. By choosing small cone angle, α (between 0.5 and three degrees), and constant, slow rotation speed, ω, the shear-stress is nearly independent of the position, except very close to the cone edge, and the velocity profile is linear. The fluid mechanical solution for this type of flow is simple, and yields, for a Newtonian fluid, a shear-stress of *(16)*:

$$\tau = \tau_w = \mu \frac{\omega}{\alpha} \quad (3)$$

The approximations of the model consider that the fluid is in intimate contact with the cone and the stationary plate, and the neglect edge effects at the outer rim of the rotating cone are negligible. A modified Reynolds number $\tilde{R}$ is used as a measure of the centrifugal forces acting on the moving fluid.

$$\tilde{R} = \frac{r^2 \omega \alpha^2}{12 \nu} \quad (4)$$

where ν is the kinematic viscosity. For $\tilde{R} < 1$, the flow is laminar, but secondary flows appear when $\tilde{R} > 1$. It is also essential that the cone axis is perpendicular to the plate; otherwise, pulsatile flow will result. Furthermore, the fluid should fill only the space between the cone and plate, otherwise an additional torque from fluid shearing can be developed *(6)*.

The main advantages of the cone-and-plate are summarized here:

1. There are no entrance and exit lengths, unlike the parallel-plate flow chamber.
2. The laminar homogeneous flow field does not depend on a hydrostatic pressure gradient.
3. Small changes in dimensions do not have a large effect on wall shear-stress.
4. Shear-stress effects on several cover slips can be examined in the same experiment.
5. The operating fluid volume is small *(16)*.

The cone-and-plate apparatus is capable of producing laminar shear-stresses between 0.01 and 100 dyn/cm².

As originally designed, the cone and plate device has a small cell:volume ratio, does not permit continuous sampling of the conditioned medium, has significant medium evaporation, which requires continuous infusion of fresh medium, and does not easily allow continuous visualization of cells during shear-stress application *(5,6,8)*.

To allow direct observation of anchorage-dependent cells, Schnittler et al. have developed a modified cone-and-plate viscometer in which the rotating cone is transparent *(9)*. Cells are visualized by phase-contrast or fluorescent optics, and the whole system is connected to a computer-controlled linear stage, in order to allow automatic recording at any point of the cell cultures. This improved rheological device consists of 12 independent units (chambers), so that a number of different conditions can be tested simultaneously during a single experiment. Another modification of the cone-and-plate device is the rheoscope, in which the cone and plate counterrotate *(17,18)*. The counter rotation of the cone and plate has the advantage that a particle at the center line between the cone and plate remains stationary to the observer and can be studied without the aid of high-speed cinematography. This device was developed to study rheological properties of blood components. If ω_{cone} and ω_{plate} are the rotation speeds of the cone and plate, respectively, then the shear-stress in the apparatus, assuming Newtonian fluid, is computed as:

$$\tau = \tau_w = \mu \frac{\left(\omega_{cone} + \omega_{plate}\right)}{2\alpha} \tag{5}$$

By increasing the angular velocity of the cone and increasing the angle between the cone and the plate, turbulent flow can be achieved with the cone-and-plate device *(19)*. Turbulent shear-stresses, achieved with this device, are on the order of 2–200 dyn/cm². Turbulence appears first at the outer boundary and spreads inward towards the cone apex; shear-stress is no longer constant across the gap. Davies et al. studied the effects of turbulent flow on endothelial cells, using a cone angle of 5 degrees. Shear-stress fluctuations were limited to 25% of the mean value *(19)*.

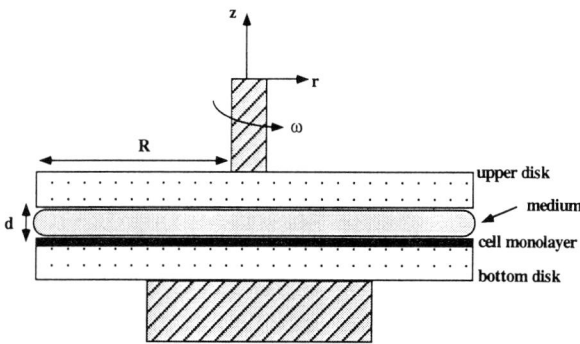

Fig. 3. Parallel-disk apparatus. Cells are growing on the bottom disk; rotation of the top disk causes the medium to flow concentrically and to produce shear-stress on cells. Shear-stress is a function of r and decreases from the perimeter to the center of the bottom disk.

2.3. Parallel-Disk Apparatus

Parallel-disk systems have the advantage, over the aforementioned systems, that if the flow pattern is well defined, shear-stress gradients can be examined on anchorage-dependent cells *(6,20,21)*. Because of the time-dependent nature of many biological processes, significant information can be obtained if cultured cells are submitted to a continuous range of shear forces simultaneously. A concurrent problem, however, is that, since cells *see* different shear-stress histories, evaluation of shear-stress effects on secreted products or intracellular messenger RNA levels is not possible. The operation of the parallel-disk system, a schematic diagram of which is shown in **Fig. 3**, is very similar to the cone-and-plate device *(20)*. Rotation of the upper disk with a constant angular velocity, ω, forces the medium between the two parallel disks to flow circumferentially, and produces a fluid shear-stress on the cells growing at the bottom disk. For a Newtonian fluid, the shear-stress is calculated as:

$$\tau_w = \mu \frac{\omega r}{d} \qquad (6)$$

where d is the vertical distance between the two disks, and r is the distance from the center of the culture dish. The model approximations include neglecting the effects of the centrifugal force in the rotating fluid, and, as a consequence, the radial and axial secondary velocity components are zero. These assumptions are valid only when the fluid velocity, as well as the Reynolds number, are small (for most applications R_e is less than 10):

$$R_e = \frac{\omega r^2}{v} \qquad (7)$$

As indicated from Eq. 6, the level of shear-stress increases from 0 at the center of the bottom disk ($r = 0$), to a maximum at the perimeter ($r = R$). The maximum reported wall shear-stress on endothelial cells growing at the bottom disk is in the range of 7–15 dyn/cm^2 *(20)*.

Parallel-disk devices are not widely used, because they have the same experimental problems as the cone-and-plate devices, and, in addition, the bulk phase shear-stress is not constant throughout the gap (d), even at low Reynolds numbers *(6)*. Furthermore, this device cannot be effectively used to study the effects of pathophysiological levels of shear-stress on endothelial cells. From **Eqs. 6** and **7**, it is possible by reducing the gap, d, to elevate shear-stress without increasing Reynolds number. However, with small gap size, nutrient supply is limited and viability problems are observed *(20)*. When the upper rotating cylinder is placed in a cylindrical container, the instrument is called disk-and-cylinder device.

One modification of the parallel-disk device is the spinning disk apparatus *(22–24)*. In this system, both disks are placed inside a cylindrical chamber full of medium buffer. The solutions for the velocity field are derived using the boundary layer theory, and the shear-stress at any position on the lower disk surface is given as *(23)*:

$$\tau_w = \left[\tau_r^2 + \tau_\phi^2 \right]^{1/2} \bigg|_{z=0} = 0.8 \mu r \left[\frac{\omega^3}{\upsilon} \right]^{1/2} \tag{8}$$

where υ is the velocity component in the circumferential direction, and τ_r, τ_ϕ are the radial and the circumferential shear-stress components. Assumptions of the model include infinite medium, infinite disk size, laminar flow, and steady-state shear field *(22,23)*. It is necessary that the boundary layer thickness be much smaller than the disk radius, in order for the edge effects to be negligible. In most applications, the boundary layer thickness is 1/100 the disk radius. Furthermore, it has been calculated that the steady state flow regime is achieved within a few revolutions of the disk. With the spinning disk system, changes in cell adhesion, cell attachment, and cell morphology in response to shear-stress have been studied *(22–24)*.

2.4. Radial Flow Device

The radial flow chamber is designed to overcome some of the mechanical problems that were encountered with the spinning disk system *(25)*. This device produces an axisymmetric flow field, providing a large shear-stress range within a given experiment *(25–27)*. As shown in **Fig. 4**, the radial flow chamber consists of two parallel stationary disks, the bottom one with a central inlet pipe, separated by a narrow spacing *(25,28)*. Tissue-culture medium is pumped at constant volumetric flow rate through the central pipe, and flows radially out

Stress Effects on Endothelial Cells

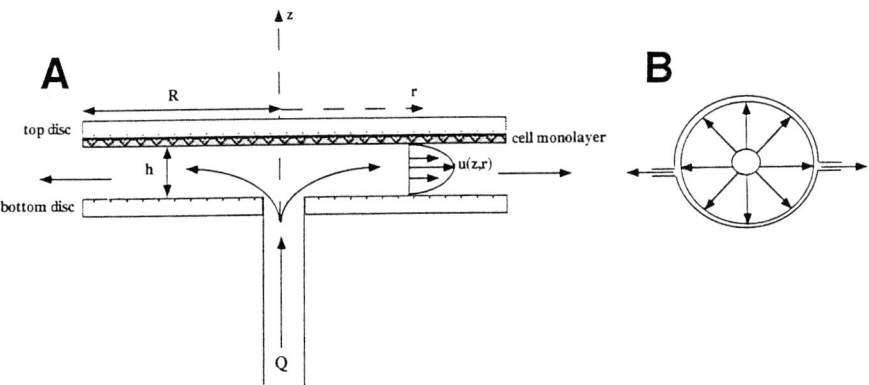

Fig. 4. **(A)** Side view of a radial flow chamber. Cells, bacteria or ligand are placed on the top disk and fluid is pumped through a central pipe. Shear-stress decreases with distance from the center of the disk. h is the distance between the two parallel disks. **(B)** Top view of the flow field. Flow cross-sectional area increases with radius.

to a collection manifold. If disks of transparent material are used, the test surface can be observed microscopically *(26)*. Fluid is restricted to flow only in the radial direction, which allows the derivation of simple equations for shear-stress gradients across the test surface. Assuming laminar axisymmetric flow and neglecting entry effects, the surface shear-stress and a local Reynolds number can be calculated as *(25)*:

$$\tau_w = \frac{3Q\mu}{\pi r h^2} \quad , \quad \mathrm{Re}_{local} = \frac{Q}{\pi r \nu} \tag{9}$$

As the cross-sectional area for flow between the two disks increases radially, shear-stress decreases radially with distance from the center. Flow may be turbulent at some small radii (except at the stagnation point), showing local Reynolds numbers above 2000, so care must be taken in the selection of the volumetric flow rates *(25)*. Most radial flow chambers are operated in the laminar flow regime ($\mathrm{Re}_{local} < 2000$, for $r > 0.05$ mm).

One advantage of this system is that operation is achieved without moving parts, which decreases problems associated with sterility, and greatly simplifies continuous monitoring of the cell responses *(25)*. This device has been mostly used to study cell adhesion/detachment to/from surfaces, and ligand–receptor bond affinity *(26,28–31)*. The principle of operation is that particles (cells, bacteria, ligand) are swept away within a circular zone around the pipe inlet, because of high shear-stress, but at the outer zone, where shear forces are lower, particles remain adherent to the test surface *(26)*. A critical radius, r_c, at the boundary between those two zones, defines the position where the particle–surface adhesive force is equal to the force exerted by the fluid to the particle. The shear-stress range examined with this device is between 0 and 70 dyn/cm^2 *(26,28)*.

A common problem of the radial flow chamber is a central area of turbulence, around the inlet, because of a sharp-edged boundary. The area and the pattern of turbulence can be manipulated by modifying the geometry of the inlet pipe/plate *(25)*. An easy alternative is to adjust the volumetric flow rate so that the critical radius is well outside the turbulent zone.

2.5. Linear Shear-Stress Flow Chamber

The linear shear-stress or tapered parallel-plate flow chamber is a modification of the parallel-plate flow chamber that is designed to generate a linear variation of shear-stress within the same flow field, without changing the flow rate or the gap height *(4,32)*. As shown in **Fig. 5**, the difference between the two flow systems is in the gasket design, which, for the linear shear-stress chamber, gives a parallel-plate flow geometry with varying width *(32)*. The idea for this design came from the fact that the velocity field of steady flow of a Newtonian fluid between two parallel-plates (also known as Hele-Shaw flow) is equivalent to the velocity field of a stagnation flow. The properties of a two-dimensional Hele-Shaw stagnation flow permit the design of a flow channel, so that the wall shear-stress is linearly distributed along the center line of the channel. The entrance end of the flow channel can be approximated by a straight line; the exit end can be formed by two perpendicular straight lines through the center of the channel. Thus, the shear-stress along the center line is calculated as *(32)*:

$$\tau_w = \frac{6\mu Q}{h^2 w_1}\left(1 - \frac{z}{L}\right) \quad (10)$$

where h is the channel height, w_1 is the entrance width, and L is the total length of the channel.

This simple flow-channel design gives a constant shear-stress gradient over the entire length of the chamber. The maximum value of shear-stress is at the entrance, and falls to zero at the exit. At the wall, and near the exit of the channel, shear-stress does not linearly decrease with distance from the entrance. Usami et al. used the linear shear-stress flow chamber to study cell–cell interactions over a wide range of shear-stress (0–45 dyn/cm^2) in a single experimental run. Other forms of a flow channel can be designed, in which the shear-stress will not be zero at the exit, by selecting a different part of the two-dimensional stagnation flow *(32)*.

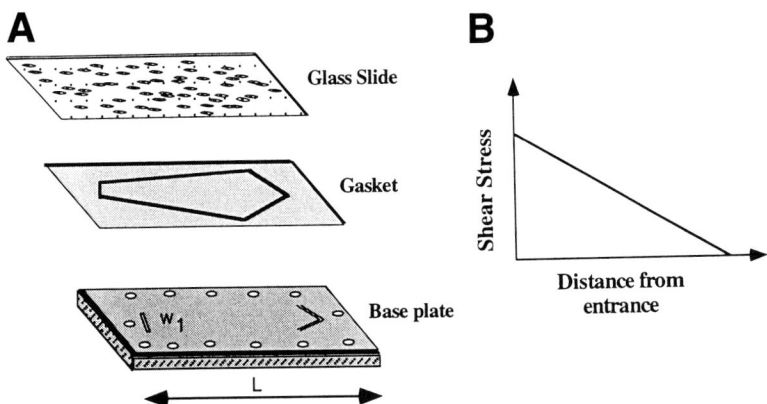

Fig. 5. **(A)** Linear shear-stress flow chamber. The only difference with the parallel-plate flow chamber is in the gasket geometry. The width between the two parallel-plates changes from the entrance to exit of the flow chamber. w_1 is the entrance width and L is the length of the chamber. **(B)** For this flow channel design, shear-stress changes from a maximum value at the entrance to zero at the exit along the center line.

2.6. Cylindrical Tube

Cylindrical tube devices have been used to study fluid-surface interactions important in thrombosis *(6,33–35)*. As indicated in **Fig. 6**, cylindrical tubes have many similarities with the parallel-plate flow chamber. For a fully developed, laminar, viscous flow in tubes, the shear-stress along the tube wall and the Reynolds number are expressed as *(6)*:

$$\tau_w = \frac{4Q\mu}{\pi R^3}, \quad \text{Re} = \frac{2Q}{\pi R \nu} \tag{11}$$

where R is the tube radius. The velocity profile is parabolic and the shear-stress is zero along the centerline and maximum along the walls. From **Eq. 11**, for a 110-μm radius R, the critical Reynolds number above which turbulent flow occurs is 2300, and the corresponding laminar shear-stress is 4600 dyn/cm². In general, higher shear-stresses can be produced in a cylindrical tube than in the parallel-plate flow chamber, but the tube must be longer than the parallel-plates *(6)*.

The cylindrical tube systems have advantages and disadvantages similar to the parallel-plate devices, since the bulk fluid shear rate and stress vary over the cross-section. At the ends of the tube nonuniform flow areas exist, because fluid flows from a reservoir to the tube, and vise versa. One of the main problems of the tube device is that microscopic monitoring is difficult because of the optical distortion caused by the curved surfaces *(6)*. This problem can be

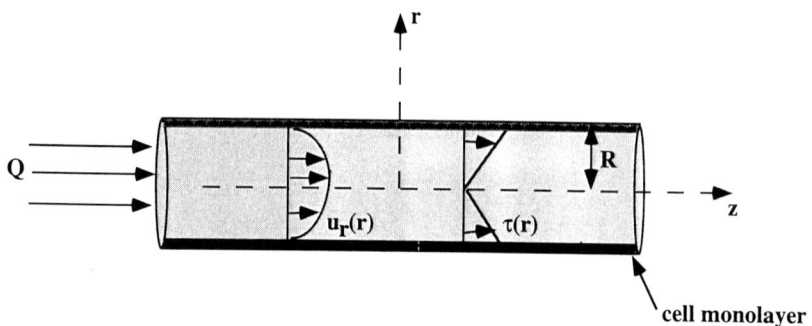

Fig. 6. Cylindrical tube. The wall velocity profile is parabolic; shear-stress varies linearly with r. Cells can be grown on the side walls of the cylinder.

partially solved by embedding the tube in a water-filled chamber. In addition, generation of confluent cell monolayers on small-diameter, closed cylindrical surfaces can be difficult.

Redmond et al. have developed a perfused transcapillary endothelial cell–smooth muscle cell co-culture system, marketed by Cellco, which mimics the architecture of the vascular wall *(35)*. They used a bundle of 230 semipermeable capillaries (cylindrical tubes with diameter 330 µm), in which endothelial cells and smooth muscle cells were seeded in the luminal and extracapillary space of the tubes, respectively. Badimon et al., developed a tubular perfusion chamber to investigate the effects of blood flow on biological materials *(33,34)*. This cylindrical chamber is flexible and can accept a variety of test surfaces (such as aortic subendothelium, collagen, and so on). The geometrical configuration of this device differs moderately from the geometry of the cylindrical tube.

2.7. Annular Flow Chamber

The annular flow chamber has been used to expose endothelium denuded vessel segments to flowing blood *(36–39)*. As shown in **Fig. 7**, inverted vessel segments (up to three) are mounted on the core of the chamber, and blood is passing through the annular spacing. Assuming laminar flow of a Newtonian fluid in an annulus, the shear rate at the vessel surface can be calculated as *(39)*:

$$\tau_w = \frac{32 g(k) Q}{\pi (d_o - d_i)^3} \ , \ g(k) = \frac{(1-k)^2}{2k(1+k)} \left[\frac{1 - k^2 - 2k^2 \ln(1/k)}{(1+k^2)\ln(1/k) + k^2 - 1} \right] \quad (12)$$

where d_0, d_i are the outside and inside diameters, respectively, $k = d_i/d_o$, and $g(k)$ is a constant. When k goes to zero, this equation reduces to the circular tube equation. As with the circular tube, the major disadvantage of the annular

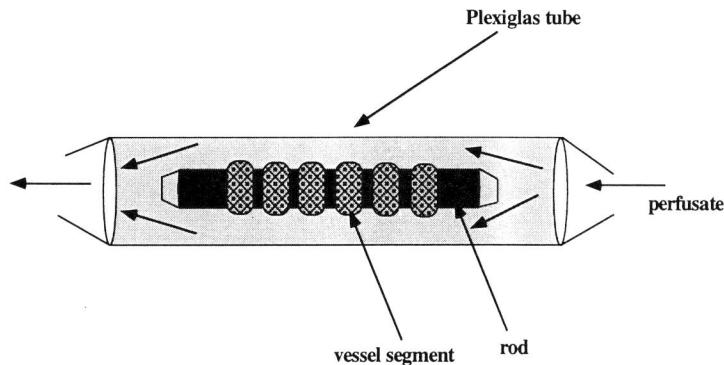

Fig. 7. Annular flow chamber. Inverted vessel segments are drawn on a rod that is placed inside a Plexiglas tube. Fluid (usually blood) enters the Plexiglas tube, and then streams along the rod, forming an annulus.

chamber is the difficulty in microscopically observing the cellular processes. Additionally, this device requires a preselection of vessels with certain dimensions, and is not easily adapted for the use of non vascular materials *(33)*. Studies with the annular chamber have been very important in developing an understanding of thrombosis, hemostasis, and the causes of various bleeding diseases.

2.8. Flow Chambers with Variable Degrees of Stenosis (or Expansion)

Devices that model flow through a stenosis (or expansion) are designed to study the response of vascular cells to changes in local fluid-dynamic conditions *(40–44)*. Such studies are very important, since disturbances in the blood flow (in areas of high shear-stress, flow stagnation, recirculation, and turbulence) can significantly contribute to processes such as platelet activation and fibrin deposition. Furthermore, near flow dividers and bifurcations, the orientation pattern of endothelial cells often changes. Flow chambers with stenosis (or expansion) are modifications of the parallel-plate flow chamber or the cylindrical tube, and contain a recess in the flow channel *(4)*. A major drawback of these devices is that they may require sophisticated computational fluid mechanics packages for the solution of the Navier-Stokes that describe fluid motion in the flow chamber.

Truskey et al. studied the effects of flow separation and recirculation on cultured endothelial cells, with the use of a sudden asymmetric expansion parallel-plate flow chamber *(42)*. Shear-stresses on the order of 80 dyn/cm^2 and shear-stress gradients as high as 2500 (dyn/cm^2)/cm, were produced in the recirculation zone of this chamber. Barstad et al. developed a parallel-plate perfusion chamber with an eccentric stenosis, to investigate thrombus formation at the apex of the stenosis *(44)*. In order to allow a gradual, rather than a step, variation in cross sectional area of the flow chamber, those investigators introduced the stenosis as an 18-mm-long planar surface with a 0.5-mm, cosine-shaped step.

3. Concluding Remarks

In this chapter, we have described some of the more popular devices used to quantitatively measure shear-stress effects on endothelial cells, and to address the experimental problems associated with them. Some of the flow chambers described here have been used principally to investigate rheological properties of blood cells, or adhesion of flowing leukocytes and platelets to surfaces. However, with minor design modifications, these devices can also be effectively used to study flow effects on endothelial cells or any other type of anchorage dependent cell. The exponentially increasing interest in understanding how fluid shear-stress regulates the structure and function of vascular cells makes necessary the development and design of improved flow-chamber technologies.

Acknowledgments

This work was supported by NIH grants HL18672 and NS23326, NASA grant NAGW-5007, Welch Foundation grant C-0938 and TATP grant 003604.

References

1. Resnick, N. and Gimbrone, M. A. (1995) Hemodynamic forces are complex regulators of endothelial gene expression. *FASEB J.* **9,** 874–882.
2. Davies, P. F. (1995) Flow-mediated endothelial mechanotransduction. *Physiol. Rev.* **75,** 519–560.
3. Patrick, C. W., Sampath, R., and McIntire, L. V. (1995) Fluid shear stress effects on vascular function, in *Biomedical Engineering Handbook* (Bronzino, J. D., ed.), CRC, Boca Raton, FL, pp. 1636–1655.
4. Slack, S. M. and Turitto, V. T. (1994) Flow chambers and their standardization for use in studies of thrombosis: on behalf of the subcommittee on Rheology of the Scientific and Standardization Committee of the ISTH. *Thromb. Haemost.* **72,** 777–781.
5. Panaro, N. J. and McIntire, L. V. (1993) Flow and shear stress effects on endothelial cell function, in *Hemodynamic Forces and Vascular Cell Biology* (Sumpio, B. E., ed.), R. G. Landers, Austin, TX, pp. 47–65.
6. Tran-Son-Tray, R. (1993) Techniques for studying the effects of physical forces on mammalian cells and measuring cell mechanical properties, in *Physical Forces and the Mammalian Cell* (Frangos, J. A., ed.), Academic, San Diego, pp. 1–59.
7. Frangos, J. A., Eskin, S. G., McIntire, L. V., and Ives, C. L. (1985) Flow effects on prostacyclin production by cultured human endothelial cells. *Science* **227,** 1477–1479.
8. Frangos, J. A., McIntire, L. V., and Eskin, S. G. (1988) Shear stress induced stimulation of mammalian cell metabolism. *Biotechnol. Bioeng.* **32,** 1053–1060.

9. Schnittler, H. J., Franke, R. P., Akbay, U., Mrowietz, C., and Drenckhahn, D. (1993) Improved in vitro rheological system for studying the effect of fluid shear stress on cultured cells. *Am. J. Physiol.* **265**, C289–C298.
10. Gopalan, P. K., Jones, D. A., McIntire, L. V., and Wayne Smith, C. (1995) Cell adhesion under hydrodynamic flow conditions, in *Current Protocols in Immunology* (Goito, R., ed.), John Wiley, New York, pp. 7. 29. 1–7. 29. 23.
11. Koslow, A. R., Stromberg, R. R., Friedman, L. I., Lutz, R. J., Hilbert, S. L., and Schuster, P. (1986) A flow system for the study of shear forces upon cultured endothelial cells. *J. Biomech. Eng.* **108**, 338–341.
12. Viggers, R. F., Wechezak, A. R., and Sauvage, L. R. (1986) An apparatus to study the response of cultured endothelium to shear stress. *J. Biomech. Eng.* **108**, 332–337.
13. Eskin, S. G., Ives, C. L., McIntire, L. V., and Navarro, L. T. (1984) Response of cultured endothelial cells to steady flow. *Microvasc. Res.* **28**, 87–94.
14. Rosenhead, L. (1963) *Laminar Boundary Layers.* Oxford University, Oxford, UK.
15. Cooke, B. M., Usami, S., Perry, I., and Nash, G. B. (1993) A simplified method for culture of endothelial cells and analysis of adhesion of blood cells under conditions of flow. *Microvasc. Res.* **45**, 33–45.
16. Dewey, C. F., Bussolari, S. R., Gimbrone, M. A., and Davies, P. F. (1981) The dynamic response of vascular endothelial cells to fluid shear stress. *J. Biomech. Eng.* **103**, 177–185.
17. Shmid-Schonbein, H., Gosen, J. V., Heinich, L., Klose, H. J., and Volger, E. (1973) A counter-rotating "rheoscope chamber" for the study of the microrheology of blood cell aggregation by microscopic observation and microphotometry. *Microvasc. Res.* **6**, 366–376.
18. Franke, R. P., Grafe, M., Schnittler, H., Seiffge, D., Mittermayer, C., and Drenckhahn, D. (1984) Induction of human endothelial stress fibers by fluid shear stress. *Nature.* **307**, 648–649.
19. Davies, P. F., Remuzzi, A., Gordon, E. J., Dewey, C. F., Jr., and Gimbrone, M. A., Jr. (1986) Turbulent fluid shear stress induces vascular endothelial cell turnover in vitro. *Proc. Natl. Acad. Sci. USA* **83**, 2114–2117.
20. Nomura, H., Ishikawa, C., Komatsuda, T., Ando, J., and Kamiya, A. (1988) A disk-type apparatus for applying fluid shear stress on cultured endothelial cells. *Biorheology.* **25**, 461–470.
21. Ando, J., Nomura, H., and Kamiya, A. (1987) The effect of fluid shear stress on the migration and proliferation of cultured endothelial cells. *Microvasc. Res.* **33**, 62–70.
22. Hochmuth, R. M., Mohandas, N., Spaeth, E. E., Williamson, J. R., Blackshear, P. L., and Johnson, D. W. (1972) Surface adhesion, deformation and detachment at low shear of red cells and white cells. *Amer. Soc. Artif. Int. Organs* **18**, 325–332.
23. Horbett, T. A., Waldburger, J. J., Ratner, B. D., and Hoffman, A. S. (1988) Cell adhesion to a series of hydrophilic-hydrophobic copolymers studied with a spinning disc apparatus. *J. Biom. Mater. Res.* **22**, 383–404.
24. Weiss, L. (1961) The measurement of cell adhesion. *Exper. Cell Res.* **Suppl. 8**, 141–153.
25. Fowler, H. W. and McKay, A. J. (1980) The measurement of microbial adhesion, in *Microbial adhesion to surfaces* (Berkey, R. C. W., Lynch, J. M., Melling, J., Rutter, P. R., and Vincent, B., eds.), John Wiley, Chichester, pp. 143–161.

26. Cozens-Roberts, C., Quinn, J. A., and Lauffenberger, D. A. (1990) Receptor-mediated adhesion phenomena. Model studies with the Radical-Flow Detachment Assay. *Biophys. J.* **58**, 107–25.
27. Kuo, S. C. and Lauffenburger, D. A. (1993) Relationship between receptor/ligand binding affinity and adhesion strength. *Biophys. J.* **65**, 2191–200.
28. Duddridge, J. E., Kent, C. A., and Laws, J. F. (1982) Effect of surface shear stress on the attachment of *Pseudomonas fluorescens* to stainless steel under defined flow conditions. *Biotechnol. Bioeng.* **24**, 153–164.
29. Cozens-Roberts, C., Lauffenburger, D. A., and Quinn, J. A. (1990) Receptor-mediated cell attachment and detachment kinetics. I. Probabilistic model and analysis. *Biophys. J.* **58**, 841–856.
30. Cozens-Roberts, C., Quinn, J. A., and Lauffenburger, D. A. (1990) Receptor-mediated cell attachment and detachment kinetics. II. Experimental model studies with the radial-flow detachment assay. *Biophys. J.* **58**, 857–872.
31. Groves, B. J. and Riley, P. A. (1987) A miniaturized parallel-plate shearing apparatus for the measurement of cell adhesion. *Cytobios* **52**, 49–62.
32. Usami, S., Chen, H. H., Zhao, Y., Chien, S., and Skalak, R. (1993) Design and construction of a linear shear stress flow chamber. *Ann. Biomed. Eng.* **21**, 77–83.
33. Badimon, L., Turitto, V., Rosemark, J. A., Badimon, J. J., and Fuster, V. (1987) Characterization of a tubular flow chamber for studying platelet interaction with biologic and prosthetic materials: deposition of indium 111-labeled platelets on collagen, subendothelium, and expanded polytetrafluoroethylene [published erratum appears in *J. Lab. Clin. Med.* **111**, 5]. *J. Lab. Clin. Med.* **110**, 706–718.
34. Badimon, L., Badimon, J. J., Turitto, V. T., and Fuster, V. (1987) Thrombosis: studies under flow conditions. *Ann. NY Acad. Sci.* **516**, 527–40.
35. Redmond, E. M., Cahill, P. A., and Sitzmann, J. V. (1995) Perfused transcapillary smooth muscle and endothelial cell co-culture-a novel in vitro model, *in Vitro Cell Dev. Biol. Anim.* **31**, 601–609.
36. Baumgartner, H. R. (1973) The role of blood flow in platelet adhesion, fibrin deposition, and formation of mural thrombi. *Microvasc. Res.* **5**, 167–179.
37. Turitto, V. T. and Baumgartner, H. R. (1975) Platelet interaction with subendothelium in a perfusion system: physical role of red blood cells. *Microvasc. Res.* **9**, 335–344.
38. Turitto, V. T., Weiss, H. J., and Baumgartner, H. R. (1980) The effect of shear rate on platelet interaction with subendothelium exposed to citrated human blood. *Microvasc. Res.* **19**, 352–265.
39. Turitto, V. T. and Baumgartner, H. R. (1979) Platelet interaction with subendothelium in flowing rabbit blood: effect of blood shear rate. *Microvasc. Res.* **17**, 38–54.
40. Lassila, R., Badimon, J. J., Vallabhajosula, S., and Badimon, L. (1990) Dynamic monitoring of platelet deposition on severely damaged vessel wall in flowing blood. Effects of different stenoses on thrombus growth. *Arteriosclerosis* **10**, 306–315.
41. Badimon, L. and Badimon, J. J. (1989) Mechanisms of arterial thrombosis in nonparallel streamlines: platelet thrombi grow on the apex of stenotic severely injured vessel wall. Experimental study in the pig model. *J. Clin. Invest.* **84**, 1134–1144.

42. Truskey, G. A., Barber, K. M., Robey, T. C., Olivier, L. A., and Combs, M. P. (1995) Characterization of a sudden expansion flow chamber to study the response of endothelium to flow recirculation. *J. Biomech. Eng.* **117,** 203–210.
43. Schoephoerster, R. T., Oynes, F., Nunez, G., Kapadvanjwala, M., and Dewanjee, M. K. (1993) Effects of local geometry and fluid dynamics on regional platelet deposition on artificial surfaces. *Arterioscler. Thromb.* **13,** 1806–1813.
44. Barstad, R. M., Roald, H. E., Cui, Y., Turitto, V. T., and Sakariassen, K. S. (1994) A perfusion chamber developed to investigate thrombus formation and shear profiles in flowing native human blood at the apex of well-defined stenoses. *Arterioscler. Thromb.* **14,** 1984–1991.

43

Quantitative Modeling of Limitations Caused by Diffusion

Athanassios Sambanis and Sanda A. Tan

1. Introduction

Transport of nutrients and metabolites in many bioartificial tissue constructs relies exclusively on diffusion, i.e., on the presence of a concentration gradient between the inside of the construct and the surrounding milieu. A quantitative evaluation of the rate of diffusional processes is thus essential for properly designing three-dimensional cell–polymer systems, and for assessing the chemical environment at various locales within the construct.

Ordinary or bulk diffusional transport is generally assumed to follow Fick's law, which is described by the equation *(1–3)*

$$N_A = -D \, \partial C_A / \partial x \qquad (1)$$

where N_A (mol/m²·s) is the molar flux of species A in the x-direction, D (m²/s) is the diffusivity coefficient, and C_A (mol/m³) is the concentration of species A. The fraction $\partial C_A / \partial x$ is thus the concentration gradient of species A in the x-direction. Bioartificial tissue constructs do not comprise a homogeneous medium for diffusion, since they generally contain pores of complex geometry and solids of different structure (e.g., biopolymers and cells). Transport through such a medium is typically described by **Eq. 1**, using an aggregate effective diffusivity, D_{eff}, based on the total cross-section of the construct (voids and solids) normal to the direction of diffusion *(3)*. In certain constructs, diffusion may not even follow Fick's law with an effective diffusivity coefficient; non-Fickian diffusion will not be considered in the context of this chapter.

Besides diffusion, transport in constructs may also occur by convection. Convective transport occurs as a result of a pressure gradient between the construct interior and the surrounding medium, and thus it is dependent on the flow field around the construct. One way to assess the presence of convective fluxes is to evaluate the transport of a solute from the medium to the construct (or vice versa), with the construct placed in different flow fields. If the rate of transport is dependent on the flow field, and not solely on the concentration gradient, then this constitutes evidence for the presence of convective fluxes.

In this chapter, we describe methods for measuring diffusivities for calcium alginate slabs and beads, and for synthetic membranes containing cells. Models of diffusion–reaction processes, which can be used to quantitatively assess the limitations caused by diffusion, are discussed. The preparation of the constructs themselves is not described in detail, since this is not the focus of this chapter. Convective transport processes are also not examined. The methods presented may be adapted to characterize diffusional limitations in other bioartificial tissue constructs, after performing any necessary modifications.

2. Materials

2.1. Construct Preparation

1. Creating an alginate construct: alginate (Keltone LV; Kelco, Chicago, IL); calcium chloride (Fisher, Norcross, GA); deionized water.
2. Creating alginate beads: alginate (Keltone LV; Kelco); calcium chloride (Fisher); syringe pump (Razel Scientific Instruments, Stamford, CT); 10-mL syringe with 18-gage needle; air tank with gas regulator (Fisher); phosphate buffer saline (PBS; Sigma, St. Louis, MO).
3. Measurement of bead sizes: microscope with grid on ocular lens (Nikon, Tokyo, Japan) or connected to image-acquisition and processing system (Perceptics, Knoxville, TN).

2.2. Specimen Preparation for Diffusion Experiment

1. Glutaraldehyde treatment and inactivation of cells: 50% glutaraldehyde (Fisher); Hanks' balanced salt solution (HBSS; Sigma).
2. Measurement of construct volume: Dulbecco's modified Eagle's medium (DMEM; Sigma); a graduated cylinder.

2.3. Diffusivity Measurement Across a Slab

1. Two reservoirs with stirrers (**Fig. 1**); copper-wire mesh with large pores (0.5 mm diameter); epoxy resin.
2. Glucose-free DMEM (Sigma); fetal bovine serum (FBS; Sigma); glucose (Fisher); or an air and a nitrogen cylinder with gas regulators.

2.4. Diffusivity Measurement Across a Hollow Fiber Membrane

1. Two reservoirs with stirrers (**Fig. 2**).
2. Hollow fiber; gas-impermeable tubing (Fisher); peristaltic pump (Cole Palmer, Vernon Hills, IL) for medium circulation.

Limitations Caused by Diffusion

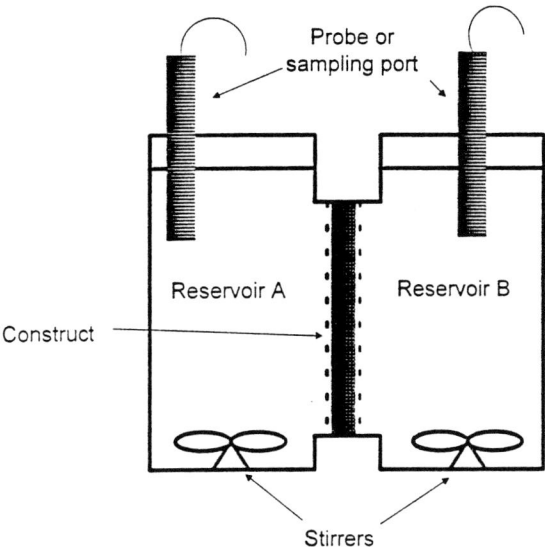

Fig. 1. Experimental setup for steady-state measurement of diffusivity across a bioartificial construct of slab or sheet geometry.

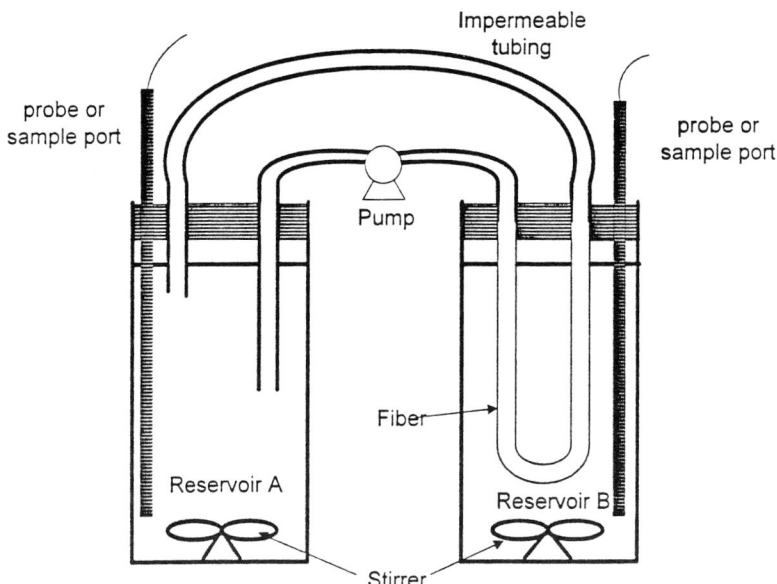

Fig. 2. Experimental setup for steady-state measurement of diffusivity across a hollow-fiber membrane.

3. Glucose-free DMEM (Sigma); FBS (Sigma); glucose (Fisher); or an air and a nitrogen cylinder with gas regulators.

2.5. Diffusivity Measurement in Alginate Beads

1. Reservoir with stirrer, e.g., a spinner flask (Corning, Corning, NY).
2. Glucose-free DMEM (Sigma); FBS (Sigma); glucose (Fisher); or ^{125}I-labeled insulin (Binax, South Portland, ME).

2.6. Analytical Techniques

1. Glucose: Trinder reagent (Sigma); spectrophotometer; cuvets.
2. Protein: Bio-Rad protein assay (Bio-Rad, Richmond CA) or BCA protein assay reagent kit (Pierce, Rockford IL); spectrophotometer; cuvets.
3. ^{125}I-labeled insulin: ^{125}I-labeled insulin (Binax); gamma counter.
4. L-glucose: L-[1-^{3}H] glucose (New England Nuclear, Boston, MA); scintillation vials and counter.
5. Oxygen: polarographic oxygen probes and amplifiers (Ingold, Wilmington, MA).

3. Methods
3.1. General Considerations

The methods described in this section for the quantitative assessment of diffusional limitations fall into two general categories: steady-state and transient measurements. Steady-state measurements involve more sophisticated experimental apparati and pose specific requirements on the geometry of constructs, but the required data processing is simpler when compared to that of transient experiments. On the other hand, transient measurements generally involve simpler experimental systems, but the processing of data for the determination of diffusion coefficients is complex. Computationally proficient investigators might find transient measurements preferable; those more experimentally inclined might opt for the steady-state techniques, provided that the construct geometry allows for their implementation.

If diffusivities are to be evaluated in constructs containing cells, the cells should be included during the measurements, since they generally affect diffusivities. Reactions should be prevented from occurring at cell sites during the experiment. This can be accomplished by inactivating the cells either before they are incorporated in the construct, or after the construct fabrication *(4)*, or by using nonmetabolizable compounds for measurements *(5)*.

Since the construct geometry may affect its pore structure, it is advisable that, whenever possible, diffusivity measurements be made on constructs having the same qualitative and quantitative geometric features as the actual ones. Furthermore, since medium components may affect the surface of constructs, diffusivity measurements should be made in media similar to those to which the constructs will eventually be exposed. For instance, adsorbed serum proteins may have a significant effect on the transport properties of the construct, so serum should be included in the medium if it will be present in the actual working environment of the bioartificial tissue.

3.2. Construct Preparation

1. Creating a slab-gel construct: A polymer solution or a mixture of polymer solution with cells is poured onto a flat surface to form a slab 1.0–5.0 mm in thickness. Solutions to cause crosslinking are then added. For an alginate gel slab, in particular, a 2% alginate solution in PBS is exposed to 1.1% $CaCl_2$ solution for approx 2 h; excess $CaCl_2$ is washed away with deionized water.
2. Creating alginate beads: Sodium alginate is dissolved in water to form 2% solution. Cells could be added to this solution if they are to be used in construct characterization. The solution is transferred to a syringe positioned on a syringe pump and connected to an 18-gage needle. Air is blown parallel to the needle at a rate necessary to obtain beads of the desired average diameter, as originally described in **ref. 6**. Droplets falling in a 1.1% $CaCl_2$ solution produce calcium alginate beads. Beads are washed with $CaCl_2$, followed by two washes with 10 mM PBS. Beads can be coated with poly-L-lysine, if this is how the actual construct is prepared *(4)*.
3. Analysis of bead sizes: Beads are sized by being placed under an inverted microscope with a grid on the ocular lens, or connected to an image analyzer. For each bead, two orthogonal diameters, D_1 and D_2, are measured, and size is represented by their geometric mean. The absolute difference of the two diameters is also recorded and compared to the geometric mean, to estimate the degree to which each bead is actually a sphere *(7)*.

3.3. Specimen Preparation for Diffusion Experiment

1. Glutaraldehyde treatment: Treat construct with 1% glutaraldehyde in HBSS for 1 h. Alternatively, cells can be inactivated by being exposed to 1% glutaraldehyde solution in HBSS for 1 h prior to fabrication of the construct, followed by washing in HBSS.
2. Measurement of construct volume: Make DMEM with 10% FBS. Pre-equilibrate the construct with medium, so that it acquires its final, constant volume. Add the construct to a graduated cylinder with medium; measure the total volume; aspirate the medium; measure the medium volume; the difference is the volume of the construct.

3.4. Steady-State Diffusivity Measurement Across a Slab or a Hollow Fiber

1. In experiments involving encapsulation, the cells are embedded within a gel matrix and/or are surrounded by a semipermeable membrane. For diffusivity measurements, a gel with entrapped cells can be fabricated in a slab or a sheet geom-

etry. The semipermeable membranes used in the bioartificial constructs could be tested in sheet forms or in hollow cylindrical shapes. This section will present the experimental setup for both geometries.

2. If cells are involved in the experiment, they are inactivated as described in **Subheading 3.3**. Alternatively, the diffusivity of a nonmetabolizable compound could be measured *(5)*. For a gel slab, the construct is placed between two chambers with the help of copper wire mesh (**Fig. 1**). The wire mesh is supported using epoxy resin. A vertical position of the slab is preferred to a horizontal position, since it reduces the chances of deformation caused by a pressure difference between the reservoirs. The reservoirs should have a volume much higher than that of the construct.

3. For a hollow fiber, a known length of the fiber is placed in reservoir B, with the ends sealed completely to impermeable tubes which lead to reservoir A. The medium from reservoir A is then pumped through the inner lumen of the hollow fiber continuously in a recirculation manner (**Fig. 2**).

4. Reservoirs A and B are filled with known, constant volumes of medium containing a high and a low (or zero) concentration of the test molecule, respectively. For glucose, the two initial concentrations could be 25 mM and 0 mM; for oxygen, the medium in one reservoir could be equilibrated with air and in the other with nitrogen. Equilibration can be achieved by stirring the medium in the presence of the appropriate gas for about 30 min. Both reservoirs must be well mixed.

5. Samples are withdrawn at known time-points for subsequent assays, or, for oxygen, dissolved oxygen probes are placed within each chamber to monitor the concentration changes *(8)*.

6. For the system of **Fig. 1**, assuming that there is negligible resistance to mass transfer from the liquid bulk to the surface of the construct, i.e., there are no significant boundary layer effects; that the partition coefficient of the test molecule between the liquid medium and the construct is 1; and that a steady-state concentration profile has been established within the slab (a good approximation when the concentrations in the reservoirs change slowly with time), then the equation describing diffusion across the construct of slab geometry is:

$$N = AD_{\text{eff}}(C_A - C_B)/L \tag{2}$$

In the above equation, N is the flux of the test molecule across the slab construct (mol/s); C_A (C_B) are the concentrations of the test molecule in reservoir A (B) (mol/m^3); A is the cross-sectional area of the construct across which diffusion occurs (m^2); D_{eff} is the effective diffusivity (m^2/s); and L is the thickness of the slab (m). **Equation 2** can be coupled with the equation describing conservation of mass of the test molecule:

$$N = -d\,(V_A C_A)/dt = d\,(V_B C_B)/dt \tag{3}$$

where V_A (V_B) is the (constant) medium volume in reservoir A (B) (m^3). **Equations 2** and **3** can be solved analytically yielding the following relationship between C_B and t:

$$\ln\{[\alpha C_{Ao} + C_{Bo} - C_B(1 + \alpha)]/[\alpha(C_{Ao} - C_{Bo})]\} = [(1 + \alpha)/\alpha](AD_{eff}/LV_B)t \quad (4)$$

where $\alpha = V_A/V_B$ is the ratio of medium volumes in the two reservoirs, and C_{Ao} (C_{Bo}) is the initial concentration of the test molecule in reservoir A (B) (mol/m^3).

Therefore, if measurements are made on C_B as a function of time, the values of the left-hand side of **Eq. 4** can be plotted against time. The slope of the resulting straight line is calculated by the least squares method, and the effective diffusivity is evaluated as

$$D_{eff} = -(slope)(\alpha LV_B)/[(1 + \alpha)A] \quad (5)$$

7. In applying **Eq. 5**, initial points collected before the establishment of a steady-state concentration profile in the slab should be discarded; i.e., the initial points of the foregoing graph are expected to not fall on a straight line, and should not be taken into account.
8. An alternative approach is to start with a slab initially empty of test molecule and to calculate the effective diffusivity by fitting to the data the transient diffusion equation

$$\partial C/\partial t = D_{eff}(\partial^2 C/\partial x^2)$$

with boundary conditions
$C(x = 0) = C_A$
$C(x = L) = C_B$
and initial condition
$C(t = 0) = 0$

In the above equations, C is the concentration of test molecule in the slab (mol/m^3) and x is the linear dimension along which diffusion occurs (m). This analysis is somewhat more involved mathematically than that of the steady-state system. Details can be found in **ref. 9**.

9. For the system of **Fig. 2**, if the same assumptions as in **step 6** apply, and, additionally, the system does not deviate significantly from the slab geometry, i.e., $R_2 - R_1$ is much smaller than R_1, where R_1 and R_2 are the inner and outer radii of the fiber, then the equations describing this system are the same as those listed for the slab construct.

If measurements are made on C_B as a function of time, the values of the left-hand side of **Eq. 4** can be plotted against time. The slope of the resulting straight line is calculated by the least squares method and the effective diffusivity is evaluated from **Eq. 5**.

3.6. Transient Measurement of Diffusivities in Beads

1. The following protocol has been employed to evaluate diffusivities of glucose and insulin in hydrogel beads containing insulin-secreting cells (4). However, the same protocol could be used, possibly with some modifications, to evaluate diffusivities of various compounds in different types of three-dimensional tissue-engineered constructs.
2. Place the beads in a stirred vessel, such as a spinner flask (Corning) containing medium free of test molecule, and having a volume 5–10× the volume of the

construct. Stir the medium, and ensure that there is movement of medium relative to the beads. Repeat this washing procedure twice to ensure a zero concentration of test molecule in the beads.
3. Transfer the beads into a spinner flask containing medium with the test molecule. The medium volume should not exceed 3× the volume of the beads. Remove medium samples for subsequent assays. Initially, samples should be removed frequently, i.e., approximately every minute; later in the experiment, the sampling frequency may be reduced.
4. Assay for the test molecule in collected samples.
5. Use the following equations, derived from **Eq. 1** and mass balances, to describe the dynamics of the system *(4)*:

$$\partial C/\partial t = D_{\text{eff}} [\partial^2 C/\partial r^2 + (2/r)(\partial C/\partial r)] \qquad (6)$$

$$V_b (dC_b/dt) = -D_{\text{eff}} A (\partial C/\partial r)_{r=R} \qquad (7)$$

Initial conditions:

$$C(t = 0, r) = C_o(r) \qquad (8)$$

$$C_b(t = 0) = C_{bo} \qquad (9)$$

Boundary conditions:

$$\partial C/\partial r (r = 0, t) = 0 \qquad (10)$$

$$C(r = R, t) = C_s(t) \qquad (11)$$

In the above equations, C is the concentration of test molecule in beads (mol/m³); t is time (s); D_{eff} is the effective diffusivity (m²/s); r is the radial position in beads (m); R is the bead radius (m); $C_o(r)$ is the initial concentration of test molecule in beads as a function of radius (mol/m³) (zero in the described experiment); C_s is the concentration of test molecule in beads at the surface; C_b is the concentration of test molecule in the liquid bulk (mol/m³); C_{bo} is the initial concentration of test molecule in the liquid bulk (mol/m³); V_b is the volume of medium surrounding the beads (m³); and A is the total surface area of beads (m²). If there are no boundary layer effects around the beads, and the partition coefficient of the test molecule between beads and liquid medium is 1, then

$$C_s = C_b \qquad (12)$$

The system of **Eqs. 7–12** may be solved numerically, as such. Alternatively, one may use its analytical solution for diffusion of solute from the liquid bulk into beads initially empty of solute, which is given by *(4,10)*

$$C_b/C_{bo} = [\alpha/(1+\alpha)] [1 + \sum_{n=1}^{\infty} 6(1+\alpha) \exp(-D_{\text{eff}} q_n^2 t/R^2)/(9 + 9\alpha + q_n^2 \alpha^2)] \qquad (13)$$

where α is the ratio of liquid volume to total bead volume and q_n is the nth root of the equation

$$\tan q_n = 3q_n/(3 + \alpha q_n) \qquad (14)$$

6. Fit **Eq. 13** to the experimental data by minimizing the sum of the squares of the differences between calculated and measured values, with the effective diffusivity, D_{eff}, being the fitted parameter. Minimization routines are available in a variety of commercially available software programs, including Microsoft Excel (Microsoft, Bothell, WA). The above procedure allows for the calculation of the effective diffusivity for transport of the test molecule from the surrounding medium into the construct.
7. To measure the effective diffusivity for transport from the construct into the surrounding medium, the construct equilibrated with medium as in **step 3** of this subheading is transferred into medium without test molecule; the medium is agitated as in **step 2**, and medium samples are collected for subsequent assays as in **step 3**.
8. Assay for the test molecule in collected samples.
9. The dynamics of the system are described by **Eqs. 6–12**, which, for diffusion of solute from beads into the liquid bulk, initially containing no solute, have the following analytical solution:

$$C_b/C_{bo} = [1/(1 + \alpha)] [1 - \sum_{n=1}^{\infty} [6\alpha(1 + \alpha) \exp(-D_{eff}q_n^2 t/R^2)/9 + 9\alpha + q_n^2\alpha^2)] \quad (15)$$

with α and q_n as defined above. Fit **Eq. 15** to the experimental data by minimizing the sum of the squares of the differences between calculated and measured values, with the effective diffusivity, D_{eff}, being the fitted parameter.

3.7. Analytical Techniques

1. Glucose concentrations are measured spectrophotometrically, using an assay kit based on the Trinder reagent (Sigma), according to manufacturer's protocols.
2. Total protein concentration in a sample is measured using a dye-binding assay (Bio-Rad) or BCA protein assay reagent kit (Pierce), according to manufacturer's protocol.
3. ^{125}I-labeled insulin concentration is measured by placing a certain sample volume in a gamma counter and measuring the counts over a certain period of time *(4)*. Radioimmunoassays on nonlabeled, insulin-related peptides yield standard errors too high for an accurate determination of diffusivities *(4)*.
4. If L-glucose is used as a nonmetabolizable compound, the concentration of its tritiated form, L-[1-^{3}H] glucose, can be measured by scintillation counting *(5)*.
5. Dissolved oxygen concentrations are measured with polarographic dissolved-oxygen probes connected to amplifiers.

3.8. Quantitative Modeling of Reaction/Diffusion Processes

The importance of diffusional resistances in the transport of nutrients and metabolites to and from constructs should be evaluated in relation to the intrinsic reaction kinetics, rather than in an absolute fashion. In other words, the same extent of diffusional resistance may be limiting the observed overall rate of a process if the intrinsic kinetics are fast, and it may not be limiting if the intrinsic kinetics are slow.

The traditional engineering approach in evaluating these effects involves calculating the value of the dimensionless Thiele modulus ϕ, which can be defined as *(3)*

$$\phi = (V/A) \, [r_2/KD_{eff}]^{1/2} \qquad (16)$$

where V is the volume of the construct (m³), A its outside surface area (m²) (so that V/A is a characteristic length), r_x is the reaction rate per unit volume in the absence of mass-transfer limitations (mol/m³·s), K is a Michaelis constant for the intrinsic reaction kinetics (mol/m³), and D_{eff} is the effective diffusivity (m²/s). If ϕ is much smaller than 1, the reaction rate is limiting and diffusional resistances are insignificant; if ϕ is much greater than 1, the overall process is limited by diffusional transport.

The intrinsic reaction kinetics, i.e., the kinetics of nutrient consumption and metabolite production by cells, can be evaluated in cultures that do not exhibit any significant mass transfer resistances, e.g., with cells cultured as monolayers. A more accurate evaluation of diffusional limitations can be obtained by combining the intrinsic kinetics with the mass transport equations in a model of the three-dimensional construct *(4,11)*. Such models can evaluate the concentration profiles of nutrients and metabolites across the construct, thus allowing an assessment of the chemical environment to which cells are exposed in various locales; these models may also be used to simulate the overall behavior exhibited by the construct in different environments *(4,11)*.

4. Notes

There are several potential pitfalls associated with the experimental setups for the quantitative measurement of diffusional resistances. Some of these pitfalls will be discussed in this section.

While making concentration measurements to calculate diffusivity across a construct, one must take into account the possibility of adsorption of proteins on the construct and on the walls of the containing vessel. One method of overcoming this problem is to pretreat the construct and vessel with the protein solution, so as to saturate all the attachment sites before the experiment. It is recommended that, prior to the experiment, a test be carried out to confirm the absence of protein loss by adsorption. The experimenter should also ensure that sample collecting devices and storage vials be pretreated and washed thoroughly, to avoid additional errors, especially when low concentrations of molecules are involved.

In experimental setups, a modified version of the actual construct may need to be used. When this is the case, care should be taken to mimic the geometric shape of the original construct, because the pore geometry may change when the three-dimensional architecture is altered. Furthermore, care should be taken to avoid deforming the construct when positioning it in the experimental apparatus and installing the monitoring equipment.

Although most proteins can be stabilized by lowering the temperature, any alteration of the environment should be avoided, and, if possible, the experiment should be carried out in an environment similar to that in which the construct will be used. This is especially important for hydrogels, which could change their physical properties, such as their pore geometry, with temperature. This also raises the question of stability of the test molecule for the duration of the experiment, as proteins tend to degrade over a long period of time. Therefore, a control should always be run alongside to account for any degradation of protein, especially in longer-term, steady-state experiments.

Last but not least, care should be taken to minimize the mass-transfer resistance caused by boundary layer effects. If increasing the stirring rate in reservoirs increases the value of the measured effective diffusivity, then this is indicative of boundary layers and/or convective transport within the construct. If boundary layer effects cannot be abolished by increasing the stirring rate, then the diffusional resistance imposed by the construct is probably too small to be of any particular significance.

References

1. Bennett, C. O. and Myers, J. E. (1974) *Momentum, Heat and Mass Transfer*, McGraw-Hill, New York, pp. 481–512.
2. Bird, R. B., Stewart, W. E., and Lightfoot, E. N. (1960) *Transport Phenomena*, John Wiley, New York, London and Sydney, pp. 495–518.
3. Blanch, H. W. and Clark, D. S. (1996) *Biochemical Engineering*, Marcel Dekker, New York, pp. 103–161.
4. Tziampazis, E. and Sambanis, A. (1995) Tissue engineering of a bioartificial pancreas: modeling the cell environment and device function. *Biotechnol. Prog.* **11,** 115–126.
5. Casciari, J. J., Sotirchos, S. V., and Sutherland, R. M. (1988) Glucose diffusivity in multicellular tumor spheroids. *Cancer Res.* **48,** 3905–3909.
6. Vorlop, K.-D. and Klein, J. (1983) New developments in the field of cell immobilization-formation of biocatalysis by ionotropic gelation, in *Enzyme Technology. III. Rotenburg Fermentation Symposium 1982* (Lefferty, R. M., ed.), Springer-Verlag, New York, pp. 219–235.
7. Papas, K. K., Constantinidis, I., and Sambanis, A. (1993) Cultivation of recombinant, insulin-secreting AtT-20 cells as free and entrapped spheroids. *Cytotechnology* **13,** 1–12.
8. Sun, Y., Furusaki S., Yamauchi, A., and Ichimura, K. (1988) Diffusivity of oxygen into carriers entrapping whole cells. *Biotech. Bioeng.* **34,** 55–58.
9. Hannoun, B. J. M. and Stephanopoulos, G. (1985) Diffusion coefficients of glucose and ethanol in cell-free and cell-occupied calcium alginate membranes. *Biotechnol. Bioeng.* **28,** 829–835.
10. Crank, J. (1975) *The Mathematics of Diffusion*, 2nd ed., Clarendon, Oxford.
11. Hannoun, B. J. M. and Stephanopoulos, G. (1990) Growth and fermentation model for alginate-entrapped *Saccharomyces cerevisiae*. *Biotechnol. Prog.* **6,** 349–356.

Index

A

Actin, hepatocyte staining, 454
Adhesion, *see* Cell adhesion;
    Leukocyte-endothelial cell
    adhesion
Agarose gel,
  Matrigel, *see* Matrigel
  neuron regeneration,
    cell entrapment, 110–111
    functionalization with laminin,
      benzophenone-based
        photochemistry in ligand
        immobilization, 105, 109,
        116
      imidazole chemistry in
        ligand immobilization,
        104, 109, 116
    gel preparation, 110
    neurite extension in hydrogels,
      imaging,
        calibration, 111
        instrumentation, 105
        length measurement, 111,
          112, 117
        spread measurement, 112
      percent of cells extending
        neurites, calculation, 112
Albumin,
  assay of spheroid production,
    247, 249–251

hepatocyte staining, 452, 454
Alcian blue, staining of alginate
  beads, 177, 183, 184
Alginate-polylysine-alginate,
  microencapsulation of
  pancreatic islet,
    air-jet microencapsulation, 474, 475
    allograft implantation and efficacy
      in diabetes models, 478
    characteristics of capsules, 471, 472
    electrostatic droplet
      microencapsulation, 475
    solution preparation, 473
    strength of capsule, optimization, 477
    xenografts,
      humans, 472
      implantation and efficacy in
        diabetes models, 478, 479
Angioplasty, *see* Coronary artery
Annular flow chamber, *see* Flow
  chamber
Arterial prosthesis,
  arterial pulsatile fluid shear and
    cycle strain analysis,
      apparatus and data collection,
        461, 462
      effects on adhesion and cell
        orientation, 462, 463
      biomechanical properties and
        computations, 463–466

fabrication of silicone rubber
  tube matrix,
  cleaning, 466
  fibronectin basement
    membrane application, 460
  materials, 458
  silicone rubber solution
    preparation, 459
  tube fabrication, 459, 460
human saphenous vein
  endothelial cell culture,
  application to matrix, 461
  isolation, 460
  materials, 458, 459
ideal criteria, 457
viability assay of cells, 462, 463
Artery, see Arterial prosthesis;
  Coronary artery
Autocrine cell loop,
  fibroblasts expressing epidermal
    growth factor receptor and
    transforming growth factor-α,
    autocrine clone creation, 144,
      145, 148
    media, 146, 147
    transforming growth factor-α,
      ligand capture determination
        with Cytosensor, 145,
        146, 150–153
      measurement in extracellular
        medium, 145, 148, 149
      plasmid construction, 144, 147,
        148
      processing in cells, 145,
        149, 150
      tetracycline effects on
        secretion, 149
  ligand capture versus escape, 143
  ligand processing, 144

B

BAL, see Bioartificial liver
Baloon angioplasty, see Coronary
  artery
β-cell, see Pancreatic islet
Bioartificial liver (BAL), see
  Minnesota Bioartificial
  Liver
Bone graft,
  evaluation in small animals,
    histological processing,
      dehydration, clearing, and
        embedding, 126, 128–130
      interpretation and
        histomorphometry, 124, 128
      materials, 123, 124
      slide preparation, 127, 130
      specimen retrieval and
        fixation, 126, 128
      staining, 127
    surgery,
      materials, 122, 123
      rabbit skull trephine defect
        model, 122, 125, 128
      transcortical bone-pin
        model, 121, 122, 124,
        125, 128
  ideal bone substitute, 134
  osteoblast culture in poly(α-
    hydroxyester) foams,
    materials, 135
    prewetting and sterilization of
      foams, 136, 137
    seeding and culture, 137–139
  replacement materials,
    limitations, 133

C

Calvarias, see Osteoblast

Index

Cartilage, *see also* Chondrocyte
  composition, 195, 197, 523
  formation with polyoxamer gel suspensions,
    chondrocyte,
      cell–polymer suspension preparation, 78–79
      preparation, 78
    injectable cartilage, 80, 82
    painting on bone, 79, 80
  mechanical deformation response,
    biochemical analysis,
      digestion, 536, 541
      DNA content assay with Hoechst 33258, 538, 539
      glycosaminoglycan assay with dimethylmethylene blue, 537, 538
      materials, 526, 527
      metabolic radiolabeling, 539, 540
    compression devices,
      mechanical spectrometer, 532, 533, 541
      post materials, 540
      static chambers, 531–533, 540, 541
    culture techniques, 531, 540
    displacement, 521–523
    dissection,
      distal ulna, 528, 529
      explantation of tissue, 528
      femoropatellar groove, 529, 540
      humoral head, 529, 540
      materials, 525, 526
    explant disk preparation,
      distal ulna, 530
      femoropatellar groove, 530, 531
      humoral head, 530, 531
    load, 521–523
    medium for culture, 525, 527, 528
    osmotic pressure, 525
    sample protocols,
      compression release, 534
      compression, 533, 534, 541
      oscillatory compression, 534, 535
    static versus dynamic compression, 524
Cell adhesion, *see also* Leukocyte–endothelial cell adhesion,
  cell–cell flow chamber analysis,
    adhesive substrate incorporation, 513, 514
    adhesive substrate preparation,
      human umbilical vein endothelial cell monolayer preparation, 512
      platelet monolayer preparation, 512, 513
      purified native adhesion receptor proteins, 511
      recombinant adhesion proteins, 512
    apparatus, 507–509
    assembly of custom polycarbonate flow-chamber apparatus, 510, 511
    attachment of cells under steady flow conditions, 516
    data presentation, 516, 517
    materials, 508–510
    static-flow versus detachment flow assays, 515, 516
    video microscopy assays of adhesive interactions, 514, 515
    wall shear stress equations, 517,

518
overview of assays, 507, 508
selectin–sialyl Lewis^x
    interactions in leukocyte rolling,
    antibody blocking and control experiments, 549, 550
    data acquisition, 548, 549
    data analysis, 550, 551
    flow chamber,
        calibration, 548
        design, 545, 546
    materials, 544, 545
    microsphere preparation,
        avidin linking, 546, 547
        biotinylated carbohydrate linking, 547
        flow cytometry analysis of carbohydrates, 547
    overview, 543, 544
    selectin substrate,
        preparation, 548
        surface density determination, 548, 551
Cellulose nitrate, enzyme microencapsulation, 318, 320, 326
Chondrocyte,
    agarose three-dimensional implantation, 200, 202, 203
    cartilage,
    compression, *see* cartilage
    formation with polyoxamer gel suspensions,
        cell isolation, 78
        cell–polymer suspension preparation, 78, 79
        injectable cartilage, 80, 82
        painting on bone, 79, 80

immortalized cells,
    adult cell isolation and culture, 175, 176, 178, 179, 187, 188
    Alcian blue staining of alginate beads, 177, 183, 184
    alginate culture system, 176, 181, 182, 188
    collagen expression in SV40 immortalization, 174, 187
    embryonic cell isolation and culture, 176, 179, 180, 188
    proteoglycans,
        metabolic radiolabeling, 177
        polyacrylamide gel electrophoresis analysis, 184, 185, 189
    retrovirus production and infection, 176, 180, 181, 188
    RNA extraction and analysis by reverse transcription-polymerase chain reaction, 177, 178, 185–187, 189
    Western blot analysis of SV40 large T antigen, 176, 177, 182, 183
isolation,
    materials, 198, 199
    plating, 201
    tissue dissection, 198–200
knee cartilage repair by implantation,
    cell culture,
        alginate suspensions, 209, 213
        medium, 211
        monolayers, 209, 213
        morphology, 211, 212
    collagen expression, 212, 213

Index 611

day-one biopsy processing, 207, 208, 212, 213
day-two biopsy processing, 208
differential potential evaluation,
  immunohistochemistry, 210, 211
  RNase protection assay, 210, 213
  materials, 205–207
  principle, 205
  processing cells from second digestion, 208, 209
primary culture,
  collagen expression, 173, 197
  medium, 199
  morphology, 197
  passaging, 201, 202
Collagen,
  contact guidance of cells, 67
  magnetic-induced alignment in tissue equivalents,
    advantages, 67, 68
    alignment, 71–73
    hardware, 69, 72
    mechanism, 68, 69
    molds, preparation, 71
    monomeric collagen solution preparation, 69–72
  microencapsulation with cells, 335, 336, 344–346
Collagen gel,
  hepatocyte culture, 232–234
  neuron regeneration, preparation and loading into guidance channels, 108, 109, 116
Collagen–glycosaminoglycan, *see* Cultured skin substitute; Nerve regeneration template; Skin regeneration template
Collagenase,
  endothelial cell isolation, 256
  hepatocyte isolation,
    liver perfusion, 238, 239, 451
    overview, 228
    pelleting, 239, 240, 451, 452
    rat surgery, 238, 448–451
    setup, 237, 238
    solutions, 448
    species variations, 228, 229
  osteoblast isolation, 297–299
Composite skin graft,
  acellular human dermis preparation,
    epidermis removal, 412, 418
    harvesting of cadaver skin, 409–412, 417–418
    materials, 409, 410
    storage of cadaver skin, 410
  grafting, 410, 411, 413–416, 418
  harvesting, 410, 411, 416–418
  keratinocyte culture on acellular human dermis,
    fibroblast feeder layer, 408
    primary culture, 408, 409, 412
    seeding, 412, 413, 418
    thickening at air–liquid interface, 413, 418
Cone and plate device, *see* Flow chamber
Coronary artery, *see also* Arterial prosthesis,
  hydrogel,
    coating application to reduce injury following angioplasty, 89

rabbit model,
   hydrogel protection against injury, 95, 96, 98, 99
   vascular injury, 91, 96, 98
vein bypass, 457
wound healing following balloon angioplasty, 85, 86
Crigler–Najjar syndrome, hepatocye microencapsulation in treatment, 317
CSS, *see* Cultured skin substitute
Cultured skin substitute (CSS),
  animal surgery,
    anesthesia, 379
    animal selection, 386
    aseptic technique, 378, 379
    bandaging of dressings, 380
    irrigation of grafts, 371
    materials, 371, 372
    postoperative care, 380, 381
    resuscitation, 380
    suturing and dressing of skin substitutes, 380
    wound preparation, 379
  cell culture,
    cryopreservation, 374, 385
    harvesting and inoculation,
      fibroblasts, 376
      keratinocytes, 376
      melanocytes, 377
    incubation schedules for skin substitute coculture, 377, 378
    maturation medium, 370, 371
    media, 367–369, 381
    primary culture of keratinocytes, melanocytes, and fibroblasts, 372, 373
    recovery of cryopreserved cells into culture, 374
    safety, 381, 382
    supplements, 367, 370
    tissue sources, 382, 383
    vessels, 371
  cell type requirements, 365, 366
  collagen–glycosaminoglycan substrate,
    fabrication, 367, 369
    inoculation frames, 369, 370
    rehydration of dry substrates, 376
  optimization of grafting deficiency, 366
  overview of preparation, 374, 376
Cylindrical tube flow chamber,
CYP1A1, *see* Ethoxyresorufin O-dealkylase

D

Dendritic cell, *see* Hematopoietic cells
Diabetes, *see* Insulin-dependent diabetes mellitus
Diffusion,
  effective diffusivity coefficient, 595, 601
  Fick's law, 595
  measurement of diffusivity,
    analyte measurements, 598, 603
    boundary layer effects, 605
    construct design and preparation, 596, 598, 599
    materials, 596, 598
    measurement in alginate beads, 598, 601–603
    measurement across hollow fiber membrane, 596, 598–

Index

600
measurement across slab, 596, 599–601
precautions, 604, 605
steady-state verus transient measurements, 598
quantitative modeling, 603, 604
Thiele modulus, 604
Dimethylmethylene blue (DMB), glycosaminoglycan assay of compressed cartilage,
Dimethylsulfoxide (DMSO), supplementation of hepatocyte cultures, 234
DMB, *see* Dimethylmethylene blue
DMSO, *see* Dimethylsulfoxide

E

EGF, *see* Epidermal growth factor
Encapsulation, *see* Macroencapsulation; Microencapsulation
Endothelial cell, *see also* Cell adhesion; Leukocyte–endothelial cell adhesion; Shear stress,
human dermal vascular cells,
culture, 261, 264, 267—268
isolation,
dissection and trypsinization, 263, 267
materials, 262, 263, 266, 267
plating, 264
selection with magnetic beads, 261, 262, 264
marker expression, 264–266
morphology, 264
human saphenous vein endothelial cell culture for arterial prosthesis,
application to matrix, 461
isolation, 460
materials, 458, 459
human umbilical vein cells, isolation,
harvesting, 256
materials, 253, 254
perfusion and collagenase digestion, 256
umbilical cord preparation and cannulation, 255
scavenger receptor staining, 257
serial propagation, 254, 255, 257
von Willebrand factor staining, 258
Epidermal growth factor (EGF), PEO star molecule as tether to glass,
advantages, 20, 21
amine-star-growth factor, preparation and purification, 22, 23, 27–31
applicability to any free amine, 21
glass surface preparation, 21, 22, 26, 27
radioiodination of growth factor, 22, 24–26
tresylation of star molecules, 21, 23, 24, 30
EROD, *see* Ethoxyresorufin O-dealkylase
Erythropoietin, microencapsulation of secreting renal cells, 317
Ethoxyresorufin O-dealkylase (EROD), assay of spheroids, 247, 250
Extracellular matrix analogs, *see also* Nerve regeneration

template; PEG-co-poly(α-hydroxy acid) copolymers; Poly(3-hydroxybutyrate-co-3-hydroxyvalerate) copolymer; Poly(L-lactic acid) foam; Polyoxamer gel; Skin regeneration template
classification of materials, 48
collagen alignment, *see* Collagen
imaging, *see* Magnetic resonance imaging
pore size, 57, 58
requirements for biodegradable polymers, 48, 75

F

Fibroblast, *see also* Ligament cell; Tenocyte,
autocrine clone creation, 144, 145, 148
media, 146, 147
transforming growth factor-α,
ligand capture determination with Cytosensor, 145, 146, 150–153
measurement in extracellular medium, 145, 148, 149
plasmid construction, 144, 147, 148
processing in cells, 145, 149, 150
tetracycline effects on secretion, 149
coculture with hepatocytes on micropatterning substrates,
fibroblast culture, 356
hepatocyte isolation and culture, 356, 357
materials, 351, 352
micropatterning on modified substrates, 357–361
troubleshooting, 361
cultured skin substitute coculture on collagen–glycosaminoglycan substrate,
cryopreservation, 374, 385
harvesting and inoculation,
fibroblasts, 376
keratinocytes, 376
melanocytes, 377
incubation schedules for skin substitute coculture, 377, 378
maturation medium, 370, 371
media, 367–369, 381
primary culture of keratinocytes, melanocytes, and fibroblasts, 372, 373
recovery of cryopreserved cells into culture, 374
safety, 381, 382
supplements, 367, 370
tissue sources, 382, 383
vessels, 371
skin equivalent culture, 394, 396
Fick's law, 595
Flow chamber,
cell adhesion analysis,
adhesive substrate incorporation, 513, 514
adhesive substrate preparation,
human umbilical vein endothelial cell monolayer preparation, 512
platelet monolayer preparation, 512, 513

purified native adhesion
receptor proteins, 511
recombinant adhesion
proteins, 512
apparatus, 507–509
assembly of custom
polycarbonate flow-chamber
apparatus, 510, 511
attachment of cells under
steady flow conditions, 516
data presentation, 516, 517
materials, 508–510
static-flow versus detachment
flow assays, 515, 516
video microscopy assays of
adhesive interactions, 514, 515
wall shear stress equations,
517, 518
leukocyte–endothelial cell
adhesion flow chamber
assay,
detachment assays, 564
endothelial monolayer
preparation, 563, 564
erythrocytes in medium,
measurement, 565, 566
leukocyte isolation, 563
materials, 554, 555
multiple flow rate capture
assays, 565
principle, 561, 563
single flow rate capture assays,
564
selectin–sialyl Lewis^x
interactions in leukocyte
rolling,
antibody blocking and control
experiments, 549, 550

data acquisition, 548—549
data analysis, 550, 551
flow chamber,
calibration, 548
design, 545, 546
materials, 544, 545
microsphere preparation,
avidin linking, 546, 547
biotinylated carbohydrate
linking, 547
flow cytometry analysis of
carbohydrates, 547
overview, 543, 544
selectin substrate,
preparation, 548
surface density
determination, 548, 551
shear stress measurement in
endothelial cells,
annular flow chamber, 588, 589
cone and plate device,
advantages, 582
principle, 580, 581
Reynolds number, 581
shear stress calculation,
581, 582
cylindrical tube flow chamber,
587, 588
linear shear stress flow
chamber, 586
parallel-disk apparatus,
limitations, 584
operation, 583
Reynolds number, 583
spinning disk modification,
584
stenosis or expansion
variation, 589, 590

wall shear stress calculation, 583, 584
parallel-plate flow chamber,
  advantages and disadvantages, 579, 580
  applications, 580
  entrance length, 579
  geometry, 578
  wall shear stress calculation, 579
radial flow chamber,
  advantages and disadvantages, 585, 586
  principle, 584—585
  Reynolds number, 585
  wall shear stress calculation, 585
Flow cytometry, hematopoietic cells, monitoring of differentiation,
  data acquisition, 282, 283, 289, 290
  data analysis, 284
  materials, 274
  staining of cells, 282, 289
Foam, see Extracellular matrix analogs; *specific foams*
Force transducer, see Muscle

G

GAG, see Glycosaminoglycan
Glycosaminoglycan (GAG), assay of compressed cartilage with dimethylmethylene blue, 537, 538

H

HEMA-MMA, see Hydroxyethylmethacrylate-methylmethacrylate

Hematopoietic cells,
  clinical applications, 271
  culture devices and setup,
    biocompatibility of materials, 276, 277
    cytokine selection, 276
    materials, 272, 284, 285
    spinner flask culture,
      cleaning and preparation, 277, 278, 287
      initiation, 278, 287
      maintenance, 278, 279, 287
    static culture initiation and maintenance, 277, 286, 287
  flow cytometry monitoring of differentiation,
    data acquisition, 282, 283, 289, 290
    data analysis, 284
    materials, 274
    staining of cells, 282, 289
  isolation,
    $CD34^+$ cell selection, 275, 276
    materials, 272
    mononuclear cell separation from whole blood, 275, 285, 286
    sources, 274, 275
  quantitative analysis,
    colony-forming cell assay, 273, 279, 280, 288, 289
    long-term culture-initiating cell assay,
      initiation, 281, 289
      materials, 274
      stroma preparation, 280, 281, 289
      time requirements, 271, 280
    total nucleated cells, 272, 273, 279, 287, 288

Index

Hemoglobin, microencapsulation as blood substitute, 315
Hepatocyte, *see also* Minnesota Bioartificial Liver,
  coculture with fibroblasts on micropatterning substrates,
    fibroblast culture, 356
    hepatocyte isolation and culture, 356, 357
    materials, 351, 352
    micropatterning on modified substrates, 357–361
    troubleshooting, 361
  cryopreservation,
    cell damage, 303, 311
    controlled rate freezing device,
      characteristics, 304
      construction, 308, 309, 311
      materials, 305, 306
    freeze-thaw protocol, 309, 311, 312
    importance, 303
    long-term culture, 304, 305, 307, 308
  implantation,
    collagen-coated microcarriers implanted into peritoneum, 436, 438–440
    considerations, 434
    injection into portal venous system, 435, 437–439, 441
    injection into spleen, 435–439, 442
    materials, 437, 438
    microencapsulated hepatocytes implanted into peritoneum, 436–438, 440, 441
    overview of implantation methods, 435
    rationale, 433, 434
    seeded hepatocytes on biodegradable polymer scaffolds, 437, 438, 440, 442
  microencapsulation,
    Crigler–Najjar syndrome treatment, 317
    implantation, 436–438, 440, 441
    preparation and encapsulation, 322–324
  mitogens, 227, 228, 230, 231
  phenotypic stability enhancement of cultures,
    collagen gel matrix supplementation, 232–234
    dimethylsulfoxide supplementation, 234
    mixed cultures with other cells, 235, 236
    spheroids, 234, 235
  primary culture,
    characteristics, 227
    gene expression patterns, 232
    growth regulation studies in sparse cultures, 230, 240
    hepatocyte growth medium system, 236, 237
    isolation of cells by collagenase perfusion,
      liver perfusion, 238, 239, 451
      overview, 228
      pelleting, 239, 240, 451, 452
      rat surgery, 238, 448–451
      setup, 237, 238
      solutions, 448

species variations, 228, 229
long-term sandwich culture,
  actin staining, 454
  albumin staining, 452, 454
  applications, 447, 448
  materials, 448–450
  plating, 452
monolayers, 229, 230, 240
spheroids,
  albumin assay, 247, 249–251
  bioreactor applications, 430
  ethoxyresorufin O-dealkylase assay, 247, 250
  formation,
    media, 246, 250
    static culture, 245, 247, 248, 250, 251
    suspension culture, 245, 246, 248, 251
  morphology and ultrastructure, 245
  phenotypic stability, 234, 235
  viability staining, 247–249
Hoechst 33258, DNA content assay of compressed cartilage, 538, 539
Hydroxyethylmethacrylate-methylmethacrylate (HEMA-MMA), microencapsulation of cells,
  cell survival and function, 336, 337
  cell suspension additives, 335, 336
  collagen coencapsulation, 335, 336, 344–346
  core solutions, 340
  Matrigel coencapsulation, 335, 336, 344, 345
  oscillating needle setup,
    large-diameter microcapsule preparation, 344
    overview, 333, 334
  physical properties of copolymers, 334, 335
  stationary needle assembly setup,
    droplet release, 331, 333, 345
    materials, 337–339
    overview, 331
    small-diameter microcapsule preparation, 342, 343
    temperature control components, 340, 341

I

IDDM, *see* Insulin-dependent diabetes mellitus
Insulin-dependent diabetes mellitus (IDDM),
  insulin therapy and onset of complications, 469
  islet transplantation, *see* Pancreatic islet
Islet transplantation, *see* Pancreatic islet

K

Keratinocyte,
  composite skin graft, cell culture on acellular human dermis,
    fibroblast feeder layer, 408
    primary culture, 408, 409, 412
    seeding, 412, 413, 418
    thickening at air–liquid interface, 413, 418
  cultured skin substitute coculture on collagen–glycosaminoglycan substrate,
    cryopreservation, 374, 385

Index

harvesting and inoculation,
   fibroblasts, 376
   keratinocytes, 376
   melanocytes, 377
incubation schedules for skin
   substitute coculture, 377, 378
maturation medium, 370, 371
media, 367–369, 381
primary culture of keratinocytes,
   melanocytes, and fibroblasts,
   372, 373
recovery of cryopreserved
   cells into culture, 374
safety, 381, 382
supplements, 367, 370
tissue sources, 382, 383
vessels, 371
epithelial sheets in skin
   replacement, 407
migration, 391, 392
re-epithelialization, 391, 392
skin equivalent culture, 398, 399, 403
submerged culture,
   fibroblast feeder layer, 394,
   396
   media, 393, 394

L

Lesch–Nyhan disease, xanthine
   oxidase microencapsulation
   in treatment, 315
Leukocyte–endothelial cell
   adhesion,
   adhesion molecules,
      expression analysis, 554, 558–
      561, 571
      types, 553
   cellular deformability assays,
      data analysis, 558

      equipment, 557
      materials, 553, 554
      micropipet aspiration, 557,
         558, 570, 571
      overview of assays, 555, 556
   in vitro flow chamber assay,
      detachment assays, 564
      endothelial monolayer
         preparation, 563, 564
      erythrocytes in medium,
         measurement, 565, 566
      leukocyte isolation, 563
      materials, 554, 555
      multiple flow rate capture
         assays, 565
      principle, 561, 563
      single flow rate capture assays,
         564
   in vivo assay,
      acute tissue preparation,
         ear microcirculation, 568
         mesentary, 567, 568
      chronic tissue preparation,
         cranial window, 569, 570
         dorsal skin-fold chamber,
            569
      data analysis, 570
      materials, 55
      principle, 566
Leukocyte rolling, selectin–sialyl
   Lewis^x interactions, flow
   chamber assay,
   antibody blocking and control
      experiments, 549–550
   data acquisition, 548, 549
   data analysis, 550, 551
   flow chamber,
      calibration, 548
      design, 545, 546

materials, 544, 545
microsphere preparation,
  avidin linking, 546, 547
  biotinylated carbohydrate linking, 547
  flow cytometry analysis of carbohydrates, 547
overview, 543, 544, 553
selectin substrate,
  preparation, 548
  surface density determination, 548, 551
Ligament cell,
isolation,
  materials, 198, 199
  plating, 201
  tissue dissection, 198–200
morphology in vivo, 197
primary culture,
  medium, 199
  passaging, 201, 202
Linear shear stress flow chamber, *see* Flow chamber
Liver, *see* Hepatocyte; Minnesota Bioartificial Liver

M

Macroencapsulation, pancreatic islets, 471
Magnetic resonance imaging (MRI), extracellular matrix analogs,
  defect detection, 40, 41, 44
  image processing and display, 38, 41, 44
  in-plane pore size and connectivity, 41, 43
  instrumentation, 37
  principle, 36, 37
  pulse sequence, 37—38

Matrigel,
  microencapsulation with cells, 335, 336, 344, 345
  neuron regeneration, Schwann cell isolation and seeding, 104, 107, 108, 116
Melanocyte, cultured skin substitute coculture on collagen–glycosaminoglycan substrate,
  cryopreservation, 374, 385
  harvesting and inoculation,
    fibroblasts, 376
    keratinocytes, 376
    melanocytes, 377
  incubation schedules for skin substitute coculture, 377, 378
  maturation medium, 370, 371
  media, 367–369, 381
  primary culture of keratinocytes, melanocytes, and fibroblasts, 372, 373
  recovery of cryopreserved cells into culture, 374
  safety, 381, 382
  supplements, 367, 370
  tissue sources, 382, 383
  vessels, 371
Microencapsulation,
cell encapsulation,
  macroporous microcapsules, 319, 320, 325, 326, 328
  materials, 318–320
  standard method, 318, 319, 322–324, 327
  two-step method for high concentration of cells, 319, 324, 325, 328
enzyme capsules,
  cellulose nitrate, 318, 320, 326

lipid–polymer membrane microcapsules retaining cofactors, 321, 322, 327
multienzyme reactions, 327
polyamide membranes, 318, 320, 321, 326, 327
erythropoietin-secreting renal cells, 317
hemoglobin blood substitute, 315
hepatocytes,
    Crigler–Najjar syndrome treatment, 317
    preparation and encapsulation, 322–324
hydroxyethylmethacrylate-methylmethacrylate capsules,
    cell survival and function, 336, 337
    cell suspension additives, 335, 336
    collagen coencapsulation, 335, 336, 344–346
    core solutions, 340
    Matrigel coencapsulation, 335, 336, 344, 345
    oscillating needle setup, large-diameter microcapsule preparation, 344
    overview, 333, 334
    physical properties of copolymers, 334, 335
    stationary needle assembly setup,
        droplet release, 331, 333, 345
        materials, 337—339
    overview, 331

small-diameter microcapsule preparation, 342, 343
temperature control components, 340, 341
immunoisolation criteria, 331
pancreatic islets, alginate-polylysine-alginate microencapsulation,
    air-jet microencapsulation, 474, 475
    allograft implantation and efficacy in diabetes models, 478
    characteristics of capsules, 471, 472
    electrostatic droplet microencapsulation, 475
    solution preparation, 473
    strength of capsule, optimization, 477
    xenografts,
        humans, 472
        implantation and efficacy in diabetes models, 478, 479
phenylalanine ammonia lyase in phenylketonuria, 315, 316
recombinant microbes, 317, 322–324
xanthine oxidase in Lesch–Nyhan disease, 315
Micropatterning, cells,
    coculture of hepatocytes and fibroblasts,
        fibroblast culture, 356
        hepatocyte isolation and culture, 356, 357
        materials, 351, 352
        micropatterning on modified substrates, 357–361

troubleshooting, 361
manipulation of cell–cell interactions, 349
microfabrication of substrates,
  approaches, 349, 350
  facilities, 352, 353, 359
  materials, 350, 351
  wafer modifications, 353, 354, 359
surface modification of substrates with adhesive proteins, 351, 354–356, 360
Minnesota Bioartificial Liver,
hepatocyte harvesting and preparation from rat,
  isolation, 427, 428
  materials, 425
  operative procedure, 426
  preoperative procedure, 426
media, 425, 426
overview, 423, 424
reactor,
  cell suspension preparation and loading, 427, 428
  design considerations, 429, 430
  hollow fiber cartridge, 425
  operation, 428, 429
  preparation, 425, 426
  rinsing, 427
speroid use, 430
MRI, *see* Magnetic resonance imaging
Muscle,
  contraction definition, 155
  data acquisition system for contraction measurements, 162
  embryogenesis, 217
  force transducers,
    performance characteristics, 158–160
    types, 157, 158
  functional assessment,
    animal handling, 164
    in situ preparations,
      apparatus, 163
      contraction measurements, 165
    in vitro preparations,
      contraction measurements, 166, 167
      muscle bath, 163
      solutions for maintenance, 163, 164
    in vivo preparations,
      apparatus, 162
      contraction measurements, 164, 165
    maximum isometric tetanic force, 167–170
    maximum power output during a single contraction, 168
    optimum muscle length, 167
  servo motors,
    performance characteristics, 160–162
    types, 160
  stimulators and optimum voltage, 156, 157, 167
  structural measurements, 156, 168, 169
  tissue culture of skeletal muscle organoids,
    characteristics of organoids, 217, 218
    immunocytochemical staining, 218, 219, 223, 224
    media, 218

Index

notched bracket construction and well attachment, 219, 220
plating of cells, 221
removal from culture wells, 221, 223
well construction,
large well, 219
materials, 218
small well, 220, 221

N

Nerve regeneration,
degeneration prevention in injury, 101, 102
failure from lack of appropriate environment, 102, 103
gel-filled nerve guidance channels,
agarose gel functionalized with laminin,
benzophenone-based photochemistry in ligand immobilization, 105, 109, 116
cell entrapment, 110, 111
collagen gel preparation and loading into guidance channels, 108, 109, 116
gel preparation, 110
imidazole chemistry in ligand immobilization, 104, 109, 116
gel loading into guidance channels, 107
Matrigel, Schwann cell isolation and seeding, 104, 107, 108, 116
nerve guide implantation,
anesthesia, 114
animal preparation, 114
closure, 115
instruments, 114
overview, 112, 113
placement, 115
rat models, 113, 114
surgical exposure, 114, 115
neurite extension in hydrogels, imaging,
calibration, 111
dorsal root ganglion length measurement, 111, 112, 117
dorsal root ganglion spread measurement, 112
instrumentation, 105
percent of cells extending neurites, calculation, 112
poly(glyceryl methacrylate)–collagen hydrogel synthesis, 107, 116
polyglyceryl methacrylate gel synthesis, 103, 104, 106, 107
polyhydroxyethylmethacrylate gel preparation, 103, 105, 106, 115
migration and patterning of neurons,
culture of chick dorsal root gaglion neurons,
cover slip culture, 488
dissection, 487, 488, 492
materials, 484, 491
substrate preparation,
acid etching, 488, 489
cleaning, 488
materials, 485–487
photoresist patterning, 489
protein patterning, 489
video microscopy,

culture chamber
  preparation, 490
  data collection and analysis, 490–493
  materials and instrumentation, 487
pathways for migration and axon cell guidance, 101
Nerve regeneration template (NRT),
  collagen–glycosaminoglycan suspension preparation, 5, 8, 9, 15
  crosslinking, sterilization, and hydration, 8, 14, 15
  matrix pore structure formation, 6, 7, 9–12, 15
  structural properties, 3–5
Neuron migration, *see* Nerve regeneration
NMR, *see* Nuclear magnetic resonance
NRT, *see* Nerve regeneration template
Nuclear magnetic resonance (NMR), *see* Magnetic resonance imaging

O

Organoid, *see* Muscle
Osteoblast,
  culture in poly(α-hydroxyester) foams,
    materials, 135
    prewetting and sterilization of foams, 136, 137
    seeding and culture, 137–139
  isolation from rat calvarias,
    calvaria harvesting, 295, 296, 299
    glass chips,
      collagenase digestion, 297–299
      osteoblast migration, 293, 294
      periosteum stripping and covering of calvaria, 296, 297, 299
      preparation, 295
      removal, 299
    materials, 294, 295, 298, 299
    primary culture heterogeneity, 293
    principle, 293, 294
    seeding of cells, 298, 300

P

Pancreatic islet,
  alginate-polylysine-alginate microencapsulation,
    air-jet microencapsulation, 474, 475
    allograft implantation and efficacy in diabetes models, 478
    characteristics of capsules, 471, 472
    electrostatic droplet microencapsulation, 475
    solution preparation, 473
    strength of capsule, optimization, 477
    xenografts,
      humans, 472
      implantation and efficacy in diabetes models, 478, 479
  culture, 474
  immune rejection of grafts,
    immunoisolation, 470, 471, 497
    macroencapsulation, 471
    minimization, 469, 470
  isolation,

Index

insulin release assay, 472, 473
mouse islets, 499, 500
pig islets, 472
rat islets, 473
planar diffusion chamber measurements,
chamber characteristics, 499
chamber volume, 501, 503
encapsulation and implantation, 500
islet isolation, 499, 500
islet number, 501–504
islet volume, 501, 503
materials, 498, 499
removal and sample preparation, 500, 502, 503
transplantation in diabetes, 469, 470
Parallel-disk apparatus, see Flow chamber
Parallel-plate flow chamber, see Flow chamber
PEG-co-poly(α-hydroxy acid) copolymers,
advantages of liquid application versus foam, 88, 89
angioplasty,
coating application to reduce injury, 89
coronary artery healing following balloon angioplasty, 86
rabbit model,
hydrogel protection against injury, 95, 96, 98, 99
vascular injury, 91, 96, 98
biocompatibility, 87, 88
macromer synthesis and structure, 87–89, 91, 92, 98
photopolymerization of hydrogels, 88, 90–92, 99

rabbit uterine horn model,
hydrogel protection against postoperative adhesions, 92–95
peritoneal injury model, 90
resorption, 88
synthesis, 89, 91, 98
water content, 86, 87
PEO star molecule, tethering of epidermal growth factor to glass,
advantages, 20, 21
amine-star-growth factor, preparation and purification, 22, 23, 27–31
applicability to any free amine, 21
glass surface preparation, 21, 22, 26, 27
radioiodination of growth factor, 22, 24–26
tresylation of star molecules, 21, 23, 24, 30
Peritoneal cavity,
rabbit uterine horn model, hydrogel protection against postoperative adhesions, 92–95
peritoneal injury model, 90
wound healing, 85, 86
PHBHV, see Poly(3-hydroxybutyrate-co-3-hydroxyvalerate)
Phenylalanine ammonia lyase, microencapsulation in phenylketonuria, 315, 316
Phenylketonuria (PKU),
phenylalanine ammonia lyase microencapsulation in treatment, 315, 316
PKU, see Phenylketonuria

Platelet, monolayer preparation for flow chamber analysis, 512, 513
PLGA foam, see Poly(DL-lactic-co-glycolic acid) foam
PLLA foam, see Poly(L-lactic acid) foam
Polyamide membranes, enzyme microencapsulation, 318, 320, 321, 326, 327
Polyethylene glycol, see PEG-co-poly(α-hydroxy acid) copolymers
Polyethylene oxide, see PEO star molecule
Polyglyceryl methacrylate gel, neuron regeneration, gel synthesis, 103, 104, 106, 107
Poly(3-hydroxybutyrate-co-3-hydroxyvalerate) (PHBHV) copolymer,
　degradation rate, 35
　magnetic resonance imaging of porosity,
　　defect detection, 40, 41, 44
　　image processing and display, 38, 41, 44
　　in-plane pore size and connectivity, 41, 43
　　instrumentation, 37
　　principle, 36, 37
　　pulse sequence, 37, 38
　porous matrx preparation, 39, 44
　purification, 39, 44
　salt particle preparation, 39
Poly(α-hydroxyesters), see Poly(L-lactic acid); Poly(DL-lactic-co-glycolic acid); Polyoxamer gel

Polyhydroxyethylmethacrylate gel, neuron regeneration, gel preparation, 103, 105, 106, 115
Poly(glyceryl methacrylate)–collagen hydrogel, neuron regeneration,
　gel loading into guidance channels, 107
　gel synthesis, 107, 11
Poly(L-lactic acid) (PLLA) foam, fabrication,
　casting, 51, 53, 54, 135–137
　drying, 52–54
　evaporation, 51
　heat-compression molding, 136–137
　leaching, 51, 52, 54
　materials, 49
　phase separation,
　　atomization, 59, 60
　　biocompatibility, 64
　　casting, naphthalene system, 60
　　casting, phenol system, 60
　　characterization, 61
　　controlled drug release, 61–63
　　inclusion of bioactive agents, 61–63
　　materials, 59
　　theory, 58, 59
　surface coating, 52, 54
　vial preparation, 50
osteoblast culture,
　materials, 135
　prewetting and sterilization of foams, 136, 137
　seeding and culture, 137–139
particulate leaching, 136

polymer solution preparation, 49, 50, 53, 54
salt particle preparation, 49, 50, 53, 54
sterilization and packing, 49, 52, 53
surface-coating solution, 49, 54
Poly(DL-lactic-co-glycolic acid) (PLGA) foam,
biodegradation rate control, 133
heat-compression molding, 136, 137
osteoblast culture,
materials, 135
prewetting and sterilization of foams, 136, 137
seeding and culture, 137–139
particulate leaching, 136
solvent-casting, 135, 136, 137
Polyoxamer gel,
cartilage tissue formation,
chondrocyte,
cell–polymer suspension preparation, 78, 79
preparation, 78
injectable cartilage, 80, 82
painting on bone, 79, 80
characteristics of polyoxamer series, 75
preparation, 76–78
thermopolymerization, 76

R

Radial flow chamber, see Flow chamber
RNA, extraction and analysis by reverse transcription-polymerase chain reaction, 177, 178, 185–187, 189
RNase protection assay, 210, 213

S

Scavenger receptor, staining in endothelial cells, 257
Selectins,
sialyl Lewis^x interactions in leukocyte rolling,
antibody blocking and control experiments, 549–550
data acquisition, 548, 549
data analysis, 550, 551
flow chamber,
calibration, 548
design, 545, 546
materials, 544, 545
microsphere preparation,
avidin linking, 546, 547
biotinylated carbohydrate linking, 547
flow cytometry analysis of carbohydrates, 547
overview, 543, 544
selectin substrate,
preparation, 548
surface density determination, 548, 551
types and ligands, 543, 544
Servo motor, see Muscle
Shear stress, see also Cell adhesion; Flow chambers,
annular flow chamber, 588, 589
cone and plate device,
advantages, 582
principle, 580, 581
Reynolds number, 581
shear stress calculation, 581, 582
cylindrical tube flow chamber, 587, 588

linear shear stress flow chamber, 586
parallel-disk apparatus,
  limitations, 584
  operation, 583
  Reynolds number, 583
  spinning disk modification, 584
  stenosis or expansion variation, 589, 590
  wall shear stress calculation, 583, 584
parallel-plate flow chamber,
  advantages and disadvantages, 579, 580
  applications, 580
  entrance length, 579
  geometry, 578
  wall shear stress calculation, 579
pathogenesis of cardiovascular disorders, 577
radial flow chamber,
  advantages and disadvantages, 585, 586
  principle, 584, 585
  Reynolds number, 585
  wall shear stress calculation, 585
Skin, see Composite skin graft; Cultured skin substitute; Keratinocyte; Skin equivalent; Skin regeneration template
Skin equivalent,
  characteristics, 391
  collagen matrix construction, 395–398
  fibroblast isolation, 394, 396
  keratinocyte growth, 398, 399, 403
  media for culture, 395
  processing for cryosectioning, 400–402
  submerged keratinocyte culture, fibroblast feeder layer, 394, 396
  media, 393, 394
  wounding model, 395, 399, 400, 403, 404
Skin regeneration template (SRT),
  clinical applications, 3
  collagen–glycosaminoglycan suspension preparation, 5, 8, 9, 15
  crosslinking, sterilization, and hydration, 7, 8, 12–16
  matrix pore structure formation, 6, 9, 15
  structural properties, 4, 5
Spheroid, see Hepatocyte
SRT, see Skin regeneration template
Stem cell, see Hematopoietic cells
Surface-bound biomolecules,
  applications, 19
  ligand density control, 20
  micropatterning substrates, modification with adhesive proteins, 351, 354–356, 360
  PEO star molecule as epidermal growth factor tether,
    advantages, 20, 21
    amine-star-growth factor, preparation and purification, 22, 23, 27–31
    applicability to any free amine, 21
    glass surface preparation, 21, 22, 26, 27

radioiodination of growth factor, 22, 24–26
tresylation of star molecules, 21, 23, 24, 30

T

Tenocyte,
  isolation,
    materials, 198, 199
    plating, 201
    tissue dissection, 198–200
  morphology in vivo, 197
  primary culture,
    medium, 199
    passaging, 201, 202
Tethered ligands, see Surface-bound biomolecules
TGF-α, see Transforming growth factor-α
Thiele modulus, 604
Transcortical bone-pin model, see Bone graft
Transforming growth factor-α (TGF-α), autocrine cell loop expressing epidermal growth factor receptor,
  autocrine clone creation, 144, 145, 148
  ligand capture determination with Cytosensor, 145, 146, 150–153
  measurement in extracellular medium, 145, 148, 149
  media, 146, 147
  plasmid construction, 144, 147, 148
  processing in cells, 145, 149, 150
  tetracycline effects on secretion, 149

Transplantation, tissue engineering alternatives, 47, 48
Trephine defect model, see Bone graft

V

Video microscopy,
  assays of adhesive interactions, 514, 515
  neurite extension in hydrogels,
    calibration, 111
    dorsal root ganglion length measurement, 111, 112, 117
    dorsal root ganglion spread measurement, 112
    instrumentation, 105
  neuron migration,
    culture chamber preparation, 490
    data collection and analysis, 490–493
    materials and instrumentation, 487
Von Willebrand factor, staining in endothelial cells, 258, 264, 265

W

Wound healing,
  coronary arteries following balloon angioplasty, 86
  peritoneal cavity, 85, 86
  skin equivalent model, 395, 399, 400, 403, 404

X

Xanthine oxidase,
  microencapsulation in Lesch–Nyhan disease, 315